T0093542

Foundations of Socio-Environmental Research

This anthology convenes 53 foundational readings that showcase the rich history of socio-environmental research from the late 1700s onwards. The introduction orients readers to the topic and how it has evolved and describes how to best use the book. The original readings are organized into six sections documenting the emergence of socio-environmental research first as a shared concern and then as a topic of specific interest to anthropology and geography; economics, sociology, and political science; ecology; ethics, religious studies, and history; and technology, energy, and materials studies. A noted scholar introduces each section, putting the readings into historical and intellectual context. The conclusion links the legacy readings to contemporary approaches to socio-environmental research and discusses how these links can enrich the reader's understanding and work. Invaluable to students, instructors, and researchers alike, this canonical reference volume illuminates underappreciated linkages across research domains and creates a shared basis for dialogue and collaboration.

William R. Burnside is a Senior Editor at *Nature Sustainability*. Previously, he was a Postdoctoral Fellow at the US National Socio-Environmental Synthesis Center (SESYNC) and an editor of environmental science textbooks. As an independent scholar, he is interested in ecology, sustainability science, and interdisciplinary research.

Simone Pulver is an Associate Professor of Environmental Studies at the University of California, Santa Barbara, where she directs the endowed Environmental Leadership Incubator. Her research investigates the role of business in sustainability.

Kathryn J. Fiorella is an Assistant Professor at Cornell University. Her research investigates how environmental change affects human health, with a focus on food systems and livelihood, food, and nutrition security.

Meghan L. Avolio is an Assistant Professor at Johns Hopkins University. She is an ecologist working in grasslands and urban environments investigating anthropogenic impacts on plant populations, communities, and ecosystems.

Steven M. Alexander is a Science Advisor at Fisheries and Oceans Canada and holds an Adjunct Assistant Professor appointment at the University of Waterloo. As an environmental social scientist, his research is centered on natural resource management, community-based conservation, and the interaction between science, policy, and practice.

Foundations of Socio-Environmental Research

Legacy Readings with Commentaries

Edited by

WILLIAM R. BURNSIDE

Independent Scholar

SIMONE PULVER

University of California, Santa Barbara

KATHRYN J. FIORELLA

Cornell University

MEGHAN L. AVOLIO

Johns Hopkins University

STEVEN M. ALEXANDER

University of Waterloo

CAMBRIDGE
UNIVERSITY PRESS

Shaftesbury Road, Cambridge CB2 8EA, United Kingdom

One Liberty Plaza, 20th Floor, New York, NY 10006, USA

477 Williamstown Road, Port Melbourne, VIC 3207, Australia

314–321, 3rd Floor, Plot 3, Splendor Forum, Jasola District Centre, New Delhi – 110025, India

103 Penang Road, #05–06/07, Visioncrest Commercial, Singapore 238467

Cambridge University Press is part of Cambridge University Press & Assessment,
a department of the University of Cambridge.

We share the University's mission to contribute to society through the pursuit of
education, learning and research at the highest international levels of excellence.

www.cambridge.org
Information on this title: www.cambridge.org/9781009177849

DOI: 10.1017/9781009177856

First published 2023

Printed in the United Kingdom by TJ Books Limited, Padstow Cornwall

A catalogue record for this publication is available from the British Library

ISBN 978-1-009-17784-9 Hardback

We dedicate this book to our families, who have supported us throughout its development and who inspire us deeply.

Contents

Contributors

Steven M. Alexander
University of Waterloo, Waterloo, ON, Canada

Meghan L. Avolio
Johns Hopkins University, Baltimore, MD, USA

Patricia Balvanera
National Autonomous University of Mexico, Morelia, Michoacán, Mexico

William R. Burnside
Independent Scholar, New York, USA

J. Baird Callicott
University of North Texas, Denton, TX, USA

Kathryn J. Fiorella
Cornell University, Ithaca, NY, USA

Marina Fischer-Kowalski
University of Natural Resources and Life Sciences, Vienna, Austria

Emilio F. Moran
Michigan State University, East Lansing, MI, USA

Richard B. Norgaard
University of California-Berkeley Emeritus, Berkeley, CA, USA

Simone Pulver
University of California, Santa Barbara, CA, USA

Richard York
University of Oregon, Eugene, OR, USA

Foreword

In 2011, the National Socio-Environmental Synthesis Center (SESYNC.org) of the University of Maryland opened its doors to the international research community by providing support and space for interdisciplinary teams to undertake projects on complex relationships between people and the natural world. To date, the work of almost 5,000 unique participants, including both academics and individuals from outside the academy, has resulted in over 800 scholarly articles and many opportunities to engage with decision-makers and stakeholders across a variety of sectors. Within a year of opening, SESYNC also launched the "Socio-Environmental Immersion Program" to support early-career scholars in undertaking interdisciplinary postdoctoral research. As I will explain, this program is part of the story behind this book project.

The Immersion Program was designed to foster an interdisciplinary disposition and an appreciation for the breadth of scholarly perspectives and traditions that can be brought to understanding human–nature relationships. My thesis shaping the structure of the Immersion Program was that a certain level of understanding of different disciplinary traditions was essential for truly integrative research on socio-environmental systems. While lexicon differences are often cited as a barrier to interdisciplinary research, there is of course much more. Namely, there can be vast differences among researchers in their epistemic stances that determine what kind of knowledge they consider valid in research. There can be different, if not divergent, methods and methodologies, as well as variation in the extent to which research is theory-based or -motivated.

With this as a backdrop, a key element of the Immersion Program was a series of highly interactive workshops involving the postdoctoral fellows and accomplished visiting scholars. Both groups were diverse, and across the ten-plus years of the Immersion Program's funding, the workshops explored a range of perspectives and methodological "lenses" through which diverse fields and sectors approached the study of socio-environmental systems. Given often high within-discipline diversity, we sought to have multiple scholarly voices in the workshops that over time spanned sociology, anthropology, history, ecology, economics, human and land science geography, political ecology, and political science. During the two- to four-day workshops, the scholars and postdoctoral fellows co-explored the assumptions, approaches, and epistemologies they brought to their work. The fellows, themselves with very different backgrounds, variously probed, sometimes challenged, and sometimes chose to selectively disengage with certain ideas put forward by the visiting scholars.

Two things were obvious from the outset of the program. First, there was no scholarly "source" that covered the range of issues associated with the diversity of socio-environmental research. Second, while some of the Immersion scholars provided a brief historical perspective on their field(s), absent was a sense of the potential linkages across disciplinary histories. Awareness of these voids persistently plagued several of those who are now editors of this book. The reading selections and new contributions in this volume are meant to fill some of that void by providing rich accountings of how thinkers from the 1700s until nearly the end of the twentieth century engaged with human–nature relationships. What this volume brings to the fore is that, while cloaked differently, the same questions we ask today have been asked a long time.

As is elegantly argued in the part commentaries, socio-environmental research is not *of* a discipline nor is it *about* a problematic. It is of many disciplines, many problematics, and many long histories. While the point of intersection is a focus on human–nature relationships, what that means, what phenomena are explored, and why ideas converge or are discordant is very much a part of how they evolved – they all have histories. Yet, as Pulver et al. (Introduction) and Burnside et al. (Conclusion) remind us, knowledge production in this field today remains fragmented, with researchers applying diverse frameworks or approaches that appeal to distinct communities or theoretical perspectives. These communities largely move forward in confined intellectual spaces not only typically absent divergent voices but absent the history and evolution of thought that has preceded them. The question remains: can deeper knowledge of these histories help move the research community in a more truly integrative direction?

One hopes the answer to that question is yes because, while polyphyletic in its origin, interest in socio-environment research today is mushrooming. While today's work is diverse, this book reveals how the origins of many threads can be traced to early ideas and voices. Even the strong "demand-side" push scholars feel today to find solutions to problems can be seen in early works. To the extent today's research involves a "recoupling of society and nature" (Norgaard, Commentary on Part III) to focus on processes that contribute to the transformation of socio-environmental relationships (York, Commentary on Part I), then perhaps this burgeoning research enterprise could help the global "we" envision paths to better and more ethical social and environmental futures.

Margaret Palmer

Director, National Socio-Environmental Synthesis Center
Distinguished University Professor, University of Maryland

Acknowledgments

We are deeply indebted to a number of individuals and institutions for their support. First, we extend a thank you to the authors of the reading selections we included. Their words and insights, some now centuries old, inspired us and continue to inform our research today.

We thank the National Socio-Environmental Synthesis Center (SESYNC). The center's postdoctoral Socio-Environmental Immersion Program created the intellectual space and community for this book to emerge. Margaret Palmer, Jon Kramer, Cynthia Wei, Kristal Jones, Elizabeth Wise, and other SESYNC staff supported our efforts throughout. Margaret's vision for the Immersion Program and Jon's guidance in formulating a SESYNC workshop supporting the book were critical.

We are indebted to many colleagues for suggestions of readings, feedback, and advice. First and foremost, this includes our commentary authors: Richard York, Emilio F. Moran, Richard B. Norgaard, Patricia Balvanera, J. Baird Callicott, and Marina Fischer-Kowalski. We thank them for their time and ideas, their insightful pieces, and their advice and patience. We are also grateful for the early work of Elizabeth Daut on this volume. We thank colleagues and visitors who participated in the Immersion Program for the many rich discussions and for vetting the earliest ideas for this project. We also thank individuals in our academic networks who generously responded to requests for suggested readings in their areas of expertise.

Generous funding came from several sources. This work was supported by SESYNC under funding received from the National Science Foundation grants DBI-1052875 and DBI-1639145. SESYNC funded and hosted a workshop for commentary authors and editors and provided substantial financial support for reprint permissions. The Morton K. Blaustein Department of Earth & Planetary Sciences at Johns Hopkins University also generously covered the cost of permission to reprint many of the included reading selections. We also thank those scholars and presses who have graciously allowed us to access and reprint readings at no cost, including the Internet Archive, Emilio F. Moran, the Institute of Development Studies, the American Economic Association and the American Economic Review, the International Labor Organization, the American Association for the Advancement of Science, the Society of Range Management, Elsevier, Gregory Cajete, Robert U. Ayres and Allen V. Kneese, and the Rauner Special Collections Library at Dartmouth College.

We thank Cambridge University Press and in particular Dominic Lewis, Aleksandra Serocka, Jenny van der Meijden, and Susan Francis for supporting our vision and helping shepherd it into print. We are also grateful to Roland Scholz and an anonymous reviewer for feedback on our book proposal and to Tom Rudel and Derek Armitage for feedback on our concluding chapter. We thank Jim Brown, Ken Conca, Charles Curtin, Courtney Lix, Dov Sax, Vernon Scarborough, and Wenfei Tong for their advice on the publishing process.

Finally, we thank family and friends who have supported us during the project. A project of this length and complexity is a marathon and labor of love, and their sustenance was vital.

This volume is the product of a fundamentally egalitarian communal effort over a long period. All editors contributed deeply and substantially throughout this time and to all aspects, and this book would not exist without our individual and collective contributions. After Burnside, who conceived the idea for this book, the reverse alphabetical order reflects our recognition of this reality and counters the alpha ordering typically used. We thank each other for the collegiality, commitment, and sacrifices along the way.

Introduction

Foundations of Socio-Environmental Research

*Simone Pulver, William R. Burnside, Steven M. Alexander,
Meghan L. Avolio, and Kathryn J. Fiorella*

In a word, after a long period of peaceful rule, Heaven-and-earth cannot stop the people from reproducing. Yet the resources with which Heaven-and-earth nourish the people are finite. (Hong Liangji (1783) *China's Population Problem*)

Man has reacted upon organized and inorganic nature, and thereby modified, if not determined, the material structure of his earthly home. (George Perkins Marsh (1864) *Man and Nature*)

[Development programmes] fail to recognise that humans, like all living things, are participants in the water cycle and can survive sustainably only through that participation. (Vandana Shiva (1988) *Staying Alive*)

What is Socio-Environmental Research?

Separated by time, geography, discipline, and identity, Hong Liangji, a political advisor to the emperor of China, George Perkins Marsh, an American scholar and naturalist, and Vandana Shiva, an Indian ecofeminist and activist, all capture the challenge and inspiration of socio-environmental research. Each analyzes the complex reciprocal relationships connecting society and environment, with the goal of intervening to protect and enhance both (Palmer 2012). **Hong Liangji's** essay was written in **1783 (Part I)** as a warning to the emperor of China's fourth Qing dynasty, at the end of a century of dramatic population growth. Hong cautioned the emperor to expect "floods, drought and pestilence [as] the means of Heaven to temper the [population] problem." Almost 100 years later, **George Perkins Marsh (1864, Part I)** argued in *Man and Nature; or, Physical Geography as Modified by Human Action* against the prevailing idea that the consequences of human action on the natural world were marginal or benign. Instead, he described the Earth as transformed by human action, from surface to atmosphere. Finally, **Vandana Shiva (1988, Part V)**, after another century of social and technological revolution, reasserts the interdependence of humans and environments. Her analysis of the large-scale damming of rivers in India showcases that human attempts to dominate nature lead to crisis and failure.

For those conducting socio-environmental research today, the history of reflection and inquiry on society–environment interrelations is a tremendous resource. However, accessing the insights of this history can be challenging. The future-focused nature of current socio-environmental research can limit recognition and engagement with its rich past. The contemporary terrain is characterized by a proliferation of conceptual frameworks (see Binder et al. 2013; Pulver et al. 2018 for reviews), each seeking to define a new approach to socio-environmental research and offering few explicit references to the shared legacies on which they build. Many foundational readings are also siloed by discipline. Over time, academic disciplines have become more specialized, each developing its own distinct theoretical frameworks and research methodologies to investigate socio-environmental concerns (Winder, 2003). Researchers are most familiar with key readings from their own disciplines, missing out on insights from other research traditions. Needed is a guide that introduces the foundations of socio-environmental research across time and discipline and links those foundations to contemporary research communities, bringing thematic coherence and historic grounding to a fragmented terrain. Our volume provides this guide. It assembles a curated set of 53 readings, spanning from the late 1700s to the early 1990s. Each reading provides insight on the interactions between society and environment and adds a new concept, dimension, or empirical approach to the further evolution of socio-environmental inquiry.

We define socio-environmental research as structured inquiry about the reciprocal relationships between society and environment. This definition highlights three common threads linking the 53 reading selections across time, place, and discipline. First, each reading emphasizes both society and environment. While various terms are used to describe the pairing that is the focus of this book – e.g., social–ecological (Berkes and Folke 1998; Ostrom 2009), nature/culture (Goldman and Schurman 2000), human and natural (Holling and Gunderson 2002; Liu et al. 2007a, b), human and environment (Scholz and Binder 2011) – the terms society and environment best capture our aims and align with the common practice of many scholars whose work is assembled here. We choose the word "society" to emphasize that human choices are influenced by the overlapping social structures in which individuals are embedded. Humans are inherently and intensely social, so human interactions with environments are channeled through various aggregations. Family, social group, class, race/ethnicity, gender, geography, politics, and culture all intersect to organize people's relations to the environment (Moran 2010). We choose the word "environment" to encompass ecosystems from micro to macro scales as well as engineered and built structures and global biophysical systems. Finally, we maintain the distinction between society and environment implied by the term socio-environmental. While humans are a part of the natural world, the level of complexity characterizing human social systems is distinctive (Burnside et al. 2012).

The second common thread is the focus on reciprocal relationships, emphasizing the interdependence of society and environment. The reading selections bring into view the causal role each plays in shaping outcomes in the other. While there is a wealth of research that engages with either the environmental basis of social life or society's impact on the environment, only a subset gives agency to both by emphasizing their interdependence. The readings introduce key dimensions of society–environment interdependence, including both the reciprocal relationships between populations and resources and the ways these relationships have been managed by societies, mediated by culture and knowledge systems, and upended by changes in technology. Some readings present society–environment interdependence as a simple relationship between two variables (e.g., population growth and resource productivity), emphasizing short-term and localized feedbacks. Others have developed global frameworks that acknowledge temporally and spatially distant interconnections and link a variety of biophysical, economic, social, and cultural trends and drivers.

Third, each reading engages in some form of structured inquiry. What constitutes structured inquiry varies across discipline and era, and we define it as suggesting researchable propositions that can be evaluated empirically or through argumentation. Our definition is purposely broad, so as to encompass reading selections not only from the Western scientific tradition but also those grounded in Indigenous knowledge, religious scholarship, and from the humanities more generally.

We characterize the selections assembled in this volume as legacy readings, denoting both their continuing influence and the forms of privilege that enabled such influence. The included reading selections have all made a lasting imprint on socio-environmental research, even if the ideas presented have inspired critique. However, this lasting influence is intertwined with the racial, gender, and class hierarchies that have determined and continue to determine which voices are pushed into the foreground. The history of scholarly scientific research is one dominated by white, male, and Western voices, and we recognize the limited representation of non-Western, non-male, and nonwhite perspectives in published socio-environmental research, historically and today. We endeavored to include a diversity of perspectives, though with limited success. Of the 53 legacy readings, seven are solo- or first-authored by women and nine by authors from the Global South.

Each of the included 53 legacy readings was carefully vetted via a multi-stage selection process. First, nominations were elicited from over 50 scholars from a range of disciplines active in socio-environmental research. For tractability, we restricted our focus primarily to selections from the biophysical science, humanities, and social science disciplines. More than 500 nominated selections were then carefully reviewed and winnowed by the editors over a four-year period. Finally, the list of legacy readings was further refined through external peer review and an internal workshop with the six leading scholars who wrote introductory commentaries for the six parts of the book. We acknowledge that some of the selected legacy readings are informed by and reify nativist, racist, and reductionist perspectives. There is debate about the continued inclusion of such readings in research and teaching. Since productive dialogue often emerges to contest such perspectives, we see their inclusion as basic to understanding the history and arc of socio-environmental inquiry.

Socio-Environmental Research in Historical Context

The two centuries that span the readings in this volume begin at the tail end of the Enlightenment and extend to the 1990s, a decade of critical evaluation of the successes and failures of environmental management through state and market institutions. During this time, research on the reciprocal interactions between human societies and their environments evolved from the observational expeditions of the past to today's data-intensive, multi-regional, interdisciplinary modeling efforts. The book is structured to acknowledge this evolution. The 53 legacy readings, spanning from the late 1700s to the mid-1990s, are organized into six parts, with an introductory commentary to each part written by a prominent scholar. The parts are sequenced to reflect the history of intellectual and political engagement with socio-environmental relationships over time and within and across disciplines.

As introduced by Richard York (Part I), reflections on the interdependence of society and environment trace back to at least the late 1700s and early 1800s, when scholars began to theorize and research the entanglements and interdependencies of societies, their natural resources, and their broader environments. In Europe, the early 1800s were characterized by the questioning of a divine order to nature and by the rise of scientific inquiry aimed at determining the fundamental laws of science governing both nature and society. Scholarship was motivated by population and resource concerns and by comparisons among world regions.

The first discipline-based research frameworks for analyzing the reciprocal relationships between society and environment emerged in geography and anthropology in the early 1900s, as highlighted by Emilio F. Moran in his commentary on Part II. Much of this research occurred in settings of colonial domination, which both enabled the extraction of information and resources and provided racialized and gendered justifications for the continued subjugation of non-Western societies. However, many geographers and anthropologists also bore witness to the disruptions caused by Western influence and documented the environmental practices of Indigenous peoples (Pels 1997; Bonnett 2003).

Parts III through VI all encompass legacy readings beginning in the post–World War II era and extending to the mid-1990s. This period was marked by the rise of industrial capitalism, which seemed to liberate society from nature's limitations, as manifest in claims of human mastery over nature, efforts to scientifically manage natural resources, and in prophecies of unlimited growth (Podeshi 2007). However, industrial capitalism also unleashed a new set of environmental ills, illustrated by the spectacular violence of environmental crises (Kahn 2007) as well as the slow violence of systemic environmental harm from new forms of chemical pollution and creeping habitat loss (Nixon 2011). In response, the twentieth century saw the growth of a range of environmental movements. Starting in the 1970s, these concerns became institutionalized in the creation of governmental agencies dedicated to environmental protection (Meyer et al. 1997). The 1970s also marked the shift to a global point of view. The images of Earth from space both embodied the capacity to collect data about global-scale ecology and evoked concerns about the finiteness of the planet (Jasanoff 2004).

During much of this time, debates about socio-environmental research were occurring mostly within disciplines. In his commentary on Part III, Richard B. Norgaard frames socio-environmental research in economics, sociology, and political science. Patricia Balvanera introduces readings from ecology in Part IV. J. Baird Callicott's commentary on Part V centers on ethical, religious, and historical approaches to socio-environmental research. Finally, Marina Fischer-Kowalski highlights themes of technology, energy, and materials in socio-environmental research in her commentary on Part VI. These disciplinary conversations began to reintegrate at the end of the twentieth century, an evolution almost taken for granted today. In the 1980s, sustainable development (WCED 1987) was put forward as the resolution to debates about both the incompatibilities of economic growth, social well-being, and environmental protection and the competing environmental priorities of the Global North (i.e., the environmental problems of affluence) and the Global South (i.e., the environmental problems of poverty). Across disciplines, many of the legacy readings from the 1980s and 1990s grapple more or less directly with the potential for sustainable development, some seeking to operationalize its vision and others offering fundamental critiques from the perspectives of race, class, gender, and geography.

While we truncate the collection of legacy readings in the mid-1990s, socio-environmental research has expanded and flourished since then. Motivated by the increasingly global nature of environmental issues, the growing research community is reflected in increasing numbers of interdisciplinary environmental journals (Yarime et al. 2010; McDonough et al. 2017) and graduate research and training programs (Vincent 2010; Vincent et al. 2013, 2017). In our concluding chapter, we examine multiple leading contemporary approaches to socio-environmental research, ranging from those that articulate specific analytic frameworks to the wider set of emerging perspectives and evolving fields that grapple with socio-environmental relationships. Like their predecessors, current research efforts both extend knowledge of socio-environmental interactions and seek to intervene to protect and enhance environmental and social systems.

Key Lineages in Socio-Environmental Research

The above history highlights the dramatic changes over the two centuries separating the oldest and most recent legacy readings in this volume. However, close engagement with the readings reveals that questions at the intersection of society and environment present an enduring challenge. Although societies now grapple with a scale of human impact that has transformed the Earth's surface (Vitousek et al. 1997; Haberl et al. 2007; Barnosky et al. 2012), crises that seem unprecedented have manifested at smaller scales in the past (Mainwaring et al. 2010). Current socio-environmental research shares much with its antecedents. Themes of interest to early scholars, including population, resources, pollution, technology, and justice, remain central to socio-environmental scholarship today. In this section, we whet readers' appetites by introducing five enduring research lineages connecting socio-environmental scholars. Our focus is both longitudinal and lateral, linking backward through time and across disciplines.

The reciprocal relationship between **human populations and their resource base** has been a core theme in socio-environmental research since the 1700s. Foundational is **Thomas Malthus'** famous treatise, *An Essay on the Principle of Population*, first published in **1798 (Part I)**. Equally foundational, though less well known, is **Hong Liangji's** essay on "China's Population Problem," published five years earlier **(1793, Part I)**. Malthus argued that while population expands geometrically (i.e., exponentially), food production can increase only arithmetically. As a result, population growth will always outstrip food production, leading to impoverishment and starvation. Likewise, Hong identified famine, disaster, and plague as natural limits to population growth. Over centuries, Malthus' hypothesis has been challenged, extended, and rediscovered. In direct opposition to Malthus and Hong's thesis, some argued that population growth leads to innovation, urbanism, and agricultural advances. These enable intensified and expanded food production, allowing societies to overcome the purported limits nature sets to population growth **(Marx 1867, Part I; Mumford 1956, Part II; Boserup 1965, Part VI)**. However, neo-Malthusian arguments resurged in the 1970s, motivated by rapid global population growth and its purported impact on the environment **(Hardin 1968, Part III**; **Ehrlich and Holdren 1971, Part VI**; **Meadows et al. 1972, Part VI)**. Although not always in the foreground, concerns of ethics and justice are deeply embedded in these debates. Malthus wrote his 1798 essay in opposition to England's Poor Laws, because he believed support for the poor would foster unsustainable population growth. Concerns about the resource impacts of growing populations were and are used to shield nativist and racist ideologies (e.g., Hardin 1968) and usually don't acknowledge the disproportionate use of resources by industrialized countries.

A closely related lineage considers societies' capacities for the **sustainable management of common-pool resources**, i.e., resources whose consumption by one prevents simultaneous consumption by another and whose use is difficult to restrict (Bromley 1991). A key feature of such resources is that the benefits of resource use accrue to the user while the costs are spread throughout the community, which, according to **Hardin's 1968 (Part III)** article, would lead to a "Tragedy of the Commons." Using a hypothetical example of shared pasture land, Hardin predicted that individual decision-making would lead to overgrazing; an outcome that could only be avoided by either government oversight or the assignment of private property rights. Prefiguring Hardin, **Gordon (1954, Part III)** analyzed fisheries as a classic case of a common-pool resource. **Ostrom (1990, Part III)** challenged Hardin by showing that local self-governed units can manage common-pool resources and have done so for generations. Her work analyzes the conditions that characterize successful common-pool resource management. One key condition is information about the resource, a theme elaborated in **Pauly's (1995, Part IV)** analysis of shifting baselines in fisheries management. These debates still resonate today in policy discussions about global fisheries (Costello et al. 2008) and climate change (Ostrom et al. 1999).

The relationship between **society and land** is a third enduring theme in socio-environmental research, offering a conceptual trove for current efforts to develop a land systems science (Verburg et al. 2015). **Semple (1911, Part II)**, an early proponent of environmental determinism, hypothesized a "land basis of society," arguing that physical environments determined social and cultural practices. That tradition is continued by those who argue that changes in the land base beget social and economic change; see, for example, **Melville's (1990, Part V)** history of environmental and social transformation of the Valle del Mezquital, Mexico. Others pushed back against environmental determinism, noting reciprocal influences between physical geography and human societies **(Humboldt 1814, Part I; Marsh 1864, Part I; Boas 1938, Part II; Mumford 1956, Part II)**. Access rules mediate the relationship between society and land, with consequences for both. Private ownership of land is central to accumulating wealth **(Ricardo 1817, Part I; Marx 1867, Part I)**, and its obverse limits accumulation of material possession **(Sahlins 1972, Part II; Sen 1981, Part III)**. A key feature of the expansion of capitalism, industrialization, and urbanization is the erosion of individuals' and communities' rights and access to land, concentrating control of resources in fewer hands and exacerbating inequality **(Marx 1867, Part I; Polanyi 1944, Part III; Blaikie and Brookfield 1987, Part II)**. This separation also undermines the lived experience of the land, causing the loss of a land ethic **(Leopold 1949; Part V)** and of knowledge of the natural world **(Gadgil et al. 1993, Part IV; Cajete 1994, Part V)**.

The role of **technology** in society–environment relationships is a fourth key research lineage **(Fisher-Kowalski and Haberl 1993, Part VI)**. Some see technology as generative, providing the means to surpass the limits imposed by nature **(Boserup 1965, Part VI)** or to help internalize the wastes of industry **(Graedel et al. 1993, Part VI)**. Readings that analyze the destructive role of technology range from **Marx's (1867, Part I)** analysis of modern agriculture robbing the soil, to feminist analyses of modern engineering's over-harvest of India's water resources **(Shiva 1988, Part V)**, to the disproportionate harm from industrial innovations borne by low-income groups and communities of color **(Bullard 1990, Part III)**. Another line of research notes the limitations of technological solutions to socio-environmental problems, predicting unforeseen consequences of technological interventions. For example, **Jevons (1865, Part I)** showed that the efficient use of resources can increase their consumption. A century later, **Beck (1986, Part VI)** argued that technological modernization generates new, manufactured risks around which society then organizes itself. **Norgaard (1994, Part III)** offers the specific example of pesticides and the emergence of pesticide tolerance. Finally, it is through technology that changes in the environment are measured, understood, and made legible **(Humboldt 1814, Part I; Moran 1981, Part II)**. In linking both society to environment and environment to society, the scale and scope of technology's influence have expanded over time. Current socio-environment research frameworks recognize technology's increasingly global consequences and the increasing permeation of both environment and society by digital technologies (Berkhout and Hertin 2004).

Our final example lineage focuses on the concept of **systems** in socio-environmental research. The idea of a system was already recognized by **Humboldt's (1814, Part I)** view of nature and society as a holistic entity, by **Darwin (1859, Part I)** in the "web of complex relations" that typify ecological systems, and by **Marx (1867, Part I)** in the idea of a metabolic exchange between society and environment. By the 1960s and 1970s, system science was emerging as an influential area of research (von Bertalanffy 1972). One important debate considered the inherent stability versus changeability of socio-environmental systems. In economics, **Daly (1974, Part III)** envisioned the idea of a steady-state or balanced economic–environmental system. In anthropology, **Rappaport (1967, Part II)** considered the role of collectively ritualized rules and norms, such as taboos, that stabilize relationships between communities and local environments, avoiding degradation and rebalancing the system. Likewise, ecologist Eugene **Odum (1969, Part IV)** advanced a "balance of nature" perspective. Shortly thereafter others challenged this equilibrium paradigm. **Meadows et al. (1972, Part VI)** predicted potential global socio-economic system collapse, while **Holling (1973, Part IV)** introduced the ideas of resilience and multiple stable system states. **Ellis and Swift (1988, Part IV)** highlighted the role of abiotic factors driving variability in system conditions and advocated for traditional pastoral management strategies as best suited for adapting to dynamic grassland systems. Taking a broader perspective, **Fischer-Kowalski and Haberl (1993, Part VI)** theorized the material exchanges between societies and environments and the social regulation of those physical exchanges in hunter-gatherer, agricultural, and industrial societies. Systems framings continue to be widely used in contemporary socio-environmental research (e.g., Turner et al. 2016; Kapsar et al. 2019; Elsawah et al. 2020).

The above lineages introduce some recognized themes in socio-environmental research, yet they are merely examples of the innumerable points of contact and tension across the legacy readings. We hope they showcase how the readings in this volume have been and can be combined and recombined to yield new insights and that readers are inspired to discover and explore new lineages and connections. As a collection, the legacy readings show that current socio-environmental research has deep roots. However, the common threads linking the selections do not point to a single perspective. Rather, multiple voices emerge – some complementary, some contradictory, and all reflecting distinct and diverse ways of knowing (i.e., ontologies) and approaches to research (i.e., epistemologies) (Moon and Blackman 2014). Collectively, the readings also introduce a range of research methods, marked by increasing sophistication over time. Qualitative methods range from ethnography to ethical inquiry to narrative and historical approaches using a range of primary sources. Quantitative methods include simple to complex surveys and models, reflecting a growing computational toolbox for analyzing social and environmental data and their interactions across space and time. Methods are often mixed, explicitly or implicitly.

A Guide to Using This Book

We welcome readers to engage with *Foundations of Socio-Environmental Research* in multiple ways. In addition to this introductory chapter, readers are guided through the volume by the six commentaries. Each describes the historical and scholarly context for the selections, orienting the reader to the socio-political forces and intellectual themes animating the conversations in which the legacy authors were engaged. Each commentary also identifies the key insights and research propositions regarding socio-environmental relationships articulated in the subset of selected readings.

For new scholars of environment and society and for those teaching in this area, this book provides a basic vocabulary and an overview of multiple research traditions that contribute to contemporary scholarship. We made accessibility a central consideration in our vetting process and, for conciseness, chose portions of some of the longer original texts. Nevertheless, some selections, especially older ones, are challenging to read. Observations and ideas are interspersed among tangentially related discussion and require an approach to reading centered on engaging closely with the text to extract key ideas. Moreover, some of the ideas presented in some reading selections may cause offense and can be difficult to read because they present hierarchical relations across social groups that have long since been rejected.

Another aim of our book is to support individuals and groups doing interdisciplinary research. The composite nature of socio-environmental systems does not respect disciplinary boundaries; socio-environmental research is inherently interdisciplinary (Palmer et al. 2016), and research from the biophysical and social sciences, engineering, humanities, and other fields each makes unique and valuable contributions. However, interdisciplinarity requires bridging differences in research perspective and purpose (Tress et al. 2005). Many of the most interesting research questions are at the interstices between disciplines. We introduce current researchers to multiple lenses and conceptualizations of relationships between society and environment. The readings selected can serve to build shared understanding across multidisciplinary teams.

Finally, this volume seeks to engage those readers who consider themselves specialists in socio-environmental research. For this group, the assembled selections provide a mix of texts that, for any given student or scholar, likely includes readings both known and unfamiliar. The reading of both can stimulate new research questions and lines of inquiry and root seemingly novel ones in forgotten soil. We challenge this community to acknowledge the historical roots of their current concerns and the continuities across centuries in key themes and relationships. Socio-environmental research encompasses a large and expanding community of scholars, institutions, funding sources, and publication outlets. Its dynamism and timeliness belie its deep roots. The climate change challenges and disease crises of today may be unprecedented in scope but not in kind. Seeing the origins and evolution of socio-environmental research along historical lineages can inspire useful clarity, productive growth, and new lines of inquiry for contemporary students and scholars.

References

Barnosky, Anthony D., Elizabeth A. Hadly, Jordi Bascompte, Eric L. Berlow, James H. Brown, Mikael Fortelius et al. "Approaching a State Shift in Earth's Biosphere." *Nature* 486, no. 7401 (2012): 52–58.

Berkes, Fikret, and Carl Folke. "Linking Social and Ecological Systems for Resilience and Sustainability." In *Linking Social and Ecological Systems: Management Practices and Social Mechanisms for Building Resilience*, edited by Fikret Berkes, Carl Folke, and Johan Colding, 1–26. Cambridge: Cambridge University Press, 1998.

Berkhout, Frans, and Julia Hertin. "De-materialising and Re-materialising: Digital Technologies and the Environment." *Futures* 36, no. 8 (2004): 903–920.

Binder, Claudia R., Jochen Hinkel, Pieter W. G. Bots, and Claudia Pahl-Wostl. "Comparison of Frameworks for Analyzing Social-Ecological Systems." *Ecology and Society* 18, no. 4 (2013): 26.

Bonnett, Alastair. "Geography As the World Discipline: Connecting Popular and Academic Geographical Imaginations." *Area* 35, no. 1 (2003): 55–63.

Bromley, Daniel W. *Environment and Economy: Property Rights and Public Policy.* Oxford: Blackwell, 1991.

Burnside, William R., James H. Brown, Oskar Burger, Marcus J. Hamilton, Melanie Moses, and Luis M. A. Bettencourt. "Human Macroecology: Linking Pattern and Process in Big-Picture Human Ecology." *Biological Reviews* 87, no. 1 (2012): 194–208.

Costello, Christopher, Steven D. Gaines, and John Lynham. "Can Catch Shares Prevent Fisheries Collapse?" *Science* 321, no. 5896 (2008): 1678–1681.

Elsawah, Sondoss, Tatiana Filatova, Anthony J. Jakeman, Albert J. Kettner, Moira L. Zellner, Ioannis N. Athanasiadis et al. "Eight Grand Challenges in Socio-Environmental Systems Modeling." *Socio-Environmental Systems Modelling* 2 (2020): 16226.

Goldman, Michael, and Rachel A. Schurman. "Closing the 'Great Divide': New Social Theory on Society and Nature." *Annual Review of Sociology* 26, no. 1 (2000): 563–584.

Haberl, Helmut, Karl Heinz Erb, Fridolin Krausmann, Veronika Gaube, Alberte Bondeau, Christoph Plutzar et al. "Quantifying and Mapping the Human Appropriation of Net Primary Production in Earth's Terrestrial Ecosystems." *Proceedings of the National Academy of Sciences* 104, no. 31 (2007): 12942–12947.

Holling, Crawford Stanley, and Lance H. Gunderson. *Panarchy: Understanding Transformations in Human and Natural Systems*. Washington, DC: Island Press, 2002.

Jasanoff, Sheila. "Heaven and Earth: The Politics of Environmental Images." *Earthly Politics: Local and Global in Environmental Governance* 31 (2004): 41–44.

Kahn, Matthew E. "Environmental Disasters As Risk Regulation Catalysts? The Role of Bhopal, Chernobyl, Exxon Valdez, Love Canal, and Three Mile Island in Shaping US Environmental Law." *Journal of Risk and Uncertainty* 35, no. 1 (2007): 17–43.

Kapsar, Kelly E., Ciara L. Hovis, Ramon F. Bicudo da Silva, Erin K. Buchholtz, Andrew K. Carlson, Yue Dou et al. "Telecoupling Research: The First Five Years." *Sustainability* 11, no. 4 (2019): 1033.

Liu, Jianguo, Thomas Dietz, Stephen R. Carpenter, Marina Alberti, Carl Folke, Emilio Moran et al. "Complexity of Coupled Human and Natural Systems." *Science* 317, no. 5844 (2007a): 1513–1516.

Liu, Jianguo, Thomas Dietz, Stephen R. Carpenter, Carl Folke, Marina Alberti, Charles L. Redman et al. "Coupled Human and Natural Systems." *Ambio: A Journal of the Human Environment* 36, no. 8 (2007b): 639–649.

Mainwaring, A. Bruce, Robert Giegengack, Claudio Vita-Finzi, and Joel S. Schwartz. *Climate Crises in Human History*. Philadelphia: American Philosophical Society, 2010.

McDonough, Kelsey, Stacy Hutchinson, Trisha Moore, and J. M. Shawn Hutchinson. "Analysis of Publication Trends in Ecosystem Services Research." *Ecosystem Services* 25 (2017): 82–88.

Meyer, John W., David John Frank, Ann Hironaka, Evan Schofer, and Nancy Brandon Tuma. "The Structuring of a World Environmental Regime, 1870–1990." *International Organization* 51, no. 4 (1997): 623–651.

Moon, Katie, and Deborah Blackman. "A Guide to Understanding Social Science Research for Natural Scientists." *Conservation Biology* 28, no. 5 (2014): 1167–1177.

Moran, Emilio F. *Environmental Social Science: Human–Environment Interactions and Sustainability*. Hoboken, NJ: John Wiley & Sons, 2010.

Nixon, Rob. *Slow Violence and the Environmentalism of the Poor*. Cambridge, MA: Harvard University Press, 2011.

Ostrom, Elinor. "A General Framework for Analyzing Sustainability of Social-Ecological Systems." *Science* 325, no. 5939 (2009): 419–422.

Ostrom, Elinor, Joanna Burger, Christopher B. Field, Richard B. Norgaard, and David Policansky. "Revisiting the Commons: Local Lessons, Global Challenges." *Science* 284, no. 5412 (1999): 278–282.

Palmer, Margaret A. "Socio-Environmental Sustainability and Actionable Science." *Bioscience* 62, no. 1 (2012): 5–6.

Palmer, Margaret A., Jonathan G. Kramer, James Boyd, and David Hawthorne. "Practices for Facilitating Interdisciplinary Synthetic Research: The National Socio-Environmental Synthesis Center (SESYNC)." *Current Opinion in Environmental Sustainability* 19 (2016): 111–122.

Pels, Peter. "The Anthropology of Colonialism: Culture, History, and the Emergence of Western Governmentality." *Annual Review of Anthropology* 26, no. 1 (1997): 163–183.

Podeschi, Christopher W. "The Culture of Nature and the Rise of Modern Environmentalism: The View through General Audience Magazines, 1945–1980." *Sociological Spectrum* 27, no. 3 (2007): 299–331.

Pulver, Simone, Nicola Ulibarri, Kathryn L. Sobocinski, Steven M. Alexander, Michelle L. Johnson, Paul F. McCord, and Jampel Dell'Angelo. "Frontiers in Socio-Environmental Research: Components, Connections, Scale and Context." *Ecology and Society* 23, no. 3 (2018): 23.

Scholz, Roland W., and Claudia R. Binder. *Environmental Literacy in Science and Society: From Knowledge to Decisions*. Cambridge: Cambridge University Press, 2011.

Tress, Bärbel, Gunther Tres, and Gary Fry. "Defining Concepts and the Process of Knowledge Production." In *From Landscape Research to Landscape Planning: Aspects of Integration, Education and Application*, edited by Bärbel Tress, Gunther Tres, Gary Fry, and Paul Opdam, 13–26. Berlin: Springer Science & Business Media, 2005.

Turner II, Billie L., Karen J. Esler, Peter Bridgewater, Joshua Tewksbury, Nadia Sitas, Brent Abrahams et al. "Socio-Environmental Systems (SES) Research: What Have We Learned and How Can We Use This Information in Future Research Programs?" *Current Opinion in Environmental Sustainability* 19 (2016): 160–168.

Verburg, Peter H., Neville Crossman, Erle C. Ellis, Andreas Heinimann, Patrick Hostert, Ole Mertz et al. "Land System Science and Sustainable Development of the Earth System: A Global Land Project Perspective." *Anthropocene* 12 (2015): 29–41.

Vincent, Shirley. *Interdisciplinary Environmental Education on the Nation's Campuses: Elements of Field Identity and Curriculum Design*. Washington, DC: National Council for Science and the Environment, 2010.

Vincent, Shirley, Stevenson Bunn, and Lilah Sloane. *Interdisciplinary Environmental and Sustainability Education on the Nation's Campuses 2012: Curriculum Design*. Washington, DC: National Council for Science and the Environment, 2013.

Vincent, Shirley, Sumedha Rao, Quiyan Fu, Katt Gu, Xiao Huang, Kaitlyn Lindaman et al. *Scope of Interdisciplinary Environmental, Sustainability, and Energy Baccalaureate and Graduate Education in the United States*. Washington, DC: National Council for Science and the Environment, 2017.

Vitousek, Peter M., Harold A. Mooney, Jane Lubchenco, and Jerry M. Melillo. "Human Domination of Earth's Ecosystems." *Science* 277, no. 5325 (1997): 494–499.

von Bertalanffy, Ludwig. "The History and Status of General Systems Theory." *Academy of Management Journal* 15, no. 4 (1972): 407–426.

Winder, Nick. "Successes and Problems When Conducting Interdisciplinary or Transdisciplinary (= Integrative) Research." In *Interdisciplinarity and Transdisciplinarity in Landscape Studies: Potential and Limitations*, edited by Bärbel Tress, Gunther Tres, Arnold van der Valk, and Gary Fry, 74–90. Wageningen: Delta Program, 2003.

World Commission on Environment and Development (WCED). *Our Common Future*. Oxford: Oxford University Press, 1987.

Yarime, Masaru, Yoshiyuki Takeda, and Yuya Kajikawa. "Towards Institutional Analysis of Sustainability Science: A Quantitative Examination of the Patterns of Research Collaboration." *Sustainability Science* 5, no. 1 (2010): 115–125.

PART I

EARLY CLASSICS OF SOCIO-ENVIRONMENTAL RESEARCH

Richard York

Over the course of the eighteenth century, the Enlightenment changed the Western world, sweeping away old certainties and clearing vast swaths of intellectual terrain. As the nineteenth century emerged, a growing commitment to reason led to new ideas and insights across all fields of human inquiry. Collected in this part are the works of eight important and influential writers from this era who have shaped socio-environmental research: Hong Liangji (1746–1809), Thomas Malthus (1766–1834), Alexander von Humboldt (1769–1859), David Ricardo (1772–1823), George Perkins Marsh (1801–1882), Charles Darwin (1809–1882), Karl Marx (1818–1883), and William Stanley Jevons (1835–1882). Most of these thinkers variously influenced or were influenced by one another. New understandings of the world developed out of the intellectual tumult to which they contributed, and they continue to affect the world in our time just as they affected it in their own.

Each of these writers is worthy of our attention, but we must reflect on why these authors were in positions to write in the first place and why their works were published and have been regularly read, while the views and ideas of others were not recorded and/or have gone unnoticed. It is the great injustice of history that we know the ideas of the social elite much more than we do those of the underclass, since typically only the privileged had access to publication or other means of transmitting their ideas. These writers are all men, which is telling about gender relationships in their time and ours. All but one are white Europeans who lived in a world built on the depredations of colonialism and slavery. No doubt, people from many cultures at different times had important insights about society and the environment, and these insights have often been lost to us through processes of exclusion or myopia. We must not lose sight of the fact that Indigenous people around the world have long recognized the systematic nature and complexity of the human–ecological condition and practiced their own versions of ecological protection and restoration (Norgaard 2019).

Why and how should we read writers from the past? The great thinkers of two centuries ago fundamentally altered our understanding of nature and what it means to be human. They asked questions that we still ask and which still defy simple answers. They struggled with ideas that continue to inspire and confound us to this day. Thus, it is important to understand the ideas of these thinkers not as mere historical curiosities but as living and dynamic bodies of thought. Although already in motion before the end of the eighteenth century, the great and terrible processes that are fundamental to our time in many regards developed their defining characteristics when these thinkers were trying to understand the rapidly changing world. The Industrial Revolution, the rise of fossil fuels, urbanization, the globalization of capitalism, and the ongoing horrors of colonialism, racism, and slavery were central features of this era. Although we have learned much over the past two centuries, in many ways the thinkers covered here had a clearer view than do we of these processes, which were then rawer and their dynamics more exposed. Also, it is often the case that early periods of intellectual movements are characterized by more diverse, wider-ranging, and more profound ideas than exist after subsequent winnowing, as fields and disciplines become more rigid as they become more rigorous (Merton 1979).

One way to engage with these thinkers is to read them as if they are in a debate with one another over major philosophical, scientific, political, and ethical questions. In following this approach, rather than discuss each of the figures considered here alone, I will introduce them relative to two interconnected thematic tensions around which many of their views and debates revolved and which resonate with debates in our time: (1) explanation via universal laws versus the value of historical thinking; and (2) the materialist challenge to romanticist views of nature. The first theme engages with questions about how much of life is determined by inalterable forces and how much potential there is for social change. The second theme highlights the intellectual and ethical tensions that emerged with the modern scientific and materialist worldview, which challenged romantic ideas that nature has intrinsic meaning and order. These two themes are central to our current concerns, where we struggle with environmental ethics and with finding new pathways to lead us out of our environmental crises. I conclude with a section that addresses the key implications of these thematic debates, which together lead us to the question of the extent to which social progress is possible in a world with biophysical constraints.

Natural Laws and Historicism

The great works of science that led to and developed throughout the Enlightenment, exemplified by the works of Isaac Newton, demonstrated the power of general, abstract physical laws to explain worldly phenomena. Although many early scientists did not dispense with the belief in a divine grand architect who designed the universe and its laws, they did move away from the idea that God regularly intervened in worldly affairs. Thus, natural laws became sufficient to explain all matters mundane. From the appreciation of the power of natural laws arose a central question with relevance across all of the sciences: Can all phenomena, including human social processes, be adequately explained by the invocation of natural laws? Historicism, an approach exemplified by the Marxian tradition, offers a response to this question. Historicism does not deny that the world is governed by natural laws but recognizes that socio-ecological relationships are the products of history and are subject to change. In our time, the historicist approach can be seen in the work of political ecologists who emphasize the importance of specific, contextual factors for understanding human–environment relationships (Peet, Robbins, and Watts 2011). Other scholars, such as the classical economists I discuss below and modern thinkers such as **Hardin** (**1968, Part III**), emphasize that general laws and principles can broadly explain the human condition. Thus, historicists emphasized that humans are free of destiny and can remake societies in many different forms. In contrast, those focused on general forces and constraints believed the social and economic processes of their times emerged from inalterable laws and, therefore, that key aspects of the human condition were immutable.

The classical economists and those that followed in their tradition, including Malthus, Ricardo, and Jevons, imitated the physical sciences in their commitment to finding universal laws, relying on abstract logical and mathematical formalisms in their efforts to develop economic and social theories. As can be seen in the selection provided here, **Thomas Malthus** (1766–1834) – an English cleric from an affluent family who studied mathematics at Jesus College, Cambridge (of which he was subsequently elected a fellow) – developed his principle of population by logical and mathematical reasoning. He recognized correctly that any population of organisms tends to grow geometrically (i.e., exponentially) if unrestrained, and that through such a process its size can become exceedingly large rapidly. He also posited that the means of subsistence (i.e., food production) can grow only arithmetically (i.e., linearly). Additionally, he assumed that the means of restraining human fertility would be not only unvirtuous but unsuccessful. Thus, he reasoned from these starting principles that population must always be restrained by rising mortality, since food production could never keep pace with the potential of population growth. Malthus made this argument in a political context, to advocate against government measures to help the poor, arguing that the suffering of the masses was dictated by natural laws and could not be alleviated in the long term by public measures.

Hong Liangji (1746–1809), a Chinese philosopher who for part of his life worked in government positions for the Qing dynasty, independently (and several years earlier) had a similar insight, arguing that food production and the provision of housing cannot keep pace with population growth and, therefore, that human numbers are likely to be limited by mortality. Hong lived during a time when China's population was growing rapidly, with concomitant social and economic challenges. Like Malthus, Hong wrote in part to make a political point: he was questioning the traditional assumption that population growth was a sign of good governance. He ends this selection by suggesting that growth was undermining the possibility of peaceful rule. The parallel of Hong's and Malthus' ideas provides a prime example of a common observation in the history of science: that discoveries are often made by multiple people working independently of one another at a similar time, but only one of the discoverers (Malthus in this case) receives widespread recognition for the discovery due to social factors rather than due to having greater insight or originality (Merton 1979).

Malthus' and Hong's arguments resonate to this day, as can be seen in ongoing debates about the role of population growth in driving environmental problems.

"Neo-Malthusians," such as Paul Ehrlich (1968) and **Garrett Hardin (1968, Part III)**, have argued that addressing population growth is of central importance in limiting environmental destruction, while some technological optimists have argued that market forces will stimulate innovations that allow humans to transcend natural limits (**Boserup 1965, Part VI**; Simon 1981; Lomborg 2001).

David Ricardo (1772–1823), who made a fortune as a businessman and subsequently became a member of the Parliament of Great Britain and Ireland, also developed his theory of rent, in the strict sense of what is paid for use of the soil in its natural condition, by reasoning from basic assumptions. Although he recognized that the productivity of land can be influenced by the work of the farmer – for example, by the "skillful rotation of crops, or the better choice of manure" (reading 4) – he had an ahistorical view, assuming that at base productivity cannot be fundamentally enhanced since it comes from the "original and indestructible powers" (reading 4) of the soil. He did not foresee the potential for human actions to degrade the environment and undermine productivity, assuming that the "gifts" of nature are free and provided in "boundless" quantities. However, influenced by Malthus, he argued that economic development and population growth will invariably lead to the cultivation of less-productive land and, therefore, drive up rents, since in his formulation rent is determined by the gap in productivity between the best and worst cultivated land. Ricardo's views anticipate modern concerns that the pressures placed on natural resources from growing production and consumption will lead to economic and political crises (e.g., Collier and Hoeffler 2005).

Writing decades after Malthus and Ricardo, when industry had eclipsed agriculture as a source of financial wealth, **William Stanley Jevons** (1835–1882), who became a professor of logic, moral philosophy, and political economy at Owens College in Manchester, England, analyzed the economy of coal and the machines it powered. He was born to an iron merchant in Liverpool and spent a stint working as a metallurgical assayer in Australia, so he was familiar with the working of industrialism. Like his classical economist forebears, he developed his theories based on logical and mathematical reasoning. Jevons wrote *The Coal Question* to address the pressing concern about the potential for Britain's economy to be hampered by a decline in coal reserves (Clark and Foster 2001). In his analysis of the history of coal use, Jevons identified the paradox for which he is famous. He noted that as the coal-powered steam engine had become more efficient, it had led not to a decline in coal consumption but to an increase. He argued that improvements in efficiency made steam engines (and other industrial equipment) less costly relative to human labor and more profitable from ever-expanding production, so industrialists would invest more in steam engines as they became more efficient. Because coal powered industrial production, coal consumption rose as technological advances improved efficiency. Recognizing that coal was finite, Jevons predicted that Britain's industrial economy would eventually collapse when coal was depleted.

The question of whether improvements in efficiency can help societies overcome ecological crises remains central in our times, as can be seen in debates over the "rebound effect," a general class of phenomena, which includes the Jevons paradox, where improvements in efficiency fail to deliver on their full promise. Advocates for using "green" technology to overcome our environmental crises suggest that efficiency is the key to conservation (Hawken, Lovins, and Hunter Lovins 1999; Gillingham et al. 2013), while scholars in the Jevons tradition highlight how improvements in efficiency are commonly associated with rising resource consumption (Polimeni et al. 2008; York and McGee 2016).

Malthus, Hong, Ricardo, and Jevons show signs of recognizing that societies are inherently dependent on natural resources and that there are limits to growth in production and consumption. However, they did not advocate for any form of conservation, taking a grimly fatalistic view that the forces they outlined were inevitable and immutable. Their views foreshadow how in our era mainstream economists discount future costs, such as the impending catastrophes driven by climate change, in favor of near-term profits (York, Clark, and Foster 2009).

Karl Marx (1818–1883), a German philosopher and social theorist who spent much of his life in exile in England (due to his radical political views), was an exemplar of historicism and materialism who offered a different perspective from the classical economists. Marx sought to explain the world without invocation of design. Whereas thinkers like Malthus saw divine intentions behind natural laws and processes, Marx saw no higher purpose but rather historical change driven by struggle in strictly material contexts. Thus, Marx admired the accomplishments of the natural sciences and accepted that our universe is governed by natural laws. However, Marx critiqued the classical economists for assuming that the regularities they identified were the necessary outcomes of invariant economic and social laws. Unlike Malthus and Ricardo, who saw the productivity of soil as largely unalterable, Marx studied the science of his day, most notably that of Justus von Liebig, to understand the conditions that led to the fertility of soil and how it could be transformed (Foster 2000).

In the first selected Marx readings, he presents some of the basic concepts of his labor theory of value, which recognizes that use-value (i.e., features of a commodity useful to humans) is produced by a combination of human labor, soil (i.e., natural resources), and instruments (i.e., technologies) of production. In this, Marx shows his fundamental materialism, which recognizes nature as central to all value. However, Marx also argued that socio-environmental relationships can take various forms and that current conditions were not unalterable. He critiques Malthus' assumption that food production can only increase arithmetically, recognizing that advances in soil science, technology, and human labor can dramatically expand agricultural productivity.

In the second reading selection, Marx highlighted how problems with food production, poverty, and soil degradation were due to the structure of capitalism rather than determined by fixed laws: "all progress in capitalistic agriculture is a progress in the art, not only of robbing the labourer, but of robbing the soil" (reading 8). He saw how social and environmental problems were interconnected and could not be adequately addressed separately, as shown in his observation that capitalism saps the two sources of all wealth, soil, and labor, through its rapacious exploitation of both. This recognition can be seen in our era in the efforts of activists and scholars to link social justice movements with radical environmentalism (Pellow 2014). More broadly, there is growing recognition in both the natural and social sciences that ruthless competition, seen as natural by the classical economists, is not the only or most important form of interaction among individuals or species. Advances in our understanding of evolutionary and ecological processes have shown that, although competition is undoubtedly a feature of the natural world, cooperation among individuals and among species is also central to survival (Nowak and Highfield 2012). Mutualisms, where species interact in interdependent ways that may advance their mutual fitness, is ubiquitous in nature and serves as a key driver of ecological–evolutionary processes (Douglas 2014).

Romanticism and Materialism

Although the recognition of natural laws was revolutionary and a key part of developing modern scientific worldviews, many thinkers, including Newton, did not dispense with a belief in divine forces but rather saw natural laws as part of an all-knowing God's design. Going beyond most other scientists and philosophers of his time, including Humboldt, Darwin dispensed with any notion of divine order or plan, accepting that there was no God, no purpose, and no inherent meaning in the universe. In this, Darwin adhered to the tenets of philosophical materialism, the view that the world can be understood in terms of itself without the invocation of supernatural forces.

Alexander von Humboldt (1769–1859) was an internationally famous scientist in his lifetime and for generations afterward. He was renowned for *Kosmos* (published in five volumes between 1845 and 1862, the last volume published posthumously based on his notes)*,* one of the most influential works of popular science ever produced. In this great work, he laid out his view of the interconnections across all scales in the natural world and the unity and harmony of nature, a view that underlies all of his work but is only dimly apparent in the one reading included here. The foundation of his fame and influence began in 1799, when he set out for South America on one of the greatest scientific journeys in history. On his five-year expedition, he traversed the continent's mountains, forests, and rivers; collected thousands of specimens of plants, animals, and minerals; documented and lamented environmental destruction (as can be seen in his contribution in this volume); conducted experiments and measurements of all sorts; studied meteorology; and marveled at the magnificence of the natural world. He was a pioneer of global physical geography who made major contributions across many scientific fields (Wulf 2015). After his return to Europe, he published 34 volumes of his travel journal.

In the selection from his journal provided here, Humboldt's keen analytical mind is on display, as he presents a well-reasoned case for how humans, through deforestation and agriculture, affected the water levels of Lake Valencia in Venezuela, by altering evaporative processes. His specific example has general implications in that it shows how human activity can affect the hydrologic cycle, the atmosphere, and the local climate. These insights are central to environmental science and foreshadow our contemporary concerns about global climate change. Thus, Humboldt was an intellectual pioneer in recognizing that society and the environment are systematically interconnected, rather than existing independently from one another. His work was not only influential in the natural sciences but was central to the development of environmentalism and environmental ethics, since his love of the natural world and advocacy for its protection permeated his work.

Charles Darwin (1809–1882), a gentleman-naturalist, came from a wealthy family that included eminent physicians, natural philosophers, industrialists, and abolitionists. Having been an unsuccessful medical student at the University of Edinburgh, he moved to Christ's

College, Cambridge, where he prepared to become a country parson, which would have allowed him time to pursue his naturalist interests. In due course, he became inspired, in part by the adventures of Humboldt, to travel the world and be a scientist.

Famously, Darwin changed the way many people thought about the natural world when he published *On the Origin of Species,* which brought his theory of evolution by natural selection to the public 20 years after his voyage around the world on HMS *Beagle* (Browne 1995, 2002). Perhaps even more radical than the specifics of the theory were the implicit philosophical underpinnings. As is demonstrated in the selection included in this volume, one of Darwin's key insights was that blind material forces of variation and differential survival can explain how species adapt to shifting local environments and change over time without requiring any higher purpose. Integral to Darwin's theory is an understanding that organisms evolve through their interactions with other organisms in an ecological context, each individual being part of the web of life. Darwin drew on Malthus' insight that populations grow exponentially and therefore have the tendency to exceed the capacity of their resource bases, leading to a struggle among individuals for finite resources. Darwin argues that individuals with characteristics that help them to succeed in this competition are more likely than others to survive and produce offspring and, thereby, shape the composition of future generations. Darwin leads us to see that as a consequence of evolutionary processes and changes in the physical environment driven by geological and other forces, ecosystems are continuously changing, never being in an ideal, foreordained order. As Darwin noted in his closing passage to *Origin*, there was grandeur in his view of life. However, it was a grandeur that emerged from conflict and randomness, not a wise and benevolent plan. Thus, Darwin's view stood in contrast to the harmony and meaning seen in nature by previous generations.

It is important to recognize that Humboldt's view of nature was of a fundamentally different sort than the one Darwin ultimately developed (Gould 2002). Humboldt was in spirit allied with romanticism, a movement that sought to glorify nature and seek inspiration from it. To Humboldt, nature had an inherent order and meaning, whereas for Darwin, order in nature only emerged from dynamic struggles among individuals. Although Darwin's materialist-evolutionary thinking won the day in the scientific community, the tension between Humboldt's romantic view of the harmony of nature and Darwin's austere view of the ultimate meaninglessness of the universe is a divide in environmentalism and ecological thinking today. Many of the concepts and metaphors used in popular environmentalism – such as the emphasis on balance and harmony in the natural world of Lovelock's (1979) Gaia hypothesis – betray romantic sympathies that stand in tension with the Darwinian recognition that natural systems are not designed and are inevitably in flux (Foster 2000). Although Humboldt highlighted processes that led to changes in the environment, unlike Darwin, he saw anthropogenic change as a perturbance of the natural order that pervaded the universe. This tension has important implications for the ethics of restoration projects, since the Darwinian view suggests that which features of ecosystems should be preserved or restored cannot be decided based on cosmic principles (Lee et al. 2014).

George Perkins Marsh (1801–1882) was a Vermont lawyer and dedicated philologist (i.e., someone who studies the forms of, historical development of, and relationships among languages), who later in life was appointed by Abraham Lincoln to serve as a diplomat for the United States in the Kingdom of Italy. A man of wide-ranging interests, he, like Marx, Humboldt, and Darwin, recognized that the Earth had been transformed over time and that humans played a growing role in shaping it (Clark and Foster 2002). His great work, *Man and Nature,* was aimed at not only understanding how humans had altered the planet but also envisioning a way to protect and restore the environment. In reading 6, he provides a foundation for environmental history. He argues that the Roman Empire contributed to its own demise by degrading the environment upon which it depended through excessive exploitation and poor management of nature's bounty. Clearly inspired by Humboldt, Marsh notes how human actions can and have altered soils, ecosystems, and the climate. However, he also notes the potential for humans to restore nature and, thereby, improve ecological conditions. Like Humboldt, Marsh shows a love of the natural world and a desire to alter how humans interact with it. His view represented a shift away from the romantic notion that nature is large and powerful and humanity small and weak, to the Industrial Age view that humans are mighty and nature frail. In this, he seems very modern, both loving the natural world and seeing it as endangered and recognizing the potential for transformations, either good or bad. Marsh's inspiration can be seen today in efforts to protect and restore ecosystems, such as in E. O. Wilson's (2016) ambitious proposal to protect half the planet.

In his concern for how societies value, or fail to value, nature and his advocacy for caution about the long-term implications of our present actions, Marsh anticipated the precautionary principle (the idea that we should avoid potentially hazardous new technologies or products until they are shown to be safe) and present-day concerns about how we should value future generations and other species (Waring 1999; O'Brien 2000; York, Clark, and Foster 2009). Marsh warns that we are "breaking up the floor and wainscoting and doors and window frames of our dwelling, for fuel to warm our bodies and seethe our pottage, and the world cannot afford to wait til the slow and sure progress of exact science has taught it a better economy" (reading 6).

Is a Better World Possible?

All of the works considered here were not only scientifically important; they also had serious political and ethical implications. Indeed, most of them were written in part to promote political ideas. In a very general sense, these writers give different answers to whether it is possible to improve the human condition. Malthus and Hong are clearly pessimists, suggesting that the problem they separately outlined, where human population will be limited by misery, is inevitable. This fit with a political stance against government support for the poor in the case of Malthus and a recognition of the limits of government intervention in the case of Hong. Ricardo, like Malthus and Hong, believed that suffering was the human lot (at least for the underclass) and that little could be done to alleviate it.

Common sense, of course, tells us that Hong and Malthus were indeed correct that geometric population growth cannot continue forever on our finite planet. However, neither of them anticipated the potential for humans to reduce their fertility by humane means. In our era, safe and effective birth control methods are both widely available and desired, and fertility rates have fallen below replacement in many regions – in many places primarily due to the free choice of individuals rather than due to coercion (although there definitely have been ugly examples of the latter) (Cohen 1995). Thus, it appears clear that the inevitable end to population growth can come about from changes in people's reproductive behaviors in new social-historical contexts rather than a gruesome crash from rising mortality.

Marsh, who decried environmental destruction, took the optimistic view that through changing socio-environmental relationships collapse was not inevitable, a debate still active today (e.g., Steffen 2007, but see Strunz et al. 2019). He was a pioneer of restoration ecology and envisioned a future in which humanity understood ecological processes and worked to live in a manner that did not undermine its means of subsistence. Humboldt, steeped in the optimism of the Enlightenment, was also a firm believer in the potential for societal improvements through progress in science and ethical thinking.

Similar to Marsh and Humboldt, Marx saw not only much that was awful in the world but also much that was good and believed firmly that a better world was possible. However, unlike many Enlightenment era thinkers, including Humboldt, he saw no natural tendency to progress. Rather, he recognized that change is unavoidable and that social improvements are possible through struggle. Nature may restrain us, but it does not determine our future; we make our own world if not in conditions of our choosing. Thus, Marx was not an unrealistic utopian but nonetheless saw hope that through collective action it was possible to change the human condition for the better. The Marxian view lets us see that many of our socio-environmental conundrums, such as the Jevons paradox, are not due to fundamental laws but rather are part of the historically particular dynamics of capitalism (Clark and Foster 2001; York and McGee 2016). Modern advocates for radical ecology (e.g., Merchant 1992, 2013) and ecological socialism (e.g., Klein 2014) follow the Marxian tradition, arguing that humans can resolve our environmental crises, but that this requires fundamental political-economic changes, contrasting with both (market-focused) technological optimists and pessimistic neo-Malthusians.

The question of whether there is hope for a better future looms over us today as the climate crisis, the unfolding mass extinction, and the ravages of infectious diseases threaten humanity. The hard-nosed recognition of limits that Hong, Malthus, Ricardo, Darwin, and Jevons dimly saw need not mean that humanity is doomed to a future of misery, although time is growing short for societies to make needed changes if we are to avoid the worst potential outcomes of our current environmental threats. Marx, Humboldt, and Marsh each envisioned the potential, through rational, scientific understanding of our world, to transform socio-environmental relationships so that we need not go over the brink.

References

Browne, Janet. *Charles Darwin: Voyaging*. Princeton: Princeton University Press, 1995.
Browne, Janet. *Charles Darwin: The Power of Place*. New York: Knopf, 2002.

Clark, Brett, and John Bellamy Foster. "William Stanley Jevons and *The Coal Question*: An Introduction to Jevons's 'Of the Economy of Fuel'." *Organization & Environment* 14, no. 1 (2001): 93–98.

Clark, Brett, and John Bellamy Foster. "George Perkins Marsh and the Transformation of Earth: An Introduction to Marsh's *Man and Nature*." *Organization & Environment* 15, no. 2 (2002): 164–169.

Cohen, Joel E. *How Many People Can the Earth Support?* New York: W.W. Norton, 1995.

Collier, Paul, and Anke Hoeffler. "Resource Rents, Governance, and Conflict." *Journal of Conflict Resolution* 49, no. 4 (2005): 625–633.

Douglas, Angela E. *The Symbiotic Habit*. Princeton: Princeton University Press, 2014.

Ehrlich, Paul R. *The Population Bomb*. San Francisco: Sierra Club Books, 1968.

Foster, John Bellamy. *Marx's Ecology: Materialism and Nature*. New York: Monthly Review Press, 2000.

Gillingham, Kenneth, Matthew J. Kotchen, David S. Rapson, and Gernot Wagner "The Rebound Effect Is Overplayed." *Nature* 493 (2013): 475–6.

Gould, Stephen Jay. "Art Meets Science in *The Heart of the Andes:* Church Paints, Humboldt Dies, Darwin Writes and Nature Blinks in the Fateful Year of 1859." In *I Have Landed: The End of a Beginning in Natural History*, edited by Stephen Jay Gould, 90–109. New York: Harmony Books, 2002.

Hawken, Paul, Amory Lovins, and L. Hunter Lovins. *Natural Capitalism: Creating the Next Industrial Revolution*. New York: Little, Brown and Company, 1999.

Klein, Naomi. *This Changes Everything: Capitalism vs. the Climate*. New York: Simon & Schuster, 2014.

Lee, Alex, Adam Pérou Hermans, and Benjamin Hale. "Restoration, Obligation, and the Baseline Problem." *Environmental Ethics* 36, no. 2 (2014): 171–186.

Lomborg, Bjorn. *The Skeptical Environmentalist: Measuring the Real State of the World*. Cambridge: Cambridge University Press, 2001.

Lovelock, James. *Gaia: A New Look at Life on Earth*. New York: Oxford University Press, 1979.

Merchant, Carolyn. *Radical Ecology: The Search for a Livable World*. London: Routledge, 1992.

Merchant, Carolyn. *Reinventing Eden: The Fate of Nature in Western Culture*. London: Routledge, 2013.

Merton, Robert K. *The Sociology of Science: Theoretical and Empirical Investigations*. Chicago: University of Chicago Press, 1979.

Norgaard, Kari M. *Salmon and Acorns Feed Our People: Colonialism, Nature, and Social Action*. New Brunswick: Rutgers University Press, 2019.

Nowak, Martin A., and Roger Highfield. *SuperCooperators: Altruism, Evolution, and Why We Need Each Other to Succeed*. New York: Free Press, 2012.

O'Brien, Mary. *Making Better Environmental Decisions: An Alternative to Risk Assessment*. Cambridge: MIT Press, 2000.

Peet, Richard, Paul Robbins, and Michael Watts, eds. *Global Political Ecology*. London: Routledge, 2011.

Pellow, David N. *Total Liberation: The Power and Promise of Animal Rights and the Radical Earth Movement*. Minneapolis: University of Minnesota Press, 2014.

Polimeni, John M., Kozo Mayumi, Mario Giampietro, and Blake Alcott. *The Jevons Paradox and the Myth of Resource Efficiency Improvements*. London: Earthscan, 2008.

Simon, Julian L. *The Ultimate Resource*. Princeton: Princeton University Press, 1981.

Steffen, Will. *Sustainability or Collapse? An Integrated History and Future of People on Earth*. Cambridge: MIT Press, 2007.

Strunz, Sebastian, Melissa Marselle, and Matthias Schröter. "Leaving the 'Sustainability or Collapse' Narrative Behind." *Sustainability Science* 14, no. 6 (2019): 1717–1728.

Waring, Marilyn. *Counting for Nothing: What Men Value and What Women Are Worth*. 2nd ed. Toronto: University of Toronto Press, 1999.

Wilson, E. O. *Half-Earth: Our Planet's Fight for Life*. New York: Liveright, 2016.

Wulf, Andrea. *The Invention of Nature: Alexander von Humboldt's New World*. New York: Knopf, 2015.

York, Richard, Brett Clark, and John Bellamy Foster. "Capitalism in Wonderland." *Monthly Review* 61, no. 1 (2009): 1–18.

York, Richard, and Julius Alexander McGee. "Understanding the Jevons Paradox." *Environmental Sociology* 2, no. 1 (2016): 77–87.

CHINA'S POPULATION PROBLEM

There has never been a people who did not delight in living under peaceful rule, nor a people not happy about living under peaceful rule that has lasted for a long time. Peaceful rule that lasts more than one hundred years is considered to have lasted a long time. But in the matter of population, it may be noted that today's population is five times as large as that of thirty years ago, ten times as large as that of sixty years ago, and not less than twenty times as large as that of one hundred years ago. Take, for example, a family that at the time of the great-great-grandfather and the great-grandfather was in possession of a ten-room house and one hundred *mou* of farmland. After the man married there were at first only the two of them; they lived in the ten-room house and upon the one hundred *mou* of land, and their resources were more than ample. Assuming that they had three sons, by the time the sons grew up, all three sons as well as the father had wives; there were a total of eight persons. Eight persons would require the help of hired servants; there would be, say, ten persons in the household. With the ten-room house and the one hundred *mou* of farmland, I believe they would have just enough space to live in and food to eat, although barely enough. In time, however, there will be grandsons, who, in turn, will marry. The aged members of the household will pass away, but there could still be more than twenty persons in the family. With more than twenty persons sharing a ten-room house and working on one hundred *mou* of farmland, I am sure that even if they eat very frugally and live in crowded quarters, their needs will not be met. Moreover, there will be great-grandchildren and great-great-grandchildren — the total number in a household will be fifty or sixty times that in the great-great-grandfather's or great-grandfather's time. For every household at the time of the great-grandfather, there will be at least ten households at the time of the great-grandson and great-great-grandson. There are families whose population has declined, but there are also lineages whose male members have greatly multiplied, compensating for the cases of decline.

Someone may say that at the time of the great-grandfather and great-great-grandfather, not all uncultivated land had been reclaimed and not all vacancies in housing available on the market had been filled. However, the amount [of available farmland and housing] has only doubled or, at the most, increased three to five times, while the population has grown ten to twenty times. Thus farmland and houses are always in short supply, while there is always a surplus of households and population. Furthermore, there are families who [have bought up or otherwise] appropriated other people's property — one person owning the houses of more than a hundred, one household occupying the farmland of a hundred households. No wonder, then, that everywhere there are people who have died from exposure to windstorm, rain, and frost, or from hunger and cold and the hardships of homelessness.

Question: Do Heaven-and-earth have a way of dealing with this situation? Answer: Heaven-and-earth's way of making adjustments lies in flood, drought, and plagues [which reduce the population]. However, people who unfortunately succumb to flood, drought, and plagues are no more than 10 or 20 percent of the total population.

Question: Do the ruler and his ministers have a way of dealing with this situation? Answer: The ruler and the ministers may make adjustments in the following ways: pursuing policies to ensure that no farmland will remain unused and that there will be no surplus labor. Migration of farmers to newly reclaimed land may be organized; heavy taxes may be reduced after a comparison is made between past and present tax rates. Extravagance in consumption may be prohibited; the wealthy household's appropriation of the property of others may be suppressed. Should there be floods, drought, and plagues, grain in the granaries may be made available, and all the funds in the government treasury may be used for relief — these are all that the ruler and his ministers can do in the way of adjustments between population and productive land.

In a word, after a long period of peaceful rule, Heaven-and-earth cannot stop the people from reproducing. Yet the resources with which Heaven-and-earth nourish the people are finite. After a period of peaceful rule, the ruler and the ministers cannot stop the people from reproducing, yet what the ruler and the ministers can do for the people is limited to the policies enumerated above. Among ten youths in a family, there are always one or two who resist being educated. Among the idle people in all the empire, how can it be expected that all will accept control from above? The housing for one person is inadequate for the needs of ten persons; how can it be sufficient for a hundred persons? The food for one person is inadequate for ten persons; how can it be sufficient for a hundred persons? This is why I am worried about peaceful rule.

[Hong, *Yiyan*, in *Juanshi geji* 1:8a–9b — KCL]

CHAP. II.

*The different ratios in which population and food in-
creafe.—The neceffary effects of thefe different ratios
of increafe.—Ofcillation produced by them in the condition of the lower claffes of fociety.—Reafons why
this ofcillation has not been fo much obferved as might
be expected.—Three propofitions on which the general
argument of the effay depends.—The different ftates
in which mankind have been known to exift propofed
to be examined with reference to thefe three propofitions.*

I SAID that population, when unchecked, increafed in a geometrical
ratio; and fubfiftence for man in an
arithmetical ratio.

Let us examine whether this pofition be juft.

I think it will be allowed, that no
ftate has hitherto exifted (at leaft that
we have any account of) where the
manners were fo pure and fimple, and
the means of fubfiftence fo abundant,
that no check whatever has exifted to
early marriages; among the lower claffes,
from a fear of not providing well for
their families; or among the higher
claffes, from a fear of lowering their condition in life. Confequently in no ftate
that we have yet known, has the power
of population been left to exert itfelf
with perfect freedom.

Whether the law of marriage be in-
ftituted, or not, the dictate of nature
and virtue, feems to be an early attachment to one woman. Suppofing a liberty
of changing in the cafe of an unfortunate choice, this liberty would not affect
population till it arofe to a height greatly
vicious; and we are now fuppofing the
exiftence of a fociety where vice is
fcarcely known.

In a ftate therefore of great equality and virtue, where pure and fimple manners prevailed, and where the means of fubfiftence were fo abundant, that no part of the fociety could have any fears about providing amply for a family, the power of population being left to exert itfelf unchecked, the increafe of the human fpecies would evidently be much greater than any increafe that has been hitherto known.

In the United States of America, where the means of fubfiftence have been more ample, the manners of the people more pure, and confequently the checks to early marriages fewer, than in any of the modern ftates of Europe, the population has been found to double itfelf in twenty-five years.

This ratio of increafe, though fhort of the utmoft power of population, yet as the refult of actual experience, we will take as our rule ; and fay,

That population, when unchecked, goes on doubling itfelf every twenty-five years, or increafes in a geometrical ratio.

Let us now take any fpot of earth, this Ifland for inftance, and fee in what ratio the fubfiftence it affords can be fuppofed to increafe. We will begin with it under its prefent ftate of cultivation.

If I allow that by the beft pofsible policy, by breaking up more land, and by great encouragements to agriculture, the produce of this Ifland may be doubled in the firft twenty-five years, I think it will be allowing as much as any perfon can well demand.

In the next twenty-five years, it is impofsible to fuppofe that the produce could be quadrupled. It would be contrary to all our knowledge of the qualities of land. The very utmoft that we can conceive, is, that the increafe in the fecond twenty-five years might equal the prefent produce. Let us then take this for our rule, though certainly far beyond the truth; and allow that by great exertion, the whole produce of the Ifland might be increafed every twenty-five years, by a quantity of fubfiftence equal to what it at prefent produces. The moft enthufiaftic fpeculator cannot fuppofe a greater increafe than this. In a few centuries it would make every acre of land in the Ifland like a garden.

Yet this ratio of increafe is evidently arithmetical.

It may be fairly faid, therefore, that the means of fubfiftence increafe in an arithmetical ratio.

. Let us now bring the effects of thefe two ratios together.

The population of the Ifland is computed to be about feven millions; and we will fuppofe the prefent produce equal to the fupport of fuch a number. In the firft twenty-five years the population would be fourteen millions; and the food being alfo doubled, the means of fubfiftence would be equal to this increafe. In the next twenty-five years the population would be twenty-eight millions; and the means of fubfiftence only equal to the fupport of twenty-one millions. In the next period, the population would be fifty-fix millions, and the means of fubfiftence juft fufficient for

half that number. And at the conclu-
fion of the firft century, the population
would be one hundred and twelve millions, and the means of fubfiftence only
equal to the fupport of thirty-five millions; which would leave a population of
feventy-feven millions totally unprovided
for.

A great emigration neceffarily implies
unhappinefs of fome kind or other in the
country that is deferted. For few per-
fons will leave their families, connections, friends, and native land, to feek
a fettlement in untried foreign climes,
without fome ftrong fubfifting caufes of
uneafinefs where they are, or the hope
of fome great advantages in the place to
which they are going.

But to make the argument more general, and lefs interrupted by the partial views of emigration, let us take the
whole earth, inftead of one fpot, and
fuppofe that the reftraints to population
were univerfally removed. If the fubfiftence for man that the earth affords
was to be increafed every twenty-five
years by a quantity equal to what the
whole world at prefent produces; this
would allow the power of production in
the earth to be abfolutely unlimited, and
its ratio of increafe much greater than
we can conceive that any pofsible exertions of mankind could make it.

Taking the population of the world at
any number, a thoufand millions, for
inftance, the human fpecies would increafe in the ratio of—1, 2, 4, 8, 16,
32, 64, 128, 256, 512, &c. and fubfiftence as—1, 2, 3, 4, 5, 6, 7, 8, 9,
10, &c. In two centuries and a quarter,
the population would be to the means of

subfiftence as 512 to 10 : in three centuries as 4096 to 13 ; and in two thoufand years the difference would be almoft
incalculable, though the produce in that
time would have increafed to an immenfe
extent.

No limits whatever are placed to the
productions of the earth ; they may increafe for ever and be greater than any
afsignable quantity ; yet ftill the power
of population being a power of a fuperior
order, the increafe of the human fpecies
can only be kept commenfurate to the
increafe of the means of fubfiftence, by
the conftant operation of the ftrong law
of necefsity acting as a check upon the
greater power.

The effects of this check remain now
to be confidered.

Among plants and animals the view of
the fubject is fimple. They are all impelled by a powerful inftinct to the increafe of their fpecies ; and this inftinct
is interrupted by no reafoning, or doubts
about providing for their offspring.
Wherever therefore there is liberty, the
power of increafe is exerted ; and the
fuperabundant effects are repreffed afterwards by want of room and nourifhment,
which is common to animals and plants ;
and among animals, by becoming the
prey of others.

The effects of this check on man are
more complicated.

Impelled to the increafe of his fpecies
by an equally powerful inftinct, reafon
interrupts his career, and afks him
whether he may not bring beings into
the world, for whom he cannot provide

the means of fubfiftence. In a ftate of equality, this would be the fimple queftion. In the prefent ftate of fociety, other confiderations occur. Will he not lower his rank in life? Will he not fubject himfelf to greater difficulties than he at prefent feels? Will he not be obliged to labour harder? and if he has a large family, will his utmoft exertions enable him to fupport them? May he not fee his offspring in rags and mifery, and clamouring for bread that he cannot give them? And may he not be reduced to the grating necefity of forfeiting his independence, and of being obliged to the fparing hand of charity for fupport?

Thefe confiderations are calculated to prevent, and certainly do prevent, a very great number in all civilized nations from purfuing the dictate of nature in an early attachment to one woman. And this reftraint almoft neceffarily, though not abfolutely fo, produces vice. Yet in all focieties, even thofe that are moft vicious, the tendency to a virtuous attachment is fo ftrong, that there is a conftant effort towards an increafe of population. This conftant effort as conftantly tends to fubject the lower claffes of the fociety to diftrefs, and to prevent any great permanent amelioration of their condition.

The way in which thefe effects are produced feems to be this.

We will fuppofe the means of fubfiftence in any country juft equal to the eafy fupport of its inhabitants. The conftant effort towards population, which is found to act even in the moft vicious focieties, increafes the number of people before the means of fubfiftence are increafed. The food therefore which before fupported feven millions, muft now be divided among feven millions and a half or eight millions. The poor confequently muft live much worfe, and many of them be reduced to fevere diftrefs. The number of labourers alfo being above the proportion of the work in the market, the price of labour muft tend toward a decreafe ; while the price of provifions would at the fame time tend to rife. The labourer therefore muft work harder to earn the fame as he did before. During this feafon of diftrefs, the difcouragements to marriage, and the difficulty of rearing a family are fo great, that population is at a ftand. In the mean time the cheapnefs of labour, the plenty of labourers, and the necefsity of an increafed induftry amongft them, encourage cultivators to employ more labour upon their land ; to turn up frefh foil, and to

manure and improve more completely what is already in tillage; till ultimately the means of fubfiftence become in the fame proportion to the population as at the period from which we fet out. The fituation of the labourer being then again tolerably comfortable, the reftraints to population are in fome degree loofened ; and the fame retrograde and progrefsive movements with refpect to happinefs are repeated.

This fort of ofcillation will not be remarked by fuperficial obfervers ; and it may be difficult even for the moft penetrating mind to calculate its periods. Yet that in all old ftates fome fuch vibration does exift ; though from various tranfverfe caufes, in a much lefs marked, and in a much more irregular manner than I have defcribed it, no reflecting man who confiders the fubject deeply can well doubt.

Many reafons occur why this ofcillation has been lefs obvious, and lefs decidedly confirmed by experience, than might naturally be expected.

One principal reafon is, that the hiftories of mankind that we poffefs, are hiftories only of the higher claffes. We have but few accounts that can be depended upon of the manners and cuftoms of that part of mankind, where thefe retrograde and progrefsive movements chiefly take place. A fatisfactory hiftory of this kind, of one people, and of one period, would require the conftant and minute attention of an obferving mind during a long life. Some of the objects of enquiry would be, in what proportion to the number of adults was the number of marriages: to what extent vicious cuftoms prevailed in confequence of the reftraints upon matrimony: what was the comparative mortality among the children of the moft diftreffed part of the community, and thofe who lived rather more at their cafe: what were the variations in the real price of labour: and what were the obfervable differences in the ftate of the lower claffes of fociety, with refpect to eafe and happinefs, at different times during a certain period.

Such a hiftory would tend greatly to elucidate the manner in which the conftant check upon population acts; and would probably prove the exiftence of the retrograde and progrefsive movements that have been mentioned; though the times of their vibration muft neceffarily be rendered irregular, from the operation of many interrupting caufes; fuch as, the introduction or failure of certain manufactures: a greater or lefs prevalent fpirit of agricultural enterprize: years of plenty, or years of fcarcity: wars and peftilence: poor laws: the invention of proceffes for fhortening labour without the proportional extenfion of the market for the commodity: and, particularly, the difference between the nominal and real price of labour; a circumftance, which has perhaps more than any other, contributed to conceal this ofcillation from common view.

It very rarely happens that the nominal price of labour univerfally falls; but we well know that it frequently remains the fame, while the nominal price of provifions has been gradually increafing. This is, in effect, a real fall in the price of labour; and during this period, the condition of the lower orders of the community muft gradually grow worfe and worfe. But the farmers and capitalifts are growing rich from the real cheapnefs of labour. Their increafed capitals enable them to employ a greater number of men. Work therefore may be plentiful; and the price of labour would confequently rife. But the want of freedom in the market of labour, which occurs more or lefs in all communities, either from parifh laws, or the more general caufe of the facility of combination among the rich, and its difficulty among the poor, operates to prevent the price of labour from rifing at the natural period, and keeps it down fome time longer; perhaps, till a year of fcarcity, when the clamour is too loud, and the necessity too apparent to be refifted.

The true caufe of the advance in the price of labour is thus concealed ; and the rich affect to grant it as an act of compafsion and favour to the poor, in confideration of a year of fcarcity ; and when plenty returns, indulge themfelves in the moft unreafonable of all complaints, that the price does not again fall ; when a little reflection would fhew them, that it muft have rifen long before, but from an unjuft confpiracy of their own.

But though the rich by unfair combinations, contribute frequently to prolong a feafon of diftrefs among the poor ; yet no pofsible form of fociety could prevent the almoft conftant action of mifery, upon a great part of mankind, if in a ftate of inequality, and upon all, if all were equal.

The theory, on which the truth of this pofition depends, appears to me fo extremely clear ; that I feel at a lofs to conjecture what part of it can be denied.

That population cannot increafe without the means of fubfiftence, is a propofition fo evident, that it needs no illuftration.

That population does invariably increafe, where there are the means of fubfiftence, the hiftory of every people that have ever exifted will abundantly prove.

And, that the fuperior power of population cannot be checked, without producing mifery or vice, the ample portion of thefe too bitter ingredients in the cup of human life, and the continuance of the phyfical caufes that feem to have produced them, bear too convincing a teftimony.

But in order more fully to afcertain the validity of thefe three propofitions, let us examine the different ftates in which mankind have been known to exift. Even a curfory review will, I think, be fufficient to convince us, that thefe propofitions are incontrovertible truths.

CHAPTER XVI.

Lake of Tacarigua.—Hot Springs of Mariara. —Town of Nueva Valencia de el Rey. — Descent toward the coasts of Porto Cabello.

THE valleys of Aragua, of which we have displayed the rich cultivation and the admirable fecundity, form a narrow basin between granitic and calcareous mountains of unequal height. On the North, they are separated by the Sierra Mariara from the seacoast; and toward the South, the chain of Guacimo and Yusma serves them as a rampart against the heated air of the steppes. Groups of hills, high enough to determine the course of the waters, close this basin on the East and West, like transverse dikes. We find these hills between the Tuy and La Victoria *, as well as on the road from

Valencia to Nirgua, and at the mountains of Torito. From this extraordinary configuration of the land, the little rivers of the valleys of Aragua form a peculiar system, and direct their course toward a basin closed on all sides. These rivers do not bear their waters to the ocean; they are collected in an interior lake, and, subject to the powerful influence of evaporation, they lose themselves, if we may use the expression, in the atmosphere. On the existence of these rivers and lakes the fertility of the soil, and the produce of cultivation in these valleys, depend. The aspect of the spot, and the experience of half a century have proved, that the level of the waters is not invariable; the waste by evaporation, and the increase from the waters running into the lake, do not uninterruptedly balance each other. The lake, being elevated one thousand feet above the neighbouring steppes of Calabozo, and one thousand three hundred and thirty-two feet above the level of the ocean, it has been suspected, that

---

* The lofty mountains of Los Teques, which give birth to the Tuy, may be looked upon as the eastern boundary of the valleys of Aragua. The level of the ground continues in fact to rise from La Victoria (269 t.) to the Hacienda de Tuy (295 t.) : but the river Tuy, turning South toward the Sierras of Guairaima and Tiara, has found an issue on the East; and it is more natural to consider as the limits of the basin of Aragua a line drawn through the sources of the streams flowing into the lake of Valencia. The charts and sections I have traced of the road from Caraccas to Nueva Valencia, and from Porto Cabello to Villa de Cura, exhibit the whole of these geological relations.

there are subterraneous communications and filtrations. The appearance of new islands, and the gradual retreat of the waters, have led to the belief, that the lake may perhaps become entirely dry. An assemblage of physical circumstances so remarkable was well fitted to fix my attention on those valleys, where the wild beauty of nature is embellished by agricultural industry, and the arts of rising civilization.

The lake of Valencia, called Tacarigua* by the Indians, exceeds in magnitude the lake of Neufchatel in Switzerland; but it's general form has more resemblance to the lake of Geneva, which is nearly at the same height above the level of the sea. The slope of the ground in the valleys of Aragua tending toward the South and the West, that part of the basin, which has remained covered with water, is the nearest to the southern chain of the mountains of Guigue, of Yusma, and of Guacimo, which stretch toward the high savannahs of Ocumare. The opposite banks of the lake of Valencia display a singular contrast; those on the South are desert, and almost uninhabited, and a screen of high mountains gives them a gloomy and monotonous aspect. The northern shore, on the contrary, is cheerful, pastoral, and decked

with the rich cultivation of the sugar-cane, coffee-tree, and cotton. Paths, bordered with cestrums, azedaracs, and other shrubs, always in flower, cross the plain, and join the scattered farms. Every house is surrounded by clumps of trees. The ceiba with it's large yellow * flowers gives a peculiar character to the landscape, mingling it's branches with those of the purple erithryna. This mixture of vivid vegetable colours contrasts with the uniform tint of an unclouded sky. In the season of drought, where the burning soil is covered with an undulating vapour, artificial irrigations preserve the verdure and fertility. Here and there the granitic rock pierces through the cultivated ground. Enormous stony masses rise abruptly in the midst of the valley. Bare and forked, they nourish a few succulent plants, which prepare mould for future ages. Often at the summit of these lonely hills a fig-tree or a clusia with fleshy leaves has fixed it's roots in the rock, and towers over the landscape. With their dead and withered branches they look like signals erected on a steep cliff. The form of these mounts betrays the secret of their ancient origin; for, when the whole of this valley was filled with water, and the waves beat at the foot of the peaks of Mariara, the Devil's Wall †, and the

---

* Fray Pedro Simon calls the lake, no doubt by mistake, Acarigua aud Tarigua. (Notic. Hist., p. 533 and 668.)

* *Carnes tollendas ;* bombax *hibiscifolius.*
† *El Rincon del Diablo.*

chain of the coast, these rocky hills were shoals or islets.

These features of a rich landscape, these contrasts between the two banks of the lake of Valencia, often reminded me of the situations of the Pays de Vaud, " where the earth, every where cultivated, and every where fertile, offers the husbandman, the shepherd, and the vinedresser, the secure fruit of their labours," while the opposite side of Chablais presents only a mountainous and half-desert country. In these distant climes, surrounded with the productions of an exotic nature, I loved to recall to mind the enchanting descriptions, with which the aspect of the Leman lake and the rocks of La Meillerie inspired a great writer. Now, while in the centre of civilized Europe, I endeavour in my turn to paint the scenes of the New World, I do not imagine I present the reader with clearer images, or more precise ideas, by comparing our landscapes with those of the equinoctial regions. It cannot be too often repeated, that Nature, under every zone, whether wild or cultivated, smiling or majestic, displays an individual character. The impressions, which she excites, are infinitely varied, like the emotions produced by works of genius, according to the age in which they were conceived, and the diversity of languages from which they derive a part of their charms. We can justly compare only what belongs to dimensions and external forms. We may institute a parallel between the colossal summit of Mount Blanc, and the mountains of Himalaya; the cascades of the Pyrenees and those of the Cordilleras: but these comparisons, useful with respect to science, fail to make known what characterises Nature in the temperate and torrid zones. On the banks of a lake, in a vast forest, at the foot of summits covered with eternal snows, it is not the simple magnitude of the objects, that penetrates us with secret admiration. What speaks to the soul, what causes such profound and various emotions, escapes our measurements, as it does the forms of language. Those who feel powerfully the charms of Nature fear to weaken their enjoyments by comparing scenes of a different character.

But it is not alone the picturesque beauties of the lake of Valencia, that have given celebrity to it's banks. This basin displays several other phenomena, the solution of which is interesting alike to physical science and to the well-being of the inhabitants. What are the causes of the diminution of the waters of the lake? Is this diminution more rapid now than in former ages? Can we presume, that an equilibrium between the waters flowing in and the loss will be shortly reëstablished? or may we apprehend, that the lake will entirely disappear?

According to astronomical observations * made at La Victoria, Hacienda de Cura, Nueva Valencia, and Guigue, the length of the lake, in it's present state, from Cagua to Guayos, is ten leagues, or twenty-eight thousand eight hundred toises. It's breadth is very unequal. If we judge from the latitudes of the mouth of the Rio Cura and the village of Guigue, it no where surpasses 2·3 leagues, or six thousand five hundred toises; most commonly it is but four or five miles. The dimensions resulting from my observations are much less, than those hitherto adopted by the natives †. It might be thought, that, to form a precise idea of the progressive diminution of the waters, it would suffice to compare the present dimensions of the lake with those attributed to it by ancient chroniclers; by Oviedo, for instance, in his History of the Province of Venezuela, published about the year 1723. This writer, in his emphatic style, gives " this interior sea, this *monstruoso cuerpo de la laguna de Valencia*," fourteen leagues in length and six in breadth. He relates,

---

* The itinerary distances from La Victoria to Cagua, as well as those from Guacara to Mocundo and to Los Guayos, were taken into consideration. Angles were taken at the island of Cura, at Cabo Blanco, and at Mocundo.

† *Depons, Voyage à La Terre Ferme*, vol. i, p. 138.

---

that at a small distance from the shore, the lead finds no bottom; and that large floating islands cover the surface of the waters, which are constantly agitated by the winds *. No importance can be attached to estimations, which, without being founded on any measurement, are expressed in leagues, *leguas*, reckoned in the colonies at three thousand, five thousand, and six thousand six hundred and fifty *varas* †. What is worthy of our attention in the works of a man, who must so often have passed over the valleys of Aragua, is the assertion, that the town of *Nueva Valencia de el Rey* was built in 1555, at the distance of half a league from the lake ‡; and that the proportion between the length of

---

* Oviedo, p. 125.

† Seamen being the first, and for a long time the only persons, who introduced into the Spanish colonies any precise ideas on the astronomical position and distances of places, it was the *legua nautica* of 6650 *varas*, or of 2854 toises, 20 in a degree, that was originally used in Mexico and South America; but this *legua nautica* has been gradually reduced to one half or one third, on account of the slowness of travelling across steep mountains, or dry and burning plains. The common people measure only time directly; and then, by arbitrary hypotheses, infer from the time the space of ground travelled over. In the course of my geographical researches, I have had frequent opportunities of examining the real value of *leagues*, by comparing the itinerary distances between points lying under the same meridian with the difference of latitudes.

the lake, and it's breadth, is as seven to three. At present the town of Valencia is separated from the lake by level ground of more than two thousand seven hundred toises, which Oviedo would no doubt have estimated as a space of a league and a half; and the length of the basin of the lake is to it's breadth as 10 to 2·3, or as 7 to 1·6. The appearance of the soil between Valencia and Guigue, the little hills that rise abruptly in the plain, East of the Cano de Cambury, and some of which (el Islote and la Isla de la Negra, or Caratapona) have even preserved the name of islands, sufficiently prove, that the waters have retired considerably since the time of Oviedo. With respect to the change in the general form of the lake, it appears to me improbable, that in the seventeenth century it's breadth was nearly the half of it's length. The situation of the granitic mountains of Mariara and of Guigue, the slope of the ground, which rises more rapidly toward the North and South than toward the East and West, are alike repugnant to this supposition.

In treating the long-discussed question of the diminution of the waters, I conceive we must distinguish the different periods, at which the sinking of their level has taken place. Wherever we examine the valleys of rivers, or the basins of lakes, we see the ancient shore at great distances. No doubt seems now to be entertained, that our rivers and lakes have undergone immense diminutions ; but many geological facts remind us also, that these great changes in the distribution of the waters have preceded all historical times ; and that for many thousand years most lakes have attained a permanent equilibrium between the produce of the water flowing in, and that of evaporation and filtration. Whenever we find this equilibrium broken, it will be more prudent to examine, whether the rupture be not owing to causes merely local, and of a very recent date, than to admit an uninterrupted diminution of the water. This reasoning is conformable to the more circumspect method of modern science. At a time when the physical history of the world, traced by the genius of some eloquent writers, borrowed all it's charms from the fictions of imagination, a new proof would have been found, in the phenomenon of which we are treating, of the contrast that these writers were fond of establishing between the two continents. To demonstrate, that America rose later than Asia and Europe from the bosom of the waters, they would have cited the lake of Tacarigua as one of those interior basins, which have not had time to become dry by the effects of a slow and gradual evaporation. I have no doubt, that, in very remote times, the whole valley, from the foot of the mountains of

Cocuyza to those of Torito and Nirgua, and from La Sierra de Mariara to the chain of Guigue, of Guacimo, and La Palma, was filled with water. Every where the form of the promontories, and their steep declivities, seem to indicate the shore of an alpine lake, similar to those of Styria and Tyrol. The same little helicites, the same valvæ, which now live in the lake of Valencia, are found in layers of three or four feet in the island, as far as Turmero and *La Concesion* near La Victoria. These facts undoubtedly prove a retreat of the waters; but nothing indicates, that this retreat has continued from that remote period to our days. The valleys of Aragua are one of the parts of Venezuela the most anciently peopled; and yet there is no mention in Oviedo, or any other old chronicler, of a sensible diminution of the lake. Ought we simply to suppose, that this phenomenon escaped their observation, at a time when the Indian population far exceeded the white, and when the banks of the lake were less inhabited? Within half a century, and particularly within these thirty years, the natural desiccation of this great basin has excited general attention. We find vast spaces of land that were formerly inundated, now dry, and already cultivated with plantains, sugar-canes, or cotton. Wherever a hut is erected on the bank of the lake, we see the shore receding from year to year. We discover islands, which, in consequence of the retreat of the waters, scarcely begin to be joined to the continent, as the rocky island of Culebra, on the side of Guigue; other islands already form promontories, as the Morro, between Guigue and Nueva Valencia, and La Cabrera, South-East of Mariara; others now rise in the islands, like scattered hills. Among these last, so easily recognised at a distance, some are only a quarter of a mile, others a league from the present shore. I shall cite as the most remarkable three granitic islands, thirty or forty toises high, on the road from Hacienda de Cura to Aguas Calientes; and at the western extremity of the lake, the Serrito de Don Pedro, Islote, and Caratapona. On visiting two islands * entirely surrounded by water, we found in the midst of brush-wood, on small flats of four, six, and even eight toises height above the surface of the lake, fine sand mixed with helicites, anciently deposited by the waters. In each of these islands may be perceived the most certain traces of the gradual sinking of the waters. But still farther, and this accident is regarded by the inhabitants as a marvellous phenomenon,

* Isla de Cura, and Cabo Blanco. The promontory of Cabrera has been connected with the shore ever since the year 1750 or 1760, by a vale, which bears the name of Portachuelo.

in 1796, three new islands appeared to the East of the island Caiguira, in the same direction as the islands Burro, Otama, and Zorro. These new islands, called by the people *los nuevos Penones,* or *las Aparecidas,* form a kind of banks, with surfaces quite flat. They rose already, in 1800, more than a foot above the *mean level of the waters.*

We observed at the beginning of this chapter, that the lake of Valencia, like the lakes of the valley of Mexico * forms the centre of a little system of rivers, none of which have any communication with the ocean. These rivers for the greater part deserve only the name of torrents, or brooks †; they are twelve or fourteen in number. The inhabitants, little acquainted with the effects of evaporation, have long imagined, that the lake has a subterranean outlet, by which a quantity of water runs out equal to that which flows in by the rivers. Some suppose, that this outlet communicates with grottoes, which they place at great depths; others admit, that the water flows through an oblique channel into the basin of the ocean. These bold hypotheses on the communication

between two neighbouring basins have presented themselves under every zone to the imagination of the vulgar, as well as to that of natural philosophers; for the latter, without confessing it, sometimes repeat popular opinions in scientific language. We hear of subterranean gulfs and outlets in the New World, as on the shores of the Caspian sea, though the lake of Tacarigua is two hundred and twenty-two toises higher, and the Caspian sea fifty-four toises lower, than the ocean; and though it is well known, that fluids find the same level, when they communicate by a lateral channel.

The changes, which the destruction of forests, the clearing of plains, and the cultivation of indigo, have produced within half a century in the quantity of water flowing in on the one hand; and on the other the evaporation of the soil, and the dryness of the atmosphere, present causes sufficiently powerful to explain the successive diminution of the lake of Valencia. I am not of the opinion of a traveller, who has visited these countries since me *, that " to set

---

* Before the opening dug by the Spaniards near Huehuetoque, and known by the name of *Desague Real.*

† The following are their names: Rios de Aragua, Turmero, Maracay, Tapatapa, Aguas Calientes, Mariara, Cura, Guacara, Guataparo, Valencia, Cano grande de Cambury, &c.

---

* Mr. *Depons (Voyage à la Terre Ferme, vol.* i, *p.* 139) adds: " The small extent of the surface of the lake" (it amounts however to 106,500,000 square toises) " renders impossible the supposition, that evaporation alone, however considerable under the tropics, could remove as much water, as the rivers furnish." In the sequel, the author himself seems to abandon " this occult cause, the hypothesis of an aperture."

the mind at rest, and for the honour of science," a subterranean issue must be admitted. By felling the trees, that cover the tops and the sides of mountains, men in every climate prepare at once two calamities for future generations ; the want of fuel, and a scarcity of water. Trees, by the nature of their perspiration, and the radiation from their leaves in a sky without clouds, surround themselves with an atmosphere constantly cool and misty. They affect the copiousness of springs, not, as was long believed, by a peculiar attraction for the vapors diffused through the air, but because, by sheltering the soil from the direct action of the Sun, they diminish the evaporation of the water produced by rain. When forests are destroyed, as they are every where in America by the European planters, with an imprudent precipitation, the springs are entirely dried up, or become less abundant. The beds of the rivers, remaining dry during a part of the year, are converted into torrents, whenever great rains fall on the heights. The sward and moss disappearing with the brush-wood from the sides of the mountains, the waters falling in rain are no longer impeded in their course : and instead of slowly augmenting the level of the rivers by progressive filtrations, they furrow during heavy showers the sides of the hills, bear down the loosened soil, and form those sudden inundations, that devas-

tate the country. Hence it results, that the destruction of forests, the want of permanent springs, and the existence of torrents, are three phenomena closely connected together. Countries that are situate in opposite hemispheres, Lombardy bordered by the chain of the Alps, and Lower Peru inclosed between the Pacific Ocean and the Cordillera of the Andes, exhibit striking proofs of the justness of this assertion *.

Till the middle of the last century, the mountains that surround the valleys of Aragua were covered with forests. Great trees of the families of mimosa, ceiba, and the fig-tree, shaded and spread coolness along the banks of the lake. The plain, then thinly inhabited, was filled with brush-wood, interspersed with trunks of scattered trees and parasite plants, envelopped with a thick sward, less capable of emitting radiant caloric than the soil that is cultivated, and therefore not sheltered from the rays of the Sun. With the destruction of trees, and the increase of the cultivation of sugar, indigo, and cotton, the springs, and all the natural supplies of the lake of Valencia, have diminished from year to year. It is difficult to form a just idea of the enormous quantity of evaporation, that takes place under the torrid zone, in a valley

* See my political Essai on New Spain, vol. i, p. 208, and the *Recherches de M. de Prony sur les Crues du Pô.*

surrounded with steep declivities, where a regular breeze and descending currents of air are felt toward evening, and the bottom of which is flat, and looks as if it were levelled by the waters. We have elsewhere remarked, that the heat, which prevails throughout the year at Cura, Guacara, Nueva Valencia, and on the borders of the lake, is the same as that which is felt at midsummer in Naples and Sicily. The mean annual temperature of the valleys of Aragua is nearly 25·5° * ; my hygrometrical observations of the month of February, taking the mean of day and night, gave 71·4° of the hair-hygrometer †. As the words great drought and great humidity have no determinate signification, and an air, that would be called very dry in the lower regions of the tropics, would be regarded as a humid air in Europe, we can judge of these relations between climates only by comparing spots placed under the same zone. Now at Cumana, where it sometimes does not rain during a whole year, and where I had the means of collecting a great number of hygrometric observations made at different hours of the day and night, the mean humidity

* 20·4° Reaumur. According to the observations of the month of February, 19·5° R.; and, at Cumana, this month is 0·7° R. below the mean temperature of the year.

† These 71·4° of apparent humidity correspond to a mean temperature of 24·3°.

of the air is 86°; corresponding to the mean temperature of 27·7° cent. Taking into account the influence of the rainy months, that is to say, estimating the difference observed in other parts of South America between the mean humidity of the dry months and that of the whole year; an annual mean humidity is obtained, for the valleys of Aragua, at farthest of 74°, the temperature being 25·5°. In this air, so hot, and at the same time so little humid, the quantity of water evaporated is enormous. The theory of Dalton estimates, under the conditions just stated, for the thickness of the sheet of water evaporated in an hour's time 0·36 *mill.* or 3·8 lines in twenty-four hours *. Assuming for the temperate zone, for instance at Paris, the mean temperature to be 10·6°, and the mean humidity 82°, we find, according to the same formulæ, 0·10 *mill.* an hour, and 1 line for twenty-four hours. If we prefer substituting for the uncertainty of these theoretical deductions the direct results of observation, we may recollect, that at Paris, and at Montmorenci, the mean annual evaporation was found by Sedileau and Cotte, to be from 32 in. 1 line to 38 in. 4 lines. Two able engineers in the South of France, Messrs. Clausade and Pin, found, that in subtracting the effects of filtrations, the waters of the canal of Languedoc, and the basin of Saint Ferréol, lose every year from 0·758 met. to 0·812 met. or from 336 to 360 lines. Mr. de Prony found nearly similar results in the Pontine marshes. The whole of these experiments, made in the latitudes of 41° and 49°, and at 10·5° and 16° of mean temperature, indicate a mean evaporation of one line, or one and three tenths, a day. Under the torrid zone, in the West India islands for instance, the effect of evaporation is four times as much, according to Le Gaux, and double according to Cassan. At Cumana in a place where the atmosphere is far more loaded with humidity than in the valley of Aragua, I have often seen evaporate, during twelve hours, in the sun, 8·8 *mill.*, in the shade 3·4 *mill.*; and I believe, that the annual produce of evaporation in the rivers near Cumana is not below one hundred and thirty inches. Experiments of this kind are extremely delicate, but what I have related will suffice to demonstrate how great must be the quantity of vapour, that rises from the lake of Valencia, and from the surrounding country, the waters of which flow into the lake. I shall have occasion elsewhere to resume this subject; for, in a work which displays the great laws of nature under different zones, we must endeavour to solve the problem of the *mean tension of the vapours* contained in the atmosphere in different

latitudes, and at different heights above the surface of the ocean.

A great number of local circumstances cause the produce of evaporation to vary; it changes as more or less shade covers the basin of the waters, with their state of motion or of repose, with their depth, and the nature and colour of their bottom : but in general evaporation depends only on three circumstances, the temperature, the tension of the vapours contained in the atmosphere, and the resistance which the air, more or less dense, more or less agitated, opposes to the diffusion of vapour. The quantity of water, that evaporates in a given spot, every thing else being equal, is proportional to the difference between the quantity of vapour, which the ambient air can contain when saturated, and the quantity, which it actually contains. Hence it follows (as Mr. Daubuisson has already observed, in subjecting my hygrometic observations to calculation), that the evaporation is not so great under the torrid zone, as might be expected from the enormous augmentation of temperature; because, in those ardent climates, the air is habitually very humid.

Since the increase of agricultural industry in the valleys of Aragua, the little rivers, that run into the lake of Valencia, can no longer be regarded as real supplies during the six months succeeding December. They remain dried up in the lower part of their course, because the planters of indigo, coffee, and sugar-canes, have made frequent drainings *(azequias)*, in order to water the ground by trenches. We may observe also, that a pretty considerable river, the Rio Pao, which rises at the entrance of the Llanos, at the foot of the range of hills called *La Galera*, heretofore mingled it's waters with those of the lake, by uniting itself with the *Cano de Cambury*, on the road from the town of Nueva Valencia to Guigue. The course of this river was then from South to North. At the end of the seventeenth century, the proprietor of a neighbouring plantation thought proper to dig at the back of the hill a new bed for the Rio Pao. He turned the river ; and, after having employed part of the water for the irrigation of his fields, he caused the rest to flow at a venture toward the South, following the declivity of the Llanos. In this new southern direction the Rio Pao, mingled with three other rivers, the Tinaco, the Guanarito, and the Chilua, falls into the Portuguesa, which is a branch of the Apure. It is a remarkable phenomenon, to observe, that by a particular disposition of the ground, and the lowering of the *ridge of division* toward the South-West, the Rio Pao separates itself from the little *system of interior rivers*, to which it originally belonged, and for a century past has communicated, through the channel of the Apure and the Oroonoko, with the ocean. What has been here effected on a small scale by the hand of man, Nature often performs, either by progressively elevating the level of the soil, or by those falls of the ground, which violent earthquakes occasion. It is probable, that, in the lapse of ages, several rivers of Soudan, and of New Holland, which are now lost in the sands, or in inland basins, will open themselves a way toward the shores of the ocean. We cannot at least doubt, that in both continents there are systems of interior rivers; which may be considered as *not entirely developed*[*]; and which communicate with each other, either in the time of great risings, or by permanent bifurcations.

The Rio Pao has scooped itself out a bed so deep and broad, that in the season of rains, when the *Cano grande de Cambury* inundates all the land to the North-West of Guigue, the waters of this *Cano*, and those of the lake of Valencia, flow back into the Rio Pao itself ; so that this river, instead of adding water to the lake, tends rather to carry it away. We see something similar in North America, where geographers have chosen to represent on their maps an imaginary chain of mountains, between

[*] *Carl Ritter, Erdkunde,* vol. i, p. 315.

the great lakes of Canada and the country of the Miamis. At the time of floods, the waters flowing into the lakes communicate with those which run into the Missisippi; and it is practicable to proceed by boats from the sources of the river St. Mary to the Wabash, as well as from the Chicago to the Illinois *. These analogous facts appear to me well worthy of the attention of hydrographers.

The land that surrounds the lake of Valencia being entirely flat and even, what I daily observed in the lakes of Mexico takes place here; a diminution of a few inches in the level of the water exposes a vast extent of ground covered with fertile mud and organic remains. In proportion as the lake retires, the planters advance toward the new shore. These natural desiccations, so important to the colonial agriculture, have been eminently considerable during the last ten years, in which all America has suffered from great droughts. Instead of marking the sinuosities of the present banks of the lake, I have advised the rich landholders in these countries, to place columns of granite in the basin itself, in order to observe from year to year the mean height of the waters. The Marquis del Toro has undertaken to put this design into execution, employing the fine granite of the Sierra de Mariara, and establishing *limnometers,* on a bottom of gneiss rock, so common in the lake of Valencia.

It is impossible to anticipate the limits, more or less narrow, to which this basin of water will one day be confined, when an equilibrium, between the streams flowing in and the produce of evaporation and filtration, shall be completely established. The idea very generally spread, that the lake will soon entirely disappear, seems to me chimerical. If in consequence of great earthquakes, or other causes equally mysterious, ten very humid years should succeed to long droughts; if the mountains should clothe themselves anew with forests, and great trees overshadow the shore and the plains of Aragua; we should more probably see the volume of the waters augment, and menace that beautiful cultivation, which now trenches on the basin of the lake.

While some of the cultivators of the valleys of Aragua fear the total disappearance of the lake, and others it's return toward the banks it has deserted, we hear the question gravely discussed at Caraccas, whether it would not be advisable, in order to give greater extent to agriculture, to conduct the waters of the lake into the Llanos, by digging a canal toward the Rio Pao. The possibility * of this enterprise

* The *dividing ridge,* namely, that which divides the waters

cannot be denied, particularly by having recourse to tunnels, or subterranean canals. The
progressive retreat of the waters has given birth
to the beautiful and luxuriant plains of Maracay, Cura, Mocundo, Guigue, and Santa Cruz
del Escoval, planted with tobacco, sugar-canes,
coffee, indigo, and cacao; but how can it be
doubted for a moment, that the lake alone
spreads fertility over this country? Deprived
of an enormous mass of vapours, which the surface of the waters sends forth daily into the
atmosphere, the valleys of Aragua would become

between the valleys of Aragua and the Llanos, lowers so
much toward the West of Guigue, as we have already
observed, that there are ravines, which conduct the waters of the Cano de Cambury, the Rio Valencia, and the
Guataparo, in the time of floods, to the Rio Pao; but it
would be easier to open a navigable canal from the lake of
Valencia to the Oroonoko, by the Pao, the Portuguesa,
and the Apure, than to dig a draining canal level with the
bottom of the lake. This bottom, according to the sounding,
and my barometric measurements, is 40 toises less than 222,
or 182 above the surface of the ocean. On the road from
Guigue to the Llanos, by the table-land of La Villa de
Cura, I found, to the South of the *dividing ridge*, and on it's
southern declivity, no point of level corresponding to the 182
toises, except near San Juan. The absolute height of this
village is 194 toises. But, I repeat, that farther toward
the West, in the country between the Cano de Cambury
and the sources of the Rio Pao, which I was not able
to visit, the point of level of the bottom of the lake is much
more toward the North.

as dry and barren as the surrounding mountains.

CHAPTER II.

IT remains however to be considered, whether the appropriation of land, and the consequent creation of rent, will occasion any variation in the relative value of commodities, independently of the quantity of labour necessary to production. In order to understand this part of the subject, we must inquire into the nature of rent, and the laws by which its rise or fall is regulated. Rent is that portion of the produce of the earth, which is paid to the landlord for the use of the original and indestructible powers of the soil. It is often however confounded with the interest and profit of capital, and in popular language the term is applied to whatever is annually paid by a no rent. Is it not however evident, that the person who paid, what he thus calls rent, paid it in consideration of the valu-

able commodity which was then standing on the land, and that he actually repaid himself with a profit, by the sale of the timber? If, indeed, after the timber was removed, any compensation were paid to the landlord for the use of the land, for the purpose of growing timber or any other produce, with a view to future demand, such compensation might justly be called rent, because it would be paid for the productive powers of the land; but in the case stated by Adam Smith, the compensation was paid for the liberty of removing and selling the timber, and not for the liberty of growing it. He speaks also of the rent of coal mines, and of stone quarries, to which the same observation applies—that the compensation given for the mine or quarry, is paid for the value of the coal or stone which can be removed from them, and has no connexion with the original and indestructible powers of the land. This is a distinction of great importance, in an inquiry concerning rent and profits; for it is found, that the laws which regulate the progress of rent, are widely different from those which regulate the progress of profits, and seldom operate in the same direction. In all improved countries, that which is annually paid to the landlord, partaking of both characters, rent and profit, is sometimes kept stationary by the effects of opposing causes, at other times advances or recedes, as one or other of these causes preponderates. In the future pages of this work, then, whenever I speak of the rent of land, I wish to be understood as speaking of that compensation, which is paid to the owner of land for the use of its original and indestructible powers.

On the first settling of a country, in which there is an abundance of rich and fertile land, a very small proportion of which is required to be cultivated for the support of the actual population, or indeed can be cultivated with the capital which the population can command, there will be no rent; for no one would pay for the use of land, when there was an abundant quantity not yet appropriated, and therefore at the disposal of whosoever might choose to cultivate it.

On the common principles of supply and demand, no rent could be paid for such land, for the reason stated, why nothing is given for the use of air and water, or for any other of the gifts of nature which exist in boundless quantity. With a given quantity of materials, and with the assistance of the pressure of the atmosphere, and the elasticity of steam, engines may perform work, and abridge human labour to a very great extent; but no charge is made for the use of these natural aids, because they are inexhaustible, and at every man's disposal. In the same manner the brewer, the distiller, the dyer, make incessant use of the air and water for the production of their commodities; but as the supply is boundless, it bears no price.* If

---

* " The earth, as we have already seen, is not the only agent of nature which has a productive power; but it is the only one, or nearly so, that one set of men take to themselves, to the exclusion of others; and of which consequently they can appropriate the benefits. The waters of rivers, and of the sea, by the power which they have of giving movement to our machines, carrying our boats, nourishing our fish, have also a productive power; the wind which turns our mills, and even the heat of the sun, work for us; but happily no one has yet been able to say: the ' wind and the sun are mine, and the service which they render must be paid for.' "—*Economie Politique, par J. B. Say,* vol. ii. p. 124.

all land had the same properties, if it were boundless in quantity, and uniform in quality, no charge could be made for its use, unless where it possessed peculiar advantages of situation. It is only then because land is of different qualities with respect to its productive powers, and because in the progress of population, land of an inferior quality, or less advantageously situated, is called into cultivation, that rent is ever paid for the use of it. When, in the progress of society, land of the second degree of fertility is taken into cultivation, rent immediately commences on that of the first quality, and the amount of that rent will depend on the difference in the quality of these two portions of land.

When land of the third quality is taken into cultivation, rent immediately commences on the second, and it is regulated as before, by the difference in their productive powers. At the same time, the rent of the first quality will rise, for that must always be above the rent of the second, by the difference between the produce which they yield with a given quantity of capital and labour. With every step in the progress of population, which

shall oblige a country to have recourse to land of a worse quality, to enable it to raise its supply of food, rent, on all the more fertile land, will rise.

Thus suppose land—No. 1, 2, 3,—to yield, with an equal employment of capital and labour, a net produce of 100, 90, and 80 quarters of corn. In a new country, where there is an abundance of fertile land compared with the population, and where therefore it is only necessary to cultivate No. 1, the whole net produce will belong to the cultivator, and will be the profits of the stock which he advances. As soon as population had so far increased as to make it necessary to cultivate No. 2, from which ninety quarters only can be obtained after supporting the labourers, rent would commence on No. 1 ; for either there must be two rates of profit on agricultural capital, or ten quarters, or the value of ten quarters must be withdrawn from the produce of No. 1, for some other purpose. Whether the proprietor of the land, or any other person, cultivated No. 1, these ten quarters would equally constitute rent ; for the cultivator of No. 2 would get the same result with his capital, whether he cultivated No. 1, paying ten quarters for rent, or continued to cultivate No. 2, paying no rent. In the same manner it might be shewn that when No. 3 is brought into cultivation, the rent of No. 2 must be ten quarters, or the value of ten quarters, whilst the rent of No. 1 would rise to twenty quarters; for the cultivator of No. 3 would have the same profits whether he paid twenty quarters for the rent of No. 1, ten quarters for the rent of No. 2, or cultivated No. 3 free of all rent.

It often, and indeed commonly happens that before No. 2, 3, 4, or 5, or the inferior lands are cultivated, capital can be employed more productively on those lands which are already in cultivation. It may perhaps be found, that by doubling the original capital employed on No. 1, though the produce will not be doubled, will not be increased by 100 quarters, it may be increased by eighty-five quarters, and that this quantity exceeds what could be obtained by employing the same capital on land, No. 3.

In such case, capital will be preferably em-

ployed on the old land, and will equally create a rent; for rent is always the difference between the produce obtained by the employment of two equal quantities of capital and labour. If with a capital of 1000*l.* a tenant obtain 100 quarters of wheat from his land, and by the employment of a second capital of 1000*l.*, he obtain a further return of eighty-five, his landlord would have the power at the expiration of his lease, of obliging him to pay fifteen quarters, or an equivalent value, for additional rent; for there cannot be two rates of profit. If he is satisfied with a diminution of fifteen quarters in the return for his second 1000*l.*, it is because no employment more profitable can be found for it. The common rate of profit would be in that proportion, and if the original tenant refused, some other person would be found willing to give all which exceeded that rate of profit to the owner of the land from which he derived it.

In this case, as well as in the other, the capital last employed pays no rent. For the greater productive powers of the first 1000*l.*, fifteen quarters is paid for rent, for the employment of the second 1000*l*. no rent whatever is paid. If a third 1000*l*. be employed on the same land, with a return of seventy-five quarters, rent will then be paid for the second 1000*l*. and will be equal to the difference between the produce of these two, or ten quarters; and at the same time the rent of the first 1000*l*. will rise from fifteen to twenty-five quarters; while the last 1000*l*. will pay no rent whatever.

If then good land existed in a quantity much more abundant than the production of food for an increasing population required, or if capital could be indefinitely employed without a diminished return on the old land, there could be no rise of rent; for rent invariably proceeds from the employment of an additional quantity of labour with a proportionally less return.

The most fertile, and most favourably situated land will be first cultivated, and the exchangeable value of its produce will be adjusted in the same manner as the exchangeable value of all other commodities, by the total quantity of labour necessary in various forms, from first to last, to produce it, and bring it to market. When land of an inferior quality is taken into cultivation, the exchangeable value of raw produce will rise, because more labour is required to produce it.

The exchangeable value of all commodities, whether they be manufactured, or the produce of the mines, or the produce of land, is always regulated, not by the less quantity of labour that will suffice for their production under circumstances highly favourable, and exclusively enjoyed by those who have peculiar facilities of production; but by the greater quantity of labour necessarily bestowed on their production by those who have no such facilities; by those who continue to produce them under the most unfavourable circumstances; meaning—by the most unfavourable circumstances, the most unfavourable under which the quantity of produce required renders it necessary to carry on the production.

Thus, in a charitable institution, where the poor are set to work with the funds of benefactors, the general prices of the commodities, which are the produce of such work, will not be governed by the peculiar facilities afforded to these workmen, but by the common, usual, and natural difficulties, which every other manufacturer will have to encounter. The manufacturer enjoying none of these facilities might indeed be driven altogether from the market, if the supply afforded by these favoured workmen were equal to all the wants of the community; but if he continued the trade, it would be only on condition that he should derive from it the usual and general rate of profits on stock; and that could only happen when his commodity sold for a price proportioned to the quantity of labour bestowed on its production.*

It is true, that on the best land, the same produce would still be obtained with the same labour as before, but its value would be enhanced in consequence of the diminished returns obtained by those who employed fresh labour and stock on the less fertile land. Notwithstanding then, that the advantages of fertile over inferior lands are in no case lost, but only transferred from the cultivator, or consumer, to the landlord, yet since more labour is required on the inferior lands, and since it is from such land only that we are enabled to furnish ourselves with the additional supply of raw produce, the comparative value of that produce will continue permanently above its former level, and make

* Has not M. Say forgotten, in the following passage, that it is the cost of production which ultimately regulates price? " The produce of labour employed on the land has this peculiar property, that it does not become more dear by becoming more scarce, because population always diminishes at the same time that food diminishes, and consequently the quantity of these products *demanded*, diminishes at the same time as the quantity supplied. Besides it is not observed that corn is more dear in those places where there is plenty of uncultivated land, than in completely cultivated countries. England and France were much more imperfectly cultivated in the middle ages than they are now; they produced much less raw produce: nevertheless from all that we can judge by a comparison with

the value of other things, corn was not sold at a dearer price. If the produce was less, so was the population; the weakness of the demand compensated the feebleness of the supply." vol. ii. 338. M. Say being impressed with the opinion that the price of commodities is regulated by the price of labour, and justly supposing that charitable institutions of all sorts tend to increase the population beyond what it otherwise would be, and therefore to lower wages, says, " I suspect that the cheapness of the goods, which come from England is partly caused by the numerous charitable institutions which exist in that country." vol. ii. 277. This is a consistent opinion in one who maintains that wages regulate price.

it exchange for more hats, cloth, shoes, &c. &c. in the production of which no such additional quantity of labour is required.

The reason then, why raw produce rises in comparative value, is because more labour is employed in the production of the last portion obtained, and not because a rent is paid to the landlord. The value of corn is regulated by the quantity of labour bestowed on its production on that quality of land, or with that portion of capital, which pays no rent. Corn is not high because a rent is paid, but a rent is paid because corn is high; and it has been justly observed, that no reduction would take place in the price of corn, although landlords should forego the whole of their rent. Such a measure would only enable some farmers to live like gentlemen, but would not diminish the quantity of labour necessary to raise raw produce on the least productive land in cultivation.

Nothing is more common than to hear of the advantages which the land possesses over every other source of useful produce, on account of the surplus which it yields in the form of rent. Yet when land is most abundant, when most productive, and most fertile, it yields no rent; and it is only when its powers decay, and less is yielded in return for labour, that a share of the original produce of the more fertile portions is set apart for rent. It is singular that this quality in the land, which should have been noticed as an imperfection, compared with the natural agents by which manufacturers are assisted, should have been pointed out as constituting its peculiar pre-eminence. If air, water, the elasticity of steam, and the pressure of the atmosphere, were of various qualities; if they could be appropriated, and each quality existed only in moderate abundance, they as well as the land would afford a rent, as the successive qualities were brought into use. With every worse quality employed, the value of the commodities in the manufacture of which they were used would rise, because equal quantities of labour would be less productive. Man would do more by the sweat of his brow, and nature perform less; and the land would be no longer pre-eminent for its limited powers.

If the surplus produce which land affords

in the form of rent be an advantage, it is desirable that, every year, the machinery newly constructed should be less efficient than the old, as that would undoubtedly give a greater exchangeable value to the goods manufactured, not only by that machinery, but by all the other machinery in the kingdom; and a rent would be paid to all those who possessed the most productive machinery.*

---

* " In agriculture too," says Adam Smith, " nature labours along with man; and though her labour costs no expense, its produce has its value, as well as that of the most expensive workman." The labour of nature is paid, not because she does much, but because she does little. In proportion as she becomes niggardly in her gifts, she exacts a greater price for her work. Where she is munificently beneficent, she always works gratis. " The labouring cattle employed in agriculture, not only occasion, like the workmen in manufactures, the reproduction of a value equal to their own consumption, or to the capital which employs them, together with its owner's profits, but of a much greater value. Over and above the capital of the farmer and all its profits, they regularly occasion the reproduction of the rent of the landlord. This rent may be considered as the produce of those powers of nature, the use of which the landlord lends to the farmer. It is greater or smaller according to the supposed extent of those powers, or in other words, according to the supposed natural or improved fertility of the land. It is the work of nature which remains, after deducting or compensating every thing which can be regarded as

The rise of rent is always the effect of the increasing wealth of the country, and of the

---

the work of man. It is seldom less than a fourth, and frequently more than a third of the whole produce. No equal quantity of productive labour employed in manufactures, can ever occasion so great a reproduction. *In them nature does nothing, man does all;* and the reproduction must always be in proportion to the strength of the agents that occasion it. The capital employed in agriculture, therefore, not only puts into motion a greater quantity of productive labour than any equal capital employed in manufactures, but in proportion too to the quantity of the productive labour which it employs, it adds a much greater value to the annual produce of the land and labour of the country, to the *real* wealth and revenue of its inhabitants. Of all the ways in which a capital can be employed, it is by far the most advantageous to the society."— Book II. chap. v. p. 15.

Does nature nothing for man in manufactures? Are the powers of wind and water, which move our machinery, and assist navigation, nothing? The pressure of the atmosphere and the elasticity of steam, which enable us to work the most stupendous engines—are they not the gifts of nature? to say nothing of the effects of the matter of heat in softening and melting metals, of the decomposition of the atmosphere in the process of dyeing and fermentation. There is not a manufacture which can be mentioned, in which nature does not give her assistance to man, and give it too, generously and gratuitously.

In remarking on the passage which I have copied from Adam Smith, Mr. Buchanan observes, " I have endeavoured to shew, in the observations on productive and unproductive

difficulty of providing food for its augmented population. It is a symptom, but it is never a cause of wealth; for wealth often increases most rapidly while rent is either stationary, or even falling. Rent increases most rapidly, as the disposable land decreases in its productive powers. Wealth increases most rapidly in those countries where the disposable land is most fertile, where importation is least restricted, and where through agricultural improvements, productions can be multiplied without any increase in the proportional quantity of labour, and where consequently the progress of rent is slow.

labour, contained in the fourth volume, that agriculture adds no more to the national stock than any other sort of industry. In dwelling on the reproduction of rent as so great an advantage to society, Dr. Smith does not reflect that rent is the effect of high price, and that what the landlord gains in this way, he gains at the expense of the community at large. There is no absolute gain to the society by the reproduction of rent; it is only one class profiting at the expense of another class. The notion of agriculture yielding a produce, and a rent in consequence, because nature concurs with human industry in the process of cultivation, is a mere fancy. It is not from the produce, but from the price at which the produce is sold, that the rent is derived; and this price is got, not because nature assists in the production, but because it is the price which suits the consumption to the supply."

If the high price of corn were the effect, and not the cause of rent, price would be proportionally influenced as rents were high or low, and rent would be a component part of price. But that corn which is produced with the greatest quantity of labour is the regulator of the price of corn, and rent does not and cannot enter in the least degree as a component part of its price. Adam Smith, therefore, cannot be correct in supposing that the original rule which regulated the exchangeable value of commodities, namely the comparative quantity of labour by which they were produced, can be at all altered by the appropriation of land and the payment of rent. Raw material enters into the composition of most commodities, but the value of that raw material as well as corn, is regulated by the productiveness of the portion of capital last employed on the land, and paying no rent; and therefore rent is not a component part of the price of commodities.

We have been hitherto considering the effects of the natural progress of wealth and population on rent, in a country in which the land is of variously productive powers; and we have seen, that with every portion of additional capital which it becomes necessary to employ on the land with a less productive return, rent would rise. It follows from the same principles, that any circumstances in the society which should make it unnecessary to employ the same amount of capital on the land, and which should therefore make the portion last employed more productive, would lower rent. Any great reduction in the capital of a country, which should materially diminish the funds destined for the maintenance of labour, would naturally have this effect. Population regulates itself by the funds which are to employ it, and therefore always increases or diminishes with the increase or diminution of capital. Every reduction of capital is therefore necessarily followed by a less effective demand for corn, by a fall of price, and by diminished cultivation. In the reverse order to that in which the accumulation of capital raises rent, will the diminution of it lower rent. Land of a less unproductive quality will be in succession relinquished, the exchangeable value of produce will fall, and land of a superior quality will be the land last cultivated, and that which will then pay no rent.

The same effects may however be produced when the wealth and population of a country are increased, if that increase is accompanied by such marked improvements in agriculture, as shall have the same effect of diminishing the necessity of cultivating the poorer lands, or of expending the same amount of capital on the cultivation of the more fertile portions.

If a million of quarters of corn be necessary for the support of a given population, and it be raised on land of the qualities of No. 1, 2, 3; and if an improvement be afterwards discovered by which it can be raised on No. 1 and 2, without employing No. 3, it is evident that the immediate effect must be a fall of rent; for No. 2, instead of No. 3, will then be cultivated without paying any rent; and the rent of No. 1, instead of being the difference between the produce of No. 3 and No. 1, will be the difference only between No. 2 and 1. With the same population, and no more, there can be no demand for any additional quantity of corn; the capital and labour employed on No. 3, will be devoted to the production of other commodities desirable to the community, and can have no effect in raising rent unless the raw material from which they are made cannot be obtained without employing capital less advantageously on the land, in which case No. 3 must again be cultivated.

It is undoubtedly true, that the fall in the relative price of raw produce, in consequence of the improvement in agriculture, or rather in consequence of less labour being bestowed on its production, would naturally lead to increased accumulation; for the profits of stock would be greatly augmented. This accumulation would lead to an increased demand for labour, to higher wages, to an increased population, to a further demand for raw produce, and to an increased cultivation. It is only, however, after the increase in the population, that rent would be as high as before; that is to say, after No. 3 was taken into cultivation. A considerable period would have elapsed, attended with a positive diminution of rent.

But improvements in agriculture are of two kinds: those which increase the productive powers of the land, and those which enable us to obtain its produce with less labour. They both lead to a fall in the price of raw produce; they both affect rent, but they do not affect it equally. If they did not occasion a fall in the price of raw produce, they would not be improvements; for it is the essential quality of an improvement to diminish the quantity of labour before required to produce a commodity; and this diminution cannot take place without a fall of its price or relative value.

The improvements which increase the productive powers of the land, are such as the more skilful rotation of crops, or the better choice of manure. These improvements absolutely enable us to obtain the same produce from a smaller quantity of land. If, by the introduction of a course of turnips, I can feed my sheep besides raising my corn, the land on which the sheep were before fed becomes unnecessary, and the same quantity of raw produce is raised by the employment of a less quantity of land. If I discover a manure which will enable me to make a piece of land produce 20 per cent. more corn, I may withdraw at least a portion of my capital from the most unproductive part of my farm. But, as I have before observed, it is not necessary that land should be thrown out of cultivation, in order to reduce rent: to produce this effect, it is sufficient that successive portions of capital are employed on the same land with different results, and that the portion which gives the least result should be withdrawn. If, by the introduction of the turnip husbandry, or by the use of a more invigorating manure, I can obtain the same produce with less capital, and without disturbing the difference between the productive powers of the successive portions of capital, I shall lower rent; for a different and more productive portion will be that which will form the standard from which every other will be reckoned. If, for example, the successive portions of capital yielded 100, 90, 80, 70; whilst I employed these four portions, my rent would be 60, or the difference between

$$
\left.
\begin{array}{l}
70 \text{ and } 100 = 30 \\
70 \text{ and } \;\,90 = 20 \\
70 \text{ and } \;\,80 = 10 \\
\overline{} \\
60
\end{array}
\right\}
\;\;
\begin{array}{l}
\text{whilst the produce} \\
\text{would be 340}
\end{array}
\;\;
\left\{
\begin{array}{r}
100 \\
90 \\
80 \\
70 \\
\overline{} \\
340
\end{array}
\right.
$$

and while I employed these portions, the rent would remain the same, although the produce of each should have an equal augmentation. If, instead of 100, 90, 80, 70, the produce should be increased to 125, 115, 105, 95, the rent would still be 60, or the difference between

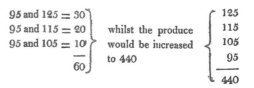

$$
\left.
\begin{array}{l}
95 \text{ and } 125 = 30 \\
95 \text{ and } 115 = 20 \\
95 \text{ and } 105 = 10 \\
\overline{} \\
60
\end{array}
\right\}
\;\;
\begin{array}{l}
\text{whilst the produce} \\
\text{would be increased} \\
\text{to 440}
\end{array}
\;\;
\left\{
\begin{array}{r}
125 \\
115 \\
105 \\
95 \\
\overline{} \\
440
\end{array}
\right.
$$

But with such an increase of produce, without an increase of demand, there could be no motive for employing so much capital on the land; one portion would be withdrawn, and consequently the last portion of capital would yield 105 instead of 95, and rent would fall to 30, or the difference between

$$
\left.
\begin{array}{l}
105 \text{ and } 125 = 20 \\
105 \text{ and } 115 = 10 \\
\overline{} \\
30
\end{array}
\right\}
\;\;
\begin{array}{l}
\text{whilst the produce would be still} \\
\text{adequate to the wants of the po-} \\
\text{pulation, for it would be 345} \\
\text{quarters, or}
\end{array}
\;\;
\left\{
\begin{array}{r}
125 \\
115 \\
105 \\
\overline{} \\
345
\end{array}
\right.
$$

the demand being only for 340 quarters. —But there are improvements which may lower the relative value of produce without lowering the corn rent, though they will lower the money rent of land. Such improvements do not increase the productive powers of the land, but they enable us to obtain its produce with less labour. They are rather directed to the formation of the capital applied to the land, than to the cultivation of the land itself. Improvements in agricultural implements, such as the plough and the threshing machine, economy in the use of horses employed in husbandry, and a better knowledge of the veterinary art, are of this nature. Less capital, which is the same thing as less labour, will be employed on the land; but to obtain the same produce, less land cannot be cultivated. Whether improvements of this kind, however, affect corn rent, must depend on the question, whether the difference between the produce obtained by the employment of different portions of capital be increased, stationary, or diminished. If four portions of capital, 50, 60, 70, 80, be employed on the land, giving each the same results, and any improvement in the

formation of such capital should enable me to withdraw 5 from each, so that they should be 45, 55, 65, and 75, no alteration would take place in the corn rent; but if the improvements were such as to enable me to make the whole saving on the largest portion of capital, that portion which is least productively employed, corn rent would immediately fall, because the difference between the capital most productive and the capital least productive would be diminished; and it is this difference which constitutes rent.

Without multiplying instances, I hope enough has been said to shew, that whatever diminishes the inequality in the produce obtained from successive portions of capital employed on the same or on new land, tends to lower rent; and that whatever increases that inequality, necessarily produces an opposite effect, and tends to raise it.

In speaking of the rent of the landlord, we have rather considered it as the proportion of the whole produce, without any reference to its exchangeable value; but since the same cause, the difficulty of production, raises the exchangeable value of raw produce, and raises also the proportion of raw produce paid to the landlord for rent, it is obvious that the landlord is doubly benefited by difficulty of production. First he obtains a greater share, and secondly the commodity in which he is paid is of greater value.*

---

* To make this obvious, and to shew the degrees in which corn and money rent will vary, let us suppose that the labour of ten men will, on land of a certain quality, obtain 180 quarters of wheat, and its value to be 4*l.* per quarter, or 720*l.*; and that the labour of ten additional men will, on the same or any other land, produce only 170 quarters in addition; wheat would rise from 4*l.* to 4*l.* 4*s.* 8*d.* for 170 : 180 : : 4*l.* : 4*l.* 4*s.* 8*d.*; or, as in the production of 170 quarters, the labour of 10 men is necessary in one case, and only of 9.44 in the other, the rise would be as 9.44 to 10, or as 4*l.* to 4*l.* 4*s.* 8*d.* If 10 men be further employed, and the return be

160,	the price will rise to	£4	10	0
150,	- - - - - - - -	4	16	0
140,	- - - - - - - -	5	2	10

Now if no rent was paid for the land which yielded 180 quarters when corn was at 4*l.* per quarter, the value of 10 quarters would be paid as rent when only 170 could be procured, which, at 4*l.* 4*s.* 8*d.* would be 42*l.* 7*s.* 6*d.*

20 qrs. when 160 were produced, which at £4 10 0 would be £90 0 0
30 qrs. . . 150 . . . . 4 16 0 . . . 144 0 0
40 qrs. . . 140 . . . . 5 2 10 . . . 205 13 4

Corn rent then would increase $\left\{\begin{matrix} 100 \\ 200 \\ 300 \\ 400 \end{matrix}\right.$ and money rent in the proportion of . . . in the proportion of $\left\{\begin{matrix} 100 \\ 212 \\ 340 \\ 485 \end{matrix}\right.$

CHAPTER III.

STRUGGLE FOR EXISTENCE.

Bears on natural selection—The term used in a wide sense—Geometrical powers of increase — Rapid increase of naturalised
animals and plants—Nature of the checks to increase—Competition universal — Effects of climate — Protection from the
number of individuals—Complex relations of all animals and
plants throughout nature—Struggle for life most severe between
individuals and varieties of the same species; often severe between species of the same genus—The relation of organism to
organism the most important of all relations.

BEFORE entering on the subject of this chapter, I must
make a few preliminary remarks, to show how the
struggle for existence bears on Natural Selection. It
has been seen in the last chapter that amongst organic
beings in a state of nature there is some individual variability; indeed I am not aware that this has ever been
disputed. It is immaterial for us whether a multitude
of doubtful forms be called species or sub-species or varieties; what rank, for instance, the two or three hundred
doubtful forms of British plants are entitled to hold, if
the existence of any well-marked varieties be admitted.
But the mere existence of individual variability and of
some few well-marked varieties, though necessary as
the foundation for the work, helps us but little in
understanding how species arise in nature. How have
all those exquisite adaptations of one part of the organisation to another part, and to the conditions of life, and
of one distinct organic being to another being, been perfected? We see these beautiful co-adaptations most
plainly in the woodpecker and missletoe; and only a
little less plainly in the humblest parasite which clings
to the hairs of a quadruped or feathers of a bird; in the
structure of the beetle which dives through the water;
in the plumed seed which is wafted by the gentlest
breeze; in short, we see beautiful adaptations everywhere and in every part of the organic world.

Again, it may be asked, how is it that varieties, which
I have called incipient species, become ultimately converted into good and distinct species, which in most
cases obviously differ from each other far more than do
the varieties of the same species? How do those groups
of species, which constitute what are called distinct
genera, and which differ from each other more than do
the species of the same genus, arise? All these results,
as we shall more fully see in the next chapter, follow
inevitably from the struggle for life. Owing to this
struggle for life, any variation, however slight and from
whatever cause proceeding, if it be in any degree profitable to an individual of any species, in its infinitely
complex relations to other organic beings and to external nature, will tend to the preservation of that individual, and will generally be inherited by its offspring.
The offspring, also, will thus have a better chance of
surviving, for, of the many individuals of any species
which are periodically born, but a small number can
survive. I have called this principle, by which each
slight variation, if useful, is preserved, by the term of
Natural Selection, in order to mark its relation to man's
power of selection. We have seen that man by selection can certainly produce great results, and can adapt
organic beings to his own uses, through the accumulation of slight but useful variations, given to him by the
hand of Nature. But Natural Selection, as we shall
hereafter see, is a power incessantly ready for action,
and is as immeasurably superior to man's feeble efforts,
as the works of Nature are to those of Art.

We will now discuss in a little more detail the struggle for existence. In my future work this subject shall be treated, as it well deserves, at much greater length. The elder De Candolle and Lyell have largely and philosophically shown that all organic beings are exposed to severe competition. In regard to plants, no one has treated this subject with more spirit and ability than W. Herbert, Dean of Manchester, evidently the result of his great horticultural knowledge. Nothing is easier than to admit in words the truth of the universal struggle for life, or more difficult—at least I have found it so—than constantly to bear this conclusion in mind. Yet unless it be thoroughly engrained in the mind, I am convinced that the whole economy of nature, with every fact on distribution, rarity, abundance, extinction, and variation, will be dimly seen or quite misunderstood. We behold the face of nature bright with gladness, we often see superabundance of food; we do not see, or we forget, that the birds which are idly singing round us mostly live on insects or seeds, and are thus constantly destroying life; or we forget how largely these songsters, or their eggs, or their nestlings, are destroyed by birds and beasts of prey; we do not always bear in mind, that though food may be now superabundant, it is not so at all seasons of each recurring year.

I should premise that I use the term Struggle for Existence in a large and metaphorical sense, including dependence of one being on another, and including (which is more important) not only the life of the individual, but success in leaving progeny. Two canine animals in a time of dearth, may be truly said to struggle with each other which shall get food and live. But a plant on the edge of a desert is said to struggle for life against the drought, though more properly it should be said to be dependent on the moisture. A plant which annually produces a thousand seeds, of which on an average only one comes to maturity, may be more truly said to struggle with the plants of the same and other kinds which already clothe the ground. The missletoe is dependent on the apple and a few other trees, but can only in a far-fetched sense be said to struggle with these trees, for if too many of these parasites grow on the same tree, it will languish and die. But several seedling missletoes, growing close together on the same branch, may more truly be said to struggle with each other. As the missletoe is disseminated by birds, its existence depends on birds; and it may metaphorically be said to struggle with other fruit-bearing plants, in order to tempt birds to devour and thus disseminate its seeds rather than those of other plants. In these several senses, which pass into each other, I use for convenience sake the general term of struggle for existence.

A struggle for existence inevitably follows from the high rate at which all organic beings tend to increase. Every being, which during its natural lifetime produces several eggs or seeds, must suffer destruction during some period of its life, and during some season or occasional year, otherwise, on the principle of geometrical increase, its numbers would quickly become so inordinately great that no country could support the product. Hence, as more individuals are produced than can possibly survive, there must in every case be a struggle for existence, either one individual with another of the same species, or with the individuals of distinct species, or with the physical conditions of life. It is the doctrine of Malthus applied with manifold force to the whole animal and vegetable kingdoms; for in this case there can be no artificial increase of food, and no prudential restraint from marriage. Although some species may be now increasing, more or less rapidly, in numbers, all cannot do so, for the world would not hold them.

There is no exception to the rule that every organic being naturally increases at so high a rate, that if not destroyed, the earth would soon be covered by the progeny of a single pair. Even slow-breeding man has doubled in twenty-five years, and at this rate, in a few thousand years, there would literally not be standing room for his progeny. Linnæus has calculated that if an annual plant produced only two seeds—and there is no plant so unproductive as this—and their seedlings next year produced two, and so on, then in twenty years there would be a million plants. The elephant is reckoned to be the slowest breeder of all known animals, and I have taken some pains to estimate its probable minimum rate of natural increase: it will be under the mark to assume that it breeds when thirty years old, and goes on breeding till ninety years old, bringing forth three pair of young in this interval; if this be so, at the end of the fifth century there would be alive fifteen million elephants, descended from the first pair.

But we have better evidence on this subject than mere theoretical calculations, namely, the numerous recorded cases of the astonishingly rapid increase of various animals in a state of nature, when circumstances have been favourable to them during two or three following seasons. Still more striking is the evidence from our domestic animals of many kinds which have run wild in several parts of the world: if the statements of the rate of increase of slow-breeding cattle and horses in South-America, and latterly in Australia, had not been well authenticated, they would have been quite incredible. So it is with plants: cases could be given of introduced plants which have become common throughout whole islands in a period of less than ten years. Several

of the plants now most numerous over the wide plains of La Plata, clothing square leagues of surface almost to the exclusion of all other plants, have been introduced from Europe; and there are plants which now range in India, as I hear from Dr. Falconer, from Cape Comorin to the Himalaya, which have been imported from America since its discovery. In such cases, and endless instances could be given, no one supposes that the fertility of these animals or plants has been suddenly and temporarily increased in any sensible degree. The obvious explanation is that the conditions of life have been very favourable, and that there has consequently been less destruction of the old and young, and that nearly all the young have been enabled to breed. In such cases the geometrical ratio of increase, the result of which never fails to be surprising, simply explains the extraordinarily rapid increase and wide diffusion of naturalised productions in their new homes.

In a state of nature almost every plant produces seed, and amongst animals there are very few which do not annually pair. Hence we may confidently assert, that all plants and animals are tending to increase at a geometrical ratio, that all would most rapidly stock every station in which they could any how exist, and that the geometrical tendency to increase must be checked by destruction at some period of life. Our familiarity with the larger domestic animals tends, I think, to mislead us: we see no great destruction falling on them, and we forget that thousands are annually slaughtered for food, and that in a state of nature an equal number would have somehow to be disposed of.

The only difference between organisms which annually produce eggs or seeds by the thousand, and those which produce extremely few, is, that the slow-breeders would require a few more years to people, under favourable conditions, a whole district, let it be ever so large. The condor lays a couple of eggs and the ostrich a score, and yet in the same country the condor may be the more numerous of the two: the Fulmar petrel lays but one egg, yet it is believed to be the most numerous bird in the world. One fly deposits hundreds of eggs, and another, like the hippobosca, a single one; but this difference does not determine how many individuals of the two species can be supported in a district. A large number of eggs is of some importance to those species, which depend on a rapidly fluctuating amount of food, for it allows them rapidly to increase in number. But the real importance of a large number of eggs or seeds is to make up for much destruction at some period of life; and this period in the great majority of cases is an early one. If an animal can in any way protect its own eggs or young, a small number may be produced, and yet the average stock be fully kept up; but if many eggs or young are destroyed, many must be produced, or the species will become extinct. It would suffice to keep up the full number of a tree, which lived on an average for a thousand years, if a single seed were produced once in a thousand years, supposing that this seed were never destroyed, and could be ensured to germinate in a fitting place. So that in all cases, the average number of any animal or plant depends only indirectly on the number of its eggs or seeds.

In looking at Nature, it is most necessary to keep the foregoing considerations always in mind—never to forget that every single organic being around us may be said to be striving to the utmost to increase in numbers; that each lives by a struggle at some period of its life; that heavy destruction inevitably falls either on the young or old, during each generation or at recurrent intervals. Lighten any check, mitigate the destruction ever so little, and the number of the species will almost instantaneously increase to any amount. The face of Nature may be compared to a yielding surface, with ten thousand sharp wedges packed close together and driven inwards by incessant blows, sometimes one wedge being struck, and then another with greater force.

What checks the natural tendency of each species to increase in number is most obscure. Look at the most vigorous species; by as much as it swarms in numbers, by so much will its tendency to increase be still further increased. We know not exactly what the checks are in even one single instance. Nor will this surprise any one who reflects how ignorant we are on this head, even in regard to mankind, so incomparably better known than any other animal. This subject has been ably treated by several authors, and I shall, in my future work, discuss some of the checks at considerable length, more especially in regard to the feral animals of South America. Here I will make only a few remarks, just to recall to the reader's mind some of the chief points. Eggs or very young animals seem generally to suffer most, but this is not invariably the case. With plants there is a vast destruction of seeds, but, from some observations which I have made, I believe that it is the seedlings which suffer most from germinating in ground already thickly stocked with other plants. Seedlings, also, are destroyed in vast numbers by various enemies; for instance, on a piece of ground three feet long and two wide, dug and cleared, and where there could be no choking from other plants, I marked all the seedlings of our native weeds as they came up, and out of the 357 no less than 295 were destroyed, chiefly by slugs and insects. If turf which has long been mown, and the case would be the same with turf closely browsed by quadrupeds, be let to grow, the more vigorous plants gradually kill the less vigorous, though fully grown, plants: thus out of twenty species growing on a little plot of turf (three feet by four) nine species perished from the other species being allowed to grow up freely.

The amount of food for each species of course gives the extreme limit to which each can increase; but very frequently it is not the obtaining food, but the serving as prey to other animals, which determines the average numbers of a species. Thus, there seems to be little doubt that the stock of partridges, grouse, and hares on any large estate depends chiefly on the destruction of vermin. If not one head of game were shot during the next twenty years in England, and, at the same time, if no vermin were destroyed, there would, in all probability, be less game than at present, although hundreds of thousands of game animals are now annually killed. On the other hand, in some cases, as with the elephant and rhinoceros, none are destroyed by beasts of prey: even the tiger in India most rarely dares to attack a young elephant protected by its dam.

Climate plays an important part in determining the average numbers of a species, and periodical seasons of extreme cold or drought, I believe to be the most effective of all checks. I estimated that the winter of 1854–55 destroyed four-fifths of the birds in my own grounds; and this is a tremendous destruction, when we remember that ten per cent. is an extraordinarily severe mortality from epidemics with man. The action of climate seems at first sight to be quite independent of the struggle for existence; but in so far as climate chiefly acts in reducing food, it brings on the most severe struggle between the individuals, whether of the same or of distinct species, which subsist on the same kind of food. Even when climate, for instance extreme cold, acts directly, it will be the least vigorous, or those which have got least food through the advancing winter, which will suffer most. When we travel from south to north, or from a damp region to a dry, we invariably see some species gradually getting rarer and rarer, and finally disappearing; and the change of climate being conspicuous, we are tempted to attribute the whole effect to its direct action. But this is a very false view: we forget that each species, even where it most abounds, is constantly suffering enormous destruction at some period of its life, from enemies or from competitors for the same place and food; and if these enemies or competitors be in the least degree favoured by any slight change of climate, they will increase in numbers, and, as each area is already fully stocked with inhabitants, the other species will decrease. When we travel southward and see a species decreasing in numbers, we may feel sure that the cause lies quite as much in other species being favoured, as in this one being hurt. So it is when we travel northward, but in a somewhat lesser degree, for the number of species of all kinds, and therefore of competitors, decreases northwards; hence in going northward, or in ascending a mountain, we far oftener meet with stunted forms, due to the *directly* injurious action of climate, than we do in proceeding southwards or in descending a mountain. When we reach the Arctic regions, or snow-capped summits, or absolute deserts, the struggle for life is almost exclusively with the elements.

That climate acts in main part indirectly by favouring other species, we may clearly see in the prodigious number of plants in our gardens which can perfectly well endure our climate, but which never become naturalised, for they cannot compete with our native plants, nor resist destruction by our native animals.

When a species, owing to highly favourable circumstances, increases inordinately in numbers in a small tract, epidemics—at least, this seems generally to occur with our game animals—often ensue: and here we have a limiting check independent of the struggle for life. But even some of these so-called epidemics appear to be due to parasitic worms, which have from some cause, possibly in part through facility of diffusion amongst the crowded animals, been disproportionably favoured: and here comes in a sort of struggle between the parasite and its prey.

On the other hand, in many cases, a large stock of individuals of the same species, relatively to the numbers of its enemies, is absolutely necessary for its preservation. Thus we can easily raise plenty of corn and rape-seed, &c., in our fields, because the seeds are in great excess compared with the number of birds which feed on them; nor can the birds, though having a superabundance of food at this one season, increase in number proportionally to the supply of seed, as their numbers are checked during winter: but any one who has tried, knows how troublesome it is to get seed from a few wheat or other such plants in a garden; I have in this case lost every single seed. This view of the necessity of a large stock of the same species for its preservation, explains, I believe, some singular facts in nature, such as that of very rare plants being sometimes extremely abundant in the few spots where they do occur; and that of some social plants being social, that is, abounding in individuals, even on the extreme confines of their range. For in such cases, we may believe, that a plant could exist only where the conditions of its life were so favourable that many could exist together, and thus save each other from utter destruction. I should add that the good effects of frequent intercrossing, and the ill effects of close interbreeding, probably come into play in some of these cases; but on this intricate subject I will not here enlarge.

Many cases are on record showing how complex and unexpected are the checks and relations between organic beings, which have to struggle together in the same country. I will give only a single instance, which, though a simple one, has interested me. In Staffordshire, on the estate of a relation where I had ample means of investigation, there was a large and extremely barren heath, which had never been touched by the hand of man; but several hundred acres of exactly the same nature had been enclosed twenty-five years previously and planted with Scotch fir. The change in the native vegetation of the planted part of the heath was most remarkable, more than is generally seen in passing from one quite different soil to another: not only the proportional numbers of the heath-plants were wholly changed, but twelve species of plants (not counting grasses and carices) flourished in the plantations, which could not be found on the heath. The effect on the insects must have been still greater, for six insectivorous birds were very common in the plantations, which were not to be seen on the heath; and the heath was frequented by two or three distinct insectivorous birds. Here we see how potent has been the effect of the introduction of a single tree, nothing whatever else having been done, with the exception that the land had been enclosed, so that cattle could not enter. But how important an element enclosure is, I plainly saw near Farnham, in Surrey. Here there are extensive heaths, with a few clumps of old Scotch firs on the distant hill-tops: within the last ten years large spaces have been enclosed, and self-sown firs are now springing up in multitudes, so close together that all cannot live.

When I ascertained that these young trees had not been sown or planted, I was so much surprised at their numbers that I went to several points of view, whence I could examine hundreds of acres of the unenclosed heath, and literally I could not see a single Scotch fir, except the old planted clumps. But on looking closely between the stems of the heath, I found a multitude of seedlings and little trees, which had been perpetually browsed down by the cattle. In one square yard, at a point some hundred yards distant from one of the old clumps, I counted thirty-two little trees; and one of them, judging from the rings of growth, had during twenty-six years tried to raise its head above the stems of the heath, and had failed. No wonder that, as soon as the land was enclosed, it became thickly clothed with vigorously growing young firs. Yet the heath was so extremely barren and so extensive that no one would ever have imagined that cattle would have so closely and effectually searched it for food.

Here we see that cattle absolutely determine the existence of the Scotch fir; but in several parts of the world insects determine the existence of cattle. Perhaps Paraguay offers the most curious instance of this; for here neither cattle nor horses nor dogs have ever run wild, though they swarm southward and northward in a feral state; and Azara and Rengger have shown that this is caused by the greater number in Paraguay of a certain fly, which lays its eggs in the navels of these animals when first born. The increase of these flies, numerous as they are, must be habitually checked by some means, probably by birds. Hence, if certain insectivorous birds (whose numbers are probably regulated by hawks or beasts of prey) were to increase in Paraguay, the flies would decrease—then cattle and horses would become feral, and this would certainly greatly alter (as indeed I have observed in parts of South America) the vegetation: this again would largely affect the insects; and this, as we just have seen in Staffordshire, the insectivorous birds, and so onwards in ever-increasing circles of complexity. We began this series by insectivorous birds, and we have ended with them. Not that in nature the relations can ever be as simple as this. Battle within battle must ever be recurring with varying success; and yet in the long-run the forces are so nicely balanced, that the face of nature remains uniform for long periods of time, though assuredly the merest trifle would often give the victory to one organic being over another. Nevertheless so profound is our ignorance, and so high our presumption, that we marvel when we hear of the extinction of an organic being; and as we do not see the cause, we invoke cataclysms to desolate the world, or invent laws on the duration of the forms of life!

I am tempted to give one more instance showing how plants and animals, most remote in the scale of nature, are bound together by a web of complex relations. I shall hereafter have occasion to show that the exotic Lobelia fulgens, in this part of England, is never visited by insects, and consequently, from its peculiar structure, never can set a seed. Many of our orchidaceous plants absolutely require the visits of moths to remove their pollen-masses and thus to fertilise them. I have, also, reason to believe that humble-bees are indispensable to the fertilisation of the heartsease (Viola tricolor), for other bees do not visit this flower. From experiments which I have tried, I have found that the visits of bees, if not indispensable, are at least highly beneficial to the fertilisation of our clovers; but humble-bees alone visit the common red clover (Trifolium pratense), as other bees cannot reach the nectar. Hence I have very little doubt, that if the whole genus of humble-bees became extinct or very rare in England, the heartsease and red clover would become very rare, or wholly disappear. The number of humble-bees in any district depends in a great degree on the number of field-mice, which destroy their combs and nests; and Mr. H. Newman, who has long attended to the habits of humble-bees, believes that "more than two-thirds of them are thus destroyed all over England." Now the number of mice is largely dependent, as every one knows, on the number of cats; and Mr. Newman says, "Near villages and small towns I have found the nests of humble-bees more numerous than elsewhere, which I attribute to the number of cats that destroy the mice." Hence it is quite credible that the presence of a feline animal in large numbers in a district might determine, through the intervention first of mice and then of bees, the frequency of certain flowers in that district!

In the case of every species, many different checks, acting at different periods of life, and during different seasons or years, probably come into play; some one check or some few being generally the most potent, but all concurring in determining the average number or even the existence of the species. In some cases it can be shown that widely-different checks act on the same species in different districts. When we look at the plants and bushes clothing an entangled bank, we are tempted to attribute their proportional numbers and kinds to what we call chance. But how false a view is this! Every one has heard that when an American forest is cut down, a very different vegetation springs up; but it has been observed that the trees now growing on the ancient Indian mounds, in the Southern United States, display the same beautiful diversity and proportion of kinds as in the surrounding virgin forests. What a struggle between the several kinds of trees

must here have gone on during long centuries, each annually scattering its seeds by the thousand; what war between insect and insect—between insects, snails, and other animals with birds and beasts of prey—all striving to increase, and all feeding on each other or on the trees or their seeds and seedlings, or on the other plants which first clothed the ground and thus checked the growth of the trees! Throw up a handful of feathers, and all must fall to the ground according to definite laws; but how simple is this problem compared to the action and reaction of the innumerable plants and animals which have determined, in the course of centuries, the proportional numbers and kinds of trees now growing on the old Indian ruins!

The dependency of one organic being on another, as of a parasite on its prey, lies generally between beings remote in the scale of nature. This is often the case with those which may strictly be said to struggle with each other for existence, as in the case of locusts and grass-feeding quadrupeds. But the struggle almost invariably will be most severe between the individuals of the same species, for they frequent the same districts, require the same food, and are exposed to the same dangers. In the case of varieties of the same species, the struggle will generally be almost equally severe, and we sometimes see the contest soon decided: for instance, if several varieties of wheat be sown together, and the mixed seed be resown, some of the varieties which best suit the soil or climate, or are naturally the most fertile, will beat the others and so yield more seed, and will consequently in a few years quite supplant the other varieties. To keep up a mixed stock of even such extremely close varieties as the variously coloured sweet-peas, they must be each year harvested separately, and the seed then mixed in due proportion, otherwise the weaker kinds will steadily decrease in numbers and disappear. So again with the varieties of sheep: it has been asserted that certain mountain-varieties will starve out other mountain-varieties, so that they cannot be kept together. The same result has followed from keeping together different varieties of the medicinal leech. It may even be doubted whether the varieties of any one of our domestic plants or animals have so exactly the same strength, habits, and constitution, that the original proportions of a mixed stock could be kept up for half a dozen generations, if they were allowed to struggle together, like beings in a state of nature, and if the seed or young were not annually sorted.

As species of the same genus have usually, though by no means invariably, some similarity in habits and constitution, and always in structure, the struggle will generally be more severe between species of the same genus, when they come into competition with each other, than between species of distinct genera. We see this in the recent extension over parts of the United States of one species of swallow having caused the decrease of another species. The recent increase of the missel-thrush in parts of Scotland has caused the decrease of the song-thrush. How frequently we hear of one species of rat taking the place of another species under the most different climates! In Russia the small Asiatic cockroach has everywhere driven before it its great congener. One species of charlock will supplant another, and so in other cases. We can dimly see why the competition should be most severe between allied forms, which fill nearly the same place in the economy of nature; but probably in no one case could we precisely say why one species has been victorious over another in the great battle of life.

A corollary of the highest importance may be deduced from the foregoing remarks, namely, that the structure of every organic being is related, in the most essential yet often hidden manner, to that of all other organic beings, with which it comes into competition for food or residence, or from which it has to escape, or on which it preys. This is obvious in the structure of the teeth and talons of the tiger; and in that of the legs and claws of the parasite which clings to the hair on the tiger's body. But in the beautifully plumed seed of the dandelion, and in the flattened and fringed legs of the water-beetle, the relation seems at first confined to the elements of air and water. Yet the advantage of plumed seeds no doubt stands in the closest relation to the land being already thickly clothed by other plants; so that the seeds may be widely distributed and fall on unoccupied ground. In the water-beetle, the structure of its legs, so well adapted for diving, allows it to compete with other aquatic insects, to hunt for its own prey, and to escape serving as prey to other animals.

The store of nutriment laid up within the seeds of many plants seems at first sight to have no sort of relation to other plants. But from the strong growth of young plants produced from such seeds (as peas and beans), when sown in the midst of long grass, I suspect that the chief use of the nutriment in the seed is to favour the growth of the young seedling, whilst struggling with other plants growing vigorously all around.

Look at a plant in the midst of its range, why does it not double or quadruple its numbers? We know that it can perfectly well withstand a little more heat or cold, dampness or dryness, for elsewhere it ranges into slightly hotter or colder, damper or drier districts. In this case we can clearly see that if we wished in imagination to give the plant the power of increasing in number, we should have to give it some advantage over its competitors, or over the animals which preyed on it. On the confines of its geographical range, a change of constitution with respect to climate would clearly be an advantage to our plant; but we have reason to believe that only a few plants or animals range so far, that they are destroyed by the rigour of the climate alone. Not until we reach the extreme confines of life, in the arctic regions or on the borders of an utter desert, will competition cease. The land may be extremely cold or dry, yet there will be competition between some few species, or between the individuals of the same species, for the warmest or dampest spots.

Hence, also, we can see that when a plant or animal is placed in a new country amongst new competitors, though the climate may be exactly the same as in its former home, yet the conditions of its life will generally be changed in an essential manner. If we wished to increase its average numbers in its new home, we should have to modify it in a different way to what we should have done in its native country; for we should have to give it some advantage over a different set of competitors or enemies.

It is good thus to try in our imagination to give any form some advantage over another. Probably in no single instance should we know what to do, so as to succeed. It will convince us of our ignorance on the mutual relations of all organic beings; a conviction as necessary, as it seems to be difficult to acquire. All that we can do, is to keep steadily in mind that each organic being is striving to increase at a geometrical ratio; that each at some period of its life, during some season of the year, during each generation or at intervals, has to struggle for life, and to suffer great destruction. When we reflect on this struggle, we may console ourselves with the full belief, that the war of nature is not incessant, that no fear is felt, that death is generally prompt, and that the vigorous, the healthy, and the happy survive and multiply.

CHAPTER I.

INTRODUCTORY.

NATURAL ADVANTAGES OF THE TERRITORY OF THE ROMAN EMPIRE—PHYS-
ICAL DECAY OF THAT TERRITORY AND OF OTHER PARTS OF THE OLD WORLD
—CAUSES OF THE DECAY—NEW SCHOOL OF GEOGRAPHERS—REACTION OF
MAN UPON NATURE—OBSERVATION OF NATURE—COSMICAL AND GEOLOGICAL
INFLUENCES—GEOGRAPHICAL INFLUENCE OF MAN—UNCERTAINTY OF OUR
METEOROLOGICAL KNOWLEDGE—MECHANICAL EFFECTS PRODUCED BY MAN
ON THE SURFACE OF THE EARTH—IMPORTANCE AND POSSIBILITY OF PHYS-
ICAL RESTORATION—STABILITY OF NATURE—RESTORATION OF DISTURBED
HARMONIES—DESTRUCTIVENESS OF MAN—PHYSICAL IMPROVEMENT—HUMAN
AND BRUTE ACTION COMPARED—FORMS AND FORMATIONS MOST LIABLE TO
PHYSICAL DEGRADATION—PHYSICAL DECAY OF NEW COUNTRIES—CORRUPT
INFLUENCE OF PRIVATE CORPORATIONS, *note*.

Natural Advantages of the Territory of the Roman Empire.

THE Roman Empire, at the period of its greatest expansion, comprised the regions of the earth most distinguished by a happy combination of physical advantages. The provinces bordering on the principal and the secondary basins of the Mediterranean enjoyed a healthfulness and an equability of climate, a fertility of soil, a variety of vegetable and mineral products, and natural facilities for the transportation and distribution of exchangeable commodities, which have not been possessed in an equal degree by any territory of like extent in the Old World or the New. The abundance of the land and of the waters adequately supplied every material want, ministered liberally to every sensuous enjoyment. Gold and silver, indeed, were not found in the profusion which has proved so baneful to the industry of lands richer in veins of the precious metals; but mines and river beds yielded them in the spare measure most favorable to stability of value in the medium of exchange, and, consequently, to the regularity of commercial transactions. The ornaments of the barbaric pride of the East, the pearl, the ruby, the sapphire, and the diamond—though not unknown to the luxury of a people whose conquests and whose wealth commanded whatever the habitable world could contribute to augment the material splendor of their social life—were scarcely native to the territory of the empire; but the comparative rarity of these gems in Europe, at somewhat earlier periods, was, perhaps, the very circumstance that led the cunning artists of classic antiquity to enrich softer stones with engravings, which invest the common onyx and carnelian with a worth surpassing, in cultivated eyes, the lustre of the most brilliant oriental jewels.

Of these manifold blessings the temperature of the air, the distribution of the rains, the relative disposition of land and water, the plenty of the sea, the composition of the soil, and the raw material of some of the arts, were wholly gratuitous gifts. Yet the spontaneous nature of Europe, of Western Asia, of Libya, neither fed nor clothed the civilized inhabitants of those provinces. Every loaf was eaten in the sweat of the brow. All must be earned by toil. But toil was nowhere else rewarded by so generous wages; for nowhere would a given amount of intelligent labor produce so abundant, and, at the same time, so varied returns of the good things of material existence. The luxuriant harvests of cereals that waved on every field from the shores of the Rhine to the banks of the Nile, the vines that festooned the hillsides of Syria, of Italy, and of Greece, the olives of Spain, the fruits of the gardens of the Hesperides, the domestic quadrupeds and fowls known in ancient rural husbandry—all these were original products of foreign climes, naturalized in new homes, and gradually ennobled by the art of man, while centuries of persevering labor were expelling the wild vegetation, and fitting the earth for the production of more generous growths.

Only for the sense of landscape beauty did unaided nature

make provision. Indeed, the very commonness of this source of refined enjoyment seems to have deprived it of half its value; and it was only in the infancy of lands where all the earth was fair, that Greek and Roman humanity had sympathy enough with the inanimate world to be alive to the charms of rural and of mountain scenery. In later generations, when the glories of the landscape had been heightened by plantation, and decorative architecture, and other forms of picturesque improvement, the poets of Greece and Rome were blinded by excess of light, and became, at last, almost insensible to beauties that now, even in their degraded state, enchant every eye, except, too often, those which a lifelong familiarity has dulled to their attractions.

Physical Decay of the Territory of the Roman Empire, and of other parts of the Old World.

If we compare the present physical condition of the countries of which I am speaking, with the descriptions that ancient historians and geographers have given of their fertility and general capability of ministering to human uses, we shall find that more than one half of their whole extent—including the provinces most celebrated for the profusion and variety of their spontaneous and their cultivated products, and for the wealth and social advancement of their inhabitants—is either deserted by civilized man and surrendered to hopeless desolation, or at least greatly reduced in both productiveness and population. Vast forests have disappeared from mountain spurs and ridges; the vegetable earth accumulated beneath the trees by the decay of leaves and fallen trunks, the soil of the alpine pastures which skirted and indented the woods, and the mould of the upland fields, are washed away; meadows, once fertilized by irrigation, are waste and unproductive, because the cisterns and reservoirs that supplied the ancient canals are broken, or the springs that fed them dried up; rivers famous in history and song have shrunk to humble brooklets; the willows that ornamented and protected the banks of the lesser watercourses are gone, and the rivulets have ceased to exist as perennial currents, because the little water that finds its way into their old channels is evaporated by the droughts of summer, or absorbed by the parched earth, before it reaches the lowlands; the beds of the brooks have widened into broad expanses of pebbles and gravel, over which, though in the hot season passed dryshod, in winter sealike torrents thunder; the entrances of navigable streams are obstructed by sandbars, and harbors, once marts of an extensive commerce, are shoaled by the deposits of the rivers at whose mouths they lie; the elevation of the beds of estuaries, and the consequently diminished velocity of the streams which flow into them, have converted thousands of leagues of shallow sea and fertile lowland into unproductive and miasmatic morasses.

Besides the direct testimony of history to the ancient fertility of the regions to which I refer—Northern Africa, the greater Arabian peninsula, Syria, Mesopotamia, Armenia, and many other provinces of Asia Minor, Greece, Sicily, and parts of even Italy and Spain—the multitude and extent of yet remaining architectural ruins, and of decayed works of internal improvement, show that at former epochs a dense population inhabited those now lonely districts. Such a population could have been sustained only by a productiveness of soil of which we at present discover but slender traces; and the abundance derived from that fertility serves to explain how large armies, like those of the ancient Persians, and of the Crusaders and the Tartars in later ages, could, without an organized commmissariat, secure adequate supplies in long marches through territories which, in our times, would scarcely afford forage for a single regiment.

It appears, then, that the fairest and fruitfulest provinces of the Roman Empire, precisely that portion of terrestrial surface, in short, which, about the commencement of the Christian era, was endowed with the greatest superiority of soil, climate, and position, which had been carried to the highest pitch of physical improvement, and which thus combined the natural and artificial conditions best fitting it for the habitation and enjoyment of a dense and highly refined and cultivated population, is now completely exhausted of its fertility, or so diminished in productiveness, as, with the exception of a few favored oases that have escaped the general ruin, to be no longer capable of affording sustenance to civilized man. If to this realm of desolation we add the now wasted and solitary soils of Persia and the remoter East, that once fed their millions with milk and honey, we shall see that a territory larger than all Europe, the abundance of which sustained in bygone centuries a population scarcely inferior to that of the whole Christian world at the present day, has been entirely withdrawn from human use, or, at best, is thinly inhabited by tribes too few in numbers, too poor in superfluous products, and too little advanced in culture and the social arts, to contribute anything to the general moral or material interests of the great commonwealth of man.

Causes of this Decay.

The decay of these once flourishing countries is partly due, no doubt, to that class of geological causes, whose action we can neither resist nor guide, and partly also to the direct violence of hostile human force; but it is, in a far greater proportion, either the result of man's ignorant disregard of the laws of nature, or an incidental consequence of war, and of civil and ecclesiastical tyranny and misrule. Next to ignorance of these laws, the primitive source, the *causa causarum*, of the acts and neglects which have blasted with sterility and physical decrepitude the noblest half of the empire of the Cæsars, is, first, the brutal and exhausting despotism which Rome herself exercised over her conquered kingdoms, and even over her Italian territory; then, the host of temporal and spiritual tyrannies which she left as her dying curse to all her wide dominion, and which, in some form of violence or of fraud, still brood over almost every soil subdued by the Roman legions.* Man can-

---

* In the Middle Ages, feudalism, and a nominal Christianity whose corruptions had converted the most beneficent of religions into the most

not struggle at once against crushing oppression and the destructive forces of inorganic nature. When both are combined against him, he succumbs after a shorter or a longer struggle, and the fields he has won from the primeval wood relapse into their original state of wild and luxuriant, but

baneful of superstitions, perpetuated every abuse of Roman tyranny, and added new oppressions and new methods of extortion to those invented by older despotisms. The burdens in question fell most heavily on the provinces that had been longest colonized by the Latin race, and these are the portions of Europe which have suffered the greatest physical degradation. "Feudalism," says Blanqui, "was a concentration of scourges. The peasant, stripped of the inheritance of his fathers, became the property of inflexible, ignorant, indolent masters; he was obliged to travel fifty leagues with their carts whenever they required it; he labored for them three days in the week, and surrendered to them half the product of his earnings during the other three; without their consent he could not change his residence, or marry. And why, indeed, should he wish to marry, when he could scarcely save enough to maintain himself? The Abbot Alcuin had twenty thousand slaves, called *serfs*, who were forever attached to the soil. This is the great cause of the rapid depopulation observed in the Middle Ages, and of the prodigious multitude of monasteries which sprang up on every side. It was doubtless a relief to such miserable men to find in the cloisters a retreat from oppression; but the human race never suffered a more cruel outrage, industry never received a wound better calculated to plunge the world again into the darkness of the rudest antiquity. It suffices to say that the prediction of the approaching end of the world, industriously spread by the rapacious monks at this time, was received without terror."—*Résumé de l'Histoire du Commerce*, p. 156.

The abbey of Saint-Germain-des-Prés, which, in the time of Charlemagne, had possessed a million of acres, was, down to the Revolution, still so wealthy, that the personal income of the abbot was 300,000 livres. The abbey of Saint-Denis was nearly as rich as that of Saint-Germain-des-Prés.—LAVERGNE, *Économie Rurale de la France*, p. 104.

Paul Louis Courier quotes from La Bruyère the following striking picture of the condition of the French peasantry in his time: "One sees certain dark, livid, naked, sunburnt, wild animals, male and female, scattered over the country and attached to the soil, which they root and turn over with indomitable perseverance. They have, as it were, an articulate voice, and when they rise to their feet, they show a human face. They are, in fact, men; they creep at night into dens, where they live on black bread, water, and roots. They spare other men the labor of ploughing,

unprofitable forest growth, or fall into that of a dry and barren wilderness.

Rome imposed on the products of agricultural labor in the
rural districts taxes which the sale of the entire harvest would
scarcely discharge; she drained them of their population by
military conscription; she impoverished the peasantry by
forced and unpaid labor on public works; she hampered
industry and internal commerce by absurd restrictions and
unwise regulations. Hence, large tracts of land were left
uncultivated, or altogether deserted, and exposed to all the
destructive forces which act with such energy on the surface
of the earth when it is deprived of those protections by which
nature originally guarded it, and for which, in well-ordered
husbandry, human ingenuity has contrived more or less efficient substitutes.* Similar abuses have tended to perpetuate
and extend these evils in later ages, and it is but recently that,
even in the most populous parts of Europe, public attention

sowing, and harvesting, and therefore deserve some small share of the
bread they have grown." "These are his own words," adds Courier;
"he is speaking of the fortunate peasants, of those who had work and
bread, and they were then the few."—*Pétition à la Chambre des Députés
pour les Villageois que l'on empêche de danser.*

Arthur Young, who travelled in France from 1787 to 1789, gives, in
the twenty-first chapter of his Travels, a frightful account of the burdens
of the rural population even at that late period. Besides the regular
governmental taxes, and a multitude of heavy fines imposed for trifling
offences, he enumerates about thirty seignorial rights, the very origin and
nature of some of which are now unknown, while those of some others,
claimed and enforced by ecclesiastical as well as by temporal lords, are as
repulsive to humanity and morality, as the worst abuses ever practised by
heathen despotism. Most of these, indeed, had been commuted for money
payments, and were levied on the peasantry as pecuniary imposts for the
benefit of prelates and lay lords, who, by virtue of their nobility, were
exempt from taxation. Who can wonder at the hostility of the French
plebeian classes toward the aristocracy in the days of the Revolution?

* The temporary depopulation of an exhausted soil may be, in some
cases, a physical, though, like fallows in agriculture, a dear-bought advantage. Under favorable circumstances, the withdrawal of man and his
flocks allows the earth to clothe itself again with forests, and in a few
generations to recover its ancient productiveness. In the Middle Ages,

has been half awakened to the necessity of restoring the disturbed harmonies of nature, whose well-balanced influences are so propitious to all her organic offspring, of repaying to our great mother the debt which the prodigality and the thriftlessness of former generations have imposed upon their successors—thus fulfilling the command of religion and of practical wisdom, to use this world as not abusing it.

New School of Geographers.

The labors of Humboldt, of Ritter, of Guyot, and their followers have given to the science of geography a more philosophical, and, at the same time, a more imaginative character than it had received from the hands of their predecessors. Perhaps the most interesting field of speculation, thrown open by the new school to the cultivators of this attractive study, is the inquiry: how far external physical conditions, and especially the configuration of the earth's surface, and the distribution, outline, and relative position of land and water, have influenced the social life and social progress of man.

Reaction of Man on Nature.

But, as we have seen, man has reacted upon organized and inorganic nature, and thereby modified, if not determined, the material structure of his earthly home. The measure of that reaction manifestly constitutes a very important element in the appreciation of the relations between mind and matter, as well as in the discussion of many purely physical problems. But though the subject has been incidentally touched upon by many geographers, and treated with much fulness of detail in regard to certain limited fields of human effort, and to certain specific effects of human action, it has not, as a whole, so far as I know, been made matter of special observation, or of his-

worn-out fields were depopulated, in many parts of the Continent, by civil and ecclesiastical tyrannies, which insisted on the surrender of the half of a loaf already too small to sustain its producer. Thus abandoned, these lands often relapsed into the forest state, and, some centuries later, were again brought under cultivation with renovated fertility.

torical research by any scientific inquirer.* Indeed, until the influence of physical geography upon human life was recognized as a distinct branch of philosophical investigation, there was no motive for the pursuit of such speculations; and it was desirable to inquire whether we have or can become the architects of our own abiding place, only when it was known how the mode of our physical, moral, and intellectual being is affected by the character of the home which Providence has appointed, and we have fashioned, for our material habitation.†

It is still too early to attempt scientific method in discussing this problem, nor is our present store of the necessary facts by any means complete enough to warrant me in promising any approach to fulness of statement respecting them. Systematic observation in relation to this subject has hardly yet begun,‡ and the scattered data which have chanced to be recorded have never been collected. It has now no place in the general scheme of physical science, and is matter of sug-

* The subject of climatic change, with and without reference to human action as a cause, has been much discussed by Moreau de Jonnes, Dureau de la Malle, Arago, Humboldt, Fuster, Gasparin, Becquerel, and many other writers in Europe, and by Noah Webster, Forry, Drake, and others in America. Fraas has endeavored to show, by the history of vegetation in Greece, not merely that clearing and cultivation have affected climate, but that change of climate has essentially modified the character of vegetable life. See his *Klima und Pflanzenwelt in der Zeit*.

 † Gods Almagt wenkte van den troon,
  En schiep elk volk een land ter woon:
  Hier vestte Zij een grondgebied,
  Dat Zij ons zelven scheppen liet.

‡ The udometric measurements of Belgrand, reported in the *Annales Forestières* for 1854, and discussed by Vallès in chap. vi of his *Études sur les Inondations*, constitute the earliest, and, in some respects, the most remarkable series known to me, of persevering and systematic observations bearing directly and exclusively upon the influence of human action on climate, or, to speak more accurately, on precipitation and natural drainage. The conclusions of Belgrand, however, and of Vallès, who adopts them, have not been generally accepted by the scientific world, and they seem to have been, in part at least, refuted by the arguments of Héricourt and the observations of Cantegril, Jeandel, and Belland. See chapter iii : *The Woods*.

gestion and speculation only, not of established and positive conclusion. At present, then, all that I can hope is to excite an interest in a topic of much economical importance, by pointing out the directions and illustrating the modes in which human action has been or may be most injurious or most beneficial in its influence upon the physical conditions of the earth we inhabit.

...

Cosmical and Geological Influences.

The revolutions of the seasons, with their alternations of temperature, and of length of day and night, the climates of different zones, and the general condition and movements of the atmosphere and the seas, depend upon causes for the most part cosmical, and, of course, wholly beyond our control. The elevation, configuration, and composition of the great masses of terrestrial surface, and the relative extent and distribution of land and water, are determined by geological influences equally remote from our jurisdiction. It would hence seem that the physical adaptation of different portions of the earth to the use and enjoyment of man is a matter so strictly belong-ing to mightier than human powers, that we can only accept geographical nature as we find her, and be content with such soils and such skies as she spontaneously offers.

Geographical Influence of Man.

But it is certain that man has done much to mould the form of the earth's surface, though we cannot always distin-guish between the results of his action and the effects of purely geological causes ; that the destruction of the forests, the drainage of lakes and marshes, and the operations of rural husbandry and industrial art have tended to produce great changes in the hygrometric, thermometric, electric, and chem-ical condition of the atmosphere, though we are not yet able to measure the force of the different elements of disturbance, or to say how far they have been compensated by each other, or by still obscurer influences; and, finally, that the myriad forms of animal and vegetable life, which covered the earth when man first entered upon the theatre of a nature whose harmonies he was destined to derange, have been, through his action, greatly changed in numerical proportion, sometimes much modified in form and product, and sometimes entirely extirpated.

The physical revolutions thus wrought by man have not all been destructive to human interests. Soils to which no nutritious vegetable was indigenous, countries which once brought forth but the fewest products suited for the sustenance and comfort of man—while the severity of their climates created and stimulated the greatest number and the most imperious urgency of physical wants—surfaces the most rugged and intractable, and least blessed with natural facilities of communication, have been made in modern times to yield and distribute all that supplies the material necessities, all that contributes to the sensuous enjoyments and conveniences of civilized life. The Scythia, the Thule, the Britain, the Germany, and the Gaul which the Roman writers describe in such forbidding terms, have been brought almost to rival the native luxuriance and easily won plenty of Southern Italy; and, while the fountains of oil and wine that refreshed old Greece and Syria and Northern Africa have almost ceased to flow, and the soils of those fair lands are turned to thirsty and inhospitable deserts, the hyperborean regions of Europe have conquered, or rather compensated, the rigors of climate, and attained to a material wealth and variety of product that, with all their natural advantages, the granaries of the ancient world can hardly be said to have enjoyed.

These changes for evil and for good have not been caused by great natural revolutions of the globe, nor are they by any means attributable wholly to the moral and physical action or inaction of the peoples, or, in all cases, even of the races that now inhabit these respective regions. They are products of a complication of conflicting or coincident forces, acting through a long series of generations; here, improvidence, wastefulness, and wanton violence; there, foresight and wisely guided persevering industry. So far, as they are purely the calculated and desired results of those simple and familiar operations of agriculture and of social life which are as universal as civilization—the removal of the forests which covered the soil required for the cultivation of edible fruits, the drying of here and there a few acres too moist for profitable husbandry, by draining off the surface waters, the substitution of domesticated and nutritious for wild and unprofitable vegetable growths, the construction of roads and canals and artificial harbors—they belong to the sphere of rural, commercial, and political economy more properly than to geography, and hence are but incidentally embraced within the range of our present inquiries, which concern physical, not financial balances. I propose to examine only the greater, more permanent, and more comprehensive mutations which man has produced, and is producing, in earth, sea, and sky, sometimes, indeed, with conscious purpose, but for the most part, as unforseen though natural consequences of acts performed for narrower and more immediate ends.

The exact measurement of the geographical changes hitherto thus effected is, as I have hinted, impracticable, and we possess, in relation to them, the means of only qualitative, not quantitative analysis. The fact of such revolutions is established partly by historical evidence, partly by analogical deduction from effects produced in our own time by operations similar in character to those which must have taken place in more or less remote ages of human action. Both sources of information are alike defective in precision; the latter, for general reasons too obvious to require specification; the former, because the facts to which it bears testimony occurred before the habit or the means of rigorously scientific observation upon any branch of physical research, and especially upon climatic changes, existed.

Importance and Possibility of Physical Restoration.

Many circumstances conspire to invest with great present interest the questions : how far man can permanently modify and ameliorate those physical conditions of terrestrial surface and climate on which his material welfare depends ; how far he can compensate, arrest, or retard the deterioration which many of his agricultural and industrial processes tend to pro-duce ; and how far he can restore fertility and salubrity to soils which his follies or his crimes have made barren or pestilential. Among these circumstances, the most prominent, perhaps, is the necessity of providing new homes for a European popula-tion which is increasing more rapidly than its means of subsist-ence, new physical comforts for classes of the people that have now become too much enlightened and have imbibed too much culture to submit to a longer deprivation of a share in the material enjoyments which the privileged ranks have hith-erto monopolized.

To supply new hives for the emigrant swarms, there are, first, the vast unoccupied prairies and forests of America, of Australia, and of many other great oceanic islands, the sparsely inhabited and still unexhausted soils of Southern and even Central Africa, and, finally, the impoverished and half-depopulated shores of the Mediterranean, and the interior of Asia Minor and the farther East. To furnish to those who shall remain after emigration shall have conveniently reduced the too dense population of many European states, those means of sensuous and of intellectual well-being which are styled " artificial wants " when demanded by the humble and the poor, but are admitted to be " necessaries " when claimed by the noble and the rich, the soil must be stimulated to its highest powers of production, and man's utmost ingenuity and energy must be tasked to renovate a nature drained, by his improvidence, of fountains which a wise economy would have made plenteous and perennial sources of beauty, health, and wealth.

In those yet virgin lands which the progress of modern discovery in both hemispheres has brought and is still bringing to the knowledge and control of civilized man, not much improvement of great physical conditions is to be looked for. The proportion of forest is indeed to be considerably reduced, superfluous waters to be drawn off, and routes of internal communication to be constructed ; but the primitive geographical and climatic features of these countries ought to be, as far as possible, retained.

...

Restoration of Disturbed Harmonies.

In reclaiming and reoccupying lands laid waste by human improvidence or malice, and abandoned by man, or occupied only by a nomade or thinly scattered population, the task of the pioneer settler is of a very different character. He is to become a co-worker with nature in the reconstruction of the damaged fabric which the negligence or the wantonness of former lodgers has rendered untenantable. He must aid her in reclothing the mountain slopes with forests and vegetable mould, thereby restoring the fountains which she provided to water them; in checking the devastating fury of torrents, and bringing back the surface drainage to its primitive narrow channels; and in drying deadly morasses by opening the natural sluices which have been choked up, and cutting new canals for drawing off their stagnant waters. He must thus, on the one hand, create new reservoirs, and, on the other, remove mischievous accumulations of moisture, thereby equalizing and regulating the sources of atmospheric humidity and of flowing water, both which are so essential to all vegetable growth, and, of course, to human and lower animal life.

Destructiveness of Man.

Man has too long forgotten that the earth was given to him for usufruct alone, not for consumption, still less for profligate waste. Nature has provided against the absolute destruction of any of her elementary matter, the raw material of her works; the thunderbolt and the tornado, the most convulsive throes of even the volcano and the earthquake, being only phenomena of decomposition and recomposition. But she has left it within the power of man irreparably to derange the combinations of inorganic matter and of organic life, which through the night of æons she had been proportioning and balancing, to prepare the earth for his habitation, when, in the fulness of time, his Creator should call him forth to enter into its possession.

Apart from the hostile influence of man, the organic and

the inorganic world are, as I have remarked, bound together
by such mutual relations and adaptations as secure, if not the
absolute' permanence and equilibrium of both, a long continuance of the established conditions of each at any given time
and place, or at least, a very slow and gradual succession of
changes in those conditions. But man is everywhere a disturbing agent. Wherever he plants his foot, the harmonies of
nature are turned to discords. The proportions and accom-
modations which insured the stability of existing arrangements are overthrown. Indigenous vegetable and animal
species are extirpated, and supplanted by others of foreign
origin, spontaneous production is forbidden or restricted, and
the face of the earth is either laid bare or covered with a new
and reluctant growth of vegetable forms, and with alien tribes
of animal life. These intentional changes and substitutions
constitute, indeed, great revolutions; but vast as is their
magnitude and importance, they are, as we shall see, insignificant in comparison with the contingent and unsought
results which have flowed from them.

The fact that, of all organic beings, man alone is to be
regarded as essentially a destructive power, and that he wields
energies to resist which, nature—that nature whom all
material life and all inorganic substance obey—is wholly
impotent, tends to prove that, though living in physical
nature, he is not of her, that he is of more exalted parentage,
and belongs to a higher order of existences than those born of
her womb and submissive to her dictates.

There are, indeed, brute destroyers, beasts and birds and
insects of prey—all animal life feeds upon, and, of course,
destroys other life,—but this destruction is balanced by compensations. It is, in fact, the very means by which the existence of one tribe of animals or of vegetables is secured against
being smothered by the encroachments of another; and the
reproductive powers of species, which serve as the food of
others, are always proportioned to the demand they are
destined to supply. Man pursues his victims with reckless
destructiveness; and, while the sacrifice of life by the lower

animals is limited by the cravings of appetite, he unsparingly
persecutes, even to extirpation, thousands of organic forms
which he cannot consume.*

* The terrible destructiveness of man is remarkably exemplified in the
chase of large mammalia and birds for single products, attended with the
entire waste of enormous quantities of flesh, and of other parts of the animal, which are capable of valuable uses. The wild cattle of South America
are slaughtered by millions for their hides and horns; the buffalo of North
America for his skin or his tongue; the elephant, the walrus, and the
narwhal for their tusks; the cetacea, and some other marine animals, for
their oil and whalebone; the ostrich and other large birds, for their
plumage. Within a few years, sheep have been killed in New England by
whole flocks, for their pelts and suet alone, the flesh being thrown away;
and it is even said that the bodies of the same quadrupeds have been used
in Australia as fuel for limekilns. What a vast amount of human nutriment, of bone, and of other animal products valuable in the arts, is thus
recklessly squandered! In nearly all these cases, the part which constitutes the motive for this wholesale destruction, and is alone saved, is
essentially of insignificant value as compared with what is thrown away.
The horns and hide of an ox are not economically worth a tenth part as
much as the entire carcass.

One of the greatest benefits to be expected from the improvements of
civilization is, that increased facilities of communication will render it possible to transport to places of consumption much valuable material that is
now wasted because the price at the nearest market will not pay freight.
The cattle slaughtered in South America for their hides would feed millions of the starving population of the Old World, if their flesh could be
economically preserved and transported across the ocean.

We are beginning to learn a better economy in dealing with the inorganic world. The utilization—or, as the Germans more happily call it,
the Verwerthung, the *beworthing*—of waste from metallurgical, chemical,
and manufacturing establishments, is among the most important results of
the application of science to industrial purposes. The incidental products
from the laboratories of manufacturing chemists often become more valuable than those for the preparation of which they were erected. The slags
from silver refineries, and even from smelting houses of the coarser metals,
have not unfrequently yielded to a second operator a better return than
the first had derived from dealing with the natural ore; and the saving of
lead carried off in the smoke of furnaces has, of itself, given a large profit
on the capital invested in the works. A few years ago, an officer of an
American mint was charged with embezzling gold committed to him for
coinage. He insisted, in his defence, that much of the metal was vola-

The earth was not, in its natural condition, completely adapted to the use of man, but only to the sustenance of wild animals and wild vegetation. These live, multiply their kind in just proportion, and attain their perfect measure of strength and beauty, without producing or requiring any change in the natural arrangements of surface, or in each other's spontaneous tendencies, except such mutual repression of excessive increase as may prevent the extirpation of one species by the encroachments of another. In short, without man, lower animal and spontaneous vegetable life would have been constant in type, distribution, and proportion, and the physical geography of the earth would have remained undisturbed for indefinite periods, and been subject to revolution only from possible, unknown cosmical causes, or from geological action.

But man, the domestic animals that serve him, the field and garden plants the products of which supply him with food and clothing, cannot subsist and rise to the full development of their higher properties, unless brute and unconscious nature be effectually combated, and, in a great degree, vanquished by human art. Hence, a certain measure of transformation of terrestrial surface, of suppression of natural, and stimulation of artificially modified productivity becomes necessary. This measure man has unfortunately exceeded. He has felled the forests whose network of fibrous roots bound the mould to the rocky skeleton of the earth ; but had he allowed here and there a belt of woodland to reproduce itself by spontaneous propagation, most of the mischiefs which his reckless destruction of the natural protection of the soil has occasioned would have been averted. He has broken up the mountain reservoirs, the percolation of whose waters through unseen channels supplied the fountains that refreshed his cattle and fertilized his fields ; but he has neglected to maintain the cisterns and the canals of irrigation which a wise antiquity

tilized and lost in refining and melting, and upon scraping the chimneys of the melting furnaces and the roofs of the adjacent houses, gold enough was found in the soot to account for no small part of the deficiency.

had constructed to neutralize the consequences of its own imprudence. While he has torn the thin glebe which confined the light earth of extensive plains, and has destroyed the fringe of semi-aquatic plants which skirted the coast and checked the drifting of the sea sand, he has failed to prevent the spreading of the dunes by clothing them with artificially propagated vegetation. He has ruthlessly warred on all the tribes of animated nature whose spoil he could convert to his own uses, and he has not protected the birds which prey on the insects most destructive to his own harvests.

Purely untutored humanity, it is true, interferes comparatively little with the arrangements of nature,* and the destruc-

---

* It is an interesting and not hitherto sufficiently noticed fact, that the domestication of the organic world, so far as it has yet been achieved, belongs, not indeed to the savage state, but to the earliest dawn of civilization, the conquest of inorganic nature almost as exclusively to the most advanced stages of artificial culture. It is familiarly known to all who have occupied themselves with the psychology and habits of the ruder races, and of persons with imperfectly developed intellects in civilized life, that although these humble tribes and individuals sacrifice, without scruple, the lives of the lower animals to the gratification of their appetites and the supply of their other physical wants, yet they nevertheless seem to cherish with brutes, and even with vegetable life, sympathies which are much more feebly felt by civilized men. The popular traditions of the simpler peoples recognize a certain community of nature between man, brute animals, and even plants; and this serves to explain why the apologue or fable, which ascribes the power of speech and the faculty of reason to birds, quadrupeds, insects, flowers, and trees, is one of the earliest forms of literary composition.

In almost every wild tribe, some particular quadruped or bird, though persecuted as a destroyer of more domestic beasts, or hunted for food, is regarded with peculiar respect, one might almost say, affection. Some of the North American aboriginal nations celebrate a propitiatory feast to the manes of the intended victim before they commence a bear hunt; and the Norwegian peasantry have not only retained an old proverb which ascribes to the same animal "*ti Mænds Styrke og tolv Mænds Vid,*" ten men's strength and twelve men's cunning; but they still pay to him something of the reverence with which ancient superstition invested him. The student of Icelandic literature will find in the saga of *Finnbogi hinn rami* a curious illustration of this feeling, in an account of a dialogue between a

tive agency of man becomes more and more energetic and unsparing as he advances in civilization, until the impoverishment, with which his exhaustion of the natural resources of the soil is threatening him, at last awakens him to the neces-

Norwegian bear and an Icelandic champion—dumb show on the part of Bruin, and chivalric words on that of Finnbogi—followed by a duel, in which the latter, who had thrown away his arms and armor in order that the combatants might meet on equal terms, was victorious. Drummond Hay's very interesting work on Morocco contains many amusing notices of a similar feeling entertained by the Moors toward the redoubtable enemy of their flocks—the lion.

This sympathy helps us to understand how it is that most if not all the domestic animals—if indeed they ever existed in a wild state—were appropriated, reclaimed and trained before men had been gathered into organized and fixed communities, that almost every known esculent plant had acquired substantially its present artificial character, and that the properties of nearly all vegetable drugs and poisons were known at the remotest period to which historical records reach. Did nature bestow upon primitive man some instinct akin to that by which she teaches the brute to select the nutritious and to reject the noxious vegetables indiscriminately mixed in forest and pasture?

This instinct, it must be admitted, is far from infallible, and, as has been hundreds of times remarked by naturalists, it is in many cases not an original faculty but an acquired and transmitted habit. It is a fact familiar to persons engaged in sheep husbandry in New England—and I have seen it confirmed by personal observation—that sheep bred where the common laurel, as it is called, *Kalmia angustifolia*, abounds, almost always avoid browsing upon the leaves of that plant, while those brought from districts where laurel is unknown, and turned into pastures where it grows, very often feed upon it and are poisoned by it. A curious acquired and hereditary instinct, of a different character, may not improperly be noticed here. I refer to that by which horses bred in provinces where quicksands are common avoid their dangers or extricate themselves from them. See BRÉMONTIER, *Mémoire sur les Dunes, Annales des Ponts et Chaussées*, 1833 : *premier sémestre*, pp. 155–157.

It is commonly said in New England, and I believe with reason, that the crows of this generation are wiser than their ancestors. Scarecrows which were effectual fifty years ago are no longer respected by the plunderers of the cornfield, and new terrors must from time to time be invented for its protection.

Civilization has added little to the number of vegetable or animal species grown in our fields or bred in our folds, while, on the contrary,

sity of preserving what is left, if not of restoring what has been wantonly wasted. The wandering savage grows no cultivated vegetable, fells no forest, and extirpates no useful plant, no noxious weed. If his skill in the chase enables him to entrap numbers of the animals on which he feeds, he compensates this loss by destroying also the lion, the tiger, the wolf, the otter, the seal, and the eagle, thus indirectly protecting the feebler quadrupeds and fish and fowls, which would otherwise become the booty of beasts and birds of prey. But with stationary life, or rather with the pastoral state, man at once commences an almost indiscriminate warfare upon all the forms of animal and vegetable existence around him, and as he advances in civilization, he gradually eradicates or transforms every spontaneous product of the soil he occupies.*

Human and Brute Action Compared.

It has been maintained by authorities as high as any known to modern science, that the action of man upon nature, though greater in *degree*, does not differ in *kind*, from

the subjugation of the inorganic forces, and the consequent extension of man's sway over, not the annual products of the earth only, but her substance and her springs of action, is almost entirely the work of highly refined and cultivated ages. The employment of the elasticity of wood and of horn, as a projectile power in the bow, is nearly universal among the rudest savages. The application of compressed air to the same purpose, in the blowpipe, is more restricted, and the use of the mechanical powers, the inclined plane, the wheel and axle, and even the wedge and lever, seems almost unknown except to civilized man. I have myself seen European peasants to whom one of the simplest applications of this latter power was a revelation.

* The difference between the relations of savage life, and of incipient civilization, to nature, is well seen in that part of the valley of the Mississippi which was once occupied by the mound builders and afterward by the far less developed Indian tribes. When the tillers of the fields which must have been cultivated to sustain the large population that once inhabited those regions perished or were driven out, the soil fell back to the normal forest state, and the savages who succeeded the more advanced race interfered very little, if at all, with the ordinary course of spontaneous nature.

that of wild animals. It appears to me to differ in essential character, because, though it is often followed by unforeseen and undesired results, yet it is nevertheless guided by a self-conscious and intelligent will aiming as often at secondary and remote as at immediate objects. The wild animal, on the other hand, acts instinctively, and, so far as we are able to perceive, always with a view to single and direct purposes. The backwoodsman and the beaver alike fell trees; the man that he may convert the forest into an olive grove that will mature its fruit only for a succeeding generation, the beaver that he may feed upon their bark or use them in the construction of his habitation. Human differs from brute action, too, in its influence upon the material world, because it is not controlled by natural compensations and balances. Natural arrangements, once disturbed by man, are not restored until he retires from the field, and leaves free scope to spontaneous recuperative energies; the wounds he inflicts upon the material creation are not healed until he withdraws the arm that gave the blow. On the other hand, I am not aware of any evidence that wild animals have ever destroyed the smallest forest, extirpated any organic species, or modified its natural character, occasioned any permanent change of terrestrial surface, or produced any disturbance of physical conditions which nature has not, of herself, repaired without the expulsion of the animal that had caused it.*

The form of geographical surface, and very probably the climate of a given country, depend much on the character of the vegetable life belonging to it. Man has, by domestication, greatly changed the habits and properties of the plants he rears; he has, by voluntary selection, immensely modified the forms and qualities of the animated creatures that serve him; and he has, at the same time, completely rooted out many forms of both vegetable and animal being.† What is there, in

---

* There is a possible—but only a possible—exception in the case of the American bison. See note on that subject in chap. iii, *post*.

† Whatever may be thought of the modification of organic species by natural selection, there is certainly no evidence that animals have exerted

the influence of brute life, that corresponds to this? We have no reason to believe that in that portion of the American continent which, though peopled by many tribes of quadruped and fowl, remained uninhabited by man, or only thinly occupied by purely savage tribes, any sensible geographical change had occurred within twenty centuries before the epoch of discovery and colonization, while, during the same period, man had changed millions of square miles, in the fairest and most fertile regions of the Old World, into the barrenest deserts.

The ravages committed by man subvert the relations and destroy the balance which nature had established between her organized and her inorganic creations; and she avenges herself upon the intruder, by letting loose upon her defaced provinces destructive energies hitherto kept in check by organic forces destined to be his best auxiliaries, but which he has unwisely dispersed and driven from the field of action. When the forest is gone, the great reservoir of moisture stored up in its vegetable mould is evaporated, and returns only in deluges of rain to wash away the parched dust into which that mould has been converted. The well-wooded and humid hills are turned to ridges of dry rock, which encumbers the low grounds and chokes the watercourses with its debris, and—except in countries favored with an equable distribution of rain through the seasons, and a moderate and regular inclination of surface—the whole earth, unless rescued by human art from the physical degradation to which it tends, becomes an assemblage of bald mountains, of barren, turfless hills, and of swampy and malarious plains. There are parts of Asia Minor, of Northern Africa, of Greece, and even of Alpine Europe, where the operation of causes set in action by man has brought the face of the earth to a desolation almost as complete as that of the moon; and though, within that brief space

upon any form of life an influence analogous to that of domestication upon plants, quadrupeds, and birds reared artificially by man; and this is as true of unforeseen as of purposely effected improvements accomplished by voluntary selection of breeding animals.

of time which we call "the historical period," they are known to have been covered with luxuriant woods, verdant pastures, and fertile meadows, they are now too far deteriorated to be reclaimable by man, nor can they become again fitted for human use, except through great geological changes, or other mysterious influences or agencies of which we have no present knowledge, and over which we have no prospective control. The earth is fast becoming an unfit home for its noblest inhabitant, and another era of equal human crime and human improvidence, and of like duration with that through which traces of that crime and that improvidence extend, would reduce it to such a condition of impoverished productiveness, of shattered surface, of climatic excess, as to threaten the depravation, barbarism, and perhaps even extinction of the species.*

...

Physical Decay of New Countries.

I have remarked that the effects of human action on the forms of the earth's surface could not always be distinguished from those resulting from geological causes, and there is also much uncertainty in respect to the precise influence of the

---

* ——"And it may be remarked that, as the world has passed through these several stages of strife to produce a Christendom, so by relaxing in the enterprises it has learnt, does it tend downwards, through inverted steps, to wildness and the waste again. Let a people give up their contest with moral evil; disregard the injustice, the ignorance, the greediness, that may prevail among them, and part more and more with the Christian ele-ment of their civilization; and in declining this battle with sin, they will inevitably get embroiled with men. Threats of war and revolution punish their unfaithfulness; and if then, instead of retracing their steps, they yield again, and are driven before the storm, the very arts they had cre-ated, the structures they had raised, the usages they had established, are swept away; 'in that very day their thoughts perish.' The portion they had reclaimed from the young earth's ruggedness is lost; and failing to stand fast against man, they finally get embroiled with nature, and are thrust down beneath her ever-living hand."—MARTINEAU's *Sermon, "The Good Soldier of Jesus Christ."*

clearing and cultivating of the ground, and of other rural operations, upon climate. It is disputed whether either the mean or the extremes of temperature, the periods of the seasons, or the amount or distribution of precipitation and of evaporation, in any country whose annals are known, have undergone any change during the historical period. It is, indeed, impossible to doubt that many of the operations of the pioneer settler tend to produce great modifications in atmospheric humidity, temperature, and electricity; but we are at present unable to determine how far one set of effects is neutralized by another, or compensated by unknown agencies. This question scientific research is inadequate to solve, for want of the necessary data; but well conducted observation, in regions now first brought under the occupation of man, combined with such historical evidence as still exists, may be expected at no distant period to throw much light on this subject.

Australia is, perhaps, the country from which we have a right to expect the fullest elucidation of these difficult and disputable problems. Its colonization did not commence until the physical sciences had become matter of almost universal attention, and is, indeed, so recent that the memory of living men embraces the principal epochs of its history; the peculiarities of its fauna, its flora, and its geology are such as to have excited for it the liveliest interest of the votaries of natural science; its mines have given its people the necessary wealth for procuring the means of instrumental observation, and the leisure required for the pursuit of scientific research; and large tracts of virgin forest and natural meadow are rapidly passing under the control of civilized man. Here, then, exist greater facilities and stronger motives for the careful study of the topics in question than have ever been found combined in any other theatre of European colonization.

In North America, the change from the natural to the artificial condition of terrestrial surface began about the period when the most important instruments of meteorological observation were invented. The first settlers in the territory now constituting the United States and the British American provinces had other things to do than to tabulate barometrical and thermometrical readings, but there remain some interesting physical records from the early days of the colonies,* and there is still an immense extent of North American soil where the industry and the folly of man have as yet produced little appreciable change. Here, too, with the present increased facilities for scientific observation, the future effects, direct and contingent, of man's labors, can be measured, and such precautions taken in those rural processes which we call improvements, as to mitigate evils, perhaps, in some degree, inseparable from every attempt to control the action of natural laws.

In order to arrive at safe conclusions, we must first obtain a more exact knowledge of the topography, and of the present superficial and climatic condition of countries where the natural surface is as yet more or less unbroken. This can only be accomplished by accurate surveys, and by a great multiplication of the points of meteorological registry,† already so

* The Travels of Dr. Dwight, president of Yale College, which embody the results of his personal observations, and of his inquiries among the early settlers, in his vacation excursions in the Northern States of the American Union, though presenting few instrumental measurements or tabulated results, are of value for the powers of observation they exhibit, and for the sound common sense with which many natural phenomena, such for instance as the formation of the river meadows, called "intervales," in New England, are explained. They present a true and interesting picture of physical conditions, many of which have long ceased to exist in the theatre of his researches, and of which few other records are extant.

† The general law of temperature is that it decreases as we ascend. But, in hilly regions, the law is reversed in cold, still weather, the cold air descending, by reason of its greater gravity, into the valleys. If there be wind enough, however, to produce a disturbance and intermixture of higher and lower atmospheric strata, this exception to the general law does not take place. These facts have long been familiar to the common people of Switzerland and of New England, but their importance has not been sufficiently taken into account in the discussion of meteorological observations. The descent of the cold air and the rise of the warm affect

numerous ; and as, moreover, considerable changes in the proportion of forest and of cultivated land, or of dry and wholly or partially submerged surface, will often take place within brief periods, it is highly desirable that the attention of observers, in whose neighborhood the clearing of the soil, or the drainage of lakes and swamps, or other great works of rural improvement, are going on or meditated, should be especially drawn not only to revolutions in atmospheric temperature and precipitation, but to the more easily ascertained and perhaps more important local changes produced by these operations in the temperature and the hygrometric state of the superficial strata of the earth, and in its spontaneous vegetable and animal products.

The rapid extension of railroads, which now everywhere keeps pace with, and sometimes even precedes, the occupation of new soil for agricultural purposes, furnishes great facilities for enlarging our knowledge of the topography of the territory they traverse, because their cuttings reveal the composition and general structure of surface, and the inclination and elevation of their lines constitute known hypsometrical sections, which give numerous points of departure for the measurement of higher and lower stations, and of course for determining the relief and depression of surface, the slope of the beds of watercourses, and many other not less important questions.*

the relative temperatures of hills and valleys to a much greater extent than has been usually supposed. A gentleman well known to me kept a thermometrical record for nearly half a century, in a New England country town, at an elevation of at least 1,500 feet above the sea. During these years his thermometer never fell lower than 26° Fahrenheit, but at the shire town of the county, situated in a basin one thousand feet lower, and ten miles distant, as well as at other points in similar positions, the mercury froze several times in the same period.

* Railroad surveys must be received with great caution where any motive exists for *cooking* them. Capitalists are shy of investments in roads with steep grades, and of course it is important to make a fair show of facilities in obtaining funds for new routes. Joint-stock companies have no souls; their managers, in general, no consciences. Cases can be cited

The geological, hydrographical, and topographical surveys, which almost every general and even local government of the civilized world is carrying on, are making yet more important contributions to our stock of geographical and general physical knowledge, and, within a comparatively short space, there will

where engineers and directors of railroads, with long grades above one hundred feet to the mile, have regularly sworn in their annual reports, for years in succession, that there were no grades upon their routes exceeding half that elevation. In fact, every person conversant with the history of these enterprises knows that in their public statements falsehood is the rule, truth the exception.

What I am about to remark is not exactly relevant to my subject; but it is hard to "get the floor" in the world's great debating society, and when a speaker who has anything to say once finds access to the public ear, he must make the most of his opportunity, without inquiring too nicely whether his observations are "in order." I shall harm no honest man by endeavoring, as I have often done elsewhere, to excite the attention of thinking and conscientious men to the dangers which threaten the great moral and even political interests of Christendom, from the unscrupulousness of the private associations that now control the monetary affairs, and regulate the transit of persons and property, in almost every civilized country. More than one American State is literally governed by unprincipled corporations, which not only defy the legislative power, but have, too often, corrupted even the administration of justice. Similar evils have become almost equally rife in England, and on the Continent; and I believe the decay of commercial morality, and I fear of the sense of all higher obligations than those of a pecuniary nature, on both sides of the Atlantic, is to be ascribed more to the influence of joint-stock banks and manufacturing and railway companies, to the workings, in short, of what is called the principle of "associate action," than to any other one cause of demoralization.

The apophthegm, "the world is governed too much," though unhappily too truly spoken of many countries—and perhaps, in some aspects, true of all—has done much mischief whenever it has been too unconditionally accepted as a political axiom. The popular apprehension of being over-governed, and, I am afraid, more emphatically the fear of being over-taxed, has had much to do with the general abandonment of certain governmental duties by the ruling powers of most modern states. It is theoretically the duty of government to provide all those public facilities of intercommunication and commerce, which are essential to the prosperity of civilized commonwealths, but which individual means are inadequate to furnish, and for the due administration of which individual guar-

be an accumulation of well established constant and historical facts, from which we can safely reason upon all the relations of action and reaction between man and external nature.

But we are, even now, breaking up the floor and wainscoting and doors and window frames of our dwelling, for fuel to warm our bodies and seethe our pottage, and the world cannot afford to wait till the slow and sure progress of exact science has taught it a better economy. Many practical lessons have been learned by the common observation of unschooled men; and the teachings of simple experience, on topics where natural philosophy has scarcely yet spoken, are not to be despised.

In these humble pages, which do not in the least aspire to rank among scientific expositions of the laws of nature, I shall

anties are insufficient. Hence public roads, canals, railroads, postal communications, the circulating medium of exchange, whether metallic or representative, armies, navies, being all matters in which the nation at large has a vastly deeper interest than any private association can have, ought legitimately to be constructed and provided only by that which is the visible personification and embodiment of the nation, namely, its legislative head. No doubt the organization and management of these institutions by government are liable, as are all things human, to great abuses. The multiplication of public placeholders, which they imply, is a serious evil. But the corruption thus engendered, foul as it is, does not strike so deep as the rottenness of private corporations; and official rank, position, and duty have, in practice, proved better securities for fidelity and pecuniary integrity in the conduct of the interests in question, than the suretyships of private corporate agents, whose bondsmen so often fail or abscond before their principal is detected.

Many theoretical statesmen have thought that voluntary associations for strictly pecuniary and industrial purposes, and for the construction and control of public works, might furnish, in democratic countries, a compensation for the small and doubtful advantages, and at the same time secure an exemption from the great and certain evils, of aristocratic institutions. The example of the American States shows that private corporations—whose rule of action is the interest of the association, not the conscience of the individual—though composed of ultra-democratic elements, may become most dangerous enemies to rational liberty, to the moral interests of the commonwealth, to the purity of legislation and of judicial action, and to the sacredness of private rights.

attempt to give the most important practical conclusions suggested by the history of man's efforts to replenish the earth and subdue it; and I shall aim to support those conclusions by such facts and illustrations only, as address themselves to the understanding of every intelligent reader, and as are to be found recorded in works capable of profitable perusal, or at least consultation, by persons who have not enjoyed a special scientific training.

INTRODUCTION.

DAY by day it becomes more obvious that the Coal we happily possess in excellent quality and abundance is the Mainspring of Modern Material Civilization. As Fuel—or the source of fire—it is the source at once of mechanical motion and of chemical change. Accordingly it is the chief agent in almost every improvement or discovery in the arts which the present age brings forth. It is to us indispensable for domestic purposes, and it has of late years been found to yield a series of organic substances, which puzzle us by their complexity, please us by their beautiful colours, and serve us by their various utility.

And as the source especially of Steam and Iron, Coal is all-powerful. This age has been called the Iron Age, and it is true that iron˙ is the material of most great novelties. By its strength, endurance, and wide range of qualities, it is fitted to be the fulcrum and lever of great works, while steam is the motive power. But coal alone can command in sufficient abundance either the iron or the steam; and coal, therefore, commands this age—the Age of Coal.

Coal, in truth, stands not beside but entirely above all other commodities. It is the material energy of the country — the universal aid — the factor in everything we do. With coal almost any feat is possible or easy; without it we are thrown back into the laborious poverty of early times.

With such facts familiarly before us, it can be no matter of surprise that year by year we make larger draughts upon a material of such myriad qualities—of such miraculous powers.

But it is at the same time impossible that men of foresight should not turn to compare with some anxiety the masses yearly drawn with the quantities known or supposed to lie within these islands.

Geologists of eminence, acquainted with the contents of our strata, and accustomed in the study of their great science to look over long periods of time with judgment and enlightenment, were long ago painfully struck by the essentially limited nature of our 'main wealth. And though others have been found to reassure the public, roundly asserting that all anticipations of exhaustion are groundless and absurd, and " may be deferred for an indefinite period," yet misgivings have constantly recurred to those really examining the question. Not long since the subject acquired new weight when prominently brought forward by Sir W. Armstrong in his Address to the British Association, at Newcastle, the very birth-place of the coal trade.

This question concerning the duration of our present cheap supplies of coal cannot but excite deep interest and anxiety wherever or whenever it is mentioned. For a little reflection will show that coal is almost the sole necessary basis of our material Power, and is that, consequently, which gives efficiency to our moral and intellectual capabilities. England's manufacturing and commercial greatness, at least, is at stake in this question, nor can we be sure that material decay may not involve us in moral and intellectual retrogression. And as there is no part of the civilized world where the life of our true and beneficent Commonwealth can be a matter of indifference, so to an Englishman who knows the grand and steadfast course his country has pursued to its present point, its future must be a matter of almost personal solicitude and affection.

The thoughtless and selfish, indeed, who fear any interference with the enjoyment of the present, will be apt to stigmatise all reasoning about the future as absurd and chimerical. But the opinions of such are closely guided by their wishes. It is true that at the best we see dimly into the future, but those who acknowledge their duty to posterity will feel impelled to use their foresight upon what facts and guiding principles we do possess. Though many data are at present wanting, or doubtful, our conclusions may be rendered so far probable as to lead to further inquiries upon a subject of such overwhelming importance. And we ought not at least to delay dispersing a set of plausible fallacies about the economy of fuel, and the discovery of substitutes for coal, which at present obscure the critical nature of the question, and are eagerly passed about among those who like to believe that we have an indefinite period of prosperity before us.

The writers who have hitherto discussed this question, being chiefly geologists, have of necessity treated it casually, and in a one-sided manner. There are several reasons why it should now receive fuller consideration. In the first place, the accomplishment of a Free Trade policy, the repeal of many laws that tended to moderate our industrial progress, and the very unusual clause in the French Treaty which secures a free export of coals, are all events tending to an indefinite increase of the consumption of coal. On the other hand, two most useful systems of Government inquiry have lately furnished us with new and accurate information bearing upon the question,—the Geological Survey now gives some degree of certainty to our estimates of the coal existing within our reach, while the returns of mineral statistics inform us very exactly of the amount of coal consumed.

Taking advantage of such information, I venture to try and shape out a first rough approximation to the probable progress of our industry and consumption of coal in a system of free industry. We of course deal only with what is probable. It is the duty of a careful writer not to reject facts or circumstances because they are only probable, but to state everything with its due weight of probability. It will be my foremost desire to discriminate certainty and doubt, knowledge and ignorance—to state those data we want as well as those we have. But I must also draw attention to principles governing this subject, which have rather the certainty of natural laws than the fickleness of statistical numbers.

It will be apparent that the first seven of the following chapters are mainly devoted to the physical data of this question, and are of an introductory character. The remaining chapters, which treat of the social and commercial aspects of the subject, constitute the more essential part of the present inquiry. It is this part of the subject which seems to me to have been too much overlooked by those who have expressed opinions concerning the duration of our coal supplies.

I have endeavoured to present a pretty complete outline of the available information in union with the arguments which the facts suggest. But such is the extent and complexity of the subject that it is impossible to notice all the bearings of fact upon fact. The chapters, therefore, have rather the character of essays treating of the more important aspects of the question; and I may here suitably devote a few words to pointing out the particular purpose of each chapter, and the bearings of one upon the other.

I commence by citing the opinions of earlier writers, who have more or less shadowed forth my conclusions; and I also quote (pp. 8–10) Mr. Hull's estimate of the coal existing in England, and adopt it as the geological datum of my arguments.

In considering the geological aspects of the question I endeavour merely to give some notion of the way in which an estimate of the existing coal is made, and of the degree of certainty attaching to it, deferring to the chapter upon Coal Mining the question of the depth to which we can follow seams of coal. It is shown that in all probability there is no precise physical limit of deep mining (pp. 40–43), but that the growing difficulties of management and extraction of coal in a very deep mine must greatly enhance its price (pp. 43–54). It is by this rise of price that gradual exhaustion will be manifested and its deplorable effects occasioned.

I naturally pass to consider whether there are yet in the cost of coal (pp. 57–67) any present signs of exhaustion; it appears that there has been no recent rise of importance (p. 61), but that at the same time the high price demanded for coals drawn from some of the deepest pits indicates the high price that must in time be demanded for even ordinary coals.

A distinct division of the inquiry comprising Chapters V. VI. and VII. treats of inventions in regard to the use of coal. It is shown that we owe almost all arts to Continental nations (pp. 68–78), except those great arts which have been called into use here, by the cheapness (pp. 80–101) and excellence of our coal. It is shown that the constant tendency of discovery is to render coal a more and more efficient agent (pp. 105–116), while there is no probability that when our coal is used up any more powerful substitute will be forthcoming (pp. 121–141). Nor can the economical use of coal reduce its consumption. On the contrary, by rendering its employment more profitable, the present demand for coal is increased and the advantage is more strongly thrown upon the side of those who will in the future have the cheapest supplies (pp. 141–5). As it is in a subsequent chapter conclusively shown that we cannot make up for a future want of coal, by importation from other countries (pp. 220–251), *it will appear that there is no reasonable prospect of any relief from a future want of the main agent of industry.* We must lose that which constitutes our peculiar energy; for considering how greatly our manufactures and navigation depend upon coal, and how vast is our consumption of it compared with other nations, it cannot be supposed we shall do without coal more than a fraction of what we do with it.

I then turn to a totally different aspect of the question, leading to some estimate of the duration of our prosperity. I first explain the natural principle, that a nation tends to multiply itself at a constant rate, so as to receive not equal additions in equal times but additions rapidly growing greater and greater (pp. 146–155).

In the chapter on Population it is incidentally pointed out that the nation, as a whole, has rapidly grown more numerous from the time when the steam-engine and other inventions involving the consumption of coal came into use (pp. 158–9). Until about 1820, however, the agricultural and manufacturing populations both increased about equally. But the former then became excessive, occasioning great pauperism (pp. 160–166), while it is only our towns and coal and iron districts which have afforded any scope for a rapid and continuous increase.

The more nearly, too, we approach industry concerned directly with coal, the more rapid and constant is the rate of growth (p. 169). The progress indeed of almost every part of our population has clearly been checked by emigration (pp. 170–2). But that this emigration is not due to pressure at home is plain from the greatly increased frequency of marriages in the last ten or fifteen years. And though this emigration temporarily checks our growth in mere numbers, it greatly promotes our welfare, and tends to induce greater future growths of population (pp. 173–180).

Attention is then drawn to the rapid and constant rate of multiplication displayed by the iron, cotton, shipping, and other great branches of our industry, the progress of which is in general quite unchecked up to the present time (pp. 191–203). The consumption of coal there is every reason to suppose has similarly been multiplying itself at a growing rate (pp. 204–6).

The present rate of increase of our coal consumption is then ascertained, and it is shown *that should the consumption multiply for rather more than a century at the same rate, the average depth of our coal-mines would be* 4,000 *feet, and the average price of coal much higher than the highest price now paid for the finest kinds of coal* (pp. 209 –215).

It is thence simply inferred that *we cannot long continue our present rate of progress.* The first check to our growing prosperity, however, must render our population excessive. Emigration may relieve it, and by exciting increased trade tend to keep up our progress, but after a time we must either sink down into poverty, adopting wholly new habits, or else witness a constant annual exodus of the youth of the country. It is further pointed out that the ultimate result will be to render labour so abundant in the United States that our iron manufactures will be underbid by the unrivalled iron and coal resources of Pennsylvania, and in a separate chapter it is shown that the crude iron manufacture will in all probability be our first loss, while it is impossible to say how much of our manufactures may not follow it.

Possible measures for checking the waste and use of coal are indeed briefly discussed, but the general conviction perhaps must force itself upon the mind, that restrictive legislation may mar but probably cannot mend or correct the natural course of industrial development. Such is a general outline of my arguments and conclusions.

When I commenced studying this question I had little thought of some of the results, and I might well hesitate at asserting things so little accordant with the unbounded confidence of the present day. But as serious misgivings do already exist, some discussion is necessary to set them at rest, or to confirm them, and perhaps to modify our views. And in entering on such a discussion, an unreserved and even an overdrawn statement of the adverse circumstances is better than weak reticence. If my conclusions are at all true they cannot too soon be recognised and kept in mind; if mistaken, I shall be among the first to rejoice at a vindication of our country's resources from all misgivings.

For my own part I am convinced that this question must before long force itself upon our attention with painful urgency. It cannot long be shirked and shelved. It must rise by degrees into the position of a great national and perhaps a party question, antithetical to that of Free Trade. There will be a Conservative Party, desirous at all cost to secure the continued and exclusive prosperity of this country as a main bulwark of the general good. On the other hand, there will be the Liberal Party, less cautious, more trustful in abstract principles and the unfettered tendencies of nature.

Bulwer, in one of his Caxtonian Essays, has described, with all his usual felicity of thought and language, the confliction of these two great parties. They have fought many battles upon this soil already, and the result as yet is that wonderful union of stability and change, of the good old and the good new, which makes the English Constitution.

But if it shall seem that this is not to last indefinitely—that some of our latest determinations of policy lead directly to the exhaustion of our main wealth—the letting down of our mainspring—I know not how to express the difficulty of the moral and political questions which will arise. Some will wish to hold to our adopted principles, and leave commerce and the consumption of coal unchecked even to the last—while others, subordinating commerce to purposes of a higher nature, will tend to the prohibition of coal exports, the restriction of trade, and the adoption of every means of sparing the fuel which makes our welfare and supports our influence upon the nations of the world.

This is a question of that almost religious importance which needs the separate study and determination of every intelligent person. And if we find that we must yield before the disposition of material wealth, which is the work of a higher Providence, we need not give way to weak discouragement concerning the future, but should rather learn to take an elevated view of our undoubted duties and opportunities in the present.

SECTION 1.—THE LABOUR-PROCESS OR THE PRODUCTION OF USE-VALUES.

THE capitalist buys labour-power in order to use it; and labour-power in use is labour itself. The purchaser of labour-power consumes it by setting the seller of it to work. By working, the latter becomes actually, what before he only was potentially, labour-power in action, a labourer. In order that his labour may reappear in a commodity, he must, before all things, expend it on something useful, on something capable of satisfying a want of some sort. Hence, what the capitalist sets the labourer to produce, is a particular use-value, a specified article. The fact that the production of use-values, or goods, is carried on under the control of a capitalist and on his behalf, does not alter the general character of that production. We shall, therefore, in the first place, have to consider the labour-process independently of the particular form it assumes under given social conditions.

Labour is, in the first place, a process in which both man and Nature participate, and in which man of his own accord starts, regulates, and controls the material re-actions between himself and Nature. He opposes himself to Nature as one of her own forces, setting in motion arms and legs, head and hands, the natural forces of his body, in order to appropriate Nature's productions in a form adapted to his own wants. By thus acting on the external world and changing it, he at the same time changes his own nature. He develops his slumbering powers and compels them to act in obedience to his sway. We are not now dealing with those primitive instinctive forms of labour that remind us of the mere animal. An immeasurable interval of time separates the state of things in which a man brings his labour-power to market for sale as a commodity, from that state in which human labour was still in its first instinctive stage. We presuppose labour in a form that stamps it as exclusively human. A spider conducts operations that resemble those of a weaver, and a bee puts to shame many an architect in the construction of her cells. But what distinguishes the worst architect from the best of bees is this, that the architect raises his structure in imagination before he erects it in reality. At the end of every labour-process, we get a result that already existed in the imagination of the labourer at its commencement. He not only effects a change of form in the material on which he works, but he also realises a purpose of his own that gives the law to his modus operandi, and to which he must subordinate his will. And this subordination is no mere momentary act. Besides the exertion of the bodily organs, the process demands that, during the whole operation, the workman's will be steadily in consonance with his purpose. This means close attention. The less he is attracted by the nature of the work, and the mode in which it is carried on, and the less, therefore, he enjoys it as something which gives play to his bodily and mental powers, the more close his attention is forced to be.

The elementary factors of the labour-process are 1, the personal activity of man, i.e., work itself, 2, the subject of that work, and 3, its instruments.

The soil (and this, economically speaking, includes water) in the virgin state in which it supplies [1] man with necessaries or

---

[1] "The earth's spontaneous productions being in small quantity, and quite independent of man, appear, as it were, to be furnished by Nature, in the same way as a small sum is given to a young man, in order to put him in a way of industry, and of making his fortune." (James Steuart: "Principles of Polit. Econ." edit. Dublin, 1770, v. I. p. 116).

the means of subsistence ready to hand, exists independently of him, and is the universal subject of human labour. All those things which labour merely separates from immediate connection with their environment, are subjects of labour spontaneously provided by Nature. Such are fish which we catch and take from their element, water, timber which we fell in the virgin forest, and ores which we extract from their veins. If, on the other hand, the subject of labour has, so to say, been filtered through previous labour, we call it raw material; such is ore already extracted and ready for washing. All raw material is the subject of labour, but not every subject of labour is raw material; it can only become so, after it has undergone some alteration by means of labour.

An instrument of labour is a thing, or a complex of things, which the labourer interposes between himself and the subject of his labour, and which serves as the conductor of his activity. He makes use of the mechanical, physical, and chemical properties of some substances in order to make other substances subservient to his aims.[1] Leaving out of consideration such ready-made means of subsistence as fruits, in gathering which a man's own limbs serve as the instruments of his labour, the first thing of which the labourer possesses himself is not the subject of labour but its instrument. Thus Nature becomes one of the organs of his activity, one that he annexes to his own bodily organs, adding stature to himself in spite of the Bible. As the earth is his original larder, so too it is his original tool house. It supplies him, for instance, with stones for throwing, grinding, pressing, cutting, &c. The earth itself is an instrument of labour, but when used as such in agriculture implies a whole series of other instruments and a comparatively high development of labour.[2] No sooner does labour undergo the

[1] "Reason is just as cunning as she is powerful. Her cunning consists principally in her mediating activity, which, by causing objects to act and re-act on each other in accordance with their own nature, in this way, without any direct interference in the process, carries out reason's intentions." (Hegel: "Encyklopädie, Erster Theil. Die Logik." Berlin, 1840, p. 382.)

[2] In his otherwise miserable work, ("Théorie de l'Econ. Polit." Paris, 1819), Ganilh enumerates in a striking manner in opposition to the "Physiocrats" the long series of previous processes necessary before agriculture properly so called can commence.

least development, than it requires specially prepared instruments. Thus in the oldest caves we find stone implements and weapons. In the earliest period of human history domesticated animals, *i.e.*, animals which have been bred for the purpose, and have undergone modifications by means of labour, play the chief part as instruments of labour along with specially prepared stones, wood, bones, and shells.[1] The use and fabrication of instruments of labour, although existing in the germ among certain species of animals, is specifically characteristic of the human labour-process, and Franklin therefore defines man as a tool-making animal. Relics of by-gone instruments of labour possess the same importance for the investigation of extinct economical forms of society, as do fossil bones for the determination of extinct species of animals. It is not the articles made, but how they are made, and by what instruments, that enables us to distinguish different economical epochs.[2] Instruments of labour not only supply a standard of the degree of development to which human labour has attained, but they are also indicators of the social conditions under which that labour is carried on. Among the instruments of labour, those of a mechanical nature, which, taken as a whole, we may call the bone and muscles of production, offer much more decided characteristics of a given epoch of production, than those which, like pipes, tubs, baskets, jars, &c., serve only to hold the materials for labour, which latter class, we may in a general way, call the vascular system of production. The latter first begins to play an important part in the chemical industries.

In a wider sense we may include among the instruments of labour, in addition to those things that are used for directly transferring labour to its subject, and which therefore, in one

[1] Turgot in his "Reflexions sur la Formation et la Distribution des Richesses" (1766) brings well into prominence the importance of domesticated animals to early civilisation.

[2] The least important commodities of all for the technological comparison of different epochs of production are articles of luxury, in the strict meaning of the term. However little our written histories up to this time notice the development of material production, which is the basis of all social life, and therefore of all real history, yet prehistoric times have been classified in accordance with the results, not of so called historical, but of materialistic investigations. These periods have been divided, to correspond with the materials from which their implements and weapons were made, viz., into the stone, the bronze, and the iron ages.

way or another, serve as conductors of activity, all such objects as are necessary for carrying on the labour-process. These do not enter directly into the process, but without them it is either impossible for it to take place at all, or possible only to a partial extent. Once more we find the earth to be a universal instrument of this sort, for it furnishes a locus standi to the labourer and a field of employment for his activity. Among instruments that are the result of previous labour and also belong to this class, we find workshops, canals, roads, and so forth.

In the labour-process, therefore, man's activity, with the help of the instruments of labour, effects an alteration, designed from the commencement, in the material worked upon. The process disappears in the product; the latter is a use-value, Nature's material adapted by a change of form to the wants of man. Labour has incorporated itself with its subject: the former is materialised, the latter transformed. That which in the labourer appeared as movement, now appears in the product as a fixed quality without motion. The blacksmith forges and the product is a forging.

If we examine the whole process from the point of view of its result, the product, it is plain that both the instruments and the subject of labour, are means of production,[1] and that the labour itself is productive labour.[2]

Though a use-value, in the form of a product, issues from the labour-process, yet other use-values, products of previous labour, enter into it as means of production. The same use-value is both the product of a previous process, and a means of production in a later process. Products are therefore not only results, but also essential conditions of labour.

With the exception of the extractive industries, in which the material for labour is provided immediately by nature, such as mining, hunting, fishing, and agriculture (so far as the

---

[1] It appears paradoxical to assert, that uncaught fish, for instance, are a means of production in the fishing industry. But hitherto no one has discovered the art of catching fish in waters that contain none.

[2] This method of determining from the standpoint of the labour-process alone, what is productive labour, is by no means directly applicable to the case of the capitalist process of production.

latter is confined to breaking up virgin soil), all branches of industry manipulate raw material, objects already filtered through labour, already products of labour. Such is seed in agriculture. Animals and plants, which we are accustomed to consider as products of nature, are in their present form, not only products of, say last year's labour, but the result of a gradual transformation, continued through many generations, under man's superintendence, and by means of his labour. But in the great majority of cases, instruments of labour show even to the most superficial observer, traces of the labour of past ages.

Raw material may either form the principal substance of a product, or it may enter into its formation only as an accessory. An accessory may be consumed by the instruments of labour, as coal under a boiler, oil by a wheel, hay by draft-horses, or it may be mixed with the raw material in order to produce some modification thereof, as chlorine into unbleached linen, coal with iron, dye-stuff with wool, or again, it may help to carry on the work itself, as in the case of the materials used for heating and lighting workshops. The distinction between principal substance and accessory vanishes in the true chemical industries, because there none of the raw material reappears, in its original composition, in the substance of the product.[1]

Every object possesses various properties, and is thus capable of being applied to different uses. One and the same product may therefore serve as raw material in very different processes. Corn, for example, is a raw material for millers, starch-manufacturers, distillers, and cattle-breeders. It also enters as raw material into its own production in the shape of seed: coal, too, is at the same time the product of, and a means of production in, coal-mining.

Again, a particular product may be used in one and the same process, both as an instrument of labour and as raw material. Take, for instance, the fattening of cattle, where the animal is the raw material, and at the same time an instrument for the production of manure.

[1] Storch calls true raw materials "matières," and accessory material "matériaux:" Cherbuliez describes accessories as "matières instrumentales."

A product, though ready for immediate consumption, may yet serve as raw material for a further product, as grapes when they become the raw material for wine. On the other hand, labour may give us its product in such a form, that we can use it only as raw material, as is the case with cotton, thread, and yarn. Such a raw material, though itself a product, may have to go through a whole series of different processes : in each of these in turn, it serves, with constantly varying form, as raw material, until the last process of the series leaves it a perfect product, ready for individual consumption, or for use as an instrument of labour.

Hence we see, that whether a use-value is to be regarded as raw material, as instrument of labour, or as product, this is determined entirely by its function in the labour process, by the position it there occupies : as this varies, so does its character.

Whenever therefore a product enters as a means of production into a new labour-process, it thereby loses its character of product, and becomes a mere factor in the process. A spinner treats spindles only as implements for spinning, and flax only as the material that he spins. Of course it is impossible to spin without material and spindles ; and therefore the existence of these things as products, at the commencement of the spinning operation, must be presumed : but in the process itself, the fact that they are products of previous labour, is a matter of utter indifference ; just as in the digestive process, it is of no importance whatever, that bread is the produce of the previous labour of the farmer, the miller, and the baker. On the contrary, it is generally by their imperfections as products, that the means of production in any process assert themselves in their character of products. A blunt knife or weak thread forcibly remind us of Mr. A., the cutler, or Mr. B., the spinner. In the finished product the labour by means of which it has acquired its useful qualities is not palpable, has apparently vanished.

A machine which does not serve the purposes of labour, is useless. In addition, it falls a prey to the destructive influence of natural forces. Iron rusts and wood rots. Yarn with which we neither weave nor knit, is cotton wasted. Living labour must seize upon these things and rouse them from their death-sleep, change them from mere possible use-values into real and effective ones. Bathed in the fire of labour, appropriated as part and parcel of labour's organism, and, as it were, made alive for the performance of their functions in the process, they are in truth consumed, but consumed with a purpose, as elementary constituents of new use-values, of new products, ever ready as means of subsistence for individual consumption, or as means of production for some new labour-process.

If then, on the one hand, finished products are not only results, but also necessary conditions, of the labour-process, on the other hand, their assumption into that process, their contact with living labour, is the sole means by which they can be made to retain their character of use-values, and be utilised.

Labour uses up its material factors, its subject and its instruments, consumes them, and is therefore a process of consumption. Such productive consumption is distinguished from individual consumption by this, that the latter uses up products, as means of subsistence for the living individual; the former, as means whereby alone, labour, the labour-power of the living individual, is enabled to act. The product, therefore, of individual consumption, is the consumer himself; the result of productive consumption, is a product distinct from the consumer.

In so far then, as its instruments and subjects are themselves products, labour consumes products in order to create products, or in other words, consumes one set of products by turning them into means of production for another set. But, just as in the beginning, the only participators in the labour-process were man and the earth, which latter exists independently of man, so even now we still employ in the process many means of production, provided directly by nature, that do not represent any combination of natural substances with human labour.

The labour process, resolved as above into its simple elementary factors, is human action with a view to the production of use-values, appropriation of natural substances to human requirements; it is the necessary condition for effecting exchange of matter between man and Nature; it is the everlasting nature-imposed condition of human existence, and therefore is independent of every social phase of that existence, or rather, is common to every such phase. It was, therefore, not necessary to represent our labourer in connexion with other labourers; man and his labour on one side, Nature and its materials on the other, sufficed. As the taste of the porridge does not tell you who grew the oats, no more does this simple process tell you of itself what are the social conditions under which it is taking place, whether under the slave-owner's brutal lash, or the anxious eye of the capitalist, whether Cincinnatus carries it on in tilling his modest farm or a savage in killing wild animals with stones.[1]

...

[1] By a wonderful feat of logical acumen, Colonel Torrens has discovered, in this stone of the savage the origin of capital. "In the first stone which he [the savage] flings at the wild animal he pursues, in the first stick that he seizes to strike down the fruit which hangs above his reach, we see the appropriation of one article for the purpose of aiding in the acquisition of another, and thus discover the origin of capital. (R. Torrens: "An Essay on the Production of Wealth," &c., pp. 70-71.)

SECTION 10.—MODERN INDUSTRY AND AGRICULTURE.

The revolution called forth by modern industry in agricul-ture, and in the social relations of agricultural producers, will be investigated later on. In this place we shall merely indi-cate a few results by way of anticipation. If the use of machinery in agriculture is for the most part free from the injurious physical effect it has on the factory operative, its action in superseding the labourers is more intense, and finds less resistance, as we shall see later in detail. In the counties of Cambridge and Suffolk, for example, the area of cultivated land has extended very much within the last 20 years (up to 1868), while in the same period the rural population has diminished, not only relatively, but absolutely. In the United States it is as yet only virtually that agricultural machines replace labourers ; in other words, they allow of the culti-vation by the farmer of a larger surface, but do not actually expel the labourers employed. In 1861 the number of persons occupied in England and Wales in the manufacture of agricul-tural machines was 1034, whilst the number of agricultural labourers employed in the use of agricultural machines and steam engines did not exceed 1205.

[1] Robert Owen, the father of Co-operative Factories and Stores, but who, as before remarked, in no way shared the illusions of his followers with regard to the bear-ing of these isolated elements of transformation, not only practically made the fac-tory system the sole foundation of his experiments, but also declared that system to be theoretically the starting point of the social revolution. Herr Vissering, Pro-fessor of Political Economy in the University of Leyden, appears to have a suspicion of this when, in his "Handboek van Praktische Staatshuishoudkunde, 1860-62," which reproduces all the platitudes of vulgar economy, he strongly supports handi- crafts against the factory system.

In the sphere of agriculture, modern industry has a more revolutionary effect than elsewhere, for this reason, that it annihilates the peasant, that bulwark of the old society, and replaces him by the wage labourer. Thus the desire for social changes, and the class antagonisms are brought to the same level in the country as in the towns. The irrational, old fashioned methods of agriculture are replaced by scientific ones. Capitalist production completely tears asunder the old bond of union which held together agriculture and manufacture in their infancy. But at the same time it creates the material conditions for a higher synthesis in the future, viz., the union of agriculture and industry on the basis of the more perfected forms they have each acquired during their temporary separation. Capitalist production, by collecting the population in great centres, and causing an ever increasing preponderance of town population, on the one hand concentrates the historical motive-power of society; on the other hand, it disturbs the circulation of matter between man and the soil, *i.e.*, prevents the return to the soil of its elements consumed by man in the form of food and clothing; it therefore violates the conditions necessary to lasting fertility of the soil. By this action it destroys at the same time the health of the town labourer and the intellectual life of the rural labourer.[1] But while upsetting the naturally grown conditions for the maintenance of that circulation of matter, it imperiously calls for its restoration as a system, as a regulating law of social production, and under a form appropriate to the full development of the human race. In agriculture as in manufacture, the transformation of production under the sway of capital, means, at the same time, the martyrdom of the producer; the instrument of labour becomes the means of enslaving, exploiting, and impoverishing the

---

[1] "You divide the people into two hostile camps of clownish boors and emasculated dwarfs. Good heavens! a nation divided into agricultural and commercial interests, calling itself sane; nay, styling itself enlightened and civilized, not only in spite of, but in consequence of this monstrous and unnatural division." (David Urquhart, l. c., p. 119.) This passage shows, at one and the same time, the strength and the weakness of that kind of criticism which knows how to judge and condemn the present, but not how to comprehend it.

labourer; the social combination and organization of labour-processes is turned into an organised mode of crushing out the workman's individual vitality, freedom, and independence. The dispersion of the rural labourers over larger areas breaks their power of resistance while concentration increases that of the town operatives. In modern agriculture, as in the urban industries, the increased productiveness and quantity of the labour set in motion are bought at the cost of laying waste and consuming by disease labour-power itself. Moreover, all progress in capitalistic agriculture is a progress in the art, not only of robbing the labourer, but of robbing the soil; all progress in increasing the fertility of the soil for a given time, is a progress towards ruining the lasting sources of that fertility. The more a country starts its development on the foundation of modern industry, like the United States, for example, the more rapid is this process of destruction.[1]

[1] See Liebig: "Die Chemie in ihrer Anwendung auf Agricultur und Physiologie, 7. Auflage, 1862," and especially the "Einleitung in die Naturgesetze des Feldbaus," in the 1st Volume. To have developed from the point of view of natural science, the negative, *i.e.*, destructive side of modern agriculture, is one of Liebig's immortal merits. His summary, too, of the history of agriculture, although not free from gross errors, contains flashes of light. It is, however, to be regretted that he ventures on such hap-hazard assertions as the following: "By greater pulverising and more frequent ploughing, the circulation of air in the interior of porous soil is aided, and the surface exposed to the action of the atmosphere is increased and renewed; but it is easily seen that the increased yield of the land cannot be proportional to the labour spent on that land, but increases in a much smaller proportion. This law," adds Liebig, "was first enunciated by John Stuart Mill in his 'Principles of Pol. Econ.', Vol. I., p. 17, as follows: 'That the produce of land increases, *cæteris paribus*, in a diminishing ratio to the increase of the labourers employed" (Mill here introduces in an erroneous form the law enunciated by Ricardo's school, for since the 'decrease of the labourers employed,' kept even pace in England with the advance of agriculture, the law discovered in, and applied to, England, could have no application to that country, at all events), "is the universal law of agricultural industry.' This is very remarkable, since Mill was ignorant of the reason for this law." (Liebig, l. c., Bd. I., p. 143 and Note.) Apart from Liebig's wrong interpretation of the word "labour," by which word he understands something quite different from what political economy does, it is, in any case, "very remarkable" that he should make Mr. John Stuart Mill the first propounder of a theory which was first published by James Anderson in A. Smith's days, and was repeated in various works down to the beginning of the 19th century; a theory which Malthus, that master in plagiarism (the whole of his population theory is a shameless plagiarism), appropriated to himself in 1815; which West developed at the same time as, and independently of, Anderson; which in the year 1817 was connected by Ricardo with the general theory of value, then made the round of the world as Ricardo's theory, and in

Capitalist production, therefore, develops technology, and the combining together of various processes into a social whole, only by sapping the original sources of all wealth—the soil and the labourer.

1820 was vulgarised by James Mill, the father of John Stuart Mill; and which, finally, was reproduced by John Stuart Mill and others, as a dogma already quite common-place, and known to every school-boy. It cannot be denied that John Stuart Mill owes his, at all events, " remarkable " authority almost entirely to such *quid-pro-quos.*

THE ROOTS OF SOCIO-ENVIRONMENTAL RESEARCH IN GEOGRAPHY AND ANTHROPOLOGY

Emilio F. Moran

The collection of papers in **Part II** of the volume, on the historical roots of socio-environmental research in geography and anthropology, deserve careful reading in the twenty-first century. Our contemporary concerns in socio-environmental research and human–environment interactions research are not new. Reading these late nineteenth- and twentieth-century papers, often citing works much earlier, and thus a more ancient concern with the balance and imbalance in our relations with the environment, is to become aware how much we have distanced ourselves from the nature that sustains us – and also how scholars have long been concerned with these issues. The non-resolution through all these years reminds us how intractable these interactions are and that we have been unable to bring about the changes required to be good stewards of nature. This continues to be a challenge today.

There has been a steady development of ideas in how anthropology and geography have treated society and environment interactions. At the outset, in the late 1800s, ideas focused on understanding many of the new societies and cultures being discovered by Europeans as they engaged in colonial travel and occupation around the world, and how artifacts and cultures related to the biogeographic characteristics of the places they found (e.g., Ratzel 1896–1898; Lowie 1937). These scholars favored views that gave the habitat primacy in explaining human behavior and deprived the population of agency. Environment determined culture. As more research was done, anthropologists and geographers found societies within the same environments that had very different cultures, thus showing the role of culture in shaping interactions with environment. **Franz Boas** (1858–1942), in the legacy paper in this volume, presents evidence against the environmental determinist view of earlier scholars by showing that primitive man (i.e., non-Western populations) was no different than dominant white European cultures. This view was further elaborated many decades later by **Marshall Sahlins** (1930–2021) in the paper in this volume, showing that hunter-gatherers were affluent and that they had low demands from their environment and lived comfortably – in contrast to contemporary urban/industrial societies wherein we have very large poor populations despite material abundance. Adjusting our standard of living to the productive conditions of the environment, and valuing distribution of existing wealth is a path to affluence, and one that may be necessary as we run out of resources on the planet. Those debates have as much relevance today as when first formulated. Environment influences what people do but does not determine it – many cultures seem to be particularly adept at constructing their environment through modifying the physical limitations that they encounter. This is a view that has become particularly important in contemporary anthropology and geography, as we discover more and more evidence for the ways in which people change the environment, rather than simply adapt to the conditions that they find in the habitat.

The debate over whether particular features of culture or society are a product of diffusion, or of independent invention, is another important theme that continues to be of interest to scholars to this day. The fact that humans migrate, and that they carry their cultural knowledge and institutions along, is not arguable, yet much ink has been spilled over this issue. How else to explain why Frederick Jackson Turner, the great historian of the American frontier, argued that the reason why American individualism flourished was because the homesteaders who went west to the frontier left behind their cultural baggage and were free to innovate and make Americans a free and innovative people? Later research showed that he was wrong about people leaving their cultural baggage behind, yet that myth survives to this day among many. Some people prefer to believe that we are always innovators, rather than adopters. In fact, we are both, depending on circumstances and the historical moment. The task is to discover which one explains a particular feature of society or culture, or if both are implicated at the same historical moment or another. The legacy article by **V. Gordon Childe** (1892–1957) in this collection is a classic discussion of how the Neolithic Revolution changed everything in our interactions with the environment. Whereas before people adjusted themselves to the productivity of the environment, just 10,000 years ago people in various places began to plant and cultivate what became domesticated plants. Gathered plants gradually began to be manipulated and selected, favoring larger grains. This attention by hunter-gatherers began to reap benefits that made cultivation ever more desirable and led to greater dependence on this mode of production than the earlier hunting and gathering. As Childe and many archeologists since then have pointed out, it was a very gradual process that coexisted with hunting and gathering but that inevitably led to greater populations, more numerous settlements, and relatively self-sufficient communities. While the article focuses on wheat and barley domestication to make its case, it does note that this Neolithic revolution happened in many other places, with other crops, and often in combination with animal domestication – but not in all cases. The picture that emerges is that people began to alter human–environment relations in a way that was richly diverse, largely independent, but in which the growing proximity of these Neolithic settlements led to some trade and exchanges of plants and animals that resulted in information being exchanged and new ways of improving productivity were found. To capture the significance of this Neolithic revolution, ecologists in introductory textbooks have often described this as a major change, from humans-with-nature to humans-against-nature.

The importance of diffusion was tackled by Boas as well. It is a contribution worth a close reading because he takes on several important concerns relevant to scholars ever since: the importance of careful inductive research, the painstaking work of historical studies to establish what is diffusion or independent invention, and why we should be cautious in coming to conclusions that seem attractive but are lacking in rigor when we put them to the careful test of historical exegesis, or critical interpretation. In the field of socio-environmental research this is particularly important today. Fortunately, one of the sub-fields of contemporary socio-environmental research, known as historical ecology (Crumley 1994; Balée 1998; Balée and Erickson 2006), grapples with the legacy of the past on the present in human–environment interactions. We call these issues "legacy effects" and they have become increasingly valuable in the research community as a way to historicize a study area (Foster and Aber 2004). Foster and Aber showed how Harvard Forest, an important biological station and part of the Long-Term Ecological Research network (LTER), had been a farm and that the forest we had been studying for decades was in fact an old-growth secondary forest, rather than a pristine one. This raises a very important issue, and a cautionary tale, about making sure that we have delved into the archives and historical record before assuming that any particular place is pristine, and that much of the Earth has had a history of transformation and landscape change that must be considered in socio-environmental analysis.

Of great importance in the field of geography is **Ellen Semple** (1864–1932). The contribution in this collection is a good example of her work, highlighting a common theme in the first four contributions included in this part of the volume (Semple, Childe, Boas, White): the importance of land (nature) to the people. Like Boas, Semple tends to view environment (nature, land) as somewhat passive in its role, but its impact shapes the response of people to its characteristics. Semple laid the ground for what has become the mantra of geography: place matters. To understand any human endeavor, whether histor-ical, sociological, or anthropological, requires us to specify the geographic conditions and characteristics. This later evolved into the rise of subareas of geography such as political geography, economic geography, and human geography. And today we can see this important dictum made concrete in the near requirement to have all our data be "spatially explicit" – made possible through the easy availability of Global Positioning System (GPS) capabilities in our phones, tablets, and GPS sensors. These ideas did not pop up suddenly in the twenty-first century. The theoretical basis for recognizing how specific geographic locations shape commerce, transportation, settlement pattern, and even culture itself was laid by the early geography scholars like Ratzel and Semple who showed in sweeping analyses why the interactions of "the land" and "the people" were the obligatory first step in any analysis in the social sciences, and certainly for many years were the obligatory first two chapters of monographs in anthropology and geography. There is an inherent critique of works that ignore the land in the analysis of society, and Semple shows why a careful discussion of land–people interactions is a requirement for deep understanding, and that this connection is fundamental to understanding social relations. A difference is drawn between how hunter-gatherers, fishers, pastoralists, and agriculturalists bond with the land, and how this bond differs in these different modes of production. This takes us to the superb article by Mumford.

Lewis Mumford's (1895–1990) contribution on the natural history of cities is one of the best in this part of the volume. It is broadly encompassing, like Childe's, on the evolution from bands, to villages, to towns, but takes one step further in the evolution into cities, and suburbia. It takes us through this evolution, defining it in terms almost contrary to those of some evolutionists or progress-oriented writers like Semple and anthropologist

Lewis H. Morgan, noting that this evolution is characterized by a gradual distancing of people from their environment. He does not assume that this evolution is progress (as Childe seems to), but sees it as simply a consequence of population growth and capacity to apply technology to harvesting a growing area to feed the population with increasing efficiency and productivity. This in turn required people to abandon what is characterized as low population and closer relations between land and people and to adopt a distancing of people from the environment through specialization. As this process moves inexorably towards cities, it puts people in ever more polluted environments, where air and water are increasingly poisoned by our economic activities, and where only elites are able to escape to the countryside and suburbs to breathe an air as fresh as past generations enjoyed in their rural villages. For an article written in the late 1950s it is remarkably sensitive to the looming crises that led to the awakening of the 1968 revolution and the 1970s environmental movement. While many viewed the 1950s as a golden age of consumerism, suburban bliss, and happy families in the United States, the article offers an early critique that all was not well and that these levels of consumption would lead to the build-up of toxic outcomes for the environment and for people who had to breathe the air and swim in ever more polluted streams. Few articles have ever been written that encompass such a grand vision of socio-environmental systems changing over millennia.

Another very influential body of work on socio-environmental systems is contributed by Julian Steward, an anthropologist, and his development of what he called cultural ecology. Steward began to formulate his ideas in the 1930s (Steward 1938) but his well-articulated theory came only in the 1950s (Steward 1955). He broke with both environmental and cultural determinism by emphasizing the use of the comparative method to test causal connections between social structure and modes of subsistence. For him it was not nature or nurture but resource utilization that was the crucial element. For a full discussion of Steward's influence see Moran (1979, 2000, 2007, and 2022). Steward used a comparative approach, dear to anthropologists since Boas at least, and he refined that sort of study by developing a unique theory that helped delimit the scope of human–environment interactions research. Whereas before, Ratzel, Semple, Alfred Kroeber, Childe, and others treated human–environment interactions in broad sweeps, Steward started a more rigorous approach: not all of the environment, but only those parts that people recognized culturally were the subject of study, and not all of human culture, but only those parts of social organization dedicated to the search for resources. This work remains influential because it poses the need to set boundaries to our studies of socio-environmental systems, and it describes how we might go about doing so.

Naturally, Steward did not have the final word about human–environment interactions. His approach worked best in hunting-gathering societies, where the direct interaction with the physical environment led to fine-tuned social outcomes. As societies moved to agricultural modes of production, the distancing of people from the land allowed for more diverse solutions, and for more imbalances. He tried to apply the cultural ecological approach to Puerto Rico, a society that is not only agricultural but connected to the US and the production of sugar for export. The approach failed to explain things as neatly as with hunter-gatherers. Bob Netting, who later became the most notable cultural ecologist of his generation, did apply it successfully in an agricultural population of Nigeria (Netting 1968), and then in the Swiss Alps (Netting 1981) – the latter more of a long-term historical analysis of population and environment than a cultural ecological study. However, for this latter approach to be doable requires detailed archival records to allow for the reconstruction of land transactions, population movement, reproductive rates, and such.

The problems in extending the cultural ecology method led to the development of what came to be known as the ecological anthropological approach, led by **Roy Rappaport** (1926–1997). Rappaport, Andrew Vayda, Clifford Geertz, and others critiqued Steward for giving primacy to social organization for subsistence and ignoring other dimensions, such as population density, religion and ritual, history, politics, and other cultural dimensions, and for giving culture a larger role than environment in shaping socio-environmental systems. Thus, they advocated putting the ecosystem at the center, and humans and their cultures as just elements within the larger ecosystem. This development happened just as the first environmental movement got going in 1968, and the adoption by anthropologists and geographers of the ecosystem concept. Rappaport's approach, as evident in the contribution in this volume, is to introduce a systems approach, coming out of systems ecology, that looks for how systems correct or maintain themselves through negative feedback, and, when needed, restructure themselves through positive feedback. It is a dynamic equilibrium approach that assumes that change will occur and that systems adjust or adapt to changing circumstances. Rappaport's (1968) work, focusing on how ritual functions as a regulator for human–environment interactions among the Tsembaga Maring in Papua New Guinea, is a classic study of system regulation. This approach took off simultaneously with the rise of ecosystem modeling, and many anthropologists and geographers translated their research into energy flow models. The interest in modeling has remained with us, although energy flow modeling gave way to other kinds of modeling approaches. Today, for example, agent-based modeling is particularly important, and reflects the growing importance of agency in the study of socio-environmental systems, and has been made possible by the computing power available to researchers right on their laptops.

The contribution of **Moran** (1946–) in this volume, and originally published as the final chapter of his book *Developing the Amazon* (1981), was influenced by such modeling during field work and resulted in a reflection on levels of analysis that are inherent to all socio-environmental research and modeling. A main insight raised by this contribution is that many of our scholarly debates can be traced back to a lack of clarity or specification about what is the level of analysis at which the researcher is operating and what is the central question of interest. Questions that want to know *how* things have come to be are best addressed by micro-level local studies where the process of decisions can be observed directly, whereas studies which want to know *why* things happen in a particular way are best addressed by focusing on larger units of analysis relating to external dynamics and structural constraints. The processes are inherently different at each scale of analysis. We see this most clearly in the difference between micro-economic theory and macro-economic theory. Those two sets of theories are inherently different and bridging them has not been possible. The same holds true in socio-environmental studies. Global Circulation Models (GCMs) and Global Atmospheric Models have important things to tell us about Earth system dynamics, but they have not been able to get at how human behavior drives these processes. For that, one needs to downscale to regional and local modeling, where we can see human decisions about energy use, consumption, reproductive behavior, and transportation choices. The contribution in this volume specifies how the difficulties in managing levels of analysis have influenced the debates over the ecology of the Amazon region but also in reference to the larger literature in geography and anthropology.

This systems perspective was criticized in the 1980s by many who saw systems perspectives as taking away from recognizing the agency of individuals in their interactions with the physical environment. The individual was given less attention in those studies, and the larger ecosystem was viewed as the greater concern, reflecting the discovery of our environmental crises leading to a strong preference for the health of the larger system. Similar concerns reappear from time to time in our contemporary situation when we overemphasize carbon emissions and carbon cycling over things like biodiversity or human agency and choice.

Problematic, too, was a tendency to reify the ecosystem concept and give it properties of a biological organism by overemphasizing energy flows and measurement of calories, ignoring time and structural change, and neglecting the role of individuals (see Moran 1990 for a review of these problems). It is important to recall that this critique took place in the context of the Reagan/Thatcher governments, which gave primacy to individuals making economic decisions to maximize their utility and to deregulation and the removal of social safety nets, and a show of less concern with system properties or the common good. Models that became dominant were inspired by economic modeling, rather than ecosystem models, and by competition, rather than system health. The work of **Marshall Sahlins** (1930–2021) in this volume presents an anthropological contrast to this trend. His work on "stone age economics" argued for the importance of economic transactions in human society from its earliest foundations in hunting and gathering societies. It is a contrast to the work of a whole generation of economists and a number of anthropological economics and economic geography scholars who seemed to suggest that *Homo economicus* was universal and to be found everywhere. Sahlins and a different group of anthropological economists led by Harold Schneider suggested that non-Western societies followed a different logic from economic man, and were driven by other values such as prestige, rank, and the common good, rather than by the rules of maximizing utility. In the following decades, attention to agency and decision-making gained ground, as scholars abandoned energy flow modeling as a major way to understand ecosystem dynamics, and instead focused more on nutrient cycling, decision-making, system complexity, and loss of biodiversity as integrative ways to understand people and resources. Among these new directions is the adoption of remotely sensed data as a basic tool in socio-environmental studies. It is difficult to imagine doing socio-environmental research today, and in the future, without the use of Earth-observing satellites capable of providing time-series data on features such as soils, vegetation, moisture, urban sprawl, night lights, ocean productivity, shifts in river courses, flooding, and many other physical changes that are hard to observe on the ground over time and space. When combined with field work it is also possible to infer with remarkable accuracy how households and communities behave in their use of resources (Goodchild 2003; Evans et al. 2005; Moran et al. 2006).

Another important and very influential direction came from geography and the leadership of **Gilbert White** (1911–2006) in the study of natural hazards. From the outset, in his dissertation, he focused on understanding how people dealt with devastating floods, and later with other natural hazards such as drought, hurricanes, and other disasters. His work inspired a generation of scholars to address these extreme socio-environmental crises, and to see them as ideal natural laboratories wherein to understand human adaptive capacities. People are tested in such crises, and how they respond as societies and cultures, and how their institutions respond to people experiencing extreme events, has been very influential in shaping theories and methods that we use in socio-environmental research to this day. A key insight is that the chief problem is that people become exposed to these flood risks when they build their homes and cities in floodplains that are cyclically exposed to floods. Instead of staying away from floodplains and flood-prone lowlands, people and their institutions propose to overcome these cyclical issues through engineering works like levees and flood-control dams, through public programs to provide relief to those who suffer from floods, and through planning mechanisms. There is a very substantial research community that can respond quickly when these events occur, that generates new approaches to responses by institutions, and which even has a dedicated program at, for example, the US National Science Foundation which focuses on extreme events. What is most striking is that in the years since 1950 when Gilbert White wrote his dissertation, efforts have rarely been directed to the obvious way to reduce risk, and that is to move settlements away from floodplains and flood-prone lowlands. Solutions such as flood insurance have simply increased the cost of floods by reducing the risk of losses and thus encouraging real-estate development in flood-prone areas.

The other important direction taking shape in this period came from the sustainable livelihoods scholars, represented in this volume by the contribution of **Robert Chambers** (1937–) and **Gordon R. Conway** (1938–). This approach puts emphasis on issues of equity, capability, and sustainability, and reminds us that our interests in sustainability are really quite recent but have grown increasingly pervasive in society and our academic discourse. We have no good examples of sustainable societies, but there is growing interest in figuring out how we might start moving towards more sustainable, equitable pathways. In getting there both equity and increasing human capabilities are necessary, yet often forgotten in the sustainable development literature, as Amartya Sen (**1981, Part III**) has often reminded us. One needs to increase human capabilities to choose one's pathway freely, rather than imposing preset pathways driven by assumptions about how economic development will lead to better lives for all. It will not, and to persist in thinking so flies against all the evidence. This paper and many others in this vein are important counters to the persistence of versions of trickle-down economic theory that has never improved the lives of the poor. Anthropologists and geographers are profoundly engaged with these issues to this day.

One of the most inspiring legacy papers in this volume comes from **Piers Blaikie** (1942–) and **Harold Brookfield** (1926–), who "invented" political ecology in their work, by emphasizing how the larger political system within which societies exist shapes the conditions of peoples worldwide. This approach is inspired by classical political economy, but with a more local focus and grounded in contemporary events and conditions. Political ecology is concerned with the political context, with core–periphery concerns raised earlier by dependency theorists (e.g., Raul Prebisch, Andre Gunder Frank) and world system theorists (e.g., Immanuel Wallerstein). Unlike these other approaches, political ecology is much less macro in its interests, and more local, more concerned with how specific populations are kept in poverty by their relations with larger political systems as they are expressed in local social and political relations. The contribution in this volume is a richly detailed exegesis of how the people in Nepal are kept poor and how poverty in turn leads to poor land management. Anthropology and geography today have very robust political ecology research communities, which make important contributions to socio-environmental systems research.

Work on socio-environmental research is not just an academic pursuit, but one that often takes place in the field, and in policy-making circles by government, NGOs, and others who seek to address the unequal access to resources of the many in order for them to have sustainable livelihoods. People with these concerns move back and forth between academia and the policy world and carry out rigorous experiments (e.g., the work of Esther Duflo at the Massachusetts Institute of Technology) to try to improve human lives by figuring out how to give them access to the resources they need. It was the Brundtland Report (WCED 1987) that first defined and set as a goal sustainable development. We have struggled ever since with how to meet this ambitious goal, with competing schools of thought on how to do so. Numerous other efforts have followed: a human dimensions of global change research agenda was articulated by the US National Research Council (1992, 1994) and was followed by the formation of an interagency committee on global change research that continues to this day within the US National Academy of Sciences. Socio-environmental systems are at the very core of the work that the National Academy of Sciences supports across the science research enterprise.

In short, socio-environmental research in the twenty-first century continues to add more refined approaches that permit analysis of global environmental changes and the underlying local and regional dynamics (e.g., Moran and Ostrom 2005; Moran 2010). This poses major, as well as new, challenges to research methods for addressing both social and environmental dimensions (Moran and Brondizio 2001; Brondizio and Moran 2013). Exploring the scholarly roots of these debates is a good way to hone one's skills to address the challenges of the future in socio-environmental systems research.

References

Balée, William. *Advances in Historical Ecology*. New York: Columbia University Press, 1998.

Balée, William, and Clark L. Erickson. *Time and Complexity in Historical Ecology: Studies of the Neotropical Lowlands*. New York: Columbia University Press, 2006.

Brondizio, Eduardo S., and Emilio F. Moran, eds. *Human–Environment Interactions: Current and Future Directions*. New York: Springer, 2013.

Crumley, Carole L., ed. *Historical Ecology: Cultural Knowledge and Changing Landscapes*. Santa Fe, CA: School of Advanced Research Press, 1994.

Evans, Tom P., Leah K. VanWey, and Emilio F. Moran. "Human–Environment Research, Spatially-explicit Data Analysis and Geographic Information Systems." In *Seeing the Forest and the Trees: Human–Environment Interactions in Forest Ecosystems*, edited by Emilio F. Moran and Elinor Ostrom, 161–185. Cambridge, MA: MIT Press, 2005.

Foster, David R., and John D. Aber, eds. *Forests in Time: The Environmental Consequences of 1,000 Years of Change in New England*. New Haven, CT: Yale University Press, 2004.

Goodchild, Michael F. "Geographic Information Science and Systems for Environmental Management." *Annual Review of Environment and Resources* 28 (2003): 493–519.

Lowie, Robert H. *The History of Ethnological Theory*. New York: Farrar & Rinehart Inc, 1937.

Moran, Emilio F. *Human Adaptability: An Introduction to Ecological Anthropology*. North Scituate, RI: Duxbury Press, 1979.

Moran, Emilio F. *Developing the Amazon*. Bloomington: Indiana University Press, 1981.

Moran, Emilio F., ed. *The Ecosystem Approach in Anthropology: From Concept to Practice*. Ann Arbor: University of Michigan Press, 1990.

Moran, Emilio F. *Human Adaptability: An Introduction to Ecological Anthropology*, 2nd ed. Boulder: Westview Press, 2000.

Moran, Emilio F. *Human Adaptability: An Introduction to Ecological Anthropology*, 3rd ed. Boulder: Westview Press, 2007.

Moran, Emilio F. 2010. *Environmental Social Science: Human–Environment Interactions and Sustainability*. Oxford: Wiley-Blackwell Publishers, 2011.

Moran, Emilio F. *Human Adaptability: An Introduction to Ecological Anthropology*, 4th ed. London: Routledge Press, 2022.

Moran, Emilio F., Ryan Adams, Bryn Bakoyéma, Fiorini T. Stefano, and Bruce Boucek. "Human Strategies for Coping with El Niño Related Drought in Amazônia." *Climatic Change* 77, no. 3 (2006): 343–61.

Moran, Emilio F., and Eduardo S. Brondizio. "Human Ecology from Space: Ecological Anthropology Engages the Study of Global Environmental Change." In *Ecology and the Sacred: Engaging the Anthropology of Roy A. Rappaport*, 64–87, edited by Ellen Messer and Michael J. Lambeck. Ann Arbor, MI: University of Michigan Press. 2001.

Moran, Emilio F., and Elinor Ostrom. *Seeing the Forest and the Trees: Human Environment Interactions in Forest Ecosystems*. Cambridge, MA: MIT Press, 2005.

National Research Council. *Global Environmental Change: Understanding the Human Dimensions*. Washington, DC: National Academies Press, 1992.

National Research Council. *Science Priorities for the Human Dimensions of Global Change*. Washington, DC: National Academies Press, 1994.

Netting, Robert M. *Hill Farmers of Nigeria: Cultural Ecology of the Kofyar of the Jos Plateau*. Seattle: University of Washington Press, 1968.

Netting, Robert M. *Balancing on an Alp: Ecological Change and Continuity in a Swiss Mountain Community*. New York: Cambridge University Press, 1981.

Rappaport, Roy. *Pigs for the Ancestors: Ritual in the Ecology of a New Guinea People*. New Haven, CT: Yale University Press, 1968.

Ratzel, Friedrich. *The History of Mankind*, translated by Arthur J. Butler. London: Macmillan, 1896–1898.

Steward, Julian H. *Basin Plateau Aboriginal Sociopolitical Groups*. Bulletin 120. Bureau of American Ethnology. Washington, DC: Smithsonian Institution, 1938.

Steward, Julian H. *Theory of Culture Change: The Methodology of Multilinear Evolution*. DeKalb: University of Illinois Press, 1955.

WCED, World Commission on Environment and Development. *Our Common Future: Report of the World Commission on Environment and Development*. Oxford: Oxford University Press, 1987.

CHAPTER III

SOCIETY AND STATE IN RELATION TO THE LAND

People and land.   EVERY clan, tribe, state or nation includes two ideas, a people and its land, the first unthinkable without the other. History, sociology, ethnology touch only the inhabited areas of the earth. These areas gain their final significance because of the people who occupy them; their local conditions of climate, soil, natural resources, physical features and geographic situation are important primarily as factors in the development of actual or possible inhabitants. A land is fully comprehended only when studied in the light of its influence upon its people, and a people cannot be understood apart from the field of its activities. More than this, human activities are fully intelligible only in relation to the various geographic conditions which have stimulated them in different parts of the world. The principles of the evolution of navigation, of agriculture, of trade, as also the theory of population, can never reach their correct and final statement, unless the data for the conclusions are drawn from every part of the world, and each fact interpreted in the light of the local conditions whence it sprang. Therefore anthropology, sociology and history should be permeated by geography.

Political geography and history   In history, the question of territory,—by which is meant mere area in contrast to specific geographic conditions— has constantly come to the front, because a state obviously involved land and boundaries, and assumed as its chief function the defence and extension of these. Therefore political geography developed early as an offshoot of history. Political science has often formulated its principles without regard to the geographic conditions of states, but as a matter of fact, the most fruitful political policies of nations have almost invariably had a geographic core. Witness the colonial policy of Holland, England, France and Portugal, the free-trade policy of England, the militantism of Germany, the whole complex question of European balance of power and the Bosporus, and the Monroe Doctrine of the United States. Dividing lines between political parties tend to follow approximately geographic lines of cleavage; and these make themselves apparent at recurring intervals of national upheaval, perhaps with centuries between, like a submarine volcanic rift. In England the southeastern plain and the northwestern uplands have been repeatedly arrayed against each other, from the Roman conquest which embraced the lowlands up to about the 500-foot contour line,[1] through the War of the Roses and the Civil War,[2] to the struggle for the repeal of the Corn Laws and the great Reform Bill of 1832.[3] Though the boundary lines have been only roughly the same and each district has contained opponents of the dominant local party, nevertheless the geographic core has been plain enough.

Political versus social geography. The land is a more conspicuous factor in the history of states than in the history of society, but not more necessary and potent. Wars, which constitute so large a part of political history, have usually aimed more or less directly at acquisition or retention of territory; they have made every petty quarrel the pretext for mulcting the weaker nation of part of its land. Political maps are therefore subject to sudden and radical alterations, as when France's name was wiped off the North American continent in 1763, or when recently Spain's sovereignty in the Western Hemisphere was obliterated. But the race stocks, languages, customs, and institutions of both France and Spain remained after the flags had departed. The reason is that society is far more deeply rooted in the land than is a state, does not expand or contract its area so readily. Society is always, in a sense, *adscripta glebae;* an expanding state which incorporates a new piece of territory inevitably incorporates its inhabitants, unless it exterminates or expels them. Yet because racial and social geography changes slowly, quietly and imperceptibly, like all those fundamental processes which we call growth, it is not so easy and obvious a task to formulate a natural law for the territorial relations of the various hunter, pastoral nomadic, agricultural, and industrial types of society as for those of the growing state.

Land basis of society.

Most systems of sociology treat man as if he were in some way detached from the earth's surface; they ignore the land basis of society. The anthropo-geographer recognizes the various social forces, economic and psychologic, which sociologists regard as the cement of societies; but he has something to add. He sees in the land occupied by a primitive tribe or a highly organized state the underlying material bond holding society together, the ultimate basis of their fundamental social activities, which are therefore derivatives from the land. He sees the common territory exercising an integrating force,—weak in primitive communities where the group has established only a few slight and temporary relations with its soil, so that this low social complex breaks up readily like its organic counterpart, the low animal organism found in an amoeba; he sees it growing stronger with every advance in civilization involving more complex relations to the land,—with settled habitations, with increased density of population, with a discriminating and highly differentiated use of the soil, with the exploitation of mineral resources, and finally with that far-reaching exchange of commodities and ideas which means the establishment of varied extra-territorial relations. Finally, the modern society or state has grown into every foot of its own soil, exploited its every geographic advantage, utilized its geographic location to enrich itself by international trade, and when possible, to absorb outlying territories by means of colonies. The broader this geographic base, the richer, more varied its resources, and the more favorable its climate to their exploitation, the more numerous and complex are the connections which the members of a social group can establish with it, and through it with each other; or in other words, the greater may be its ultimate historical significance. The polar regions and the subtropical deserts, on the other hand, permit man to form only few and intermittent relations with any one spot, restrict economic methods to the lower stages of development, produce only the small, weak, loosely organized horde, which never evolves into a state so long as it remains in that retarding environment.

**Morgan's
Societas.**

Man in his larger activities, as opposed to his mere physiological or psychological processes, cannot be studied apart from the land which he inhabits. Whether we consider him singly or in a group—family, clan, tribe or state—we must always consider him or his group in relation to a piece of land. The ancient Irish sept, Highland clan, Russian mir, Cherokee hill-town, Bedouin tribe, and the ancient Helvetian canton, like the political state of history, have meant always a group of people and a bit of land. The first presupposes the second. In all cases the form and size of the social group, the nature of its activities, the trend and limit of its development will be strongly influenced by the size and nature of its habitat. The land basis is always present, in spite of Morgan's artificial distinction between a theoretically landless *societas*, held together only by the bond of common blood, and the political *civitas* based upon land.[4] Though primitive society found its conscious bond in common blood, nevertheless the land bond was always there, and it gradually asserted its fundamental character with the evolution of society.

...

Fishing tribes have their chief occupation determined by their habitats, which are found along well stocked rivers, lakes, or coastal fishing grounds. Conditions here encourage an early adoption of sedentary life, discourage wandering except for short periods, and facilitate the introduction of agriculture wherever conditions of climate and soil permit. Hence these fisher folk develop relatively large and permanent social groups, as testified by the ancient lake-villages of Switzerland, based upon a concentrated food-supply resulting from a systematic and often varied exploitation of the local resources. The coöperation and submission to a leader necessary in pelagic fishing often. gives the preliminary training for higher political organization.[14] All the primitive stocks of the Brazilian Indians, except the mountain Ges, are fishermen and agriculturists; hence their annual migrations are kept within narrow limits. Each linguistic group occupies a fixed and relatively well defined district.[15] Stanley found along the Congo large permanent villages of the natives, who were engaged in fishing and tilling the fruitful soil, but knew little about the country ten miles back from the river. These two generous means of subsistence are everywhere combined in Polynesia, Micronesia and Melanesia; there they are associated with dense populations and often with advanced political organization, as we find it in the feudal monarchy of Tonga and the savage Fiji Islands.[16] Fisher tribes, therefore, get an early impulse forward in civilization;[17] and even where conditions do not permit the upward step to agriculture, these tribes have permanent relations with their land, form stable social groups, and often utilize their location on a natural highway to develop systematic trade. For instance, on the northwest coast of British Columbia and Southern Alaska, the Haida, Tlingit and Tsimshean Indians have portioned out all the land about their seaboard villages among the separate families or households as hunting, fishing, and berrying grounds. These are regarded as private property and are handed down from generation to generation. If they are used by anyone other than the owner, the privilege must be paid for. Every salmon stream has its proprietor, whose summer camp can be seen set up at the point where the run of the fish is greatest. Combined with this private property in land there is a brisk trade up and down the coast; and a tendency toward feudalism in the village communities, owing to the association of power and social distinction with wealth and property in land.[18]

Land
bond in
pastoral
societies.

Among pastoral nomads, among whom a systematic use of their territory begins to appear, and therefore a more definite relation between land and people, we find a more distinct notion than among wandering hunters of territorial ownership, the right of communal use, and the distinct obligation of common defense. Hence the social bond is drawn closer. The nomad identifies himself with a certain district, which belongs to his tribe by tradition or conquest, and has its clearly defined boundaries. Here he roams between its summer and winter pastures, possibly one hundred and fifty miles apart, visits its small arable patches in the spring for his limited agricultural ventures, and returns to them in the fall to reap their meager harvest. Its springs, streams, or wells assume enhanced value, are things to be fought for, owing to the prevailing aridity of summer; while ownership of a certain tract of desert or grassland carries with it a certain right in the bordering settled district as an area of plunder.[19]

The Kara-Kirghis stock, who have been located since the sixteenth century on Lake Issik-Kul, long ago portioned out the land among the separate families, and determined their limits by natural features of the landscape.[20] Sven Hedin found on the Tarim River poles set up to mark the boundary between the Shah-yar and Kuchar tribal pastures.[21] John de Plano Carpini, traveling over southern Russia in 1246, immediately after the Tartar conquest, found that the Dnieper, Don, Volga and Ural rivers were all boundaries between domains of the various millionaries or thousands, into which the Tartar horde was organized.[22] The population of this vast country was distributed according to the different degrees of fertility and the size of the pastoral groups.[23] Volney observed the same distinction in the distribution of the Bedouins of Syria. He found the barren cantons held by small, widely scattered tribes, as in the Desert of Suez; but the cultivable cantons, like the Hauran and the Pachalic of Aleppo, closely dotted by the encampments of the pastoral owners.[24]

The large range of territory held by a nomadic tribe is all successively occupied in the course of a year, but each part only for a short period of time. A pastoral use of even a good district necessitates a move of five or ten miles every few weeks. The whole, large as it may be, is absolutely necessary for the annual support of the tribe. Hence any outside encroachment upon their territory calls for the united resistance of the tribe. This joint or social action is dictated by their common interest in pastures and herds. The social administration embodied in the apportionment of pastures among the families or clans grows out of the systematic use of their territory, which represents a closer relation between land and people than is found among purely hunting tribes. Overcrowding by men or livestock, on the other hand, puts a strain upon the social bond. When Abraham and Lot, typical nomads, returned from Egypt to Canaan with their large flocks and herds, rivalry for the pastures occasioned conflicts among their shepherds, so the two sheiks decided to separate. Abraham took the hill pastures of Judea, and Lot the plains of Jordan near the settled district of Sodom.[25]

Strength of the land bond in the state. The geographic basis of a state embodies a whole complex of physical conditions which may influence its historical development. The most potent of these are its size and zonal location; its situation, whether continental or insular, inland or maritime, on the open ocean or an enclosed sea; its boundaries, whether drawn by sea, mountain, desert or the faint demarking line of a river; its forested mountains, grassy plains, and arable lowlands; its climate and drainage system; finally its equipment with plant and animal life, whether indigenous or imported, and its mineral resources. When a state has taken advantage of all its natural conditions, the land becomes a constituent part of the state,[27] modifying the people which inhabit it, modified by them in turn, till the connection between the two becomes so strong by reciprocal interaction, that the people cannot be understood apart from their land. Any attempt to divide them theoretically reduces the social or political body to a cadaver, valuable for the study of structural anatomy after the method of Herbert Spencer, but throwing little light upon the vital processes.

Weak land tenure of hunting and pastoral tribes. A people who makes only a transitory or superficial use of its land has upon it no permanent or secure hold. The power to hold is measured by the power to use; hence the weak tenure of hunting and pastoral tribes. Between their scattered encampments at any given time are wide interstices, inviting occupation by any settlers who know how to make better use of the soil. This explains the easy intrusion of the English colonists into the sparsely tenanted territory of the Indians, of the agricultural Chinese into the pasture lands of the Mongols beyond the Great Wall, of the American pioneers into the hunting grounds of the Hudson Bay Company in the disputed Oregon country.[28] The frail bonds which unite these lower societies to their soil are easily ruptured and the people themselves dislodged, while their land is appropriated by the intruder. But who could ever conceive of dislodging the Chinese or the close-packed millions of India? A modern state with a given population on a wide area is more vulnerable than another of like population more closely distributed; but the former has the advantage of a reserve territory for future growth.[29] This was the case of Kursachsen and Brandenburg in the sixteenth century, and of the United States throughout its history. But beside the danger of inherent weakness before attack, a condition of relative underpopulation always threatens a retardation of development. Easy-going man needs the prod of a pressing population. [Compare maps pages 8 and 103 for examples.]

Land and food supply. Food is the urgent and recurrent need of individuals and of society. It dictates their activities in relation to their land at every stage of economic development, fixes the locality of the encampment or village, and determines the size of the territory from which sustenance is drawn. The length of residence in one place depends upon whether the springs of its food supply are perennial or intermittent, while the abundance of their flow determines how large a population a given piece of land can support.

Hunter and fisher folk, relying almost exclusively upon what their land produces of itself, need a large area and derive from it only an irregular food supply, which in winter diminishes to the verge of famine. The transition to the pastoral stage has meant the substitution of an artificial for a natural basis of subsistence, and therewith a change which more than any other one thing has inaugurated the advance from savagery to civilization.[30] From the standpoint of economics, the forward stride has consisted in the application of capital in the form of flocks and herds to the task of feeding the wandering horde;[31] from the standpoint of alimentation, in the guarantee of a more reliable and generally more nutritious food supply, which enables population to grow more steadily and rapidly; from the standpoint of geography, in the marked reduction in the per capita amount of land necessary to yield an adequate and stable food supply. Pastoral nomadism can support in a given district of average quality from ten to twenty times as many souls as can the chase; but in this respect is surpassed from twenty to thirtyfold by the more productive agriculture. While the subsistence of a nomad requires 100 to 200 acres of land, for that of a skillful farmer from 1 to 2 acres suffice.[32] In contrast, the land of the Indians living in the Hudson Bay Territory in 1857 averaged 10 square miles per capita; that of the Indians in the United States in 1825, subsidized moreover by the government, $1\frac{1}{4}$ square miles.[33]

With transition to the sedentary life of agriculture, society makes a further gain over nomadism in the closer integration of its social units, due to permanent residence in larger and more complex groups; in the continuous release of labor from the task of mere food-getting for higher activities, resulting especially in the rapid evolution of the home; and finally in the more elaborate organization in the use of the land, leading to economic differentiation of different localities and to a rapid increase in the population supported by a given area, so that the land becomes the dominant cohesive force in society. [See maps pages 8 and 9.]

Migratory agriculture. Agriculture is adopted at first on a small scale as an adjunct to the chase or herding. It tends therefore to partake of the same extensive and nomadic character[34] as these other methods of gaining subsistence, and only gradually becomes sedentary and intensive. Such was the superficial, migratory tillage of most American Indians, shifting with the village in the wake of the retreating game or in search of fresh unexhausted soil. Such is the agriculture of the primitive Korkus in the Mahadeo Hills in Central India. They clear a forested slope by burning, rake over the ashes in which they sow their grain, and reap a fairly good crop in the fertilized soil. The second year the clearing yields a reduced product and the third year is abandoned. When the hamlet of five or six families has exhausted all the land about it, it moves to a new spot to repeat the process.[35]

The same superficial, extensive tillage, with abandonment of fields every few years, prevails in the Tartar districts of the Russian steppes, as it did among the cattle-raising Germans at the beginning of their history. Tacitus says of them, *Arva per annos mutant et superest ager*,[36] commenting at the same time upon their abundance of land and their reluctance to till. Where nomadism is made imperative by aridity, the agriculture which accompanies it tends to become fixed, owing to the few localities blessed with an irrigating stream to moisten the soil. These spots, generally selected for the winter residence, have their soil enriched, moreover, by the long stay of the herd and thus avoid exhaustion.[37] Often, however, in enclosed basins the salinity of the irrigating streams in their lower course ruins the fields after one or two crops, and necessitates a constant shifting of the cultivated patches; hence agriculture remains subsidiary to the yield of the pastures. This condition and effect is conspicuous along the termini of the streams draining the northern slope of the Kuen Lun into the Tarim basin. [38]

Theory of progress from the standpoint of geography. The study of history is always, from one standpoint, a study of progress. Yet after all the century-long investigation of the history of every people working out its destiny in its given environment, struggling against the difficulties of its habitat, progressing when it overcame them and retrograding when it failed, advancing when it made the most of its opportunities and declining when it made less or succumbed to an invader armed with better economic or political methods to exploit the land, it is amazing how little the land, in which all activities finally root, has been taken into account in the discussion of progress. Nevertheless, for a theory of progress it offers a solid basis. From the standpoint of the land social and political organizations, in successive stages of development, embrace ever increasing areas, and make them support ever denser populations; and in this concentration of population and intensification of economic development they assume ever higher forms. It does not suffice that a people, in order to progress, should extend and multiply only its local relations to its land. This would eventuate in arrested development, such as Japan showed at the time of Perry's visit. The ideal basis of progress is the expansion of the world relations of a people, the extension of its field of activity and sphere of influence far beyond the limits of its own territory, by which it exchanges commodities and ideas with various countries of the world. Universal history shows us that, as the geographical horizon of the known world has widened from gray antiquity to the present, societies and states have expanded their territorial and economic scope; that they have grown not only in the number of their square miles and in the geographical range of their international intercourse, but in national efficiency, power, and permanence, and especially in that intellectual force which feeds upon the nutritious food of wide comparisons. Every great movement which has widened the geographical outlook of a people, such as the Crusades in the Middle Ages, or the colonization of the Americas, has applied an intellectual and economic stimulus. The expanding field of advancing history has therefore been an essential concomitant and at the same time a driving force in the progress of every people and of the world.

Man's increasing dependence upon nature. Since progress in civilization involves an increasing exploitation of natural advantages and the development of closer relations between a land and its people, it is an erroneous idea that man tends to emancipate himself more and more from the control of the natural conditions forming at once the foundation and environment of his activities. On the contrary, he multiplies his dependencies upon nature;[46] but while increasing their sum total, he diminishes the force of each. There lies the gist of the matter. As his bonds become more numerous, they become also more elastic. Civilization has lengthened his leash and padded his collar, so that it does not gall; but the leash is never slipped. The Delaware Indians depended upon the forests alone for fuel. A citizen of Pennsylvania, occupying the former Delaware tract, has the choice of wood, hard or soft coal, coke, petroleum, natural gas, or manufactured gas. Does this mean emancipation? By no means. For while fuel was a necessity to the Indian only for warmth and cooking, and incidentally for the pleasureable excitement of burning an enemy at the stake, it enters into the manufacture of almost every article that the Pennsylvanian uses in his daily life. His dependence upon nature has become more far-reaching, though less conspicuous and especially less arbitrary.

Increase in kind and amount. These dependencies increase enormously both in variety and amount. Great Britain, with its twenty thousand merchant ships aggregating over ten million tons, and its immense import and export trade, finds its harbors vastly more important to-day for the national welfare than in Cromwell's time, when they were used by a scanty mercantile fleet. Since the generation of electricity by water-power and its application to industry, the plunging falls of the Scandinavian Mountains, of the Alps of Switzerland, France, and Italy, of the

Southern Appalachians and the Cascade Range, are geographical features representing new and unsuspected forms of national capital, and therefore new bonds between land and people in these localities. Russia since 1844 has built 35,572
miles (57,374 kilometers) of railroad in her European territory, and thereby derived a new benefit from her level plains,
which so facilitate the construction and cheap operation of
railroads, that they have become in this aspect alone a new
feature in her national economy. On the other hand, the
galling restrictions of Russia's meager and strategically confined coasts, which tie her hand in any wide maritime policy,
work a greater hardship to-day than they did a hundred years
ago, since her growing population creates a more insistent
demand for international trade. In contrast to Russia, Norway, with its paucity of arable soil and of other natural resources, finds its long indented coastline and the coast-bred
seamanship of its people a progressively important national
asset. Hence as ocean-carriers the Norwegians have developed a merchant marine nearly half as large again as that of
Russia and Finland combined—1,569,646 tons[47] as against
1,084,165 tons.

This growing dependence of a civilized people upon its land
is characterized by intelligence and self-help. Man forms a
partnership with nature, contributing brains and labor, while
she provides the capital or raw material in ever more abundant
and varied forms. As a result of this coöperation, held by the
terms of the contract, he secures a better living than the savage
who, like a mendicant, accepts what nature is pleased to dole
out, and lives under the tyranny of her caprices.

NOTES TO CHAPTER III

1. H. J. Mackinder, Britain and the British Seas, p. 196. London, 1904.

2. Gardner, Atlas of English History, Map 29. New York, 1905.

3. Hereford George, Historical Geography of Great Britain, pp. 58-60. London, 1904.

4. Lewis Morgan, Ancient Society, p. 62. New York, 1878.

5. Franklin H. Giddings, Elements of Sociology, p. 247. New York, 1902.

6. Schoolcraft, The Indian Tribes of the United States, Vol. I, pp. 198-200, 224. Philadelphia, 1853.

7. *Ibid.*, Vol. I, pp. 231-232, 241.

8. Roosevelt, The Winning of the West, Vol. I, pp. 70-73, 88. New York, 1895.

9. McGee and Thomas, Prehistoric North America, pp. 392-393, 408, Vol. XIX, of *History of North America*, edited by Francis W. Thorpe, Philadelphia, 1905. *Eleventh Census Report on the Indians*, p. 51. Washington, 1894.

10. Hans Helmolt, History of the World, Vol. II, pp. 249-250. New York, 1902-1906.

11. Spencer and Gillen, Northern Tribes of Central Australia, pp. 13-15. London, 1904.

12. Ratzel, History of Mankind, Vol. I, p. 126. London, 1896-1898.

13. Roscher, *National-Oekonomik des Ackerbaues*, p. 24. Stuttgart, 1888.

14. Ratzel, History of Mankind, Vol. I, p. 131. London, 1896-1898.

15. Paul Ehrenreich, *Die Einteilung und Verbreitung der Völkerstämme Brasiliens*, Peterman's *Geographische Mittheilungen*, Vol. XXXVII, p. 85. Gotha, 1891.

16. Roscher, *National-Oekonomik des Ackerbaues*, p. 26, Note 5. Stuttgart, 1888.

17. *Ibid.*, p. 27.

18. Albert Niblack, The Coast Indians of Southern Alaska and Northern British Columbia, pp. 298-299, 304, 337-339. Washington, 1888.

19. Ratzel, History of Mankind, Vol. III, p. 173. London, 1896-1898.

20. *Ibid.*, Vol. III, pp. 173-174.

21. Sven Hedin, Central Asia and Tibet, Vol. I, p. 184. New York and London, 1903.

22. John de Plano Carpini, Journey in 1246, p. 130. *Hakluyt Society*, London, 1904.

23. Journey of William de Rubruquis in 1253, p. 188. *Hakluyt Society*, London, 1903.

24. Volney, quoted in Malthus, Principles of Population, Chap. VII, p. 60. London, 1878.

25. Genesis, Chap. XIII, 1-12.

26. Herbert Spencer, Principles of Sociology, Vol. I, p. 457. New York.

27. Heinrich von Treitschke, *Politik*, Vol. I, pp. 202-204. Leipzig, 1897.

28. E. C. Semple, American History and Its Geographic Conditions, pp. 206-207. Boston, 1903.

29. Roscher, *Grundlagen des National-Oekonomie*, Book VI. *Bevolkerung*, p. 694, Note 5. Stuttgart, 1886.

30. Edward John Payne, History of the New World Called America, Vol. I, p. 303-313. Oxford and New York, 1892.

31. Roscher, *National-Oekonomik des Ackerbaues*, pp. 31, 52. Stuttgart, 1888.

32. *Ibid.*, p. 56, Note 5.

33. For these and other averages, Sir John Lubbock, Prehistoric Times, pp. 593-595. New York, 1872.

34. Roscher, *National-Oekonomik des Ackerbaues*, pp. 79-80, p. 81, Note 7. Stuttgart, 1888. William I. Thomas, Source Book for Social Origins, pp. 96-112. Chicago, 1909.

35. Capt. J. Forsyth, The Highlands of Central India, pp. 101-107, 168. London, 1889.

36. Tacitus, *Germania*, III.

37. Roscher, *National-Oekonomik des Ackerbaues*, p. 32, Note 15 on p. 36. Stuttgart, 1888.

38. E. Huntington, The Pulse of Asia, pp. 202, 203, 212, 213, 236-237. Boston, 1907.

39. Sheldon Jackson, Introduction of Domesticated Reindeer into Alaska, pp. 20, 25-29, 127-129. Washington, 1894.

40. Quoted in Alexander von Humboldt, Aspects of Nature in Different Lands, pp. 62, 139. Philadelphia, 1849.

41. Edward John Payne, History of the New World Called America, Vol. I, pp. 311-321, 333-354, 364-366. New York, 1892.

42. Prescott, Conquest of Peru, Vol. I, p. 47. New York, 1848.

43. McGee and Thomas, Prehistoric North America, Vol. XIX, pp. 151-161, of *The History of North America*, edited by Francis W. Thorpe, Philadelphia, 1905.

44. Ratzel, *Anthropo-geographie*, Vol. II, pp. 264-265.

45. Malthus, Principles of Population, Chapters V and VII. London, 1878.

46. Nathaniel Shaler, Nature and Man in America, pp. 147-151. W. Z. Ripley, Races of Europe, Chap. I, New York, 1899.

47. Justus Perthes, *Taschen-Atlas*, pp. 44, 47. Gotha, 1910.

CHAPTER V

THE NEOLITHIC REVOLUTION

THROUGHOUT the vast eras of the Ice Ages man had made no fundamental change in his attitude to external Nature. He had remained content to take what he could get, though he had vastly improved his methods of getting and had learned discrimination in what he took. Soon after the end of the Ice Age man's attitude (or rather that of a few communities) to his environment underwent a radical change fraught with revolutionary consequences for the whole species. In absolute figure the era since the Ice Age is a trifling fraction of the total time during which men or man-like creatures have been active on the earth. Fifteen thousand years is a generous estimate of the post-glacial period, as against a conservative figure of 250,000 years for the preceding era. Yet in the last twentieth of his history man has begun to control Nature, or has at least succeeded in controlling her by co-operating with her.

The steps by which man's control was made effective have been gradual, their effects cumulative. But among them we may distinguish some which, judged by the standards explained in Chapter I, stand out as revolutions. The first revolution that transformed human economy gave man control over his own food supply. Man began to plant, cultivate, and improve by selection edible grasses, roots, and trees. And he succeeded in taming and firmly attaching to his person certain species of animal in return for the fodder he was able to offer, the protection he could afford, and the forethought he could exercise. The two steps are closely related. Many authorities now hold that cultivation is everywhere older than stock-breeding. Others, notably the German historical school, believe that, while some human groups were beginning to cultivate plants, other groups were domesticating animals. Very few still contend that a stage of pastoralism universally preceded cultivation. For purposes of exposition we shall here adopt the first theory. Even to-day many tribes of cultivators survive who possess no domestic animals. In Central Europe and Western China, where mixed farming has been for centuries the prevailing economy, the oldest peasants revealed by the archæologist's spade relied very little, if at all, on domestic animals, but lived on agricultural produce and perhaps a little game.

Quite a large variety of plants are capable of providing a staple diet under cultivation. Rice, wheat, barley, millet, maize, yams, sweet potatoes, respectively support considerable populations even to-day. But in the civilizations which have contributed most directly and most generously to the building up of the cultural heritage we enjoy, wheat and barley lie at the foundations of the economy. These two cereals offer, in fact, exceptional advantages. The food they yield is highly nutritious, the grains can easily be stored, the return is relatively high, and, above all, the labour involved in cultivation is not too absorbing. The preparation of the fields and sowing certainly demand a considerable effort; some weeding and watching are requisite while the crop is ripening ; harvest demands intensive exertion by the whole community. But these efforts are seasonal. Before and after sowing come intervals in which the fields need practically no attention. The grain-grower enjoys substantial spells of leisure, during which he can devote himself to other occupations. The rice-grower, on the other hand, enjoys no such respite. His toil need perhaps never be so intensive as that demanded by the grain harvest, but it is more continuous.

As the historic civilizations of the Mediterranean basin, Hither Asia, and India were built upon cereals, we shall confine our attention to the economies based upon wheat and barley. The histories of these have been much more extensively studied than those of other cultivated plants, and may be briefly indicated.

Both wheats and barleys are domesticated forms of wild grasses. But in each case cultivation, the deliberate selection for seed purposes of the best plants, and conscious or accidental crossings of varieties have produced grains far larger and more nutritive than any wild grass-seeds. Two wild grasses ancestral to wheat are known—dinkel and wild emmer. Both grow wild in mountainous countries, the former in the Balkans, the Crimea, Asia Minor, and the Caucasus, emmer further south in Palestine and perhaps in Persia.

The present distribution may, of course, be deceptive ; climate has changed greatly since the time when cultivation began, and plant geography is dependent upon climate. Indeed, Vavilov, arguing from different premises, has proposed Afghanistan and North-western China as the original centres of wheat-growing. In any case, wild dinkel is the parent of a small,

unsatisfying wheat, extensively cultivated in Central Europe in prehistoric times and still grown in Asia Minor. A far superior grain may be got from the cultivation of emmer (*Triticcum dicoccum*). Emmer seems to have been the oldest wheat cultivated in Egypt, in Asia Minor, and in Western Europe, and is often grown to-day. But the majority of the modern bread wheats belong to a third variety (*Triticum vulgare*), to which no wild ancestor is known. They may result from crosses between emmer and some unknown grass. The oldest wheat-grains found in Mesopotamia, Turkestan, Persia, and India belong to this group.

The wild ancestors of barley are also mountain grasses. They have been reported from Marmarica in North Africa, Palestine, Asia Minor, Transcaucasia, Persia, Afghanistan, and Turkestan. Vavilov's methods would point to primary centres of barley cultivation in Abyssinia and South-eastern Asia. The questions where cultivation started and whether in one centre or in several are still undecided. Because sickles have recently been found in cave-dwellings in Palestine, accompanied by a set of tools appropriate to a food-gathering economy rather than to the culture normally associated with the first revolution, it is argued that cereal cultivation started in or near Palestine. But it is not impossible that these cave-dwellers (termed Natufians) were just a backward tribe who had adopted some elements of culture from more progressive cultivators elsewhere, but had not yet thoroughly reorganized their economy.

As a revolution the introduction of a food-producing economy should affect the lives of all concerned, so as to be reflected in the population curve. Of course, no " vital statistics " have been recorded to prove that the expected increase of population did occur. But it is easy to see that it should. The community of food-gatherers had been restricted in size by the food supplies available—the actual number of game animals, fish, edible roots, and berries growing in its territory. No human effort could augment these supplies, whatever magicians might say. Indeed, improvements in the technique or intensification of hunting and collecting beyond a certain point would result in the progressive extermination of the game and an absolute diminution of supplies. And, in practice, hunting populations appear to be nicely adjusted to the resources at their disposal. Cultivation at once breaks down the limits thus imposed. To increase the food supply it is only necessary to sow more seed, to bring more land under tillage. If there are more mouths to feed, there will also be more hands to till the fields.

Again, children become economically useful. To hunters children are liable to be a burden. They have to be nourished for years before they can begin to contribute to the family larder effectively. But quite young toddlers can help in weeding fields and scaring off birds or trespassing beasts. If there are sheep and cattle, boys and girls can mind them. *A priori*, then, the probability that the new economy was accompanied by an increase of population is very high. That population did really expand quite fast seems to be established by archæology. Only so can we explain the apparent suddenness with which peasant communities sprang up in regions previously deserted or tenanted only by very sparse groups of collectors.

Round the lake that once filled the Fayum depression

the number of Old Stone Age tools is certainly imposing. But they have to be spread over so many thousand years that the population they attest may be exiguous. Then quite abruptly the shore of a somewhat shrunken lake is found to be fringed with a chain of populous hamlets, all seemingly contemporary and devoted to farming. The Nile valley from the First Cataract down to Cairo is quickly lined with a chain of flourishing peasant villages, all seeming to start about the same time and all developing steadily down to 3000 B.C. Or take the forest plains of Northern Europe. After the Ice Age we find there scattered settlements of hunters and fishers along the coasts, on the shores of lagoons, and on sandy patches in the forest. The relics collected from such sites should probably be spread over a couple of thousand years, and so are only compatible with a tiny population. But then, within the space of a few centuries, Denmark first, and thereafter Southern Sweden, North Germany, and Holland, become dotted with tombs built of gigantic stones. It must have taken a considerable force to build such burial-places, and in fact some contain as many as 200 skeletons. The growth of the population must then have been rapid. It is true that in this case the first farmers, who were also the architects of the big stone graves, are supposed to have been immigrants. But as they are supposed to have come by boat from Spain round Orkney and across the North Sea, the actual immigrant population cannot have been very large. The multitude implied by the tombs must have resulted from the fecundity of a few immigrant families and that of the older hunters who had joined with these in exploiting the agricultural resources of the virgin north. Finally, the human skeletons assigned to the New Stone Age in Europe alone are several hundred times more numerous than those from the whole of the Old Stone Age. Yet the New Stone Age in Europe lasted at the outside 2000 years—less than one hundredth of the time assigned to the Old !

It would be tedious to pile up evidence; its implications are clear. It was only after the first revolution—but immediately thereafter—that our species really began to multiply at all fast. Certain other implications and consequences of this first or " neolithic " revolution can be considered later. It is desirable at this point to enter a caveat.

The adoption of cultivation must not be confused with the adoption of a sedentary life. It has been customary to contrast the settled life of the cultivator with the nomadic existence of the " homeless hunter." The contrast is quite fictitious. Last century the hunting and fishing tribes of the Pacific coasts of Canada possessed permanent villages of substantial, ornate, and almost luxurious wooden houses. The Magdalenians of France during the Ice Age certainly occupied the same cave over several generations. On the other hand, certain methods of cultivation impose a sort of nomadism upon their practitioners. To many peasants in Asia, Africa, and South America, even to-day, cultivation means simply clearing a patch of scrub or jungle, digging it up with a hoe or just a stick, sowing it, and then reaping the crop. The plot is not fallowed, still less manured, but just re-sown next year. Of course, under such conditions the yield declines conspicuously after a couple of seasons. Thereupon another plot is cleared, and the process repeated till that too is exhausted. Quite soon all the available land close to the settlement has been cropped to exhaustion. When that has happened, the people move away and start afresh elsewhere.

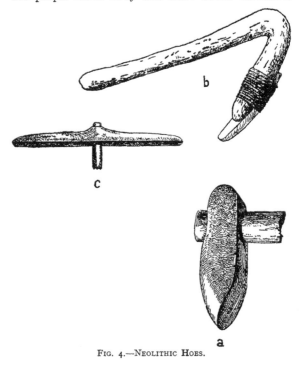

FIG. 4.—NEOLITHIC HOES.

Their household goods are simple enough to be easily transported. The houses themselves are flimsy hovels, probably grown foul by prolonged occupation, and can readily be replaced.

What has just been described is the most primitive form of cultivation, often termed hoe-culture or garden-culture. Nature soon posed a question to the first cultivators—the problem of soil exhaustion. The easiest way of dealing with the issue was to dodge it and move away. Actually that solution is perfectly satisfactory so long as there is plenty of cultivable land and the cultivator is content to do without such luxuries and refinements as impede migration. It was, of course, a nuisance to have to clear a new bit of forest every few years, but that was surely less trouble than thinking out a new solution. In any case this sort of cultivation prevailed throughout Europe north of the Alps in prehistoric times. It may have survived among some German tribes down to the beginning of our era; for the geographer Strabo remarks upon their readiness to shift their settlements. It is still practised to-day, for example, among the rice-growing Nagas in Assam, among the Boro in the Amazon basin, and even by grain-growers in the Sudan. Yet it is a wasteful method, and ultimately limits the population, since suitable land is nowhere unrestricted.

If nomadic garden-culture is the most primitive type of cultivation, it is not quite the simplest nor certainly the oldest. Throughout the great belt of countries now arid or desert, between the temperate forests of the north and the jungles of the tropics, the most suitable land for cultivation is often found on the alluvial soils deposited where intermittent torrents flow out from the hills on to the plains, and in the valleys of rivers that periodically overflow their banks. In that arid zone the muddy flood plains beside the great rivers and the silts, spreading out like a fan at the mouth of a torrent's gorge, form a welcome contrast to the unfertile sands or sterile rocks of the desert. And on them the waters remaining from the floods take the place of the uncertain rains in providing the moisture needed for the germination and ripening of crops. And so in the Eastern Sudan the Hadendoa scattered millet seeds on the wet mud left by the Nile flood every autumn and just waited for the crop to sprout. Whenever a thunderstorm on Mount Sinai has brought down the Wady el Arish in flood, the desert Arabs hasten to sow barley grains in the freshly deposited silt and gather a welcome crop.

Now under such conditions the floods, thus utilized, not only water the crop, they create fresh soil. The flood waters are yellow and muddy from the sediments which they have collected in their impetuous passage through the hills. As they spread out and flow more gently, the suspended mud is deposited as a deep silt on the inundated lands. The silt contains the chemical constituents that last year's crop had taken from the soil, which is thus renewed and refertilized. Under conditions of natural irrigation, the cultivator need not be nomadic. He can go on cultivating the same patches year after year provided they are flooded between each harvest.

Now the method of cultivation just described is possible precisely in the regions where wild ancestors of wheat and barley are probably native. Perry has cogently argued that irrigation is the oldest method of grain-growing. In particular, conditions in the Nile Valley would have been exceptionally favourable to the deliberate cultivation of cereals. The Nile, swollen by the monsoon rains on the Abyssinian plateau, overflows its banks with remarkable regularity every autumn. The flood arrives at a convenient season when the heat is no longer so intense as to scorch up young shoots. And so, Perry suggests, the reliable and timely Nile flood first prompted men to plant seeds deliberately and let them grow. Food-gatherers must have utilized grains of wild wheat and barley for food long before they started cultivating them. Handfuls of such seeds, scattered on the wet silt of the Nile's flood, would be the direct ancestors of all cultivated cereals. And natural irrigation would be the prototype of all systems of cultivation.

Perry's plausible and consistent account of the Egyptian origin of agriculture is, of course, just a theory supported by even less direct evidence than the Palestinian origin mentioned on p. 77. At the time of the oldest agricultural settlements in the Nile Valley the rainfall in Hither Asia and North Africa was more generous than to-day, so that irrigation was by no means the sole method of getting crops to grow. The idea of cultivating cereals undoubtedly spread very rapidly; North Syria, Iraq, and the Persian plateau are studded with the ruins of agricultural villages going back nearly, if not quite, as far as the oldest of such villages in Egypt. Migratory garden-culture would explain this rapid diffusion quite simply. It is not so easy to see how a system, developed under the exceptional conditions of the Nile Valley, should have been transplanted to Persia and Mesopotamia, with their very different and less favourable circumstances. As to Europe, it is highly probable that the idea of cultivation and the cereals cultivated were first introduced by hoe-cultivators spreading from North Africa over Western Europe and by others expanding from the Danube basin into Belgium and Germany; for the wild ancestors of wheat and barley are not to be expected north of the Balkans.

On the other hand, cultivation in Egypt was not so simple. In its natural state the Nile Valley must have been a chain of swamps, choked with reedy jungle that sheltered hippopotami and other troublesome beasts. To make it cultivable the swamps must be drained and cleared and their dangerous denizens driven out. Such an undertaking was possible only to a well-organized community of some size and equipped with efficient tools. On the whole it looks as if cultivation dependent on the Nile flood would be later than simple hoe-cultivation and derived therefrom. It is really unprofitable to speculate how, where, and when the cultivation of cereals began. It is perhaps slightly less futile to inquire how the primary form of food-production was completed and transformed into mixed farming, assuming the theory enunciated on p. 75.

In practically all the oldest food-producing settlements examined by archæologists in Europe, Hither Asia, and North Africa, the basic industry is mixed farming; in addition to the cultivation of cereals, animals are bred for food. This economy is characteristic of the "neolithic" stage wherever such exist. The food animals kept were not very varied: horned cattle, sheep, goats, and swine. Comparatively few species—fowls are the most important—have been added to the farmyards in subsequent periods or other countries. Horned cattle require rather rich grass, but can live on well-watered steppes, in naturally irrigated valleys, and even in forests that are not too dense. Pigs like swamps or woodlands; sheep and goats can thrive under dry, though not absolutely desert, conditions, and both are at home in hilly and mountainous country. Wild goats probably once ranged all along the mountains that divide Eurasia lengthwise, perhaps from the Pyrenees, or at least from the Balkans eastwards to the Himalaya. Wild sheep lived along the same chains, but in three distinct varieties. The mouflon survives in the Mediterranean islands and the hill country of Hither Asia from Turkey to Western Persia; east of the mouflon, in Turkestan, Afghanistan, and the Punjab, is the home of the urial; still farther east, in the mountains of Central Asia, lives the argal. In Africa no wild sheep is known. The oldest Egyptian sheep belong to the urial stock, as do the oldest European flocks; but the mouflon is represented side by side with the urial on early monuments from Mesopotamia. The reader will note that ancestors of our farmyard animals lived wild in most of the regions that seem likely to have comprised the cradle of grain-growing. But the absence of wild sheep from Africa makes Egypt unlikely as a starting point of mixed farming.

As already remarked, the period when the food-producing economy became established was one of climatic crises adversely affecting precisely that zone of arid sub-tropical countries where the earliest farmers appear, and where the wild ancestors of cultivated cereals and domestic animals actually lived. The melting of the European ice-sheets and the contraction of the high pressures or anticyclones over them involved a northward shift in the normal path of the rain-bearing depressions from the Atlantic. The showers that had watered North Africa and Arabia were deflected over Europe. Desiccation set in. Of course the process was not sudden or catastrophic. At first and for long, the sole harbinger would be the greater

severity and longer duration of periodical droughts. But quite a small reduction in the rainfall would work a devastating change in countries that were always relatively dry. It would mean the difference between continuous grasslands and sandy deserts interrupted by occasional oases.

A number of animals that could live comfortably with a twelve-inch annual rainfall would become a surplus population if the precipitation diminished by a couple of inches for two or three years on end. To get food and water, the grass-eaters would have to congregate round a diminishing number of springs and streams—in oases. There they would be more exposed than ever to the attacks of beasts of prey—lions, leopards, and wolves—that would also gravitate to the oases for water. And they would be brought up against man too; for the same causes would force even hunters to frequent the springs and valleys. The huntsman and his prey thus find themselves united in an effort to circumvent the dreadful power of drought. But if the hunter is also a cultivator, he will have something to offer the famished beasts: the stubble of his freshly reaped fields will afford the best grazing in the oasis. Once the grains are garnered, the cultivator can tolerate half-starved mouflons or wild oxen trespassing upon his garden plots. Such will be too weak to run away, too thin to be worth killing for food. Instead, man can study their habits, drive off the lions and wolves that would prey upon them, and perhaps even offer them some surplus grain from his stores. The beasts, for their part, will grow tame and accustomed to man's proximity.

Hunters to-day, and doubtless in prehistoric times, have been accustomed to make pets of young wild animals for ritual ends or just for fun. Man has allowed the dog to frequent his camp in return for the offal of his prey and refuse from his feasts. Under the conditions of incipient desiccation the cultivator has the chance of attaching to his ménage not only isolated young beasts, but the remnants of complete flocks or herds, comprising animals of both sexes and all ages. If he just realizes the advantage of having a group of such half-tamed beasts hanging round the fringes of his settlement as a reserve of game easily caught, he will be on the way to domestication.

Next he must exercise restraint and discrimination in using this reserve of meat. He must refrain from frightening the beasts unnecessarily or killing the youngest and tamest. Once he begins to kill only the shyest and least amenable bulls or rams, he will have started selective breeding, eliminating untractable brutes, and consequently favouring the more docile. But he must also use his new opportunities of studying the life of the beasts at close range. He will thus learn about the processes of reproduction, the animals' needs of food and water. He must act upon his knowledge. Instead of merely driving the herd away when the time comes round for sowing his plots again, he must follow the beasts, guide them to suitable pastures and water, and continue to ward off predatory carnivora. It can thus be imagined how with lapse of time a flock or a herd should have been bred that was not only tame, but actually dependent upon man.

That result could happen only provided the peculiar climatic conditions continued long enough, and suitable animals were haunting human settlements. No doubt experiments were tried with various species; herds of antelopes and gazelles were kept by the Egyptians about 3000 B.C. These and other unknown experiments were fruitless. Luckily, cattle, sheep, goats, and pigs were included in the wild fauna of the desiccated regions in Asia. These did become firmly attached to man and ready to follow him.

At first the tame or domesticated beast would presumably be regarded only as a potential source of meat, an easily accessible sort of game. Other uses would be discovered later. It might be noticed that crops flourished best on plots that had been grazed over. Ultimately the value of dung as a fertilizer would be realized. The process of milking can only have been discovered when men had had ample opportunity of studying at close quarters the suckling of calves and lambs and kids. But once the trick was grasped, milk would become a second staple. It could be obtained without killing the beast, without touching your capital. Selection would again be applied. The best milkers would be spared, and their young reared in preference to other calves, lambs, or kids. Still later the hair of sheep or goats would win appreciation. It could be treated by processes, perhaps originally applied to plant fibres, and woven into cloth or else beaten into felt. Wool is entirely the artificial product of selective breeding. On wild sheep it is merely a down between the hairs. It was still unknown to the Egyptians even after 3000 B.C. But in Mesopotamia sheep were being bred for their wool before that date. The harnessing of animals to bear burdens or draw ploughs and vehicles is a late adaptation, and will be considered among the steps leading up to the second revolution in human economy (p. 137).

The minimal characteristics of simple cultivation

have already been considered. But these must now be pictured as combined with stock-breeding if we are to understand the basic economy revealed in neolithic settlements in North Africa, Hither Asia, and Europe. If the number of animals kept remains quite small, the account already given will hold good : the animals will be put to graze on the stubble after the harvest and at other seasons on natural pastures round the settlement. Beyond telling off a few youths to look after the herd, the communal economy can be left as already described. But as soon as the flocks exceed a low limit, special provision may have to be made for them. Trees and scrub may be burned off to make room for grass. In a river valley it may be thought worth while to clear or irrigate special meadows to serve as pasture for cattle. Crops may be deliberately grown, harvested, and conserved to serve exclusively as fodder. Or the animals may be driven far afield to find pastures in the dry season. In Mediterranean lands, Persia, and Asia Minor there is good summer grazing on the hills which in winter are snow-clad. And so sheep and cattle are driven up to hill pastures in the spring. And now a regular company of the village's inhabitants must accompany the herds to ward off wild beasts, to milk the cows and ewes. The herders must generally take with them supplies of grain and other equipment. In some cases the fraction of the community that migrate with their gear to the summer pastures is quite small. But in hot and dry countries, like Persia, parts of the Eastern Sudan, and in the north-western Himalayas, the bulk of the community abandons its village in the stifling valley and accompanies the herds to the cooler hills. Only a few stay behind to look after the fields and dwellings.

From this it is no far cry to a purely pastoral economy in which cultivation plays a negligible rôle. Pure pastoral nomadism is familiar, and is illustrated by several peoples in the Old World; the Bedouin of Arabia and Mongolian tribes of Central Asia are the best-known examples. How old such a mode of life may be is uncertain. Pastoralists are not likely to leave many vestiges by which the archæologist could recognize their presence. They tend to use vessels of leather and basketry instead of pots, to live in tents instead of in excavated shelters or huts supported by stout timber posts or walls of stone or brick. Leather vessels and baskets have as a rule no chance of surviving; tents need not even leave deep post-holes to mark where they once stood. (Though wood decays, modern archæology can recognize the hole made by a post five thousand years ago.)

The failure to recognize prehistoric settlement sites or groups of relics belonging to pure pastoralists is not in itself any proof that such did not exist. To that extent the postulate of the " historical school," that pure pastoralism and pure hoe-culture were originally practised independently by separate peoples and that mixed farming resulted from their subsequent fusion, is irrefutable. Yet Forde has recently emphasized the instability of pure pastoralism. Many typical pastoral tribes to-day, like the patriarchs in Genesis, actually cultivate grain, though in an incidental and rather casual manner. If they grow no grain themselves, pastoral nomads are almost always economically dependent upon settled peasant villages. The cultivators may be tributaries or serfs to the pastoralists, but they are essential to their subsistence.

Whatever its origin, stock-breeding gave man control

over his own food supply in the same way as cultivation did. In mixed farming it becomes an equal partner in the food-producing economy. But just as the term "cultivation" covers many distinct modes of gaining a livelihood, so the single phrase "mixed farming" marks an equal disparity and diversity. The several different modes of cultivation may be combined in varying degrees with distinct attitudes to the livestock. The diversity of the permutations and combinations has just been suggested. The multiplicity of concrete applications of the food-producing economy must never be forgotten.

It must be remembered, too, that food-production does not at once supersede food-gathering. If to-day hunting is only a ritual sport and game is a luxury for the rich, fishing is still a great industry, contributing directly to everybody's diet. At first hunting, fowling, fishing, the collection of fruits, snails, and grubs continued to be essential activities in the food-quest of any food-producing group. Grain and milk began as mere supplements to a diet of game, fish, berries, nuts, and ants' eggs. Probably at first cultivation was an incidental activity of the women while their lords were engaged in the really serious business of the chase. Only slowly did it win the status of an independent and ultimately predominant industry. When the archæological record first reveals neolithic communities in Egypt and Britain, survivals from the food-gathering régime clearly stand on an equal footing with grain-growing and stock-breeding. Only subsequently does their economic importance decline. After the second revolution, hunting and fowling have become, as with us, ritual sports, or else, like fishing, specialized industries practised by groups within the community or by independent societies, economically dependent upon an agricultural civilization.

...

10. THE INTERPRETATIONS OF CULTURE

...

When we base our study on these observations, it appears that serious objections may be made against the assumption of the occurrence of a general sequence of cultural stages among all the races of man; that rather we recognize both a tendency of diverse customs and beliefs to converge towards similar forms, and a development of customs in divergent directions. In order to interpret correctly these similarities in form, it is necessary to investigate their historical development; and only when the historical development in different areas is the same, will it be admissible to consider the phenomena in question as equivalent. From this point of view the facts of cultural contact assume a new importance (see p. 169).

Culture has also been interpreted in other ways. Geographers try to explain forms of culture as a necessary result of geographical environment.

It is not difficult to illustrate the important influence of geographical environment. The whole economic life of man is limited by the resources of the country in which he lives. The location of villages and their size depends upon the available food-supply; communication upon available trails or waterways. Environmental influences are evident in the territorial limits of tribes and peoples; seasonal changes of food-supply may condition seasonal migrations. The variety of habitations used by tribes of different areas demonstrate its influence. The snow house of the Eskimo, the bark wigwam of the Indian, the cave dwelling of the tribes of the desert, may serve as illustrations of the way in which in accordance with the available materials protection against exposure is attained. Scarcity of food may condition a nomadic life, and the necessity of carrying household goods on the back favors the use of skin receptacles and baskets as substitutes for pottery.

The special forms of utensils may be modified by geographic conditions. Thus the complex bow of the Eskimo which is related to Asiatic forms takes a peculiar form owing to the lack of long, elastic material for bow staves. Even in the more complex forms of the mental life, the influence of environment may be found; as in nature myths explaining the activity of volcanoes or the presence of curious land forms, or in beliefs and customs relating to the local characterization of the seasons.

However, geographical conditions have only the power to modify culture. By themselves they are not creative. This is clearest wherever the nature of the country limits the development of culture. A tribe, living without foreign trade in a given environment is limited to the resources of his home country. The Eskimo has no vegetable food supplies to speak of; the Polynesian who lives on an atoll has no stone and no skins of large mammals; the people of the desert have no rivers furnishing fish or offering means of travel. These self-evident limitations are often of great importance.

It is another question whether external conditions are the immediate cause of new inventions. We can understand that a fertile soil will induce an agricultural people whose numbers are increasing rapidly to improve its technique of agriculture, but not, how it could be the cause of the invention of agriculture. However rich in ore a country may be, it does not create techniques of handling metals; however rich in animals that might be domesticated, it will not lead to the development of herding if the people are entirely unfamiliar with the uses of domesticated animals.

If we should claim that geographical environment is the sole determinant that acts upon the mind assumed to be the same in all races of mankind, we should be necessarily led to the conclusion that the same environment will produce the same cultural results everywhere.

This is obviously not true, for often the forms of cultures of peoples living in the same kind of environment show marked differences. I do not need to illustrate this by comparing the American settler with the North American Indian, or the successive races of people that have settled in England, and have developed from the Stone Age to the modern English. It may, however, be desirable to show that among primitive tribes, geographical environment alone does not by any means determine the type of culture. Proof of this may be found in the mode of life of the hunting and fishing Eskimo and the reindeer-breeding Chukchee (Bogoras, Boas 3); the African pastoral Hottentot and the hunting Bushmen in their older, wider distribution (Schultze); the Negrito and the Malay of southeastern Asia (Martin).

Environment always acts upon a preexisting culture, not on an hypothetical cultureless group. Therefore it is important only insofar as it limits or favors activities. It may even be shown that ancient customs, that may have been in harmony with a certain type of environment, tend to survive under new conditions, where they are of disadvantage rather than of advantage to the people. An example of this kind, taken from our own civilization, is our failure to utilize unfamiliar kinds of food that may be found in newly settled countries. Another example is presented by the reindeer-breeding Chukchee, who carry about in their nomadic life a tent of most complicated structure, which corresponds in its type to the older permanent house of the coast dwellers, and contrasts in the most marked way with the simplicity and light weight of the Eskimo tent.[1] Even among the Eskimo, who have so marvelously well succeeded in adapting themselves to their geographical environment, customs like the taboo on

[1] Bogoras, pp. 177 et seq.; Boas 3: p. 551.

the promiscuous use of caribou and seal prevent the fullest use of the opportunities offered by the country.

Thus it would seem that environment has an important effect upon the customs and beliefs of man, but only insofar as it helps to determine the special forms of customs and beliefs. These are, however, based primarily on cultural conditions, which in themselves are due to other causes.

At this point the students of anthropo-geography who attempt to explain the whole cultural development on the basis of geographical environmental conditions are wont to claim that these causes themselves are founded on earlier conditions, in which they have originated under the stress of environment. This claim is inadmissible because the investigation of every single cultural feature demonstrates that the influence of environment brings about a certain degree of adjustment between environment and social life, but that a complete explanation of the prevailing conditions, based on the action of environment alone, is never possible. We must remember, that, no matter how great an influence we may ascribe to environment, that influence can become active only by being exerted upon the mind; so that the characteristics of the mind must enter into the resultant forms of social activity. It is just as little conceivable that mental life can be explained satisfactorily by environment alone, as that environment can be explained by the influence of the people upon nature, which, as we all know, has brought about changes of watercourses, the destruction of forests and changes of fauna. In other words, it seems entirely arbitrary to disregard the part that psychical or social elements play in determining the forms of activities and beliefs which occur with great frequency all over the world.

...

CHAPTER I

A COMPREHENSIVE VIEW OF THE FLOOD PROBLEM

The Flood Problem in the United States

Every year receding flood waters in one or more sections of the United States expose muddy plains where people were poorly prepared to meet the overflow. Small-town shopkeepers digging their goods out of Ohio River silt; Alabama farmers collecting their scattered and broken possessions; and New England manufacturers taking inventory in water-soaked warehouses, testify to the dislocating effects of floods and to the unsatisfactory adjustment which man has made to them in many valleys. For the most part, floods in the United States leave in their wake a dreary scene of impaired health, damaged property, and disrupted economic life.

The effects of floods are not everywhere disastrous, however, or even disturbing to the economy. Each year ebbing flood waters also reveal plains in which a relatively satisfactory arrangement of human occupance has taken place. Pittsburgh merchants returning to stores which, because of adequate preparations, suffered only minor losses; Montana ranchers appraising the increased yields of hay to be obtained because of fresh deposits of moisture; and New Orleans citizens carrying on their business behind a levee withstanding a flood crest high above the streets, illustrate wise adjustments to flood hazard.

It has become common in scientific as well as popular literature to consider floods as great natural adversaries which man seeks persistently to over-power. According to this view, floods always are watery marauders which do no good, and against which society wages a bitter battle. The price of victory is the cost of engineering works necessary to confine the flood crest; the price of defeat is a continuing chain of flood disasters. This simple and prevailing view neglects in large measure the possible feasibility of other forms of adjustment, of which the Pittsburgh and Montana cases are examples.

Floods are "acts of God," but flood losses are largely
acts of man. Human encroachment upon the flood plains of rivers
accounts for the high annual toll of flood losses. Although in
a few drainage areas the frequency and magnitude of floods have
increased as a result of exploitative use of the up-stream lands,
the flood menace elsewhere has changed but little while man has
moved into the natural paths of flooded rivers or has restricted
the channels so as to heighten normal flood crests. Moreover,
floods may be beneficial as well as harmful, and even where they
are completely harmful there are remedies other than physical
structures built to afford protection. Recognizing these facts,
flood-plain occupance cannot be considered realistically as a
matter solely of man against the marauder.

Dealing with floods in all their capricious and violent
aspects is a problem in part of adjusting human occupance to
the flood-plain environment so as to utilize most effectively
the natural resources of the plain, and, at the same time, of
applying feasible and practicable measures for minimizing the
detrimental impacts of floods. This problem in the United States
involves at least 35,000,000 acres of land known to be subject
to flood. A large part of that land is not cultivated, but the
cultivated portions are among the more productive agricultural
resources of the nation. Of the 59 cities in the United States
having a population of more than 150,000 in 1940, 19 or more suffer at times from high water. Eight of them (Springfield, Hart-
ford, Pittsburgh, Cincinnati, Louisville, Kansas City, Denver,
and Los Angeles) have serious flood hazards in highly important
sections. In addition, two cities--Dayton and New Orleans--
occupy land which has been protected fully from flood. Although
most of the densely settled flood plains are in the Northeastern
Manufacturing Belt and along the Lower Mississippi River, economically important encroachments have been made upon flood plains
in all sections of the United States. For the nation as a whole,
the mean annual property loss resulting from floods certainly is
more than $75,000,000 and probably exceeds $95,000,000. The toll
in human life is approximately 83 deaths annually. For the heavy
damages to health and to productive activity no measuring units
are available.

Purpose and Method of Analysis

Because of the great diversity in flood conditions and in flood plains and their occupance, it is impracticable to formulate more than a few generalizations with respect to flood problems in the United States. Solutions to such problems can be developed effectively only by examining the environmental and social conditions in each locality having a flood problem. No attempt is made in this dissertation to make that kind of an examination, locality by locality. It is believed, however, that specific local problems could be appraised more fully, and that better solutions could be found for them if a broader and essentially geographical approach to the flood problem were to be adopted. Such an approach would take account of all relevant factors affecting the use of flood plains, would consider all feasible adjustments to the conditions involved, and would be practical in application.

The remainder of this chapter outlines the points of view which have dominated public action in dealing with the flood problem in the United States, and suggests a more nearly comprehensive approach meeting the foregoing requirements as to breadth and practicability. Succeeding chapters attempt to show the validity and implications of that approach. Chapter II defines the concepts of flood, flood plain, and flood-plain occupance. Chapter III points out the chief factors--natural and social-- which have been important in the occupation of American flood plains. The range of human adjustment to the flood hazard is described in Chapter IV. Finally Chapter V states the conclusions of the investigation, and suggests ways of applying them to public policy affecting the flood problem, and to geographical research.

These findings are the results of an examination of the available literature on flood problems in the United States, comprising chiefly the reports of the U. S. Corps of Engineers and the U. S. Department of Agriculture on their flood-control surveys; reports of state and municipal engineering surveys; bulletins on floods prepared by the U. S. Geological Survey and the U. S. Weather Bureau; geographical studies of flood plains; and relevant statements in technical and trade journals. They also reflect a large body of unpublished material which the author was privileged to review while associated with the National Resources

Planning Board and its predecessors. As the major findings began
to take shape from the review of the literature on floods, they
were tested by reconnaissance studies of flood plains selected
for their diversity of conditions and lying within the Potomac,
Delaware, Upper Ohio, and Los Angeles basins.

Three Public Approaches to the Flood Problem

Public action with respect to floods in the United States
has emerged from three streams of thought, each reflecting a distinct social technique, and each fostered by a separate professional group.

The engineer has approached the problem by inquiring "Is
flood protection warranted?" He has utilized levees, dams, flood-
ways, channel improvements, and similar engineering devices to
curb flood flows. The public welfare official has sought to determine "How best to alleviate flood distress?" He has relied
upon soup kitchens, rescue boats, emergency grants, rehabilitation
loans and like measures to cushion the social effects of flood.
The property owner has been aided somewhat by the meteorologist
who, asking "When will the next flood occur and how high will it
be," has made forecasts that enable public officials and property
owners alike to evacuate some of their goods and to prepare in
other ways for the on-coming flood. Each approach has helped to
reduce flood losses and to increase the utility of flood-plain
resources. Each has developed fruitful methods of coping with
floods. These three approaches, either singly or in combination,
do not point, however, to solutions of the flood problem which
promise maximum use of all flood plains with minimum social costs.

Engineering

The traditional public attack upon the flood problem in
the United States has been to determine whether or not the flood
plain under consideration could and should be protected from
floods. This is the engineering approach. Under it the cost of
building protective works is calculated, the prospective benefits
from protection are estimated, and if the benefits seem to exceed
the cost, the work is recommended for construction. This has
been the prevailing Federal policy since 1917, and it has dominated
the flood-protection work of state and county agencies before and
since.

<u>Major trends in public policy</u>.--In non-Federal as in Federal flood protection from 1850 to 1941 the emphasis was upon curbing flood waters by engineering. Alternative forms of adjustment were not taken into account in drafting state legislation or in district plans. Evaluation of benefits received somewhat more detailed treatment, however, in state than in Federal statutory directive policies, and this might well be expected inasmuch as state work required direct assessment of a part of the costs on the benefiting property owners and public bodies. Such assessments became relatively less frequent, of course, with the adoption of the new Federal policy in 1936.

Among the first reactions to the Flood Control Act of 1936 was the authorization of state agencies to cooperate in financing and reviewing the plans for projects adopted by the Congress. New York State created a special Flood Control Commission for that purpose in 1936,[1] and in the same year Pennsylvania appropriated funds for use by its Department of Forests and Waters.[2] The Flood Control Act of 1938 brought most of those state activities to an abrupt halt. Cooperation in reservoir construction was abandoned, except in New York where the state continued to acquire land on behalf of the Federal government and where the remaining cooperation in levees and channel improvements could be handled readily by municipalities and counties without the participation of state agencies. Illinois and Pennsylvania also continue to give some financial assistance to local communities for flood protection. During the period 1936-1938 there were the beginnings of interstate agencies to facilitate cooperation with the Federal government in the financing of reservoir programs for the Merrimack, Connecticut, and Ohio basins. The 1938 act eliminated any need for cooperation, and relegated state participation to the role of criticizing and promoting Federal reservoir plans.

During the 90 years that have passed since large-scale flood-protective operations first began in the United States there has been a progressive widening of interest in the engineering

Engineering (New York: McGraw-Hill Book Co., 1925).
     Bernard A. Etcheverry, <u>Land Drainage and Flood Protection</u> (New York: McGraw-Hill Book Co., 1931).

[1] <u>New York Laws of 1936</u>, c. 16.

[2] <u>Pennsylvania Laws of 1936</u>, Act. No. 46.

structures to control or abate floods. The early interest in
channel improvement and in levees was supplemented by the development of diversion structures and detention reservoirs, and, more
recently, by multiple-purpose storage reservoirs. Attention also
has been given to the possibilities of land-management measures,
and of permanent shifts in the use of flood plains. While the
number of engineering works has increased, the area of interest
in flood protection also has grown. Not only has survey work
spread far from the banks of the Lower Mississippi, but the scope
of individual surveys has been broadened to include consideration
of all possible types of engineering works in association with
works for the use or control of water for other purposes. Drainage basins now are being treated as study units, and, within them,
water conservation programs are being developed in an effort to
serve all feasible human purposes.

Increasing use of large-scale levee structures and reservoirs has led to the assumption by the Federal government of large
responsibility for planning and construction. For 67 years after
1850, county, municipal, and other intrastate organizations were
mainly responsible for flood-protection. Beginning in 1917 the
Federal activities expanded slowly until 1936, when a national
policy of Federal aid was established. That policy at first encouraged state and interstate organizations for flood protection, but
in 1938 the trend was reversed and Federal responsibility was extended to encompass all protection programs involving reservoirs.

The early intrastate efforts laid heavy stress upon identification of flood-protection benefits and assessment of costs,
this interest reaching its peak during the rapid drainage and levee
district expansion of the first two decades of the twentieth century. Federal surveys prior to the 1930's paid only slight attention to benefit and cost analysis. Beginning in the 1930's much
more precise studies of those factors were made, and the trend is
now in the direction of even greater detail and precision.

Notwithstanding these and many other changes in policy
governing public aid to flood protection, the basic approach has
remained essentially the same. "Is flood protection warranted,"
continues to be the prevailing question.

During the entire period under discussion, public action
has unfolded in response to gradually widening realization of the
flood menace, but it has been translated into legislation only at

irregular intervals determined for the most part by hydrologic
conditions. Each flood of national importance since 1915 has
given rise to a change in Federal flood-protection policy: each
major change in policy prior to 1941 has been made within a year
or two after a great flood. National catastrophes have led to
insistent demands for national action, and the timing of the legislative process has been set by the tempo of destructive floods.

Forecasting

Except in the few areas where complete protection has been
afforded by engineering works, dwellers of flood plains require
as early a warning as possible of the coming of floods in order
that people and property can be evacuated, and in order that other
precautions can be taken against flood losses. The U. S. Weather
Bureau and its predecessors established the first flood forecasting
service along the Lower Mississippi River in 1870, and received
regular appropriations from the Congress for a separate service
after 1891.[1]

Inasmuch as the early forecasts were based upon the stages
of rivers upstream from the reaches subject to flood, the system
was applied most extensively and yielded best results along the
lower reaches of large rivers such as the Mississippi, Ohio, Red,
and Arkansas. By 1936, the Weather Bureau had organized more than
70 centers from which forecasts of river stage were released. The
forecasts utilized rainfall as well as stage data, and involved
numerous empirical formulae developed by the Bureau's forecasters
in the light of previous flood experiences.

In 1937 the Bureau, in cooperation with the Commonwealth
of Pennsylvania and the U. S. Geological Survey, launched a new
type of forecasting for the Susquehanna Basin, drawing upon tele-
phonic reports on weather conditions and upon the unit-hydrograph
method of estimating stream flow so as to build up the expected
flow for any desired point. The Bureau has since expanded such
methods to a few other large streams, and recently has laid plans
to expedite service to the headwater reaches where heretofore, because of small drainage area, high rate of runoff concentration,

---

[1] U. S. Signal Office, Annual Report of the Chief Signal
Officer, 1891 (Washington: Government Printing Office, 1892),
pp. 13-16.

and probability of intense precipitation, the Bureau has not been able to make forecasts.

Prior to 1937 the system grew slowly, and made only a few advances in forecasting techniques. It was understaffed, and it received little public attention except in time of flood. In some areas the War Department developed its own independent and superior service.

The Bureau's forecasts, even on large streams, have not always been successful, and there are instances, such as along the Lower Ohio River during the flood of January, 1937, when the forecasted peaks were as much as 6 feet below the actual peaks occurring 24 hours later. At Pittsburgh during the March, 1936, flood the crest one morning was 9 feet above that predicted on the preceding day, although that prediction was for a stage exceeding any previously recorded stage.

The forecasting system has reduced flood losses materially, and has prevented much human distress, although, as will be shown in Chapter IV, it is difficult to make a close estimate of its value. On many flood plains it is an essential part of human adjustment to flood hazard. It is an integral phase of the operation of flood-control reservoirs. At no time has the Bureau tried, however, to stimulate more effective use of the forecasts in preventing flood losses, and it has not even made detailed studies to determine the degree to which forecasts have been applied.

The history of the forecasting effort reveals an early development of stage-forecasting methods which led to a long and somewhat unimaginative period of forecasting on the lower reaches of large rivers, followed, in 1937, by rapid improvement in methods of forecasting, with particular reference to smaller drainage areas upstream.

Public Relief

The third main line of public action consists of relieving human distress resulting from floods. From the earliest damaging floods, community assistance has been given to those who suffer impairment of health or damage to property. Prior to the organization of a national Red Cross, relief from flood disasters commonly was handled by groups of interested citizens. For example,

after the Pittsburgh and Allegheny flood of July 26, 1874, which
caused the death of more than 119 persons, a special committee
was appointed by the mayors of the two cities to reduce the distress. That group raised $63,000 from local donations, and ex-
pended the entire amount in cash payments to deserving sufferers,
in purchase of merchandise, in burials, and in a deposit for the
benefit of children made orphans by the flood.[1] It prided itself
upon not having to request outside aid and upon meeting all urgent
demands for help. Federal assistance seems to have been given
only in the assignment of troops to guard the scene of the disaster. Similarly, after the great Miami River flood of 1913, the
people of Dayton, Ohio, raised a large fund to be used by the Dayton Citizens' Relief Committee for the aid of their fellow towns-
men who had suffered losses, the fund later being given over to
the Red Cross for administration.[2] These amounts now appear small
by comparison with expenditures currently made for flood relief
by Federal agencies.

The Federal government began in 1874 to make direct grants
of food and shelter to flood sufferers, and appropriations were
made for those purposes following each great flood thereafter, as
shown in Table 3. The funds were disbursed chiefly by the Quar-
termaster Corps of the War Department, and were administered for
the most part by local and state agencies and by the Red Cross.

As the annual toll of flood losses increased and as the
American Red Cross enlarged and extended its field staff, its national and chapter organizations became the center of public-
relief work. The Red Cross assisted in caring for the victims of
a few of the serious floods in the 1920's and then during the
great Mississippi River flood of 1927 it took over in large measure the work of administering aid by Federal as well as non-Federal
agencies.[3] While other agencies such as the Corps of Engineers

---

[1]Report of the Citizens' Executive Relief Committee of the
Cities of Pittsburgh and Allegheny for the Relief of the Sufferers
by the Flood of July 26, 1874 (Allegheny: Ogden and Vance, 1876),
pp. 1-61.

[2]Bock, op. cit., pp. 11-14.

[3]The Congressional charter of organization approved by
the Act of January 5, 1905 (33 U.S. Statutes 599) and amended by
the Act of February 27, 1917 (39 U.S. Statutes 946), charges the
American National Red Cross "To continue and carry on a system of

TABLE 3 - Continued

Authorizing Act		Provision
Reference	Date	
		and flood stricken areas of south-western states; $6,000,000 appropriated in deficiency act of March 4, 1929 (U. S. 1635).
46 U.S. 84	3/12/30	Appropriates $1,660,000 to aid state of Alabama in construction of roads, etc., damaged by floods in 1929.
46 U.S. 386	5/27/30	Appropriates $506,067 to aid state of Georgia in construction of roads, etc., damaged by floods in 1929.
46 U.S. 489	6/2/30	Appropriates $805,561 to aid state of South Carolina in construction of roads, etc., damaged by floods in 1929.
46 U.S. 829	6/28/30	Relieves state of Vermont from accountability for certain Federal property lost, etc., in connection with relief work incident to flood of November, 1927.

and the State Departments of Health defended the levee lines or guarded against epidemic, the Red Cross, under the general direction of Herbert Hoover as Secretary of Commerce, assumed the principal role of guiding emergency evacuation and of feeding, clothing, housing and providing medical services for the refugees. For the next ten years the Red Cross continued as the major national agency dealing with flood distress.

In 1929 and 1930 Congress appropriated funds for use by states in repairing flood losses (see Table 3). These appropriations were the first to place the Federal government squarely in the position of rehabilitating flood sufferers. After 1933 the newly-organized Federal relief agencies began to extend aid for such purposes, and during the Ohio River flood of 1937 a large

National and international relief in time of peace and to apply the same in mitigating the sufferings caused by pestilence, famine, fire, flood, and other National calamities, and to devise and carry on measures for preventing the same."

part of the total relief expenditures was made by those agencies
directly (see Chapter IV).

After 1927 it became the common practice of municipal and
state governments and of the Red Cross to provide for flood refugees while away from their homes, and to lend aid to those who
needed cash or goods in order to regain their previous level of
living. According to need, Red Cross grants of funds were used
for repairs, food, clothing, and other essentials. Once returned
to his home, a flood refugee received aid in cleanup, repairs,
and rehabilitation only in so far as he could not finance those
operations himself. The work-relief agencies and the Civilian
Conservation Corps contributed additional help by participating
in the cleanup activities and by repairing publicly-owned structures such as bridges and roads. Direct grants or assignments of
work relief were made to flood sufferers in the lower income
brackets. In 1934 the Reconstruction Finance Corporation received
its first authorization to make loans for repairs of damages from
floods and other catastrophes, and in 1937, the Disaster Loan
Corporation was established to extend credit to individuals or
corporations requiring help in rehabilitation.[1] These sources of
aid, combined with the authority which the Congress gradually gave
to the Corps of Engineers to repair levees and other flood-protection structures damaged by floods, covered a much wider range of
flood losses than that encompassed by the Red Cross relief program. By 1941, the total public relief program, as described in
detail in Chapter IV, had expanded to cover, by grant or loan,
most of the serious losses from flood.

The Private Approach to the Flood Problem

In considering the action taken by public agencies in
building flood-protective works, in forecasting floods, and in
cushioning the harmful effects of floods, it should not be forgotten that flood losses have been reduced substantially by the
action of individuals and corporations. Where public action to

[1] 48 U.S. Statutes 589. Act of April 13, 1934. This was
later extended through 1936.
        50 U.S. Statutes 19. Act of February 11, 1937. The ac-
tivities of the Corporation were extended through 1940 by the Act
of March 4, 1939.

cope with floods is deferred or is not economically feasible, it
remains for the property owners and other residents to deal with
floods as best they can. It probably is neither unjust nor exaggerated to suggest that most such residents have made no special
effort to meet or remedy the flood hazard. There are no national
statistics to support this suggestion, but a review of the efforts
of many flood-plain residents to adjust their activities to floods
indicates that once the flood plain is occupied, other adjustments
to floods are slow in coming and are the exception in so far as
residential and commercial occupance is concerned.

   Aside from making emergency preparations when an imminent
flood threatens, flood-plain occupants have a choice of readjust-
ing their patterns of land use, of altering buildings and other
physical structures, and of laying aside adequate reserves or insurance against future catastrophes. Such readjustments have been
made successfully in one or more sections of the United States,
but they are untried for the most part, and they do not command
the attention of Federal or national organizations. While the
Federal government has devoted its efforts to engineering, hydrologic, and welfare activities affecting floods, property owners
and residents have been left to themselves to find other practicable ways of reducing flood losses. In this effort they have
had no systematic aid from Federal agencies.

   In a few instances, city planners have sought to direct
urban growth away from areas subject to flood hazard. Several
states, as already indicated, have curbed, by use of the police
power, undue encroachment upon stream channels. Certain industries, such as the railway and electric-power industries, have
given special attention to techniques of reducing flood losses.
Early but abortive attempts have been made to insure against flood
losses. At least one Federal loan agency--the Federal Housing
Administration--has discouraged building of new housing facilities
in flood plains. Otherwise, flood-plain dwellers have been left
to themselves to deal with the flood problem as best they can.

Summary of Prevailing National Policy

   The policy declared by the Congress in the Flood Control
Act of 1936, as amended, represents one segment only of the total
national policy relating to the flood problem. Taking into account all phases of public action and inaction, the policy in es-

sence is one of protecting the occupants of flood plains against
floods, of aiding them when they suffer flood losses, and of
encouraging more intensive use of flood plains. By providing
plans and all or at least half of the cost of protective works,
the Federal government, under the policy established in 1936 and
1938, reduces the flood hazard for the present occupants and
stimulates new occupants to venture into some flood plains that
otherwise might have remained unsettled or sparsely settled. Even
though no protection is provided or planned, the Federal forecasting system tends to encourage continued use of flood plains by
reducing the expectancy of loss and discomfort from flood disasters. Public relief is now so widespread that the threat of flood,
while not pleasant, has lost many of its ominous qualities. If a
community wishes to relocate outside of a flood plain, Federal
help is given to the extent that funds might otherwise have been
expended on local-protective works, but if a flood-plain occupant
wishes to rehabilitate a relatively profitable business or desirable residence in the old location after a flood he may obtain
Federal aid for that purpose.

At the same time, the occupants are themselves concerned
in an important degree with reducing flood losses by emergency
removal and by changes in land use and structures. Except in so
far as the forecasting system promotes emergency removal, the prevailing public policy is largely neutral; it neither encourages
nor discourages such activities.

Obviously, the flood plains of the United States will not
be permanently evacuated and returned to nature merely because of
the annual bill for their occupancy, which now approaches
$95,000,000. Neither will they be occupied as intensely as con-
sistent with other relevant physical and cultural conditions
solely because, irrespective of cost, suitable engineering and
land-use devices can be developed to curb or prevent floods. No
general rule can be established as to the most satisfactory arrangement of land occupance in relation to local stream regimen
and flood-plain conditions. In some instances, profound modifications in the stream regimen or channel have been necessary, and
in other instances the cultural forms and patterns have been adjusted delicately to the earlier landscape. By and large, a
fairly harmonious combination has been developed. Wherever the
adjustments are not satisfactory, as attested by crippling flood

losses, wherever a regressive occupance obtains, or wherever the
flood-plain resources are not used as fully as practicable, a readjustment may be in the public interest. This, it has been
shown, is the central flood problem: how best to readjust land
occupance and flood-plain phenomena in harmonious relationship?

Outline of a Geographical Approach

From the three converging streams of public action with
respect to the flood problem, and from corollary fields of action,
such as land-use planning, we may draw an approach to this problem
more comprehensive than any one of them. It is a view which considers all possible alternatives for reducing or preventing flood
losses; one which assesses the suitability of flood-protective
works along with measures to abate floods, to evacuate people and
property before them, to minimize their damaging effects, to repair the losses caused by them, and to build up financial reserves
against their coming. It is a view which takes account of all
relevant benefits and costs. It analyzes the factors affecting
the success of possible uses of a flood-plain. It seeks to find
a use of the flood plain which yields maximum returns to society
with minimum social costs, and it promotes that use.

Unless the major factors affecting flood-plain use are
appraised, there can be no assurance that the recommended use is
beneficial. Unless all possible forms of adjustment to floods are
canvassed, the less expensive ones cannot be selected with certainty. Unless the analysis leads to practicable forms of readjustment, there is little purpose in examining these possibilities.

Analyses of this character have not been made in the past,
and even the need for them has been stated only in general terms.

Marsh, while primarily interested in the prevention and
protection phases of the flood problem, appears to have recognized
these propositions in his discussion of floods in 1898.[1] McGee
called attention in 1891 to several possible adjustments and noted
with a tinge of pessimism that, "As population has increased, men
have not only failed to devise means for suppressing or for

---

[1]George P. Marsh, The Earth as Modified by Human Action--
A Last Revision of "Man and Nature" (New York: Charles Scribners
Sons, 1898), pp. 472-474, 498.

escaping this evil, but have, with singular short-sightedness, rushed into its chosen paths."[1] Semple in 1911 described several types of riverine adjustments but did not analyze the problem of reducing flood losses.[2] Russell merely noted some of the factors affecting the occupance of flood plains.[3] J. Russell Smith called attention to the need for a different attack upon the Mississippi River problem following the flood of 1927[4] and various editorial writers[5] and public agencies[6] suggested after the 1936 and 1937 floods that a broader approach was desirable, but their suggestions have not found wide acceptance in practice. Today there are no studies or programs which meet the requirements outlined above.

This geographical approach to the flood problem appears to be more nearly national in scope, and more nearly sound from a social standpoint than the approaches which dominate prevailing public policy. The remainder of the dissertation states the evidence in support of this approach and shows its implications in public policy and in geographical research.

---

[1]W. J. McGee, "The Flood Plains of Rivers," _Forum_, XI (1891), 221-234.

[2]Ellen Churchill Semple, _Influence of Geographic Environment_ (New York: Henry Holt and Company, 1911), pp. 322-327, 363-370.

[3]J. Russell Smith, "Plan or Perish," _Survey_, LVIII (1927), 370-377.

[4]I. C. Russell, _Rivers of North America_ (New York: G. P. Putman's Sons, 1898), p. 114.

[5]"A Modest Proposal for Flood Control," _New Republic_, May 19, 1937, p. 34.

[6]New York, Division of State Planning, _A Common Sense View of the Flood Problem_, Bulletin No. 28 (Albany, May, 1937). A similar view is taken by Allen Hazen in _Flood Flows: A Study of Frequencies and Magnitudes_ (New York: John Wiley and Sons, 1930), pp. 2-3, 177-179.

The Natural History of Urbanization

LEWIS MUMFORD*

THE EMERGENCE OF THE CITY

The natural history of urbanization has not yet been written, for only a small part of the preliminary work has been done. The literature of the city itself, until a half-century ago, was barren to the point of nonexistence; and even now the ecologists of the city, dealing too largely with a late and limited aspect of urbanism, have hardly staked out the ground that is to be covered. Our present purpose, accordingly, is to make use of such studies as have so far been made in order to ask more pointed questions and so, incidentally, to indicate further fields of profitable study.

Whether one looks at the city morphologically or functionally, one cannot understand its development without taking in its relationship to earlier forms of cohabitation that go back to non-human species. One must remember not only the obvious homologies of the anthill and the beehive but also the nature of fixed seasonal habitations in protected sites, like the breeding grounds of many species of birds.

* Mr. Mumford since 1952 has been Professor of City Planning at the University of Pennsylvania, Philadelphia. He is a fellow of the American Academy of Arts and Sciences, a member of the American Philosophical Society, an honorary associate of the Royal Institute of British Architects, and an honorary fellow of Stanford University. His publications include: *Technics and Civilization*, 1934; *The Culture of Cities*, 1938; and *The Conduct of Life*, 1951.

Though permanent villages date only from Neolithic times, the habit of resorting to caves for the collective performance of magical ceremonies seems to date back to an earlier period; and whole communities, living in caves and hollowed-out walls of rock, have survived in widely scattered areas down to the present. The outline of the city as both an outward form and an inward pattern of life might be found in such ancient assemblages. Whatever the aboriginal impetus, the tendency toward formal cohabitation and fixed residence gave rise, in Neolithic times, to the ancestral form of the city: the village, a collective utility brought forth by the new agricultural economy. Lacking the size and complexity of the city, the village nevertheless exhibits its essential features: the encircling mound or palisade, setting it off from the fields; permanent shelters; storage pits and bins, with refuse dumps and burial grounds recording silently past time and spent energy. At this early stage, at least, Mark Jefferson's observation (1931) holds true: urban and rural, city and country, are one thing, not two things.

Though the number of families per acre in a village is greater than the number per square mile under a pastoral economy, such settlements bring with them no serious disturbance in the natural environment; indeed, the relation may even be favorable for building up the soil and increasing its natural productivity. Archeological explorers in Alaska have been able to detect early settlements by noting the greenness of the vegetation around the otherwise submerged village sites, probably due to the enrichment of the soil from the nitrogenous human and animal waste accumulated near by. Early cities, as we find them in Mesopotamia and Egypt, maintain the symbiotic relation with agriculture that we find in the village. In countries like China, still governed by the principles of village economy, even contemporary cities with high population density, such as Keyes describes (1951), exhibit the same reciprocal relations: "The most concentrated highly developed agriculture is just outside the walls of cities." King estimated (1927) that each million city dwellers in China account for more than 13,000 pounds of nitrogen, 2,700 pounds of phosphorous, and almost 4,500 pounds of potassium in the daily night soil returned to the land. Brunhes' description (1920) of cities under "unproductive occupation of the soil" does not altogether hold for the earliest types or, as I shall show, for the latest types of city.

The emergence of the city from the village was made possible by the improvements in plant cultivation and stock-breeding that came with Neolithic culture; in particular, the cultivation of the hard grains that could be produced in abundance and kept over from year to year without spoiling. This new form of food not merely offered insurance against starvation in the lean years, as was recorded in the famous story of Joseph in Egypt, but likewise made it possible to breed and support a bigger population not committed to food-raising. From the standpoint of their basic nutrition, one may speak of wheat cities, rye cities, rice cities, and maize cities, to characterize their chief source of energy; and it should be remembered that no other source was so important until the coal seams of Saxony and England were opened. With the surplus of manpower available as Neolithic man escaped from a subsistence economy, it was possible to draw a larger number of people into other forms of work and service: administration, the mechanical arts, warfare, systematic thought, and religion. So the once-scattered population of Neolithic times, dwelling in hamlets of from ten to fifty houses (Childe, 1954), was concentrated into "cities," ruled and regimented on a different plan. These early cities bore many marks of their village origins, for they were still in essence agricultural towns: the main source of their food supply was in the land around them; and, until the means of transport had greatly improved and a system of centralized control had developed, they could not grow beyond the limit of their local water supply and their local food sources.

This early association of urban growth with food production governed the relation of the city to its neighboring land far longer than many observers now realize. Though grains were transported long distances (even as special food accessories like salt had circulated in earlier times), cities like Rome, which drew mainly on the distant granaries of Africa and the Near East—to say nothing of the oyster beds of Colchester in England—were exceptions down to the nineteenth century. As late as fifty years ago large portions of the fruits and vegetables consumed in New York and Paris came from nearby market gardens, sometimes on soils greatly enriched, if not almost manufactured, with urban refuse, as Kropotkin pointed out in *Fields, Factories, and Workshops* (1899). This means that one of the chief determinants of large-scale urbanization has been nearness to fertile agricultural land; yet, paradoxically, the growth of most cities has been achieved by covering over and removing from cultivation the very land—often, indeed, the richest alluvial soils—whose existence at the beginning made their growth possible. The tendency of cities to grow along rivers or near accessible harbors was furthered not alone by the need for easy transportation but by the need to draw on aquatic sources of food to supplement those produced by the soil. This rich and varied diet may itself have contributed to the vital energy of city dwellers as contrasted with the more sluggish ways of hinterlanders and perhaps may also have partly offset the bad effect of close quarters in spreading communicable diseases. While modern means of transport have equalized these advantages, they have not yet hastened the migration of urban populations to upland sites on poorer soils, though often these present more salubrious climates and better living conditions.

The village and the small country town are historic constants. One of the outstanding facts about urbanization is that, while the urban population of the globe in 1930 numbered around 415,-000,000 souls, or about a fifth of the total population, the remaining four-fifths still lived under conditions approximating that of the Neolithic economy (Sorre, 1952). In countries as densely peopled as India, as late as 1939, according to the *Statesman's Yearbook,* less than 10 per cent of the total population lived in cities. These "Neolithic" conditions include the utilization of organic sources of energy, vegetable and animal, the use of a local supply of drinking water, the continuous cultivation of land within walking distance of the village, the partial use of human dung along with that of animals for fertilizer, a low concentration of inorganic refuse, like glass and metals, and an absence of air pollution. In many parts of the world, village settlements, far from encroaching on ara-

ble land, occupy barren hill sites of little use for agriculture; the stony outcrop of an Italian hill town involves only a slightly more symmetrical arrangement of the original rock strata. The chief weakness of these settlements, particularly in parts of the world long cultivated, notably in Spain, Greece, or China, is due to the peasant's begrudging the land needed for forest cover; he thus tends, by overtillage, to promote erosion and to create a further imbalance among the bird, insect, and plant populations. But, just as the early village economy was indebted to the astronomical calendar produced in the temple cities for the timely planting of their crops, so the present development of ecological knowledge, which has led to increasing concern and care for the woodland preserves in highly urbanized countries, may in time counteract the otherwise destructive effects of earlier stages in urban settlement.

URBAN SYMBIOSIS AND DOMINANCE

With the first growth of urban populations in ancient Mesopotamia, the symbiotic relations that originally held between village and land were not greatly altered. "The city," as Childe (1942, p. 94) describes its earliest manifestations, "is girt with a brick wall and a fosse, within the shelter of which man found for the first time a world of his own, relatively secure from the immediate pressure of raw, external nature. It stands out in an artificial landscape of gardens, fields, and pastures, created out of reed swamp and desert by the collective activity of preceding generations in building dykes and digging canals." Though these cities represented "a new magnitude in human settlement," the populations of Lagash, Umma, and Khafaje are "reliably estimated to have been 19,000, 16,000, and 12,000 respectively during the third millennium." The Levitical cities described in the Bible, confirmed by modern excavations of Gezer, had a town area of about 22 acres, with pasture land, permanently reserved, amounting to about 300 acres (Osborn, 1946). More than four thousand years later, as late as the sixteenth century, the characteristic size of the city in western Europe ranged from 2,000 to 20,000 people; it was only in the seventeenth century that cities of more than 100,000 began to multiply. In both the Near East in ancient times and in western Europe in the Middle Ages, cities prudently retained some portion of the land within their walls for gardens and the harboring of animals for food in case of military siege. Even the vast domains of Babylon must not mislead us into looking upon it as comparable in density to modern London. A map drawn in 1895 by Arthur Schneider, and republished by Hassert (1907), shows that Babylon covered an area big enough to contain Rome, Tarentum, Syracuse, Athens, Ephesus, Thebes, Jerusalem, Carthage, Sparta, Alexandria, and Tyre, together with almost as much open space between these cities as they occupied in their own right. Even in Herodotus' time, Babylon had many of the aspects of an overgrown village.

The Neolithic economy appears to have been a co-operative one. The concentration upon plant cultivation in small neighborly communities, never with a sufficient surplus of food or power to promote too much arrogance in man's relation with other men or with nature, established a natural balance between fields and settlements. In Europe, as Élisée Reclus long ago noted, country towns and villages tended to spread evenly, as far as topography allowed, about the space of a day's walk apart. With the introduction of metallurgy, during the succeeding period of urbanization, came technological specialization, caste differentiation, and heightened temptations to aggression; and with this began a dis-

regard for the welfare of the community as a whole and, in particular, a tendency to ignore the city's dependence upon its local resources. Excess of manpower abetted an excessive belief in the power of man—a belief deepened, no doubt, by the efficacy of the new edged weapons and armor in giving control to aggressive minorities who took the law into their own hands. With the development of long-distance trading, numerical calculation, and coinage, this urban civilization tended to throw off its original sense of limits and to regard all forms of wealth as purchasable by trade or procurable by a demonstration of military power. What could not be grown or produced in the local region could be, by theft or exchange, obtained elsewhere. In time this urban economy made the mistake of applying the pragmatic standards of the market place to the environment itself: the process began of building over the interior open spaces and building out over the surrounding land.

Until modern times the extension of a city's walls marked its growth as surely as does each additional ring of a tree. The wall had perhaps a formative role in the transformation of the village into the city; when made of heavy, permanent materials, surrounded by a moat, it gave the city a means of protection the little village could not afford. Not merely was it capable of military defense, but the city, through its surplus population, could muster enough manpower to hold against a large army of attackers. The earliest meaning of "town" is an inclosed or fortified place. The village that, because of its defensible site, offered protection against predators of all kinds would in times of peril attract families from more exposed areas and so, with a larger, mixed population, would turn into a city. Thus the temple citadel would add to its original population and, even after the danger had passed, would retain some of those who sought shelter and so become a city. In Greece, at least, the city comes into existence, historically, as such a synoecism.

But the morphological difference between the village and the city is not simply the result of the latter's superior site or of the fact that its geographic situation enables it to draw on a wider area for resources, foods, and men and in turn to export their products to a larger market, though both are facts conducive to population growth and economic expansion. What distinguish city from village are mainly two facts. The first of these is the presence of an organized social core, around which the whole structure of the community coheres. If this nucleation may begin in the village stage, as remains of temples seem to indicate, there is a general shift of household occupations and rituals into specialized collective institutions, part of the intensified social division of labor brought in with civilization itself. But, from the standpoint of the city's relation to the earth, the important point to notice is that, in this social core or nucleus, the sharpest departures from the daily habits and the physical structure of the village take place. Thus the temple, unlike the hut, will be built of permanent materials, with solid stone walls, often plated with precious stones or roofed with rare timber taken from a distant quarry or forest, all conceived on a colossal scale, while the majority of dwelling houses will still be built of clay and reed, or wattle and daub, on the old village pattern. While the temple area will be paved, the streets and alleys of the rest of the city will remain unpaved. As late as imperial Rome, pavement will be introduced first into the Forum, while most of the arteries remain uncovered, to become sloughs of mud in rainy weather. Here too, in the urban palace, as early as Akkad, such technological innovations as baths, toilets, and drains will appear—innovations that remain far beyond the reach of the urban populations-at-large until modern times.

Along with this bold aesthetic transformation of the outward environment, another tendency distinguishes the city from the village—a tendency to loosen the bonds that connect its inhabitants with nature and to transform, eliminate, or replace its earth-bound aspects, covering the natural site with an artificial environment that enhances the dominance of man and encourages an illusion of complete independence from nature. The first age of the "urban revolution," to use Childe's term, had little extrahuman power and few machines. Its technological heritage, once it had learned to smelt copper and iron, was in every sense a static one; and its major skills, weaving aside, were concentrated on fashioning utensils and utilities (pots, jars, vats, bins) and on building great collective works (dams, irrigation systems, buildings, roads, baths) and, finally, cities themselves. Having learned to employ fire of relatively high intensity to glaze and smelt ores, these early civilizations offset its danger by creating a fireproof environment. The importance of this fact, once papyrus and paper were in use, can hardly be overestimated. In this general transformation from the transient to the fixed, from fragile and temporary structures to durable buildings, proof against wind, weather, and fire, early man emancipated himself likewise from the fluctuations and irregularities of nature. Each of the utilities that characterized the new urban form—the wall, the durable shelter, the arcade, the paved way, the reservoir, the aqueduct, the sewer—lessened the impact of nature and increased the dominance of man. That fact was revealed in the very silhouette of the city, as the traveler beheld it from a distance. Standing out in the vegetation-clad landscape, the city became an inverted oasis of stone or clay. The paved road, a man-made desert that speeds traffic and makes it largely independent of the weather and the seasons; the irrigation ditch, a man-made river system that releases the farmer from irregularities of seasonal rainfall; the water main, an artificial brook that turns the parched environment of the city into an oasis; the pyramid, an artificial mountain that serves as symbolic reminder of man's desire for permanence and continuity—all these inventions record the displacement of natural conditions with a collective artifact of urban origin.

Physical security and social continuity were the two great contributions of the city. Under those conditions every kind of conflict and challenge became possible without disrupting the social order, and part of this new animus was directed into a struggle with the forces of nature. By serving as a secure base of operations, a seat of law and government, a repository of deeds and contracts, and a marshaling yard for manpower, the city was able to engage in long-distance activities. Operating through trade, taxation, mining, military assault, and road-building, which made it possible to organize and deploy thousands of men, the city proceeded to make large-scale transformations of the environment, impossible for groups of smaller size to achieve. Through its storage, canalization, and irrigation, the city, from its earliest emergence in the Near East, justified its existence, for it freed the community from the caprices and violences of nature—though no little part of that gift was nullified by the further effect of subjecting the community more abjectly to the caprices and violences of men.

URBAN DISPLACEMENT OF NATURE

Unfortunately, as the disintegration of one civilization after another re-minds us, the displacement of nature in the city rested, in part, upon an illusion—or, indeed, a series of illusions—as to the nature of man and his institutions: the illusions of self-sufficiency and independence and of the possibility of physical continuity without conscious renewal. Under the protective mantle of the city, seemingly so permanent, these illusions encouraged habits of predation or parasitism that eventually undermined the whole social and economic structure, after having worked ruin in the surrounding landscape and even in far-distant regions. Many elements supplied by nature, necessary for both health and mental balance, were lacking in the city. Medicine, as practiced by the Hippocratic School in the great retreats, like that at Kos, concerned with airs, waters, and places, seems at an early age to have employed in therapy natural elements that were depleted or out of balance even in the relatively small Aegean cities of the fifth century B.C., though their ruling classes spent no small part of their leisure in the exercise of the body. Through the ages the standard prescription for most urban illnesses—and perhaps as effective as more specific remedies—is retreat to some little village by seacoast or mountain—that is, restoration to a pre-urban natural environment. In times of plague the retreat repeatedly has taken on the aspects of a rout. Though man has become the dominant species in every region where the city has taken hold, partly because of the knowledge and the system of public controls over both man and nature he exercises there, he has yet to safeguard that position by acknowledging his sustained and inescapable dependence upon all his biological partners. With the ecological implications of this fact, I shall deal later.

Probably no city in antiquity had a population of much more than a million inhabitants, not even Rome; and, except in China, there were no later Romes until the nineteenth century. But, long before a million population is reached, most cities come to a critical point in their development. That occurs when the city is no longer in symbiotic relationship with its surrounding land; when further growth overtaxes local resources, like water, and makes them precarious; when, in order to continue its growth, a city must reach beyond its immediate limits for water, for fuel, for building materials, and for raw materials used in manufacture; and, above all, when its internal birth rate becomes inadequate to provide enough manpower to replace, if not to augment, its population. This stage has been reached in different civilizations at different periods. Up to this point, when the city has come to the limits of sustenance in its own territory, growth takes place by colonization, as in a beehive. After this point, growth takes place, in defiance of natural limitations, by a more intensive occupation of the land and by encroachment into the surrounding areas, with the subjugation by law or naked force of rival growing cities bidding for the same resources.

Most of the characteristics of this second form of urban growth can be observed in the history of Rome. Here the facts are better documented than they are for most ancient cities; and the effects upon the landscape have remained so visible that they suggested to George Perkins Marsh (1864, 1874) the principal lines of his investigation of *The Earth as Modified by Human Action*. Rome of the Seven Hills is an acropolis type of city, formed by a cluster of villages united for defense; and the plain of the Tiber was the original seat of their agriculture. The surplus population of this region conquered first the neighboring territories of the Etruscans and then those of more distant lands. By systematic expropriation, Rome brought wheat, olive oil, dried fish, and pottery back to the original site to sustain its growing population. To facilitate the movement of its legions and speed up the processes of administration, it carved roads through the landscape with triumphant disregard of the nature of the terrain. These roads and viaducts went hand in hand with similar works of engineering, the aqueducts and reservoirs necessary to bring water to Rome. By short-circuiting the flow of water from mountainside to sea, the city monopolized for its special uses a considerable amount of the runoff; and, to offset some of the effects of metropolitan overcrowding, it created a cult of the public bath that in turn imposed a heavy drain upon the fuel supplied by the near-by forest areas. The advance of technology, with central hot-air heating, characteristically hastened the process of deforestation, as was later to happen in the glass- and ironmaking and shipbuilding industries of northern Europe and to be repeated today in the heavy industrial demand for cellulose. Meanwhile, the sewers of Rome, connected to public toilets, polluted the Tiber without returning the precious mineral contents to the soil, though even in imperial Rome dung farmers still collected most of the night soil from the great tenements of the proletariat. At this stage the symbiotic relation turns into a parasitic one; the cycle of imbalance begins, and the mere massing of the demand in a single center results in denudations and desiccations elsewhere. The more complete the urbanization, the more definite is the release from natural limitations; the more highly the city seems developed as an independent entity, the more fatal are the consequences for the territory it dominates. This series of changes characterizes the growth of cities in every civilization: the transformation of eopolis into megalopolis. If the process wrought damage to the earth even in the ancient world, when cities as big as Rome, Carthage, and Alexandria were the exception rather than the rule, we have good reason to examine carefully the probable consequences of the present wave of urbanization.

MODERN FORCES OF EXPANSION

Let me sum up the observations so far made with respect to the natural history of cities. In the first stage of urbanization the number and size of cities varied with the amount and productivity of the agricultural land available. Cities were confined mainly to the valleys and flood plains, like the Nile, the Fertile Crescent, the Indus, and the Hwang Ho. Increase of population in any one city was therefore limited. The second stage of urbanization began with the development of large-scale river and sea transport and the introduction of roads for chariots and carts. In this new economy the village and the country town maintained the environmental balance of the first stage; but, with the production of grain and oil in surpluses that permitted export, a specialization in agriculture set in and, along with this, a specialization in trade and industry, supplementing the religious and political specialization that dominated the first stage. Both these forms of specialization enabled the city to expand in population beyond the limits of its agricultural hinterland; and, in certain cases, notably in the Greek city of Megalopolis, the population in smaller centers was deliberately removed to a single big center—a conscious reproduction of a process that was taking place less deliberately in other cities. At this stage the city grew by draining away its resources and manpower from the countryside without returning any equivalent goods. Along with this went a de-

structive use of natural resources for industrial purposes, with increased concentration on mining and smelting.

The third stage of urbanization does not make its appearance until the nineteenth century, and it is only now beginning to reach its full expansion, performance, and influence. If the first stage is one of urban balance and cooperation, and the second is one of partial urban dominance within a still mainly agricultural framework, behind both is an economy that was forced to address the largest part of its manpower toward cultivating the land and improving the whole landscape for human use. The actual amount of land dedicated to urban uses was limited, if only because the population was also limited. This entire situation has altered radically during the last three centuries by reason of a series of related changes. The first is that world population has been growing steadily since the seventeenth century, when the beginnings of reasonable statistical estimates, or at least tolerable guesses, can first be made. According to the Woytinskys (1953), the average rate of population increase appears to have gone up steadily: 2.7 per cent from 1650 to 1700; 3.2 per cent in the first half of the eighteenth century and 4.5 per cent in the second half; 5.3 per cent from 1800 to 1850; 6.5 per cent from 1850 to 1900; and 8.3 per cent from 1900 to 1950. As the Woytinskys themselves remark, these averages should not be taken too seriously; yet there is a high probability that an acceleration has taken place and hardly any doubt whatever that the world population has doubled during the last century, while the manpower needed to maintain agricultural productivity in mechanized countries has decreased.

By itself this expansion might mean no more than that the less populated parts of the earth would presently acquire densities comparable to those of

India and China, with a great part of the increase forced to undertake intensive cultivation of the land. But this increase did not take place by itself; it was accompanied by a series of profound technological changes which transformed the classic "age of utilities" into the present "age of the machine" and a predominantly agricultural civilization into an urban one—or possibly a suburban one. These two factors, technical improvement and population growth, have been interacting since at least the sixteenth century, for it was the improvement in the sailing ship and the art of navigation that opened up the almost virginal territory of the New World. The resulting increase of food supply, in terms of added tillage, was further augmented by New World crops like maize and the potato. Meanwhile, the increased production of energy foods—vegetable oils, animal fats, and sugar cane and sugar beet—not merely helped support a larger population but in turn, through the supply of fat, turned soap from a courtly luxury to a household necessity; and this major contribution to hygiene—public and personal—probably did more to lower the death rate than any other single factor. From the beginning of the nineteenth century the surplus population made it possible for old cities to expand and new cities to be founded. As Webber long ago pointed out (1899), the rate was even faster in Germany in the second half of the nineteenth century than it was in the United States.

This wave of urbanization was not, as is sometimes thought, chiefly dependent upon the steam engine or upon improvements in local transportation. The fact is that the number of cities above the 100,000 mark had increased in the seventeenth century, well before the steam engine or the power loom had been invented. London passed the million mark in population by 1810, before it had a mechanical means of transportation or the beginning of an adequate water supply (in parts of London piped water was turned on only twice a week). But a marked change, nevertheless, took place in urban growth during the nineteenth century.

At this moment the four natural limits on the growth of cities were thrown off: the nutritional limit of an adequate food and water supply; the military limit of protective walls and fortifications; the traffic limit set by slow-moving agents of reliable transportation like the canalboat; and the power limit to regular production imposed by the limited number of water-power sites and the feebleness of the other prime movers—horse and wind power. In the new industrial city these limits ceased to hold. While up to this time growth was confined to commercial cities favorably situated at the merging point of two or more diverse regions with complementary resources and skills, urban development now went on in places that had easy access to the coal measures, the iron-ore beds, and the limestone quarries. Pottery towns, cotton towns, woolen towns, and steel towns, no longer held down in size, flourished wherever the tracks for steam locomotives could be laid and the steam engine established as a source of power. The only limitation on the spread and multiplication of towns under this regime was the disability of the steam locomotive to operate efficiently on grades of more than 2 per cent. Whereas the water power and wind power of the eotechnic period had tended to distribute industry in the coastal cities of high winds or along fast-running upland streams, coal power tended to group industry in the valleys near the mine pits or along the railroad lines that constituted a continuation of the mine and the mining environment (Mumford, 1934). Industry, like agriculture, competes for the heavy lowland soils. As for the railroad itself, it is one of the greatest devourers of land and transformers of landscape. The marshaling yards of its great urban terminals put large areas out of urban or agricultural use.

GROWTH OF THE CONURBATION

Up to the middle of the nineteenth century, water-power sites, the seats of earlier industrial improvements, continued to attract industries into mill villages; but, with the coming of the railroad, industries grouped together in cities in order to take advantage of the surplus labor that accumulated there. From this time on, whole districts, such as Elberfeld-Barmen, Lille-Roubaix, the Black Country, and the Delaware Valley, become urbanized, and the limits of city growth are reached only when one city, by its conversion of farmland into building lots, coalesces with another city engaged in the same process. Growth of this kind, automatic and unregulated, a result of the railroad and the factory, had never been possible before; but now the agents of mechanization not merely created their own environment but set a new pattern for the growth of already existing great cities. Looking at Bartholomew's population map of Britain early in the present century, Patrick Geddes discovered (1915) that urbanization had taken a new form: urban areas, hitherto distinct, both as political units and as topographic features, had in fact flowed together and formed dense population masses on a scale far greater than any of the big cities of the past, forming a new configuration as different as the city itself was from its rural prototypes. He called this new kind of urban grouping the "conurbation." This new urban tissue was less differentiated than the old. It presented an impoverished institutional life; it showed fewer signs of social nucleation; and it tended to increase in size, block by block, avenue by avenue, "develop-

ment" by "development," without any individuality of form and, most remarkable of all, without any quantitative limits (West Midland Group, 1948).

This concentration of industry had marked effects upon the entire environment. The new source of power—coal; the new industrial processes, massed in the new steelworks and coke ovens; the new chemical plants for manufacturing chlorine, sulfuric acid, and hundreds of other potentially noxious compounds —all poured their waste products into the air and waters on a scale that made it impossible for the local environment to absorb them as it might have absorbed the effluvia of a village industry or the organic waste of a tannery or a slaughter-house. Streams hitherto well stocked with fish, salubrious for bathing, and even potable became poisonous sewers; while the fall of soot, chemical dust, silica, and steel particles choked vegetation in what open ground remained and left their deposits in human lungs. The effects of this pollution, and the possibility of more radical and irretrievable pollution to come through the use of atomic reactors, are dealt with in chapters that follow. Here the point to mark is that it was a natural penalty of over-concentration. The very ubiquity of the new type of city, coupled with its density, increases, for example, the threat of a lethal fog from chemicals normally in the air, such as wiped out over five thousand lives in a single week in London in 1952; a mass exodus by cars, at the low speed imposed by a heavy fog, would itself add to the deadly gases already in the air.

The extension of the industrial conurbation not merely brings with it the obliteration of the life-sustaining natural environment but actually creates, as substitute, a definitely antiorganic environment; and even where, in the interstices of this urban development, land remains unoccupied, it progressively ceases to be of use for either agriculture or recreation. The removal of the topsoil, or its effacement by buildings and slag piles, brings on no temporary denudation; it results in deserts that, even if every effort suggested by science were made, might take centuries to redeem for human occupancy, to say nothing of more organic forms of cultivation. Though the conurbation came into existence through the dense industrial occupation of a whole region rather than through the overgrowth of a single dominant city, the two types overlap. In England, Birmingham itself, though the center of congeries of smaller towns, has passed the million mark, to become the second city in Britain. By offering a big local market, the great conurbations, in addition to attracting the consumption trades and industries, have brought in petroleum refineries, chemical plants, and steelworks, which gravitate to the cheaper land on the edge of metropolitan areas. This tends to create industrial defilement at the point where Sir John Evelyn, in 1661 in his pamphlet *Fumifugium* (1933), proposed to create a protective green belt, filled with aromatic shrubs, to purify the already noisome air of London. This extension of the area of industrial pollution into the very land that the overgrown city needs for mass recreation—accessible to sunlight, to usable ocean, river front, and woodland—likewise lessens the advantage of the only form of temporary escape left: retreat to the suburb.

From the very nature of the city as a market, a workshop, and a place of civic assemblage, there is a direct relation between its growth and the growth of transportation systems, though, in the case of seaways and airways, the latter may be visible only in the increase of harbor facilities and storehouses. In general, one may say that, the heavier the urbanization, the heav-

ier the transportation network, not merely within but without. From ancient Rome to recent times, the fifteen-foot roadway remained the outsize. But, with the eighteenth century, land transportation takes a new turn. In 1861, Wilhelm Heinrich Riehl noted it (1935) in the change from the rural highroads of the old town economy to the new *Landstrasse*, planned in more systematic fashion by the new bureaucracy—wider by three feet, more heavily paved, and often lined with trees, as in the beautiful highway lined with ancient lindens between Lübeck and Travemunde. With the coming of railroad transportation, the width of the new kind of permanent way again increased; the railroad made fresh demands for large areas of flat, low-lying land to serve as marshaling yards, adjacent to the city or even cutting a great wedge through it. The economy of the water-level route again turned to a non-agricultural use of precisely the land that was often the most fertile available and spoiled even its recreational value. With the introduction of the motorcar, even secondary roads demanded pavement, and arterial roads both widened and multiplied, with the result that around great metropolises six-, seven-, and eight-lane highways with two-hundred-foot rights of way have become increasingly common. They are further complicated by great traffic circles or clover-leaf patterns of overpass and underpass to permit the continuous flow of traffic at intersections, however wasteful of land these junctions may be. In the case of parkways planned to follow the ridges, like the Taconic State Parkway in New York State, the land given over to the road may be of minor value either for agricultural or for civic use; but where the highway engineer ignores the contours, follows the valleys, and cuts through hills to maintain his level, the motorway may be an active agent both in eroding the soil and in disrupting the habitat. The yielding of water navigation to land transport has aggravated this damage; and every further congestion of population leads to still more highway-building of a permanent and costly kind to accommodate the mass week-end exit of motorists. Thus the city, by its incontinent and uncontrolled growth, not merely sterilizes the land it immediately needs but vastly increases the total area of sterilization far beyond its boundaries.

THE SUBURBAN OVERSPILL

At this point we are confronted with two special phenomena known only in embryonic form in other urban cultures: the production of a new kind of urban tissue, in the open pattern of the suburb, and the further development of a mass transportation by means of self-propelled, individual vehicles, trucks, and motorcars. The first change, the result of seeking an environment free from noise, dirt, and overcrowding of the city, actually antedated the means that made it possible on a mass scale. In London this suburban movement began as early as Elizabethan times as a reaction against the over-building and overcrowding that had then taken place in the center of the city; and at the end of the eighteenth century a similar exodus occurred among merchants who could afford a private coach to take them into the city. With increased facilities of transportation offered by the public coach and the railroad, this suburban movement became more common through the nineteenth century, as witness the growth of St. John's Wood, Richmond, and Hampstead in London, of Chestnut Hill and Germantown in Philadelphia, and of the Hudson River suburbs in New York. But, up to 1920, it was mainly the upper-income groups that could afford the luxury of sunlight, fresh air, gardens, open spaces, and ac-

cess to the open country. The new open-type plan, with houses set in gardens, at densities of from two houses to ten or twelve per acre, had long been characteristic of American country towns, most notably those of New England; indeed, this open pattern dominated west of the Alleghenies. But this standard now became universalized in the upper-class suburb, though its economic base lay outside the area the suburb occupied and from the beginning demanded a heavy sacrifice of man-hours in commuting to the distant metropolis. The low cost of suburban land and the possibility of economizing on local utilities like roads and sewers encouraged luxurious standards of space and gave those who could afford to escape a superior biological environment and perhaps, if Thorndyke is correct (1939), a superior social one. The initiative of a few farsighted industrialists, like Lever (Port Sunlight, 1887) and Cadbury (Bournville, 1895), proved that similar standards could be applied to building working-class quarters when land was sufficiently cheap.

Since 1920 the spread of private motor vehicles has completed the work of enlarging potential suburban territory, an expansion already well begun in the 1900's by interurban electric transit. The exodus to suburbia has taken in wave after wave of city dwellers, at lower and lower income levels, seeking to escape the congested and disordered environment of the big city. This removal from the city has not been accompanied by any equivalent decentralization of industry; rather it has served to sustain an antiquated pattern of concentration. The pattern of population distribution around great cities has been the product, not of social foresight for public ends, but mainly of private initiative for private profit, though it could not have taken place on its present scale in America without a vast public investment in highways, expressways, bridges, and tunnels. The result of this uncontrolled spread of the suburb has been to nullify the very purposes that brought the movement into existence.

But suburban agglomeration cannot be treated as a fact in itself; it carries with it, through the demands of the motorcar, both for private transportation and for the movement of goods, an enormous increase in paved roads, which eat into the surviving agricultural and wilderness areas and permanently sterilize ever larger quantities of land. The filling-up of marshes, the coverage of rich soils with buildings, the felling of woodlands, the clogging of local brooks and streams, and the abandonment of local springs and wells were all secondary disturbances of the early type of metropolis, even when it reached a population of a million people. When Rome was surrounded by the Aurelian wall in A.D. 274, it covered, according to Carcopino (1940), a little more than 5 square miles. The present area of Greater London is about a hundred and thirty times as great as this, while it is roughly six hundred and fifty times as great as the area, namely, 677 acres, surrounded by its wall in the Middle Ages. The metropolitan area of New York is even more widespread; it covers something like 2,514 square miles; and already a good case could be made out for treating a wide coastal strip from Boston to Washington as one continuous conurbation, geographically speaking (see Fig. 43, pp. 38–39). This difference in magnitude between every earlier type of urban development and that characterizing our own age is critical. What is more, as population increases, the percentage of the population in cities increases, too, and the ratio of those going into metropolitan areas is even higher. Even in England, though the amount of land occupied by cities, "built-over land," is low (2.2 per cent) in proportion to the

entire land area of the British Isles, this is more than half the area of "first-class" land available for agriculture and is a tenth of the "good land" available, according to Sir L. Dudley Stamp's classification (1952). Since requirements for manufacture and urban development are for accessible, graded land, these demands conflict with the needs of the farmer; they compete for the same good soils, and only government intervention in England, since 1932, has saved this misuse of valuable agricultural land.

Under modern technical conditions the open pattern of the residential suburb is not confined to domestic needs alone. The demand for large land areas characterizes modern factory organization, with its horizontally ordered assembly lines, housed in spreading one-story structures, and, above all, airports for long-distance flights, whose demand for landing lanes and approaches on the order of miles has increased with the size and speed of planes. In addition, the noise of planes, especially jets, sterilizes even larger areas of land for residential use as both hazardous to life and dangerous to health. There are many urban regions, like that tapped by the main-line railroads from Newark, New Jersey, to Wilmington, Delaware, where urban tissue has either displaced the land or so completely modified its rural uses as to give the whole area the character of a semi-urban desert. Add to this, in every conurbation, the ever larger quantity of land needed for collective reservoir systems, sewage works, and garbage-disposal plants as dispersed local facilities fall out of use.

As a result of population increase and urban centralization, one further demand for land, unfortunately a cumulative one, must be noted: the expansion of urban cemeteries in all cultures that maintain, as most "Christian" nations do, the Paleolithic habit of earth burial. This has resulted in the migration of the burying ground from the center to the outskirts of metropolitan areas, where vast cemeteries serve, indeed, as temporary suburban parks, until they become a wilderness of stone monuments. Unless the custom of periodically emptying out these cemeteries, as was done in London and Paris with the bones in old churchyards, takes hold, or until cremation replaces burial, the demand for open spaces for the dead threatens to crowd the quarters of the living on a scale impossible to conceive in earlier urban cultures.

URBAN-RURAL BALANCE

Whereas the area of the biggest cities, before the nineteenth century, could be measured in hundreds of acres, the areas of our new conurbations must now be measured in thousands of square miles. This is a new fact in the history of human settlement. Within a century the economy of the Western world has shifted from a rural base, harboring a few big cities and thousands of villages and small towns, to a metropolitan base whose urban spread not merely has engulfed and assimilated the small units, once isolated and self-contained, as the amoeba engulfs its particles of food, but is fast absorbing the rural hinterland and threatening to wipe out many natural elements favorable to life which in earlier stages balanced off against depletions in the urban environment. From this, even more critical results follow. Already, New York and Philadelphia, which are fast coalescing into a single conurbation along the main-line railroads and the New Jersey Turnpike, find themselves competing for the same water supply, as Los Angeles competes with the whole state of Arizona. Thus, though modern technology has escaped from the limitations of a purely local supply of water, the massing of population makes demands that, even apart

from excessive costs (which rise steadily as distance increases), put a definable limit to the possibilities of further urbanization. Water shortages may indeed limit the present distribution long before food shortages bring population growth to an end.

This situation calls for a new approach to the whole problem of urban settlement. Having thrown off natural controls and limitations, modern man must replace them with an at least equally effective man-made pattern. Though alternative proposals may be left to that portion of this volume dealing with the future, one new approach has fifty years of experience behind it and may properly be dealt with under the head of history. In the last decade of the nineteenth century two projects came forth relating to the need, already visible by then, to achieve a different balance among cities, industries, and natural regions from that which had been created by either the old rural economy, the free town economy, or the new metropolitan economy. The first of these suggestions was the work of the geographer Peter Kropotkin. His book *Fields, Factories, and Workshops* (1899) dealt with the alteration in the scale of technically efficient enterprise made possible by the invention of the electric motor. The other book, *Tomorrow,* published in 1898 by Howard, embodied a proposal to counteract the centralization of the great metropolis by reintroducing the method of colonization to take care of its further growth. Howard proposed to build relatively self-contained, balanced communities, supported by their local industry, with a permanent population, of limited number and density, on land surrounded by a swath of open country dedicated to agriculture, recreation, and rural occupation. Howard's proposal recognized the biological and social grounds, along with the psychological pressures, that underlay the current movement to suburbia. It recognized the social needs that were causing an exodus from rural regions or drab, one-industry towns into the big city. Without disparaging such real advantages as the concentrated activities and institutions of the city offered, Howard proposed to bring about a marriage between town and country. The new kind of city he called the "garden city," not so much because of its internal open spaces, which would approach a sound suburban standard, but more because it was set in a permanent rural environment.

Besides invoking the Aristotelian ideas of balance and limits, Howard's greatest contribution in conceiving this new garden city was provision for making the surrounding agricultural area an integral part of the city's form. His invention of a horizontal retaining wall, or green belt, immune to urban building, was a public device for limiting lateral growth and maintaining the urban-rural balance. In the course of twenty years two such balanced communities, Letchworth (1903) and Welwyn (1919), were experimentally founded by private enterprise in England. The soundness of the garden-city principle was recognized in the Barlow report (1940) on the decentralization of industry. Thanks to World War II, the idea of building such towns on a great scale, to drain off population from the overcrowded urban centers, took hold. This resulted in the New Towns Act of 1947, which provided for the creation of a series of new towns, fourteen in all, in Britain. This open pattern of town-building, with the towns themselves dispersed through the countryside and surrounded by permanent rural reserves, does a minimum damage to the basic ecological fabric. To the extent that their low residential density, of twelve to fourteen houses per acre, gives individual small gardens to almost every family, these towns not merely maintain a balanced micro-environment but actually grow garden produce whose value is higher than that produced when the land was used for extensive farming or grazing (Block, 1954).

On the basis of the garden-city principle, Stein (1951) and others have put forth the possibility of establishing a new type of city by integrating a group of communities into an organized design that would have the facilities of a metropolis without its congestion and loss of form. The basis for this kind of grouping was laid down in the survey of the state of New York made by the Commission of Housing and Regional Planning, of which Stein was chairman, and was published with Henry Wright in 1926. Wright, the planning adviser, here pointed out that the area of settlement was no longer the crowded terminal metropolitan areas of the railroad period but that electric power and motor transportation had opened up a wide belt on each side of the railroad trunk lines, equally favorable for industry, agriculture, and urban settlement. The most fertile soil and the most valuable geological deposits were almost entirely in the areas below the thousand-foot level; and, in planning for new urban settlement, the reservation of forest areas for water catchment and recreation, for lumber, and for electric power was important. Instead of treating the city as an intrusive element in a landscape that would finally be defaced or obliterated by the city's growth, this new approach suggested the necessity of creating a permanent rural-urban balance. In the regional city, as Stein conceived it, organization would take the place of mere agglomeration and, in doing so, would create a reciprocal relation between city and country that would not be overthrown by further population growth (Mumford, 1925, 1938; MacKaye, 1928; Stein, 1951).

With this statement of the problems raised for us today by the natural history of urbanization, our survey comes to an end. The blind forces of urbanization, flowing along the lines of least resistance, show no aptitude for creating an urban and industrial pattern that will be stable, self-sustaining, and self-renewing. On the contrary, as congestion thickens and expansion widens, both the urban and the rural landscape undergo defacement and degradation, while unprofitable investments in the remedies for congestion, such as more superhighways and more distant reservoirs of water, increase the economic burden and serve only to promote more of the blight and disorder they seek to palliate. But however difficult it is to reverse unsound procedures that offer a temporary answer and immediate—

often excessive—financial rewards, we now have a prospect of concrete alternatives already in existence in England and partly established in a different fashion by the regional planning authority for the highly urbanized Ruhr Valley in Germany. With these examples before us, we have at least a hint of the future task of urbanization: the re-establishment, in a more complex unity, with a full use of the resources of modern science and techniques, of the ecological balance that originally prevailed between city and country in the primitive stages of urbanization. Neither the blotting-out of the landscape nor the disappearance of the city is the climax stage of urbanization. Rather, it is the farsighted and provident balancing of city populations and regional resources so as to maintain in a state of high development all the elements—social, economic, and agricultural—necessary for their common life.

REFERENCES

BARLOW, ANTHONY M.
 1940 *Royal Commission on Distribution of Industrial Population Report.* London: H.M. Stationery Office. 320 pp.

BLOCK, GEOFFREY D. M.
 1954 *The Spread of Towns.* London: Conservative Political Centre. 57 pp.

BRUNHES, JEAN
 1920 *Human Geography, an Attempt at a Positive Classification: Principles and Examples.* 2d ed. Chicago: Rand McNally & Co. 648 pp.

CARCOPINO, JEROME
 1940 *Daily Life in Ancient Rome: The People and the City at the Height of the Empire.* New Haven, Conn.: Yale University Press. 342 pp.

CHILDE, V. GORDON
 1942 *What Happened in History.* Harmondsworth: Penguin Books. 288 pp.
 1954 "Early Forms of Society," pp. 38–57 in SINGER, CHARLES; HOLMYARD, E. J.; and HALL, A. R. (eds.),

A History of Technology. Oxford: Clarendon Press. 827 pp.

EVELYN, JOHN
 1933 *Fumifugium: Or the Inconvenience of the Aer and Smoake of London Dissipated.* Reprint of 1661 pamphlet. London: Oxford University Press. 49 pp.

GEDDES, PATRICK
 1915 *Cities in Evolution: An Introduction to the Town Planning Movement and to the Study of Civics.* London: Williams & Norgate. 409 pp. (Rev. ed. by JAQUELINE TYRWHITT and ARTHUR GEDDES. London: Williams & Norgate, 1949. 241 pp.)

HASSERT, KURT
 1907 *Die Städte: Geographisch Betrachtet.* Leipzig: B. G. Teubner. 137 pp.

HOWARD, EBENEZER
 1898 *To-morrow: A Peaceful Path to Real Reform.* London: Swann, Sonnenschein & Co. 176 pp.

1902 *Garden Cities of To-morrow.* London: Swann, Sonnenschein & Co. 167 pp.

1945 *Garden Cities of To-morrow.* With a Preface by F. J. OSBORN and an Introduction by LEWIS MUMFORD. London: Faber & Faber. 168 pp.

JEFFERSON, MARK

1931 "Distribution of the World's City Folks: A Study in Comparative Civilization," *Geographical Review,* XXI, No. 3, 446–65.

KEYES, FENTON

1951 "Urbanism and Population Distribution in China," *American Journal of Sociology,* LVI, No. 6, 519–27.

KING, F. H.

1927 *Farmers of Forty Centuries.* New York: Harcourt, Brace & Co. 379 pp.

KROPOTKIN, PETER

1899 *Fields, Factories, and Workshops.* New York: G. P. Putnam & Sons. 477 pp.

MACKAYE, BENTON

1928 *The New Exploration: A Philosophy of Regional Planning.* New York: Harcourt, Brace & Co. 235 pp.

MARSH, GEORGE P.

1864 *Man and Nature.* London: Sampson, Low & Son. 577 pp.

1874 *The Earth as Modified by Human Action: A New Edition of "Man and Nature."* New York: Scribner, Armstrong & Co. 656 pp.

1885 *The Earth as Modified by Human Action: A Last Revision of "Man and Nature."* New York: Charles Scribner's Sons. 629 pp. (Note that in this last edition of *Man and Nature,* Marsh refers for the first time, in a long footnote [p. 473], under the heading "Inundations and Torrents," to the influence of large urban masses on climate, particularly heat and precipitation—an anticipation of present-day studies.) Last printing in 1907.

MUMFORD, LEWIS

1934 *Technics and Civilization.* New York: Harcourt, Brace & Co. 495 pp.

1938 *The Culture of Cities.* New York: Harcourt, Brace & Co. 586 pp.

MUMFORD, LEWIS (ed.)

1925 "Regional Planning Number," *Survey Graphic,* LIV, No. 3, 128–208.

OSBORN, F. J.

1946 *Green-Belt Cities: The British Contribution.* London: Faber & Faber. 191 pp.

RIEHL, WILHELM HEINRICH

1935 *Die Naturgeschichte des Deutschen Volkes.* Reprint of 1861 edition. Leipzig: Alfred Kröner Verlag. 407 pp.

SCHNEIDER, ARTHUR

1895 "Stadtumfänge in Altertum und Gegenwart," *Geographische Zeitschrift,* I, 676–79.

SORRE, MAX

1952 *Les Fondements de la géographie humaine.* 3 vols. Paris: Librairie Armand Colin.

STAMP, L. DUDLEY

1948 *The Land of Britain: Its Use and Misuse.* London: Longman's Green. 570 pp.

1952 *Land for Tomorrow.* New York: American Geographical Society; Bloomington, Ind.: Indiana University Press. 230 pp.

STEIN, CLARENCE S.

1951 *Toward New Towns for America.* Chicago: Public Administration Service. 245 pp.

STEIN, CLARENCE S., and WRIGHT, HENRY

1926 *Report of the Commission of Housing and Regional Planning to Governor Alfred E. Smith, May 7, 1926.* (New York State Document.) Albany. 82 pp.

THORNDYKE, EDWARD LEE

1939 *Your City.* New York: Harcourt, Brace & Co. 204 pp.

WEBBER, ADNA FERRIN

1899 *The Growth of Cities in the Nineteenth Century: A Study in Statistics.* New York: Macmillan Co. 495 pp.

WEST MIDLAND GROUP

1948 *Conurbation: A Planning Survey of Birmingham and the Black Country.* London: Architectural Press. 288 pp.

WOYTINSKY, W. S. and E. C.

1953 *World Population and Production.* New York: Twentieth Century Fund. 1,268 pp.

Ritual Regulation of Environmental Relations Among a New Guinea People[1]

Roy A. Rappaport
University of Michigan

Most functional studies of religious behavior in anthropology have as an analytic goal the elucidation of events, processes, or relationships occurring within a social unit of some sort. The social unit is not always well defined, but in some cases it appears to be a church, that is, a group of people who entertain similar beliefs about the universe, or a congregation, a group of people who participate together in the performance of religious rituals. There have been exceptions. Thus Vayda, Leeds, and Smith (1961) and O. K. Moore (1957) have clearly perceived that the functions of religious ritual are not necessarily confined within the boundaries of a congregation or even a church. By and large, however, I believe that the following statement by Homans (1941: 172) represents fairly the dominant line of anthropological thought concerning the functions of religious ritual:

> Ritual actions do not produce a practical result on the external world—that is one of the reasons why we call them ritual. But to make this statement is not to say that ritual has no function. Its function is not related to the world external to the society but to the internal constitution of the society. It gives the members of the society confidence, it dispels their anxieties, it disciplines their social organization.

No argument will be raised here against the sociological and psychological functions imputed by Homans, and many others before him, to ritual. They seem to me to be plausible. Nevertheless, in some cases at least, ritual does produce, in Homans' terms, "a practical result on the world" external not only to the social unit composed of those who participate together in ritual performances but also to the larger unit composed of those who entertain similar beliefs concerning the universe. The material presented here will show that the ritual cycles of the Tsembaga, and of other local territorial groups of Maring speakers living in the New Guinea interior, play an important part in regulating the relationships of these groups with both the nonhuman components of their immediate environments and the human components of their less immediate environments, that is, with other similar territorial groups. To be more specific, this regulation helps to maintain the biotic communities existing within their territories, redistributes land among people and people over land, and limits the frequency of fighting. In the absence of authoritative political statuses or offices, the ritual cycle likewise provides a means for mobilizing allies when warfare may be undertaken. It also provides a mechanism for redistributing local pig surpluses in the form of pork throughout a large regional population while helping to assure the local population of a supply of pork when its members are most in need of high quality protein.

Religious ritual may be defined, for the purposes of this paper, as the prescribed performance of conventionalized acts manifestly directed toward the involvement of nonempirical or supernatural agencies in the affairs of the actors. While this definition relies upon the formal characteristics of the performances and upon the motives for undertaking them, attention will be focused upon the empirical effects of ritual performances and sequences of ritual performances. The religious rituals to be discussed are regarded as neither more nor less than part of the behavioral repertoire employed by an aggregate of organisms in adjusting to its environment.

The data upon which this paper is based were collected during fourteen months of field work among the Tsembaga, one of about twenty local groups of Maring speakers living in the Simbai and Jimi Valleys of the Bismarck Range in the Territory of New Guinea. The size of Maring local groups varies from a little over 100 to 900. The Tsembaga, who in 1963 numbered 204 persons, are located on the south wall of the Simbai Valley. The country in which they live differs from the true highlands in being lower, generally more rugged, and more heavily forested. Tsembaga territory rises, within a total surface area of 3.2 square miles, from an elevation of 2,200 feet at the Simbai river to 7,200 feet at the ridge crest. Gardens are cut in the secondary forests up to between 5,000 and 5,400 feet, above which the area remains in primary forest. Rainfall reaches 150 inches per year.

The Tsembaga have come into contact with the outside world only recently; the first government patrol to penetrate their territory arrived in 1954. They were considered uncontrolled by the Australian government until 1962, and they remain unmissionized to this day.

The 204 Tsembaga are distributed among five putatively patrilineal clans, which are, in turn, organized into more inclusive groupings on two hierarchical levels below that of the total local group.[2] Internal political structure is highly egalitarian. There are no hereditary or elected chiefs, nor are there even "big men" who can regularly coerce or command the support of their clansmen or co-residents in economic or forceful enterprises.

It is convenient to regard the Tsembaga as a population in the ecological sense, that is, as one of the components of a system of trophic exchanges taking place within a bounded area. Tsembaga territory and the biotic community existing upon it may be conveniently viewed as an ecosystem. While it would be permissible arbitrarily to designate the Tsembaga as a population and their territory with its biota as an ecosystem, there are also nonarbitrary reasons for doing so. An ecosystem is a system of material exchanges, and the Tsembaga maintain against other human groups exclusive access to the resources within their territorial borders. Conversely, it is from this territory alone that the Tsembaga ordinarily derive all of their food-stuffs and most of the other materials they require for survival. Less anthropocentrically, it may be justified to regard Tsembaga territory with its biota as an ecosystem in view of the rather localized nature of cyclical material exchanges in tropical rainforests.

As they are involved with the nonhuman biotic community within their territory in a set of trophic exchanges, so do they participate in other material relationships with other human groups external to their territory. Genetic materials are exchanged with other groups, and certain crucial items, such as stone axes, were in past obtained from the outside. Furthermore, in the area occupied by the Maring speakers, more than one local group is usually involved in any process, either peaceful or warlike, through which people are redistributed over land and land redistributed among people.

The concept of the ecosystem, though it provides a convenient frame for the analysis of interspecific trophic exchanges taking place within limited geographical areas, does not comfortably accommodate intraspecific exchanges taking place over wider geographic areas. Some sort of geographic population model would be more useful for the analysis of the relationship of the local ecological population to the larger regional population of which it is a part, but we lack even a set of appropriate terms for such a model. Suffice it here to note that the relations of the Tsembaga to the total of other local human populations in their vicinity are similar to the relations of local aggregates of other animals to the totality of their species occupying broader and more or less continuous regions. This larger, more inclusive aggregate may resemble what geneticists mean by the term population, that is, an aggregate of interbreeding organisms persisting through an indefinite number of generations and either living or capable of living in isolation from similar aggregates of the same species. This is the unit which survives through long periods of time while its local ecological (*sensu stricto*) subunits, the units more or less independently involved in interspecific trophic exchanges such as the Tsembaga, are ephemeral.

Since it has been asserted that the ritual cycles of the Tsembaga regulate relationships within what may be regarded as a complex system, it is necessary, before proceeding to the ritual cycle itself, to describe briefly, and where possible in quantitative terms, some aspects of the place of the Tsembaga in this system.

The Tsembaga are bush-fallowing horticulturalists. Staples include a range of root crops, taro (*Colocasia*) and sweet potatoes being most important, yams and manioc less so. In addition, a great variety of greens are raised, some of which are rich in protein. Sugar cane and some tree crops, particularly *Pandanus conoideus,* are also important.

All gardens are mixed, many of them containing all of the major root crops and many greens. Two named garden types are, however, distinguished by the crops which predominate in them. "Taro-yam gardens" were found to produce, on the basis of daily harvest records kept on entire gardens for close to one year, about 5,300,000 calories[3] per acre during their harvesting lives of 18 to 24 months; 85 per cent of their yield is harvested between 24 and 76 weeks after planting. "Sugar-sweet potato gardens" produce about 4,600,000 calories per acre during their harvesting lives, 91 per cent being taken between 24 and 76 weeks after planting. I estimated that approximately 310,000 calories per acre is expended on cutting, fencing, planting, maintaining, harvesting, and walking to and from taro-yam gardens. Sugar-sweet potato gardens required an expenditure of approximately 290,000 calories per acre.[4] These energy ratios, approximately 17:1 on taro-yam gardens and 16:1 on sugar-sweet potato gardens, compare favorably with figures reported for swidden cultivation in other regions.[5]

Intake is high in comparison with the reported dietaries of other New Guinea populations. On the basis of daily consumption records kept for ten months on four households numbering in total sixteen persons, I estimated the average daily intake of adult males to be approximately 2,600 calories, and that of adult females to be around 2,200 calories. It may be mentioned here that the Tsembaga are small and short statured. Adult males average 101 pounds in weight and approximately 58.5 inches in height; the corresponding averages for adult females are 85 pounds and 54.5 inches.[6]

Although 99 per cent by weight of the food consumed is vegetable, the protein intake is high by New Guinea standards. The daily protein consumption of adult males from vegetable sources was estimated to be between 43 and 55 grams, of adult females 36 to 48 grams. Even with an adjustment for vegetable sources, these values are slightly in excess of the recently published WHO/FAO daily requirements (Food and Agriculture Organization of the United Nations 1964). The same is true of the younger age categories, although soft and discolored hair, a symptom of protein deficiency, was noted in a few children. The WHO/FAO protein requirements do not include a large "margin for safety" or allowance for stress; and, although no clinical assessments were undertaken, it may be suggested that the Tsembaga achieve nitrogen balance at a low level. In other words, their protein intake is probably marginal.

Measurements of all gardens made during 1962 and of some gardens made during 1963 indicate that, to support the human population, between .15 and .19 acres are put into cultivation per capita per year. Fallows range from 8 to 45 years. The area in secondary forest comprises approximately 1,000 acres, only 30 to 50 of which are in cultivation at any time. Assuming calories to be the limiting factor, and assuming an unchanging population structure, the territory could support—with no reduction in lengths of fallow and without cutting into the virgin forest from which the Tsembaga extract many important items—between 290 and 397 people if the pig population remained minimal. The size of the pig herd, however, fluctuates widely. Taking Maring pig husbandry procedures into consideration, I have estimated the human carrying capacity of the Tsembaga territory at between 270 and 320 people.

Because the timing of the ritual cycle is bound up with the demography of the pig herd, the place of the pig in Tsembaga adaptation must be examined.

First, being omnivorous, pigs keep residential areas free of garbage and human feces. Second, limited numbers of pigs rooting in secondary growth may help to hasten the development of that growth. The Tsembaga usually permit pigs to enter their gardens one and a half to two years after planting, by which time second-growth trees are well established there. The Tsembaga practice selective weeding; from the time the garden is planted, herbaceous species are removed, but tree species are allowed to remain. By the time cropping is discontinued and the pigs are let in, some of the trees in the garden are already ten to fifteen feet tall. These well-established trees are relatively impervious to damage by the pigs, which, in rooting for seeds and remaining tubers, eliminate many seeds and seedlings that, if allowed to develop, would provide some competition for the established trees. Moreover, in some Maring-speaking areas swiddens are planted twice, although this is not the case with the Tsembaga. After the first crop is almost exhausted, pigs are penned in the garden, where their rooting eliminates weeds and softens the ground, making the task of planting for a second time easier. The pigs, in other words, are used as cultivating machines.

Small numbers of pigs are easy to keep. They run free during the day and return home at night to receive their ration of garbage and substandard tubers, particularly sweet potatoes. Supplying the latter requires little extra work, for the substandard tubers are taken from the ground in the course of harvesting the daily ration for humans. Daily consumption records kept over a period of some months show that the ration of tubers received by the pigs approximates in weight that consumed by adult humans, i.e., a little less than three pounds per day per pig.

If the pig herd grows large, however, the substandard tubers incidentally obtained in the course of harvesting for human needs become insufficient, and it becomes necessary to harvest especially for pigs. In other words, people must work for the pigs and perhaps even supply them with food fit for human consumption. Thus, as Vayda, Leeds, and Smith (1961: 71) have pointed out, there can be too many pigs for a given community.

This also holds true of the sanitary and cultivating services rendered by pigs. A small number of pigs is sufficient to keep residential areas clean, to suppress superfluous seedlings in abandoned gardens, and to soften the soil in gardens scheduled for second plantings. A larger herd, on the other hand, may be troublesome; the larger the number of pigs, the greater the possibility of their invasion of producing gardens, with concomitant damage not only to crops and young secondary growth but also to the relations between the pig owners and garden owners.

All male pigs are castrated at approximately three months of age, for boars, people say, are dangerous and do not grow as large as barrows. Pregnancies, therefore, are always the result of unions of domestic sows with feral males. Fecundity is thus only a fraction of its potential. During one twelve-month period only fourteen litters resulted out of a potential 99 or more pregnancies. Farrowing generally takes place in the forest, and mortality of the young is high. Only 32 of the offspring of the above-mentioned fourteen pregnancies were alive six months after birth. This number is barely sufficient to replace the number of adult animals which would have died or been killed during most years without pig festivals.

The Tsembaga almost never kill domestic pigs outside of ritual contexts. In ordinary times, when there is no pig festival in progress, these rituals are almost always associated with misfortunes or emergencies, notably warfare, illness, injury, or death. Rules state not only the contexts in which pigs are to be ritually slaughtered, but also who may partake of the flesh of the sacrificial animals. During warfare it is only the men participating in the fighting who eat the pork. In cases of illness or injury, it is only the victim and certain near relatives, particularly his co-resident agnates and spouses, who do so.

It is reasonable to assume that misfortune and emergency are likely to induce in the organisms experiencing them a complex of physiological changes known collectively as "stress." Physiological stress reactions occur not only in organisms which are infected with disease or traumatized, but also in those experiencing rage or fear (Houssay *et al.* 1955: 1096), or even prolonged anxiety (National Research Council 1963: 53). One important aspect of stress is the increased catabolization of protein (Houssay *et al.* 1955: 451; National Research Council 1963: 49), with a net loss of nitrogen from the tissues (Houssay *et al.* 1955: 450). This is a serious matter for organisms with a marginal protein intake. Antibody production is low (Berg 1948: 311), healing is slow (Large and Johnston 1948: 352), and a variety of symptoms of a serious nature are likely to develop (Lund and Levenson 1948: 349; Zintel 1964: 1043). The status of a protein-depleted animal, however, may be significantly improved in a relatively short period of time by the intake of high quality protein, and high protein diets are therefore routinely prescribed for surgical patients and those suffering from infectious diseases (Burton 1959: 231; Lund and Levenson 1948: 350; Elman 1951: 85ff; Zintel 1964: 1043ff).

It is precisely when they are undergoing physiological stress that the Tsembaga kill and consume their pigs, and it should be noted that they limit the consumption to those likely to be experiencing stress most profoundly. The Tsembaga, of course, know nothing of physiological stress. Native theories of the etiology and treatment of disease and injury implicate various categories of spirits to whom sacrifices must be made. Nevertheless, the behavior which is appropriate in terms of native understandings is also appropriate to the actual situation confronting the actors.

We may now outline in the barest of terms the Tsembaga ritual cycle. Space does not permit a description of its ideological correlates. It must suffice to note that Tsembaga do not necessarily perceive all of the empirical effects which the anthropologist sees to flow from their ritual behavior. Such empirical consequences as they may perceive, moreover, are not central to their rationalizations of the performances. The Tsembaga say that they perform the rituals in order to rearrange their relationships with the supernatural world. We may only reiterate here that behavior undertaken in reference to their "cognized environment"—an environment which includes as very important elements the spirits of ancestors—seems appropriate in their "operational environment," the material environment specified by the anthropologist through operations of observation, including measurement.

Since the rituals are arranged in a cycle, description may commence at any point. The operation of the cycle becomes clearest if we begin with the rituals performed during warfare. Opponents in all cases occupy adjacent territories, in almost all cases on the same valley wall. After hostilities have broken out, each side performs certain rituals which place the opposing side in the formal category of "enemy." A number of taboos prevail while hostilities continue. These include prohibitions on sexual intercourse and on the ingestion of certain things—food prepared by women, food grown on the lower portion of the territory, marsupials, eels, and, while actually on the fighting ground, any liquid whatsoever.

One ritual practice associated with fighting which may have some physiological consequences deserves mention. Immediately before proceeding to the fighting ground, the warriors eat heavily salted pig fat. The ingestion of salt, coupled with the taboo on drinking, has the effect of shortening the fighting day, particularly since the Maring prefer to fight only on bright sunny days. When everyone gets unbearably thirsty, according to informants, fighting is broken off.

There may formerly have been other effects if the native salt contained sodium (the production of salt was discontinued some years previous to the field work, and no samples were obtained). The Maring diet seems to be deficient in sodium. The ingestion of large amounts of sodium just prior to fighting would have permitted the warriors to sweat normally without a lowering of blood volume and consequent weakness during the course of the fighting. The pork belly ingested with the salt would have provided them with a new burst of energy two hours or so after the commencement of the engagement. After fighting was finished for the day, lean pork was consumed, offsetting, at least to some extent, the nitrogen loss associated with the stressful fighting (personal communications from F. Dunn, W. Mac-Farlane, and J. Sabine, 1965).

Fighting could continue sporadically for weeks. Occasionally it terminated in the rout of one of the antagonistic groups, whose survivors would take refuge with kinsmen elsewhere. In such instances, the victors would lay waste their opponents' groves and gardens, slaughter their pigs, and burn their houses. They would not, however, immediately annex the territory of the vanquished. The Maring say that they never take over the territory of an enemy for, even if it has been abandoned, the spirits of their ancestors remain to guard it against interlopers. Most fights, however, terminated in truces between the antagonists.

With the termination of hostilities a group which has not been driven off its territory performs a ritual called "planting the *rumbim.*" Every man puts his hand on the ritual plant, *rumbim* (*Cordyline fruticosa* (L.), A. Chev; *C. terminalis,* Kunth), as it is planted in the ground. The ancestors are addressed, in effect, as follows:

We thank you for helping us in the fight and permitting us to remain on our territory. We place our souls in this *rumbim* as we plant it on our ground. We ask you to care for this *rumbim*. We will kill pigs for you now, but they are few. In the future, when we have many pigs, we shall again give you pork and uproot the *rumbim* and stage a *kaiko* (pig festival). But until there are sufficient pigs to repay you the *rumbim* will remain in the ground.

This ritual is accompanied by the wholesale slaughter of pigs. Only juveniles remain alive. All adult and adolescent animals are killed, cooked, and dedicated to the ancestors. Some are consumed by the local group, but most are distributed to allies who assisted in the fight.

Some of the taboos which the group suffered during the time of fighting are abrogated by this ritual. Sexual intercourse is now permitted, liquids may be taken at any time, and food from any part of the territory may be eaten. But the group is still in debt to its allies and ancestors. People say it is still the time of the *bamp ku,* or "fighting stones," which are actual objects used in the rituals associated with warfare. Although the fighting ceases when *rumbim* is planted, the concomitant obligations, debts to allies and ancestors, remain outstanding; and the fighting stones may not be put away until these obligations are fulfilled. The time of the fighting stones is a time of debt and danger which lasts until the *rumbim* is uprooted and a pig festival (*kaiko*) is staged.

Certain taboos persist during the time of the fighting stones. Marsupials, regarded as the pigs of the ancestors of the high ground, may not be trapped until the debt to their masters has been repaid. Eels, the "pigs of the ancestors of the low ground," may neither be caught nor consumed. Prohibitions on all intercourse with the enemy come into force. One may not touch, talk to, or even look at a member of the enemy group, nor set foot on enemy ground. Even more important, a group may not attack another group while its ritual plant remains in the ground, for it has not yet fully rewarded its ancestors and allies for their assistance in the last fight. Until the debts to them have been paid, further assistance from them will not be forthcoming. A kind of "truce of god" thus prevails until the *rumbim* is uprooted and a *kaiko* completed.

To uproot the *rumbim* requires sufficient pigs. How many pigs are sufficient, and how long does it take to acquire them? The Tsembaga say that, if a place is "good," this can take as little as five years; but if a place is "bad," it may require ten years or longer. A bad place is one in which misfortunes are frequent and where, therefore, ritual demands for the killing of pigs arise frequently. A good place is one where such demands are infrequent. In a good place, the increase of the pig herd exceeds the ongoing ritual demands, and the herd grows rapidly. Sooner or later the substandard tubers incidentally obtained while harvesting become insufficient to feed the herd, and additional acreage must be put into production specifically for the pigs.

The work involved in caring for a large pig herd can be extremely burdensome. The Tsembaga herd just prior to the pig festival of 1962-63, when it numbered 169 animals, was receiving 54 per cent of all of the sweet potatoes and 82 per cent of all of the manioc harvested. These comprised 35.9 per cent by weight of all root crops harvested. This figure is consistent with the difference between the amount of land under cultivation just previous to the pig festival, when the herd was at maximum size, and that immediately afterwards, when the pig herd was at minimum size. The former was 36.1 per cent in excess of the latter.

I have estimated, on the basis of acreage yield and energy expenditure figures, that about 45,000 calories per year are expended in caring for one pig 120-150 pounds in size. It is upon women that most of the burden of pig keeping falls. If, from a woman's daily intake of about 2,200 calories, 950 calories are allowed for basal metabolism, a woman has only 1,250 calories a day available for all her activities, which include gardening for her family, child care, and cooking, as well as tending pigs. It is clear that no woman can feed many pigs; only a few had as many as four in their care at the commencement of the festival; and it is not surprising that agitation to uproot the *rumbim* and stage the *kaiko* starts with the wives of the owners of large numbers of pigs.

A large herd is not only burdensome as far as energy expenditure is concerned; it becomes increasingly a nuisance as it expands. The more numerous pigs become, the more frequently are gardens invaded by them. Such events result in serious disturbances of local tranquillity. The garden owner often shoots, or attempts to shoot, the offending pig; and the pig owner commonly retorts by shooting, or attempting to shoot, either the garden owner, his wife, or one of his pigs. As more and more such events occur, the settlement, nucleated when the herd was small, disperses as people try to put as much distance as possible between their pigs and other people's gardens and between their gardens and other people's pigs. Occasionally this reaches its logical conclusion, and people begin to leave the territory, taking up residence with kinsmen in other local populations.

The number of pigs sufficient to become intolerable to the Tsembaga was below the capacity of the territory to carry pigs. I have estimated that, if the size and structure of the human population remained constant at the 1962-1963 level, a pig population of 140 to 240 animals averaging 100 to 150 pounds in size could be maintained perpetually by the Tsembaga without necessarily inducing environmental degradation. Since the size of the herd fluctuates, even higher cyclical maxima could be achieved. The level of toleration, however, is likely always to be below the carrying capacity, since the destructive capacity of the pigs is dependent upon the population density of both people and pigs, rather than upon population size. The denser the human population, the fewer pigs will be required to disrupt social life. If the carrying capacity is exceeded, it is likely to be exceeded by people and not by pigs.

The *kaiko* or pig festival, which commences with the planting of stakes at the boundary and the uprooting of the *rumbim,* is thus triggered by either the additional work attendant upon feeding pigs or the destructive capacity of the pigs themselves. It may be said, then, that there are sufficient pigs to stage the *kaiko* when the relationship of pigs to people changes from one of mutualism to one of parasitism or competition.

A short time prior to the uprooting of the *rumbim,* stakes are planted at the boundary. If the enemy has continued to occupy its territory, the stakes are planted at the boundary which existed before the fight. If, on the other hand, the enemy has abandoned its territory, the victors may plant their stakes at a new boundary which encompasses areas previously occupied by the enemy. The Maring say, to be sure, that they never take land belonging to an enemy, but this land is regarded as vacant, since no *rumbim* was planted on it after the last fight. We may state here a rule of land redistribution in terms of the ritual cycle: *If one of a pair of antagonistic groups is able to uproot its rumbim before its opponents can plant their rumbim, it may occupy the latter's territory.*

Not only have the vanquished abandoned their territory; it is assumed that it has also been abandoned by their ancestors as well. The surviving members of the erstwhile enemy group have by this time resided with other groups for a number of years, and most if not all of them have already had occasion to sacrifice pigs to their ancestors at their new residences. In so doing they have invited these spirits to settle at the new locations of the living, where they will in the future receive sacrifices. Ancestors of vanquished groups thus relinquish their guardianship over the territory, making it available to victorious groups. Meanwhile, the *de facto* membership of the living in the groups with which they have taken refuge is converted eventually into *de jure* membership. Sooner or later the groups with which they have taken up residence will have occasion to plant *rumbim,* and the refugees, as co-residents, will participate, thus ritually validating their connection to the new territory and the new group. A rule of population redistribution may thus be stated in terms of ritual cycles: *A man becomes a member of a territorial group by participating with it in the planting of rumbim.*

The uprooting of the *rumbim* follows shortly after the planting of stakes at the boundary. On this particular occasion the Tsembaga killed 32 pigs out of their herd of 169. Much of the pork was distributed to allies and affines outside of the local group.

The taboo on trapping marsupials was also terminated at this time. Information is lacking concerning the population dynamics of the local marsupials, but it may well be that the taboo which had prevailed since the last fight—that against taking them in traps—had conserved a fauna which might otherwise have become extinct.

The *kaiko* continues for about a year, during which period friendly groups are entertained from time to time. The guests receive presents of vegetable foods, and the hosts and male guests dance together throughout the night.

These events may be regarded as analogous to aspects of the social behavior of many nonhuman animals. First of all, they include massed epigamic, or courtship, displays (Wynne-Edwards 1962: 17). Young women are presented with samples of the eligible males of local groups with which they may not otherwise have had the opportunity to become familiar. The context, moreover, permits the young women to discriminate amongst this sample in terms of both endurance (signaled by how vigorously and how long a man dances) and wealth (signaled by the richness of a man's shell and feather finery).

More importantly, the massed dancing at these events may be regarded as epideictic display, communicating to the participants information concerning the size or density of the group (Wynne-Edwards 1962: 16). In many species such displays take place as a prelude to actions which adjust group size or density, and such is the case among the Maring. The massed dancing of the visitors at a *kaiko* entertainment communicates to the hosts, while the *rumbim* truce is still in force, information concerning the amount of support they may expect from the visitors in the bellicose enterprises that they are likely to embark upon soon after the termination of the pig festival.

Among the Maring there are no chiefs or other political authorities capable of commanding the support of a body of followers, and the decision to assist another group in warfare rests with each individual male. Allies are not recruited by appealing for help to other local groups as such. Rather, each member of the groups primarily involved in the hostilities appeals to his cognatic and affinal kinsmen in other local groups. These men, in turn, urge other of their co-residents and kinsmen to "help them fight." The channels through which invitations to dance are extended are precisely those through which appeals for military support are issued. The invitations go not from group to group, but from kinsman to kinsman, the recipients of invitations urging their co-residents to "help them dance."

Invitations to dance do more than exercise the channels through which allies are recruited; they provide a means for judging their effectiveness. Dancing and fighting are regarded as in some sense equivalent. This equivalence is expressed in the similarity of some pre-fight and pre-dance rituals, and the Maring say that those who come to dance come to fight. The size of a visiting dancing contingent is consequently taken as a measure of the size of the contingent of warriors whose assistance may be expected in the next round of warfare.

In the morning the dancing ground turns into a trading ground. The items most frequently exchanged include axes, bird plumes, shell ornaments, an occasional baby pig, and, in former times, native salt. The *kaiko* thus facilitates trade by providing a market-like setting in which large numbers of traders can assemble. It likewise facilitates the movement of two critical items, salt and axes, by creating a demand for the bird plumes which may be exchanged for them.

The *kaiko* concludes with major pig sacrifices. On this particular occasion the Tsembaga butchered 105 adult and adolescent pigs, leaving only 60 juveniles and neonates alive. The survival of an additional fifteen adolescents and adults was only temporary, for they were scheduled as imminent victims. The pork yielded by the Tsembaga slaughter was estimated to weigh between 7,000 and 8,500 pounds, of which between 4,500 and 6,000 pounds were distributed to members of other local groups in 163 separate presentations. An estimated 2,000 to 3,000 people in seventeen local groups were the beneficiaries of the redistribution. The presentations, it should be mentioned, were not confined to pork. Sixteen Tsembaga men presented bridewealth or child-wealth, consisting largely of axes and shells, to their affines at this time.

The *kaiko* terminates on the day of the pig slaughter with the public presentation of salted pig belly to allies of the last fight. Presentations are made through the window in a high ceremonial fence built specially for the occasion at one end of the dance ground. The name of each honored man is announced to the assembled multitude as he charges to the window to receive his hero's portion. The fence is then ritually torn down, and the fighting stones are put away. The pig festival and the ritual cycle have been completed, demonstrating, it may be suggested, the ecological and economic competence of the local population. The local population would now be free, if it were not for the presence of the government, to attack its enemy again, secure in the knowledge that the assistance of allies and ancestors would be forthcoming because they have received pork and the obligations to them have been fulfilled.

Usually fighting did break out again very soon after the completion of the ritual cycle. If peace still prevailed when the ceremonial fence had rotted completely—a process said to take about three years, a little longer than the length of time required to raise a pig to maximum size—*rumbim* was planted as if there had been a fight, and all adult and adolescent pigs were killed. When the pig herd was large enough so that the *rumbim* could be uprooted, peace could be made with former enemies if they were also able to dig out their *rumbim*. To put this in formal terms: *If a pair of antagonistic groups proceeds through two ritual cycles without resumption of hostilities their enmity may be terminated.*

The relations of the Tsembaga with their environment have been analyzed as a complex system composed of two subsystems. What may be called the "local subsystem" has been derived from the relations of the Tsembaga with the nonhuman components of their immediate or territorial environment. It corresponds to the ecosystem in which the Tsembaga participate. A second subsystem, one which corresponds to the larger regional population of which the Tsembaga are one of the constituent units and which may be designated as the "regional subsystem," has been derived from the relations of the Tsembaga with neighboring local populations similar to themselves.

It has been argued that rituals, arranged in repetitive sequences, regulate relations both within each of the subsystems and within the larger complex system as a whole. The timing of the ritual cycle is largely dependent upon changes in the states of the components of the local subsystem. But the *kaiko*, which is the culmination of the ritual cycle, does more than reverse changes which have taken place within the local subsystem. Its occurrence also affects relations among the components of the regional subsystem. During its performance, obligations to other local populations are fulfilled, support for future military enterprises is rallied, and land from which enemies have earlier been driven is occupied. Its completion, furthermore, permits the local population to initiate warfare again. Conversely, warfare is terminated by rituals which preclude the reinitiation of warfare until the state of the local subsystem is again such that a *kaiko* may be staged and completed. Ritual among the Tsembaga and other Maring, in short, operates as both transducer, "translating" changes in the state of one subsystem into information which can effect changes in a second subsystem, and homeostat, maintaining a number of variables which in sum comprise the total system within ranges of viability. To repeat an earlier assertion, the operation of ritual among the

Tsembaga and other Maring helps to maintain an undegraded environment, limits fighting to frequencies which do not endanger the existence of the regional population, adjusts man-land ratios, facilitates trade, distributes local surpluses of pig throughout the regional population in the form of pork, and assures people of high quality protein when they are most in need of it.

Religious rituals and the supernatural orders toward which they are directed cannot be assumed *a priori* to be mere epiphenomena. Ritual may, and doubtless frequently does, do nothing more than validate and intensify the relationships which integrate the social unit, or symbolize the relationships which bind the social unit to its environment. But the interpretation of such presumably *sapiens*-specific phenomena as religious ritual within a framework which will also accommodate the behavior of other species shows, I think, that religious ritual may do much more than symbolize, validate, and intensify relationships. Indeed, it would not be improper to refer to the Tsembaga and the other entities with which they share their territory as a "ritually regulated ecosystem," and to the Tsembaga and their human neighbors as a "ritually regulated population."

NOTES

1. The field work upon which this paper is based was supported by a grant from the National Science Foundation, under which Professor A. P. Vayda was principal investigator. Personal support was received by the author from the National Institutes of Health. Earlier versions of this paper were presented at the 1964 annual meeting of the American Anthropological Association in Detroit, and before a Columbia University seminar on Ecological Systems and Cultural Evolution. I have received valuable suggestions from Alexander Alland, Jacques Barrau, William Clarke, Paul Collins, C. Glen King, Marvin Harris, Margaret Mead, M. J. Meggitt, Ann Rappaport, John Street, Marjorie Whiting, Cherry Vayda, A. P. Vayda and many others, but I take full responsibility for the analysis presented herewith.

2. The social organization of the Tsembaga will be described in detail elsewhere.

3. Because the length of time in the field precluded the possibility of maintaining harvest records on single gardens from planting through abandonment, figures were based, in the case of both "taro-yam" and "sugar-sweet potato" gardens, on three separate gardens planted in successive years. Conversions from the gross weight to the caloric value of yields were made by reference to the literature. The sources used are listed in Rappaport (1966: Appendix VIII)

4. Rough time and motion studies of each of the tasks involved in making, maintaining, harvesting, and walking to and from gardens were undertaken. Conversion to energy expenditure values was accomplished by reference to energy expenditure tables prepared by Hipsley and Kirk (1965: 43) on the basis of gas exchange measurements made during the performance of garden tasks by the Chimbu people of the New Guinea highlands.

5. Marvin Harris, in an unpublished paper, estimates the ratio of energy return to energy input ratio on Dyak (Borneo) rice swiddens at 10:1. His estimates of energy ratios on Tepotzlan (Meso-America) swiddens range from 13:1 on poor land to 29:1 on the best land.

6. Heights may be inaccurate. Many men wear their hair in large coiffures hardened with pandanus grease, and it was necessary in some instances to estimate the location of the top of the skull.

BIBLIOGRAPHY

Berg, C. 1948. Protein Deficiency and Its Relation to Nutritional Anemia, Hypo-proteinemia, Nutritional Edema, and Resistance to Infection. Protein and Amino Acids in Nutrition, ed. M. Sahyun, pp. 290-317. New York.

Burton, B. T., ed. 1959. The Heinz Handbook of Nutrition. New York.

Elman, R. 1951. Surgical Care. New York.

Food and Agriculture Organization of the United Nations. 1964. Protein: At the Heart of the World Food Problem. World Food Problems 5. Rome.

Hipsley, E., and N. Kirk. 1965. Studies of the Dietary Intake and Energy Expenditure of New Guineans. South Pacific Commission, Technical Paper 147. Noumea.

Homans, G. C. 1941. Anxiety and Ritual: The Theories of Malinowski and Radcliffe-Brown. American Anthropologist 43: 164-172.

Houssay, B. A., et al. 1955. Human Physiology. 2nd edit. New York.

Large, A., and C. G. Johnston. 1948. Proteins as Related to Burns. Proteins and Amino Acids in Nutrition, ed. M. Sahyun, pp. 386-396. New York.

Lund, C. G., and S. M. Levenson. 1948. Protein Nutrition in Surgical Patients. Proteins and Amino Acids in Nutrition, ed. M. Sahyun, pp. 349-363. New York.

Moore, O. K. 1957. Divination—a New Perspective. American Anthropologist 59: 69-74.

National Research Council. 1963. Evaluation of Protein Quality. National Academy of Sciences—National Research Council Publication 1100. Washington.

Rappaport, R. A. 1966. Ritual in the Ecology of a New Guinea People. Unpublished doctoral dissertation, Columbia University.

Vayda, A. P., A. Leeds, and D. B. Smith. 1961. The Place of Pigs in Melanesian Subsistence. Proceedings of the 1961 Annual Spring Meeting of the American Ethnological Society, ed. V. E. Garfield, pp. 69-77. Seattle.

Wayne-Edwards, V. C. 1962. Animal Dispersion in Relation to Social Behaviour. Edinburgh and London.

Zintel, Harold A. 1964. Nutrition in the Care of the Surgical Patient. Modern Nutrition in Health and Disease, ed. M. G. Wohl and R. S. Goodhart, pp. 1043-1064. Third edit. Philadelphia.

The Original Affluent Society

If economics is the dismal science, the study of hunting and gathering economies must be its most advanced branch. Almost universally committed to the proposition that life was hard in the paleolithic, our textbooks compete to convey a sense of impending doom, leaving one to wonder not only how hunters managed to live, but whether, after all, this was living? The specter of starvation stalks the stalker through these pages. His technical incompetence is said to enjoin continuous work just to survive, affording him neither respite nor surplus, hence not even the "leisure" to "build culture." Even so, for all his efforts, the hunter pulls the lowest grades in thermodynamics—less energy/capita/year than any other mode of production. And in treatises on economic development he is condemned to play the role of bad example: the so-called "subsistence economy."

The traditional wisdom is always refractory. One is forced to oppose it polemically, to phrase the necessary revisions dialectically: in fact, this was, when you come to examine it, the original affluent society. Paradoxical, that phrasing leads to another useful and unexpected conclusion. By the common understanding, an affluent society is one in which all the people's material wants are easily satisfied. To assert that the hunters are affluent is to deny then that the human condition is an ordained tragedy, with man the prisoner at hard labor of a perpetual disparity between his unlimited wants and his insufficient means.

For there are two possible courses to affluence. Wants may be "easily satisfied" either by producing much or desiring little. The familiar conception, the Galbraithean way, makes assumptions peculiarly appropriate to market economies: that man's wants are great, not to say infinite, whereas his means are limited, although improvable: thus, the gap between means and ends can be narrowed by industrial productivity, at least to the point that "urgent goods" become plentiful. But there is also a Zen road to affluence, departing from premises somewhat different from our own: that human material wants are finite and few, and technical means unchanging but on the whole adequate. Adopting the Zen strategy, a people can enjoy an unparalleled material plenty—with a low standard of living.

That, I think, describes the hunters. And it helps explain some of their more curious economic behavior: their "prodigality" for example—the inclination to consume at once all stocks on hand, as if they had it made. Free from market obsessions of scarcity, hunters' economic propensities may be more consistently predicated on abundance than our own. Destutt de Tracy, "fish-blooded bourgeois doctrinaire" though he might have been, at least compelled Marx's agreement on the observation that "in poor nations the people are comfortable," whereas in rich nations "they are generally poor."

This is not to deny that a preagricultural economy operates under serious constraints, but only to insist, on the evidence from modern hunters and gatherers, that a successful accomodation is usually made. After taking up the evidence, I shall return in the end to the real difficulties of hunting-gathering economy, none of which are correctly specified in current formulas of paleolithic poverty.

...

Rethinking Hunters and Gatherers

Constantly under pressure of want, and yet, by travelling, easily able to supply their wants, their lives lack neither excitement or pleasure (Smyth, 1878, vol. 1, p. 123).

Clearly, the hunting-gathering economy has to be revaluated, both as to its true accomplishments and its true limitations. The procedural fault of the received wisdom was to read from the material circumstances to the economic structure, deducing the absolute difficulty of such a life from its absolute poverty. But always the cultural design improvises dialectics on its relationship to nature. Without escaping the ecological constraints, culture would negate them, so that at once the system shows the impress of natural conditions and the originality of a social response—in their poverty, abundance.

What are the real handicaps of the hunting-gathering *praxis?* Not "low productivity of labor," if existing examples mean anything. But the economy is seriously afflicted by the *imminence of diminishing returns.* Beginning in subsistence and spreading from there to every sector, an initial success seems only to develop the probability that further efforts will yield smaller benefits. This describes the typical curve of food-getting within a particular locale. A modest number of people usually sooner than later reduce the food resources within convenient range of camp. Thereafter, they may stay on only by absorbing an increase in real costs or a decline in real returns: rise in costs if the people choose to search farther and farther afield, decline in returns if they are satisfied to live on the shorter supplies or inferior foods in easier reach. The solution, of course, is to go somewhere else. Thus the first and decisive contingency of hunting-gathering: it requires movement to maintain production on advantageous terms.

But this movement, more or less frequent in different circumstances, more or less distant, merely transposes to other spheres of production the same diminishing returns of which it is born. The manufacture of tools, clothing, utensils, or ornaments, however easily done, becomes senseless when these begin to be more of a burden than a comfort. Utility falls quickly at the margin of portability. The construction of substantial houses likewise becomes absurd if they must soon be abandoned. Hence the hunter's very ascetic conceptions of material welfare: an interest only in minimal equipment, if that; a valuation of smaller things over bigger; a disinterest in acquiring two or more of most goods; and the like. Ecological pressure assumes a rare form of concreteness when it has to be shouldered. If the gross product is trimmed down in comparison with other economies, it is not the hunter's productivity that is at fault, but his mobility.

Almost the same thing can be said of the demographic constraints of hunting-gathering. The same policy of *débarassment* is in play on the level of people, describable in similar terms and ascribable to similar causes. The terms are, cold-bloodedly: diminishing returns at the margin of portability, minimum necessary equipment, elimination of duplicates, and so forth—that is to say, infanticide, senilicide, sexual continence for the duration of the nursing period, etc., practices for which many food-collecting peoples are well known. The presumption that such devices are due to an inability to support more people is probably true—if "support" is understood in the sense of carrying them rather than feeding them. The people eliminated, as hunters sometimes sadly tell, are precisely those who cannot effectively transport themselves, who would hinder the movement of family and camp. Hunters may be obliged to handle people and goods in parallel ways, the draconic population policy an expression of the same ecology as the ascetic economy. More, these tactics of demographic restraint again form part of a larger policy for counteracting diminishing returns in subsistence. A local group becomes vulnerable to diminishing returns—so to a greater velocity of movement, or else to fission—in proportion to its size (other things equal). Insofar as the people would keep the advantage in local production, and maintain a certain physical and social stability, their Malthusian practices are just cruelly consistent. Modern hunters and gatherers, working their notably inferior environments, pass most of the year in very small groups widely spaced out. But rather than the sign of underproduction, the wages of poverty, this demographic pattern is better understood as the cost of living well.

Hunting and gathering has all the strengths of its weaknesses. Periodic movement and restraint in wealth and population are at once imperatives of the economic practice and creative adaptations, the kinds of necessities of which virtues are made. Precisely in such a framework, affluence becomes possible. Mobility and moderation put hunters' ends within range of their technical means. An undeveloped mode of production is thus rendered highly effective. The hunter's life is not as difficult as it looks from the outside. In some ways the economy reflects dire ecology, but it is also a complete inversion.

Reports on hunters and gatherers of the ethnological present— specifically on those in marginal environments—suggest a mean of three to five hours per adult worker per day in food production. Hunters keep banker's hours, notably less than modern industrial workers (unionized), who would surely settle for a 21–35 hour week. An interesting comparison is also posed by recent studies of labor costs among agriculturalists of neolithic type. For example, the average adult Hanunoo, man or woman, spends 1,200 hours per year in swidden cultivation (Conklin, 1957, p. 151); which is to say, a mean of three hours twenty minutes per day. Yet this figure does not include food gathering, animal raising, cooking and other direct subsistence efforts of these Philippine tribesmen. Comparable data are beginning to appear in reports on other primitive agriculturalists from many parts of the world. The conclusion is put conservatively when put negatively: hunters and gatherers need not work longer getting food than do primitive cultivators. Extrapolating from ethnography to prehistory, one may say as much for the neolithic as John Stuart Mill said of all labor-saving devices, that never was one invented that saved anyone a minute's labor. The neolithic saw no particular improvement over the paleolithic in the amount of time required per capita for the production of subsistence; probably, with the advent of agriculture, people had to work harder.

There is nothing either to the convention that hunters and gatherers can enjoy little leisure from tasks of sheer survival. By this, the evolutionary inadequacies of the paleolithic are customarily explained, while for the provision of leisure the neolithic is roundly congratulated. But the traditional formulas might be truer if reversed: the amount of work (per capita) increases with the evolution of culture, and the amount of leisure decreases. Hunters' subsistence labors are characteristically intermittent, a day on and a day off, and modern hunters at least tend to employ their time off in such activities as daytime sleep. In the tropical habitats occupied by many of these existing hunters, plant collecting is more reliable than hunting itself. Therefore, the women, who do the collecting, work rather more regularly than the men, and provide the greater part of the food supply. Man's work is often done. On the other hand, it is likely to be highly erratic, unpredictably required; if men lack leisure, it is then in the Enlightenment sense rather than the literal. When Condorcet attributed the hunter's unprogressive condition to want of "the leisure in which he can indulge in thought and enrich his understanding with new combinations of ideas," he also recognized that the economy was a "necessary cycle of extreme activity and total idleness." Apparently what the hunter needed was the *assured* leisure of an aristocratic *philosophe.*

Hunters and gatherers maintain a sanguine view of their economic state despite the hardships they sometimes know. It may be that they sometimes know hardships because of the sanguine views they maintain of their economic state. Perhaps their confidence only encourages prodigality to the extent the camp falls casualty to the first untoward circumstance. In alleging this is an affluent economy, therefore, I do not deny that certain hunters have moments of difficulty. Some do find it "almost inconceivable" for a man to die of hunger, or even to fail to satisfy his hunger for more than a day or two (Woodburn, 1968, p. 52). But others, especially certain very peripheral hunters spread out in small groups across an environment of extremes, are exposed periodically to the kind of inclemency that interdicts travel or access to game. They suffer—although perhaps only fractionally, the shortage affecting particular immobilized families rather than the society as a whole (cf. Gusinde, 1961, pp. 306-307).

Still, granting this vulnerability, and allowing the most poorly situated modern hunters into comparison, it would be difficult to prove that privation is distinctly characteristic of the hunter-gatherers. Food shortage is not the indicative property of this mode of production as opposed to others; it does not mark off hunters and gatherers as a class or a general evolutionary stage. Lowie asks:

> But what of the herders on a simple plane whose maintenance is periodically jeopardized by plagues—who, like some Lapp bands of the nineteenth century were obliged to fall back on fishing? What of the primitive peasants who clear and till without compensation of the soil, exhaust one plot and pass on to the next, and are threatened with famine at every drought? Are they any more in control of misfortune caused by natural conditions than the hunter-gatherer? (1938, p. 286)

Above all, what about the world today? One-third to one-half of humanity are said to go to bed hungry every night. In the Old Stone Age the fraction must have been much smaller. *This* is the era of hunger unprecedented. Now, in the time of the greatest technical power, is starvation an institution. Reverse another venerable formula: the amount of hunger increases relatively and absolutely with the evolution of culture.

This paradox is my whole point. Hunters and gatherers have by force of circumstances an objectively low standard of living. But taken as their *objective*, and given their adequate means of production, all the people's material wants usually can be easily satisfied. The evolution of economy has known, then, two contradictory movements: enriching but at the same time impoverishing, appropriating in relation to nature but expropriating in relation to man. The progressive aspect is, of course, technological. It has been celebrated in many ways: as an increase in the amount of need-serving goods and services, an increase in the amount of energy harnessed to the service of culture, an increase in productivity, an increase in division of labor, and increased freedom from environmental control. Taken in a certain sense, the last is especially useful for understanding the earliest stages of technical advance. Agriculture not only raised society above the distribution of natural food resources, it allowed neolithic communities to maintain high degrees of social order where the requirements of human existence were absent from the natural order. Enough food could be harvested in some seasons to sustain the people while no food would grow at all; the consequent stability of social life was critical for its material enlargement. Culture went on then from triumph to triumph, in a kind of progressive contravention of the biological law of the minimum, until it proved it could support human life in outer space—where even gravity and oxygen were naturally lacking.

Other men were dying of hunger in the market places of Asia. It has been an evolution of structures as well as technologies, and in that respect like the mythical road where for every step the traveller advances his destination recedes by two. The structures have been political as well as economic, of power as well as property. They developed first within societies, increasingly now between societies. No doubt these structures have been functional, necessary organizations of the technical development, but within the communities they have thus helped to enrich they would discriminate in the distribution of wealth and differentiate in the style of life. The world's most primitive people have few possessions, *but they are not poor.* Poverty is not a certain small amount of goods, nor is it just a relation between means and ends; above all it is a relation between people. Poverty is a social status. As such it is the invention of civilization. It has grown with civilization, at once as an invidious distinction between classes and more importantly as a tributary relation—that can render agrarian peasants more susceptible to natural catastrophes than any winter camp of Alaskan Eskimo.

All the preceding discussion takes the liberty of reading modern hunters historically, as an evolutionary base line. This liberty should not be lightly granted. Are marginal hunters such as the Bushmen of the Kalahari any more representative of the paleolithic condition than the Indians of California or the Northwest Coast? Perhaps not. Perhaps also Bushmen of the Kalahari are not even representative of marginal hunters. The great majority of surviving hunter-gatherers lead a life curiously decapitated and extremely lazy by comparison with the other few. The other few are very different. The Murngin, for example: "The first impression that any stranger must receive in a fully functioning group in Eastern Arnhem Land is of industry. . .

And he must be impressed with the fact that with the exception of very young children . . . there is no idleness" (Thomson, 1949a, pp. 33-34). There is nothing to indicate that the problems of livelihood are more difficult for these people than for other hunters (cf. Thomson, 1949b). The incentives of their unusual industry lie elsewhere: in "an elaborate and exacting ceremonial life," specifically in an elaborate ceremonial exchange cycle that bestows prestige on craftsmanship and trade (Thomson, 1949a, pp. 26, 28, 34 f, 87 passim). Most other hunters have no such concerns. Their existence is comparatively colorless, fixed singularly on eating with gusto and digesting at leisure. The cultural orientation is not Dionysian or Apollonian, but "gastric," as Julian Steward said of the Shoshoni. Then again it may be Dionysian, that is, Bacchanalian: "Eating among the Savages is like drinking among the drunkards of Europe. Those dry and ever-thirsty souls would willingly end their lives in a tub of malmsey, and the Savages in a pot full of meat; those over there talk only of drinking, and these here only of eating" (LeJeune, 1897, p. 249).

It is as if the superstructures of these societies had been eroded, leaving only the bare subsistence rock, and since production itself is readily accomplished, the people have plenty of time to perch there and talk about it. I must raise the possibility that the ethnography of hunters and gatherers is largely a record of incomplete cultures. Fragile cycles of ritual and exchange may have disappeared without trace, lost in the earliest stages of colonialism, when the intergroup relations they mediated were attacked and confounded. If so, the "original" affluent society will have to be rethought again for its originality, and the evolutionary schemes once more revised. Still this much history can always be rescued from existing hunters: the "economic problem" is easily solvable by paleolithic techniques. But then, it was not until culture neared the height of its material achievements that it erected a shrine to the Unattainable: *Infinite Needs*.

Levels of Analysis in Amazonian Research: Theory, Method, and Policy Implications

11

The preceding chapters have raised a number of recurrent dilemmas present in Amazonian studies: is the Amazon fragile or not? Are its soils fertile or sterile? Can societies in the region advance beyond the level of tropical forest cultures without destroying the habitat? Was Transamazon Highway colonization a failure or a success? What will the future hold for the region? These questions are not simply provocative speculations. Despite a not insignificant volume of research on each of these topics, there is evidence to support positions on both sides of these questions. After examining the available research, it is my contention that some of the heated debates about the Amazon's habitability and potential reflect a tendency to generalize about processes in one level of analysis from data and research carried out at another level. In this chapter I explore the implications of this tendency for understanding human ecological interactions in the Amazon Basin generally, and Transamazon colonization in particular.

The great bulk of Amazonian research, whether anthropological, agronomic, or ecological, has been highly site specific. There is an absence of systematic coverage of habitat types, of the impact of various types of technologies per habitat type, and of representative aggregate data for major social, ecological, and economic indicators. Perhaps the most systematic sampling carried out in the Amazon has been by agronomists (Sombroeck 1966; Falesi 1972; IPEAN 1974). But even these surveys restricted themselves to sites within close proximity of the Amazon River and the Transamazon Highway respectively, thereby leaving out many habitat types. The heated debate in anthropology over the availability of animal protein in the Amazon and its social and cultural consequences is based chiefly on data from three sites (Chagnon 1968; Holmberg 1969; Siskind 1973), none of which had hunting as the main focus of their research design. As evidence has accumulated in this past decade contrary views have emerged, but all studies, again, have been site specific and do not permit aggregation, since systematic sampling of habitat types has not yet taken place. Despite the lack of systematic sampling applied to soils of the Amazon and to animal biomass productivity per habitat type, there has been no lack of theorizing about the role of proteins in tropical forest societies or about the consequences of poor soils on aboriginal societies or others to follow. Are these debates based on an appropriate sampling procedure capable of addressing the macro-level questions that were posed at the beginning of this chapter?

Much of the available data is simply not comparable but, rather, distinct to particular habitats and levels of technology. Whereas it is normal to seek to understand one level in terms of the other, such a task may not be appropriate. The question before us is, can site-specific studies (micro-level) be the basis of region-wide statements and analyses (macro-level)? A moment's reflection will tell us that extrapolations from one level to another cannot work. In sociology this sliding between levels is known as the "ecological fallacy," wherein statements about individuals are derived from aggregate data (Robinson 1950). It is generally understood that micro- and macro-levels of analysis have distinct goals and answer different questions. Economics long ago recognized this fact formally by distinguishing between macro- and micro-economics. Macro-level studies rely on aggregate data from a broad and representative sample of the universe in question. Micro-level studies rely on careful observation of individuals in a population in order to understand the internal dynamics of the population.

Each level possesses a specificity that is remarkably unique to it, and each level permits the investigator to focus on different types of questions. Questions concerned with *how* certain processes take place are best handled by micro-level studies. Questions which, on the other hand, ask *why* things happen in a given manner call for consideration of external relations and structural constraints. How people are organized into production units calls for site-specific methods. Why they are organized into individual households rather than communal ones for producing goods can only be tackled by a comparative approach heavily reliant on historical, economic, and political analyses of a large and representative sample of data. The macro-study is not only bigger than the sum of the micro-studies, it is structurally different.

Geographers have been particularly aware of the scale problem in reference to trying to comprehend a large region while only studying small areas within it. McCarthy et al. (1956:16) noted that "every change in scale will bring about the statement of a new problem and there is no basis for assuming that associations existing at one scale will also exist at another." The caveats of the 1950s have given way to calls for integration of the macro- and micro-levels of analysis (Beer 1968; Dogan and Rokkam 1969), but the dilemma of bringing together what are different processes remains unresolved.

This concern is part of the larger question of how scientists can delimit their field of study, and what the implications of setting those particular bounds are to the relevance or completeness of the analysis. That the bounds that one sets to one's investigation define the scope and relevance of one's conclusions has long been recognized. What is less frequently noted is that the conclusions at each of these levels are distinct and, at the same time, each one is relevant to a complete understanding of human behavior. The results of a community study may not be generalized to a whole society, but the internal structure of a community is relevant to understanding how a community is affected by larger external forces (Epstein 1964:102). At a level of analysis broader than the community it might appear that the larger external forces shape the life of local communities in relatively homogeneous ways (Watson 1964:155; Blok 1974; Schneider and Schneider 1976). The focus on the community, on the other hand, shows individuals responding positively to actually change the external forces themselves (Bennett 1967), so that communities also affect external forces. Only limited explicit attention has been given to this issue of level shifting (Hagget 1965).

A system under study must be bounded. The thorny issue of how one area of social life within a larger whole can be isolated in order to permit systematic study has been discussed at great length in a volume edited by Gluckman (1964). The authors in that volume concluded that the chief concern was how to bound the system without omitting anything relevant or important. They concluded that a social anthropologist is justified, and can be most productive, if he/she restricts the scope to what they call micro-sociology and leaves aggregate phenomena to macro-sociologists, economists, and political scientists. While such an allocation of research interests has been productive for all the disciplines involved, it has not facilitated study of how local and larger social units dynamically relate to each other. I believe that by thinking about levels of analysis along disciplinary lines it has been possible to ignore some important questions: What is the relationship between types of data and a given level of analysis? and What does one learn about human adaptation from each level of analysis?

It has not been sufficiently recognized that each level's scope obscures relationships observable at other levels, particularly as one moves from local research to regional or national levels (Devons and Gluckman 1964: 211). The greater the scope of the level, the less are details of group and individual behavior and ideology analytically recognized. There are, for example, significant differences in the mean demographic behavior of small and large populations because of greater variation in the former. The scale of the researched area also affects the analytic results in significant ways. The Amazon is a perfect case in point. Confusion over the area in question is common. Some writers, as we have here, use the term to refer to the Amazon Basin, the drainage area of the Amazon River and its tributaries. Others use the legal definition used in Brazil for development planning (*Amazonia legal*), which includes large areas of savannas in the central plateau. Others define it by rainfall parameters. Since population distribution is usually uneven, small areal units have a wider range and variance of population distribution than larger units such as countries or large regions like the Amazon. Likewise, since mobility occurs mainly over short distances, small areal units are critical to microdemographics but appear as insignificant in aggregated data as compared to natural changes in population (J. Clarke 1976). Thus the relative significance of migration and natural change depends more on the size of the area studied (level of analysis) than on real demographic differences. The relationship of the particular to the general is a scale-linkage problem that remains incompletely resolved (cf. Hagget 1967).

This lack of recognition is evident in the frequency with which analysts use micro-level data to address regional or national issues and the even more frequent use of aggregate data to design programs to be implemented at the local level. What has been missing from both research and policymaking is the conscious acknowledgement that each level addresses very different kinds of processes and that generalizations should confine themselves to the level in which the research took place.

Ellen considers the issue of "analytical closure or boundedness" as central to both social and ecological studies. He argues that the most important consideration in bounding a system at either the local or the regional level is whether or not the system is able to reproduce itself within that unit (1979:10). The elegant notion of "system" or "ecosystem" that has gained popularity since the 1960s tends to cloud over the difficulty of bounding an analytical unit. Anyone familiar with the ecosystem concept will readily recognize that the ecosystem is a flexible unit which is bounded by the needs of the researcher. It seldom, if ever, has characteristics which facilitate replicability, nor is it obvious on reading a study what the breadth of the study was—Was it an intensive study of a household, a village, a valley, a geoeconomic region, a nation, or a biome? The positive uses of the ecosystem concept are many and I, in no way, wish to denigrate the many ways that this concept serves to emphasize the interconnectedness of the living and nonliving components of the biosphere. What I wish to point out, however, is that reference to the ecosystem as one's unit of analysis has produced results that address important human ecological relations within a single level—not to the whole of the human adjustments.

Geertz was the first anthropologist to argue for the usefulness of the ecosystem as a unit of analysis in social cultural anthropology. In *Agricultural Involution* (1963) he tested Steward's emphasis on subsistence and found it wanting. He showed by a broad use of historical records that

Indonesia's agricultural patterns could be understood only in terms of the economic restrictions of the Dutch colonial authorities. In fact, Geertz used the *region* of Indonesia as his ecosystem unit. He identified two contrasting agricultural systems and discovered their roots in the historical development of Indonesia's colonial economy.

In contrast, Rappaport's study of ritual and ecology in the New Guinea Highlands (1968) defined the ecosystem unit in terms of the material exchanges of a *local population*. This difference is not surprising given the flexibility of the ecosystem concept in biology. But how can one compare studies which deal with ecosystemic interactions at different levels? In the last chapter of *Pigs for the Ancestors* (1968) Rappaport acknowledged that a local population engages in material and nonmaterial exchanges with *other local populations* which, in the aggregate, can be called "regional populations." These, he suggests, are likely to be more appropriate units of analysis for long-range evolutionary studies, given the ephemeral quality of local populations (1968:226). Unfortunately, other anthropologists did not follow up on this insight into the differences between relatively synchronic micro-level studies and diachronic macro-level approaches.

From study of the varied uses of the ecosystem unit by Rappaport and Geertz, and from the examination of debates surrounding the human occupation of the Amazon Basin, I began to pursue the possible association between levels of analysis and major points of disagreement in Amazonian studies. Anthropologists have long used local communities as their fundamental units of study wherein a cultural or ethnographic method could be applied (Steward 1950:21). Most scholars have been quite aware that individual communities are part of larger wholes, but such functional interdependencies have seldom become a part of the analysis itself.

It has been common in ecological anthropology to carry out studies in local communities in order to facilitate the quantification of variables or to study populations before the full impact of the modern world reaches them (Rappaport 1968; Nietschmann 1973; Waddell 1972; Moran 1975; Baker and Little 1976). While all these researchers recognize the value of addressing larger population and ecological units, they have chosen to limit the scope of the investigation for the sake of precise data gathering. In such studies, the community is seen as a closed system for the purposes of analysis. While such a device has undoubtedly provided great insight into internal functional patterns, it has tended to neglect the relations of the community to the outside world and the impact of the outside world upon the community's structure and function. Localized studies provide insight into family structure, subsistence strategies, labor inputs, health and nutritional status, flow of energy, socialization, and cultural institutions. Studies at this level cannot address issues of social evolution, explain changes in the economic structure of society, patterns of economic development, or political economy. These issues can be addressed only with a different type of research method which emphasizes historical, geographical, economic, and political change over time.

Ecological anthropological studies will always require investigation of the way local populations interact with other components in the local ecosystem. The case for an emphasis on micro-level study in ecology has been convincingly argued by Brookfield (1970:20), who pointed out that an adaptive system can best be studied at this level because such a system model "acquires the closest orthomorphism with empirical fact." Nevertheless, regional analyses add a very different and much needed insight into the processes of human adaptation. A regional study emphasizes historical and economic factors and considers many local-level phenomena as secondary to the historical forces at play (cf. Braudel 1973; Smith, Thomas 1959; Bloch 1966). It is notable that the regional analyses of both Geertz (1963) and Bennett (1969) predominantly use historical factors in explanation. Bennett defined the region of the North American Northern Plains in terms of its historical unity (1969:26). He was able to explain the adaptive strategies of four distinct ethnic groups in terms of differential access to resources, differential access to power loci, and social/cultural differences. Thus, while he was able to flesh out the social/cultural details of the population through local interviews and by studying ethnic interactions, an understanding of the operative historical and economic forces required aggregate data from social and economic history. Such a choice of level can come only from a recognition of the appropriate level at which certain questions can be addressed. Julian Steward's difficulties in achieving his goals in his Puerto Rico study can be traced to a failure to shift from his micro-level analysis of a group's "culture core" to macro-level analysis of the Puerto Rican political economy. Steward's earlier micro-level studies had successfully generated sophisticated analyses of the internal structure of patrilineal bands and their articulation with selected habitat features. The Puerto Rico study, however, needed to move beyond the study of specific human/habitat interactions and move towards a dynamic model of structural transformations and relate the articulation of Puerto Rican communities to the World System through which many community features could be understood through time.

The relevance of the above considerations about levels of analysis and the scale of sampling appropriate to answering given questions helps shed light on some of the major questions asked about human occupation of the Amazon. An examination of the levels at which generalizations have been made will serve both to suggest the articulation between levels and to identify some of the major gaps in our current knowledge about the area. Of special interest will be the results from studies of soils, use of other environmental resources, healthfulness of the region, provision of agricultural inputs to colonization, and the analysis of the constraints to development.

Levels of Analysis and Amazonian Soils

The problem of scale emerges at the outset as one of the fundamental problems in the ability to differentiate between Amazonian soils. Most maps available are at a scale of from 1:100,000 to 1:500,000. These macro-scale maps show the soils of the Amazon to be primarily oxisols (latosols), with a small area of inceptisols (alluvial soils) along the flood plain (National Academy of Science 1972). If a connection is attempted between these soils and the human use of them, as is often the case, discussions will emphasize that these soils are problematic; their utilization is restricted to long fallow swidden agriculture with shifting of fields every two to three years due to rapid declines in fertility caused by the loss of the limited nutrients made available after the burn. A great deal of the prevalent pessimism about the potential of the Amazon as a settlement area is based on this level of analysis.

But how representative is the available data? Does it provide an adequate enough representative sample of the whole basin to permit macro-level generalizations about the region's potential? Reliance, until recently, on the simple dichotomy between the flood plain (2 percent) and the terra firme (98 percent) is at a level of generality not likely to generate systematic scaling of the Amazonian regional system, and suggests that the two areas are more similar than is likely to be the case. Given the difficulties of approaching the problem of local and regional variation, some researchers have suggested the use of "nested sampling" (Hagget 1967:175). Such an approach serves to sample a region in a systematic manner by sampling a number of units at different scale levels: region, sub-region, district, sub-district, and localities. Studies done elsewhere have noted that variability increases with movement down to the micro-scale units in the sampling process. For the Amazon there is no available soil mapping at a scale that permits observation of specific soils, except for a few isolated localities (Furley 1979; Ranzani 1978). Is the absence of such detailed micro-scaled maps critical?

When one changes the scale from the Amazon as a whole to specific sub-regions, the homogeneity evident at the regional level yields to increased variability. Instead of two soil types, three to five become common. Not only is there increased detail in visible soil types, but even the areal extent of soil types may be misjudged (Ranzani 1978). A technologically sophisticated aerial survey of the Amazon using sideways-looking radar (RADAM 1974) at a scale of 1:100,000 observed that the dominant soil type in the sub-region of Marabá was that of ultisols. A localized study by Ranzani in Marabá (1978), at a scale of 1:10,000, concluded that oxisols constituted 65 percent, entisols 22 percent, and ultisols *only* 13 percent of the soils in the area in question. Scale is important when variability is present.

Whereas maps at a scale of 1:100,000 to 1:500,000 may be useful in matters of geologic history and geomorphology and answering general questions about the relationship between soils and biotic productivity, speciation, and climate, they are of little use in addressing questions about the human use of resources, the social organization and structure of human communities, and their adaptive strategies. The general unavailability of micro-scale mapping has been a major obstacle to the study of human ecological processes in the Amazon and may explain the use of macro-scale maps by investigators. Nevertheless, the difficulties resulting from generalizations based on macro-scale maps need to be recognized.

The problem lies in the different amounts of information that can be included in a soil map. Maps in orders of 1:20,000 and up are not useful for land management. A planner or a researcher using such a map will assume that all the soils labeled with a particular name will have the same characteristics, as does a "typical profile." Yet, soils within the same order, say oxisols, and within close proximity to each other, may still differ significantly in effective cation exchange capacity, in exchangeable aluminum, in pH, in water retention capacity, and in nutrient levels in different horizons (Buol and Couto 1978:71). A macro-scale map might suggest that soil users can move from place to place with a uniform land management approach and can expect similar results everywhere. This has been, in fact, a dominant viewpoint in anthropological writings about the Amazon tropical rain forest peoples. Given the lack of micro-scale studies of a sufficient number of areas by systematic sampling, it has been easy for investigators to dismiss variations as "non-representative" and to accept the macro-scale data. Such a decision is incorrect from the point of view of geographical sampling (Duncan, Cuzzort, and Duncan 1961) and its analytical implications.

From the point of view of policy, reliance on macro-scale maps has had serious consequences. The decision to focus colonization along the Transamazon Highway in Altamira was based on the identification of medium to high fertility alfisols, which appeared to dominate in the region cut by the road. Colonists were placed on all available lots as they arrived, since soils appeared homogeneously good, and the same crops were promoted. It was not until the colonists were settled on their land that micro-level soil sampling was carried out by Moran (1975a), Smith (1976a), and Fearnside (1978) in the Altamira area, and by Ranzani (1978) and Smith (1976a) in the Marabá region. These investigators discovered that the soils of the area are a patchwork, with radical differences in nearly every kilometer. Thus, the soils of Altamira were highly variable, with the medium to high fertility alfisols making up only 8 percent of the total soils and being scattered in small patches.

The recognition that land use planning can only take place at the micro-level could have led to a different pattern of occupation, a less homogeneous effort at agricultural extension, reduced likelihood of loan defaults due merely to location on soils too poor for farming, and reduction of many other problems that affected the performance of farmers in the Transamazon and which were discussed earlier in this book. Macro-level land planning led to the promotion of three cereal crops throughout the region and later to cattle or plantation crops.

Not only was planning affected by the inappropriate macro-perspective to land planning, but so was the evaluation of farmer performance. Whereas some farmers familiar with Amazonian micro-variability refused to go along with the practices promoted by the government and obtained good yields from their diversified agricultural operations, the use of aggregate production data, rather than farm management surveys, hid the differential performance of farmers and led to a reduction of support to the whole small-farming population. Instead of identifying the strategies that worked, as a good farm management survey would have done, the aggregate analysis provided no details about what management practices worked, but only stated that the output did not meet expectations set before the project began. At the evaluation stage, the agencies involved used inappropriate quantitative tools to measure farmers' productivity, to identify limiting factors, and to correct actions. All that the aggregate analysis measured was the output of the sector in response to inputs provided. The analysis did not show, *and could not show*, that the inputs were not timed to the needs of farmers, that institutional performance was a constraint in itself, and that the technological inputs were in part responsible for the low yields in two out of the three years measured. The micro-level analysis of Vila Roxa farming carried out earlier in this book brought out the variability in farmers' responses to available soils and natural resources. I have shown here and elsewhere (Moran 1979b) that the caboclo population had precise knowledge of forest resources and soil types, and that they got better results than did farmers following practices promoted by the planners. Their use of the region's resources is more complete, more rational, and more efficient than that of outsiders.

Levels of Analysis and the Amazon's Healthfulness

Most ecological studies which address problems of populations in the humid tropics inevitably cite poor health conditions and health status as serious constraints on the well-being of these populations. Indeed, a number of potential health problems are present in the tropics, and without adequate control they can become serious obstacles to human adaptability. However, my micro-level analysis has shown that farmers lost fewer working days due to illness than they lost in the process of obtaining credit and other dealings with government institutions. In fact, they lost no more days due to illness (11.5 days) than the standard number of "sick days" currently allotted to American (U.S.) state government employees (12 days per year).

Several of the most common health problems in Vila Roxa and elsewhere in the Transamazon resulted from the low economic status of the population, and are traditionally reduced with an increase in purchasing power. The most severe problems required some modification in cultural understanding: better knowledge of how vegetation falls when cut (trauma) and better knowledge of how the malarial plasmodium behaves in the body (to reduce the incidence of insufficient dosages of medication). Health problems, then, were not a significant deterrent to productivity, although they can be if the income of the population declines and nutritional intake is reduced. The difficulties in farm productivity, as we have seen, are in no small part a result of the mismatch between macro-level policy and evaluation and the micro-level realities of farm households managing very diverse production factors, which are unrecognized by those providing the inputs.

Levels of Analysis and the Provision of Inputs

Macro-level analyses had often cited that in the Amazon and elsewhere the lack of sufficient credit was a major obstacle to increases in food production. In the Amazon Basin, analysts have added that the archaic system of aviamento, wherein traders controlled the supply of goods to the Amazon interior and long-term credit was extended at exorbitant rates, was fundamental to the permanence of the region in a state of underdevelopment. The macro-level solution to this situation was to provide credit at favorable rates through the normal channels of the Bank of Brazil. In making such a policy decision, planners failed to take into consideration: (a) the structure of social relations, (b) the costs of monitoring the credit worthiness of a population in a rain forest region, (c) the differential experience of farmers with bank credit, and (d) the traditional forms of allocating cash inputs—all micro-level constraints on the use of capital resources. Nor did the planning process allow for the imperfections in the administration of inputs by government agencies at the local level. Credit institutions were unable to release funds in accordance with the agricultural schedule of the specific areas in question. Extension agents were unable to monitor the progress of farm work, yet continued to require elaborate procedures designed to monitor credit worthiness. The aggregated data showed a high rate of credit default; however, the reasons for such defaults could only come from farm management surveys—micro-level studies at the level of the individual farm, which were not part of the monitoring process in the colonization scheme.

Micro-level analysis of the cost of credit has shown earlier that costs have been unreasonably high. Cost of loans was not the 7 percent per annum face cost, but was 50 percent due to lost labor time in obtaining release of funds. Moreover, lack of previous experience in using agricultural credit led to misallocation. Clientilistic farmers tended to consume their loans rather than apply them to the intensive use of limited areas, as did brokers. Micro-level analyses might have led to the creation of education programs geared at improved management of capital resources.

The credit structure created reflected stereotypes about "modernization" and the realities of the Brazilian economic system. Manioc and other root crops were associated with the Northeast and the Amazon Basin, both viewed in Brasilia as economically-backward areas. In order to modernize, planners saw the change of crops as a vehicle to socio-economic change. In addition, the growing urban populations of Brazil consume large amounts of rice, corn, and beans. By giving credit for the cultivation of these three crops they expected that they would not only be modernizing Amazonian agriculture, but also be providing a product with high demand in national markets. Although such a view reflects considerable naiveté about how to get agriculture moving, it does not really tell us why the credit schemes failed.

The problem can also be understood if one analyzes the influence of external bureaucratic factors within the micro-context. Before the creation of a small farm sector in the Amazon, the banks had preferred to give large loans to ranchers. The shift in policy in 1970 was forced upon the banks from above. This formal change in policy, however, could not legislate the informal behavior of credit personnel. Bank managers did not treat all farmers equally, as the presidential fiat might suggest and as planning might have it. Rather, they gave better credit deals to farmers capable of interactive behavior which they recognized as equal to theirs (e.g., entrepreneurial farmers, "having a future"), and those who had cattle ranching as an objective. The administrative costs of establishing a few cattle enterprises is less for the bank and the government than are the costs of providing numerous small loans (Katzman 1976). Schemes to produce manioc flour on a commercial scale did not receive much attention, despite the scarcity of manioc flour in the region and evidence in local markets that its price per kilo was four times that of rice.

Credit institutions were also not attuned to the agronomic constraints to cereal production in some parts of the Amazon Basin. While cereals can be grown in the Amazon Basin, they are more susceptible to pests and diseases, and require soils of higher initial fertility than do root crops.

Though cereals could be grown well on alfisols, they would not fare so well on oxisols and ultisols. They also usually require substantial fertilizer inputs after the first or second year of cultivation, but levels of fertilizer input for tropical soils have as yet to be worked out per crop and per soil type (Tropical Soils Research Program 1976:137).

The aggregated data from the banks and the government food wholesaler, CIBRAZEM, showed that low amounts of rice, corn, and beans had been marketed by Transamazon farmers. These low production levels were attributed to the low level of technology in use and the lack of entrepreneurial spirit among the farmers. It was a case, as Wood and Schmink (1979) have reminded us, of blaming the victim. The negative evaluation of farmer performance comes as no surprise when one considers the inappropriateness of the data. Input/output data for a sector created a mere three years before could have hardly yielded results capable of explaining poor performance. Only micro-level farm management investigations could have gotten to those factors.

Macro- and Micro-Approaches to Development

Cardoso and Müller (1977), Bunker (1978:23), and Wood and Schmink (1979) attribute the shift in policy from small farming to large-scale development in 1974 to contradictions in the Brazilian authoritarian system. On the one hand, the regime seeks legitimation through the execution of socially-beneficial schemes, and on the other hand, yields to pressures from the dominant capitalist sector, which supports its continued existence. These internal contradictions reflect the ambivalence of a regime seeking permanence but aware that its benefits are inequitably distributed. These contradictions, in turn, negate possible benefits that might be derived from the socially-oriented projects undertaken. The lack of institutional cooperation, the high cost of access to government services, and the imposition of unfamiliar regulations and inappropriate crop priorities have directly contributed to the low levels of performance by farmers, which has justified the reduction of incentives to the small farm sector. This conclusion is entirely justified.

However, such an analysis addresses only macro-level structural constraints and neglects other considerations of, at least, comparable significance in understanding the process. While I will not deny the structural contradictions in the process of Transamazon colonization, I have shown in this book that nearly 40 percent of the small farmers in a micro-level study had achieved broker or rural manager status and respectable levels of income and productivity; had assumed religious and economic leadership in their communities; and are expanding their landholdings, cattle herds, and productive capacities. Emphasis on the exploitation of rural labor by external capitalism homogenizes the colonist population into an undifferentiated peasantry. In fact, some of these same farmers became, not just tools of capitalism, but capitalists themselves.

Emphasis on the macro-level would deny that 40 percent of the Vila Roxa population their real accomplishments in the face of the multilevel constraints posed by institutions, interest groups in the authoritarian state, and the complex ecological variability from place to place in the Amazon. It is often forgotten that most farmers are still on their land (80 percent), that new ones arrive each day from throughout Brazil to claim tracts, and that levels of production have increased more often than they have decreased.

It is all too common when discussing development or human adaptation from a macro-perspective to forget that at that level it is the population which "develops" or "adapts," not the individuals. While it may be true that the human species has shown remarkable adaptability to the varied environments of the earth, this is not equivalent to saying that local populations or individuals show comparable adaptability. Indeed, the concept of adaptation should always specify "adaptive for whom?" since what is adaptive to an individual may spell ruin for the group and yet benefit the biotic environment due to reduced pressure (Ellen 1979). Rappaport, in the final chapter of *Pigs for the Ancestors* (1968), noted that local populations were highly ephemeral, which is not to say that the species is maladaptive. I have shown elsewhere (Moran 1975) that local populations hold on to familiar strategies of resource use even in radically different environments, sometimes with disastrous consequences (cf. Laughlin and Brady 1978). New ecological and economic structures allow some individuals to develop appropriate strategies, while others will continue to rely on familiar strategies that satisfy minimum survival requirements. Some individuals and groups in a local population will adapt more than others to changing environments. While no one has yet dared to quantify such a proposition, I dare say that there are probably more failures than successes in the process of human adaptation to new environmental settings. While such a prospect appears gloomy, it also represents the necessary process of adaptation—in which individuals possessing a variable set of qualities and resources are differentially capable of responding to changes in their environment. The alternative to this process would be to have a homogeneous population whose inventory of strategies might lack solutions to as yet unforeseen circumstances.

Human adaptation and social differentiation do not occur in a vacuum. This process of social reproduction reflects the adaptation of the Transamazon population to local habitat (micro), to the economic and structural realities of the Brazilian nation, and to the ability to function within the social field provided by a colonization setting and a rain forest environment (macro). About 40 percent of the population of Vila Roxa succeeded by making these and other appropriate adjustments to both micro- and macro-level constraints. This 40 percent has differentiated itself from the other 60 percent by assuming the status of rural managers, a status that was neither inherited nor achieved by inhumane exploitation of the 60 percent majority. In fact, it was the 60 percent which began to refer to the other 40 percent as "patron" and to reestablish the hierarchical order to which they were adapted in dependency-breeding plantations. That as many as 40 percent of the population have been able to adjust despite relatively lowly beginnings is a remarkable achievement for any population. For instance, only about 18 percent were able to achieve comparable control over their economic activities in Marabá (Velho 1976: 208). These differences may be due to ecological constraints, greater structural conflicts between colonists and large-scale developers, and differences in the make-up of the population itself (Velho 1972; Smith 1976a; Gall 1978; Fearnside 1978; Wood and Schmink 1979).

The development of the Amazon is constrained by a centralized approach to planning and implementation. This approach did not begin in 1964, but has deep historical roots (Roett 1972). The priorities set for the development of the Amazon were aggregate in character: to improve Brazil's foreign exchange balance, to promote national integration, and to reduce social tensions in the Northeast. In none of the planning documents is there mention of the project in terms of micro-level processes. The performance of Brazil's agricultural sector has kept up with the rapid increase in population and its demand for food only through the opening up of new lands (Schuh 1970), not through improved performance of the sector. To some degree this is the result of the economics of frontier agriculture vis-à-vis post-frontier intensification (Margolis 1979). It is also a result of the patrimonial or authoritarian structure of Brazil's bureaucracy, which has been incapable of using anything other than aggregate inputs for policy-making.

The choice made in 1974 to turn to large-scale development took place because of the structure of the Brazilian bureaucracy and the aggregate inputs that serve to formulate its policies. Hirschman has pointed out that planners tend to be biased against programs that involve technological uncertainties and prefer to avoid projects that involve dealing with large numbers of people (1967:39–44). The problem is, thus, general to all bureaucratic structures. In Brazil's case that structure is also remarkably centralized (i.e., authoritarian), which makes the structure of decision-making even less amenable to inputs from micro-level studies. The more centralized the structure of decision-making, the less able it is to process complex information incorporating the variability present in any areally extensive system. The result is a structure of decision-making which is insensitive to micro-level variability and has a tendency to homogenization of both environmental and social variables. Economists have noted that the result of economic policies in Brazil since 1964 has been to increase the gap between income groups, to a more evident distinction between haves and have-nots. Ecologists have also noted a tendency to treat the Amazon as a forest which can be cleared with equivalent results anywhere. The question which is central to the future of the Amazon region is whether or not the structure of the Brazilian bureaucracy is capable of adjusting its policies to include inputs from specific sites to optimize productivity and conservation per site. The implications of such a structural change will be explored in a future publication (Moran n.d.).

Articulating Levels of Analysis

The analytical articulation of micro- with macro-levels of analysis is still in a conceptual and methodological developmental stage. Responsible for the current state of ambiguity are the disciplinary lines followed by most investigators and a tendency to work within a given level, to the exclusion of the other. Nevertheless, scientists are often expected to address questions and make generalizations that reach beyond the bounds of the level at which they work. It is in overextending the natural limits of a given level that problems have arisen in the interpretation of processes such as the potential of the Amazon, the productivity and management of the soils, the obstacles to colonization, and the balance between conservation and productivity.

For the purposes of field research, it is seldom practical to try to investigate more than one level. But as the debates that have been reviewed earlier would suggest, to mix levels leads to unproductive debates, which obscure the complex processes studied. Levels are hierarchically structured and exhibit both vertical and horizontal interactions. Most research has focused upon horizontal interactions within a given level. Vertical hierarchical organization has received much less attention and remains problematic both theoretically and methodologically.

Minimally, the first requirement to overcome the current dilemma is the recognition of the distinctiveness of levels of analysis. Once they are recognized, the differences between levels pose little difficulty for specific empirical studies with limited and clear objectives. A study of a sample of groups in a rural community makes it possible to generalize about group structure, although not about the structure of peasantries. To arrive at the latter type of generalization requires systematic comparisons of a fair number of different groups in different rural areas—ideally, a representative sample of them. In short, the research design must be adapted to the level of organization to be explained, and explanations must be confined to that level. Each explanation nests within the other level and operates within the general constraints set by the other level. Thus, while each hypothesis is restricted to a particular level of analysis, structural and functional aspects of ecosystems are affected by processes at other levels. Each hierarchical level adds a layer to our understanding of the total human adaptational situation. Levels, while distinct, are continuous and not dichotomous (Goldscheider 1971:39–44). What this distinction implies is that we must confront this methodological and sampling problem in the hopes of clarifying explanations of the human adaptive process. While we should all aspire to integrate macro- and micro-level explanations, this integration cannot be achieved by mixing levels between the data-gathering and the interpretational stages. Synthesis can only result from preliminary separation of micro- and macro-analyses. Only after the level-specific processes have been interpreted can we hope for a reintegration of the levels.

Amazonian studies stand to benefit a great deal from the clear separation between levels of analysis. The micro-level adjustments studied by ecologically-oriented social scientists need to be clarified in light of the modifications of these adjustments resulting from macro-level processes, such as the capitalization of the countryside, the creation of dependence on formal organizations, and the extraction of surplus to fuel the Brazilian and international urban-industrial complex. Most anthropological studies see the impact of these macro-level processes upon local communities as a serious loss. Indeed, viewing local communities as affected by and affecting regional, national, and international processes changes many of the considerations that individuals must evaluate in making decisions, and the potential for individual loss is probably increased. It is highly doubtful that local communities, even in "isolated areas of the Amazon," have ever been truly "closed" and therefore naturally bounded as units of analysis. The bounding is a product of the types of questions being asked by the investigator. Amazonian community studies have yielded useful analyses about social organization, cultural values, and subsistence systems in specific human communities. Less successful have been efforts to connect these micro-level studies to regional-level questions. Regional-level approaches have successfully addressed questions about the conflict between the goals of national and international entities and local populations, between production and conservation, and between development at a regional scale and decreases in living standards by individuals.

Amazonian Potential

The next decade will further clarify the question of the potential of the Amazon. There is increased recognition that the region is very heterogeneous and probably varies a great deal in fragility and/or resiliency from place to place. We already know that white-sand/black-water river watersheds are particularly susceptible to desertification. The soils are extremely variable throughout the basin and demand site-specific strategies of utilization. Clearly, areas with low initial soil fertility should be protected from predatory forms of exploitation, and reserves should be created to prevent a breakdown in the closed nutrient cycle of the forest. Macro-level approaches, whether through region-wide colonization or the promotion of development poles (e.g., Polamazonia), are too macro- to permit the development of the necessary site-specific strategies of resource use. What is needed is a nested approach to resource use, as discussed earlier: building up systematic sampling of sub-regions, districts, and localities so that information flows from specific sites through each level of the hierarchy in order to permit adjustment to variation. Continued study of human communities under a variety of conditions and levels of integration to the regional, national, and international community is needed to better understand the balance between social development and environmental conservation in the Amazon.

The current state of knowledge is too fragmentary and the forms of exploitation too insensitive to micro-level variation to assert that the limiting factor to Amazonian development is everywhere environmental, rather than structural and resulting from inappropriate macro-level policies. The success of Transamazon colonization was mixed. Using government projections as a standard against which to measure success, the colonization proved a failure. Such projections, however, were optimistic in the extreme and failed to take into consideration the common obstacles faced by a population adjusting to an unfamiliar habitat, the difficulty of delivering inputs according to a site-specific agricultural schedule, and the insensitivity of the planning policies in addressing local needs. At the local level it is possible to see numerous success stories of individuals who, with limited means, achieved remarkable results, and who provide insight into the strategies that may work in creating productive agricultural systems.

Predicting the future of the Amazon does not seem a profitable exercise. The danger of continued deforestation without adequate knowledge of its consequences to specific areas is real and worrisome. On the other hand, the pace of research has quickened and useful results have in some cases been incorporated into conservation plans. The continued decimation of aboriginal peoples is a loss of major consequence, but efforts to guarantee them rights to land will probably meet with increased success as the international community pressures Brazil into obeying its own Constitution. The Amazon will continue to be exploited, sometimes poorly, sometimes disastrously, and in time, perhaps, wisely. The monolithic developmental strategies used until now may give way to the needed flexible strategies capable of incorporating diversity of local conditions into the execution of projects designed to exploit the land, provide jobs and income, and achieve the multiple goals pursued by citizens and nation alike. These multiple goals will not be achieved easily. They can be made a part of policy only when there is change in the bureaucratic structures around which Brazil is organized as a nation, when micro-level information can be processed by those same structures, and when decisions reflect the existence of many Amazons, not just of a homogeneous one.

2   Approaches to the study of land degradation

Piers Blaikie and Harold Brookfield

1   Chains of explanation

We have described our approach to the explanation of land degradation in any specific area as regional political ecology, and essentially the approach follows a chain of explanation. It starts with the land managers and their direct relations with the land (crop rotations, fuelwood use, stocking densities, capital investments and so on). Then the next link concerns their relations with each other, other land users, and groups in the wider society who affect them in any way, which in turn determines land management. The state and the world economy consitute the last links in the chain. Clearly then, explanations will be highly conjunctural, although relying on theoretical bases drawn from natural and social science. In this context we examine the major 'single hypothesis' approaches to land degradation. After all, there *are* discernible patterns of social change and land degradation, and models which would claim a degree of universality. The first is the explanation of land degradation in terms of population pressure and is the main concern of a number of influential theories. The second is very much more limited and different in character and explains degradation in terms of maladaptions and ignorance of land managers themselves – the problem lies uniquely with them. This chapter examines each of these models in turn in sections 2 and 3, and finishes with a case study which illustrates their performances.

2   Population and land degradation

2.1   Attribution and generalization

Any attempt to find the cause of land degradation is somewhat akin to a 'whodunnit', except that no criminal will ultimately confess, and Hercule Poirot is unable to assemble the suspects on a Nile steamer or in the dining car of a snowbound Orient Express for the final confrontation. The analogy is an apt one. Murders are generally easier to identify than land degradation; but guilt is often shared in different degrees between different people (e.g. assassin, accessory, etc.), as in each case of land degradation. However, any *general* statement about the causes of land degradation is of a very different order from the usual 'whodunnit', except perhaps in the case of the Orient Express, where *all* the suspects were found guilty! For the purpose of the analysis of land degradation such a statement may be true but is not very useful. At the other extreme, single general hypotheses of guilt do not get us very far either.

Perhaps the most common such hypothesis is that which attributes degradation to pressure of population on resources (which we shall henceforth call PPR), and therefore to growth in the numbers and density of population on the face of the earth. It does indeed seem self-evident that growth in numbers will cause land to be used more heavily; and that as the *per capita* area of arable and grazing land grows smaller, the sheer necessity of production will force farmers to use land in disregard of the long-term consequences. Yet, very severe degradation can occur in the total absence of PPR. Periods of population decline have often been periods of severe damage to the land.

2.2 Malthus rides again

Only a minority, among whom we are not included, would continue in the 1980s to believe that the rapid increase in the earth's population, an increase now heavily concentrated in the less-developed countries, is no matter for serious concern. A steady decline in the amount of arable land per head of the world's population means that this land will be required to produce more in order to provide the food and industrial raw materials needed by a population anticipated to grow by at least half its 1975 number during the limited space of the last quarter of this century. In the less-developed countries it is projected that the *per capita* area of arable land will have declined from just under 0.5 ha in 1955 to under 0.2 ha by the year 2000 (Council on Environmental Quality 1982: 403).

In modelling the effects on the land of population growth, however, the assumptions are based on inadequately measured and understood current trends often stated with great conviction.

Eckholm, for example, writes:

> Whatever the root causes of suicidal land treatment and rapid population growth . . . and the causes of both are numerous and complex . . . in nearly every instance the rise in human numbers is the immediate catalyst of deteriorating food-production systems. (1976: 18)

Ehrlich and Ehrlich more baldly say that 'an area must be considered overpopulated . . . if the activities of the population are leading to a steady deterioration of the environment' (1970: 201). Yet they go on to say that while Australia may be underpopulated 'the "frontier philosophy" is more rampant in Australia than in the United States in terms of environmental deterioration and agricultural overexploitation'.

The Ehrlichs' confusion is helpful, for it at once challenges their own simplification. The terms 'underpopulation' and 'overpopulation', like PPR itself, imply that there must exist a critical population density at which none of these conditions obtain. This critical population is often described as the 'carrying capacity' of the land, a notion which applies to human populations a principle that is well-established among animal populations. Once animal numbers exceed the available food resources, they undergo a severe decline through mortality or other Malthusian checks. At least in so far as human populations depend on their land for their livelihood, similar conditions should apply. Or so the argument runs.

Efforts to employ the carrying-capacity concept continue to be made, notwithstanding criticisms. A notable effort is that made on a pan-tropical scale by the Food and Agriculture Organization of the United Nations (FAO) in association with the International Institute for Applied Systems Analysis (FAO 1982; Higgins *et al*. 1984). Here countries were subdivided on the basis of the major soil types and the length-of-growing season periods. Calculations of production of a range of crops were made under conditions of 'low input', 'intermediate input' and 'high input' – these being in all cases conditions of modern technology and fertilizers, not of labour. Assessment was made of country-level ability to supply the projected year-2000 population, with encouraging results except in the arid regions of Africa and Bangladesh, provided that 'high levels of farming technology' are adopted. This is an interesting exercise, but its assumptions carry a degree of unreality, and the assumption of access by the majority of the agrarian population to high levels of modern farming technology is perhaps the most unreal of all.

If carrying capacity changes with each turn in the course of socioeconomic evolution, each new technological input or new crop introduction, and can also vary markedly according to the bounty or otherwise of rainfall in a given year, of what use is the concept? Writing of Malthus' original essay, Peacock has argued that 'like any other theory, the theory of population must be regarded as a conditional hypothesis' (in Glass 1953: 66).

2.3   Malthus unhorsed?: The Boserup hypothesis

The objections to the PPR explanation of land degradation introduced above do not constitute a refutation. An alternative hypothesis was required, and was provided by Boserup (1965, 1981). Boserup proposed that the neo-Malthusian view of population capacity as dependent on resources and the state of technology was erroneous. She gathered evidence to show that output from a given area responds far more generously to additional inputs of labour than the neo-Malthusians suppose, even under pre-industrial conditions. She then proposed that 'the growth of population is a major determinant of technological change in agriculture' (Boserup 1965: 56). Population becomes the independent variable, and the dependent variables become agro-technology, the intensiveness of labour inputs and hence the capacity of the system to support people.

Boserup's conclusions were policy conclusions for development strategy,

more explicit in a later summary (Boserup 1970). As such, however,
were largely disregarded. The prevailing view among development
ists of whatever background was that the modern biogenetic, chemical
rganizational agricultural revolution, rather than labour intensification,
l solve the problems if they were to be solved. A more gloomy view,
to the neo-Malthusian approach, would be that of Cassen:

ile the Boserup theory may have some validity in the broad sweep of
tory ... there is no reason to believe the argument is of general validity
oday's developing countries. Cases of the opposite effects are not hard
find: over-exploitation of land, overgrazing of pasture, man-made
sion and so forth (1976: 807).

is an important ambiguity in Boserup's hypothesis which concerns the
ation process itself. Her model may be likened to a toothpaste tube –
ation growth applies pressure on the tube, and somehow, in an
ined way, squeezes out agricultural innovation at the other end.
ver, as Cassen points out above, there are many contrary examples.
appears at the other end of the tube is often not innovation but
dation. Why? One of a variety of explanations provided in this book is
ck of access to productive resources on the part of the cultivator, and
intimately linked to the class nature of most land management (see also
er 6).

at is not to deny that PPR *is* often an important and reinforcing link
lucing this access to sectors of an agrarian population. This argument
lready been made by one of the authors (Blaikie 1985a: 18, 107), and
only briefly be recapitulated here. There is a wide variety of land
gement practices which can be adopted and there is a good deal of
ion in the amount of resources required for each practice on the part of
dividual cultivator or pastoralist. Agronomic methods of management
ually less demanding than soil conservation works, but even the former
re some spare capacity of labour, nutrients, land or capital. Crop residue
poration is one effective conservation technique but requires that the
r can 'do without' residues which may be important for other purposes
fuel, roofing for houses or fodder for livestock). Cover crops require
labour and seed, while a soil-conserving reorganization of crop planting
may demand a reallocation or increase in labour demand. These
vations apply *a fortiori* to chemical fertilizers, tree planting or the
ruction of grassed waterways. Also, state-sponsored research and
opment in conservation requiring relatively plentiful and locally
ble resources tend to be neglected in favour of more paying concerns
as the development of commercial crops on large farms on uneroded
(Beets 1982; Belshaw 1979; Richards 1985). Thus PPR may well
ce degradation and fail to produce agricultural innovation. Nonethe-
lack of access to productive resources must itself not be promoted as a

By attempting to isolate population as a single causal variable Boserup comes close to the very neo-Malthusians whom she criticizes. For, while neo-Malthusians believe that there are ultimate limits to the capacity of the land to support population without famine, damage or both, Boserup merely converts these limits into launching pads without successfully demonstrating that this conversion can always be made in all environments, or can continue indefinitely. Indeed, the evidence of diminishing returns, stagnation in rural wages and increased hours of work in both of Boserup's books, and especially in the second, suggests that Malthus climbs slowly back into the saddle as the Boserup sequence advances. At least within the domain of pre-industrial and early industrial agriculture which is her preferred ground, Boserup emerges more as a corrector of Malthus than as his refutor.

Cassen (1976) has drawn attention to the impact of population growth in some countries (notably in South Asia) upon the composition of resource costs required to produce basic consumption goods. The argument runs as follows. As long as population growth creates a demand in basic consumption items which can be met by the application of labour alone to land, the impact of population growth is not deleterious and may well run the course which Boserup argues (the creation of landesque capital, clearing and cultivation of new land and so on). Mao Zedong's adage of 'with every mouth comes a pair of hands' may hold. However, when increased aggregate demand for consumption items has to be met by expensive purchased inputs using scarce foreign exchange, then in a sense these are 'wasted' in keeping alive a growing population without increasing surpluses and savings. In an important sense, the reassuring model of Boserup is also stood on its head, and PPR may again assert degrading tendencies upon the land. Whether the 'Green Revolution' restores Boserup rather than Malthus to a standing position, and if so for how long, is a matter of contemporary controversy.

2.4   Innovation and intensification

The question really comes back to land management in the course of achieving production. We must once again pause to clarify the meaning of terms. The term 'intensification' is much used in literature about the Boserup model to mean the adoption of production systems which gain more output, averaged over time, from a given unit of land. Thus, elimination of the fallow year from the European three-field system and its replacement by fodder crops would be intensification. So also, however, would be the addition of labour inputs to wet rice cultivation to squeeze more and more production from the same field, as described by Geertz (1963).

Properly speaking, however, intensification means the addition of inputs up to, or beyond, the economic margin where application of further inputs will not increase total productivity; in the case of agriculture these are measured against constant land, and in pre-industrial agriculture we mean mainly inputs of labour, plus livestock. When, however, there is a change in the manner in which factors of production are used the inputs are applied in qualitatively new ways; a new curve of intensification is created. Such qualitative changes are innovations.

Brookfield (1984a) sought to distinguish between innovation and simple production intensification by means of the example of the West Indian sugar industry under slavery and subsequently. A production system which involved intensive land management with slave labour, and one effect of which may have been to save the production base from rapid degradation, was introduced for gain by seventeenth-century entrepreneurs. Requiring heavy inputs of labour, from 4000 to 8700 person-hours/year/ha, for both production and management, this system created a high density of population, the effect of which, even for a century after slavery was abolished, was to inhibit the innovation of new farming practices. Labour was cheap and abundant; furthermore, a large population required to be supported (Goveia 1969). Only the modern introduction of cane-breeding and fertilizers finally increased production per worker and so created a labour shortage. The old situation had endured some 200 years, with the effect that:

> Population-based theory is turned on its head in this case: given the social conditions of production, pressure of population on resources became a disincentive to innovate. Intensification and innovation became alternatives in a situation in which the means to innovate existed and were known, but conditions produced by intensification led the landholders to resist adoption. (Brookfield 1984a: 32)

In section 4 of this chapter we shall examine material from Nepal, where a high PPR is interpreted as both the cause of degradation, and also as the means by which degradation is managed and contained. Without anticipating this examination, it is clear that PPR can be seen both as creating a need to exploit resources in environmentally sensitive areas in such a way as to expose them to damage, and also as providing the means of a labour-intensive management system which seeks to contain the consequences. There is no reason, even on *prima facie* grounds, why both cannot be true.

2.5 Taking a position on population and degradation

The growth of human numbers, and the growth of numbers among their livestock, can undoubtedly create stress. It requires the extension of interference into new areas, and the subjection of these areas to the high levels of damage that follow initial interference. It requires the occupation of sites of lower resilience and higher sensitivity, for which existing management practices may be inadequate. Since the expansion is likely to be carried out largely by those displaced from older areas by poverty, or by other pressures of social or political origin, the new land has to be managed by those with the fewest resources to devote or divert to its management. Here we see the three definitions of margin and marginalization (economic, ecological and political economic) combining in a downward spiral. An increase in damage to the land is an inevitable consequence, at least for a time, and PPR and lack of access to the means to innovate go hand in hand.

High PPR may also create stresses within existing systems with well-tried management practices. As the margin of subsistence grows narrower, so the pressure to maximize short-term production will grow stronger. The need to innovate will grow, but the means with which to innovate will be lacking. Wealthier landholders whose own resources are not gravely threatened by the 'downstream' effects of degradation on the land of their poorer neighbours, may welcome the growing abundance of cheap labour and see no need to embark on larger innovations which might be of benefit to all. While they may have the means to innovate, they may not see it as being to their advantage to do so. Grinding poverty is a poor environment for good management, and a favourable one for degradation. But grinding poverty is not only brought about by PPR, though it may well be exacerbated by it, as in the following case study of Nepal.

Except in the presence of a situation such as outlined above, however, a high PPR also provides abundant labour with which to undertake intensive management. Where known and tried innovations are available, but require a high labour input, they are more likely to be both undertaken and maintained under conditions of high population density. The long-enduring ecosystem stability in Roman and Byzantine Palestine, and after that in Lebanon, noted in chapter 7, existed under conditions of very high rural population density. Most of modern Java, despite the serious erosion that takes place in headwater areas and on land of high environmental sensitivity that is unsuited to irrigated terracing, exemplifies the high productivity obtainable under intensive management with extremely high densities of population. The peril is that such systems require abundance of labour not only for their establishment but also for their maintenance. If some of that labour is withdrawn, as by an increase in off-farm employment opportunities, or by emigration, or by the demands on male labour generated by the state for corvée work, or for war, the consequences can be disastrous. The created system itself is one of high sensitivity, although it is resilient as long as the necessary inputs for its maintenance are available.

Where land is abundant, the need to conserve it may not be apparent. It is only after major damage has occurred, as in North America in the 1930s, that the need to halt and reverse damage comes to be perceived. Shifting cultivators in the humid tropics are often regarded as a prime example of destructive land users because of their practices, made possible by low levels of PPR. This may be to malign them. None the less, some of the major modern examples of land degradation are in areas of low to medium population density, rather than in areas of high density.

For the present, then, we adopt an open approach to the relation of population pressure to land degradation. Degradation can occur under rising PPR, under declining PPR, and without PPR. We do not accept that population pressure leads inevitably to land degradation, even though it may almost inevitably lead to extreme poverty when it occurs in underdeveloped, mainly rural, countries. The question of why management fails, or breaks down, is not answered so simply. Population is certainly one factor in the situation, and the present rapid growth of rural populations in many parts of the world makes it, in association with other causes, a critical factor. But 'in association with other causes' is the essential part of that statement, for the other causes can themselves be sufficient. PPR is something that can operate on both sides, contributing to degradation, and aiding management and repair. In general and theoretical terms, then, Hercule Poirot remains with no proven case against his prime suspect. Unlike the situation in a 'whodunnit', however, this has become obvious at an early stage in the narrative.

3 Behavioural questions and their context

3.1 Them and us

For the natural or physical scientist who diagnoses the immediate causes of a specific problem of land degradation, the debate about PPR is of use only in so far as (s)he seeks general explanation; his or her more immediate concern is with the means of introducing or enforcing protective measures, and (s)he is often aware of considerable resistance on the part of farmers and rural communities in general. Indeed it has been claimed by Blaikie (1985a) that most conservation policies introduced by governments fail, although there are significant exceptions. Major schemes with substantial funding for compensation to cultivators, and/or with the political means of mass mobilization, have a better chance of achieving results. Models for this exist even in the nineteenth century, as, for example, in major efforts to introduce large-scale land management changes in the French Alps in the 1860s (Henin 1979). In modern China, the efforts of thousands of workers could be mobilized to plant shelter-belts against wind erosion, fix dunes and terrace hillsides in the loess regions, though apparently without significant success, as Smil argues in chapter 11. But the lesser task of obtaining the co-operation of farmers in works of protection and drainage, often involving restrictions on land use and the creation of some landesque capital at the farm level, is seldom achieved without a great deal of persuasion and example, and not infrequently fails.

For a long time it was the fashion to decry the 'stupidity' or the 'conservatism' or the 'uncaring idleness' of such farmers, or to stress their 'ignorance'. A minority of farmers, like a minority in any walk of life, are certainly stupid and many more are conservative, but conservatism does not necessarily arise only from an unwillingness to change. Where there is a known set of practices and behavioural responses, it is thus much easier for the farmer to adhere to an established pattern than to make changes, as Kirkby (1973) showed in Mexico. Changes may only be forced when a major 'discontinuity' becomes apparent; yields have fallen alarmingly; more land has been lost than can simply be rationalized away; the alternative to the risk of doing something new is the seeming certainty of losing everything.

However, where farmers are peasants, in contact with a larger economy, and are subjected to pressures to change from government, different conditions apply. Bailey (1966) wrote a classic interpretation of the 'peasant view of the bad life' which is particularly relevant to the assumption by 'outsiders' that peasants are traditional, conservative and stupid, when these outsiders are confronted by yet another failure to bring about effective land management. To the peasant the government is seen as the 'enemy' and its representatives have a completely different cognitive map from that of the peasant. Intervention by government to bring about 'better' land management is frequently met by suspicion and non-comprehension. In the multiplex and non-specialized relationships of a peasant society, specialized interventions such as forest protection or pastoral regulations seem incomprehensible and quite incompatible with the moral economy of peasant society. Often the peasant views the future as the 'round of time' rather than the 'arrow of time'. The farmer allocates resources on the assumption that next year will be, more or less, like this year and is seen against a round of time – so many years before an ox is replaced, two or three years before the house needs rethatching, and so on. Not so a conservation officer, who must persuade the target-group of tangible benefits through innovations, and try to set in motion a definite change within a specific time period.

An alternative explanation for conservatism of land managers is an economic one, and rests on the often-observed risk-aversion behaviour of peasant and other farmers. Living and working in an environment of risk and uncertainty in which it is not possible to predict with confidence that such-and-such a set of inputs will yield such-and-such a return, farmers are reluctant to embark on new practices which might increase risk. Poorer farmers, it is argued, adopt what Lipton (1968) termed a 'survival algorithm'. Lacking the resources to weather failure, they both suffer greater risk and are more inclined to behave in a risk-averting manner than wealthier farmers. Innovative behaviour involves risk and uncertainty. The rich farmers are better able to bear its risks, and so stand to gain more from its benefits.

'Ignorance', 'stupidity' and 'conservatism' imply that there *is* a choice, but people are too ignorant, stupid and conservative to make the right one: that provided by governments or international 'experts'. Economic constraints caused by uncertainty and risk, compounded by onerous relations of production, perhaps too by PPR, provide a more clearly defined economic map of what is possible and what is not. Unfeasible government plans can all too easily be laid at the door of unappreciative farmers and pastoralists. At least economic explanations give some rationality to peasant behaviour and partially take the lid from the black box of ignorance and conservatism.

3.2   Ignorance and perception

Ignorance in a non-pejorative sense is another matter altogether. Ignorance of subtle changes in the quality of the soil is not only possible, but widespread. There is a number of degrading processes which show little immediate effect until a threshold of resilience is past, or until some untoward event exposes an increase in sensitivity. Leaching and pan-formation within the soil may operate in this way, as may changes in structure and chemical status. A striking example has recently emerged in eastern Australia where the planting of clover to upgrade pastures by nitrogen fixation has greatly improved capability since it began some fifty years ago. Surplus nitrogen gradually acidifies the soil in depth, releasing plant-toxic trace elements (aluminium and manganese); plants root to shallower depth to avoid them and become more sensitive to drought. The resilience of the system has a threshold beyond which acidification causes capability to suffer a sharp decline (Williams 1980; Bromfield *et al.* 1983; CSIRO 1985). Not only is none of this perceptible to the land manager until the limits of resilience are reached, but it has not even been perceived by soil scientists until lately.

It is equally possible for land managers to remain quite ignorant of the effects of low rates of erosion, where these exceed still lower rates of soil formation, until a critical level of accumulated loss is reached, and this can even take centuries (Hurni and Messerli 1981; Hurni 1983). The effect of loss of organic matter in restricting the ability of crops to respond to inorganic fertilizers is not likely to become apparent until the latter are applied. Both ignorance of degradation and its perception are functions of the rate and accumulated degree of degradation, as well as of the intelligence of the land manager.

A succession of good years can delay the perception of degradation to a critical extent, or at least facilitate an optimistic ignorance of the real consequences of observable changes. Without doubt, this has been an important factor in the 'degradation crises' that have struck many areas of new settlement during the past century. Newly settled farmers and graziers, who have used the natural 'capital' of long periods of 'rest' on land never before ploughed or grazed since the present soil–vegetation complexes were formed, had little means of knowing that the land they worked was of high sensitivity and low resilience until disaster – usually a drought – exposed the consequences of a period of heavy use. Writing just before the disasters of the 1930s, Webb (1931) showed how the perception of the Great Plains of the United States evolved rapidly during the nineteenth century, to the extent of a belief that occupation of the land increased rainfall and could ban the spectre of drought. In Australia, Meinig (1962), Heathcote (1965, 1969) and Williams (1974, 1979) have shown how the hazards of the semi-arid regions were first ignored, then harshly recognized and later only partly accepted.

Given the apparent success of remedies and adaptations a false sense of security could quickly become re-established. However, 'ignorance' shades from real to wilful, and may arise in part from other causes, such as a strong market imperative, a need to occupy new land for cash cropping or because of PPR elsewhere, or an ethos which believes that 'man' (and here we use 'man' rather than 'people') can and even must 'master nature'. This ethos, strikingly analysed by Passmore (1974), is at the root of much of the 'ignorance' so strongly complained of by those who work for a better management of the land. We present a stark illustration in chapter 11.

Still further removed from involuntary ignorance is the speculative abuse of land by commercial ranching and farming corporations and individuals, and by logging contractors. Here ignorance is not an accurate term since whether these land users know of the costs they inflict on future users of the land, and on present users elsewhere, through externalizing their costs, is beside the point. Wholesale disappearance of forests is the result (Plumwood and Routley 1982; Myers 1985), while the devastation of large tracts of agricultural land through now-abandoned commercial enterprises (Dinham and Hines 1983) is undoubtedly the result of calculated human agency and not of ignorance, nor stupidity.

4 The erosion problem in crowded Nepal – crisis of environment or crisis of explanation?

4.1 The environmental situation

Nepal is a classic area for the study of land degradation; we have already referred to it in chapter 1, and shall do so again at several points in this book. With an immensely varied environment, including the world's highest mountains, a strip of the Gangetic plain, and the high-altitude desert of the trans-Himalaya, Nepal is among the world's least 'developed' countries with a high and rising density of population on its limited areas of arable land (see figure 2.1). Rural population densities reach over 1500 per km^2 of cultivated land, or 15 to the ha, and there are districts in the middle hills with even higher densities. This is similar to densities in central Java and is 50 per cent higher than Bangladesh (Strout 1983).

In the main agricultural areas of the middle hills or *Pahad* almost all arable land is terraced. Irrigated land (*khet*), whether fed by rivers or from springs, grows rice and winter crops of wheat and potato wherever possible; dry land (*pakho*) is also terraced, but with a slope from almost level to as much as 25° or occasionally more. These dry lands grow mostly maize, often with finger millet as a relay crop which is transplanted from a seed bed under the growing maize. Unterraced land, used mainly for pasture (*charan*) but with a few cultivated patches, now occupies only the steepest slopes and ridgetops, but was in former times much more extensive and the site of shifting and semi-permanent cultivation among the mixed forests (*Quercus* spp., *Castanopsis* spp., *Pinus roxburghii*), and *Rhododendron arboreum* that used to

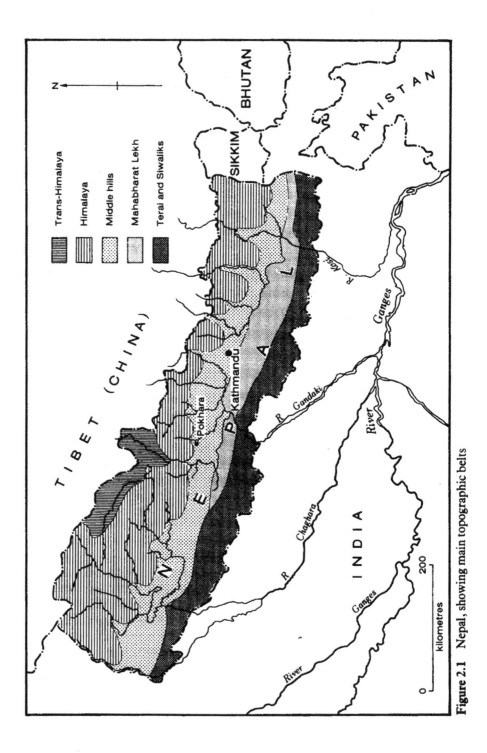

Figure 2.1   Nepal, showing main topographic belts

cover these hills (Burkill 1910; Mahat 1985). Cultivation and the collection of firewood were not the only reasons for clearance; charcoal-making for the metal industries of Nepal continued to be a major cause of depredation until early in the present century, though this has now diminished, and trees are also lopped to provide leaf-fodder for livestock (Bajracharya 1983). Livestock is a very important element in Nepalese farming systems (Axinn and Axinn 1983), and the numbers of livestock per human inhabitant are among the highest in the developing world.

The growth in population has eliminated the forest from large areas and has replaced it by cultivation. Commentators agree that virtually all land capable of being terraced has now been taken up in the middle hills (e.g. Caplan 1970: 6; Mahat 1985), so that forest boundaries are not now much in retreat. The degraded condition of the remaining forests is indeed now attributed to historical rather than to current practices (Mahat 1985; Nepal–Australia Forestry Project 1985). Yet this relative stability is only recent. In 1928 the government policy of replacing forest by human cultivation wherever possible was praised by one observer, who felt that

> this policy must be pursued for many years before there need be the slightest grounds for fearing that sufficient forest will not remain. For in the temperate zone [the middle hills] it is certain that cultivation can never occupy more than one-third of the total area. . . . Perhaps in the valley of Katmandu (*sic*) and its vicinity a condition has been reached in which it would be wise to call a halt. . . . But elsewhere the day on which restriction of cultivation need become a question for consideration is still far off. (Collier 1928/1976: 253)

From ancient times the State owned all land, and as it was the principal source of revenue to rulers no cultivable land should be allowed to lie idle (Stiller 1975; Regmi 1976). These principles were sustained and developed by the rulers of unified Nepal after 1768 so that the first king himself directed that all land convertible into fields should be reclaimed, and if homesteads were built on such land they should be moved (Regmi 1978). Peasants paid the state half the produce of the land in tax, or rent, and later this was paid to officials and others who received grants of land (and its income) in lieu of salary or as reward for service. Even so, only the irrigated *khet* land seems to have had firm definition until modern times, and in a village near Pokhara, Macfarlane (1976: 52, 87) noted that the *pakho* land was not shared out individually until about 1940; before that 'patches of jungle were slashed and burnt by those who had the labour'. By the late 1960s, however, only a few rocky and steep patches remained to be cleared; 'the limits of maize cultivation had been reached'. On the other hand, there have been no significant changes in areas of agricultural land and forest since at least 1900 in a densely peopled area at Thokarpa, east of Kathmandu, despite large population increases (Mahat 1985: 232).

Exposure of the slopes has serious consequences for runoff under the torrential monsoon rain, especially at the beginning of the monsoon when there is little ground cover after the long dry season. Heavy erosion of the sloping dry terraces is a result, especially as the risers of these terraces are cut back each dry season to add new mineral phosphorus and soil to the terraces, which are thus widened and progressively flattened through time; each season a proportion of these terraces is rendered useless, and the land reverts to uncropped *pakho*; each season, however, new or reclaimed terraces are created on the *pakho* land.[1] Slope-wash, gullying and especially mass movement occur on all slopes, and mass movement also affects the lower lying *khet* terraces, which become overcharged with water and saturated in depth. In occasional flash floods, the best valley bottom *khet* is sometimes washed out by heavily laden streams which widen their channels. As much as 50 per cent of the eroded material may be carried into the lowlands, causing wide channel braiding, choking rivers and adding up to a metre of soil per 20 years to adjacent fields; sediment loads as high as 25,000 ppm are regularly recorded in the major rivers.

The distribution and nature of erosion are, however, important. Dry terraces may lose only 0.4–1.6 mm/year, with corresponding losses of organic matter, nitrogen, phosphorus and potassium, and some of this is redeposited downslope, especially in irrigated terraces which can gain soil and nutrients. Much higher losses are experienced under degraded forest without surface protection from small shrubs, and on grazing land. In one small watershed gross estimated losses ranged from only 2 t/ha/year (tonnes per hectare per year) from the irrigated terraces to 20 t/ha/year from grazing land. However, while top-soil lost from this 63 ha catchment *and* carried beyond its bounds totalled only 220 t (3.86 t/ha from 57 ha of the area), the total loss was 1320 t/year, the balance being derived from mass wasting on the remaining 6 ha (Carson 1985). Caine and Mool (1982) calculated annual mass wasting losses to be as high as 13 t/ha from another catchment but derived from only 1 per cent of its area. Moreover, much of the landslip damage is repaired, and much land that appeared irreparably damaged when mapped in 1979–80 (Kienholz *et al.* 1983) was totally reclaimed under terraces, wet or dry, three or four years later (Ives 1984; 1985). Our own observations, using the 'geomorphic damage map' of the Ives and Messerli (1981) Mountain Hazards Mapping Project in the field, confirm that specific areas mapped as debris flow in 1979–80 were again wholly terraced in 1984 (Brookfield 1984b).

The importance of accelerated erosion in Nepal has been the subject of a great deal of uncertainty (Thompson and Warburton 1985a, b). A variety of

[1] Most of the modern literature cited in this chapter refers to dry terraces as *bari*, thus distinguished from the irrigated *khet* and the unterraced *pakho*. The Nepalese geographer Dr Harka Gurung (personal communication) advises that this is incorrect, being based on the writings of Kathmandu scholars who are unfamiliar with the languages of most of the farmers. *Bari* are infield dry terraces, heavily manured and mulched and often fenced; *pakho* is outfield, whether terraced or not. Gurung's advice is followed in this chapter. The same usage also seems to be followed by Regmi (1971).

commentators have been convinced that Nepal is an ecological catastrophe. A single doom-laden quotation from a wide choice will suffice:

> Population growth in the context of a traditional agrarian technology is forcing farmers onto even steeper slopes, slopes unfit for sustained farming even with the astonishingly elaborate terracing practised there. Meanwhile, villagers must roam farther and farther from their homes to gather fodder and firewood, thus surrounding villages with a widening circle of denuded hillsides. (Eckholm 1976: 77)

However, a detailed and quantified analysis of the extent of degradation has produced a more qualified interpretation. A recent FAO study stated that the results of their analysis suggest that past descriptions of conditions in Nepal have exaggerated the erosion problems, and go on to say that the 3 per cent figure of the total land surface classified under poor and very poor watershed conditions is not a serious erosion problem (FAO 1980: 5). Interestingly, they attribute part of this possible misconception to the effects of road-biased rural tourism, according with Chambers' (1984) apposite diagnosis of misconceptions of foreign and urban-based observers. There is an often-visited panorama which almost all visitors see as they travel west from Kathmandu by road towards Pokhara, at the point of leaving the Kathmandu Valley. From the lip of the valley westward and northward lies a rather desolate scene of considerable erosion, deforestation and degraded pastures. The higher parts of this same area were clad in heavily lopped *Quercus semecarpifolia* scrub when Kirkpatrick traversed it in 1793 (Burkill 1910). This scene finds its way into numerous articles and publications on the ecological crisis of Nepal, but is not representative of the country as a whole. Indeed, much of the agricultural land is in surprisingly good condition (FAO 1980: 7). The worst land erosion is on shifting cultivation land in the Mahabharat Lekh, and in the arid trans-Himalaya, where mass wasting and wind erosion carry away all soil from steep, unprotected slopes. In general, it seems that *charan* land including abandoned cultivation patches and pastures suffer from most serious erosion, particularly sheet erosion and loss of topsoil. *Pakho* land is the next most seriously affected followed by *khet* (Carson 1985: 7). However, mass wasting and catastrophic events such as large-scale slope failures are probably *not* due to human interference (Carson 1985; Ramsay 1985). Hence the most noticeable and dramatic forms of erosion, remarked on by rural tourists, are probably not caused by Nepalese farmers at all.

4.2   How Nepalese farmers cope

At a superficial level, the Nepalese problem has appeared to many observers, local as well as foreign, to be a classic example of the effect of PPR, specifically through deforestation and the dry terracing of steeply sloping land in a sensitive environment. Eckholm (1976) suggests that Nepal will slide away into the Ganges by the year 2000, and has no doubt that PPR is the principal culprit. A secondary culprit, however, is the Nepalese farmer of whom the Asian Development Bank (1982 II: 34) was highly critical. The practice of building unbunded, outward-sloping dry terraces was attacked with particular severity, and inward-sloping bench terraces were urged as a means of checking runoff. Water management was also regarded as primitive, a view shared by two local agricultural scientists (Nepali and Regmi 1981) who wrote that 'the technology of water management is scanty if not absent'. Under pressure of declining resources *per capita* cattle holdings are declining, and moreover the technology of composting manure is primitive, so that up to 52 per cent of the nitrogen and up to 80 per cent of the phosphorus are oxidized. Over 90 per cent of the fodder consumed by the animals does no more than keep them alive, leaving only 10 per cent for yield of 'economic products' (Asian Development Bank 1982). All this is regarded as very inefficient.

Others differ, at least to parts of this deluge of complaint. Axinn and Axinn (1983) analyse the flows of energy within the farm system, and stress the vital significance of 'keeping the animals alive' to plough the land and manure it, both 'economic products' of their existence. Ives (1985) notes that dry terraces slope outward to avoid waterlogging of dry crops, and to prevent accumulation and penetration of water which would cause landslipping; the absence of a bund is a deliberate measure to *ensure* runoff. Like ourselves, he is impressed by the skill of Nepalese levelling, terrace construction and water management, which includes extensive systems designed to enlarge the command area of irrigation flows. Gurung (personal communication) argues that it is the capital needed to build irrigation systems that is 'scanty', not the technology. Against the view of the critics is the glowing tribute paid by Cool:

> Personal observation suggests that it may require up to twenty years to fully transform an afforested hillside into a relatively stable irrigated terrace. The enterprise is marked with difficulty, setbacks and occasional failure. Yet what stands out is the skill and energy that goes into their design and execution and how successful the hill farmer is in maintaining and improving his terraced fields year after year, generation after generation. Flooding, landslips, goats and cattle, and occasional earth-quakes are taken in stride. With only hand tools and simple bullock-drawn ploughs, but with enormous fortitude, the mountain farmer rebuilds, reploughs, reseeds and survives. (1983: 7)

Much greater detail concerning the manner in which farmers of the Kolpu Khola area manage the specific hazards of their environment is provided by Johnson, Olson and Manandhar (1982) in a paper written within the 'natural-hazards school' paradigm, but illuminated by use of ethnoscientific method (Conklin 1954) and by a degree of social awareness not common among this school of research. Methods of maintenance, repair and damage reduction are described, both those used by individual farmers and those known to them but beyond their means. The problems are well understood, although the supernatural forms part of the folk explanation of sudden and unpredictable terrace collapse. Maintenance is a regular and time-consuming task. When damage actually occurs the problem becomes one of resources. Thus:

> Farmers evaluate the options and, often, must choose the less effective one which is, however, the one within their means. Timing is a crucial factor in this decision. Constraints and limited resources may lead the farmer to postpone taking preventive measures in the face of warning signs. This may result in rapid deterioration or destruction of the endangered field, or it may allow time for the accumulation of resources for complete repair. (Johnson, Olson and Manandhar 1982: 84)

All farmers, however, are willing to experience temporary loss, even of long duration, in order to reduce the risk of greater loss. Thus farmers may cut irrigation off from endangered *khet* land and use it as lower yielding dry terrace, or even let it lie waste until consolidation is achieved. Even more drastic is the deliberate diversion of erosive flows of water on to land threatened with slumping to wash it away before new terraces are built to entrap new soil and rebuild irrigated fields, which then take some years to recover full capability. On the other hand, loose temporary terraces are sometimes made on the *pakho* land in order to obtain a little extra production at extreme risk. Differences in resources between farmers may lead to differences in net damage suffered by the poor and the wealthy, so that 'the overall effect of "random" landslides and floods may result in increased disparities between rich and poor' (Johnson, Olson and Manandhar 1982: 188).

The larger socioeconomic problems which underlie both differences in the quality of land management, and in the impact of damage, thus emerge. Some of these are discussed in a wider context in chapter 6, and a detailed study is offered by Blaikie, Cameron and Seddon (1980). Bajracharaya (1983) also places deforestation in such a context. Gurung (1982) argues forcefully that 'the basic problem of the Himalayan region . . . is not ecological but the low level of development'. However, development of some types is not necessarily beneficial to the land. Khanal (1981) offers convincing demonstration of the effects of labour migration on land management in another part of the middle hills. While 90 per cent of the population of Nepal still depends on the land for its major sustenance, by one source only some 64 per cent of rural income is farm income (National Planning Commission, Nepal 1978), though by another the proportion is 85 per cent, 60 per cent from crops and 25 per cent from livestock (Nepal Rastra Bank, cited in Nepal–Australia Forestry Project 1985). Already some two-thirds of the agricultural production of Nepal comes from the lowlands (or *terai*) along the Indian border, where only one-third of the people live; a major shift in population is taking place (Goldstein, Ross and Schuler 1983), and the effect of a substantial reduction in labour inputs into the intensive farming and land-management systems of the middle hills could be more devastating than the increase in PPR that has taken place in historical and recent times. Some shortage of labourers and livestock hands is already felt in some areas (Gurung personal communication).

There is a further question of significance to be derived from the Nepalese case. It has been widely assumed that severe land degradation is a recent problem, but this is mainly because the problem has only been recognized since foreigners began to enter freely and move around the country after 1950. Logically, one would suppose that earlier shifting cultivation and wood cutting for charcoal-making on a large scale would have done more damage than the intensive terracing that has covered the same hills under rising PPR in modern times. Maize and potatoes were introduced into Nepal in the eighteenth century (Regmi 1978), and the effect of these crops, which offer poor ground cover in the early stages of growth, must surely have been to increase erosion before terracing was widely established.

Certainly, the massive transportation of material from the Himalayas and the middle hills into the Ganges plain is not new. The Kosi river, which drains eastern Nepal, has shifted its course 120 km to the west (i.e. up-Ganges) since 1736, and flooding has been experienced since ancient times (Carson 1985). The contribution of deforestation is disputed in the light of this evidence, and in any case forest destruction in Nepal has been in progress for several centuries (Mahat 1985). There is some slender evidence that the Ganges itself may have become more shoal-ridden during the first half of the nineteenth century, necessitating the replacement of early steamboats by shallower draught vessels (Headrick 1981: 22) but catastrophic events (such as glacial or debris dam-bursts) have been bringing vast quantities of sediment down into the lowlands for millennia.

In these circumstances, the effect of rising PPR in Nepal appears double-edged. Terracing has probably diminished erosion; deforestation and the creation of larger areas of grazing land have probably increased it. It is by no means clear that the balance has been towards greater degradation. Nor is it clear that the pessimist's vision, caricatured by Ives as desertification of the Himalayas, 'the devastation of the Ganges and Brahmaputra plains and a major opportunity for Dutch polder engineers in the Bay of Bengal' (1985: 428) will come about under present practices. 'Development' of a type which would involve major reduction of available rural labour in the middle hills, or the introduction of machinery and of large but untried works conceived by engineers without understanding of the dynamic montane environment, might however be more conducive to this result.

4.3   The 'chain of explanation' in the middle hills

The precise role of PPR still needs to be explored further: in terms of the interpretation of whether accelerated erosion *is* a major and widespread phenomenon; whether it has recently become worse; and also in terms of the causal relationship between population growth and erosion. A more comprehensive frame of analysis must be provided or, to use the metaphor with which this chapter started, the chain of explanation must be followed back to broader socioeconomic links.

One of the most fundamental relationships in evaluating the impact of PPR is that between population growth, extension of cultivation and the productivity of land and labour. Between 1970–1 and 1980–1 the increase in population was greater than the increase in cropped area in both highlands and lowlands (Gurung forthçoming), and in the highlands most of the increase was due to extension of double-cropping particularly involving wheat cultivation on *khet* (Mahat 1985). However, inadequate data indicate that food-crop production declined by 0.5 per cent between 1970–2 and 1980–2 in the middle hills but increased by 9.6 per cent in the lowlands, against area increases of 11.6 and 15.1 per cent respectively (Gurung forthcoming). Forest clearance in the *terai* has now reached the limits of good land, and a weakening situation of irrigated rice production probably reflects the extension of cultivation on to less fertile land.

Mahat (1985) describes the double-cropping, inter-cropping and relay-cropping systems in use in a part of the middle hills east of Kathmandu. Livestock manure is the basis of production, and in the hills there is one 'livestock unit' per head of population, mainly cattle and buffalo (41 and 22 per cent at weightings of 1.0 and 1.5) plus goats and pigs for meat. However, in Nepal as a whole, the annual increase in cattle population between 1966–7 and 1979–80 was only 0.12 per cent while the buffalo population declined by 1.3 per cent (Rajbhandari and Shah 1981). The consequences for production are probably better illustrated by a representative interview with a farmer than by unreliable statistics:

> Some thirty years ago we still produced enough grain to allow us to exchange surplus for necessary daily goods, which we could not get from our farming. Of the grains harvested, one third was exchanged.... While the good farmers who have enough cattle and do very intensive cultivation can still increase their yields, this is not the general trend. In a *khet* (irrigated) field where we sowed 4 *mana* of seed we used to get 1 *muri* of paddy; now we need an area with 8 *mana* of seed to get 1 *muri*. Our wheat used to have big ears and long halms and we filled six baskets a day, nowadays it is sometimes only one or two. In many houses there is no longer enough food. For some, the harvest grains are sufficient for only three to four months a year. (Banister and Thapa 1980: 90)

The small size of corn-cobs on stacks outside houses in the middle hills is a matter of observation, though whether or not any trend is present is impossible to determine except by farmers' recall.

Intensive questionnaire work in west-central Nepal undertaken by Blaikie, Cameron and Seddon in 1973–4 and 1980 (Blaikie, Cameron and Seddon 1979, 1980) indicated that farmers were in fact well aware of decline in yields on old-established fields, particularly dry fields which supported maize and millet. The main problem is one of a reduced availability of plant nutrients, which come predominantly from composting of forest products and involve a 'transference of fertility' from the forest to arable land (Blaikie 1985b). Cattle are stall-fed much of the time and are fed forest-litter, tree loppings and hand-cut grass, all of which are gathered daily. The manure, together with leaf material from animal bedding, is then applied to fields. As the population rises, the increased demand for food crops is met by heavier lopping, which thins and destroys the forest. Mahat's (1985) informants at Thokarpa, east of Kathmandu, held that the greatest forest degradation occurred between 1951 and 1962 when there was a partial breakdown in central government administration, but that controls have been restored under the local government system established in the latter year.

Fuel needs also reduce forest cover. Indeed, wood fuel demands exceed supply by 2.3:1 in the middle hills and by 4:1 in the drier far west. One source estimates that all accessible forests will be eliminated within twenty years (Asian Development Bank 1982: 12), but other information cited here sheds doubt on this estimate and indeed all estimates are fraught with uncertainty (Thompson and Warburton 1985a). Crop yields decline when the forest-to-arable ratio is upset, and whatever the direct effect of deforestation on erosion it certainly has an effect on the capability of the arable lands. The inability of most Nepalese households to make good this 'energy crisis' by importing chemical fertilizers, and kerosene or other alternative fuel for cooking is an essential part of the explanation.

So we must pursue the question of PPR and degradation to the next link in the chain of explanation. The problem, as Harka Gurung argues, could be conceived as a lack of development. Simply, incomes are so low that few can *afford* chemical fertilizer, or any capital investment other than that generated entirely by their own labour. Half the households of a rural household survey undertaken in 1973–4 (Blaikie, Cameron and Seddon 1980) sold less than Rs. 250 (*c.* £10) of agricultural, pastoral and handicraft produce per year. Most budgets were balanced to within a few rupees, or were in deficit (Blaikie in Seddon, Blaikie and Cameron 1979: 69). Also chemical fertilizers do not produce a satisfactory return on unirrigated crops and, therefore, are not usually applied to summer maize and millet, but are largely limited to winter wheat, and winter maize at low altitudes, to industrial crops and sugar cane in the *terai*, and to vegetable growing by larger farmers in the Kathmandu Valley. Thus, as traditional forms of energy use start to fail because of population growth and degradation, the peasantry is, and becomes more, unable to transform itself and substitute new and imported forms of energy for agriculture, livestock rearing and fuel. It also becomes less and less able to harness water for more irrigation, especially as the replacement of local smelting of iron and copper by imported metal has reduced the population of miners whose winter off-season work used to be the building of irrigation systems for villagers (H. Gurung personal communication).

However, rural households *have* responded to PPR in other ways. They have migrated out of the hills altogether (so that the *terai* now holds over 40 per cent of the population of Nepal, according to the 1981 Census). They continue to extend cultivation and upgrade land (from dry *pakho* to terraced *pakho* and to *khet*). As the Land Resources Mapping Project has shown, a number of new crop rotations have been introduced, particularly on south-facing slopes and lower elevations, involving relay-cropping, inter-cropping, the selective introduction of quick maturing varieties, zero-tillage cultivation allowing broadcast wheat to be sown before the harvesting of *padi*-rice in *khet* fields, and so on. However, the endeavours and ingenuity of Nepalese farmers can in many areas of the middle hills and mountains barely cope with the increasing challenges being heaped upon them. However, lest the erroneous impression arises that PPR *is* the ultimate culprit, the question must then to be asked, *why* is the majority of the Nepalese rural population so poor?

The last link in the chain of explanation is vital, but it is only explored in outline here to avoid anticipating the argument of chapter 6. Nepal has had a long history of political independence, but also of quite important economic relations with British and later independent India. The independent state, however, taxed the farmers heavily and placed heavy demands on their labour for corvée work; in some areas landlordism developed as state officials were allocated areas of land and people to exploit in lieu of salary (Regmi 1971, 1978). Until very recent times the state remained unchanged in its antique and quasi-feudal form and kept out modernizing reforms in education, forms of representative government as well as productive capitalism in industry and agriculture (Blaikie, Cameron and Seddon 1980). Nepal remains landlocked and dependent upon India, both politically and economically. Such surpluses as there are in Nepal tend to be used in merchanting, smuggling, real estate and speculative purchase of land. Attempts at manufacture are undermined by cheaper products from India, and the 'leaky' frontier allows grain from the surplus-producing *terai* to flow south to India rather than to the food-deficient hills. The state itself has had great difficulty in outgrowing its quasi-feudal and extractive nature to meet the daunting challenges of development at the present time (Shaha 1975; Blaikie and Seddon 1978).

4.4   Crisis in explanation?

The explanation starts with changes in the intrinsic properties of the Nepalese landscape and ends with the problems of the Nepalese state and its relations with India. Each link of the explanation is firmly closed around the next, but initially the direction is away from the environment and towards more general 'development problems' of the agrarian economy and the Nepalese state. It leads back to land degradation because a vicious circle of links has been created, leading from degradation to underdevelopment and back again to degradation.

The palliation or gradual reversal of the problem of poverty requires a series of 'techno-political' decisions. However, there is a considerable degree of indeterminacy between the 'levers of policy' and intended results. For example, we cannot state with certainty that more commercial economic opportunities for Nepalese farmers (e.g. growing fruit and other crops for which they have a comparative advantage) would *necessarily* improve land management and reduce degradation. As the discussion of the land-management decision-making process in chapter 4 will suggest, there are many *other* exits from the vicious circle just described, aside from more effective land management. The more the explanation is linked to the social and political economic context, the more unpredictable and less direct the impact of environmental policy becomes. On the other hand the more directly environmental and technical the explanation, the less it is able to account for human agency in the management of land. The knack in explanation must lie in the ability to grasp a few strategic variables that both relate closely together in a causal manner, and which are relatively sensitive to change. In that way the most promising policy variables and paths of social change can be identified.

In the Nepalese case, the need to distinguish between the natural processes of a high-energy environment and the effects of human interference – and further to distinguish between the harmful and the positive elements of that interference – is of paramount importance. As we shall see again and again in this book, not enough is known of the modern history of the land surface to establish this distinction, although there is certainly enough evidence to discard the generalizations of even a decade ago. What is certain is that the management efforts of the farmers of the middle hills are under stress of such an order as to threaten the basis of their livelihood. More specifically, the stresses are felt most keenly on the overgrazed forest-fringe areas and on the upland dry-terraced fields. Here, poverty is the basic cause of poor management, and the consequence of poor management is deepening poverty.

Sustainable rural livelihoods: practical concepts for the 21st century

Robert Chambers and Gordon R. Conway

December 1991

The purpose of this paper is to provoke discussion by exploring and elaborating the concept of sustainable livelihoods. It is based normatively on the ideas of capability, equity, and sustainability, each of which is both end and means.

In the 21st century livelihoods will be needed by perhaps two or three times the present human population. A livelihood comprises people, their capabilities and their means of living, including food, income and assets. Tangible assets are resources and stores, and intangible assets are claims and access. A livelihood is environmentally sustainable when it maintains or enhances the local and global assets on which livelihoods depend, and has net beneficial effects on other livelihoods. A livelihood is socially sustainable which can cope with and recover from stress and shocks, and provide for future generations.

For policy and practice, new concepts and analysis are needed. Future generations will vastly outnumber us but are not represented in our decision-making. Current and conventional analysis both undervalues future livelihoods and is pessimistic. Ways can be sought to multiply livelihoods by increasing resource-use intensity and the diversity and complexity of small-farming livelihood systems, and by small-scale economic synergy. Net sustainable livelihood effects and intensity are concepts which deserve to be tested. They entail weighing factors which include environmental and social sustainability, and net effects through competition and externalities.

The objective of sustainable livelihoods for all provides a focus for anticipating the 21st century, and points to priorities for policy and research. For policy, implications include personal environmental balance sheets for the better off, and for the poorer, policies and actions to enhance capabilities, improve equity, and increase social sustainability. For research, key questions are better understanding of (a) conditions for low human fertility, (b) intensity, complexity and diversity in small-farming systems, (c) the livelihood-intensity of local economies, and (d) factors influencing migration. Practical development and testing of concepts and methods are indicated.

For the reader, there is a challenge to examine this paper from the perspective of a person alive in a hundred years' time, and then to do better than the authors have done.

This publication was originally published by the Institute of Development Studies (IDS). Access at https://www.ids.ac.uk/publications/sustainable-rural-livelihoods-practical-concepts-for-the-21st-century/.

1 Context, conventions and concepts

1.1 Change and uncertainty

In almost every domain of human life, change is accelerating. This is true wherever we look, in the ecological, economic, intellectual, political, professional, psychological, social or technological aspects of our lives. Along with other changes, human aspirations are growing at an accelerating rate, not least because of the rapidity of technological change in access to information. It is not just that change is fast; it is getting faster and faster. In this unprecedented context, two aspects stand out. First, the conventional or normal concepts, values, methods and behaviour prevalent in professions are liable to lag further and further behind the frontiers. Second, future conditions become harder and harder to predict. In this flux and future uncertainty, we can expect (though who can be sure?) that change will continue to accelerate, that much professionalism will continue to be behind the times, and that we will continue to be out of date and wrong in our anticipation of the future.

One prediction, though, seems reasonably secure: short of a holocaust, pandemic, or sequence of massive disasters, the human population of the 21st century will be much larger than it is now. Although world population growth is decelerating, the population may well rise at least to somewhere in the range of 10 to 15 billion, or two or three times the 5 billion mark passed in the late 1980s.

The burden of this growth will fall largely on the poorer countries. In the current projections for the 36 year period 1989 to 2025, both the populations of low income countries, and those of middle income countries, are expected to rise by more than three quarters (for these and other estimates see WDR 1991: 254–5). If we take low income countries alone, population would rise by 2.3 billion, from 2.9 to 5.2 billion, and in 2025 still be rising. In sub Saharan Africa, (SSA), population would treble in the next 40 years.

The prospects for the future of cities in developing countries are particularly grim (see for example Sadik 1991; Schiffer 1991). Growth rates are reported often between 8 and 9 per cent per annum and cities of over 20 million may become common. The term mega-city is now being followed by hyper-city. In one estimate, by 2025, the inhabitants of cities in developing countries will total nearly 4 billion (Schiffer 1991). But the prospects are grim, too, for the rural poor. In round figures for SSA, even if the urban population of about 130 million were five times larger by 2025, at 650 million, the rural population would still by current estimates almost double, from 330 to about 650 million.

The implications for urban and rural development strategies are profound. When so many millions are already trapped in totally unacceptable poverty, it would be massively difficult simply to enable just them alone to gain adequate and decent levels of living; but when the huge anticipated population increases of the future are added, the prospect is daunting indeed. The challenge posed is both practical and analytical. What sorts of concepts and analysis could help us, the human race, meet this challenge better?

While much of the evidence and arguments in this paper apply to urban conditions, our main focus is rural. This is for two reasons. First, the needs of the rural poor are likely to get even less attention in the future. Visible misery, articulate aspiration and political organisation and influence in the cities may combine to concentrate resources in urban areas. Second, at the margin the larger the number of people who can live decently in rural areas, so the less will be the human pressure and misery of the towns. There are shades, subtleties and exceptions, but the major empirical and normative equation from which we start is to seek ways for most rural areas in developing countries to support many more people.

This conflicts with recognition that in low and middle income countries, the exploitation of rural resources is already often unsustainable, and least sustainable in those regions, countries and zones with the lowest urbanisation, the highest population growth rates, and the most vulnerable rural environments. Any strategy for environment and development for the 21st century which is concerned with people, equity and sustainability has, then, to confront the question of how a vastly larger number of people can gain at least basically decent rural livelihoods in a manner which can be sustained, many of them in environments which are fragile and marginal.

In looking for clues and answers to that question, a starting point is to examine some aspects of conventional analysis in the social sciences.

1.2 Defects of conventional professional analysis

In social science disciplines and professions, the context of accelerating change and uncertainty is often confronted by a conventional conservatism in concepts, values, methods and behaviour. In some universities, teachers and textbooks persist in conditioning students with the standard routines and reflexes of earlier times. Three modes of thinking in development teaching and analysis have proved singularly resistant to change: production thinking, employment thinking, and poverty-line thinking.

1.2.1 Production thinking

Problems defined variously as 'hunger', undernutrition, malnutrition, and famine are in this mode seen as problems of production, of producing enough food. There is, though, overwhelming evidence (see e.g. Sen 1981 for the seminal work) that these are much more problems of entitlements, of being able to command food supplies, than of production or supply.

1.2.2 Employment thinking

The problems of the poor are seen as lack of employment, leading to the prescription of generating large numbers of new 'workplaces' (e.g. Schumacher 1973). The ideal is full employment, in which everyone has a 'job'. But this misfits much rural reality, in which people seek to put together a living through multifarious activities.

1.2.3 Poverty-line thinking

Deprivation is defined in terms of a single continuum, the poverty line, which is measured in terms of incomes (especially wages or salaries) or consumption. The aim then is to enable more people to rise above the line, and fewer to sink below it. But deprivation and wellbeing, as poor rural people perceive them, have many dimensions which do not correspond with this measure (see e.g. Jodha 1988).

These three modes of analysis share two defects: an industrialised country imprint; and reductionism for ease of measurement. Production, employment and cash income as indicators of wellbeing are industrial preoccupations; and all three are also amenable to measurement along single scales – amount produced (whether tonnes of steel or tonnes of foodgrains), numbers employed in jobs, and earnings or wages in a weekly or monthly pay-packet. They have narrow conceptual bases thought out and designed in central places, and applied top-down to elicit data that fit into preset boxes. These concepts and measures, generated in urban conditions and for professional convenience, do not fit or capture the complex and diverse realities of most rural life. They account for the failure of much conventional analysis to pick up or show the plural priorities of the rural poor and their many and varied strategies to obtain a living.

1.3 Fundamentals: capability, equity and sustainability

In the 1970s and 1980s, priorities and prescriptions for development changed rapidly. In some fields, such as neo-classical economics, theory itself generated change. In others, theory and concepts lagged behind practice and experience. Gaps appeared most marked where linkages were weak between objectives and methods, or between different disciplines. But disciplinary reductionism – the limiting of values, concepts and methods to the narrow concerns of a single academic and professional discipline – has been increasingly challenged. Gaps and cross-linkages between ecology, economics and other social sciences offer scope and need for practical concepts. The question is whether concepts can be found which are useful both analytically, to generate insight and hypotheses for research, and practically, as a focus and tool for decision-making.

From the flux and debate of the past few years, we have taken three concepts variously found in or evolved in the social and biological sciences, and which have increasingly commanded consensus. (In an earlier draft, we included diversity, but this is more controversial, and the mutual reinforcement with the other three is less strong and more ambiguous than between the three themselves.) Each concept is represented by a single word. Each has two sides, normative and descriptive. Used normatively, each states a desirable goal or criterion for evaluation; and used descriptively, each can be empirically observed or in principle measured. The three concepts are capability, equity, and sustainability.

In proposing these three concepts, we, like others, are trapped in implicit paternalism. Faced with diversity and change in human conditions, values and aspirations, no search for universal concepts can fully escape top-down generalisation and prescription. So capability, equity, and sustainability are 'our' concepts, not 'theirs'. They are justified only as a stage in a constant struggle of questioning, doubt, dialogue and self-criticism, in which we try to see what is right and practicable, and what fits 'their' conditions and priorities, and those of humankind as a whole. In these, and other concepts, there can and should be nothing final.

The three concepts of capability, equity and sustainability are linked. Each is also both end and means: that is to say, each is seen as good in itself, as an end; and each is also seen as a means to good ends, to the extent that it can support the others. Linked together, capability, equity and sustainability present a framework or paradigm for development thinking which is both normative and practical. However, like other concepts of what is good, they are not always or necessarily mutually supporting (for example equity in access to a resource by no means assures sustainable resource use without appropriate and effective institutions for resource management and exploitation). The search has to be for ways in which these three concepts, as objectives, can be combined so that in practice conflict is low and mutual support high.

1.3.1 Capability

The word capability has been used by Amartya Sen (Sen 1984, 1987; Dreze and Sen 1989) to refer to being able to perform certain basic functionings, to what a person is capable of doing and being. It includes, for example, to be adequately nourished, to be comfortably clothed, to avoid escapable morbidity and preventable mortality, to lead a life without shame, to be able to visit and entertain one's friends, to keep track of what is going on and what others are talking about (Sen 1987:18; Dreze and Sen 1990: 11). Quality of life is seen in terms of valued activities and the ability to choose and perform those activities. The word capability has, thus a wide span, and being democratically defined, has diverse specific meanings for different people in different places, including the many criteria of wellbeing of poor people themselves (for examples of which see Jodha 1988).

Within the generality of Sen's use of capability, there is a subset of livelihood capabilities that include being able to cope with stress and shocks, and being able to find and make use of livelihood opportunities. Such capabilities are not just reactive, being able to respond to adverse changes in conditions; they are also proactive and dynamically adaptable. They include gaining access to and using services and information, exercising foresight, experimenting and innovating, competing and collaborating with others, and exploiting new conditions and resources.

1.3.2 Equity

In conventional terms, equity can be measured in terms of relative income distribution. But we use the word more broadly, to imply a less unequal distribution of assets, capabilities and opportunities and especially enhancement of those of the most deprived. It includes an end to discrimination against women, against minorities, and against all who are weak, and an end to urban and rural poverty and deprivation.

1.3.3 Sustainability

In development prose, 'sustainable' has replaced 'integrated' as a versatile synonym for 'good'. Few, if any, dissent from the view that development should now be sustainable. There are, though, many meanings and interpretations of the term (Lele 1991). Environmentally, sustainability refers to the new global concerns with pollution, global warming, deforestation, the overexploitation of non-renewable resources and physical degradation. It has become orthodox, verbally if not in behaviour, to take a long-term view, to have a sense of the global village with finite resources threatened by wasteful and polluting consumption on the one hand, and by rapid growth of population on the other. In common parlance, sustainability connotes self-sufficiency and an implicit ideology of long-term self-restraint and self-reliance. It is used to refer to life styles which touch the earth lightly; to organic agriculture with low external inputs; to institutions which can raise their own revenue; to processes which are self-supporting without subsidy. Socially, in the livelihood context, we will use sustainability in a more focused manner to mean the ability to maintain and improve livelihoods while maintaining or enhancing the local and global assets and capabilities on which livelihoods depend.

2 Livelihoods

2.1 Sustainable livelihoods as an integrating concept

Capabilities, equity, and sustainability combine in the concept of sustainable livelihoods. A livelihood in its simplest sense is a means of gaining a living. Capabilities are both an end and means of livelihood: a livelihood provides the support for the enhancement and exercise of capabilities (an end); and capabilities (a means) enable a livelihood to be gained. Equity is both an end and a means: any minimum definition of equity must include adequate and decent livelihoods for all (an end); and equity in assets and access are preconditions (means) for gaining adequate and decent livelihoods. Sustainability, too, is both end and means: sustainable stewardship of resources is a value (or end) in itself; and it provides conditions (a means) for livelihoods to be sustained for future generations.

A concept of sustainable livelihoods was put forward in the report of an Advisory Panel of the World Commission on Environment and Development. In calling for a new analysis, it proposed sustainable livelihood security as an integrating concept, and made it central to its report (WCED 1987a: 2–5). The definition was as follows:

> Livelihood is defined as adequate stocks and flows of food and cash to meet basic needs. Security refers to secure ownership of, or access to, resources and income-earning activities, including reserves and assets to offset risk, ease shocks and meet contingencies. Sustainable refers to the maintenance or enhancement of resource productivity on a long-term basis. A household may be enabled to gain sustainable livelihood security in many ways – through ownership of land, livestock or trees; rights to grazing, fishing, hunting or gathering; through stable employment with adequate remuneration; or through varied repertoires of activities.

The Panel argued that this was an integrating concept, since sustainable livelihood security was a precondition for a stable human population, a prerequisite for good husbandry and sustainable management, and a means of reversing or restraining destabilising processes, especially rural to urban migration. Sustainable livelihoods were seen as a means of serving the objectives of both equity and sustainability. From our perspective, sustainable livelihoods also provide the resources and conditions for the enhancement and exercise of capabilities.

Modifying the WCED panel definition, we propose the following working definition of sustainable livelihoods:

*   a livelihood comprises the capabilities, assets (stores, resources, claims and access) and activities required for a means of living: a livelihood is sustainable which can cope with and recover from stress and shocks, maintain or enhance its capabilities and assets, and provide sustainable livelihood opportunities for the next generation; and which contributes net benefits to other livelihoods at the local and global levels and in the short and long term.

2.2 Determinants of livelihood

There are numerous initial determinants of livelihood strategy. Many livelihoods are largely predetermined by accident of birth. Livelihoods of this sort may be ascriptive: in village India, children may be born into a caste with an assigned role as potters, shepherds, or washerpeople. Gender as socially defined is also a pervasive ascriptive determinant of livelihood activities. Or not necessarily ascriptively, a person may be born, socialised and apprenticed into an inherited livelihood – as a cultivator with land and tools, a pastoralist with animals, a forest dweller with trees, a fisherperson with boat and tackle, or a shopkeeper with shop and stock; and each of these may in turn create a new household or households in the same occupation.

Many livelihoods are also less singular or predetermined. Some people improvise livelihoods with degrees of desperation, what they do being largely determined by the social, economic and ecological environment in which they find themselves. A person or household may also choose a livelihood, especially through education and migration. Those who are better off usually have a wider choice than those who are worse off, and a wider choice is usually generated by economic growth. In a future of accelerating change, adaptable capabilities to exploit new opportunities may be both more needed and more prevalent.

2.3 The nature of human livelihoods

The simple definition of a livelihood as a means of securing a living summarises a reality which comes into focus as being complex as its parts are found and named, and its structure unravelled.

The definition of a livelihood can be at different hierarchical levels. The most commonly used descriptively is the household, usually meaning the human group which shares the same hearth for cooking. In adopting this level here, it is important to recognise an individual or intrahousehold level, in which the wellbeing and access of some household members, and especially women and children, may be inferior to that of others, especially men; and also the broader levels of the extended family, the social group, and the community. These levels are widely significant, but for the sake of brevity and clarity, we will here use the household as the unit of analysis.

In our provisional anatomy of a household livelihood, we postulate four categories of parts:

People	their livelihood capabilities	)
		)repertoire
Activities	what they do	)
Assets	tangible (resources and stores)	)
	and intangible (claims and access)	) portfolio
	which provide material and social	)
	means	)
Gains or outputs	a living, what they gain from what they do	

The core of a livelihood can be expressed as a living, and the main components and relationships presented as in Figure 2.1.

Of these four, the most complex is the portfolio of tangible and intangible assets.

In approaching the portfolio, Swift (1989) provides a good starting point. In his analysis of human vulnerability and responses to famine, he distinguished three classes of asset – investments, stores and claims. We accept and expand his broad use of 'asset' but shift the emphasis to include 'normal' living as well as survival in crisis. Adopting his division of assets into tangible and intangible, we separate out stores and resources as tangible, and claims and access as intangible. While these are large categories, putting them together avoids problems of overlap, since stores are often also resources and vice versa, and claims require access if they are to have any value. The two groups can be outlined as follows:

Stores and resources: These are tangible assets commanded by a household. **Stores** include food stocks, stores of value such as gold, jewellery and woven textiles, and cash savings in banks of thrift and credit schemes.

Figure 2.1 Components and Flows in a Livelihood

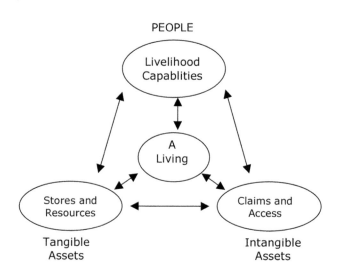

Resources include land, water, trees, and livestock; and farm equipment, tools, and domestic utensils. Assets are often both stores and resources, as with livestock, trees and savings.

Claims and access: These are intangible assets of a household. **Claims** are demands and appeals which can be made for material, moral or other practical support or access. The support may take many forms, such as food, implements, loans, gifts, or work. Claims are often made at times of stress or shock, or when other contingencies arise. Claims may be made on individuals or agencies, on relatives, neighbours, patrons, chiefs, social groups or communities, or on NGOs, government or the international community, including programmes for drought relief, or poverty alleviation. They are based on combinations of right, precedent, social convention, moral obligation and power. **Access** is the opportunity in practice to use a resource, store or service or to obtain information, material, technology, employment, food or income. Services here include transport, education, health, shops and markets. Information includes extension services, radio, television and newspapers. Technology includes techniques of cultivation and new seeds. Employment and other income-earning activities include rights to common property resources (CPRs) such as fuelwood or grazing on state or communal lands.

Out of these tangible and intangible assets people construct and contrive a living, using physical labour, skills, knowledge, and creativity. Skills and knowledge may be acquired within the household, passed on from generation to generation as indigenous technical knowledge, or through apprenticeship, or more formally through education or extension services, or through experiment and innovation.

Rural livelihoods, themselves, comprise one, or more often several, activities. These can include cultivation, herding, hunting, gathering, reciprocal or wage labour, trading and hawking, artisanal work such as weaving and carving, processing, providing services in transport, fetching and carrying and the like, begging, and theft. They variously provide food, cash, and other goods to satisfy a wide variety of human needs. Some of these outputs are consumed immediately, and others go into short or long-term stores, to be consumed later or to be invested in other assets.

As Swift (1989) points out, such investments occur when production leads to a surplus beyond immediate consumption requirements. Investments are made in enhancing or acquiring resources, in establishing claims, in gaining access, and in improving capabilities. Resources may be enhanced through investing labour as in terracing to improve the stock of soil, or through investing money in a cart to take produce to market. Claims may be established by investing in a marriage or by giving presents. Access to information may be obtained by investment in a radio or in education. Capabilities may be enhanced again through investment in (useful) education and training, and in apprenticeship. The results of successful investments are an added variety or quality of assets and/or capabilities which can be used for further production or in responding to future contingencies and threats to survival.

In addition to direct and physical benefits, adequate and decent livelihoods can and often do have other good effects. They can improve capabilities in the broader sense of the term by providing conditions and opportunities for widening choices, diminishing powerlessness, promoting self-respect, reinforcing cultural and moral values, and in other ways improving the quality of living and experience.

3 Sustainability

The sustainability of livelihoods raises many questions. These fall into two groups: whether a livelihood is sustainable environmentally, in its effects on local and global resources and other assets; and whether it is sustainable socially, that is, able to cope with stress and shocks, and retain its ability to continue and improve. Sustainability is thus a function of how assets and capabilities are utilised, maintained and enhanced so as to preserve livelihoods.

Environmental sustainability concerns the **external** impact of livelihoods on other livelihoods; social sustainability concerns their **internal** capacity to withstand outside pressures.

3.1 Environmental sustainability

Most conventional thinking equates sustainability with preservation or enhancement of the productive resource base, particularly for future generations. This can be separated into two levels.

The first level is local. The question here is whether livelihood activities maintain and enhance, or deplete and degrade, the local natural resource base. This is the familiar focus on visible aspects of sustainability. On the negative side, livelihood activities may contribute to desertification, deforestation, soil erosion, declining water tables, salinisation and the like. On the positive side, livelihood activities can improve productivity of renewable resources like air and river water, soil, organic soil fertility, and trees.

The second level is global. The question here is whether, environmentally, livelihood activities make a net positive or negative contribution to the long-term environmental sustainability of other livelihoods. This is the now familiar, but less visible, focus on issues such as pollution, greenhouse gases and global warming, the ozone layer, the irreversible use of the world's store of non-renewable resources, and the use of sinks (such as the sea for carbon dioxide) for pollution emissions (Agarwal and Narain 1991).

To this thinking on sustainability which is concerned with tangible assets, we would add the notion of preservation or enhancement of intangible assets. Livelihood activities can be regarded as environmentally unsustainable if they have a net negative effect on the claims and access needed by others. Claims and access can be diminished in several ways, including by law, by force, or by bureaucratic barriers. Examples of negative effects on claims and access to resources at the local level are their erosion or loss through appropriation and exclusion by the powerful. The livelihoods of the powerful gain, but there are net losses.

At the global level, livelihoods are threatened by international trade and other agreements that reduce claims and access to global markets for livelihood products and to global common properties, for example to ocean fisheries. The pervasive links between the global and the local levels (Davies and Leach 1991) are important and easily overlooked.

In this paper, we are concerned mainly with the local level, and mainly in the South. It is, though, imperative to point out that globally, the least environmentally sustainable livelihoods are those of the rich, mainly in the North. Any per capita calculation of the net source and sink demands made by the rich

– the people of countries of the North, and the rich of the South – would show their livelihoods to be far less sustainable in the global dimension than those of the poor.

Locally, the main challenge is to enhance the sustainable livelihood- intensity of resource use, especially in the rural areas of the South. Globally, the main challenge is to reduce the unsustainability of livelihoods, especially in the urban areas of the North.

3.2 Social sustainability

In terms of equity, the environmental sustainability of livelihoods has to be complemented by the social sustainability of all livelihoods. Social sustainability refers to whether a human unit (individual, household or family) can not only gain but maintain an adequate and decent livelihood. This has two dimensions, one negative, one positive. The negative dimension is reactive, coping with stress and shocks; and the positive dimension is proactive, enhancing and exercising capabilities in adapting to, exploiting and creating change, and in assuring continuity.

3.2.1 Coping with stress and shocks

The livelihoods and survival of human individuals, households, groups and communities are vulnerable to stresses and shocks. Vulnerability here has two aspects: external, the stresses and shocks to which they are subject; and internal, the capacity to cope (IDS 1989). Stresses are pressures which are typically continuous and cumulative, predictable and distressing, such as seasonal shortages, rising populations or declining resources, while shocks are impacts which are typically sudden, unpredictable, and traumatic, such as fires, floods and epidemics (Conway 1987; Conway and Barbier 1990). Any definition of livelihood sustainability has to include the ability to avoid, or more usually to withstand and recover from, such stresses and shocks.

Examples of livelihood stresses which build up gradually are: declining labour work available; declining real wages; declining yields on soils which degrade through salinisation, acidity or erosion; declining common property resources, and having to go further and spend longer for less, for fuel, fodder, grazing or water; declining water tables; declining rainfall; population pressures on resources leading to declining farm size and declining returns to labour; ecological change leading to lower bio-economic productivity; indebtedness; physical disabilities like river blindness, the effects of which build up gradually affecting the whole household (Evans 1989); and the domestic cycle with its periods of high ratios of dependents to active adults.

Regularly occurring stresses arise from cycles which are either diurnal (midday and afternoon heat, mosquitoes in the evening and at night, cold and difficulty seeing at night …) or seasonal. For the sustainability of livelihoods, seasonal stresses are more significant than diurnal. They have physical, biological, and socio-economic dimensions (Chambers, Longhurst and Pacey 1981; Chen 1991; Gill 1991) which often interlock at bad times of the year.

Examples of shocks affecting whole communities include wars, persecutions and civil violence, droughts, storms, floods, fires, famines, landslips, epidemics of crop pests or of animal or human illness, and the collapse of a market. Examples of shocks affecting individuals and households include accidents and sudden sickness; the death of a family member or of a valued animal; loss of assets through theft, fire or other disaster; and loss of a job.

Human and household strategies for coping with stress and shocks have a substantial literature (for some recent sources see the journal Disasters passim; Rahmato 1987; Corbett 1988; IDS 1989; de Waal 1989; Agarwal 1990; Gill 1991; Chen 1991). The strategies consist of mixes of the following:

- **stint**: reduce current consumption; shift to lower quality foods; draw on energy stored in the body
- **hoard**: accumulate and store food and other assets
- **protect**: preserve and protect the asset base for recovery and reestablishment of the livelihood
- **deplete**: draw upon household stores of food; pledge or sell assets
- **diversify**: seek new sources of food – wild foods, gleanings, wild animals, foods stored by rats and other animals (Beck 1991); diversify work activities and sources of income. especially in off-seasons
- **claim**: make claims on relatives, neighbours, patrons, the community, NGOs, the government, the international community, variously by calling in debts, appealing to reciprocity and good will, begging, and political action
- **move**: disperse family members, livestock (Rahmato 1987), and assets; and/or migrate.

Sustainable livelihoods are those that can avoid or resist such stresses and shocks and/or that are resilient and able to bounce back. Households' portfolio of tangible (stores and resources) and intangible (claims and access) assets can be understood as partly chosen by design to reduce vulnerability and to enable the household to survive stress and shocks with minimum risk of threat to the future livelihood. (A substantial though scattered literature shows tenacity on the part of the poor in protecting and hanging onto their productive assets in difficult times). Similarly, the repertoire of activities of household members can be interpreted partly as designed to spread risk.

Security is a basic dimension in livelihood sustainability. Assets can be vulnerable. Stores of grain can stolen, or destroyed by floods, fire or pests. Households can be deprived of their resources or their resource rights. Claims may be lost, as with death of a relative on whom a claim could have been made. Even access may disappear, as with government action to withdraw a bus service to market, or to close of a school or health centre.

Reducing vulnerability has two dimensions. The first is external through public action – to reduce external stress and shocks through flood prevention, disaster preparedness, off-season public works to provide employment, prophylaxis against diseases, and the like. The second is internal through private action, in which a household adds to its portfolio of assets and repertoire of responses so that it can respond more effectively and with less loss.

3.2.2 Dynamic livelihood capabilities

Social sustainability of a livelihood also depends on positive and dynamic competence, the ability to perceive, predict, adapt to, and exploit changes in the physical, social and economic environment. This aspect of sustainability has been recognised in agriculture in the work and writing of Roland Bunch (1985; 1988; 1989). In this approach, small farmers are enabled to improve their own experimentation, to conduct their own extension, and to organise to manage and exploit links with the wider economy. Awareness, experimental innovation, and adaptability contribute to dynamic capabilities. Through these, a farm family's livelihood can become more sustainable in uncertain and changing conditions where markets and prices fluctuate, and where old opportunities shrink and new ones appear and expand.

3.2.3 Intergenerational sustainability

The social sustainability of a livelihood also involves maintaining and enhancing capabilities for future generations. This intergenerational sustainability can be direct or indirect.

In its direct form, intergenerational sustainability takes the form of the inheritance of assets and/or skills: land or the tools of a trade are passed on to the next generation; skills and knowledge are transmitted from parents to children through family apprenticeship.

In its indirect form, intergenerational sustainability is achieved through children moving to other places or into other occupations. There they find or create new livelihoods which may be the same or different from those of the earlier generation. To enhance this form of sustainability, households often invest in education and the acquisition by children of skills other than those available within the household.

As rural populations rise, farm sizes diminish, and change accelerates, so dynamic livelihood capability and inter-generational sustainability become more critical.

4 Practical analysis

To translate concepts of livelihoods and sustainability into fair and efficient policies requires appropriate analytical orientations and methods. Three of these will be examined: valuing future livelihoods; enhancing livelihood-intensity; and estimating livelihood effects.

4.1 Valuing future livelihoods

Planning for future livelihoods implies the placing of a value on the future. The Brundtland Report (WCED 1987:8) emphasised this in its much-quoted definition of sustainable development as meeting the needs of the present without compromising the ability of future generations to meet their own needs, and thereby raised the issue of equity for future generations.

In practice, future generations and their livelihoods are undervalued in decision-making for four reasons:

- **Innumeracy** – the failure to recognise the numbers involved. Short of a massive global calamity or failure of human organisation, we who are alive today will be vastly outnumbered by future generations. If the population of the world were to stabilise at, say, 15 billion, with a life expectancy of 75, then future people would number 20 billion per century. Over a millenium, 200 billion people would live. At only 5 billion, we who are alive today would then be a minuscule minority, outnumbered by future people 4 to 1 over any one century, 40 to 1 over a thousand years, and 400 to 1 over ten thousand (roughly the equivalent of the time since agriculture began).

- **Undemocratic democracy** – the lack of democratic representation of future people. Future generations have no votes. Their interests can only be represented through the exercise of our imagination, altruism, restraint, and responsible stewardship.

- **Discounting** – devaluing the future. Economists discount the future in order to maximise net present value; politicians discount the future in order to win votes at elections which occur every five years or so; and businessmen discount the future in order to make profits now, and to repay interest on loans.

    The implications in economics are well known but deserve underlining. Discounting as taught and practised in conventional social cost-benefit analysis gives lower present values to benefits and costs the further away they are in the future. The manner in which this appears to justify development decisions with bad long-term environmental effects is not in dispute (see for example Pearce *et al* 1989). A cost of 50 million pounds in 100 years time has present values of only 69,016 pounds when discounted at 2 per cent, only 3,802 pounds at 5 per cent, and only 36 pounds at 10 per cent (Pearce *et al* 1989:136). Future benefits too are given similar low values. If we were starting from scratch, and were commissioned to devise a mode of valuation to penalise future generations, it is discounting that we would be driven to invent.

- **Uncertainty** – inability to predict the future. Futurologists have often been spectacularly wrong in their past predictions. Faced with accelerating ecological, technological and social change, we have few grounds for supposing that many current predictions will be better. To take only some of the more obvious examples, huge unknowns include the potentials and impacts of biotechnology, of nuclear fusion, and of the social impacts of the new mass communications.

Innumeracy, undemocratic democracy, discounting and uncertainty are bad reasons for taking a short-term view. Human ingenuity has a capacity to rationalise short-sighted selfishness as inadvertent altruism, as manifest in neoclassical economic theory. The political and professional challenge here is to offset the temptations of short-term advantage and to forego short-term benefits, for the sake of those future people who we can expect to outnumber us so astronomically.

Contrary to the values of normal democracy and discounting, intergenerational equity requires setting a higher value on future sustainable livelihoods than on present ones. This is to the extent that pressure of livelihoods on resources can be assumed to be more intense in the future. If there will be greater pressure and competition for livelihoods, then an adequate and sustainable livelihood will be harder to achieve, and will therefore be worth more than one at present.

This can be illustrated by hypothetical development decisions, and their anticipated effects on numbers of acceptable livelihoods.

Table 4.1 Hypothetical development decisions and livelihood trajectories

Number of livelihoods in

	10 years	50 years	100 years	150 years
Needed	50	100	150	150
Project A	100	50	25	10
Project B	75	75	75	75
Project C	25	50	100	100
Project D	25	50	100	150

Project A could be a canal irrigation system which generates temporary livelihoods during construction, and then later suffers declining productivity from silting and salinisation.

Project B could be small-scale lift irrigation with a good water table where investment takes place rapidly and production is then sustained.

Project C could be a diverse hardwood forestry project with slow-growing trees owned by communities or individuals who defer harvesting.

Project D could be progressive colonisation and intensification of agriculture in a swamp or in a newly forming river delta area, in each case with a land frontier.

Many considerations impinge on choices between projects like these, including externalities, the livelihood effects of reinvestment of capital generated, and uncertainties about technological change, about outcomes, and about livelihood trajectories. All the same, the table serves to illustrate the conflicting interests of generations. Normal innumeracy, undemocratic democracy, discounting and uncertainty prefer project A, or project B as a second best. Future generations would prefer project D, but that is now a rare case based on an atypical land frontier. So they would settle for C; and be appalled that we could ever have been so selfish and short-sighted as to have chosen A or B.

4.2 Enhancing livelihood-intensity

4.2.1 Professional pessimism

Recent writing about intergenerational equity has tended to be modest in its definition of sustainability, and so of our responsibilities for the future. This is reflected in the following quotations (cited from Pearce *et al* 1989: 173–185):

> … sustainability might be redefined in terms of a requirement that the use of resources today should not reduce real incomes in the future...sustainability ought to mean that a given stock of resources - trees, soil quality, water and so on – should not decline.
>
> (Markandya and Pearce 1988)

> The sustainability criterion suggests that, at a minimum, future generations should be left no worse off than current generations.
>
> (Tietenberg 1984)

> We summarise the necessary conditions [for sustainable development] as 'constancy of the natural capital stock.' More strictly, the requirement is for non-negative changes in the stock of natural resources such as soil and soil quality, ground surface waters and their quality, land biomass, water biomass, and the waste assimilation capacity of receiving environment.
>
> (Pearce, Barbier and Markandya 1988)

> In principle, such an optimal [sustainable growth] policy would seek to maintain 'an acceptable' rate of growth in per capita real incomes without depleting the national capital asset stock or the natural environmental asset stock.
>
> (Turner 1988)

These statements use negative syntax to express limited, even defensive objectives. The phrases 'should not reduce', 'should not decline', 'at a minimum...left no worse off', 'non-negative changes', 'non-declining', 'without depleting' – suggest a holding operation or rearguard action rather than an offensive. If economics is the dismal science, and environmentalism doleful, their hybrid can appear morbidly depressed.

Part of the problem is the concept of capital stocks and wealth. A constant stock notion taken on its own bequeaths future more numerous generations with less per capita. If the world population were to stabilise at 15 billion, then with constant stocks, future people would have only one third per capita of what we have today. The problem here lies in the concept of stock or wealth used by economists and environmentalists. They tend to think of environmental resources as a fixed quantity which is being used up. This is true for some resources but not others. It is true for non-renewable fossil fuels and for minerals. In the right conditions, it has not been true for many small farming systems which, in many parts of the world, have enhanced their natural resources stock and their wealth.

We do not minimise the massive problems posed by population growth, human greed, and depletion of the global resource base. But negative approaches tend to be self-fulfilling: if the focus is only on problems, opportunities are easy to miss. Moreover, current poverty and future populations present such an enormous and acute challenge that it is irresponsible not to explore more positive approaches. The question has to be how, from the finite resources of the globe, to generate a vastly larger number of decent, adequate and secure livelihoods which are also sustainable and self-enhancing.

4.2.2 Practical optimism

A more optimistic view emerges from studying what poor people do. There are no grounds for facile optimism. Nevertheless, in terms of resource productivity, and of livelihood-intensity, the actual practices of poor and small farmers and also of some who are landless suggest greater sustainable livelihood potentials than most concerned professionals have recognised. These potentials are found in two dimensions: in enhanced intensity and productivity of resource use; and in small-scale economic synergy.

a   Resource-use intensity and productivity

The livelihood potentials of resource use have been habitually underestimated, as for example by the often hysterical population and resources literature of the 1940s and 1950s. The underestimate has been made in two domains.

First, in small-scale farming there is better understanding now that for the complex, diverse and risk-prone agriculture of much of the South (outside the generally flat and irrigated green revolution areas), bio-economic productivity is enhanced and stabilised not by simplifying with high-input packages, but by complicating and diversifying with multiple interlinking enterprises. Mixed cropping, agroforestry, aquaculture, cut-and-carry stall feeding of livestock, the creation and protection of microenvironments which concentrate soil, water and nutrients (Chambers 1990), and intensified highly diverse home gardening (Nunez 1984; Cheatle and Njoroge 1991; Soemarwoto and Conway in press) are labour- and livelihood-intensive responses to risk and to rising population to land ratios. Many of the complications add synergistically to the nutrient flows internal to the farming system. Increasing heterogeneity will also diversify activities and outputs, and may then provide opportunities for more kinds of livelihood.

Second, and again echoing the findings of ecological analogies, degraded resources quite often present immense livelihood potential. Paradoxically, degradation has often protected resources for the poor. Because land is degraded – deforested, eroded, waterlogged, bare from grazing, flooded, or unsustainably cropped – it has low value. But again and again, when management practices are changed, remarkable bio-economic potential is realised (see e.g. Conroy and Litvinoff 1988, especially Bunch 1988 and Mishra and Sarin 1988). Some of the potentials are in growing trees: in India, some 69 million hectares of degraded lands could be growing trees to produce annual biomass increments dramatically greater than those current (Chambers, Saxena and Shah 1989: 39–49), with at least ten-fold increases in the production of most minor forest products. Since these are livelihood-intensive, the degraded Forest land of India (36 out of the 69 million hectares) could provide sustainable livelihoods to millions more poor people.

b   Small-scale economic synergy

The scale and significance in rural areas of activities other than a household directly farming its own land have been underperceived. One reason has been the reductionism of large-scale questionnaire survey methodology which tends to the recording of short and simple responses. But much evidence of a more social anthropological nature indicates that high proportions of incomes of the poor, even of those with land, derive from sources other than direct farming (see e.g. Chuta and Liedholm 1979 for an early review; also Harriss 1989). A study in rural Bangladesh (Magor and Orr 1990) found that only 37 per cent of reported household incomes came directly from agriculture, with 44 per cent from labour (including agricultural labour), and 19 from business, service and other sources. The proportions from direct agriculture rose sharply with landholding size. The implication is that, though at a low economic level in this case, many opportunities can be generated and exploited locally. When this occurs, there can be a synergy of recirculation of income.

To optimise that synergy raises questions of the market, of urban produce, of scale and cost of technology, of bureaucratic hassles through licensing and restrictions, and of local power structure. To optimise the synergy of recirculation is an issue which should perhaps be confronted more as a central concern, especially as more areas of the world will experience the population pressures now current in Bangladesh, and those in Bangladesh will intensify even more. A general hypothesis can be that recirculation through local purchases and provision of goods and services will be more livelihood-intensive than their import from outside.

Though we would argue that the scope for intensifying and complicating farming systems, and for small-scale synergy, is considerable in most environments, there are bio-economic limits. Just as there may be thresholds in population to land ratios for the adoption of more intensive technologies (Boserup 1965), so there must be thresholds beyond which immiseration is almost inevitable. It is not easy, for example, to be sanguine about the prospects of adequate and decent livelihoods for the rural poorer of Bangladesh in 2025, when the national population which stood at 111 million in 1989 is estimated to have risen by 77 per cent to 196 million (WDR 1991: 254). Restraining population growth has been, and will remain, one of the most vital and difficult preconditions for adequate and sustainable livelihoods for all in the long-term; and especially in those regions and countries where even with present populations they are so far from being achieved.

4.3 Net sustainable livelihoods

Concepts of wellbeing or deprivation have often been determined by their measureability. Convenience of measuring income or consumption has reinforced the definitions of deprivation as poverty, and of poverty as low income or low consumption. The ideas of employment, a job and a workplace are reinforced by the relative ease with which these can be identified and counted, especially in urban and industrial contexts; as professionals, we define as significant whatever we capture and can count in our crude and standard nets; and this misses much of the transient, mobile, dispersed and diverse livelihood activities of the rural poor.

In contrast, livelihoods, and sustainable livelihoods (SLs), are concepts which have evolved more from open-ended fieldwork than from the closed concerns of surveys and statistics. As we have seen, the empirical reality which they seek to encompass is not simple. To recapitulate, as narrowly defined, sustainable livelihoods include not just income and consumption, but ability to handle stress and shocks, and to satisfy basic needs; as defined more broadly, they include environmental sustainability, and good effects on others' livelihoods. SLs have many dimensions and multiple causality. They take different forms for different people in different environments. Not surprisingly they are not easy to measure or estimate; and any attempt to reduce measurement to a single scale or indicator risks doing violence to precisely the complexity and diversity which many rural livelihoods manifest – in themselves, in their relationships with the physical environment, and with each other.

In the practical world, though, criteria and ways of thinking are needed which can be used to make judgements about what to do and how to do it. The measurement fallacy has partly been to suppose that it is always necessary to know 'how much?' when often it is enough to know 'more' or 'less' or a trend. Also, an evaluative concept which conflates several criteria is usable once assessments of orders of magnitude, of relative values, and of trends are accepted as useful and usable for decision-making. As Carruthers and Clayton (1978) have indicated, decisions can be based on matrices which list estimated values for several criteria. Judgement can then be used to give weightings.

On these lines, the composite criterion we propose is net sustainable livelihoods (net SLs). This is a measure of the number of environmentally and socially sustainable livelihoods that provide an adequate living in a context less their negative effects on the benefits and sustainability of the totality of other livelihoods anywhere. Based on this, two practical concepts are net SL effects, and net SL intensity.

Net SL effects are the net adequate and sustainable livelihoods generated and supported by a livelihood itself, or by an enterprise, project, programme or policy, or by a resource use or locality, or by a social, economic or political grouping or system.

Net SL-intensity relates net adequate and sustainable livelihoods as numerator to the denominator of another livelihood, or an enterprise, project, programme or policy, or a resource use or locality, or a social, economic or political grouping or system. The more the net sustainable livelihoods supported, the higher the net SL-intensity.

Assessing net SL effects can be considered under three headings: environmental sustainability; social sustainability; and net effects.

4.3.1 Environmental sustainability effects

Assessing sustainability of livelihoods requires judgements about an increasingly unpredictable future. But the need to use judgement, and difficulty in using it, are not good reasons for holding off when the alternatives to a best guess are a less than best guess, based on easier but less fitting criteria. To the contrary, such difficulties and the need for the explicit exercise of judgement explain past neglect, and make it more important than ever to make good efforts now. Rather than trying to measure sustainability, we propose suggestive indicators, and for policy purposes, assessments of trends in these indicators.

Any given livelihood can in principle be assessed for its environmental sustainability. For the people of the rich North and the rich of the South, global rather than local environmental sustainability is more important. Here a measure of net per capita demand on the environment for its source and sink functions should be feasible. A do-it-yourself manual for the rich to assess their own net demands would seem overdue. For poor people in the South, a similar calculus is possible, but on equity grounds less important since their per capita net demand is lower.

For rural people in the South, local environmental sustainability is more relevant, pointing to the local livelihood resource base. This picks up physical and biological aspects of sustainability, and especially farming systems, common property resources, and raw materials needed for livelihoods. Some measures here will be those commonly used in the physical and biological sciences for assessing effects such as soil erosion, deforestation, and salinisation, and also the sustainable augmentation of productivity and livelihood-intensity of resource use.

The sustainable livelihood focus shifts some physical and biological criteria from their normal values. To illustrate, in livelihood effects, secondary forest can score better than primary forest: it generally has higher primary productivity and also higher livelihood-intensity. An example is 'degraded' sal forest in India which, as it regenerates, produces a rich harvest of non-timber products, including silkworms, fruits for oil, fodder grass, and leaves that are made into plates (Malhotra and Poffenberger 1990). Or again, in West Africa, conservationists are concerned about the loss of rainforest to farmbush fallow, a product of local rotational agricultural practices. Yet farmbush is far superior to forest in livelihood terms, as the main source of cultivable fertile land and of a wide range of non-timber forest products (pers comm. Leach M.).

The resource base has also to be that of a locality as a whole, not of just one part of it. For example, a slope may erode, but farmers may be trapping the silt lower down to make new fields. From a technical physical angle, this gets bad marks for sustainability; from a livelihood angle, it gets good marks, by concentrating soil, water and nutrients in stable and more highly productive micro-environments which provide more livelihoods which are also more stable and sustainable.

Rather more difficult to assess are the impacts of livelihood activity on the intangible assets of other livelihoods. Again there are dichotomies between rich and poor, global and local, that parallel the impacts on tangible assets. Appropriation of physical and biological resources by the rich is usually accompanied by denials of claims and access to resources by the poor, and these need to be included in the environmental sustainability calculus.

For a locality, community, or project, suggestive indicators are the status and trends in:

- migration (and off-season opportunities)
- rights and access to land, water, trees, and common property resources, and the security of those rights
- local non-farm income
- formal employment.

4.3.2 Social sustainability effects

Livelihood capability is a key part of the social sustainability of a livelihood. Indicators, though, are not obvious. Education has no simple correlation with capability; for some people education is enabling, enhancing capability; and for others it is disabling, diminishing capability, as with the trained distaste for farming of many educated young rural men. Health and physical competence are clearer: but even here, some of the handicapped find skills and niches which assure adequate livelihoods. Skills themselves are either difficult to measure or liable to mislead: old skills become obsolete, and newly acquired skills in training programmes can prove useless compared with apprenticeship. Important as livelihood capability is, it eludes easy counting, which may be one reason why its day as a concept used by professionals has yet to come.

Assets are easier to assess. The sustainability of a livelihood can be linked with its net assets. Of these, tangible assets are easier to identify and estimate. A useful and usable indicator of vulnerability and security, may be the net asset position of a household, defined as the value of its realisable tangible assets less liabilities such as debts.

Intergenerational social sustainability is hard to assess but making the effort does matter. Many trends – in increasing numbers of people, in smaller farm units, in higher dependence on off-farm income, in numbers of livelihoods from labouring, in continuing rural-rural and rural-urban migration – have mainly (though not universally or only) negative implications. Best judgements have to be made about these if present policies are to serve future people.

4.3.3 Net effects

Net effects brings together both sustainability and livelihoods. For net effects to be fully positive four conditions have to be met: first, environmental sustainability, and second, social sustainability, as both discussed above; then third, adequacy of livelihood, which presents problems of definition which require a separate paper; and fourth, net livelihoods. This last, concerning the net effects on the sum of sustainable livelihoods, raises additional questions which point to competition, and to externalities.

a   Competition

Many livelihoods are in competition. Proximate adverse effects of competition apply especially to the poor. A livelihood may be sustainable in itself but may weaken or destroy others. Where many compete for few opportunities, as in oversupplied labour markets, each livelihood diminishes others. Or the success of one enterprise is at the cost of others. A prospering shopkeeper drives rivals to ruin. A landowner buys up the land of distressed neighbours. A household appropriates common land and deprives the poor of pasture and wild foods. Net SL effects have to deduct adverse effects on other livelihoods and on their sustainability.

b   Externalities

Livelihoods and life styles have many indirect effects or externalities. Adverse externalities are marked in the net demands of the richer on the global environment. Let us suppose that an intervention improves the living standards of people who already have adequate (though not environmentally sustainable) livelihoods, say a majority in an industrialised country. If that improvement increases their net demand on the environment, the net SL-intensity of the intervention will be negative: for there will have been no gain in adequate livelihoods, and a loss in global sustainability.

In summary, then, interventions which make the adequate but unsustainable livelihoods of the rich sustainable, or which make the inadequate but sustainable livelihoods of the poor adequate, should have positive SL-intensity. But, and depending on weightings, the most positive SL-intensity should be where the unsustainable and inadequate livelihoods of the poor become both sustainable and adequate.

To conclude, the assessment of net SL effects, like the measurement of poverty, is a problem for professionals rather than for poor rural people. It is not simple to do, but useful for the questions it raises. It becomes easier the less one tries to measure, and the more one relies on commonsense, judgement, relative rather than absolute values, and trends and directions of change.

The key to assessment, here, will often be the local experts, the rural people themselves. Under the rubrics of rapid rural appraisal (RRA) and participatory rural appraisal (PRA), a battery of participatory methods has been developed which enable them to do more of the analysis themselves. Their criteria vary, fitting and reflecting local conditions and aspirations. Adequate and sustainable livelihoods are a common aspiration of the poor. Professional assessments will always be needed; but the more poor rural people themselves play a part in making assessments, the more they will be empowered; and the more policies and practice support their priorities, so the more they will be enabled to achieve for themselves the sorts of sustainable livelihoods they want and need.

5 Conclusion: for the 21st century and beyond

In this paper we have not tried to cover every aspect of sustainable livelihoods. Everything is connected with everything else, and the reader will have no difficulty in listing external factors like international debt, transnational enterprises, defence expenditures, international terms of trade, agricultural pricing and much else that could have been included. We have narrowed attention largely to the household, and especially the poorer rural rather than urban household in the South, but we have also taken a long-term perspective. There are implications for policy, for research, and for the reader.

5.1 Policy implications

Policy implications follow from the long-term view of an extended future for the human race, of many more people having to live on the earth. Here we will select a few of the most salient.

For the richer, the priority is to change their life styles to make lower demands on the environment. If the rich make lower demands, more is left for the poor and for future generations. Pricing and taxation policies could contribute to reducing environmental demand. A world-wide campaign for awareness and abstinence is also implied. One part could be personal livelihood environmental balance sheets. These would show the scale of personal debt to the environment and to future generations. They would apply much more to the rich than the poor, showing individually how much we are drawing on our common global capital.

For the poorer, who are our main concern here, and especially the rural poorer, the three concepts with which we started apply.

5.1.1 Enhancing capability

Livelihood capability in a context of change and unpredictability requires being adaptable, versatile, quick to change, well-informed, and able to exploit diverse, complicating and changing resources and opportunities. There are practical implications for the provision of enabling infrastructure and services, including:

- education for livelihood-linked capability
- health, both preventive and curative to prevent permanent disability
- bigger and better baskets of choices for agriculture, and support for farmers' experiments
- transport, communications and information services (about rights, markets, prices, skills...)
- flexible credit for new small enterprises

5.1.2 Improving equity

Giving priority to the capabilities, assets and access of the poorer, including minorities and women. Practical implications for these groups include:

- redistribution of tangible assets, especially land, and land to the tiller
- secure rights to land, water, trees and other resources, and secure inheritance for children
- protection and management of common property resources and equitable rights of access for the poorer
- enhancing the intensity and productivity of resource use, and exploiting small-scale economic synergy
- rights and effective access to services, especially education, health and credit
- removing restrictions which impoverish and weaken the poor

5.1.3 Increasing social sustainability

Reducing vulnerability by restraining external stress, minimising shocks, and providing safety nets, so that poor people do not become poorer. Practical measures are many (see also IDS 1989) and include:

- peace and equitable law and order
- disaster prevention
- counterseasonal strategies to provide food, income and work for the poorer at bad times of the year
- prompt support in bad years, and high prices for tangible assets people sell in distress
- health services that are accessible and effective in bad seasons, including treatment for accidents
- conditions for lower fertility

Among the poorer, these three thrusts are mutually supporting.

5.2 Research implications

Any list of research implications **for the rich** should be headed by estimating the net environmental demands of their livelihoods, and devising individual environmental balance sheets to provoke awareness and action to reduce demands.

For the rural poorer, our main concern, we see four sets of key research questions. None is new; but we suggest they are critical and demand renewed attention:

5.2.1 Population

Understanding better why people want to have more or fewer children, and what needs to be done to create conditions in which they will want and be able to have fewer, especially learning lessons from societies, groups and rural conditions in which small family size has been preferred.

5.2.2 Intensity, complexity and diversity

In agriculture: understanding better the sequences and conditions for intensification, complication and diversification, including synergy, sequences, gains from microenvironments, bioeconomic limits to livelihoods, and methods and scope for farmers' own analysis and innovation.

5.2.3 Livelihood-intensity of local economic circulation and non-farm activities

Understanding better the synergy of local economic circulation, and the dynamics and livelihood-intensity of local economies, their links with outside markets, and how locally based livelihoods can be increased and sustained.

5.2.4 Sustainable rural livelihoods and migration

Understanding better how more people can want and be able to continue to gain their livelihoods in rural areas, how to prevent distress migration to urban areas, and how to support voluntary reversals of rural-urban migration.

Finally, straddling between practice and research, are questions of **practical development and testing of concepts and methods**. These include:

- assessing the environmental and socio-economic conditions under which livelihood -intensity is maximised
- testing and developing SL-intensity and net SL effects as practical concepts in policy-formulation and project cycles
- exploring the utility of SL criteria as complements or alternatives to discounting, and how to offset the normal view from theÿpresent by taking the view backwards from the future

5.3 Implications for the reader

In this paper we have tried to open up and explore concepts, analogies and relationships to fit future human needs. In doing so we have rejected some conventional professional wisdom. In the spirit of exploration, we have also allowed ourselves the liberty of speculation, of seeing where a line of thinking would lead. In consequence, the paper raises more questions than it answers. It also puts forward combinations of working concepts, categories and hypotheses for testing for practical utility.

The reader can judge the concept of sustainable livelihoods from a contemporary view. We would prefer it to be judged by a view from the future, let us say, from a hundred years' time. If humankind has not annihilated itself, and if no major disaster has befallen, there will perhaps be between 10 and 20 billion people alive then. How would we, if we were them, standing there, a hundred years hence, judge this paper and its concepts? What would we, from that future vantage point, say that it missed? What, with what we knew then, would we wish had been written here?

We leave those questions as challenges, for readers to take that stance, and to do better than we have done.

Glossary

For the convenience of readers, and especially for those who skim, dip and miss bits, here are the meanings given to a number of terms in the text.

- access: opportunity in practice to use a resource, store or service, or to obtain information, material, technology, employment, food or income.
  see pages: i, 4–9, 11, 19–20, 22–23

- assets: resources and stores (tangible assets), and claims and access (intangible assets), which a person or household commands and can use towards a livelihood.
  see pages: i, 4–9, 11–12, 19–20, 22–23

- capabilities: what a person or household is capable of doing and being (after Sen). Livelihood capabilities comprise the ability to gain a livelihood, including abilities to cope with stress and shocks, to be dynamically adaptable, and to explore and exploit opportunities.
  see pages: i, 4–10, 12, 22

- claims: demands and appeals which can be made for material, moral or other practical support or access. Claims are based on combinations of right, precedent, social convention, moral obligation, and power.
  see pages: i, 6–9, 11, 19

- environmental sustainability: referring to a livelihood, maintaining or enhancing physical livelihood potentials locally and globally, and having net beneficial effects on assets and opportunities for other livelihoods in the short and long term.
  see pages: 9–10, 18–20

- livelihood: a means of living, and the capabilities, assets and activities required for it.
  see pages: i, 2, 4–24

- net sustainable livelihood effect refers to the net adequate and sustainable livelihoods generated and supported by a livelihood itself, or by an enterprise, project, policy or programme, or by a resource or locality, or by a social, economic of political grouping or system.
  see page: i

- shocks are impacts which are typically sudden, unpredictable, and traumatic, such as fires, floods, storms, epidemics, thefts, civil disorder, and wars. Contrast stresses.
  see pages: i, 4–6, 9–11, 18, 23

- social sustainability referring to a livelihood, the ability of a human unit (individual, household or family) to cope with and recover from stresses and shocks, to adapt to and exploit changes in its physical, social and economic environment, and to maintain and enhance capabilities for future generations.

  see pages: i, 9–10, 12, 18, 20, 23

- stresses are pressures which are typically cumulative, predictable, and variously continuous or cyclical, such as seasonal shortages, rising populations, declining soil fertility, and air pollution. Contrast shocks.

  see pages: 10–11

- sustainable livelihoods: a livelihood is sustainable which can cope with and recover from stress and shocks, maintain or enhance its capabilities and assets, and provide sustainable livelihood opportunities for the next generation; and which contributes net benefits to other livelihoods at the local and global levels and in the short and long term.

  see pages: i, 5–6, 9, 11, 13, 17–18, 20–22, 24

References

Agarwal, A. and Narain, S., 1991, Global Warming in an Unequal World: A case of environmental colonialism, Centre for Science and Environment, New Delhi

Agarwal, B., 1989, 'Social security and the family in rural India', DEP 21, The Development Economics Research Programme, London School of Economics, London, September

Beck, T., 1989, 'Survival strategies and power among the poorest in a West Bengal village', IDS Bulletin Vol 20 No 2, pp 23–32

—— 1991, 'Poverty and Power: survival of the poorest in three villages of West Bengal, India', PhD thesis submitted to the University of London, February

Boserup, E., 1965, The Conditions of Agricultural Growth: the economics of agrarian change under population pressure, George Allen and Unwin, London

Bunch, R.., 1985, Two Ears of Corn: a guide to people-centered agricultural improvement, World Neighbours, 5116 North Portland, Oklahoma City, Oklahoma 73112

—— 1988, 'Guinope integrated development program, Honduras', in Czech Conroy and Miles Litvinoff (eds) The Greening of Aid: Sustainable Livelihoods in Practice, Earthscan, London, pp 40–44

—— 1989, 'Encouraging farmers' experiments' in Robert Chambers, Arnold Pacey and Lori Ann Thrupp (eds) Farmer First: Farmer Innovation and Agricultural Research, Intermediate Technology Publications, London, pp 55–61

Carruthers, I. and Clayton, E., 1978, 'Ex post evaluation of agricultural projects – its implication for planning', Journal of Agricultural Economics (UK), Vol 28 No 3

Chambers, R.., 1990, 'Microenvironments unobserved', Gatekeeper Series No 22, International Institute for Environment and Development, 3 Endsleigh Street, London, WC1H ODD

Chambers, R.., Longhurst, R. and Pacey, A. (eds), 1981, Seasonal Dimensions to Rural Poverty, Frances Pinter, London

Chambers, R,., Pacey, A. and Thrupp, L-A. (eds), 1989, Farmer First: Farmer Innovation and Agricultural Research, Intermediate Technology Publications, London

Chambers, R., Saxena, N.C. and Shah, T., 1989, To the Hands of the Poor: Water and Trees, Oxford and IBH, 66 Janpath, New Delhi 110001, and Intermediate Technology Publications, London

Cheatle, R.J. and Njoroge J., 1991, 'Smallholder adoption of composting and double digging: an NGO approach to environmental conservation', paper for the workshop Environment and the Poor: Soil and Water Management for Sustainable Smallholder Development, the Aberdares, Arusha and Nairobi, 31 May to 12 June 1991

Chen, M. A., 1991, Coping with Seasonality and Drought, Sage Publications, New Delhi; Newbury Park, California and London

Chuta, E. and Liedholm, C., 1979, 'Rural non-farm employment: a review of the state of the art', Rural Development Paper No 4, Michigan State University, East Lansing

Conroy, C. and Miles, L. (eds), 1988, The Greening of Aid: Sustainable Livelihoods in Practice, Earthscan, London

Conway, G.R., 1987, 'The properties of agroecosystems', Agricultural Systems No 24, pp 95–117

Conway and Barbier, E.B., 1990, After the Green Revolution: sustainable agriculture for development, Earthscan, London

Corbett, J., 1988, 'Famine and household coping strategies', World Development, Vol 16 No 9, pp 1099–1112

Davies, S. and Leach, M., 1991, 'Globalism versus villagism: national and international issues in food security and the environment', IDS Bulletin Vol 22 No 3, July, pp 43–50

de Waal, A., 1989, Famine that Kills: Darfur, Sudan 1984–1985, Clarendon Press, Oxford

Dréze, J. and Sen, A., 1989, 'Public action for social security: foundations and strategy', DEP No 20, The Development Economics Research programme, London School of Economics, London. August

Evans, T., 1989, 'The impact of permanent disability on rural households: river blindness in Guinea', IDS Bulletin No 20 No 2, pp 41–48

Gill, G.J., 1991, Seasonality and Agriculture in the Developing World: A Problem of the Poor and the Powerless, Cambridge University Press, Cambridge

Harriss, J., 1989, 'Knowing About Rural Economic Change: problems arising from a comparison of the results of 'macro' and 'micro' research in Tamil Nadu', in Pranab Bardhan (ed.) Conversations between Economists and Anthropologists, Oxford University Press, pp 137–173

Hartmann, B. and Boyce, J., 1983, A Quiet Violence: View from a Bangladesh Village, Zed Press, London

IDS, 1989, 'Vulnerability: how the poor cope', IDS Bulletin Vol 20 No 2

Jodha, N.S., 1988, 'Poverty in India: a minority view', Economic and Political Weekly, Special Number, November pp 2421–8

Lele, S.M., 1991, 'Sustainable development: a critical review' World Development Vol 19 No 6, pp 607–621

Magor, N. and Orr, A., 1990, 'Reducing vulnerability in Bangladesh', paper for the 1990 Asian Farming Systems Research and Extension Symposium, Bangkok, 19–22 November

Malhotra, K.C. and Poffenberger, M., 1990, Forest Regeneration through Community Protection: The West Bengal experience, Ford Foundation, New Delhi

Mishra, P.R. and Sarin, M., 1988, 'Social security through social fencing, Sukhomajri and Nada, North India', in Czech Conroy and Miles Litvinoff (eds) The Greening of Aid: Sustainable Livelihoods in Practice, Earthscan, London, pp 22–28

Nunez, V.K., 1984, Household Gardens: Theoretical Considerations on an Old Survival Strategy, International Potato Center, Lima, Peru

Pearce D., Markandya, A. and Barbier, E.B., 1989, Blueprint for a Green Economy, Earthscan, London

Rahmato, D., 1988, 'Peasant survival strategies', in Angela Penrose (ed) Beyond the Famine: an Examination of Issues behind Famine in Ethiopia, International Institute for Relief and Development, Food for the Hungry International, Geneva

Sadik, N., 1991, 'Confronting the challenge of tomorrow's cities – today', Development Forum Vol 19 No 2 March–April p 3

Schiffer, R.L., 1991, 'Pioneers in uncharted territory' Development Forum, Vol 19 No 2, March-April pp 3 and 21

Schumacher, E.F., 1973, Small is Beautiful: a Study of Economics as if People Mattered, Abacus Edition (1974) published by the Penguin Group

Sen, A., 1981, Poverty and Famines: an Essay on Entitlement and Famines, Clarendon Press, Oxford

—— 1984, Resources, Values and Development, Basil Blackwell, Oxford

—— 1987, The Standard of Living, The Tanner Lectures, Clare Hall, Cambridge 1985, Cambridge University Press, Cambridge

Soemarwoto, O. and Conway, G.R. in press, 'The Javanese home garden', Journal of Farming Systems Research and Extension

Swift, J., 1989, 'Why are rural people vulnerable to famine?', IDS Bulletin Vol 20 No 2

WCED, 1987a, Food 2000: Global Policies for Sustainable Agriculture, a Report of the Advisory Panel on Food Security, Agriculture, Forestry and Environment to the World Commission on Environment and Development, Zed Books Ltd, London and New Jersey

—— 1987b, Our Common Future, The Report of the World Commission on Environment and Development, Oxford University Press, Oxford, New York

WDR, 1991, World Development Report 1991, Oxford University Press for the World Bank

PART III

SOCIO-ENVIRONMENTAL RESEARCH IN ECONOMICS, SOCIOLOGY, AND POLITICAL SCIENCE

Richard B. Norgaard

Karl Polanyi opens the readings in this part with two very powerful sentences:

> What we call land is an element of nature inextricably interwoven with man's institutions. To isolate it and form a market out of it was perhaps the weirdest of all undertakings of our ancestors. (reading 19)

All of the readings in this part grapple with how people and nature have become separated and need to be interwoven again. To put these readings in context, we need to ask how nature and human institutions became unwoven in the first place.

For the vast majority of human history, knowledge of the couplings between people and nature informed how people organized themselves to survive and thrive in their environment. Water was a community resource. Forests were shared for hunting wildlife, gathering fuel, and felling trees for lumber. Pastures were grazed in common, too. Traditional knowledge guided forest and pasture practices, while social norms kept human use, individually and collectively, within what communities understood to be the bounds of nature. Their understanding was sometimes wrong, but when population levels were low, people could ameliorate disaster by extending their activities further from their village or, in extreme situations, changing locations.

Over time, social and environmental systems became decoupled in the modern Western mind and institutions through a combination of mutually reinforcing transitions in England and Europe that were then spread around the globe.

First, beginning with the Renaissance, there has been the Western belief that humankind can progressively better control nature through modern science and technology. The decoupling of society and environment, of what was inextricably interwoven, has been an objective of modernity. Particular technologies based on narrow science, especially fossil fuel technologies, massively transformed people's relation to nature with the equally massive, delayed costs of climate change. Modern science focused ever more sharply on particular threads, while the big picture, the complex interweaving of the threads between people and nature, became increasingly blurry. Meanwhile the hope for human progress became anchored to an ever-growing economy. This book identifies how parts of the blurred interweaving have been brought into focus again and have helped facilitate the development of coupled socio-environmental systems analysis.

Second, society reorganized from feudal systems to market systems. Feudal estates, for all their limitations, were much more self-sufficient than comparable regions today. Serfs engaged with nature in multiple ways. As markets and trade arose, spurred in 1776 by Adam Smith's rationalization of markets in *The Wealth of Nations*, people specialized their roles and became interdependent with each other over greater distances. The combustion of fossil fuels also facilitated trade and specialization, expanding the reach of the market and eventually growing the number of people who lived around industrial centers. Within a few centuries, few people interacted directly with nature, and those who did specialized in farming a few crops and knew less about other crops. Historically, people grew up learning about nature and society directly. Now, for many, nature is modeled, monitored, analyzed, and disseminated in books and on computer screens.

Third, another key transition is the idea that land could be divided up into separate parcels and owned and managed by individuals. A recent incarnation is that nature consists of parts we can think of as natural capital that might also be owned and managed, though perhaps by a community. But pieces of land and parts of nature are all interconnected, and they need to be tightly reweaved with diverse social institutions. Private property enables the interconnection through markets, with individuals choosing to do what is best for them. A plethora of regulations on economic behavior are modest efforts to reweave people and nature in order to offset the unexpected social and environmental consequences of new technologies that had transformed the inextricable interweaving. The readings in this part all grapple with the consequences of having replaced historic ways of socially organizing around environmental complexities with simpler modern notions of progress, property, individualism, and markets.

The first reading selection, by economic historian **Karl Polanyi** (1886–1964), reflects on the transition from feudal relations between people and land in England and Europe to private property relations with land uses determined by market incentives. Born in Vienna, Polanyi moved to England in 1933 to escape fascism and subsequently lived in the United States and Canada. In *The Great Transformation: The Political and Economic Origins of Our Time* (1944), Polanyi describes a critical transition in how Western people understood the natural world and subsequently related to it. In his description, society and environment were "inextricably interwoven" in feudal systems, with commons playing a central role. With modest variations around the globe and changing over time, feudal systems survived for up to a millennium and beyond, indeed even well into the twentieth century (Linklater 2013). Serfs, the people who worked the land in feudal systems, were left with little more than that necessary for survival, but they did have a modicum of shelter, access to particular parcels of farmland, shared access to pasture and woodland commons, a church, and community mores that provided some individual security. Polanyi tells how serfs were severed from the land as land became private and managed in response to markets. Severed also from shelter and community, because these were inextricably interwoven with land, serfs were transformed into seasonal agricultural wage laborers or driven to the city to find work in the emerging commercial and industrial economy. Land use decisions became a matter of maximizing the return on land in response to market signals, with grim consequences.

Writing as Nazi bombs were falling on London, Polanyi suggested how breakdowns in markets for land and labor before and during the Great Depression contributed to the rise of Hitler's fascism, while comparable breakdowns in the ways land was inextricably interwoven with institutions in the Soviet Union contributed to the rise of Stalin and authoritarianism. Polanyi's solution was a re-embedding of markets for land and labor in social institutions. His book is a classic work on the emergence of the market economy.

Taking a narrower focus, economist **H. Scott Gordon's** (1924–2019) analysis of the economics of a hypothetical fishery in 1954 can also be read as an effort to reconnect people and nature. Gordon's work extends that of Harold Hotelling (1931), who postulated that market competition was sufficient to set the price and rate of extraction for the optimal depletion of exhaustible resources like oil and timber, by identifying the institutional requirements for sustainably managing a depletable, open access fishery in a market economy. Gordon refers to the fishery as a common-pool resource rather than an open-access resource, but the fishery is unlike the traditional commons described by Karl Polanyi and Elinor Ostrom. Gordon posits that there are no social norms or governing regulations limiting how much an individual fisherman is allowed to catch. He then shows that if fishing is not regulated, fish stocks will become depleted as each fisher maximizes their own profits, and all fishers suffer. With regulation, the economically optimum level of fishing can be reached. Free markets need rules, and Gordon showed why rules were necessary for an open-access, single species fishery. He left the more difficult problems of multiple fish species and changing ocean conditions to others. While Gordon's work falls short of a full synthesis of social and environmental systems, it highlights the need for a recoupling, an idea that resonated in fisheries science 40 years later (**Pauly 1995, Part IV**).

Drawing on work on common-pool resources from neoclassical economics, human ecologist **Garrett Hardin** (1915–2003) wrote one of the most cited papers of all time. In "Tragedy of the Commons" published in *Science* in 1968, Hardin presents a Malthusian argument (**Malthus 1798, Part I**) on how people globally do not limit their populations and are destroying nature – air, water, and the biosphere – because much of nature is commonly available to all. Hardin stresses how commons, because they are freely available, are overused. Hardin calls the overpopulation problem and associated environmental problems "no technical solution problems." This argument had a stunning impact on natural scientists, in part because Hardin was addressing a social issue in language that was clear to them. Social scientists, on the other hand, quickly pointed out, and still do, that Hardin was a naïve Malthusian who mistook an open access resource for a commons. As a consequence, social scientists downplayed Hardin's environmental concerns. Hardin had not read Karl Polanyi or Karl Marx (**1867, Part I**), for example, and seriously misinterpreted the history of commons and the role they played before the transition to a market economy (Cox 1985; Basarto and Ostrom 2009). Nevertheless, this paper has been on the syllabus of environmental studies courses around the world for decades. It continues to be cited, and Hardin's historically incorrect concept of the commons has now become the dominant understanding of the term, showcasing the legacy effect of his idea (Amaral 2013).

"The Tragedy of the Commons" is a good example among many of unnecessary disciplinary isolation, if not disciplinary arrogance. Hardin could have sought the knowledge of social scientists only buildings away on the Santa Barbara campus of the University of California and written a more powerful and insightful paper that would have united biological and social scientists. As we read this paper more than a half century later, we see again how many important things Hardin had to say that could have been said historically and institutionally correctly and thereby more deeply, bringing social scientists along too, had he humbly sought a coauthor from the social sciences who complemented his strengths in biology (Feeny et al. 1990).

Economist **Herman E. Daly** (1938–) also challenged the narrow perspective of his discipline. In that process, Daly helped found ecological economics, a transdiscipline that more fully synthesizes across economics and ecology by addressing the interdependence and coevolution of human economies and natural ecosystems. Within that field, Daly is famous for his arguments for a no growth or steady-state economy (1973) that operates within the finite limits of nature, reflecting his training in economics with Nicholas Georgescu-Roegen, author of *The Entropy Law and the Economic Process* (1971). Like Hardin, Daly saw nature as an open access resource with scientifically definable thresholds within which the economy must work. Arresting growth is a first step along the way to learning how to live within nature. A similar view was also advanced by Kenneth Boulding, who had earlier written "The Economics of the Coming Spaceship Earth" (1966), and by Donella and Dennis Meadows, Jorgen Randers, and William H. Behrens III in the Club of Rome study on *The Limits to Growth* (**Meadows et al. 1972, Part VI**). These critiques of economic growth, accented by the 1973 petroleum crisis documenting the economy's dependence on oil, likely helped Daly land this 1974 paper in the flagship journal of the American Economic Association. A little over a decade later, the World Bank hired Daly to help it incorporate resources and environment into its thinking about sustainable development, but the bank proved too entrenched in conventional thinking to incorporate Daly's ecological economics.

Economists were not alone in their efforts to reintroduce the environment into their discipline. **William R. Catton Jr.** (1926–2019) and **Riley E. Dunlap** (1943–) were instrumental in defining the now established sub-discipline of environmental sociology with our selected article written in 1978. Before that time, the discipline had long ignored environmental factors as relevant to social systems. It was not always so, however. Sociologist Edward Alsworth Ross' 1927 demographic analysis *Standing Room Only* is based on a solid understanding of land and agricultural systems. The second chapter of my 1964 college textbook *Handbook of Modern Sociology* is titled "Social Organization and the Ecosystem" (Duncan 1964). Catton and Dunlap's manifesto is informed by the modern environmental movement that arose in the 1960s, and they thereby reclaimed the importance of the environment in sociology. Their article juxtaposes two modes of sociological thinking about society and environment: the human exceptionalism paradigm, which views society as unlimited by nature, versus the new environmental paradigm, which "accepts 'environmental' variables as meaningful for sociological investigation" (see also Freudenburg et al. 1995). Notably, their article demonstrates the reciprocal relations between society and environment using the example of social inequality and environmental harm. They project that economic inequality leads to greater resource exploitation, which in turn exacerbates inequality. Since its foundational years, environmental sociology has taken an increasingly expansive approach to society and environment, engaging with other branches of sociology and other disciplines.

The recoupling of society and environment in and across disciplines was not only an intellectual effort but also an effort inspired by daunting societal challenges, as reflected in a chapter from **Amartya Sen's** (1933–) book *Poverty and Famines*. Born in the state of Bengal in then British India, Sen trained in economics and mathematics in both India and the United Kingdom. In 1998, he won the economics prize in honor of Alfred Nobel for arguing that the advance and spread of human capabilities should be central to our understanding of freedom, well-being, and economic development. In *Poverty and Famines*, Sen documents the limits of the core argument of Malthus (**1798, Part I**), arguing that environmental factors such as droughts that affect food availability are not sufficient to explain why famines occur. Sen shows in case after case, of which our reading centers on the Sahel region in North Africa, that the key factor is a society's entitlement system. The distribution of assets and access to land, the economic security provided, and even the entitlement to food per se, combined with the dynamics of markets, affect who and how many are deprived of food. Droughts and other environmental factors are still important to understanding famines, but Sen shows that a more complete understanding results from examining the interactions between social and environmental systems. Sen's insights on the multiple interdependencies of society and environment in food systems are mirrored, if not always acknowledged, in contemporary food systems research.

Sociologist **Robert D. Bullard** (1946–) is widely recognized as one of the first scholars in the field of environmental justice, now so commonly referred to that it goes simply by its initials, EJ. Bullard started his career in urban sociology and housing. Switching his focus from who lived where to what was dumped where, Bullard first documented the racial inequalities of waste dumping in Houston (Bullard 1983). In 1987, the United Church of Christ's Commission on Racial Justice under the leadership of Reverend Benjamin Chavis, Jr., held a conference on environmental justice and issued a report on "Toxic Wastes and Race." This brought EJ to national attention in the US. The chosen reading in this book is the second chapter of Bullard's classic book *Dumping in Dixie: Race, Class, and Environmental Quality* (1990). The reading lays out the history of industrialization, urban development, and waste dumping in the American Southeast and how even, indeed sometimes especially, regulated dumping affects people of color and the poor far more than the rich.

Environmental justice became a well-recognized field within environmental sociology and spread across many disciplines, too. Bullard stressed the injustices to Black Americans, and the field expanded to all people who have been marginalized in the development process around the globe. The income and racial disparities between those who have caused climate change and those who suffer its consequences, including future generations, is surely the biggest EJ issue of our time and has spurred the field of climate justice. The atmosphere has been an open access dump to which anyone could emit greenhouse gases, but in fact the rich did so first and did so rapidly, pushing the natural system past thresholds before the poor and people of color even got started. Closing the open access of the atmosphere and making it into a managed commons has proven especially difficult because the benefits and costs have been so unfairly distributed.

For political scientist **Elinor Ostrom** (1933–2012), the core concern was recognition of the prevalence and longevity of community-based institutions for managing natural resource systems. She won the 2009 economics Nobel Prize for showing that traditional common property institutions for the management of pastures, forests, and fisheries still exist not only in developing countries but also in Europe, the United States, and Japan. More importantly, contrary to Hardin 1968, commons institutions continue to work effectively, protecting water, land, and fisheries to their users' satisfaction. Commonly managed property resources (CPRs in Ostrom's writings) should not be privatized à la Hardin – and as advocated by most economists as well – without comparing the pros and cons of property vs. commons institutions. CPRs can be superior to private property for attaining both individual and social goals. Operating under the will of its users, common property can be used well and in the interests of all, incorporating most of the benefits or costs individual users impose on each other. Ostrom was the first woman, and now, in 2022, one of but two women to have won the economics Nobel Prize. Ostrom and her husband Vincent Ostrom were trained in political science and considered themselves theorists of how institutions work, not economists. Most economists did not even know who Ostrom was or why she deserved the Nobel Prize, suggesting again that more scholars could at least be scanning the horizons of their disciplines. Ostrom founded and led a school of CPR scholars through regular week-long workshops through which she taught and inspired hundreds of scholars who subsequently incorporated her reasoning and approach in their own research (Amaral 2013).

The selection by **Sharachchandra Lélé** (1962–) zeroes in on the friction between the rhetoric of sustainable development and its application. Trained in engineering, systems analysis, and the natural sciences in India and rapidly learning social theory for his PhD at University of California, Berkeley, Lélé was an emerging ecological economist in the late 1980s. At that time, the idea that development should be environmentally sustainable gained global credence with the publication of the Brundtland Commission's *Our Common Future* report (WCED 1987). "Sustainable development" very quickly became a buzzword in the development literature of academia and international development agencies; indeed, none were advocating "unsustainable" development. Sustainable development posits the possibility of win–win–win solutions that simultaneously enable economic growth, protect the environment, and promote social justice. Strategies to address air pollution provide a case in point. In the 1990s, economists Grossman and Kreuger (1995) showed that urban air pollution initially rose with development and then declined with more development. They concluded that the solution to reducing pollution, which they equated with sustainability, was more development. The pattern of growth and decline of urban air pollution with increasing national income was labelled the environmental Kuznets curve and the pattern has been found for other types of environmental degradation. It reaped instant favor among economists and intense scrutiny and deep criticism among ecological economists and environmental scientists for ignoring the offshoring of pollution from more to less developed regions, for ignoring pollutants like greenhouse emissions and solid waste that do not follow the pattern, and for ignoring the role politics and other aspects of social systems play in controlling pollution (Stern 2004).

Lélé was intensely aware that the sustainable development frameworks being proposed presumed knowledge of the linkages between social and natural systems that were both highly complex and virtually unknown. He was irritated by the presumptiveness of it all and concerned that implementing ill-informed sustainable development would do more damage than good. Lélé saw sustainable development as a coupled systems problem. The boundary and connectivity issues Lélé raised in the early years of sustainable development became the focus of much subsequent work, and yet they remain largely unsettled.

The final legacy reading in this part is a chapter from my 1994 book *Development Betrayed* (**Norgaard 1994**). Born in 1943, I grew up loving the natural sciences as a student but realized early that people were the problem in socio-environmental systems, in particular their economic beliefs. In the early 1960s, I came to fully understand this when I saw the Glen Canyon of the Colorado River, a place I knew and loved, drown behind a dam built to produce electricity. I earned a PhD in economics from the University of Chicago by observing, like an ethnographer, the core beliefs and central narratives of economists. I learned evolutionary ecology early through interdisciplinary research on the biological control of agricultural pests and by coming to know the Amazon rainforest. Bringing the two disciplines together, I am among the founders of ecological economics.

My reading from *Development Betrayed* (1994) suggests how the processes of economic development might be understood and pursued in environmentally sustainable ways. It is informed by my experience as a member of a 19-person, Brazilian-Amazon planning team. I observed well-trained planners argue that the next development plan, after 500 years of development failure by Westerners in the Amazon (see **Moran 1981, Part II**), was going to work. I argued that it would be better to promote diverse development experiments and try to learn from them what to do next. The development of Europe and North America had not been planned, it had coevolved. My book elaborates a socio-environmental framework based on change through the coevolution of knowledge, values, organizational, technological, and natural systems. I used this framework to suggest how social and environmental systems affect each other, both through cause and effect and by natural selection of characteristics that better fit those in the overall changing system. I designed the framework to be more explicit on the social side. I might have also divided the environment into the geosphere and biosphere, for example. However, the number of connections, i.e., arrows, nearly doubles with the addition of simply one more system. The diagram is already at the limit of comfortable comprehension. If the globe were a patchwork quilt of coevolving multiple, diverse socio-environmental systems, each patch could learn from the experiments underway in other patches. A coevolving patchwork quilt of cultures would be more resilient to a new stressor than a single globalized system dominated by market or any particular pattern of thinking. This coevolutionary way of thinking gives more credence to traditional knowledges and helps us see the consequences of the dominance of modern ways of thinking. My reading shows that not only is coupled socio-environmental systems thinking important, but that the systemic framing used is also critical.

The decoupling of societies from nature, launched by the Enlightenment and the Industrial Revolution, resulted in dramatic economic growth largely driven by fossil fuels, with a vast array of adverse effects on both societies and environments. It upended systems of shared use of natural resources, traditionally referred to as "commons," which encompassed the environmental resource and the combination of traditional knowledge and social norms that regulated its shared use. The selections reviewed here all seek a basis for the recoupling of society and environment. One key connecting strand is a focus on the management possibilities of commons as resource systems. Viewed through a modern lens, the term has been reinterpreted to mean a resource that can be accessed by everyone without restriction. The term "open access resource" is more appropriate for the typical use of the term commons today. Gordon, Hardin, and Ostrom present evolving viewpoints on how patterns of resource access affect resource viability, which in turn influences resource governance. These debates resonate today as concerns like climate change, declining fish stocks, and biodiversity loss are framed as "global commons" challenges. Another key strand across readings is the re-embedding proposed by Polanyi. But while Polanyi advocated for the re-embedding of markets in society, Daly, Sen, Catton and Dunlap, Lélé, and Norgaard theorize a re-embedding of society and economy within natural systems and propose knowledge systems that support learning and experimentation under conditions of uncertainty. Finally, a central contribution of the selections here is to spotlight the inequalities in access to resources, consumption of resources, exposure to pollution, and responsibility for environmental improvement across race, class, and geography (e.g., Polanyi, Bullard, Catton and Dunlap, Ostrom, Lélé).

References

Amaral, Eduardo. "Ostrom, Hardin and the Commons: A Critical Appreciation and a Revisionist View." *Environmental Science and Policy* 36 (2013): 11–23.

Basarto, Xavier, and Elinor Ostrom. "The Core Challenges of Moving beyond Garrett Hardin." *Journal of Natural Resources Policy Research* 1, no. 3 (2009): 255–259.

Boulding, Kenneth E. "The Economics of the Coming Spaceship Earth." In *Environmental Quality in a Growing Economy*, edited by Henry E. Jarrett, 3–14. Baltimore: Johns Hopkins University Press, 1966.

Bullard, Robert D. "Solid Waste Sites and the Black Houston Community." *Sociological Inquiry* 53, no. 2–3 (1983): 273–288.

Cox, Susan Jane Buck. "No Tragedy of the Commons." *Environmental Ethics*, 7, no. 1 (1985): 49–61.

Daly, Herman E. *Toward a Steady-State Economy*. San Francisco: W.H. Freeman, 1973.

Duncan, Otis Dudley. "Chapter 2: Social Organization and the Ecosystem." In *Handbook of Modern Sociology,* edited by Robert E. L. Farris. Chicago: Rand McNally, 1964.

Feeny, David, Fikret Birkes, Bonnie J. McKay, and James M. Acheson. "The Tragedy of the Commons 22 Years Later." *Human Ecology* 18, no. 1 (1990): 1–19.

Freudenburg, William R., Susan Frickel, and Robert Gramling. "Beyond the Nature/ Society Divide: Learning to Think about a Mountain." *Sociological Forum* 10, no. 3 (1995): 361–392.

Georgescu-Roegen, Nicholas. *The Entropy Law and the Economic Process*. Cambridge, MA: Harvard University Press, 1971.

Grossman, Gene M., and Alan B. Krueger. "Economic Growth and the Environment." *Quarterly Journal of Economics* 110, no. 2 (1995): 353–377.

Hotelling, Harold. "The Economics of Exhaustible Resources." *Journal of Political Economy* 39, no. 2 (1931): 137–175.

Linklater, Andro. *Owning the Earth: The Transforming History of Land Ownership.* London: Bloomsbury, 2013.

Ross, Edward Alsworth. *Standing Room Only.* New York: The Century Co, 1927.

Stern, David I. "The Rise and Fall of the Environmental Kuznets Curve." *World Development* 32, no. 8 (2004): 1419–1439.

WCED, World Commission on Environment and Development. *Our Common Future: Report of the World Commission on Environment and Development.* Oxford: Oxford University Press, 1987.

15

MARKET AND NATURE

WHAT WE CALL LAND is an element of nature inextricably interwoven with man's institutions. To isolate it and form a market out of it was perhaps the weirdest of all undertakings of our ancestors.

Traditionally, land and labor are not separated; labor forms part of life, land remains part of nature, life and nature form an articulate whole. Land is thus tied up with the organizations of kinship, neighborhood, craft, and creed—with tribe and temple, village, gild, and church. One Big Market, on the other hand, is an arrangement of economic life which includes markets for the factors of production. Since these factors happen to be indistinguishable from the elements of human institutions, man and nature, it can be readily seen that market economy involves a society the institutions of which are subordinated to the requirements of the market mechanism.

The proposition is as utopian in respect to land as in respect to labor. The economic function is but one of many, vital functions of land. It invests man's life with stability; it is the site of his habitation; it is a condition of his physical safety; it is the landscape and the seasons. We might as well imagine his being born without hands and feet as carrying on his life without land. And yet to separate land from man and organize society in such a way as to satisfy the requirements of a real-estate market was a vital part of the utopian concept of a market economy.

Again, it is in the field of modern colonization that the true significance of such a venture becomes manifest. Whether the colonist needs land as a site for the sake of the wealth buried in it, or whether he merely wishes to constrain the native to produce a surplus of food and raw materials, is often irrelevant; nor does it make much difference whether the native works under the direct supervision of the colonist or only under some form of indirect compulsion, for in every and any case the social and cultural system of native life must be first shattered.

There is close analogy between the colonial situation today and that of Western Europe a century or two ago. But the mobilization of land which' in exotic regions may be compressed into a few years or decades may have taken as many centuries in Western Europe.

The challenge came from the growth of other than purely commercial forms of capitalism. There was, starting in England with the Tudors, argricultural capitalism with its need for an individualized treatment of the land, including conversions and enclosures. There was industrial capitalism which—in France as in England—was primarily rural and needed sites for its mills and laborers' settlements, since the beginning of the eighteenth century. Most powerful of all, though affecting more the use of the land than its ownership, there was the rise of industrial towns with their need for practically unlimited food and raw material supplies in the nineteenth century.

Superficially, there was little likeness in the responses to these challenges, yet they were stages in the subordination of the surface of the planet to the needs of an industrial society. The first stage was the commercialization of the soil, mobilizing the feudal revenue of the land. The second was the forcing up of the production of food and organic raw materials to serve the needs of a rapidly growing industrial population on a national scale. The third was the extension of such a system of surplus production to overseas and colonial territories. With this last step land and its produce were finally fitted into the scheme of a self-regulating world market.

Commercialization of the soil was only another name for the liquidation of feudalism which started in Western urban centers as well as in England in the fourteenth century and was concluded some five hundred years later in the course of the European revolutions, when the remnants of villeinage were abolished. To detach man from the soil meant the dissolution of the body economic into its elements so that each element could fit into that part of the system where it was most useful. The new system was first established alongside the old which it tried to assimilate and absorb, by securing a grip on such soil as was still bound up in precapitalistic ties. The feudal sequestration of the land was abolished. "The aim was the elimination of all claims on the part of neighborhood or kinship organizations, especially those of virile aristocratic stock, as well as of the Church—claims, which exempted land from commerce or mortgage." [1] Some of this was achieved by individual force and violence, some by revolution from

[1] Brinkmann, C., "Das soziale System des Kapitalismus," *Grundriss der Sozial-ökonomik,* 1924.

above or below, some by war and conquest, some by legislative action, some by administrative pressure, some by spontaneous small-scale action of private persons over long stretches of time. Whether the dislocation was swiftly healed or whether it caused an open wound in the body social depended primarily on the measures taken to regulate the process. Powerful factors of change and adjustment were introduced by the governments themselves. Secularization of Church lands, for instance, was one of the fundaments of the modern state up to the time of the Italian *Risorgimento* and, incidentally, one of the chief means of the ordered transference of land into the hands of private individuals.

The biggest single steps were taken by the French Revolution and by the Benthamite reforms of the 1830's and 1840's. "The condition most favorable to the prosperity of agriculture exists," wrote Bentham, "when there are no entails, no unalienable endowments, no common lands, no right of redemptions, no tithes. . . ." Such freedom in dealing with property, and especially property in land, formed an essential part of the Benthamite conception of individual liberty. To extend this freedom in one way or another was the aim and effect of legislation such as the Prescriptions Acts, the Inheritance Act, the Fines and Recoveries Act, the Real Property Act, the general Enclosure Act of 1801 and its successors,[2] as well as the Copyhold Acts from 1841 up to 1926. In France and much of the Continent the Code Napoléon instituted middle-class forms of property, making land a commerciable good and making mortgage a private civil contract.

The second step, overlapping the first, was the subordination of land to the needs of a swiftly expanding urban population. Although the soil cannot be physically mobilized, its produce can, if transportation facilities and the law permit. *"Thus the mobility of goods to some extent compensates the lack of interregional mobility of the factors*; or (what is really the same thing) trade mitigates the disadvantages of the unsuitable geographical distribution of the productive facilities."[3] Such a notion was entirely foreign to the traditional outlook. "Neither with the ancients, nor during the early Middle Ages—this should be emphatically asserted—were the goods of every day life regularly bought and sold."[4] Surpluses of grain were supposed to

[2] Dicey, A. V., *op. cit.*, p. 226.
[3] Ohlin, B., *Interregional and International Trade*, 1935, p. 42.
[4] Bücher, K., *Entstehung der Volkswirtschaft*, 1904. Cf. also Penrose, E. F., *Population Theories and Their Application*, 1934, quotes Longfield, 1834, for the first mention of the idea that movements of commodities may be regarded as substitutes for movements of the factors of production.

provision the neighborhood, especially the local town; corn markets up to the fifteenth century had a strictly local organization. But the growth of towns induced landlords to produce primarily for sale on the market and—in England—the growth of the metropolis compelled authorities to loosen the restrictions on the corn trade and allow it to become regional, though never national.

Eventually agglomeration of the population in the industrial towns of the second half of the eighteenth century changed the situation completely—first on a national, then on a world scale.

To effect this change was the true meaning of free trade. The mobilization of the produce of the land was extended from the neighboring countryside to tropical and subtropical regions—the industrial-agricultural division of labor was applied to the planet. As a result, peoples of distant zones were drawn into the vortex of change the origins of which were obscure to them, while the European nations became dependent for their everyday activities upon a not yet ensured integration of the life of mankind. With free trade the new and tremendous hazards of planetary interdependence sprang into being.

The scope of social defense against all-round dislocation was as broad as the front of attack. Though common law and legislation speeded up change at times, at others they slowed it down. However, common law and statute law were not necessarily acting in the same direction at any given time.

In the advent of the labor market common law played mainly a positive part—the commodity theory of labor was first stated emphatically not by economists but by lawyers. On the issue of labor combinations and the law of conspiracy, too, the common law favored a free labor market, though this meant restricting the freedom of association of organized workers.

But, in respect to land, the common law shifted its role from encouraging change to opposing it. During the sixteenth and seventeenth centuries, more often than not common law insisted on the owner's right to improve his land profitably even if this involved grave dislocation in habitations and employment. On the Continent this process of mobilization involved, as we know, the reception of Roman law, while in England common law held its own and succeeded in bridging the gap between restricted medieval property rights and modern individual property without sacrificing the principle of judge-made law vital to constitutional liberty. Since the eighteenth century, on the other hand, common law in land acted as a conserver of the past in the face of modernizing legislation. But eventually, the Benthamites had their way, and, between 1830 and 1860, freedom of contract was extended to the land. This powerful trend was reversed only in the 1870's when legislation altered its course radically. The "collectivist" period had begun.

The inertia of the common law was deliberately enhanced by statutes expressly passed in order to protect the habitations and occupations of the rural classes against the effects of freedom of contract. A comprehensive effort was launched to ensure some degree of health and salubrity in the housing of the poor, providing them with allotments, giving them a chance to escape from the slums and to breathe the fresh air of nature, the "gentleman's park." Wretched Irish tenants and London slum dwellers were rescued from the grip of the laws of the market by legislative acts designed to protect their habitations against the juggernaut, improvement. On the Continent it was mainly statute law and administrative action that saved the tenant, the peasant, the agricultural laborer from the most violent effects of urbanization. Prussian conservatives such as Rodbertus, whose Junker socialism influenced Marx, were blood brothers to the Tory-Democrats of England.

Presently, the problem of protection arose in regard to the agricultural populations of whole countries and continents. International free trade, if unchecked, must necessarily eliminate ever-larger compact bodies of agricultural producers.[5] This inevitable process of destruction was very much aggravated by the inherent discontinuity in the development of modern means of transportation, which are too expensive to be extended into new regions of the planet unless the prize to be gained is high. Once the great investments involved in the building of steamships and railroads came to fruition, whole continents were opened up and an avalanche of grain descended upon unhappy Europe. This was contrary to classical prognostication. Ricardo had erected it into an axiom that the most fertile land was settled first. This was turned to scorn in a spectacular manner when the railways found more fertile land in the antipodes. Central Europe, facing utter destruction of its rural society, was forced to protect its peasantry by introducing corn laws.

But if the organized states of Europe could protect themselves

[5] Borkenau, F., *The Totalitarian Enemy*, 1939, Chapter "Towards Collectivism."

against the backwash of international free trade, the politically unorganized colonial peoples could not. The revolt against imperialism was mainly an attempt on the part of exotic peoples to achieve the political status necessary to shelter themselves from the social dislocations caused by European trade policies. The protection that the white man could easily secure for himself, through the sovereign status of his communities was out of reach of the colored man as long as he lacked the prerequisite, political government.

The trading classes sponsored the demand for mobilization of the land. Cobden set the landlords of England aghast with his discovery that farming was "business" and that those who were broke must clear out. The working classes were won over to free trade as soon as it became apparent that it made food cheaper. Trade unions became the bastions of anti-agrarianism and revolutionary socialism branded the peasantry of the world an indiscriminate mass of reactionaries. International division of labor was doubtless a progressive creed; and its opponents were often recruited from amongst those whose judgment was vitiated by vested interests or lack of natural intelligence. The few independent and disinterested minds who discovered the fallacies of unrestricted free trade were too few to make any impression.

And yet their consequences were no less real for not being consciously recognized. In effect, the great influence wielded by landed interests in Western Europe and the survival of feudal forms of life in Central and Eastern Europe during the nineteenth century are readily explained by the vital protective function of these forces in retarding the mobilization of the land. The question was often raised: what enabled the feudal aristocracy of the Continent to maintain their sway in the middle-class state once they had shed the military, judicial, and administrative functions to which they owed their ascendency? The theory of "survivals" was sometimes adduced as an explanation, according to which functionless institutions or traits may continue to exist by virtue of inertia. Yet it would be truer to say that no institution ever survives its function—when it appears to do so, it is because it serves in some other function, or functions, which *need not include the original one*. Thus feudalism and landed conservatism retained their strength as long as they served a purpose that happened to be that of restricting the disastrous effects of the mobilization of land. By this time it had been forgotten by free traders that land formed part of the territory of the country, and that the territorial character of sovereignty was not merely a result of sentimental associations, but of massive facts, including economic ones. "In contrast to the nomadic peoples, the cultivator commits himself to improvements *fixed in a particular place.* Without such improvements human life must remain elementary, and little removed from that of animals. And how large a role have these fixtures played in human history! It is they, the cleared and culti-vated lands, the houses, and the other buildings, the means of com-munication, the multifarious plant necessary for production, including industry and mining, all the permanent and immovable improvements that tie a human community to the locality where it is. They cannot be improvised, but must be built up gradually by generations of patient effort, and the community cannot afford to sacrifice them and start afresh elsewhere. Hence that *territorial* character of sovereignty, which permeates our political conceptions." [6] For a century these obvious truths were ridiculed.

The economic argument could be easily expanded so as to include the conditions of safety and security attached to the integrity of the soil and its resources—such as the vigor and stamina of the population, the abundance of food supplies, the amount and character of defense materials, even the climate of the country which might suffer from the denudation of forests, from erosions and dust bowls, all of which, ultimately, depend upon the factor land, yet none of which respond to the supply-and-demand mechanism of the market. Given a system entirely dependent upon market functions for the safeguarding of its existential needs, confidence will naturally turn to such forces outside the market system which are capable of ensuring common interests jeopardized by that system. Such a view is in keeping with our appre-ciation of the true sources of class influence: instead of trying to ex-plain developments that run counter to the general trend of the time by the (unexplained) influence of reactionary classes, we prefer to explain the influence of such classes by the fact that they, even though inci-dentally, stand for developments only seemingly contrary to the general interest of the community. That their own interests are often all too well served by such a policy offers only another illustration of the truth that classes manage to profit disproportionately from these services that they may render to the commonalty.

An instance was offered by Speenhamland. The squire who ruled the village struck upon a way of slowing down the rise in rural wages and the threatening dislocation of the traditional structure of village

[6] Hawtrey, R. G., *The Economic Problem,* 1933.

life. In the long run, the method chosen was bound to have the most nefarious results. Yet the squires would not have been able to maintain their practices, unless by doing so they had assisted the country as a whole to meet the ground swell of the Industrial Revolution.

On the continent of Europe, again, agrarian protectionism was a necessity. But the most active intellectual forces of the age were engaged in an adventure which happened to shift their angle of vision so as to hide from them the true significance of the agrarian plight. Under the circumstances, a group able to represent the endangered rural interests could gain an influence out of proportion to their numbers. The protectionist countermovement actually succeeded in stabilizing the European countryside and in weakening that drift towards the towns which was the scourge of the time. Reaction was the beneficiary of a socially useful function which it happened to perform. The identical function which allowed reactionary classes in Europe to make play with traditional sentiments in their fight for agrarian tariffs was responsible in America about a half century later for the success of the TVA and other progressive social techniques. The same needs of society which benefited democracy in the New World strengthened the influence of the aristocracy in the Old.

Opposition to mobilization of the land was the sociological background of that struggle between liberalism and reaction that made up the political history of Continental Europe in the nineteenth century. In this struggle, the military and the higher clergy were allies of the landed classes, who had almost completely lost their more immediate functions in society. These classes were now available for any reactionary solution of the *impasse* to which market economy and its corollary, constitutional government, threatened to lead since they were not bound by tradition and ideology to public liberties and parliamentary rule.

Briefly, economic liberalism was wedded to the liberal state, while landed interests were not—this was the source of their permanent political significance on the Continent, which produced the crosscurrents of Prussian politics under Bismarck, fed clerical and militarist *revanche* in France, ensured court influence for the feudal aristocracy in the Hapsburg empire, made Church and Army the guardians of crumbling thrones. Since the connection outlasted the critical two generations once laid down by John Maynard Keynes as the practical alternative to eternity, land and landed property were now credited with a congenital bias for reaction. Eighteenth century England with its Tory free traders and agrarian pioneers was as forgotten as the Tudor engrossers and their revolutionary methods of making money from the land; the Physiocratic landlords of France and Germany with their enthusiasm for free trade were obliterated in the public mind by the modern prejudice of the everlasting backwardness of the rural scene. Herbert Spencer, with whom one generation sufficed as a sample of eternity, simply identified militarism with reaction. The social and technological adaptability recently shown by the Nipponese, the Russian, or the Nazi army would have been inconceivable to him.

Such thoughts were narrowly time-bound. The stupendous industrial achievements of market economy had been bought at the price of great harm to the substance of society. The feudal classes found therein an occasion to retrieve some of their lost prestige by turning advocates of the virtues of the land and its cultivators. In literary romanticism Nature had made its alliance with the Past; in the agrarian movement of the nineteenth century feudalism was trying not unsuccessfully to recover its past by presenting itself as the guardian of man's natural habitat, the soil. If the danger had not been genuine, the stratagem could not have worked.

But Army and Church gained prestige also by being available for the "defense of law and order," which now became highly vulnerable, while the ruling middle class was not fitted to ensure this requirement of the new economy. The market system was more allergic to rioting than any other economic system we know. Tudor governments relied on riots to call attention to local complaints; a few ringleaders might be hanged, otherwise no harm was done. The rise of the financial market meant a complete break with such an attitude; after 1797 rioting ceases to be a popular feature of London life, its place is gradually taken by meetings at which, at least in principle, the hands are counted which otherwise would be raining blows.[7] The Prussian king who proclaimed that to keep the peace was the subject's first and foremost duty, became famous for this paradox; yet very soon it was a commonplace. In the nineteenth century breaches of the peace, if committed by armed crowds, were deemed an incipient rebellion and an acute danger to the state; stocks collapsed and there was no bottom

[7] Trevelyan, G. M., *History of England*, 1926, p. 533. "England under Walpole, was still an aristocracy, tempered by rioting." Hannah More's "repository" song, "The Riot" was written "in ninety-five, a year of scarcity and alarm"—it was the year of Speenhamland. Cf. *The Repository Tracts*, Vol. I, New York, 1835. Also *The Library*, 1940, fourth series, Vol. XX, p. 295, on "Cheap Repository Tracts (1795–98)."

in prices. A shooting affray in the streets of the metropolis might destroy a substantial part of the nominal national capital. And yet the middle classes were unsoldierly; popular democracy prided itself on making the masses vocal; and, on the Continent, the bourgeoisie still clung to the recollections of its revolutionary youth when it had itself faced a tyrannic aristocracy on the barricades. Eventually, the peasantry, least contaminated by the liberal virus, were reckoned the only stratum that would stand in their persons "for law and order." One of the functions of reaction was understood to be to keep the working classes in their place, so that markets should not be thrown into panic. Though this service was only very infrequently required, the availability of the peasantry as the defenders of property rights was an asset to the agrarian camp.

The history of the 1920's would be otherwise inexplicable. When, in Central Europe, the social structure broke down under the strain of war and defeat, the working class alone was available for the task of keeping things going. Everywhere power was thrust upon the trade unions and Social Democratic parties: Austria, Hungary, even Germany, were declared republics although no active republican party had ever been known to exist in any of these countries before. But hardly had the acute danger of dissolution passed and the services of the trade unions become superfluous than the middle classes tried to exclude the working classes from all influence on public life. This is known as the counterrevolutionary phase of the postwar period. Actually, there was never any serious danger of a Communist regime since the workers were organized in parties and unions actively hostile to the Communists. (Hungary had a Bolshevik episode literally forced upon the country when defense against French invasion left no alternative to the nation.) The peril was not Bolshevism, but disregard of the rules of market economy on the part of trade unions and working-class parties, in an emergency. For under a market economy otherwise harmless interruptions of public order and trading habits might constitute a lethal threat [8] since they could cause the breakdown of the economic regime upon which society depended for its daily bread. This explained the remarkable shift in some countries from a supposedly imminent dictatorship of the industrial workers to the actual dictatorship of the peasantry. Right through the twenties the peasantry deter-

[8] Hayes, C., *A Generation of Materialism, 1870–1890*, remarks that "most of the individual states, at least in Western and Central Europe, now possessed a seemingly superlative internal stability."

mined economic policy in a number of states in which they normally played but a modest role. They now happened to be the only class available to maintain law and order in the modern high-strung sense of the term.

The fierce agrarianism of postwar Europe was a side light on the preferential treatment accorded to the peasant class for political reasons. From the Lappo movement in Finland to the Austrian *Heimwehr* the peasants proved the champions of market economy; this made them politically indispensable. The scarcity of food in the first postwar years to which their ascendency was sometimes credited had little to do with this. Austria, for instance, in order to benefit the peasants financially, had to lower her food standards by maintaining duties for grain, though she was heavily dependent upon imports for her food requirements. But the peasant interest had to be safeguarded at all cost even though agrarian protectionism might mean misery to the town dwellers and an unreasonably high cost of production to the exporting industries. The formerly uninfluential class of peasants gained in this manner an ascendency quite disproportionate to their economic importance. Fear of Bolshevism was the force which made their political position impregnable. And yet that fear, as we saw, was not fear of a working-class dictatorship—nothing faintly similar was on the horizon—but rather the dread of a paralysis of market economy, unless all forces were eliminated from the political scene that, under duress, might set aside the rules of the market game. As long as the peasants were the only class able to eliminate these forces, their prestige stood high and they could hold the urban middle class in ransom. As soon as the consolidation of the power of the state and—even before that—the forming of the urban lower middle class into storm troops by the fascists, freed the bourgeoisie from dependence upon the peasantry, the latter's prestige was quickly deflated. Once the "internal enemy" in town and factory had been neutralized or subdued, the peasantry was relegated to its former modest position in industrial society.

The big landowners' influence did not share in this eclipse. A more constant factor worked in their favor—the increasing military importance of agricultural self-sufficiency. The Great War had brought the basic strategic facts home to the public, and thoughtless reliance on the world market gave way to a panicky hoarding of food-producing capacity. The "reagrarianization" of Central Europe started by the Bolshevik scare was completed in the sign of autarchy. Besides the argument of the "internal enemy" there was now the argument of the "external enemy." Liberal economists, as usual, saw merely a romantic aberration induced by unsound economic doctrines, where in reality towering political events were awakening even the simplest minds to the irrelevance of economic considerations in the face of the approaching dissolution of the international system. Geneva continued its futile attempts to convince the peoples that they were hoarding against imaginary perils, and that if only all acted in unison free trade could be restored and would benefit all. In the curiously credulous atmosphere of the time many took for granted that the solution of the economic problem (whatever that may mean) would not only assuage the threat of war but actually avert that threat forever. A hundred years' peace had created an insurmountable wall of illusions which hid the facts. The writers of that period excelled in lack of realism. The nation-state was deemed a parochial prejudice by A. J. Toynbee, sovereignty a ridiculous illusion by Ludwig von Mises, war a mistaken calculation in business by Norman Angell. Awareness of the essential nature of the problems of politics sank to an unprecedented low point.

Free trade which, in 1846, had been fought and won on the Corn Laws, was eighty years later fought over again and this time lost on the same issue. The problem of autarchy haunted market economy from the start. Accordingly, economic liberals exorcised the specter of war and naïvely based their case on the assumption of an indestructible market economy. It went unnoticed that their arguments merely showed how great was the peril of a people which relied for its safety on an institution as frail as the self-regulating market. The autarchy movement of the twenties was essentially prophetic: it pointed to the need for adjustment to the vanishing of an order. The Great War had shown up the danger and men acted accordingly; but since they acted ten years later, the connection between cause and effect was discounted as unreasonable. "Why protect oneself against passed dangers?" was the comment of many contemporaries. This faulty logic befogged not only an understanding of autarchy but, even more important, that of fascism. Actually, both were explained by the fact that, once the common mind has received the impress of a danger, fear remains latent, as long as its causes are not removed.

We claimed that the nations of Europe never overcame the shock of the war experience which unexpectedly confronted them with the perils of interdependence. In vain was trade resumed, in vain did swarms of international conferences display the idylls of peace, and dozens of governments declare for the principle of freedom of trade—no people could forget that unless they owned their food and raw material sources themselves or were certain of military access to them, neither sound currency nor unassailable credit would rescue them from helplessness. Nothing could be more logical than the consistency with which this fundamental consideration shaped the policy of communities. The source of the peril was not removed. Why then expect fear to subside?

A similar fallacy tricked those critics of fascism—they formed the great majority—who described fascism as a freak devoid of all political *ratio*. Mussolini, it was said, claimed to have averted Bolshevism in Italy, while statistics proved that for more than a year before the March on Rome the strike wave had subsided. Armed workers, it was conceded, occupied the factories in 1921. But was that a reason for disarming them in 1923, when they had long climbed down again from the walls where they had mounted guard? Hitler claimed he had saved Germany from Bolshevism. But could it not be shown that the flood of unemployment which preceded his chancellorship had ebbed away before his rise to power? To claim that he averted that which no longer existed when he came, as was argued, was contrary to the law of cause and effect, which must also hold in politics.

Actually, in Germany as in Italy, the story of the immediate postwar period proved that Bolshevism had not the slightest chance of success. But it also showed conclusively that in an emergency the working class, its trade unions and parties, might disregard the rules of the market which established freedom of contract and the sanctity of private property as absolutes—a possibility which must have the most deleterious effects on society, discouraging investments, preventing the accumulation of capital, keeping wages on an unremunerative level, endangering the currency, undermining foreign credit, weakening confidence and paralyzing enterprise. Not the illusionary danger of a communist revolution, but the undeniable fact that the working classes were in the position to force possibly ruinous interventions, was the source of the latent fear which, at a crucial juncture, burst forth in the fascist panic.

The dangers to man and nature cannot be neatly separated. The reactions of the working class and the peasantry to market economy both led to protectionism, the former mainly in the form of social legislation and factory laws, the latter in agrarian tariffs and land laws. Yet there was this important difference: in an emergency, the farmers and peasants of Europe defended the market system, which the working-class policies endangered. While the crisis of the inherently unstable system was brought on by both wings of the protectionist movement, the social strata connected with the land were inclined to compromise with the market system, while the broad class of labor did not shrink from breaking its rules and challenging it outright.

THE ECONOMIC THEORY OF A COMMON-PROPERTY RESOURCE: THE FISHERY[1]

H. SCOTT GORDON

Carleton College, Ottawa, Ontario

I. INTRODUCTION

THE chief aim of this paper is to examine the economic theory of natural resource utilization as it pertains to the fishing industry. It will appear, I hope, that most of the problems associated with the words "conservation" or "depletion" or "overexploitation" in the fishery are, in reality, manifestations of the fact that the natural resources of the sea yield no economic rent. Fishery resources are unusual in the fact of their common-property nature; but they are not unique, and similar problems are encountered in other cases of common-property resource industries, such as petroleum production, hunting and trapping, etc. Although the theory presented in the following pages is worked out in terms of the fishing industry, it is, I believe, applicable generally to all cases where natural resources are owned in common and exploited under conditions of individualistic competition.

II. BIOLOGICAL FACTORS AND THEORIES

The great bulk of the research that has been done on the primary production phase of the fishing industry has so far been in the field of biology. Owing to the lack of theoretical economic research,[2] biologists have been forced to extend the scope of their own thought into the economic sphere and in some cases have penetrated quite deeply, despite the lack of the analytical tools of economic theory.[3] Many others, who have paid no specific attention to the economic aspects of the problem have nevertheless recognized that the ultimate question is not the ecology of life in the sea as such, but man's use of these resources for his own (economic) purposes. Dr. Martin D. Burkenroad, for example, began a recent article on fishery management with a section on "Fishery Management as Political Economy," saying that "the Management of fisheries is intended for the benefit of man, not fish; therefore effect of management upon fishstocks cannot be regarded as beneficial *per se*."[4] The

[1] I want to express my indebtedness to the Canadian Department of Fisheries for assistance and co-operation in making this study; also to Professor M. C. Urquhart, of Queen's University, Kingston, Ontario, for mathematical assistance with the last section of the paper and to the Economists' Summer Study Group at Queen's for affording opportunity for research and discussion.

[2] The single exception that I know is G. M. Gerhardsen, "Production Economics in Fisheries," *Revista de economia* (Lisbon), March, 1952.

[3] Especially remarkable efforts in this sense are Robert A. Nesbit, "Fishery Management" ("U.S. Fish and Wildlife Service, Special Scientific Reports," No. 18 [Chicago, 1943]) (mimeographed), and Harden F. Taylor, *Survey of Marine Fisheries of North Carolina* (Chapel Hill, 1951); also R. J. H. Beverton, "Some Observations on the Principles of Fishery Regulation," *Journal du conseil permanent international pour l'exploration de la mer* (Copenhagen), Vol. XIX, No. 1 (May, 1953); and M. D. Burkenroad, "Some Principles of Marine Fishery Biology," *Publications of the Institute of Marine Science* (University of Texas), Vol. II, No. 1 (September, 1951).

[4] "Theory and Practice of Marine Fishery Management," *Journal du conseil permanent international pour l'exploration de la mer*, Vol. XVIII, No. 3 (January, 1953).

great Russian marine biology theorist, T. I. Baranoff, referred to his work as "bionomics" or "bio-economics," although he made little explicit reference to economic factors.[5] In the same way, A. G. Huntsman, reporting in 1944 on the work of the Fisheries Research Board of Canada, defined the problem of fisheries depletion in economic terms: "Where the take in proportion to the effort fails to yield a satisfactory living to the fisherman";[6] and a later paper by the same author contains, as an incidental statement, the essence of the economic optimum solution without, apparently, any recognition of its significance.[7] Upon the occasion of its fiftieth anniversary in 1952, the International Council for the Exploration of the Sea published a *Rapport Jubilaire*, consisting of a series of papers summarizing progress in various fields of fisheries research. The paper by Michael Graham on "Overfishing and Optimum Fishing," by its emphatic recognition of the economic criterion, would lead one to think that the economic aspects of the question had been extensively examined during the last half-century. But such is not the case. Virtually no specific research into the economics of fishery resource utilization has been undertaken. The present state

of knowledge is that a great deal is known about the biology of the various commercial species but little about the economic characteristics of the fishing industry.

The most vivid thread that runs through the biological literature is the effort to determine the effect of fishing on the stock of fish in the sea. This discussion has had a very distinct practical orientation, being part of the effort to design regulative policies of a "conservation" nature. To the layman the problem appears to be dominated by a few facts of overriding importance. The first of these is the prodigious reproductive potential of most fish species. The adult female cod, for example, lays millions of eggs at each spawn. The egg that hatches and ultimately reaches maturity is the great exception rather than the rule. The various herrings (Clupeidae) are the most plentiful of the commercial species, accounting for close to half the world's total catch, as well as providing food for many other sea species. Yet herring are among the smallest spawners, laying a mere hundred thousand eggs a season, which, themselves, are eaten in large quantity by other species. Even in inclosed waters the survival and reproductive powers of fish appear to be very great. In 1939 the Fisheries Research Board of Canada deliberately tried to kill all the fish in one small lake by poisoning the water. Two years later more than ninety thousand fish were found in the lake, including only about six hundred old enough to have escaped the poisoning.

The picture one gets of life in the sea is one of constant predation of one species on another, each species living on a narrow margin of food supply. It reminds the economist of the Malthusian law of population; for, unlike man, the

[5] Two of Baranoff's most important papers— "On the Question of the Biological Basis of Fisheries" (1918) and "On the Question of the Dynamics of the Fishing Industry" (1925)—have been translated by W. E. Ricker, now of the Fisheries Research Board of Canada (Nanaimo, B.C.), and issued in mimeographed form.

[6] "Fishery Depletion," *Science*, XCIX (1944), 534.

[7] "The highest take is not necessarily the best. The take should be increased only as long as the extra cost is offset by the added revenue from sales" (A. G. Huntsman, "Research on Use and Increase of Fish Stocks," *Proceedings of the United Nations Scientific Conference on the Conservation and Utilization of Resources* [Lake Success, 1949]).

fish has no power to alter the conditions of his environment and consequently cannot progress. In fact, Malthus and his law are frequently mentioned in the biological literature. One's first reaction is to declare that environmental factors are so much more important than commercial fishing that man has no effect on the population of the sea at all. One of the continuing investigations made by fisheries biologists is the determination of the age distribution of catches. This is possible because fish continue to grow in size with age, and seasonal changes are reflected in certain hard parts of their bodies in much the same manner as one finds growth-rings in a tree. The study of these age distributions shows that commercial catches are heavily affected by good and bad brood years. A good brood year, one favorable to the hatching of eggs and the survival of fry, has its effect on future catches, and one can discern the dominating importance of that brood year in the commercial catches of succeeding years.[8] Large broods, however, do not appear to depend on large numbers of adult spawners, and this lends support to the belief that the fish population is entirely unaffected by the activity of man.

There is, however, important evidence to the contrary. World Wars I and II, during which fishing was sharply curtailed in European waters, were followed by indications of a significant growth in fish populations. Fish-marking experiments, of which there have been a great number, indicate that fishing is a major cause of fish mortality in developed fisheries. The introduction of restrictive laws has often been followed by an increase in fish populations, although the evidence on this point is capable of other interpretations which will be noted later.

General opinion among fisheries biologists appears to have had something of a cyclical pattern. During the latter part of the last century, the Scottish fisheries biologist, W. C. MacIntosh,[9] and the great Darwinian, T. H. Huxley, argued strongly against all restrictive measures on the basis of the inexhaustible nature of the fishery resources of the sea. As Huxley put it in 1883: "The cod fishery, the herring fishery, the pilchard fishery, the mackerel fishery, and probably all the great sea fisheries, are inexhaustible: that is to say that nothing we do seriously affects the number of fish. And any attempt to regulate these fisheries seems consequently, from the nature of the case, to be useless."[10] As a matter of fact, there was at this time relatively little restriction of fishing in European waters. Following the Royal Commission of 1866, England had repealed a host of restrictive laws. The development of steam-powered trawling in the 1880's, which enormously increased man's predatory capacity, and the marked improvement of the trawl method in 1923 turned the pendulum, and throughout the interwar years discussion centered on the problem of "overfishing" and "depletion." This was accompanied by a considerable growth of restrictive regula-

---

[8] One example of a very general phenomenon: 1904 was such a successful brood year for Norwegian herrings that the 1904 year class continued to outweigh all others in importance in the catch from 1907 through to 1919. The 1904 class was some thirty times as numerous as other year classes during the period (Johan Hjort, "Fluctuations in the Great Fisheries of Northern Europe," *Rapports et procès-verbaux, Conseil permanent international pour l'exploration de la mer*, Vol. XX [1914]; see also E. S. Russell, *The Overfishing Problem* [Cambridge, 1942], p. 57).

[9] See his *Resources of the Sea* published in 1899.

[10] Quoted in M. Graham, *The Fish Gate* (London, 1943), p. 111; see also T. H. Huxley, "The Herring," *Nature* (London), 1881.

tions.[11] Only recently has the pendulum begun to reverse again, and there has lately been expressed in biological quarters a high degree of skepticism concerning the efficacy of restrictive measures, and the Huxleyian faith in the inexhaustibility of the sea has once again begun to find advocates. In 1951 Dr. Harden F. Taylor summarized the overall position of world fisheries in the following words:

Such statistics of world fisheries as are available suggest that while particular species have fluctuated in abundance, the *yield of the sea fisheries as a whole or of any considerable region has not only been sustained, but has generally increased with increasing human populations*, and there is as yet no sign that they will not continue to do so. No single species so far as we know has ever become extinct, and no regional fishery in the world has ever been exhausted.[12]

In formulating governmental policy, biologists appear to have had a hard struggle (not always successful) to avoid oversimplification of the problem. One of the crudest arguments to have had some support is known as the "propagation theory," associated with the name of the English biologist, E. W. L. Holt.[13] Holt advanced the proposition that legal size limits should be established at a level that would permit every individual of the species in question to spawn at least once. This suggestion was effectively demolished by the age-distribution studies whose results have been noted above. Moreover, some fisheries, such as the "sardine" fishery of the Canadian Atlantic Coast, are specifically for *immature* fish. The history of this particular fishery shows no evidence whatever that

the landings have been in any degree reduced by the practice of taking very large quantities of fish of prespawning age year after year.

The state of uncertainty in biological quarters around the turn of the century is perhaps indicated by the fact that Holt's propagation theory was advanced concurrently with its diametric opposite: "the thinning theory" of the Danish biologist, C. G. J. Petersen.[14] The latter argued that the fish may be too plentiful for the available food and that thinning out the young by fishing would enable the remainder to grow more rapidly. Petersen supported his theory with the results of transplanting experiments which showed that the fish transplanted to a new habitat frequently grew much more rapidly than before. But this is equivalent to arguing that the reason why rabbits multiplied so rapidly when introduced to Australia is because there were no rabbits already there with which they had to compete for food. Such an explanation would neglect all the other elements of importance in a natural ecology. In point of fact, in so far as food alone is concerned, thinning a cod population, say by half, would not double the food supply of the remaining individuals; for there are other species, perhaps not commercially valuable, that use the same food as the cod.

Dr. Burkenroad's comment, quoted earlier, that the purpose of practical policy is the benefit of man, not fish, was not gratuitous, for the argument has at times been advanced that commercial fishing should crop the resource in such a way as to leave the stocks of fish in the sea completely unchanged. Baranoff was largely responsible for destroying this

[11] See H. Scott Gordon, "The Trawler Question in the United Kingdom and Canada," *Dalhousie Review*, summer, 1951.

[12] Taylor, *op. cit.*, p. 314 (Dr. Taylor's italics).

[13] See E. W. L. Holt, "An Examination of the Grimsby Trawl Fishery," *Journal of the Marine Biological Association* (Plymouth), 1895.

[14] See C. G. J. Petersen, "What Is Overfishing?" *Journal of the Marine Biological Association* (Plymouth), 1900–1903.

approach, showing most elegantly that a commercial fishery cannot fail to diminish the fish stock. His general conclusion is worth quoting, for it states clearly not only his own position but the error of earlier thinking:

> As we see, a picture is obtained which diverges radically from the hypothesis which has been favoured almost down to the present time, namely that the natural reserve of fish is an inviolable capital, of which the fishing industry must use only the interest, not touching the capital at all. Our theory says, on the contrary, that a fishery and a natural reserve of fish are incompatible, and that the exploitable stock of fish is a changeable quantity, which depends on the intensity of the fishery. The more fish we take from a body of water, the smaller is the basic stock remaining in it; and the less fish we take, the greater is the basic stock, approximating to the natural stock when the fishery approaches zero. Such is the nature of the matter.[15]

The general conception of a fisheries ecology would appear to make such a conclusion inevitable. If a species were in ecological equilibrium before the commencement of commercial fishing, man's intrusion would have the same effect as any other predator; and that can only mean that the species population would reach a new equilibrium at a lower level of abundance, the divergence of the new equilibrium from the old depending on the degree of man's predatory effort and effectiveness.

The term "fisheries management" has been much in vogue in recent years, being taken to express a more subtle approach to the fisheries problem than the older terms "depletion" and "conservation." Briefly, it focuses attention on the quantity of fish caught, taking as the human objective of commercial fishing the derivation of the largest sustainable catch. This approach is often hailed in the biological literature as the "new theory" or the "modern formulation" of the fisheries problem.[16] Its limitations, however, are very serious, and, indeed, the new approach comes very little closer to treating the fisheries problem as one of human utilization of natural resources than did the older, more primitive, theories. Focusing attention on the maximization of the catch neglects entirely the inputs of other factors of production which are used up in fishing and must be accounted for as costs. There are many references to such ultimate economic considerations in the biological literature but no analytical integration of the economic factors. In fact, the very conception of a *net economic yield* has scarcely made any appearance at all. On the whole, biologists tend to treat the fisherman as an exogenous element in their analytical model, and the behavior of fishermen is not made into an integrated element of a general and systematic "bionomic" theory. In the case of the fishing industry the large numbers of fishermen permit valid behavioristic generalization of their activities along the lines of the standard economic theory of production. The following section attempts to apply that theory to the fishing industry and to demonstrate that the "overfishing problem" has its roots in the economic organization of the industry.

III. ECONOMIC THEORY OF THE FISHERY

In the analysis which follows, the theory of optimum utilization of fishery re-

[15] T. I. Baranoff, "On the Question of the Dynamics of the Fishing Industry," p. 5 (mimeographed).

[16] See, e.g., R. E. Foerster, "Prospects for Managing Our Fisheries," *Bulletin of the Bingham Oceanographic Collection* (New Haven), May, 1948; E. S. Russell, "Some Theoretical Considerations on the Overfishing Problem," *Journal du conseil permanent international pour l'exploration de la mer*, 1931, and *The Overfishing Problem*, Lecture IV.

sources and the reasons for its frustration in practice are developed for a typical demersal fish. Demersal, or bottom-dwelling fishes, such as cod, haddock, and similar species and the various flat-fishes, are relatively nonmigratory in character. They live and feed on shallow continental shelves where the continual mixing of cold water maintains the availability of those nutrient salts which form the fundamental basis of marine-food chains. The various feeding grounds are separated by deep-water channels which constitute barriers to the movement of these species; and in some cases the fish of different banks can be differentiated morphologically, having varying numbers of vertebrae or some such distinguishing characteristic. The significance of this fact is that each fishing ground can be treated as unique, in the same sense as can a piece of land, possessing, at the very least, one characteristic not shared by any other piece: that is, location.

(Other species, such as herring, mackerel, and similar pelagic or surface dwellers, migrate over very large distances, and it is necessary to treat the resource of an entire geographic region as one. The conclusions arrived at below are applicable to such fisheries, but the method of analysis employed is not formally applicable. The same is true of species that migrate to and from fresh water and the lake fishes proper.)

We can define the optimum degree of utilization of any particular fishing ground as that which maximizes the net economic yield, the difference between total cost, on the one hand, and total receipts (or total value production), on the other.[17] Total cost and total production can each be expressed as a function of the degree of fishing intensity or, as the biologists put it, "fishing effort," so that a simple maximization solution is possible. Total cost will be a linear function of fishing effort, if we assume no fishing-induced effects on factor prices, which is reasonable for any particular regional fishery.

The production function—the relationship between fishing effort and total value produced—requires some special attention. If we were to follow the usual presentation of economic theory, we should argue that this function would be positive but, after a point, would rise at a diminishing rate because of the law of diminishing returns. This would not mean that the fish population has been reduced, for the law refers only to the *proportions* of factors to one another, and a fixed fish population, together with an increasing intensity of effort, would be assumed to show the typical sigmoid pattern of yield. However, in what follows it will be assumed that the law of diminishing returns in this pure sense is inoperative in the fishing industry. (The reasons will be advanced at a later point in this paper.) We shall assume that, as fishing effort expands, the catch of fish increases at a diminishing rate but that it does so because of the effect of catch upon the fish population.[18] So far as the argument of the next few pages is concerned, all that is formally necessary is to assume that, as fishing intensity increases, catch will grow at a diminishing rate. Whether this reflects the pure law of diminishing returns or the reduction

[17] Expressed in these terms, this appears to be the monopoly maximum, but it coincides with the social optimum under the conditions employed in the analysis, as will be indicated below.

[18] Throughout this paper the conception of fish population that is employed is one of *weight* rather than *numbers*. A good deal of the biological theory has been an effort to combine growth factors and numbers factors into weight sums. The following analysis will neglect the fact that, for some species, fish of different sizes bring different unit prices.

of population by fishing, or both, is of no particular importance. The point at issue will, however, take on more significance in Section IV and will be examined there.

Our analysis can be simplified if we retain the ordinary production function instead of converting it to cost curves, as is usually done in the theory of the firm. Let us further assume that the functional relationship between average production (production-per-unit-of-fishing-effort) and the quantity of fishing effort is uniformly linear. This does not distort the

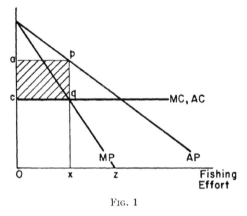

FIG. 1

results unduly, and it permits the analysis to be presented more simply and in graphic terms that are already quite familiar.

In Figure 1 the optimum intensity of utilization of a particular fishing ground is shown. The curves AP and MP represent, respectively, the average productivity and marginal productivity of fishing effort. The relationship between them is the same as that between average revenue and marginal revenue in imperfect competition theory, and MP bisects any horizontal between the ordinate and AP. Since the costs of fishing supplies, etc., are assumed to be unaffected by the amount of fishing effort, marginal cost and average cost are identical and

constant, as shown by the curve MC, AC.[19] These costs are assumed to include an opportunity income for the fishermen, the income that could be earned in other comparable employments. Then Ox is the optimum intensity of effort on this fishing ground, and the resource will, at this level of exploitation, provide the maximum net economic yield indicated by the shaded area $apqc$. The maximum sustained physical yield that the biologists speak of will be attained when marginal productivity of fishing effort is zero, at Oz of fishing intensity in the chart shown. Thus, as one might expect, the optimum economic fishing intensity is less than that which would produce the maximum sustained physical yield.

The area $apqc$ in Figure 1 can be regarded as the rent yielded by the fishery resource. Under the given conditions, Ox is the best rate of exploitation for the fishing ground in question, and the rent reflects the productivity of that ground, not any artificial market limitation. The rent here corresponds to the extra productivity yielded in agriculture by soils of better quality or location than those on the margin of cultivation, which may produce an opportunity income but no more. In short, Figure 1 shows the determination of the intensive margin of utilization on an intramarginal fishing ground.

We now come to the point that is of greatest theoretical importance in understanding the primary production phase of the fishing industry and in distinguishing it from agriculture. In the sea fish-

[19] Throughout this analysis, fixed costs are neglected. The general conclusions reached would not be appreciably altered, I think, by their inclusion, though the presentation would be greatly complicated. Moreover, in the fishing industry the most substantial portion of fixed cost—wharves, harbors, etc.—is borne by government and does not enter into the cost calculations of the operators.

eries the natural resource is not private property; hence the rent it may yield is not capable of being appropriated by anyone. The individual fisherman has no legal title to a section of ocean bottom. Each fisherman is more or less free to fish wherever he pleases. The result is a pattern of competition among fishermen which culminates in the dissipation of the rent of the intramarginal grounds. This can be most clearly seen through an analysis of the relationship between the fishermen are free to fish on whichever ground they please, it is clear that this is not an equilibrium allocation of fishing effort in the sense of connoting stability. A fisherman starting from port and deciding whether to go to ground *1* or *2* does not care for *marginal* productivity but for *average* productivity, for it is the latter that indicates where the greater total yield may be obtained. If fishing effort were allocated in the optimum fashion, as shown in Figure 2, with *Ox* on

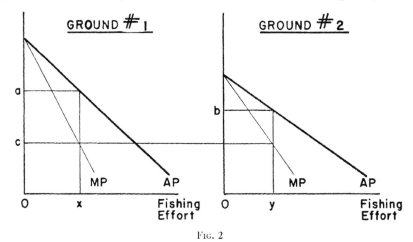

Fig. 2

intensive margin and the extensive margin of resource exploitation in fisheries.

In Figure 2, two fishing grounds of different fertility (or location) are shown. Any given amount of fishing effort devoted to ground *2* will yield a smaller total (and therefore average) product than if devoted to *1*. The maximization problem is now a question of the allocation of fishing effort between grounds *1* and *2*. The optimum is, of course, where the marginal productivities are equal on both grounds. In Figure 2, fishing effort of *Ox* on *1* and *Oy* on *2* would maximize the total net yield of *Ox* + *Oy* effort if marginal cost were equal to *Oc*. But if under such circumstances the individual *1*, and *Oy* on *2*, this would be a disequilibrium situation. Each fisherman could expect to get an average catch of *Oa* on *1* but only *Ob* on *2*. Therefore, fishermen would shift from *2* to *1*. Stable equilibrium would not be reached until the average productivity of both grounds was equal. If we now imagine a continuous gradation of fishing grounds, the extensive margin would be on that ground which yielded nothing more than outlaid costs plus opportunity income—in short, the one on which average productivity and average cost were equal. But, since average cost is the same for all grounds and the average productivity of all grounds is also brought to equality by the free and competitive nature of fishing, this means that the intramarginal grounds also yield no rent. It is entirely possible that some grounds would be exploited at a level of *negative* marginal productivity. What happens is that the rent which the intramarginal grounds are capable of yielding is dissipated through misallocation of fishing effort.

This is why fishermen are not wealthy, despite the fact that the fishery resources of the sea are the richest and most indestructible available to man. By and large, the only fisherman who becomes rich is one who makes a lucky catch or one who participates in a fishery that is put under a form of social control that turns the open resource into property rights.

Up to this point, the remuneration of fishermen has been accounted for as an opportunity-cost income comparable to earnings attainable in other industries. In point of fact, fishermen typically earn less than most others, even in much less hazardous occupations or in those requiring less skill. There is no effective reason why the competition among fishermen described above must stop at the point where opportunity incomes are yielded. It may be and is in many cases carried much further. Two factors prevent an equilibration of fishermen's incomes with those of other members of society. The first is the great immobility of fishermen. Living often in isolated communities, with little knowledge of conditions or opportunities elsewhere; educationally and often romantically tied to the sea; and lacking the savings necessary to provide a "stake," the fisherman is one of the least mobile of occupational groups. But, second, there is in the spirit of every fisherman the hope of the "lucky catch." As those who know fishermen well have often testified, they

are gamblers and incurably optimistic. As a consequence, they will work for less than the going wage.[20]

The theory advanced above is substantiated by important developments in the fishing industry. For example, practically all control measures have, in the past, been designed by biologists, with sole attention paid to the production side of the problem and none to the cost side. The result has been a wide-open door for the frustration of the purposes of such measures. The Pacific halibut fishery, for example, is often hailed as a great achievement in modern fisheries management. Under international agreement between the United States and Canada, a fixed-catch limit was established during the early thirties. Since then, catch-per-unit-effort indexes, as usually interpreted, show a significant rise in the fish population. W. F. Thompson, the pioneer of the Pacific halibut management program, noted recently that "it has often been said that the halibut regulation presents the only definite case of sustained improvement of an overfished deep-sea fishery. This, I believe, is true and the fact should lend special importance to the principles which have been deliberately used to obtain this improvement."[21] Actually, careful study of the statistics indicates that the estimated recovery of halibut stocks could not have been due principally to the control measures, for the average catch was, in fact, greater during the recovery years than during the years of

[20] "The gambling instinct of the men makes many of them work for less remuneration than they would accept as a weekly wage, because there is always the possibility of a good catch and a financial windfall" (Graham, *op. cit.*, p. 86).

[21] W. F. Thompson, "Condition of Stocks of Halibut in the Pacific," *Journal du conseil permanent international pour l'exploration de la mer*, Vol. XVIII, No. 2 (August, 1952).

decline. The total amount of fish taken was only a small fraction of the estimated population reduction for the years prior to regulation.[22] Natural factors seem to be mainly responsible for the observed change in population, and the institution of control regulations almost a coincidence. Such coincidences are not uncommon in the history of fisheries policy, but they may be easily explained. If a long-term cyclical fluctuation is taking place in a commercially valuable species, controls will likely be instituted when fishing yields have fallen very low and the clamor of fishermen is great; but it is then, of course, that stocks are about due to recover in any case. The "success" of conservation measures may be due fully as much to the sociological foundations of public policy as to the policy's effect on the fish. Indeed, Burkenroad argues that biological statistics in general may be called into question on these grounds. Governments sponsor biological research when the catches are disappointing. If there are long-term cyclical fluctuations in fish populations, as some think, it is hardly to be wondered why biologists frequently discover that the sea is being depleted, only to change their collective opinion a decade or so later.

Quite aside from the *biological* argument on the Pacific halibut case, there is no clear-cut evidence that halibut fishermen were made relatively more prosperous by the control measures. Whether or not the recovery of the halibut stocks was due to natural factors or to the catch limit, the potential net yield this could have meant has been dissipated through a rise in fishing costs. Since the method of control was to halt fishing when the limit had been reached, this created a

great incentive on the part of each fisherman to get the fish before his competitors. During the last twenty years, fishermen have invested in more, larger, and faster boats in a competitive race for fish. In 1933 the fishing season was more than six months long. In 1952 it took just twenty-six days to catch the legal limit in the area from Willapa Harbor to Cape Spencer, and sixty days in the Alaska region. What has been happening is a rise in the average cost of fishing effort, allowing no gap between average production and average cost to appear, and hence no rent.[23]

Essentially the same phenomenon is observable in the Canadian Atlantic Coast lobster-conservation program. The method of control here is by seasonal closure. The result has been a steady growth in the number of lobster traps set

---

[22] See M. D. Burkenroad, "Fluctuations in Abundance of Pacific Halibut," *Bulletin of the Bingham Oceanographic Collection*, May, 1948.

[23] The economic significance of the reduction in season length which followed upon the catch limitation imposed in the Pacific halibut fishery has not been fully appreciated. E.g., Michael Graham said in summary of the program in 1943: "The result has been that it now takes only five months to catch the quantity of halibut that formerly needed nine. This, *of course*, has meant profit, where there was none before" (*op. cit.*, p. 156; my italics). Yet, even when biologists have grasped the economic import of the halibut program and its results, they appear reluctant to declare against it. E.g., W. E. Ricker: "This method of regulation does not necessarily make for more profitable fishing and certainly puts no effective brake on waste of effort, since an unlimited number of boats is free to join the fleet and compete during the short period that fishing is open. However, the stock is protected, and yield approximates to a maximum if quotas are wisely set; as biologists, perhaps we are not required to think any further. Some claim that any mixing into the economics of the matter might prejudice the desirable biological consequences of regulation by quotas" ("Production and Utilization of Fish Population," in a Symposium on Dynamics of Production in Aquatic Populations, Ecological Society of America, *Ecological Monographs*, XVI [October, 1946], 385). What such "desirable biological consequences" might be, is hard to conceive. Since the regulatory policies are made by man, surely it is necessary they be evaluated in terms of human, not piscatorial, objectives.

by each fisherman. Virtually all available lobsters are now caught each year within the season, but at much greater cost in gear and supplies. At a fairly conservative estimate, the same quantity of lobsters could be caught with half the present number of traps. In a few places the fishermen have banded together into a local monopoly, preventing entry and controlling their own operations. By this means, the amount of fishing gear has been greatly reduced and incomes considerably improved.

That the plight of fishermen and the inefficiency of fisheries production stems from the common-property nature of the resources of the sea is further corroborated by the fact that one finds similar patterns of exploitation and similar problems in other cases of open resources. Perhaps the most obvious is hunting and trapping. Unlike fishes, the biotic potential of land animals is low enough for the species to be destroyed. Uncontrolled hunting means that animals will be killed for any short-range human reason, great or small: for food or simply for fun. Thus the buffalo of the western plains was destroyed to satisfy the most trivial desires of the white man, against which the long-term food needs of the aboriginal population counted as nothing. Even in the most civilized communities, conservation authorities have discovered that a bag-limit *per man* is necessary if complete destruction is to be avoided.

The results of anthropological investigation of modes of land tenure among primitive peoples render some further support to this thesis. In accordance with an evolutionary concept of cultural comparison, the older anthropological study was prone to regard resource tenure in common, with unrestricted exploitation, as a "lower" stage of development comparative with private and group

property rights. However, more complete annals of primitive cultures reveal common tenure to be quite rare, even in hunting and gathering societies. Property rights in some form predominate by far, and, most important, their existence may be easily explained in terms of the necessity for orderly exploitation and conservation of the resource. Environmental conditions make necessary some vehicle which will prevent the resources of the community at large from being destroyed by excessive exploitation. Private or group land tenure accomplishes this end in an easily understandable fashion.[24] Significantly, land tenure is found to be "common" only in those cases where the hunting resource is migratory over such large areas that it cannot be regarded as husbandable by the society. In cases of group tenure where the numbers of the group are large, there is still the necessity of co-ordinating the practices of exploitation, in agricultural, as well as in hunting or gathering, economies. Thus, for example, Malinowski reported that among the Trobriand Islanders one of the fundamental principles of land tenure is the co-ordination of the productive activities of the gardeners by the person possessing magical leadership in the group.[25] Speaking generally, we may say that stable primitive cultures appear to have discovered the dangers of common-property tenure and to have de-

[24] See Frank G. Speck, "Land Ownership among Hunting Peoples in Primitive America and the World's Marginal Areas," *Proceedings of the 22nd International Congress of Americanists* (Rome, 1926), II, 323–32.

[25] B. Malinowski, *Coral Gardens and Their Magic*, Vol. I, chaps. xi and xii. Malinowski sees this as further evidence of the importance of magic in the culture rather than as a means of co-ordinating productive activity; but his discussion of the practice makes it clear that the latter is, to use Malinowski's own concept, the "function" of the institution of magical leadership, at least in this connection.

veloped measures to protect their resources. Or, if a more Darwinian explanation be preferred, we may say that only
those primitive cultures have survived
which succeeded in developing such institutions.

Another case, from a very different
industry, is that of petroleum production. Although the individual petroleum
producer may acquire undisputed lease
or ownership of the particular plot of
land upon which his well is drilled, he
shares, in most cases, a common pool of
oil with other drillers. There is, consequently, set up the same kind of competitive race as is found in the fishing
industry, with attending overexpansion
of productive facilities and gross wastage
of the resource. In the United States,
efforts to regulate a chaotic situation in
oil production began as early as 1915.
Production practices, number of wells,
and even output quotas were set by governmental authority; but it was not until
the federal "Hot Oil" Act of 1935 and
the development of interstate agreements
that the final loophole (bootlegging) was
closed through regulation of interstate
commerce in oil.

Perhaps the most interesting similar
case is the use of common pasture in the
medieval manorial economy. Where the
ownership of animals was private but the
resource on which they fed was common
(and limited), it was necessary to regulate the use of common pasture in order
to prevent each man from competing and
conflicting with his neighbors in an effort
to utilize more of the pasture for his own
animals. Thus the manor developed its
elaborate rules regulating the use of the
common pasture, or "stinting" the common: limitations on the number of animals, hours of pasturing, etc., designed
to prevent the abuses of excessive individualistic competition.[26]

There appears, then, to be some truth
in the conservative dictum that everybody's property is nobody's property.
Wealth that is free for all is valued by
none because he who is foolhardy enough
to wait for its proper time of use will only
find that it has been taken by another.
The blade of grass that the manorial
cowherd leaves behind is valueless to
him, for tomorrow it may be eaten by
another's animal; the oil left under the
earth is valueless to the driller, for another may legally take it; the fish in the
sea are valueless to the fisherman, because there is no assurance that they
will be there for him tomorrow if they
are left behind today. A factor of production that is valued at nothing in the business calculations of its users will yield
nothing in income. Common-property
natural resources are free goods for the
individual and scarce goods for society.
Under unregulated private exploitation,
they can yield no rent; that can be accomplished only by methods which make
them private property or public (government) property, in either case subject to
a unified directing power.

IV. THE BIONOMIC EQUILIBRIUM OF
THE FISHING INDUSTRY

The work of biological theory in the
fishing industry is, basically, an effort to
delineate the ecological system in which
a particular fish population is found. In
the main, the species that have been extensively studied are those which are
subject to commercial exploitation. This
is due not only to the fact that funds are
forthcoming for such research but also
because the activity of commercial fishing vessels provides the largest body of
data upon which the biologist may work.

[26] See P. Vinogradoff, *The Growth of the Manor*
[London, 1905], chap. iv; E. Lipson, *The Economic
History of England* [London, 1949], I, 72.

Despite this, however, the ecosystem of the fisheries biologist is typically one that excludes man. Or, rather, man is regarded as an exogenous factor, having influence on the biological ecosystem through his removal of fish from the sea, but the activities of man are themselves not regarded as behaviorized or determined by the other elements of a system of mutual interdependence. The large number of independent fishermen who exploit fish populations of commercial importance makes it possible to treat man as a behavior element in a larger, "bionomic," ecology, if we can find the rules which relate his behavior to the other elements of the system. Similarly, in their treatment of the principles of fisheries management, biologists have overlooked essential elements of the problem by setting maximum physical landings as the objective of management, thereby neglecting the economic factor of input cost.

An analysis of the bionomic equilibrium of the fishing industry may, then, be approached in terms of two problems. The first is to explain the nature of the equilibrium of the industry as it occurs in the state of uncontrolled or unmanaged exploitation of a common-property resource. The second is to indicate the nature of a socially optimum manner of exploitation, which is, presumably, what governmental management policy aims to achieve or promote. These two problems will be discussed in the remaining pages.

In the preceding section it was shown that the equilibrium condition of uncontrolled exploitation is such that the net yield (total value landings *minus* total cost) is zero. The "bionomic ecosystem" of the fishing industry, as we might call it, can then be expressed in terms of four variables and four equations. Let P represent the population of the particular fish species on the particular fishing bank in question; L the total quantity taken or "landed" by man, measured in value terms; E the intensity of fishing or the quantity of "fishing effort" expended; and C the total cost of making such effort. The system, then, is as follows:

$$P = P(L),\qquad(1)$$
$$L = L(P, E),\qquad(2)$$
$$C = C(E),\qquad(3)$$
$$C = L.\qquad(4)$$

Equation (4) is the equilibrium condition of an uncontrolled fishery.

The functional relations stated in equations (1), (2), and (3) may be graphically presented as shown in Figure 3. Segment *1* shows the fish population as a simple negative function of landings. In segment *2* a map of landings functions is drawn. Thus, for example, if population were P_3, effort of Oe would produce Ol of fish. For each given level of population, a larger fishing effort will result in larger landings. Each population contour is, then, a production function for a given population level. The linearity of these contours indicates that the law of diminishing returns is not operative, nor are any landings-induced price effects assumed to affect the value landings graphed on the vertical axis. These assumptions are made in order to produce the simplest determinate solution; yet each is reasonable in itself. The assumption of a fixed product price is reasonable, since our analysis deals with one fishing ground, not the fishery as a whole. The cost function represented in equation (3) and graphed in segment *3* of Figure 3 is not really necessary to the determination, but its inclusion makes the matter somewhat clearer. Fixed prices of input factors—"fishing effort"—is assumed, which is reasonable again on the assumption that a small part of the total fishery is being analyzed.

Starting with the first segment, we see that a postulated catch of Ol connotes an equilibrium population in the biological ecosystem of Op. Suppose this population to be represented by the contour P_3 of segment 2. Then, given P_3, Oe is the effort required to catch the postulated landings Ol. This quantity of effort involves a total cost of Oc, as shown in segment 3 of the graph. In full bionomic

found. If the case were represented by C and L_1, the fishery would contract to zero; if by C and L_2, it would undergo an infinite expansion. Stable equilibrium requires that either the cost or the landings function be nonlinear. This condition is fulfilled by the assumption that population is reduced by fishing (eq. [1] above). The equilibrium is therefore as shown in Figure 5. Now Oe represents a fully stable equilibrium intensity of fishing.

The analysis of the conditions of stable equilibrium raises some points of general theoretical interest. In the foregoing we

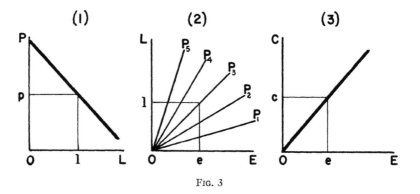

FIG. 3

equilibrium, $C = L$, and if the particular values Oc and Ol shown are not equal, other quantities of all four variables, L, P, E, and C, are required, involving movements of these variables through the functional system shown. The operative movement is, of course, in fishing effort, E. It is the equilibrating variable in the system.

The equilibrium equality of landings (L) and cost (C), however, must be a position of stability, and $L = C$ is a necessary, though not in itself sufficient, condition for stability in the ecosystem. This is shown by Figure 4. If effort-cost and effort-landings functions were both linear, no stable equilibrium could be

have assumed that stability results from the effect of fishing on the fish population. In the standard analysis of economic theory, we should have employed the law of diminishing returns to produce a landings function of the necessary shape. Market factors might also have been so employed; a larger supply of fish, forthcoming from greater fishing effort, would reduce unit price and thereby produce a landings function with the necessary negative second derivative. Similarly, greater fishing intensity might raise the unit costs of factors, producing a cost function with a positive second derivative. Any one of these three— population effects, law of diminishing returns, or market effects—is alone sufficient to produce stable equilibrium in the ecosystem.

As to the law of diminishing returns, it has not been accepted per se by fisheries biologists. It is, in fact, a principle that becomes quite slippery when one applies it to the case of fisheries. Indicative of this is the fact that Alfred Marshall, in whose *Principles* one can find extremely little formal error, misinterprets the application of the law of dimin-

estingly enough, his various criticisms of the indexes were generally accepted, with the significant exception of this one point. More recently, A. G. Huntsman warned his colleagues in fisheries biology that "[there] may be a decrease in the take-per-unit-of-effort without any decrease in the total take or in the fish population. . . . This may mean that there has been an increase in fishermen rather than a decrease in fish."[29] While these statements run in terms of average

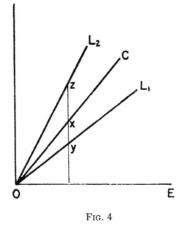

Fɪɢ. 4

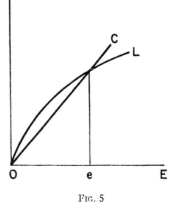

Fɪɢ. 5

ishing returns to the fishing industry, arguing, in effect, that the law exerts its influence through the reducing effect of fishing on the fish population.[27] There have been some interesting expressions of the law or, rather, its essential varying-proportions-of-factors aspect, in the biological literature. H. M. Kyle, a German biologist, included it in 1928 among a number of reasons why catch-per-unit-of-fishing-effort indexes are not adequate measures of population change.[28] Inter-

rather than marginal yield, their underlying reasoning clearly appears to be that of the law of diminishing returns. The point has had little influence in biological circles, however, and when, two years ago, I advanced it, as Kyle and Huntsman had done, in criticism of the standard biological method of estimating population change, it received pretty short shrift.

[28] "Die Statistik der Seefischerei Nordeuropas," *Handbuch der Seefischerei Nordeuropas* (Stuttgart, 1928).

[29] A. G. Huntsman, "Fishing and Assessing Populations," *Bulletin of the Bingham Oceanographic Collection* (New Haven), May, 1948.

[27] See H. Scott Gordon, "On a Misinterpretation of the Law of Diminishing Returns in Alfred Marshall's *Principles,*" *Canadian Journal of Economics and Political Science,* February, 1952.

In point of fact, the law of diminishing returns is much more difficult to sustain in the case of fisheries than in agriculture or industry. The "proof" one finds in standard theory is not empirical, although the results of empirical experiments in agriculture are frequently adduced as subsidiary corroboration. The main weight of the law, however, rests on a *reductio ad absurdum*. One can easily demonstrate that, were it not for the law of diminishing returns, all the world's food could be grown on one acre of land. Reality is markedly different, and it is because the law serves to render this reality intelligible to the logical mind, or, as we might say, "explains" it, that it occupies such a firm place in the body of economic theory. In fisheries, however, the pattern of reality can easily be explained on other grounds. In the case at least of developed demersal fisheries, it cannot be denied that the fish population is reduced by fishing, and this relationship serves perfectly well to explain why an infinitely expansible production is not possible from a fixed fishing area. The other basis on which the law of diminishing returns is usually advanced in economic theory is the prima facie plausibility of the principle as such; but here, again, it is hard to grasp any similar reasoning in fisheries. In the typical agricultural illustration, for example, we may argue that the fourth harrowing or the fourth weeding, say, has a lower marginal productivity than the third. Such an assertion brings ready acceptance because it concerns a process with a zero productive limit. It is apparent that, ultimately, the land would be completely broken up or the weeds completely eliminated if harrowing or weeding were done in ever larger amounts. The law of diminishing returns signifies simply that

such a zero limit is *gradually approached*, all of which appears to be quite acceptable on prima facie grounds. There is nothing comparable to this in fisheries at all, for there is no "cultivation" in the same sense of the term, except, of course, in such cases as oyster culture or pond rearing of fish, which are much more akin to farming than to typical sea fisheries.

In the biological literature the point has, I think, been well thought through, though the discussion does not revolve around the "law of diminishing returns" by that name. It is related rather to the fisheries biologist's problem of the interpretation of catch-per-unit-of-fishing-effort statistics. The essence of the law is usually eliminated by the assumption that there is no "competition" among units of fishing gear—that is, that the ratio of gear to fishing area and/or fish population is small. In some cases, corrections have been made by the use of the compound-interest formula where some competition among gear units is considered to exist.[30] Such corrections, however, appear to be based on the idea of an increasing catch-population ratio rather than an increasing effort-population ratio. The latter would be as the law of diminishing returns would have it; the idea lying behind the former is that the total population in existence represents the maximum that can be caught, and, since this maximum would be gradually approached, the ratio of catch to population has some bearing on the efficiency of fishing gear. It is, then, just an aspect of the population-reduction effect. Similarly, it has been pointed out that, since fish are recruited into the

[30] See, e.g., W. F. Thompson and F. H. Bell, *Biological Statistics of the Pacific Halibut Fishery*, No. 2: *Effect of Changes in Intensity upon Total Yield and Yield per Unit of Gear: Report of the International Fisheries Commission* (Seattle, 1934).

catchable stock in a seasonal fashion, one can expect the catch-per-unit-effort to fall as the fishing season progresses, at least in those fisheries where a substantial proportion of the stock is taken annually. Seasonal averaging is therefore necessary in using the catch-effort sta-

the fishery, nor is there any prima facie ground for its acceptance.

Let us now consider the exploitation of a fishing ground under unified control, in which case the equilibrium condition is the maximization of net financial yield, $L - C$.

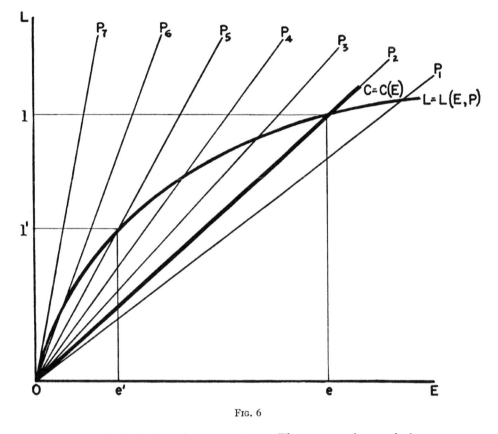

Fig. 6

tistics as population indexes from year to year. This again is a population-reduction effect, not the law of diminishing returns. In general, there seems to be no reason for departing from the approach of the fisheries biologist on this point. The law of diminishing returns is not necessary to explain the conditions of stable equilibrium in a static model of

The map of population contours graphed in segment 2 of Figure 3 may be superimposed upon the total-landings and total-cost functions graphed in Figure 5. The result is as shown in Figure 6. In the system of interrelationships we have to consider, population changes affect, and are in turn affected by, the amount of fish landed. The map of population contours does not include this roundabout effect that a population change has upon itself. The curve labeled L, however, is a landings function which accounts for the fact that larger landings reduce the population, and this is why it is shown to have a steadily diminishing slope. We may regard the landings function as moving progressively to lower population contours P_7, P_6, P_5, etc., as total landings increase in magnitude. As a consequence, while each population contour represents many hypothetical combinations of E, L, and P, only one such combination on each is actually compatible in this system of interrelationships. This combination is the point on any contour where that contour is met by the landings function L. Thus the curve labeled L may be regarded as tracing out a series of combinations of E, L, and P which are compatible with one another in the system.

The total-cost function may be drawn as shown, with total cost, C, measured in terms of landings, which the vertical axis represents.[31] This is a linear function of effort as shown. The optimum intensity of fishing effort is that which maximizes $L - C$. This is the monopoly solution; but, since we are considering only a single fishing ground, no price effects are introduced, and the social optimum coincides with maximum monopoly revenue. In this case we are maximizing the yield of a natural resource, not a privileged position, as in standard monopoly theory. The rent here is a social surplus yielded by the resource, not in any part due to artificial scarcity, as is monopoly profit or rent.

If the optimum fishing intensity is that which maximizes $L - C$, this is seen to

be the position where the slope of the landings function equals the slope of the cost function in Figure 6. Thus the optimum fishing intensity is Oe' of fishing effort. This will yield Ol' of landings, and the species population will be in continuing stable equilibrium at a level indicated by P_5.

The equilibrium resulting from uncontrolled competitive fishing, where the rent is dissipated, can also be seen in Figure 6. This, being where $C = L$, is at Oe of effort and Ol of landings, and at a stable population level of P_2. As can be clearly seen, the uncontrolled equilibrium means a higher expenditure of effort, higher fish landings, and a lower continuing fish population than the optimum equilibrium.

Algebraically, the bionomic ecosystem may be set out in terms of the optimum solution as follows. The species population in equilibrium is a linear function of the amount of fish taken from the sea:

$$P = a - bL . \qquad (1)$$

In this function, a may be described as the "natural population" of the species—the equilibrium level it would attain if not commercially fished. All natural factors, such as water temperatures, food supplies, natural predators, etc., which affect the population are, for the purposes of the system analyzed, locked up in a. The magnitude of a is the vertical intercept of the population function graphed in segment 1 of Figure 3. The slope of this function is b, which may be described as the "depletion coefficient," since it indicates the effect of catch on population. The landings function is such that no landings are forthcoming with either zero effort or zero population; therefore,

$$L = cEP . \qquad (2)$$

[31] More correctly, perhaps, C and L are both measured in money terms.

The parameter c in this equation is the technical coefficient of production or, as we may call it simply, the "production coefficient." Total cost is a function of the amount of fishing effort.

$$C = qE .$$

The optimum condition is that the total net receipts must be maximized, that is,

$$L - C \text{ to be maximized} .$$

Since q has been assumed constant and equal to unity (i.e., effort is counted in "dollars-worth" units), we may write $L - E$ to be maximized. Let this be represented by R:

$$R = L - E , \qquad (3)$$

$$\frac{dR}{dE} = 0 . \qquad (4)$$

The four numbered equations constitute the system when in optimality equilibrium. In order to find this optimum, the landings junction (2) may be rewritten, with the aid of equation (1), as:

$$L = cE (a - bL) .$$

From this we have at once

$$L (1 + cEb) = cEa ,$$

$$L = \frac{caE}{1 + cbE} .$$

To find the optimum intensity of effort, we have, from equation (3):

$$\frac{dR}{dE} = \frac{dL}{dE} - \frac{dE}{dE}$$

$$= \frac{(1 + cbE)(ca) - caE(cb)}{(1 + cbE)^2} - 1 ,$$

$$= \frac{ca}{(1 + cbE)^2} - 1 ;$$

for a maximum, this must be set equal to zero; hence,

$$ca = (1 + cbE)^2 ,$$

$$1 + cbE = \pm \sqrt{ca} ,$$

$$E = \frac{-1 \pm \sqrt{ca}}{cb} .$$

For positive E,

$$E = \frac{\sqrt{ca} - 1}{cb} .$$

This result indicates that the effect on optimum effort of a change in the production coefficient is uncertain, a rise in c calling for a rise in E in some cases and a fall in E in others, depending on the magnitude of the change in c. The effects of changes in the natural population and depletion coefficient are, however, clear, a rise (fall) in a calling for a rise (fall) in E, while a rise (fall) in b means a fall (rise) in E.

The Tragedy of the Commons

The population problem has no technical solution;
it requires a fundamental extension in morality.

Garrett Hardin

At the end of a thoughtful article on the future of nuclear war, Wiesner and York (*1*) concluded that: "Both sides in the arms race are . . . confronted by the dilemma of steadily increasing military power and steadily decreasing national security. *It is our considered professional judgment that this dilemma has no technical solution.* If the great powers continue to look for solutions in the area of science and technology only, the result will be to worsen the situation."

I would like to focus your attention not on the subject of the article (national security in a nuclear world) but on the kind of conclusion they reached, namely that there is no technical solution to the problem. An implicit and almost universal assumption of discussions published in professional and semipopular scientific journals is that the problem under discussion has a technical solution. A technical solution may be defined as one that requires a change only in the techniques of the natural sciences, demanding little or nothing in the way of change in human values or ideas of morality.

In our day (though not in earlier times) technical solutions are always welcome. Because of previous failures in prophecy, it takes courage to assert that a desired technical solution is not possible. Wiesner and York exhibited this courage; publishing in a science journal, they insisted that the solution to the problem was not to be found in the natural sciences. They cautiously qualified their statement with the phrase, "It is our considered professional judgment. . . ." Whether they were right or not is not the concern of the present article. Rather, the concern here is with the important concept of a class of human problems which can be called "no technical solution problems," and, more specifically, with the identification and discussion of one of these.

It is easy to show that the class is not a null class. Recall the game of tick-tack-toe. Consider the problem, "How can I win the game of tick-tack-toe?" It is well known that I cannot, if I assume (in keeping with the conventions of game theory) that my opponent understands the game perfectly. Put another way, there is no "technical solution" to the problem. I can win only by giving a radical meaning to the word "win." I can hit my opponent over the head; or I can drug him; or I can falsify the records. Every way in which I "win" involves, in some sense, an abandonment of the game, as we intuitively understand it. (I can also, of course, openly abandon the game—refuse to play it. This is what most adults do.)

The class of "No technical solution problems" has members. My thesis is that the "population problem," as conventionally conceived, is a member of this class. How it is conventionally conceived needs some comment. It is fair to say that most people who anguish over the population problem are trying to find a way to avoid the evils of overpopulation without relinquishing any of the privileges they now enjoy. They think that farming the seas or developing new strains of wheat will solve the problem—technologically. I try to show here that the solution they seek cannot be found. The population problem cannot be solved in a technical way, any more than can the problem of winning the game of tick-tack-toe.

What Shall We Maximize?

Population, as Malthus said, naturally tends to grow "geometrically," or, as we would now say, exponentially. In a finite world this means that the per capita share of the world's goods must steadily decrease. Is ours a finite world?

A fair defense can be put forward for the view that the world is infinite; or that we do not know that it is not. But, in terms of the practical problems that we must face in the next few generations with the foreseeable technology, it is clear that we will greatly increase human misery if we do not, during the immediate future, assume that the world available to the terrestrial human population is finite. "Space" is no escape (*2*).

A finite world can support only a finite population; therefore, population growth must eventually equal zero. (The case of perpetual wide fluctuations above and below zero is a trivial variant that need not be discussed.) When this condition is met, what will be the situation of mankind? Specifically, can Bentham's goal of "the greatest good for the greatest number" be realized?

No—for two reasons, each sufficient by itself. The first is a theoretical one. It is not mathematically possible to maximize for two (or more) variables at the same time. This was clearly stated by von Neumann and Morgenstern (*3*), but the principle is implicit in the theory of partial differential equations, dating back at least to D'Alembert (1717–1783).

The second reason springs directly from biological facts. To live, any organism must have a source of energy (for example, food). This energy is utilized for two purposes: mere maintenance and work. For man, maintenance of life requires about 1600 kilocalories a day ("maintenance calories"). Anything that he does over and above merely staying alive will be defined as work, and is supported by "work calories" which he takes in. Work calories are used not only for what we call work in common speech; they are also required for all forms of enjoyment, from swimming and automobile racing to playing music and writing poetry. If our goal is to maximize population it is obvious what we must do: We must make the work calories per person approach as close to zero as possible. No gourmet meals, no vacations, no sports, no music, no literature, no art. . . . I think that everyone will grant, without

The author is professor of biology, University of California, Santa Barbara. This article is based on a presidential address presented before the meeting of the Pacific Division of the American Association for the Advancement of Science at Utah State University, Logan, 25 June 1968.

From Hardin, G. (1968). The Tragedy of the Commons. Science 162: 1243–1248. Reprinted with permission from AAAS.

argument or proof, that maximizing population does not maximize goods. Bentham's goal is impossible.

In reaching this conclusion I have made the usual assumption that it is the acquisition of energy that is the problem. The appearance of atomic energy has led some to question this assumption. However, given an infinite source of energy, population growth still produces an inescapable problem. The problem of the acquisition of energy is replaced by the problem of its dissipation, as J. H. Fremlin has so wittily shown (4). The arithmetic signs in the analysis are, as it were, reversed; but Bentham's goal is still unobtainable.

The optimum population is, then, less than the maximum. The difficulty of defining the optimum is enormous; so far as I know, no one has seriously tackled this problem. Reaching an acceptable and stable solution will surely require more than one generation of hard analytical work—and much persuasion.

We want the maximum good per person; but what is good? To one person it is wilderness, to another it is ski lodges for thousands. To one it is estuaries to nourish ducks for hunters to shoot; to another it is factory land. Comparing one good with another is, we usually say, impossible because goods are incommensurable. Incommensurables cannot be compared.

Theoretically this may be true; but in real life incommensurables *are* commensurable. Only a criterion of judgment and a system of weighting are needed. In nature the criterion is survival. Is it better for a species to be small and hideable, or large and powerful? Natural selection commensurates the incommensurables. The compromise achieved depends on a natural weighting of the values of the variables.

Man must imitate this process. There is no doubt that in fact he already does, but unconsciously. It is when the hidden decisions are made explicit that the arguments begin. The problem for the years ahead is to work out an acceptable theory of weighting. Synergistic effects, nonlinear variation, and difficulties in discounting the future make the intellectual problem difficult, but not (in principle) insoluble.

Has any cultural group solved this practical problem at the present time, even on an intuitive level? One simple fact proves that none has: there is no prosperous population in the world today that has, and has had for some time, a growth rate of zero. Any people that has intuitively identified its optimum point will soon reach it, after which its growth rate becomes and remains zero.

Of course, a positive growth rate might be taken as evidence that a population is below its optimum. However, by any reasonable standards, the most rapidly growing populations on earth today are (in general) the most miserable. This association (which need not be invariable) casts doubt on the optimistic assumption that the positive growth rate of a population is evidence that it has yet to reach its optimum.

We can make little progress in working toward optimum poulation size until we explicitly exorcize the spirit of Adam Smith in the field of practical demography. In economic affairs, *The Wealth of Nations* (1776) popularized the "invisible hand," the idea that an individual who "intends only his own gain," is, as it were, "led by an invisible hand to promote . . . the public interest" (5). Adam Smith did not assert that this was invariably true, and perhaps neither did any of his followers. But he contributed to a dominant tendency of thought that has ever since interfered with positive action based on rational analysis, namely, the tendency to assume that decisions reached individually will, in fact, be the best decisions for an entire society. If this assumption is correct it justifies the continuance of our present policy of laissez-faire in reproduction. If it is correct we can assume that men will control their individual fecundity so as to produce the optimum population. If the assumption is not correct, we need to reexamine our individual freedoms to see which ones are defensible.

Tragedy of Freedom in a Commons

The rebuttal to the invisible hand in population control is to be found in a scenario first sketched in a little-known pamphlet (6) in 1833 by a mathematical amateur named William Forster Lloyd (1794–1852). We may well call it "the tragedy of the commons," using the word "tragedy" as the philosopher Whitehead used it (7): "The essence of dramatic tragedy is not unhappiness. It resides in the solemnity of the remorseless working of things." He then goes on to say, "This inevitableness of destiny can only be illustrated in terms of human life by incidents which in fact involve unhappiness. For it is only by them that the futility of escape can be made evident in the drama."

The tragedy of the commons develops in this way. Picture a pasture open to all. It is to be expected that each herdsman will try to keep as many cattle as possible on the commons. Such an arrangement may work reasonably satisfactorily for centuries because tribal wars, poaching, and disease keep the numbers of both man and beast well below the carrying capacity of the land. Finally, however, comes the day of reckoning, that is, the day when the long-desired goal of social stability becomes a reality. At this point, the inherent logic of the commons remorselessly generates tragedy.

As a rational being, each herdsman seeks to maximize his gain. Explicitly or implicitly, more or less consciously, he asks, "What is the utility *to me* of adding one more animal to my herd?" This utility has one negative and one positive component.

1) The positive component is a function of the increment of one animal. Since the herdsman receives all the proceeds from the sale of the additional animal, the positive utility is nearly +1.

2) The negative component is a function of the additional overgrazing created by one more animal. Since, however, the effects of overgrazing are shared by all the herdsmen, the negative utility for any particular decision-making herdsman is only a fraction of −1.

Adding together the component partial utilities, the rational herdsman concludes that the only sensible course for him to pursue is to add another animal to his herd. And another; and another. . . . But this is the conclusion reached by each and every rational herdsman sharing a commons. Therein is the tragedy. Each man is locked into a system that compels him to increase his herd without limit—in a world that is limited. Ruin is the destination toward which all men rush, each pursuing his own best interest in a society that believes in the freedom of the commons. Freedom in a commons brings ruin to all.

Some would say that this is a platitude. Would that it were! In a sense, it was learned thousands of years ago, but natural selection favors the forces of psychological denial (8). The individual benefits as an individual from his ability to deny the truth even though society as a whole, of which he is a part, suffers.

Education can counteract the natural tendency to do the wrong thing, but the inexorable succession of generations requires that the basis for this knowledge be constantly refreshed.

A simple incident that occurred a few years ago in Leominster, Massachusetts, shows how perishable the knowledge is. During the Christmas shopping season the parking meters downtown were covered with plastic bags that bore tags reading: "Do not open until after Christmas. Free parking courtesy of the mayor and city council." In other words, facing the prospect of an increased demand for already scarce space, the city fathers reinstituted the system of the commons. (Cynically, we suspect that they gained more votes than they lost by this retrogressive act.)

In an approximate way, the logic of the commons has been understood for a long time, perhaps since the discovery of agriculture or the invention of private property in real estate. But it is understood mostly only in special cases which are not sufficiently generalized. Even at this late date, cattlemen leasing national land on the western ranges demonstrate no more than an ambivalent understanding, in constantly pressuring federal authorities to increase the head count to the point where overgrazing produces erosion and weed-dominance. Likewise, the oceans of the world continue to suffer from the survival of the philosophy of the commons. Maritime nations still respond automatically to the shibboleth of the "freedom of the seas." Professing to believe in the "inexhaustible resources of the oceans," they bring species after species of fish and whales closer to extinction (9).

The National Parks present another instance of the working out of the tragedy of the commons. At present, they are open to all, without limit. The parks themselves are limited in extent—there is only one Yosemite Valley—whereas population seems to grow without limit. The values that visitors seek in the parks are steadily eroded. Plainly, we must soon cease to treat the parks as commons or they will be of no value to anyone.

What shall we do? We have several options. We might sell them off as private property. We might keep them as public property, but allocate the right to enter them. The allocation might be on the basis of wealth, by the use of an auction system. It might be on the basis of merit, as defined by some agreed-upon standards. It might be by lottery. Or it might be on a first-come, first-served basis, administered to long queues. These, I think, are all the reasonable possibilities. They are all objectionable. But we must choose—or acquiesce in the destruction of the commons that we call our National Parks.

Pollution

In a reverse way, the tragedy of the commons reappears in problems of pollution. Here it is not a question of taking something out of the commons, but of putting something in—sewage, or chemical, radioactive, and heat wastes into water; noxious and dangerous fumes into the air; and distracting and unpleasant advertising signs into the line of sight. The calculations of utility are much the same as before. The rational man finds that his share of the cost of the wastes he discharges into the commons is less than the cost of purifying his wastes before releasing them. Since this is true for everyone, we are locked into a system of "fouling our own nest," so long as we behave only as independent, rational, free-enterprisers.

The tragedy of the commons as a food basket is averted by private property, or something formally like it. But the air and waters surrounding us cannot readily be fenced, and so the tragedy of the commons as a cesspool must be prevented by different means, by coercive laws or taxing devices that make it cheaper for the polluter to treat his pollutants than to discharge them untreated. We have not progressed as far with the solution of this problem as we have with the first. Indeed, our particular concept of private property, which deters us from exhausting the positive resources of the earth, favors pollution. The owner of a factory on the bank of a stream—whose property extends to the middle of the stream—often has difficulty seeing why it is not his natural right to muddy the waters flowing past his door. The law, always behind the times, requires elaborate stitching and fitting to adapt it to this newly perceived aspect of the commons.

The pollution problem is a consequence of population. It did not much matter how a lonely American frontiersman disposed of his waste. "Flowing water purifies itself every 10 miles," my grandfather used to say, and the myth was near enough to the truth when he was a boy, for there were not too many people. But as population became denser, the natural chemical and biological recycling processes became overloaded, calling for a redefinition of property rights.

How To Legislate Temperance?

Analysis of the pollution problem as a function of population density uncovers a not generally recognized principle of morality, namely: *the morality of an act is a function of the state of the system at the time it is performed* (10). Using the commons as a cesspool does not harm the general public under frontier conditions, because there is no public; the same behavior in a metropolis is unbearable. A hundred and fifty years ago a plainsman could kill an American bison, cut out only the tongue for his dinner, and discard the rest of the animal. He was not in any important sense being wasteful. Today, with only a few thousand bison left, we would be appalled at such behavior.

In passing, it is worth noting that the morality of an act cannot be determined from a photograph. One does not know whether a man killing an elephant or setting fire to the grassland is harming others until one knows the total system in which his act appears. "One picture is worth a thousand words," said an ancient Chinese; but it may take 10,000 words to validate it. It is as tempting to ecologists as it is to reformers in general to try to persuade others by way of the photographic shortcut. But the essense of an argument cannot be photographed: it must be presented rationally—in words.

That morality is system-sensitive escaped the attention of most codifiers of ethics in the past. "Thou shalt not . . ." is the form of traditional ethical directives which make no allowance for particular circumstances. The laws of our society follow the pattern of ancient ethics, and therefore are poorly suited to governing a complex, crowded, changeable world. Our epicyclic solution is to augment statutory law with administrative law. Since it is practically impossible to spell out all the conditions under which it is safe to burn trash in the back yard or to run an automobile without smog-control, by law we delegate the details to bureaus. The result is administrative law, which is rightly feared for an ancient reason—*Quis custodiet ipsos custodes?*—"Who shall watch the watchers themselves?" John Adams said that we must have "a government of laws and not men." Bureau administrators, trying to evaluate the morality of acts in the total system, are singularly liable to corruption, producing a government by men, not laws.

Prohibition is easy to legislate (though not necessarily to enforce); but how do we legislate temperance? Experience indicates that it can be accomplished best through the mediation of administrative law. We limit possibilities unnecessarily if we suppose that the sentiment of *Quis custodiet* denies us the use of administrative law. We should rather retain the phrase as a perpetual reminder of fearful dangers we cannot avoid. The great challenge facing us now is to invent the corrective feedbacks that are needed to keep custodians honest. We must find ways to legitimate the needed authority of both the custodians and the corrective feedbacks.

Freedom To Breed Is Intolerable

The tragedy of the commons is involved in population problems in another way. In a world governed solely by the principle of "dog eat dog"—if indeed there ever was such a world—how many children a family had would not be a matter of public concern. Parents who bred too exuberantly would leave fewer descendants, not more, because they would be unable to care adequately for their children. David Lack and others have found that such a negative feedback demonstrably controls the fecundity of birds (*11*). But men are not birds, and have not acted like them for millenniums, at least.

If each human family were dependent only on its own resources; *if* the children of improvident parents starved to death; *if*, thus, overbreeding brought its own "punishment" to the germ line— *then* there would be no public interest in controlling the breeding of families. But our society is deeply committed to the welfare state (*12*), and hence is confronted with another aspect of the tragedy of the commons.

In a welfare state, how shall we deal with the family, the religion, the race, or the class (or indeed any distinguishable and cohesive group) that adopts overbreeding as a policy to secure its own aggrandizement (*13*)? To couple the concept of freedom to breed with the belief that everyone born has an equal right to the commons is to lock the world into a tragic course of action.

Unfortunately this is just the course of action that is being pursued by the United Nations. In late 1967, some 30 nations agreed to the following (*14*):

The Universal Declaration of Human Rights describes the family as the natural and fundamental unit of society. It follows that any choice and decision with regard to the size of the family must irrevocably rest with the family itself, and cannot be made by anyone else.

It is painful to have to deny categorically the validity of this right; denying it, one feels as uncomfortable as a resident of Salem, Massachusetts, who denied the reality of witches in the 17th century. At the present time, in liberal quarters, something like a taboo acts to inhibit criticism of the United Nations. There is a feeling that the United Nations is "our last and best hope," that we shouldn't find fault with it; we shouldn't play into the hands of the archconservatives. However, let us not forget what Robert Louis Stevenson said: "The truth that is suppressed by friends is the readiest weapon of the enemy." If we love the truth we must openly deny the validity of the Universal Declaration of Human Rights, even though it is promoted by the United Nations. We should also join with Kingsley Davis (*15*) in attempting to get Planned Parenthood-World Population to see the error of its ways in embracing the same tragic ideal.

Conscience Is Self-Eliminating

It is a mistake to think that we can control the breeding of mankind in the long run by an appeal to conscience. Charles Galton Darwin made this point when he spoke on the centennial of the publication of his grandfather's great book. The argument is straightforward and Darwinian.

People vary. Confronted with appeals to limit breeding, some people will undoubtedly respond to the plea more than others. Those who have more children will produce a larger fraction of the next generation than those with more susceptible consciences. The difference will be accentuated, generation by generation.

In C. G. Darwin's words: "It may well be that it would take hundreds of generations for the progenitive instinct to develop in this way, but if it should do so, nature would have taken her revenge, and the variety *Homo contra-*cipiens* would become extinct and would be replaced by the variety *Homo progenitivus*" (*16*).

The argument assumes that conscience or the desire for children (no matter which) is hereditary—but hereditary only in the most general formal sense. The result will be the same whether the attitude is transmitted through germ cells, or exosomatically, to use A. J. Lotka's term. (If one denies the latter possibility as well as the former, then what's the point of education?) The argument has here been stated in the context of the population problem, but it applies equally well to any instance in which society appeals to an individual exploiting a commons to restrain himself for the general good—by means of his conscience. To make such an appeal is to set up a selective system that works toward the elimination of conscience from the race.

Pathogenic Effects of Conscience

The long-term disadvantage of an appeal to conscience should be enough to condemn it; but has serious short-term disadvantages as well. If we ask a man who is exploiting a commons to desist "in the name of conscience," what are we saying to him? What does he hear?—not only at the moment but also in the wee small hours of the night when, half asleep, he remembers not merely the words we used but also the nonverbal communication cues we gave him unawares? Sooner or later, consciously or subconsciously, he senses that he has received two communications, and that they are contradictory: (i) (intended communication) "If you don't do as we ask, we will openly condemn you for not acting like a responsible citizen"; (ii) (the unintended communication) "If you *do* behave as we ask, we will secretly condemn you for a simpleton who can be shamed into standing aside while the rest of us exploit the commons."

Everyman then is caught in what Bateson has called a "double bind." Bateson and his co-workers have made a plausible case for viewing the double bind as an important causative factor in the genesis of schizophrenia (*17*). The double bind may not always be so damaging, but it always endangers the mental health of anyone to whom it is applied. "A bad conscience," said Nietzsche, "is a kind of illness."

To conjure up a conscience in others

is tempting to anyone who wishes to extend his control beyond the legal limits. Leaders at the highest level succumb to this temptation. Has any President during the past generation failed to call on labor unions to moderate voluntarily their demands for higher wages, or to steel companies to honor voluntary guidelines on prices? I can recall none. The rhetoric used on such occasions is designed to produce feelings of guilt in noncooperators.

For centuries it was assumed without proof that guilt was a valuable, perhaps even an indispensable, ingredient of the civilized life. Now, in this post-Freudian world, we doubt it.

Paul Goodman speaks from the modern point of view when he says: "No good has ever come from feeling guilty, neither intelligence, policy, nor compassion. The guilty do not pay attention to the object but only to themselves, and not even to their own interests, which might make sense, but to their anxieties" (18).

One does not have to be a professional psychiatrist to see the consequences of anxiety. We in the Western world are just emerging from a dreadful two-centuries-long Dark Ages of Eros that was sustained partly by prohibition laws, but perhaps more effectively by the anxiety-generating mechanisms of education. Alex Comfort has told the story well in The Anxiety Makers (19); it is not a pretty one.

Since proof is difficult, we may even concede that the results of anxiety may sometimes, from certain points of view, be desirable. The larger question we should ask is whether, as a matter of policy, we should ever encourage the use of a technique the tendency (if not the intention) of which is psychologically pathogenic. We hear much talk these days of responsible parenthood; the coupled words are incorporated into the titles of some organizations devoted to birth control. Some people have proposed massive propaganda campaigns to instill responsibility into the nation's (or the world's) breeders. But what is the meaning of the word responsibility in this context? Is it not merely a synonym for the word conscience? When we use the word responsibility in the absence of substantial sanctions are we not trying to browbeat a free man in a commons into acting against his own interest? Responsibility is a verbal counterfeit for a substantial quid pro quo. It is an attempt to get something for nothing.

If the word responsibility is to be used at all, I suggest that it be in the sense Charles Frankel uses it (20). "Responsibility," says this philosopher, "is the product of definite social arrangements." Notice that Frankel calls for social arrangements—not propaganda.

Mutual Coercion

Mutually Agreed upon

The social arrangements that produce responsibility are arrangements that create coercion, of some sort. Consider bank-robbing. The man who takes money from a bank acts as if the bank were a commons. How do we prevent such action? Certainly not by trying to control his behavior solely by a verbal appeal to his sense of responsibility. Rather than rely on propaganda we follow Frankel's lead and insist that a bank is not a commons; we seek the definite social arrangements that will keep it from becoming a commons. That we thereby infringe on the freedom of would-be robbers we neither deny nor regret.

The morality of bank-robbing is particularly easy to understand because we accept complete prohibition of this activity. We are willing to say "Thou shalt not rob banks," without providing for exceptions. But temperance also can be created by coercion. Taxing is a good coercive device. To keep downtown shoppers temperate in their use of parking space we introduce parking meters for short periods, and traffic fines for longer ones. We need not actually forbid a citizen to park as long as he wants to; we need merely make it increasingly expensive for him to do so. Not prohibition, but carefully biased options are what we offer him. A Madison Avenue man might call this persuasion; I prefer the greater candor of the word coercion.

Coercion is a dirty word to most liberals now, but it need not forever be so. As with the four-letter words, its dirtiness can be cleansed away by exposure to the light, by saying it over and over without apology or embarrassment. To many, the word coercion implies arbitrary decisions of distant and irresponsible bureaucrats; but this is not a necessary part of its meaning. The only kind of coercion I recommend is mutual coercion, mutually agreed upon by the majority of the people affected.

To say that we mutually agree to coercion is not to say that we are required to enjoy it, or even to pretend we enjoy it. Who enjoys taxes? We all grumble about them. But we accept compulsory taxes because we recognize that voluntary taxes would favor the conscienceless. We institute and (grumblingly) support taxes and other coercive devices to escape the horror of the commons.

An alternative to the commons need not be perfectly just to be preferable. With real estate and other material goods, the alternative we have chosen is the institution of private property coupled with legal inheritance. Is this system perfectly just? As a genetically trained biologist I deny that it is. It seems to me that, if there are to be differences in individual inheritance, legal possession should be perfectly correlated with biological inheritance—that those who are biologically more fit to be the custodians of property and power should legally inherit more. But genetic recombination continually makes a mockery of the doctrine of "like father, like son" implicit in our laws of legal inheritance. An idiot can inherit millions, and a trust fund can keep his estate intact. We must admit that our legal system of private property plus inheritance is unjust—but we put up with it because we are not convinced, at the moment, that anyone has invented a better system. The alternative of the commons is too horrifying to contemplate. Injustice is preferable to total ruin.

It is one of the peculiarities of the warfare between reform and the status quo that it is thoughtlessly governed by a double standard. Whenever a reform measure is proposed it is often defeated when its opponents triumphantly discover a flaw in it. As Kingsley Davis has pointed out (21), worshippers of the status quo sometimes imply that no reform is possible without unanimous agreement, an implication contrary to historical fact. As nearly as I can make out, automatic rejection of proposed reforms is based on one of two unconscious assumptions: (i) that the status quo is perfect; or (ii) that the choice we face is between reform and no action; if the proposed reform is imperfect, we presumably should take no action at all, while we wait for a perfect proposal.

But we can never do nothing. That which we have done for thousands of years is also action. It also produces evils. Once we are aware that the status quo is action, we can then compare its discoverable advantages and disadvantages with the predicted advantages and disadvantages of the proposed reform, discounting as best we can for our lack of experience. On the basis of such a comparison, we can make a rational decision which will not involve the unworkable assumption that only perfect systems are tolerable.

Recognition of Necessity

Perhaps the simplest summary of this analysis of man's population problems is this: the commons, if justifiable at all, is justifiable only under conditions of low-population density. As the human population has increased, the commons has had to be abandoned in one aspect after another.

First we abandoned the commons in food gathering, enclosing farm land and restricting pastures and hunting and fishing areas. These restrictions are still not complete throughout the world.

Somewhat later we saw that the commons as a place for waste disposal would also have to be abandoned. Restrictions on the disposal of domestic sewage are widely accepted in the Western world; we are still struggling to close the commons to pollution by automobiles, factories, insecticide sprayers, fertilizing operations, and atomic energy installations.

In a still more embryonic state is our recognition of the evils of the commons in matters of pleasure. There is almost no restriction on the propagation of sound waves in the public medium. The shopping public is assaulted with mindless music, without its consent. Our

government is paying out billions of dollars to create supersonic transport which will disturb 50,000 people for every one person who is whisked from coast to coast 3 hours faster. Advertisers muddy the airwaves of radio and television and pollute the view of travelers. We are a long way from outlawing the commons in matters of pleasure. Is this because our Puritan inheritance makes us view pleasure as something of a sin, and pain (that is, the pollution of advertising) as the sign of virtue?

Every new enclosure of the commons involves the infringement of somebody's personal liberty. Infringements made in the distant past are accepted because no contemporary complains of a loss. It is the newly proposed infringements that we vigorously oppose; cries of "rights" and "freedom" fill the air. But what does "freedom" mean? When men mutually agreed to pass laws against robbing, mankind became more free, not less so. Individuals locked into the logic of the commons are free only to bring on universal ruin; once they see the necessity of mutual coercion, they become free to pursue other goals. I believe it was Hegel who said, "Freedom is the recognition of necessity."

The most important aspect of necessity that we must now recognize, is the necessity of abandoning the commons in breeding. No technical solution can rescue us from the misery of overpopulation. Freedom to breed will bring ruin to all. At the moment, to avoid hard decisions many of us are tempted to propagandize for conscience and responsible parenthood. The temptation must be resisted, because an appeal to independently acting con-

sciences selects for the disappearance of all conscience in the long run, and an increase in anxiety in the short.

The only way we can preserve and nurture other and more precious freedoms is by relinquishing the freedom to breed, and that very soon. "Freedom is the recognition of necessity"—and it is the role of education to reveal to all the necessity of abandoning the freedom to breed. Only so, can we put an end to this aspect of the tragedy of the commons.

References

1. J. B. Wiesner and H. F. York, *Sci. Amer.* **211** (No. 4), 27 (1964).
2. G. Hardin, *J. Hered.* **50**, 68 (1959); S. von Hoernor, *Science* **137**, 18 (1962).
3. J. von Neumann and O. Morgenstern, *Theory of Games and Economic Behavior* (Princeton Univ. Press, Princeton, N.J., 1947), p. 11.
4. J. H. Fremlin, *New Sci.*, No. 415 (1964), p. 285.
5. A. Smith, *The Wealth of Nations* (Modern Library, New York, 1937), p. 423.
6. W. F. Lloyd, *Two Lectures on the Checks to Population* (Oxford Univ. Press, Oxford, England, 1833), reprinted (in part) in *Population, Evolution, and Birth Control*, G. Hardin, Ed. (Freeman, San Francisco, 1964), p. 37.
7. A. N. Whitehead, *Science and the Modern World* (Mentor, New York, 1948), p. 17.
8. G. Hardin, Ed. *Population, Evolution, and Birth Control* (Freeman, San Francisco, 1964), p. 56.
9. S. McVay, *Sci. Amer.* **216** (No. 8), 13 (1966).
10. J. Fletcher, *Situation Ethics* (Westminster, Philadelphia, 1966).
11. D. Lack, *The Natural Regulation of Animal Numbers* (Clarendon Press, Oxford, 1954).
12. H. Girvetz, *From Wealth to Welfare* (Stanford Univ. Press, Stanford, Calif., 1950).
13. G. Hardin, *Perspec. Biol. Med.* **6**, 366 (1963).
14. U. Thant, *Int. Planned Parenthood News*, No. 168 (February 1968), p. 3.
15. K. Davis, *Science* **158**, 730 (1967).
16. S. Tax, Ed., *Evolution after Darwin* (Univ. of Chicago Press, Chicago, 1960), vol. 2, p. 469.
17. G. Bateson, D. D. Jackson, J. Haley, J. Weakland, *Behav. Sci.* **1**, 251 (1956).
18. P. Goodman, *New York Rev. Books* **10**(8), 22 (23 May 1968).
19. A. Comfort, *The Anxiety Makers* (Nelson, London, 1967).
20. C. Frankel, *The Case for Modern Man* (Harper, New York, 1955), p. 203.
21. J. D. Roslansky, *Genetics and the Future of Man* (Appleton-Century-Crofts, New York, 1966), p. 177.

THE WORLD DYNAMICS OF
ECONOMIC GROWTH

The Economics of the Steady State

By HERMAN E. DALY*

> But if your theory is found to be against the second law of thermodynamics, I can give you no hope; there is nothing for it but to collapse in deepest humiliation.
>
> *Sir Arthur Eddington*

My title is somewhat pretentious since at present this "new economics" consists only of a definition of a steady-state economy, some arguments for its necessity and desirability, and some disciplined speculations on its appropriate institutions and the problem of transition, each of which will be briefly discussed below.

I. What is a Steady-State Economy?

A steady-state economy is defined by constant stocks of physical wealth (artifacts) and a constant population, each maintained at some chosen, desirable level by a low rate of throughput—i.e., by low birth rates equal to low death rates and by low physical production rates equal to low physical depreciation rates, so that longevity of people and durability of physical stocks are high. The throughput flow, viewed as the cost of maintaining the stocks, begins with the extraction (depletion) of low entropy resources at the input end, and terminates with an equal quantity of high entropy waste (pollution) at the output end. The throughput is the inevitable cost of maintaining the stocks of people and artifacts and should be minimized subject to the maintenance of a

* Professor of economics, Louisiana State University.

chosen level of stocks (Kenneth E. Boulding).

The services (want satisfaction) yielded by the stocks of artifacts (and people) are the ultimate benefit of economic activity, and the throughput is the ultimate cost. The stock of physical wealth is an accumulated flow of throughput, and thus in the final analysis is a cost. Ultimate efficiency is the ratio of service to throughput. But to yield a service, the throughput flow must be first accumulated into stocks even if of short duration. It is the existence of a table or a doctor at a point in time that yields services, not their gradual depreciation nor the productive process by which they are replaced. Stocks are intermediate magnitudes that yield services and require throughput for maintenance and replacement. This may be expressed in the equation:

$$(1) \quad \frac{\text{Ultimate}}{\text{Efficiency}} = \overset{(1)}{\frac{\text{Service}}{\text{Throughput}}} = \overset{(2)}{\frac{\text{Service}}{\text{Stock}}} \times \overset{(3)}{\frac{\text{Stock}}{\text{Throughput}}}$$

Since by definition stocks are constant at a level corresponding to some concept of sufficiency or maturity, progress in the steady state consists in increasing ultimate efficiency (ratio 1) in two ways: by maintaining the stock with less throughput (increase ratio 3 or "maintenance efficiency") and by getting more service per unit of time from the same stock (increase ratio 2 or "service efficiency"). The laws of thermodynamics provide a theoretical limit to the improvement of maintenance efficiency. Whether there is any theoretical limit to increase in service efficiency resulting from the limits of the human stomach and nervous system is less clear, but in my opinion likely.

Over short periods of time the throughput cost of maintaining the constant stock may decrease due to improvements in maintenance efficiency, but over the long run it must increase because as better grade (lower entropy) sources of raw materials are used up, it will be necessary to process ever larger amounts of materials using ever more energy and capital equipment to get the same quantity of needed mineral. Thus a steady-state economy, as here defined, does not imply constant throughput, much less static technology, nor does it imply eternal life for the economic system. It is simply a strategy for good stewardship, for maintaining our spaceship and permitting it to die of old age rather than from the cancer of growthmania. It is basically an extension of the demographers' model of a stationary population to include the populations of physical artifacts, and the fundamental idea is found in John Stuart Mill's discussion of the stationary state of classical economics.

The term "economic growth" conventionally refers to an increase in the flow of "real *GNP*," which is a value *index* of the *physical* flow of throughput. The (measurable) throughput is, in turn, an index of (unmeasurable) service only if ratios 2 and 3 (or their product) are constant or increasing. This may have been the case in the past, but for the future it is doubtful. As the growing throughput pushes against biophysical limits, it provokes a decline in service efficiency (more of the stock must be devoted to the defensive use of repairing life support systems that formerly provided free services). Also, since our institutions are geared to a continually increasing throughput, we may willingly lower maintenance efficiency for the sake of per-

mitting a larger throughput (e.g. planned obsolescence and fashion). If someone wants to redefine "economic growth" as an increase in nonmaterial services and then argue that it can and should grow forever, he is free to do so. But this hardly constitutes a refutation of the steady-state economy, which is defined in terms of measurable physical stocks, not unmeasurable psychic fluxes.

Nor are the levels at which the stocks of people and artifacts are maintained necessarily frozen for all eternity. As a result of technical and moral evolution it may become both possible and desirable to grow or to decline to a different level. But then growth or decline would be seen as a temporary transition from one steady state to another and not as the norm for a healthy economy. Technical and moral change would lead growth rather than being blindly pushed down the path of least resistance by the growth juggernaut.

At what point should growth in stocks and maximization of production flow give way to stock maintenance and the minimization of the production flow? There are a large number of steady-state levels of stocks to choose from, and such a choice is a difficult problem of ecology and ethics. But our inability to define the optimum level does not mean that we will not someday discover that we have grown beyond it. It is more important to learn to be stable at existing or nearby levels than to know in advance which level is optimal. Knowledge of the latter without the former merely allows us to recognize and wave goodbye to the optimum as we grow through it. Besides, the optimum may well be a broad plateau within which one place is as good as another as long as we don't go too near the edge.

The radical change implied by a steady state is evident from the foregoing, and from W. W. Rostow's characterization of our present economy of high mass consumption, to which all countries unrealistically aspire, as one "in which compound interest becomes built, as it were, into our habits and institutions" (p. 7). This built-in exponential growth and its unfortunate consequences constitute the theme of the much-maligned little book, *Limits to Growth*, by D. H. Meadows et al. Before discussing the radical departure of "deinstitutionalizing" compound interest or at least uncoupling it from all physical dimensions, we must consider whether such a change is really necessary and/or desirable.

II. The Necessity and Desirability of the Steady State

Our economy is a subsystem of the earth, and the earth is apparently a steady-state open system. The subsystem cannot grow beyond the frontiers of the total system and, if it is not to disrupt the functioning of the latter, must at some much earlier point conform to the steady-state mode. The technocratic project of redesigning the world (substituting technosphere for ecosphere) so as to allow for indefinite economic growth is a bit of hubris that has received the insufficiently pejorative label of "growthmania."

The conceptual roots of growthmania are to be found in the orthodox doctrines of "relative scarcity" and "absolute wants." Relative (or "Ricardian") scarcity refers to the scarcity of a particular resource relative to another resource or to a lower quality of the same resource. Absolute (or "Malthusian") scarcity refers to the scarcity of all resources in general, relative to population and per capita consumption levels. The solution to relative scarcity is substitution. Absolute scarcity assumes that all economical substitutions are made so that the total burden of absolute scarcity is minimized but still exists and may still increase. Even an efficiently borne burden can become too heavy. Substitu-

tion is always of one form of low entropy matter-energy for another. There is no substitute for low entropy itself, and low entropy is scarce, both in its terrestrial source (finite stocks of concentrated fossil fuels and minerals) and in its solar source (a fixed rate of inflow of solar energy). (See Georgescu-Roegen.) Both the human economy and the nonhuman part of the biosphere depend on the same limited budget of low entropy and on the allocative pattern which that budget has evolved over millennia. The entropy of the human sector is reduced and kept low by the continual importation of low entropy from, and exportation of high entropy to, the nonhuman sector (Daly, 1968). If too much low entropy is diverted to economic growth in the human sector, or if too many evolutionary allocative patterns are disrupted in the process of diversion, then the complex life support systems of the biosphere will begin to fail. Growth in population and per capita consumption result in increasing absolute scarcity, which is manifested in the increasing prevalence of "external costs"—i.e., the system becomes more generally sensitive to particular interferences as the web of general interdependence is stretched ever tighter by growth in the populations of people and artifacts and the resulting stress on the entropy budget.

Orthodox economic theory has assumed that all scarcity is relative: "Nature imposes particular scarcities, not an inescapable general scarcity" (Harold J. Barnett and Chandler Morse, p. 11). Therefore the answer to scarcity is always substitution, and since relative price changes induce substitution, the policy recommendation is "internalization of externalities," usually via pollution taxes. The following statement is representative of orthodox complacency: " . . . the problem of environmental pollution is a simple matter of correcting a minor resource misallocation by means of pollution charges . . . " (Wilfred Beckerman, p. 327). But price rigging by itself is ineffective in coping with increasing absolute scarcity since its mode of operation is only to induce substitution. What substitute is there for resources in general, for low entropy? How is it possible to raise the *relative* price of *all* resources? Attempts to do so result in inflation rather than substitution.

A similar distinction between absolute and relative wants has also been obscured by orthodox economics. Following Keynes we may define absolute wants as those we feel independent of the situation of our fellow human beings. Relative wants are those that we feel only if their satisfaction makes us feel superior to our fellows. The importance of this distinction is that only relative wants are infinite and that relative wants cannot be universally satisfied by growth because the relative satisfactions of the elite are cancelled as growth raises the general level. This effect can be avoided, and often is, by allowing growth to increase inequality so that the relatively well off become relatively better off. But it is quite impossible for everyone to become better off relative to everyone else. In spite of this extremely important distinction, orthodox theory assumes that wants *in general* are insatiable and extends to all wants the dignity of absolute status—i.e., the satisfaction of relative and absolute wants is considered equally legitimate and equally capable of satisfaction in the aggregate by means of economic growth. The assumption of equal legitimacy is a value judgment (though it is treated by many economists as the avoidance of a value judgment), and the assumption of equal capability of satisfaction is either a logical error or an implicit acceptance of a value judgment in favor of increasing inequality.

The implication of the dogmas of the relativity of all scarcity and the absolute-

ness of all wants is growthmania. If there is no absolute scarcity to limit the *possibility* of growth (infinite substitutability of relatively abundant for relatively scarce resources) and no merely relative wants to limit the *desirability* or efficacy of growth (wants in general are infinite and all wants are equally worthy and capable of satisfaction by growth), then "growth forever and the more the better" is the logical consequence. It is also the *reductio ad absurdum* that exposes the growth orthodoxy as a rigorous exercise in wishful thinking, as a theory that is against the second law of thermodynamics as well as against common sense. It is simply a brute fact that there is such a thing as absolute scarcity and such a thing as relative wants. Furthermore, these latter categories eventually become dominant at the margin as growth continues. The implication of absolute scarcity and relative wants is the opposite of growthmania, namely, the steady state.

At this point the growthmaniacs usually make a burnt offering to the god of technology: surely economic growth can continue indefinitely because technology will continue to "grow exponentially" as it has in the past. This elaborately misses the point. The alleged "exponential growth" of technology is not directly measurable and is only inferred from the permissive role that it has played in making possible the measured exponential growth in the physical magnitudes of production, depletion and pollution (i.e., the throughput). Such technical progress is more a part of the problem than the solution. What must be appealed to is a *qualitative change* in the direction of technical progress, not a continuation of alleged quantitative trends. The institutions to be discussed in the next section seek to induce just such a change toward resource-saving technology and patterns of living, and to a greater reliance on solar energy and renewable resources. But we can be fairly certain that no new technology will abolish absolute scarcity because the laws of thermodynamics apply to all possible technologies. No one can be absolutely certain that we will not some day discover perpetual motion and how to create and destroy matter and energy. But the reasonable assumption for economists is that this is an unlikely prospect and that while technology will continue to pull rabbits out of hats, it will not pull an elephant out of a hat—much less an infinite series of ever-larger elephants!

But the ideology of growth continues to transcend the ordinary logic of elementary economics. Growth is the basis of national power and prestige. Growth offers the prospect of prosperity for all with sacrifice by none. It is a substitute for redistribution. The present sins of poverty and injustice will be washed away in a future sea of abundance, vouchsafed by the amazing grace of compound interest. This evasion, common to both capitalism and communism, was never totally honest. It is now increasingly exposed as absurd.

III. Speculations on the Steady State

The first design principle disciplining our speculations on institutions is to provide the necessary social control with a minimum sacrifice of personal freedom, to provide macro stability while allowing for micro variability, to combine the macro static with the micro dynamic. A second design principle, closely related to the first, is to maintain considerable slack between the actual environmental load and the maximum carrying capacity. The closer the actual approaches the maximum, the more rigorous, finely tuned, and micro oriented our controls will have to be. We lack the knowledge and ability to assume detailed central control of the spaceship, even if such were desirable, so therefore we should leave it on "automatic

pilot" as it has been for eons. But the automatic pilot only works if the actual load is small relative to the maximum. A third design principle, important for making the transition, is to start from existing initial conditions rather than an imaginary "clean slate," and a fourth is to build in the ability to tighten constraints gradually. Minimum faith is placed in our ability to plan a detailed blueprint for a new society. Maximum faith is placed in the basic regenerative powers of life and in the possibility of moral growth, once the root physical process of degeneration (unlimited growth) is arrested.

The kinds of institutions required follow directly from the definition. We need (1) an institution for stabilizing population, (2) an institution for stabilizing physical wealth and keeping throughput below ecological limits, and, less obviously but most importantly, (3) an institution limiting the degree of inequality in the distribution of the constant stocks among the constant population since growth can no longer be appealed to as the answer to poverty.

What specific institutions can perform these functions and are most in harmony with the general design principles discussed above? Elsewhere I have outlined a model and can here only briefly describe it (Daly, 1973, 1974a). The model builds on the existing institutions of private property and the price system and is thus fundamentally conservative. But it extends these institutions to areas previously not included: control of aggregate births (marketable birth license plan as first proposed by Boulding) and control of aggregate depletion of basic resources (depletion quotas auctioned by the government). Extending the market, under the discipline of aggregate quotas, to these vital areas is necessary to deal with increasing absolute scarcity since, as argued above, price controls deal only with relative scarcity. Quantitative limits are set with reference to ecological and ethical criteria, and the price system is then allowed, by auction and exchange, to allocate depletion quotas and birth quotas efficiently. The throughput is controlled at its input (depletion) rather than at the pollution end because physical control is easier at the point of lower entropy. Orthodox economics suggests price controls at the output end (pollution taxes), while steady-state economics suggests quantitative controls at the input end (depletion quotas).

With more vital areas of life officially subject to the discipline of the price system, it will become more urgent to establish the institutional preconditions of free and mutually beneficial exchange, namely, to limit the degree of inequality in the distribution of income and wealth and to limit the monopoly power of corporations. A distributist institution establishing a minimum income and a maximum income and wealth would go a long way toward achieving that end, while leaving room for differential reward and incentives within reasonable limits. There might be one set of limits for individuals, one for families, and one for corporations. Natural monopolies should be publicly owned and operated.

Birth quotas, depletion quotas, and distributive limits can all be varied continuously and applied with any degree of gradualism desired. Moreover, all three control points are price system parameters, and altering them does not interfere with the static allocative efficiency of the market. Externalities involving ecological, demographic, and distributive issues are "externalized" by means of quotas rather than "internalized" in rigged market prices. Yet the effect is much the same in that prices rise to reflect previously unaccounted dimensions of scarcity, and prices become a safer guide to market decisions.

The net advantage of the quota scheme is that it limits aggregate throughput, whereas price controls merely alter through-put composition, providing a useful fine-tuning supplement to quotas, but not a substitute. The higher resource prices resulting from limited depletion would have the dynamic effect of inducing resource-saving technology and a shift to greater dependence on solar energy and renewable resources. The receipts of the depletion quota auction could help finance the minimum income. The marketable birth license plan would also have an equalizing effect on per capita income distribution.

Such institutional change is obviously not on the political agenda for 1974. Nor should it be since it is speculative, has not had the benefit of widespread professional criticism, and thus may contain terrible mistakes. But mistakes will not be discovered and better ideas will not be offered unless economists awake from the dogmatic slumber of growthmania induced by the soporific doctrines of relative scarcity and absolute wants and put the steady-state paradigm on the agenda for academic debate.

REFERENCES

H. J. Barnett and C. Morse, *Scarcity and Growth*, Baltimore 1963.

W. Beckerman, "Economists, Scientists, and Environmental Catastrophe," *Oxford Econ. Papers*, Nov. 1972, *24*, 327.

F. H. Bormann and W. R. Burch, eds., *Growth, Limits and the Quality of Life*, San Francisco 1974 (forthcoming).

K. E. Boulding, *Economics as a Science*, New York 1970.

J. Culbertson, *Economic Development: An Ecological Approach*, New York 1971.

H. E. Daly, "On Economics as a Life Science," *J. Polit. Econ.*, May/June 1968, *76*, 392–406.

———, "In Defense of a Steady-State Economy," *Amer. J. Agri. Econ.*, Dec. 1972, *54*, 945–954.

———, ed., *Toward a Steady-State Economy*, San Francisco 1973.

———, "Long Run Environmental Constraints and Trade-Offs Between Human and Artifact Populations," *Internat. Populat. Conf.*, IUSSP, Liege 1973, *3*, 453–460.

———, "A Model for a Steady-State Economy" in Bormann and Burch, 1974a.

———, "Steady-State Economics Versus Growthmania: A Critique of the Orthodox Conceptions of Growth, Wants, Scarcity, and Efficiency," *Policy Scien.* (forthcoming) Summer 1974b.

Editors of *The Ecologist*, "A Blueprint for Survival," *The Ecolo.*, Jan. 1972.

N. Georgescu-Roegen, *The Entropy Law and the Economic Process*, Cambridge, Mass. 1971.

D. H. Meadows et al., *The Limits to Growth*, New York 1972.

D. L. Meadows and D. H. Meadows, eds., *Toward Global Equilibrium: Collected Papers*, Cambridge, Mass., 1973.

W. Ophuls, *Prologue to a Political Theory of the Steady State*, (unpublished) Ph.D. dissertation in political science, Yale University, 1973.

W. W. Rostow, *The Stages of Economic Growth*, New York 1960.

"The No-Growth Society," *Daedalus*, American Academy of Arts and Sciences Proceedings, *102*, Fall 1973.

ENVIRONMENTAL SOCIOLOGY: A NEW PARADIGM*

WILLIAM R. CATTON, JR.
RILEY E. DUNLAP
Washington State University

The American Sociologist 1978, Vol. 13 (February):41–49

Ostensibly diverse and competing theoretical perspectives in sociology are alike in their shared anthropocentrism. From any of these perspectives, therefore, much contemporary and future social experience has to seem anomalous. Environmental sociologists attempt to understand recent societal changes by means of a nonanthropocentric paradigm. Because ecosystem constraints now pose serious problems both for human societies and for sociology, three assumptions quite different from the prevalent Human Exceptionalism Paradigm (HEP) have become essential. They form a New Environmental Paradigm (NEP). Sociologists who accept this New Environmental Paradigm have no difficulty appreciating the sociological relevance of variables traditionally excluded from sociology. The core of environmental sociology is, in fact, study of interactions between environment and society. Recent work by NEP-oriented sociologists on issues pertaining to social stratification exemplifies the utility of this paradigm.

Sociology appears to have reached an impasse. Efforts of sociologists to assimilate into their favorite theories some of the astounding events that have shaped human societies within the last generation have sometimes contributed more to the fragmentation of the sociological community than to the convincing explanation of social facts. But as Thomas Kuhn (1962:76) has shown, such an impasse

* The authors contributed equally to the preparation of this paper, and are listed alphabetically. At various times we have had the benefit of stimulating discussions, for which we are grateful, with the following colleagues or graduate students: Don A. Dillman, Viktor Gecas, Dennis L. Peck, Kenneth R. Tremblay, Jr., Kent D. Van Liere, John M. Wardwell, and Robert L. Wisniewski. We are especially indebted to Don Dillman for his critical reading of an earlier draft of this paper. Dunlap's contribution to the paper was supported by Project 0158, Department of Rural Sociology, Washington State University, and this is Scientific Paper No. 4933, College of Agriculture Research Center, Washington State University, Pullman, WA 99164.

often signifies "that an occasion for re-tooling has arrived."

The rise of environmental problems, and especially apprehensions about "limits to growth," signalled sharp departures from the exuberant expectations most sociologists had shared with the general public. Environmental problems and constraints contributed to the general uneasiness in American society brought about by events in the sixties. Sociologists, no less than other thinking people, are still grappling with the dramatic shift from the calmer fifties, when the American dreams of social progress, upward mobility, and societal stability seemed secure.

In 1976 the American Sociological Association, following precedents set a few years earlier in the Rural Sociological Society and in the Society for the Study of Social Problems, established a new "Section on Environmental Sociology."[1] In this paper we shall try to account for the development of environmental sociology by showing how it represents an attempt to understand recent societal changes that are difficult to comprehend from traditional sociological perspectives. We contend that, rather than simply representing the rise of another speciality within the discipline, the emergence of environmental sociology reflects the development of a new paradigm, and that this paradigm may help to extricate us from the impasse referred to above.

The "New Environmental Paradigm" (NEP) implicit in environmental sociology is, of course, only one among several current candidates to replace or amend the increasingly obsolescent set of "domain assumptions" which have defined the nature of social reality for most sociologists. Environmental sociologists, no less than the advocates of the very different alternatives Gouldner (1970) has described, are attempting to come to grips with a changed "sense of what is real." Further, we believe the NEP may contribute to a better understanding of contemporary and future social conditions than is possible with previous sociological perspectives. To illustrate the power of this paradigm to

shed new light on important sociological issues, we shall briefly describe some recent NEP-based examinations of problems in stratification. But first we must contrast the old and new sets of assumptions.

The "Human Exceptionalism Paradigm"

The numerous competing theoretical perspectives in contemporary sociology—e.g., functionalism, symbolic interactionism, ethnomethodology, conflict theory, Marxism, and so forth—are prone to exaggerate their differences from each other. They purport to be paradigms in their own right, and are often taken as such (see, e.g., Denisoff, *et al.*, 1974, and Ritzer, 1975). But they have also been construed simply as competing "pre-paradigmatic" perspectives (Friedrichs, 1972). We maintain that their apparent diversity is not as important as the fundamental anthropocentrism underlying *all* of them.

This mutual anthropocentrism is part of a basic sociological worldview (Klausner, 1971:10–11). We call *that* worldview the "Human Exceptionalism Paradigm" (HEP). We contend that acceptance of the assumptions of the HEP has made it difficult for most sociologists, regardless of their preferred orientation, to deal meaningfully with the social implications of ecological problems and constraints. Thus, the HEP has become increasingly obstructive of sociological efforts to comprehend contemporary and future social experience.

The HEP comprises several assumptions that have either been challenged by recent additions to knowledge, or have had their optimistic implications contradicted by events of the seventies. Accepted explicitly or implicitly by all existing theoretical persuasions, they include:

1. Humans are unique among the earth's creatures, for they have culture.
2. Culture can vary almost infinitely and can change much more rapidly than biological traits.
3. Thus, many human differences are socially induced rather than inborn, they can be socially altered, and in-

---

[1] In the late sixties a Natural Resources Research Group was formed in the RSS, and in 1973 SSSP established an Environmental Problems Division.

convenient differences can be eliminated.

4. Thus, also, cultural accumulation means that progress can continue without limit, making all social problems ultimately soluble.

Sociological acceptance of such an optimistic worldview was no doubt fostered by prevalence of the doctrine of progress in Western culture, where academic sociology was spawned and nurtured. It was under the American branch of Western culture that sociology flourished most fully, and it has been clear to foreign analysts of American life, from Tocqueville to Laski, that most Americans (until recently) ardently believed that the present was better than the past and the future would improve upon the present. Sociologists could easily share that conviction when natural resources were still so plentiful that limits to progress remained unseen. The historian, David Potter (1954:141), tried to alert his colleagues to some of the unstated and unexamined assumptions shaping their studies; his words have equal relevance for sociologists: "The factor of abundance, which we first discovered as an environmental condition and which we then converted by technological change into a cultural as well as a physical force, has . . . influenced all aspects of American life in a fundamental way."[2]

Not only have sociologists been too unmindful of the fact that our society derived special qualities from past abundance; the heritage of abundance has made it difficult for most sociologists to perceive the possibility of an era of uncontrived scarcity. For example, ecological concepts such as "carrying capacity" are alien to the vocabularies of most sociologists (Catton, 1976a; 1976b), yet disregard for this concept has been tantamount to assuming an environment's carrying capacity is always enlargeable as needed—thus denying the possibility of scarcity.

Neglect of the ecosystem-dependence of human society has been evident in sociological literature on economic development (e.g., Horowitz, 1972), which has simply not recognized biogeochemical limits to material progress. And renewed sociological attention to a theory of societal evolution (e.g., Parsons, 1977) has seldom paid much attention to the resource base that is subjected to "more efficient" exploitation as societies become more differentiated internally and are thereby "adaptively upgraded." In such literature, the word "environment" refers almost entirely to a society's "symbolic environment" (cultural systems) or "social environment" (environing social systems).[3]

It is the habit of neglecting laws of other sciences (such as the Principle of Entropy and the Law of Conservation of Energy)[4]—as if human actions were unaffected by them—that enables so distinguished a sociologist as Daniel Bell (1973: 465) to assert that the question before humanity is "not subsistence but standard of living, not biology, but sociology," to insist that basic needs "are satiable, and the possibility of abundance is real," to impute "apocalyptic hysteria" to "the ecology movement," and to regard it as trite rather than questionable to expect "compound interest" growth to continue for another hundred years. Likewise, this neglect permits Amos Hawley (1975:8–9) to write that "there are no known limits to the improvement of technology" and the population pressure on nonagricultural resources is neither "currently being felt or likely to be felt in the early future." Such views reflect a staunch commitment to the HEP.

Environmental Sociology and the "New Environmental Paradigm"

When public apprehension began to be aroused concerning newly visible environmental problems, the scientists

---

[2] For an early warning that this exuberance-producing force could be temporry, see Sumner (1896). Few twentieth century sociologists have taken the warning seriously.

[3] Even sociological human ecologists have limited their attention primarily to the *social* or *spatial* environment, rather than the *physical* environment (see Michelson, 1976:17–23), reflecting their adherence to the HEP.

[4] See Miller (1972) for a lucid discussion of these laws.

who functioned as opinion leaders were not sociologists. They included such individuals as Rachel Carson, Barry Commoner, Paul Ehrlich and Garrett Hardin—biologists. Leadership in highlighting the precariousness of the human condition was mostly forfeited by sociologists, because until recently, most of us had been socialized into a worldview that makes it difficult to recognize the reality and full significance of the environmental problems and constraints we now confront. Due to our acceptance of the HEP, our discipline has focused on humans to the neglect of habitat; consideration of our *social* environment has crowded out consideration of our physical circumstances (Michelson, 1976:17). Further, we have had unreserved faith that equilibrium between population and resources could and would be reached in noncatastrophic ways, since technology and organization would mediate the relations between a growing population and its earthly habitat (see, e.g., Hawley, 1975).

But, stimulated by troubling events, some sociologists began to read such works as Carson (1962), Commoner (1971), Ehrlich and Ehrlich (1970), and Hardin (1968), and began to shed the blinders of the HEP. As long-held assumptions began to lose their power over our perceptions, we began to recognize that the reality of ecological constraints posed serious problems for human societies *and* for the discipline of sociology (see, e.g., Burch, 1971). It began to appear that, in order to make sense of the world, it was necessary to rethink the traditional Durkheimian norm of sociological purity— i.e., that social facts can be explained *only* by linking them to other *social* facts. The gradual result of such rethinking has been the development of environmental sociology.

Environmental sociology is clearly still in its formative years. At the turn of the decade rising concern with "environment" as a social problem led to numerous studies of public attitudes toward environmental issues and of the "Environmental Movement" (see Albrecht and Mauss, 1975). A coalition gradually developed between sociologists

with such interests and sociologists with a range of other concerns—including rather established interests such as the "built" environment, natural hazards, resource management and outdoor recreation, as well as newer interests such as "social impact assessment" (mandated by the National Environmental Policy Act of 1969). After the energy crisis of 1973, numerous sociologists (including many with prior interests in one or more of the above areas) began to investigate the effects of energy shortages in particular, and resource constraints in general, on society: the stratification system, the political order, the family, and so on. (For an indication of the range of interests held by environmental sociologists see Dunlap, 1975, and Manderscheid, 1977; for reviews of the literature see Dunlap and Catton, forthcoming, and Humphrey and Buttel, 1976.)

These diverse interests are linked into an increasingly distinguishable specialty known as environmental sociology by the acceptance of "environmental" variables as meaningful for sociological investigation. Conceptions of "environment" range from the "manmade" (or "built") environment to the "natural" environment, with an array of "human-altered" environments—e.g., air, water, noise and visual pollution—in between. In fact, *the study of interaction between the environment and society is the core of environmental sociology*, as advocated several years ago by Schnaiberg (1972).[5] This involves studying the effects of the environment on society (e.g., resource abundance or scarcity on stratification) and the effects of society on the environment (e.g., the contributions of differing economic systems to environmental degradation).[6]

The study of such interaction rests on the recognition that sociologists can no longer afford to ignore the environment in

_____

[5] This does not mean that environmental sociologists focus only on bi-variate relationships between social and environmental variables, as illustrated by Schnaiberg's (1975) "societal-environmental dialectic" (to be discussed below).
[6] For an alternative and narrower view of the domain of environmental sociology see Zeisel (1975: Chap. 1).

their investigations, and this in turn appears to depend on at least tacit acceptance of a set of assumptions quite different from those of the HEP. From the writings of several environmental sociologists (e.g., Anderson, 1976; Burch, 1971, 1976; Buttel, 1976; Catton, 1976a, 1976b; Morrison, 1976; Schnaiberg, 1972, 1975) it is possible to extract a set of assumptions about the nature of social reality which stand in stark contrast to the HEP. We call this set of assumptions the "New Environmental Paradigm" or NEP (see Dunlap and Van Liere, 1977 for a broader usage of the term, referring to emerging public beliefs):

1. Human beings are but one species among the many that are interdependently involved in the biotic communities that shape our social life.
2. Intricate linkages of cause and effect and feedback in the web of nature produce many unintended consequences from purposive human action.
3. The world is finite, so there are potent physical and biological limits constraining economic growth, social progress, and other societal phenomena.

Environmental Facts and Social Facts

Sociologists who adhere to the NEP readily accept as factual the opening sentences of the lead article (by a perceptive economist) in a recent issue of *Social Science Quarterly* devoted to "Society and Scarcity": "We have inherited, occupy, and will bequeath a world of scarcity: resources are not adequate to provide all of everything we want. It is a world, therefore, of limitations, constraints, and conflict, requiring the bearing of costs and calling for communal coordination" (Allen, 1976:263). Persistent adherents of the HEP, on the other hand, accustomed to relying on endless and generally benign technological and organizational breakthroughs, could be expected to discount such a statement as a mere manifestation of the naive presumption that the "state of the arts" is fixed (see, e.g., Hawley, 1975:6–7).

Likewise, sociologists who have been

converted to the assumptions of the NEP have no difficulty appreciating the sociological relevance of the following fact: the $36 billion it now costs annually to import oil to supplement depleted American supplies is partially defrayed by exporting $23 billion worth of agricultural products—grown at the cost of enormous soil erosion (van Bavel, 1977). Environmental sociologists expect momentous social change if soil or oil, or both are depleted. But sociologists still bound by the HEP would probably ignore such topics, holding that oil and soil are irrelevant variables for sociologists. However, we believe that only by taking into account such factors as declining energy resources can sociologists continue to understand and explain "social facts." We will attempt to demonstrate this by examining some work by NEP-oriented sociologists in one of the areas they have begun to examine—social stratification.

Usefulness of the NEP: Recent Work in Social Stratification

The bulk of existing work in stratification appears to rest on the Human Exceptionalism Paradigm, as it ". . . does not adequately consider the context of resource constraints or lack thereof in which the stratification system operates . . ." (Morrison, 1973:83). We will therefore describe recent work in the area by environmental sociologists, in an effort to illustrate the insights into stratification processes provided by the NEP. We will limit the discussion to three topics: the current decline in living conditions experienced by many Americans; contemporary and likely future cleavages in our stratification system; and the problematic prospects for ending self-perpetuating poverty.

Recent decline in standard of living: A majority of Americans are concerned about their economic situation (Strumpel, 1976:23), and in *Food, Shelter and the American Dream*, Aronowitz (1974) exemplifies the growing awareness that *something* is not going according to expectation—that old ideals of societal progress, increasing prosperity and material comfort, and individual and intergenerational mobility for *all* segments of society are *not* being realized (also see Anderson, 1976:1–3). Yet, even a "critical sociologist" such as Aronowitz seems impeded by the HEP in attempting to understand these changes. He views recent shortages in food, gasoline, heating oil, and so on, entirely as the result of "manipulations" by large national and supranational corporations, and is skeptical of the idea that resource scarcities may be real. Thus, his solution to the decline in the American standard of living would apparently be solely political—reduce the power of large corporations.

Although many environmental sociologists would not deny that oil companies have benefited from energy shortages, their acceptance of the NEP leads to a different explanation of recent economic trends. Schnaiberg (1975:6–8), for instance, has explicated a very useful "societal-environmental dialectic." Given the *thesis* that "economic expansion is a social desideratum" and the *antithesis* that "ecological disruption is a necessary consequence of economic expansion," a dialectic emerges with the acceptance of the proposition that "ecological disruption is harmful to human society." Schnaiberg notes three alternative *syntheses* of the dialectic: (1) an *economic synthesis* which ignores ecological disruptions and attempts to maximize growth; (2) a *managed scarcity synthesis*[7] which deals with the most obvious and pernicious consequences of resource-utilization by imposing controls over selected industries and resources; and (3) an *ecological synthesis* in which "substantial control over both production and effective demand for goods" is used to minimize ecological disruptions and maintain a "sustained yield" of resources. Schnaiberg (1975:9–10) argues that the synthesis adopted will be influenced by the basic economic structure of a society, with "regressive" (inequality-magnifying)

societies most likely to maintain the "economic" synthesis and "progressive" (equality-fostering) societies least resistive to the "ecological" synthesis.[8] Not surprisingly, therefore, the U.S., with its "non-redistributive" economy, has increasingly opted for "managed scarcity" as the solution to environmental and resource problems.[9]

Managed scarcity involves, for example, combating ecological disruptions by forcing industries to abate pollution, with resultant costs passed along to consumers via higher prices, and combating resource shortages via higher taxes (and thus higher consumer prices) on the scarce resources. There is growing recognition of the highly regressive impacts of both mechanisms (Morrison, 1977; Schnaiberg, 1975), and thus governmental reliance on "managed scarcity" to cope with pollution and resource shortages at least partly accounts for the worsening economic plight of the middle-, working, and especially lower-classes—a plight in which adequate food and shelter are often difficult to obtain. Unfortunately, these economic woes cannot simply be corrected by returning to the economic synthesis. The serious health threats posed by pollutants, the potentially devastating changes in the ecosystem wrought by unbridled economic and technological growth (e.g., destruction of the protective ozone layer, alteration of atmospheric temperature), and the undeniable reality of impending shortages in crucial resources such as oil, all make reversion to the traditional economic synthesis impossible in the long run (see, e.g., Anderson, 1976; Miller, 1972). Of course, as Morrison (1976) has noted, the pressures to return to this synthesis are great, and understanding them

[7] Schnaiberg's term was "planned scarcity," but because adherents of the HEP might suppose that phrase referred to scarcity *caused* by planners—rather than scarcities (and their costs) *allocated* by planners—we prefer to speak of "managed scarcity."

[8] Thus, e.g., among industrialized nations Sweden appears to have come closest to the ecological synthesis, while China represents the closest approximation to it by any developing nation (Anderson, 1976:242–251). In contrast, highly regressive developing nations such as Brazil seem strongly committed to the economic synthesis.

[9] The U.S. economy is "non-redistributive" because overall patterns of inequality (i.e., relative shares of wealth) have been altered very little by growth, even though all strata have improved their lot via growth (Schnaiberg, 1975:9; Zeitlin, 1977: Part 2).

provides insights into contemporary and future economic cleavages.

Cleavages within the stratification system: Schnaiberg's ecological synthesis amounts to what others have termed a "stationary" or "steady-state" society, and it is widely agreed that such a society would need to be far more egalitarian than the contemporary U.S. (Anderson, 1976:58–61; Daly, 1973:168–170).[10] Achieving the necessary redistribution would be very difficult, and opposition to it would be likely to result in serious, but unstable cleavages within the stratification system. In the long run, as environmental constraints become more obvious, ecologically aware "haves" are likely to opt for increased emphasis on managed scarcity to cope with them. The results would be disastrous for the "have nots," as slowed growth and higher prices would reverse the traditional trend in the U.S. in which *all* segments of society have improved their material condition—not because they obtained a larger slice of the "pie," but because the pie kept growing (Anderson, 1976:28–33; Morrison, 1976). Slowed growth *without* increased redistribution will result in real (as well as relative) deprivation among the "have nots," making class conflict more likely than ever before.[11] As Morrison (1976:299) notes, "Class antagonisms that are soothed by general economic growth tend to emerge as more genuine class conflicts when growth slows or ceases." Thus, in the long run the NEP suggests that Marx's predictions about class conflict may become more accurate, although for reasons Marx could not have foreseen.

In the short run, however, a very different possibility seems likely. The societal pressures resulting from managed scarcity are such that large portions of *both* "haves" and "have nots" will push for a reversion to the economic (growth) synthesis. In fact, Morrison (1973) has predicted the emergence of a Dahrendorfian (i.e., non-Marxian) cleavage: "growthists vs. nongrowthists," with *all* those highly dependent upon industrial growth (workers *and* owners) coalescing to oppose environmentalists (who typically hold positions—in the professions, government, education, for example—less directly dependent on growth). The staunch labor union support for growth, and the successful efforts of industry to win the support of labor and the poor in battles against environmentalists, both suggest the emergence of this coalition. Somewhat ironically, therefore, support for continued economic growth has united capitalists and the "left" (used broadly to include most labor unions, advocates for the poor, and academic Marxists). Not only does this support reveal the extent to which most of the left has abandoned hopes for real *redistribution* in favor of getting a "fair share" of a growing pie, but it also reveals a misunderstanding of the distribution of costs and benefits of traditional economic growth.

The "Culture of Poverty" solidified: Sociologists guided by the NEP have not only questioned the supposed universal benefits of growth, but they consistently point to the generally neglected "costs" of growth—costs which tend to be very regressive (Anderson, 1976:30–31; Schnaiberg, 1975:19). Thus, it is increasingly recognized that the workplace and inner city often constitute serious health hazards, and that there is generally a strong inverse relationship between SES and exposure to environmental pollution (Schnaiberg, 1975:19). Further, in his study of the SES-air pollution relationship, Burch (1976:314) has gone so far as to suggest that, "Each of these pollutants when ingested at certain modest levels over continuing periods, is likely to be an important influence upon one's ability to persist in the struggle for improvement of social position . . . These exposures, like nutritional deficiencies, seem one mechanism by which class inequalities are

---

[10] For example, wasteful consumption due to excess wealth and economic growth stemming from the investment of excess capital (for profit) would need to be halted, as would pressure for economic growth stemming from the unmet needs of the lower strata (see, e.g., Anderson, 1976:58–61; Daly, 1973:168–171).

[11] Managed scarcity and slowed growth are also likely to exacerbate tensions between developed and developing nations. This leads NEP-oriented sociologists (e.g., Anderson, 1976:258–269; Morrison, 1976) to see the future of international development quite differently than sociologists bound by the HEP (see, e.g., Horowitz, 1972).

reinforced." This leads him to suggest that efforts to eradicate poverty which do not take into account the debilitating impact of environmental insults are likely to fail.

Conclusion

We have attempted to illustrate the utility of the NEP by focusing on issues concerning stratification, for we believe this is one of many aspects of society that will be significantly affected by ecological constraints. As noted above, in the short run we expect tremendous pressure for reverting to the economic growth synthesis, for such a strategy seeks to alleviate societal tensions at the expense of the environment. Of course, the NEP implies that such a strategy cannot continue indefinitely (and the evidence seems to support this—see, e.g., Miller, 1972). Thus we are faced with the necessity of choosing between managed scarcity and an ecological synthesis.[12] The deleterious effects of the former are already becoming obvious; they help account for the trends described by Aronowitz and others. However, the achievement of a truly ecological synthesis will require achieving a steady-state society, a very difficult goal. As students of social organization, sociologists should play a vital role in delineating the characteristics of such a society, feasible procedures for attaining it, and their probable social costs. (See Anderson, 1976 for a preliminary effort.) Until sociology extricates itself from the Human Exceptionalism Paradigm, however, such a task will be impossible.

REFERENCES

Albrecht, Stan L. and Armand L. Mauss
  1975   "The environment as a social problem." Pp. 556–605 in A. L. Mauss, Social Problems as Social Movements. Philadelphia: Lippincott.

---

[12] A point implicit in our discussion is worth making explicit: the NEP suggests that resource scarcities are unavoidable, but—as Schnaiberg's work indicates—societies may react to them in a variety of ways. Thus, as Schnaiberg (1975:17) suggests, sociologists should begin to examine the social impacts (especially distributional impacts) of *alternative* responses to scarcity.

Allen, William R.
  1976   "Scarcity and order: The Hobbesian problem and the Humean resolution." Social Science Quarterly 57:263–275.
Anderson, Charles H.
  1976   The Sociology of Survival: Social Problems of Growth. Homewood, Illinois: Dorsey.
Aronowitz, Stanley
  1974   Food, Shelter and the American Dream. New York: Seabury Press.
Bell, Daniel
  1973   The Coming of Post-Industrial Society. New York: Basic Books.
Burch, William R., Jr.
  1971   Daydreams and Nightmares: A Sociological Essay on the American Environment. New York: Harper and Row.
  1976   "The peregrine falcon and the urban poor: Some sociological interrelations." Pp. 308–316 in P. J. Richerson and J. McEvoy III (eds.), Human Ecology: An Environmental Approach. North Scituate, Mass.: Duxbury.
Buttel, Frederick H.
  1976   "Social science and the environment: Competing theories." Social Science Quarterly 57:307–323.
Carson, Rachel
  1962   Silent Spring. Boston: Houghton-Mifflin.
Catton, William R., Jr.
  1976a "Toward prevention of obsolescence in Sociology." Sociological Focus 9:89–98.
  1976b "Why the future isn't what it used to be (and how it could be made worse than it has to be)." Social Science Quarterly 57:276–291.
Commoner, Barry
  1971   The Closing Circle. New York: Knopf.
Daly, Herman E.
  1973   "The steady-state economy: Toward a political economy of biophysical equilibrium and moral growth." Pp. 149–174 in H.E. Daly (ed.), Toward a Steady-State Economy. San Francisco: W. H. Freeman.
Denisoff, R. Serge, Orel Callahan, and Mark H. Levine (eds.)
  1974   Theories and Paradigms in Contemporary Sociology. Itasca, Illinois: Peacock.
Dunlap, Riley E. (ed.)
  1975   Directory of Environmental Sociologists. Pullman: Washington State University, College of Agriculture Research Center, Circular No. 586.
Dunlap, Riley E. and William R. Catton, Jr.
  forth- "Environmental sociology." Annual Recoming view of Sociology. Palo Alto, Calif.: Annual Reviews, Inc.
Dunlap, Riley E. and Kent D. Van Liere
  1977   "The 'new environmental paradigm': A proposed measuring instrument and preliminary results." Paper presented at the Annual Meeting of the American Sociological Association, Chicago.
Ehrlich, Paul R. and Anne H. Ehrlich
  1970   Population, Resources, Environment. San Francisco: W. H. Freeman.

Friedrichs, Robert W.
1972   A Sociology of Sociology. New York: Free Press.
Gouldner, Alvin W.
1970   The Coming Crisis of Western Sociology. New York: Basic Books.
Hardin, Garrett
1968   "The tragedy of the commons." Science 162:1243–1248.
Hawley, Amos H. (ed.)
1975   Man and Environment. New York: New York Times Company.
Horowitz, Irving L.
1972   Three Worlds of Development: The Theory and Practice of International Stratification. 2nd ed. New York: Oxford University Press.
Humphrey, Craig R. and Frederick H. Buttel
1976   "New directions in environmental sociology." Paper presented at the Annual Meeting of the Society for the Study of Social Problems, New York.
Klausner, Samuel Z.
1971   On Man in His Environment. San Francisco: Jossey-Bass.
Kuhn, Thomas S.
1962   The Structure of Scientific Revolutions. Chicago: University of Chicago Press.
Manderscheid, Ronald W. (ed.)
1977   Annotated Directory of Members: Ad Hoc Committee on Housing and Physical Environment. Adelphi, Maryland: Mental Health Study Center, NIMH.
Michelson, William H.
1976   Man and His Urban Environment. 2nd ed. Reading, Mass.: Addison-Wesley.
Miller, G. Tyler, Jr.
1972   Replenish the Earth: A Primer in Human Ecology. Belmont, Calif.: Wadsworth.
Morrison, Denton E.
1973   "The environmental movement: Conflict dynamics." Journal of Voluntary Action Research 2:74–85.
1976   "Growth, environment, equity and scarcity." Social Science Quarterly 57:292–306.
1977   "Equity impacts of some major energy alternatives." Paper presented at the Annual Meeting of the American Sociological Association, Chicago.
Parsons, Talcott
1977   The Evolution of Societies (ed. by Jackson Toby). Englewood Cliffs, NJ: Prentice-Hall.
Potter, David M.
1954   People of Plenty. Chicago: University of Chicago Press.
Ritzer, George
1975   Sociology: A Multiple Paradigm Science. Boston: Allyn and Bacon.
Schnaiberg, Allan
1972   "Environmental sociology and the division of labor." Department of Sociology, Northwestern University, mimeograph.
1975   "Social syntheses of the societal-environmental dialectic: The role of distributional impacts." Social Science Quarterly 56:5–20.
Strumpel, Burkhard (ed.)
1976   Economic Means for Human Needs. Ann Arbor: Institute for Social Research, University of Michigan.
Sumner, William Graham
1896   "Earth hunger or the philosophy of land grabbing." Pp. 31–64 in A. G. Keller (ed.), Earth Hunger and Other Essays. New Haven: Yale University Press, 1913.
van Bavel, Cornelius H. M.
1977   "Soil and oil." Science 197:213.
Zeisel, John
1975   Sociology and Architectural Design. New York: Russell Sage Foundation.
Zeitlin, Maurice (ed.)
1977   American Society, Inc. 2nd ed. Chicago: Rand McNally.

Received 8/24/77   Accepted 10/25/77

Chapter 8

Drought and Famine in the Sahel

8.1 THE SAHEL, THE DROUGHT, AND THE FAMINE

The name Sahel is derived from an Arabic word meaning 'shore' or 'border'. The Sahel refers to the border of the world's largest tropical desert: the Sahara. It is, in fact, the fringe of the desert, lying between the desert and the tropical rain forests of Africa. But within this general conception of the Sahel, a great many alternative specifications of it can be found in the vast literature on the Sahel produced by geographers and climatologists. It is useful to begin by sorting out the different approaches, if only to avoid possible confusion later in the analysis of the Sahelian drought.

(1) *The ecological definitions* The Sahel can be defined as the 'dry zone', comprising the 'arid' zone (with average rainfall per year less than 100 mm, or 4 in.) and the 'semi-arid' zone (with rainfall between 100 and 500 mm, or between 4 and 20 in.), on the southern fringe of the Sahara, and this coincides with a 'tropical steppe vegetation belt'.[1] It runs across the broadest part of Africa from the Atlantic to the Red Sea. The Sahel is sometimes defined not as the entire dry zone immediately south of the Sahara, but as only the semi-arid zone there. While covering less than the whole dry zone in this part of the world, it too runs from the Atlantic to the Red Sea.[2]

(2) *The politico-ecological definitions* On this view, the Sahel is defined as the 'semi-arid vegetation belt' in six West African countries, viz. Mauritania, Senegal, Mali, Upper Volta, Niger, and Chad.[3] Alternatively, and more broadly, the Sahel could refer to the 'dry zone' in these six countries.

[1] Harrison Church (1973), p. 62; see also Harrison Church (1961).
[2] See Winstanley (1976), p. 189. Winstanley's specification of rainfall for the semi-arid region is also on the *higher* side, viz. 'between 200 and 600 mm', so that not merely the arid zone but also regions at the northern end of the semi-arid zone as defined by Harrison Church (1973) is excluded in this view of the Sahel.
[3] See Matlock and Cockrum (1976) and Swift (1977b). Sometimes Gambia and Cape Verde are added to this list of six countries.

---

(3) *The political definitions* The word Sahel could be used, as it has been by the international news media, simply to refer to these six *countries* in West Africa (Mauritania, Senegal, Mali, Upper Volta, Niger, and Chad), affected by the recent drought.[4]

There is no point in spilling blood over the choice of the definition that should be used, but one must be careful to distinguish between the different senses of the word Sahel to be found in the literature, since not a little confusion has arisen from vagueness owing to this plethora of definitions—sometimes used implicitly rather than explicitly.

In this work the Sahel in the purely political characterization will be called 'Sahelian countries', covering Mauritania, Senegal, Mali, Upper Volta, Niger, and Chad. The politico-ecological definition will be used in the form of calling the dry Sahel region in the Sahelian countries as 'the dry Sahel region'. Finally, the purely ecological definition applied to the whole of Africa (without noting political divisions) will be captured by the expression 'the Sahel belt of Africa'.

I turn now to the drought in question. The reference is to the period of low rainfall during 1968–73. There is some controversy as to whether 1968 or even 1969 were, in fact, years of drought. It has been argued that 'the 1968–69 rains were more or less normal; they only appeared low by comparison with the pluvian 1960s, which detracted attention from the cyclical nature of rainfall in that area'.[5] It appears that 'the Sahel and Sudan zones probably received more rain between 1956 and 1965 than at any time in this century', and, it has been argued, after that 'just less than average rainfall' appeared as 'drought'.[6] But there can be little doubt that the drought period involved a very considerable shift in the rainfall isohyets in a southerly direction, making an arid zone out of parts of semi-arid regions and turning non-dry regions into semi-arid ones. And this is so not only in contrast with the immediate pluvial past, but also compared with earlier record.[7] While it is certainly not correct to say that 'the Sahel has never had a drought like this one',[8] the shift in the rainfall pattern

[4] See the map of Sahel countries on p. 130.
[5] Wiseberg (1976), p. 122, quoting the views of an MIT group of climatologists.
[6] Matlock and Cockrum (1976), p. 238.
[7] See Winstanley (1976), Bradley (1977), Schove (1977).
[8] Brown and Anderson (1976), p. 162; it must, however, be said in fairness to the authors that the drought *was* very unusual in terms of its *impact*, though not as a drought as

was very substantial, and the contrast with the rainy 1960s was particularly sharp.

The problem applied not merely to the dry Sahel region in the Sahelian countries, but to the whole of the Sahel belt of Africa, covering also parts of Sudan and Ethiopia. Indeed, there were clear links in the drought pattern associated with the Ethiopian famine studied in the last chapter and the drought affecting the Sahelian countries.

It is also worth noting that the short-fall of rain compared with normal was more severe further north within the Sahelian region; i.e., areas that normally receive less rainfall anyway suffered a bigger relative short-fall as well.[9] This is not altogether unusual, since there is some evidence that as we move north in this part of Africa not only does the mean rainfall level fall, but the coefficient of variation rises substantially.[10]

The pastoral and agricultural economy of the Sahel region was severely affected by the drought.[11] While, as we shall presently see, there were factors other than the drought in the causation of the famine, it would be stupid to pretend that the drought was not seriously destructive. The peak year of suffering seems to have been 1973, and the drought waned only with the good rains of 1974. The animal loss altogether was estimated, according to some calculations, to be as high as 'some 40 percent to 60 percent',[12] but there is a good deal of disputation on these quantitative magnitudes, and much variation from region to region. There is, however, very little scope for doubting the severity of the suffering that accompanied the drought, and the famine conditions that developed during this period. Of the six countries, Mauritania, Mali, and Niger seem to have been hit harder than the other countries.

Already by 1969 there were reports of 'prolonged drought across West Africa'.[13] The situation got worse as the drought progressed, and by the spring of 1972 the United Nations World

such. On the comparative evidence from earlier periods, see the technical papers cited above, and also Dalby and Harrison Church (1973), pp. 13–16, 29–45.

[9] See Bradley (1977), p. 50, Figure 2.

[10] See Winstanley (1976), p. 197, Figure 8.6.

[11] See Shear and Stacy (1976), Winstanley (1976), and Matlock and Cockrum (1976). In Section 8.2 of this chapter the output situation is reviewed.

[12] El-Khawas (1976), p. 77. See also Winstanley (1976) and Matlock and Cockrum (1976).

[13] See Sheets and Morris (1976), p. 36.

Food Programme noted that drought in the Sahelian countries had become 'endemic', requiring that 'special treatment' be given to the region in providing emergency food aid. By September that year the FAO had identified a coming 'disaster', with 'an acute emergency situation developing in large areas due to exceptionally poor harvests in the Sahel'. The level of starvation was by then, it appears, much greater than in the preceding four years of drought, and 'children and the elderly had already begun to succumb'.[14]

The peak year of the famine in the Sahelian countries was 1973, the starvation having by then gathered momentum in a cumulative process of destitution and deprivation.[15] The number of famine deaths during that year was estimated to be around 100,000.[16] But there is a good deal of disputation about the mortality estimates,[17] and rather little direct evidence on which a firm estimate can be based. There is also much debate on the extent to which the famine unleashed the forces of epidemic in the Sahelian countries. The threat of epidemics was widely noted, and reports of flaring up of diseases like measles did come through, but it has been argued that the epidemics were rather mild and to a great extent were confined to the relief camps where infections could spread fast.[18] Certainly, the epidemic flare-up was nothing in comparison with what was observed in some earlier famines, e.g. the Bengal famine of 1943 (see Appendix D below).

The relief operations, though slow to start with, were on quite a massive scale, and the provision of food, medicine, and shelter was helped by a good deal of international co-operation. The efficiency of the relief operations remains a highly disputed topic. Some have seen in such operations the clue to 'how disaster was avoided',[19] and have assigned a good deal of credit to inter-national efforts as well as to efforts within the countries

---

[14] Sheets and Morris (1976), p. 38. The authors provide a blow-by-blow account of the international recognition of the Sahelian disaster and a sharp critique of the delay in responding to it.

[15] See Newman (1975).

[16] Center for Disease Control (1973), and Kloth (1974). See also Imperato (1976).

[17] See Seaman, Holt, Rivers, and Murlis (1973), Imperato (1976), and Caldwell (1977), pp. 94–5.

[18] See Imperato (1976), pp. 295–7, Sheets and Morris (1976), pp. 61–3, and Seaman, Holt, Rivers, and Murlis (1973), p. 7.

[19] See Imperato (1976) and Caldwell (1977).

themselves. Others have emphasized the sluggishness, the chaos, and the discrimination between different communities of victims as evidence of unpardonable mismanagement.[20]

8.2 FAD VIS-À-VIS ENTITLEMENTS

Was the Sahelian famine caused by a decline in food availability? The *per capita* food output did go down quite substantially. The figures of food availability per head in the six Sahelian countries as presented by the FAO are given in Table 8.1. There is much less of a decline in the food consumption per head judging by the FAO figures on calorie and protein consumption per head, as presented in Table 8.2. But, apart from Senegal, the other countries had a decline in food consumption per head as well, so that the FAD hypothesis does not stand rejected on the basis of these data.

TABLE 8.1
Net Food Output per Head (Index)

	1961–5	1968	1969	1970	1971	1972	1973	1974	1975
Chad	100	90	87	84	84	69	63	69	71
Mali	100	93	99	97	90	73	61	65	80
Mauritania	100	98	100	98	92	80	67	68	68
Niger	100	92	99	88	88	81	57	65	60
Senegal	100	81	87	66	89	56	67	88	107
Upper Volta	100	106	103	102	93	85	76	84	90

Source: FAO Production Yearbook 1976, Table 6.

TABLE 8.2
Calorie Consumption per head in Sahel countries

	1961–5	1972	1973	1974
Chad	100	76	73	76
Mali	100	86	86	88
Mauritania	100	91	94	95
Niger	100	86	89	85
Senegal	100	93	104	107
Upper Volta	100	86	85	96

Source: Calculated from data given in Table 97 of FAO Production Yearbook 1976.

[20] See Sheets and Morris (1976), El-Khawas (1976) and Wiseberg (1976).

It is also worth emphasizing that the FAD view is not rejected even by the otherwise important observation that nearly all the Sahelian countries had enough food within their borders to prevent starvation had the food been divided equally, and that 'throughout the whole Sahel [the Sahelian countries] in every year from 1968 to 1972 the per caput supply [of cereals] exceeded this figure [FAO/WHO recommended food-intake per person] comfortably'.[21] It appears that an FAO survey 'documented that every Sahelian country, with the possible exception of mineral-rich Mauritania, actually produced enough grain to feed its total population even during the worst drought year'.[22] This is obviously relevant in emphasizing the importance of unequal distribution in starvation as well as giving us an insight into the type of information that FAO looked into in understanding the famine, but it does not, of course, have any bearing on the correctness of the FAD view of famines. The FAD approach is concerned primarily not with the *adequacy* of over-all food supply, but with its *decline* compared with past experience. In particular, the FAD claim is not that famines occur if and only if the average food supply per head is insufficient compared with some nutritional norm, but that famines occur if and only if there is a sharp *decline* in the average food availability per head. And, it could be argued, such a decline did take place for the Sahelian countries as a group during the drought of 1969–73. If the FAD approach to famines were to seek refuge in some comforting bosom, it probably couldn't do better in the modern world than choose the Sahelian famine: the food availability did go down, and—yes—there was a famine!

Despite this, I would argue that, even for the Sahelian famine, the FAD approach delivers rather little. First, in the peak year of famine, 1973, the decline in food availability per head was rather small even in comparison with the pluvial early 1960s for Mali, Mauritania, Niger, and Upper Volta (less than a 15 per cent decline of calorie availability per head), and none at all for Senegal. While the decline was much sharper for Chad (27 per cent), Chad was one of the less affected countries in the region,

[21] Marnham (1977), p. 17. See also Lappé and Collins (1977, 1978), and Lofchie (1975).

[22] Lappé and Collins (1977), p. 2, and footnote 6, quoting a letter from Dr. M. Ganzin, Director, Food Policy and Nutrition Division, FAO, Rome, a major authority on the Sahelian economy.

the famine having been most severe in Mauritania, Mali, and Niger, and possibly Upper Volta.[23] But none of these latter countries had a very sharp decline in food availability per head (Table 8.2). It could, however, be argued that in terms of food *output* per head (Table 8.1), Niger and Mali did have the biggest decline in the famine year of 1973, with the indices standing respectively at 57 and 61. But Senegal, which was less affected, had an index value of 56 for 1972, the lowest index value for any country for any year. As we move away from the gross factual statements to a bit more detailed information, the FAD analysis starts limping straightway.

Second, the rationale of the FAD approach, concentrating as it does on aggregate supply, rests in ignoring distributional changes. But there is clear evidence that dramatic shifts in the distribution of purchasing power were taking place in the drought years in the Sahelian countries, mainly between the dry Sahel regions in these countries and the rest of the regions. The drought affected the Sahel area rather than the savanna, in which the vast majority of the population of Sahelian countries live. The relief camps even in the south were full of people who had migrated from the Sahelian north into less dry areas in the south. The famine victims were almost exclusively the nomadic pastoral population from the Sahel region (including nomads, semi-nomads, and transhumants) and the sedentary population living in the Sahel region (agriculturists, fishermen, etc.).[24] Rather than looking at aggregate statistics like those of food availability per head, it would clearly make much more sense to look at the economic conditions of these groups of people in understanding the Sahelian famine.

Third, we indeed have direct evidence of the decline in income and purchasing power of pastoralists and agriculturists living in the Sahelian region. The destruction of crops and the death of animals in these parts of the Sahelian countries were very substantial.[25] While the crops affected were often of foodgrains and the animals do supply edible products, most sections of the Sahelian population rely also on trade for food.[26] The severe

[23] See Center for Disease Control (1973) and Kloth (1974).

[24] See Sheets and Morris (1976), Kloth (1974), Copans *et al.* (1975), Lofchie (1975), and Imperato (1976).

[25] See Winstanley (1976) and Matlock and Cockrum (1976).

[26] The role of the commercialization in heightening the impact of the drought on the

decline in agricultural and pastoral output in these particular regions thus meant a sharp reduction in the ability of the affected people to command food, whether from one's own output or through exchange. The situation is somewhat comparable to the picture we already found in Ethiopia during the drought and famine there (see Chapter 7 above), which could be seen as a related phenomenon in another part of the Sahel belt of Africa.

It is worth mentioning in this context that, from the point of view of the suffering of the individual agriculturist, it matters rather little whether the crop destroyed happens to be a food crop which is consumed directly, or a cash crop which is sold to buy food. In either case the person's entitlement to food collapses. It is this collapse that directly relates to his starvation (and that of his family) rather than some remote aggregate statistics about food supply per head. The same applies to the pastoralist who lives by selling animals and animal products; and, like the Afar or Somali pastoralists in Ethiopia, the Sahel pastoralists also rely substantially on trading animals and animal products for other goods, including grains, which provide cheaper calories.[27]

Thus, despite superficial plausibility, the FAD approach throws rather little light on the Sahelian famine. It is not my contention that the FAD *always* makes wrong predictions. (If it did, then the FAD approach could have provided a good basis of prediction, *applied in reverse*!) Predictions based on FAD may sometimes prove right (as in gross statements about the over-all Sahelian famine), sometimes wrong (as in the Bengal famine or in the Ethiopian famine). So the first point is that it isn't much of a predictor to rely on. But the more important point is that, even when its prediction happens to come out right about the existence of a famine in a broad area, it provides little guidance about the character of the famine—*who* died, *where*, and *why*?

8.3 DESTITUTION AND ENTITLEMENT

Of the total population of some 25 million in the Sahelian countries in 1974, about 10 per cent can be described as nomadic.[28] The definition includes semi-nomads and transhum-

---

Sahelian economy has been particularly emphasized by Comité Information Sahel (1974). See also Meillassoux (1974), Raynaut (1977), and Berry, Campbell and Emker (1977). Also Imperato (1976), pp. 285–6.

[27] See, among others, Seaman, Holt, Rivers, and Murlis (1973), and Haaland (1977).

[28] See Caldwell (1975).

ants in addition to pure nomads. At the time the drought hit, the population of the dry Sahel region in these countries was roughly 5 to 6 million.[29] So the nomads were no more than half the population of the dry Sahel region itself, and quite a small proportion of the total population of the Sahelian countries. But the share of the nomads in the mortality induced by the famine and in nutritional deficiency were both remarkably high,[30] and even among the victims of the famine there is some evidence of 'a shocking contrast between the nutritional state of sedentary victims of the drought and the deep starvation of the nomads'.[31]

The destitution of the nomadic pastoralist seems to have followed a process not dissimilar to the fate of the Ethiopian herdsmen of the Afar or Somali communities, studied in the last chapter. The animal loss in the dry Sahel area varied between 20 to 100 per cent depending on the region,[32] and this was compounded, as in Ethiopia, by the 'rapid destocking of animals in the Sahel' coinciding with 'a sharp drop in prices'.[33] The reasons for the fall in price seem to have been similar to those encountered in analysing the Ethiopian famine (see Chapter 7). A pastoralist depends on consuming cereals part of the year as a cheaper source of calories,[34] and may become more—and not less—dependent on exchanging animals for cereals in the straitened circumstances caused by the loss of animals.[35] The fixed money obligations arising from taxes, etc., that have to be met makes the pastoralist inclined to sell more rather than less, when the relative prices of animals fall.[36] The emaciation of animals would provide a more direct explanation of the sharp fall in prices. Whatever the cause, the statistics of the loss of animals understate the magnitude of the economic decline sustained by the pastoralist.

[29] See Imperato (1976), p. 285, and Marnham (1977), p. 7.

[30] See Kloth (1974), Sheets and Morris (1976), and Imperato (1976).

[31] Sheets and Morris (1976), p. 53.

[32] See Glantz (1976), pp. 77–8, 199–200, and FAO *Production Yearbook*, 1974, Tables 107–10.

[33] Club du Sahel (1977), p. 55.

[34] See Chapter 7 and also the references cited in note 27 in this chapter.

[35] The sale of 'an unusual proportion of their animals' was thus 'one of the pastoralists' means of defence' (Caldwell, 1977, p. 96).

[36] This notion leads naturally to the possibility of a backward-bending supply curve, which has been frequently mentioned. But since this is only one factor among many, such a backward bend may not occur in the *total* supply curve. The position also depends on the precise circumstances facing the pastoralist (see Haaland, 1977), and there is some empirical evidence against the backward bend (see Khalifa and Simpson, 1972).

The pastoralists migrated down south in quite large numbers during the Sahelian famine, and this movement of people and animals was one way of reducing the impact of the famine—indeed, according to one view was this group's 'chief defence'.[37] Better feeding conditions for the animals as well as opportunities for wage labour in the south permitted some relief, and the relief camps set up there also took a good many of the pastoralists from the north. There is some evidence, however, of discriminating treatment against the pastoral nomadic people in the relief camps, and a firm suggestion that the Sahelian governments were closely tied to (and more responsive to the needs of) the majority sedentary communities.[38]

Another group severely affected by the famine was the sedentary agriculturists from the dry Sahel region. While the position of the dry Sahel pastoralist was rather similar to that of the Afar and Somali nomads in the Ethiopian famine, that of the dry Sahel agriculturist could be compared with the predicament of the Wollo cultivator. His output was down, and whether he lived by eating his own product or by selling it in the market, his entitlement to food underwent a severe decline. For many cash crops, e.g. groundnuts, the impact of the drought seems to have been no less than on the total quantity of food output as such (see Table 8.3). The food output in the country in which he lived might have gone down by a moderate ratio, but his command over the food that was there slipped drastically. The extent of the decline varied depending on the precise region in question since the drought was far from uniform,[39] but for some of the dry Sahel agriculturists it was a case of straightforward ruin. The problem was made more difficult by the monetary obligations of taxation,[40] which had to be paid despite the drought, and by the loss of job opportunity arising from what we have been calling 'derived deprivation'.[41]

[37] Caldwell (1977), p. 96.

[38] See Sheets and Morris (1976).

[39] See Bradley (1977), pp. 39–40.

[40] It has been argued that the obligation to pay taxes in monetary terms has been an important reason for Sahelian farmers for moving from food crops to cash crops (see Berry, Campbell and Emker, 1977, p. 86). See also Comité Information Sahel (1974).

[41] See Chapters 1, 6, and 7. In the drought year the Sahelian labourer sought employment further away from home than normal. See, for example, Faulkingham (1977), Tables 17.7 and 17.8; there is a remarkable increase in the number of labourers from this Hausa village in Niger going to distant Kano (360 km), Lagos (910 km) and Abidjan (1,420 km) for wage employment in such occupations as being 'water career'.

TABLE 8.3
Food Output Compared with Output of Groundnuts in shell

	1966–8	1969	1970	1971	1972	1973	1974
Senegal							
Food	100	101	78	108	70	84	94
Groundnut	100	88	65	110	65	85	95
Niger							
Food	100	99	97	89	88	64	80
Groundnut	100	92	79	82	93	29	64
Mali							
Food	100	107	101	108	86	68	88
Groundnut	100	97	125	121	119	79	151
Chad							
Food	100	104	103	102	80	72	76
Groundnut	100	119	119	78	47	52	52

Source: Calculated from Tables 7 and 43 in *FAO Production Yearbook 1975*, changing the base of the index to the average for 1966, 1967, and 1968, with unchanged weights.

8.4 SOME POLICY ISSUES

Droughts may not be avoidable, but their effects can be. 'The weather and climate modification schemes proposed since 1900 for the regions in West and Central Africa surrounding and including the Sahara Desert'[42] are worthy fields of investigation. I expect some day one might indeed grow rice or catch fish in the Sahara. But while science marches on slowly, some means have to be found for freeing the Sahelian population from vulnerability to droughts and the prospect of famines. I end this chapter with brief comments on this issue.

One approach is to argue that the problem is not as bad as it might look from the perspective of the recent drought. One could say that droughts come and go, and they don't come all that frequently either: in this century the serious ones for the Sahel countries have been 1910–14, 1941–2, 1969–73. Is there a real need for special action? While smugness—like virtue—is its own reward, it is not a reward on which the Sahelian population can bank. Periodic famines may not be as bad as perpetual starvation, but one can scarcely find that situation acceptable.

Another approach is to argue that there is no real need to make the dry Sahel region economically non-vulnerable, since people

[42] Glantz and Parton (1976). See also Franke and Chasin (1979).

can emigrate. It has been argued that 'the inevitably limited
funds available for development can be more effectively
employed' in the moister area south of the dry Sahelian region,
and that 'relevant governments should consider the feasibility of
a strategic withdrawal from areas of very low and unreliable
rainfall'.[43] This approach is further supported by the observation
that many nomads interviewed in the south during and immediately after the drought years indicated a willingness to stay
on in the south.

Depopulation as a solution to the problem of the Sahelian
region overlooks a number of important factors. First, it seems
that the bulk of the population from the Sahelian north who
migrated south have, in fact, returned north again. Even if many
had stayed on, that would have still left the Sahelian countries
with the problem of what to do with the remaining dry Sahel
population. It is not easy to change the way of life of a large
community, nor is there any reason to suppose that the Sahelian
northerner typically finds the economic opportunities in the
south to be superior to those ruling in the north in the normal
years. The problem is one of fluctuation of economic circumstances between wet and dry years rather than of a collapse of the
economic potential of the north in general.

Second, it does seem economically most wasteful to give up the
use of the resources available for agriculture, pastoralism, and
fishing in the north and to crowd the southern savanna even more
than it is already crowded. The southerner typically did not
welcome the northern invasion, and there is good reason to think
that large-scale permanent immigration would indeed worsen
the economic conditions of the southern population.

Finally, the Sahelian countries do get many things from the
economic production in the north, both of pastoral products as
well as of cash crops. The Sahelian north is no 'basket case', and
while a severe drought may make the northerner dependent on
the southerner for his survival, the southerner normally gets
various commodities and foreign exchange from the production
taking place in the north.

It is necessary, therefore, to think of solutions of the Sahelian
problem without choosing the drastic simplicity of depopulating

[43] This view is attributed to 'Professor Hodder of the School of Oriental and African
Studies and others' by David Dalby in Dalby and Harrison Church (1973), p. 21.

the Sahel region. Since the source of the problem is variability rather than a secular decline, it is tempting to think in terms of insurance arrangements. The first question then is: what is to be insured? If FAD had been a good theory of famines and total food availability a good guide to starvation, then the thing to insure would have been the over-all supply of food in these countries. Some system of international insurance could be used to make resources available for import of food from abroad when domestic production of food fails and when the earnings of foreign exchange go down, making it difficult to import food by using currently earned exchange. While the Sahelian countries have had a declining trend of food output per head over the last decade or so, the trend rates have been much influenced by the sequence of drought years. Also, there have been some substantial shifts from food crops to cash crops, which reduce the food output per head but not necessarily the country's ability to buy food, since cash crops can be sold and exchanged for food in the international market. On the FAD approach, therefore, the problem may look simple enough: it is a case of occasional fall in food supply per head, which could be made good by an international insurance mechanism, so that enough food is available within the country.

But FAD isn't a very reliable guide. Despite its superficial plausibility in explaining the Sahelian famine, it delivers rather little even in this case, as we saw in the last section. There is need for concern not merely with how to get food into the country, but also with the way it could be commanded by the affected population. It is necessary to devise ways by which the population most affected by drought and economic difficulty can have the ability to obtain food through economic mechanisms. This contrast is particularly important in the context of recent developments such as growth of cash cropping, which have added to the country's over-all earning power but seem to have led to a decline of the exchange entitlement of particular sections of the population.

Thus, the insurance arrangements have to deal with command over food at the family level, and not merely at the national level. Private insurance of families of nomadic pastoralists and other groups in the dry Sahel region isn't an easy task, and indeed, such insurance rarely exists in any poor and backward country. Thus, the public sector would have to play a major role in the task of guaranteeing food to the vulnerable sections. Famine relief operations are, of course, insurance systems in some sense, but if a long-run solution to the problem of vulnerability has to be found, then clearly a less *ad hoc* system would have to be devised. Various alternative ways of doing this, varying from guaranteed social security benefits in the form of income supplementation to employment guarantee schemes, can be considered in this context.

But social security is not the only aspect of the problem. It is important to recognize the long-run changes in the Sahelian economy that have made the population of the dry Sahel region more vulnerable to droughts. One factor that has been noted is the increased vulnerability arising from the growing commericialization of the Sahelian economy.[44] As the Sahelian population has become more dependent on cash crops, there has been an increased dependence on markets for meeting its food requirements, and this has tended to supplement the variability arising from climatic fluctuations.[45] In the light of the analysis of variation of entitlement relations with which this book has been much concerned (see Chapters 1, 5, 6, and 7), this point may not need very much elaboration here. Compared with the farmer or the pastoralist who lives on what he grows and is thus vulnerable only to variations of his own output (arising from climatic considerations and other influences), the grower of cash crops, or the pastoralist heavily dependent on selling animal products, is vulnerable both to output flucutations and to shifts in marketability of commodities and in exchange rates.[46] The worker employed in wage-based farms or other occupations is, of course, particularly vulnerable, since employment fluctuations owing to climatic shifts, or to other factors such as 'derived deprivation', can be very sharp indeed. While commercialization may have

[44] See Comité Information Sahel (1974).
[45] See Meillassoux (1974) and Comité Information Sahel (1974). See also Ball (1976) and Berry, Campbell, and Emker (1977).
[46] While price and quantity fluctuations can operate in opposite directions, thereby tempering the effects of each other, this is far from guaranteed. Indeed, as was seen in the last chapter and also in Section 8.3 above, the pastoralists often have to face a price decline along with a reduction in quantities. Similarly, a decline in agricultural output could lead to a reduction of wage employment, thereby leading to a collapse of the 'effective price' of labour power for those who are fired.

opened up new economic opportunities, it has also tended to increase the vulnerability of the Sahel population.

Commercialization has also had other worrying effects. The 'traditional symbiotic relationship between nomadic livestock and the crops' may have been disrupted in some regions by the growth of cash crops with a different seasonal rhythm (e.g., cotton being harvested later than the traditional food crops).[47] The livestock eating the post-harvest stubs from the agricultural fields, and in its turn fertilizing the fields by providing dung, may have fitted in nicely with traditional agriculture, but have often not co-ordinated at all well with cash crops. Furthermore, when traditional grazing land has been taken over for commercial farming, the pastoral population has, of course, directly suffered from a decline of resources.

Another long-run trend that seems to have been important is the partial breakdown of the traditional methods of insurance.[48] The political division of the dry Sahel region has put arbitary constraints on pastoral movements, reducing the scope for anti-drought responses. The practice of storing animals on the hoof as insurance seems to have become more expensive because of taxes. 'Fall-back hunting' has become almost impossible because of reduction of wild animals in the region.[49] The collapse of the traditional methods of fighting economic problems arising from periodical droughts may have played an important part in making the dry Sahel region more vulnerable to draught in recent years than it need have been. On some of these changes corrective policy actions are worth considering, but many of these developments are difficult to reverse.

Additional vulnerability has also arisen from over-grazing and an increase in the size of livestock population in the dry Sahelian region. To some extent, this was the result of the pluvial 1960s, preceding the drought,[50] and the problem would be different for

[47] See Norton (1976), pp. 260–1. Also Swift (1977b), pp. 171–3.

[48] See Swift (1977a, 1977b). See also Copans *et al.* (1975), Haaland (1977), and Caldwell (1977).

[49] Another long-run factor working in the same direction is the decline of trading possibilities across the Sahara, especially the disruption of the traditional Tuareg caravan trade, first by French interference and then by competition from traders with motorized transport (Baier, 1974; cited by Berry, Campbell and Emker, 1977, pp. 87–8).

[50] See Norton (1976), p. 261.

quite some time after the drought. But the nature of the nomadic economy has built-in forces operating in that direction. Since the pastures are held communally and animals owned privately, there is a conflict of economic rationale in the package, which becomes relevant when pasture land gets short in supply. Having additional animals for grazing adds to families' incomes, and while this might lead to loss of grass cover and erosion, and thus to reduced productivity for the pastoralists as a group, the loss to the individual family from the latter may be a good deal less than its gain from having additional animals. Thus a conflict of the type of the so-called 'prisoners' dilemma'[51] is inherent in the situation.

The problem is further compounded by the fact that the animals, aside from adding to the family's usual income, also serve as insurance, as was noted before, so that the tendency to enlarge one's herd, causing over-grazing, tends to be stimulated as uncertainty grows. And the over-grazing, in its turn, adds to the uncertainty, by denuding the grass cover and helping desert formation.

Contrary to what is sometimes said, the tendency of the pastoralist to have a large stock, contributing to over-grazing, does not indicate anything in the least foolish or shortsighted from his personal point of view. Policy formulation can hardly be helped by the failure to recognize this important fact.

The complexity of the problem was recently suggested by an official from a major donor country who had interviewed a Fulani herder in northern Upper Volta in the spring of 1973. He reported that the farmer, asked how he had been affected by the recent drought, said he had 100 head of cattle and had lost 50. The farmer continued, 'Next time I will have 200,' implying that by starting with twice as many he would save the 100 cattle he wants. Yet the land's carrying capacity is such that he will still have only 50 cattle, but his loss will have been much greater.[52]

As a piece of economic reasoning, this is, of course, sheer rubbish, as the Fulani herder must have seen straightaway, unless mesmerized by respect for the 'major donor country'. If the Fulani herder had begun with 200 rather than 100 animals, he would not have ended up with the same 50, and in fact in large

---

[51] For the basic Prisoners' Dilemma model, see Luce and Raiffa (1958), and conflicts of this kind can be found in many economic situations (see Sen, 1967a).

[52] Glantz (1976), p. 7.

pastoral communities he would have a good chance of retaining the same proportion, viz. 50 per cent, thereby ending up with the planned 100. The proportion of herd loss does depend on the *total* stock of animals, given the availability of grazing land, but the influence of one herder's individual herd size on the total animal stock will be very small indeed. So it isn't that he himself loses by starting off with 200 rather than 100 head of cattle, but that he thus contributes a little bit to the general reduction of survival possibility of animals in the whole region, and if many herders do the same then the survival ratio will indeed come down significantly. But the individual herder can hardly undo this general problem by keeping his own herd small unilaterally. The herder is sensible enough within his sphere of control, which is his own stock, but the totality of these individually sensible actions produces a social crisis.

In tackling this aspect of the problem, several alternative approaches are possible, varying from incentive schemes using taxes and subsidies, or regulations governing the size of the herds, to the formation of pastoral co-operatives.[53] The problem arises from private entitlement being positively responsive to one's private ownership of animals while being negatively responsive to ownership by others, and from each herder having direct control only over his own stock. The object of institutional change will be to eliminate the conflicts that arise from this dichotomy.

The food problem of the population in the dry Sahel region depends crucially on this set of institutional factors affecting food entitlement through production and exchange, and there is scope for action here. As discussed earlier, there is also need for a mechanism for directly tackling the problem of vulnerability through public institutions guaranteeing food entitlement. The last category includes not merely distribution of food when the problem becomes acute, but also more permanent arrangements for entitlement through social security and employment protection. What is needed is not ensuring food availability, but guaranteeing food entitlement.

---

[53] Various institutional reforms have been discussed in recent years. See Comité Information Sahel (1974), Copans *et al.* (1975), Widstrand (1975), Dahl and Hjort (1976, 1979), Glantz (1976), Norton (1976), Rapp, Le Houérou, and Lundholm (1976), Swift (1977a, 1977b), and Toupet (1977), among many other important contributions.

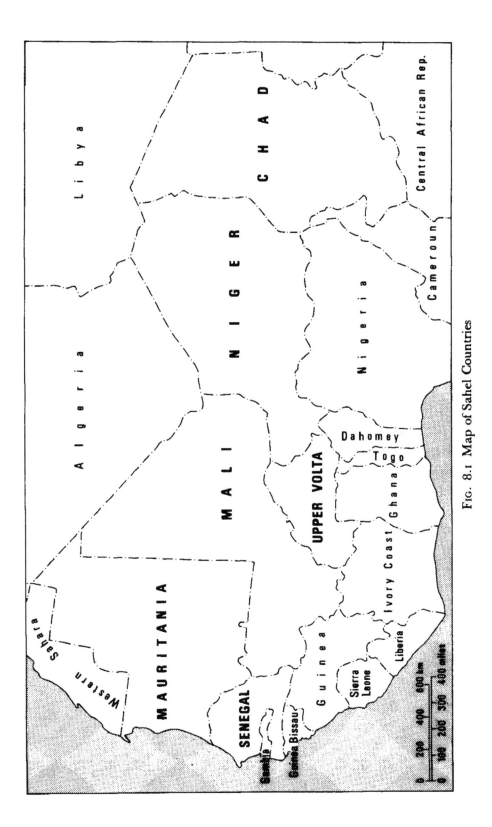

Fig. 8.1 Map of Sahel Countries

CHAPTER   TWO

◆

Race, Class, and the Politics of Place

The southern United States, with its unique history and its plantation-economy legacy, presents an excellent opportunity for exploring the environment-development dialectic, residence-production conflict, and residual impact of the de facto industrial policy (i.e., "any job is better than no job") on the region's ecology. The South during the 1950s and 1960s was the center of social upheavals and the civil rights movement. The 1970s and early 1980s catapulted the region into the national limelight again, but for different reasons. The South in this latter period was undergoing a number of dramatic demographic, economic, and ecological changes. It had become a major growth center.

Growth in the region during the 1970s was stimulated by a number of factors. They included (1) a climate pleasant enough to attract workers from other regions and the "underemployed" workforce already in the region, (2) weak labor unions and strong right-to-work laws, (3) cheap labor and cheap land, (4) attractive areas for new industries, i.e., electronics, federal defense, and aerospace contracting, (5) aggressive self-promotion and booster campaigns, and (6) lenient environmental regulations.[1] Beginning in the mid-1970s, the South was transformed from a "net exporter of people to a powerful human magnet."[2] The region had a number of factors it promoted as important for a "good business climate," including "low business taxes, a good infrastructure of municipal services, vigorous law enforce-ment, an eager and docile labor force, and a minimum of business regulations."[3]

The rise of the South intensified land-use conflicts revolving around "use value" (neighborhood interests) and "exchange value" (business interests). Government and business elites became primary players in affecting land-use decisions and growth potentialities. The "growth machine," thus, sometimes pitted neighborhood interests against the interests of industrial expansion. However, economic boosters could usually count on their promise of jobs as an efficient strategy of neutralizing local opposition to growth projects. Harvey Molotch emphasized the importance of jobs as a selling point in growth machine politics:

> Perhaps the key ideological prop for the growth machine, especially in terms of sustaining support from the working-class majority, is the claim that growth "makes jobs." This claim is aggressively promulgated by developers, builders, and chambers of commerce; it becomes part of the statesman talk of editorialists and political officials. Such people do not speak of growth as useful to profits—rather, they speak of it as necessary for making jobs.[4]

Competition intensified as communities attempted to expand their work force and lure new industries away from other locations. There was a "clear preference for clean industries that require highly skilled workers over dirty industries that use unskilled workers."[5] Many communities could not afford to be choosy. Those communities that failed to penetrate the clean industry market were left with a choice of dirty industry or no industry at all. These disparities typify the changing industrial pattern in the South.

Before moving to the next section, we need to delineate the boundaries of the South. We have chosen to use the U.S. Bureau of the Census South Region, sixteen states and the District of Columbia, as the study area (see Figure 2.1). The South has the largest population of any region in the country. More than 75.4 million inhabitants, nearly one-third of the nation's population, lived in the South in 1980.[6] All of the southern states experienced a net in-migration during the 1970s. The South, during the 1970s and 1980s, also grew at a faster rate than the

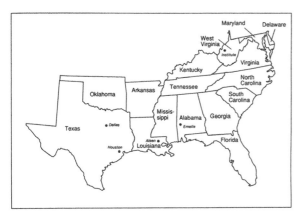

FIGURE 2.1   Location of industrial facilities in the study area

nation as a whole—a factor that had important economic, political, and ecological implications.

The South also has the largest concentration of blacks in the country. In 1980, more than 14 million blacks lived in the region. Blacks were nearly one-fifth of the region's population. In the 1970s the region's black population increased by nearly 18 percent. In 1980, six of the southern states had black populations that exceeded 20 percent (35.2 percent of the population in Mississippi, 30.4 percent in South Carolina, 29.4 percent in Louisiana, 26.8 percent in Georgia, 25.6 percent in Alabama, and 22.4 percent in North Carolina).

Consequences of Uneven Development

The South has gone to great lengths to shed its image as a socially and economically "backward" region. However, slick public relations and image management campaigns have not been able to hide decades of neglect and underdevelopment. Many of the old problems remain, while new problems were created as a direct result of the new growth. Migrants to urban

areas and incumbent residents who had marginal skills generally found themselves in the growing unemployment lines.[7] Individuals who do not have the requisite education often become part of the region's expanding underclass.

The South's new prosperity was mainly confined to metropolitan areas. Growth in the urban South heightened status differences between rich and poor and between blacks and whites. Poverty coexisted amid affluence. Poverty, however, represented a source of cheap labor. The large pool of docile and nonunionized labor was part of the so-called "good business climate."[8]

William Falk and Thomas Lyson described the uneven economic development and plight of rural southerners in their book *High Tech, Low Tech, No Tech.* The authors wrote:

> Not all citizens have benefited from the upturn in the southern economy. In fact, many may not have benefited at all. Blacks, women, and people living in rural areas have, in varying degrees, received little or none of the job opportunities and economic affluence that has washed over the region. The quality of life and opportunity for improvement for these "people left behind" have remained essentially unchanged over the last fifty years.[9]

Development disparities are heightened by business policies that direct jobs away from minority communities through the systematic avoidance of urban ghettos and rural blackbelt counties. The blackbelt represents geopolitical power (or the potential for empowerment). It also represents the epitome of American apartheid with its rigid segregation practices, second-class status for blacks, and staunch white resistance to black majority rule. Falk and Lyson studied 147 southern blackbelt counties (a band of counties with 40 percent or more black population extending from North Carolina to Louisiana) and discovered these areas lagging far behind other counties in the region, partly because of the concentration of unskilled, poorly educated workers. The authors summed up their findings by writing:

> If the SMSA counties are seen as the "pride of the South," the Black Belt can be viewed as the "Sunbelt's stepchild." The

industrial growth and development that has washed over the region has left the 147 Black Belt counties with a residue of slow growth and stagnant and declining industries. . . . High tech industries have virtually ignored the Black Belt. . . . In short, by any yardstick of industrial development, the Black Belt remains mired in the backwater of the southern economy.[10]

The persistent problem of uneven development and economic disparities caused many writers to challenge the existence of a "New" South. Chet Fuller, a black journalist, traveled across the region in the late 1970s and discovered that "the much touted progress of some southern cities is more illusion than reality."[11] The region was portrayed as booming with industrial growth and expanding employment opportunities that were once closed to the black masses. The New South was promoted as a changed land where blacks could now share in the American Dream. Fuller argued that "power has not changed hands in the South, not from white hands to black hands."[12] What is "new" about an area where blacks are systematically denied access to jobs, housing, and other residential amenities?

Black communities still suffer from institutionalized discrimination. Discriminatory practices occur at various levels of government and affect the location of polling places, municipal landfills, and toxic-waste dumps. Discrimination, thus, involves a "process of defending one group's privilege gained at the expense of another."[13] Black communities and their inhabitants must defend themselves against hostile external forces that shape land-use decisions and environmental policies.

Why focus on the South? The South has always been home for a significant share of the black population. More than 90 percent of black Americans lived in the southern states at the turn of the century. A little more than one-half (53 percent) of all blacks were living in the region in 1980, the same percentage as in 1970.[14] In an effort to improve their lives, millions of rural blacks migrated from the South to other regions of the nation. From the mid-1940s to the late 1960s, nearly 4.5 million more blacks left the South than migrated to it. Beginning in the mid-1970s, however, the number of blacks moving into the South exceeded the number departing for other regions of the country. For the period 1975–1980, over 415,000 blacks moved

into the South, while 220,000 left the region (or a net in-migration of 195,000 blacks), thereby reversing the longstanding black exodus. More than 411,000 blacks migrated to the South during the 1980–1985 period while 324,000 moved out of the region, a net in-migration of 87,000 blacks.[15]

As industry and jobs relocated to the region, job seekers followed. More than 17 million new jobs were added in the South between 1960 and 1985, compared to 11 million jobs added in the West, and a combined total of 13 million jobs added in the Midwest and Northeast.[16] The challenges that the South must face rest with how its resources—housing, jobs, public services, political representation, etc.—are shared with blacks who historically have not gotten their fair share of the region's amenities. The major reason for this discrepancy has been the location preferences of businesses. Industries that relocated to the South generally built new factories where they could find surplus white labor and "avoided places with a high ratio of poor and unskilled blacks."[17] The plight of millions of blacks has been exacerbated by the combination of economic recession (and depression-like conditions in many black communities), federal budget cuts, growing tension among individuals competing for limited jobs and other scarce resources, and the federal retreat on enforcement of civil rights and antidiscrimination laws.[18]

The social climate of the South was changed dramatically by the civil rights movement. Some gains were also made in the political arena. Most of these gains were made after the passage of the Voting Rights Act of 1965. There were 1,469 black elected officials in 1970, 4,912 in 1980, and 5,654 in 1984.[19] The number of black officeholders increased to 6,681 in 1987. There were twenty-three blacks in the U.S. Congress in 1989. This number represented only 5.3 percent of the 435 members of the U.S. House of Representatives. There were no blacks in the U.S. Senate.

Only four blacks from the Deep South were serving in Congress (Harold Ford of Memphis, Mickey Leland of Houston, John Lewis of Atlanta, and Michael Espy of Yazoo City, Mississippi) in 1989. Espy and Lewis were first elected in 1986. Espy became the first black elected to Congress from Mississippi since Reconstruction. Although some 53 percent of the nation's blacks live in the South, 62 percent of the black elected officials were found in the region.[20] In spite of the progress that has been made since the civil rights movement of the 1960s and 1970s, blacks remain underrepresented as political officeholders.[21] They are also underrepresented in policy-making boards and commissions, including industrial and environmental regulatory bodies. The interests of all-white industrial boards, zoning commissions, and governmental regulatory bodies may run counter to those of the black community. Until these policy-setting institutions are made more inclusive, we are likely to find an intensification of locational conflicts and charges of racial discrimination.

Endangered Environs

Millions of urban and rural blacks are physically trapped in inner cities while the job centers, especially for white-collar and service occupations, are moving to the suburbs. Housing discrimination, residential segregation, and limited public transportation severely limit the access of urban blacks to the job-rich suburbs. Many black workers must settle for nearby manufacturing jobs—the ones that have not moved out of the inner city. Because of their proximity to polluting industries, black communities have the most to gain from effective environmental enforcement mechanisms.[22]

Unlike their white counterparts, black communities do not have a long history of dealing with environmental problems. Blacks were involved in civil rights activities during the height of the environmental movement, roughly during the late 1960s and early 1970s. Many social justice activists saw the environmental movement as a smoke screen to divert attention and resources away from the important issue of the day—white racism. On the other hand, the key environmental issues of this period (e.g., wildlife and wilderness preservation, energy and resource conservation, and regulation of industrial polluters) were not high priority items on the civil rights agenda.

Social justice, political empowerment, equal education, fair employment practices, and open housing were major goals of social justice advocates. It was one thing to talk about "saving trees" and a whole different story when one talked about "saving low-income housing" for the poor. As a course of action, black communities usually sided with those who took an active role on the housing issue. Because eviction and displacement are fairly common in black communities (particularly for inner-city residents), decent and affordable housing became a more salient issue than the traditional environmental issues. Similarly, unemployment and poverty were more pressing social problems for African Americans than any of the issues voiced by middle-class environmentalists.

In their desperate attempt to improve the economic conditions of their constituents, many civil rights advocates, business leaders, and political officials directed their energies toward bringing jobs to their communities by relaxing enforcement of pollution standards and environmental regulations and sometimes looking the other way when violations were discovered. In many instances, the creation of jobs resulted in health risks to workers and residents of the surrounding communities.

Industrial policies remained paternalistic toward those who were less well-off. Polluting industries were brought into poor communities with little input from local community leaders.[23] When questions were raised by concerned citizens, the argument of jobs for local residents was used to quell dissent. Environmental risks were offered as unavoidable trade-offs for jobs and a broadened tax base in economically depressed communities. Jobs were real; environmental risks were unknown. This scenario proved to be the de facto industrial policy in "poverty pockets" and job-hungry communities around the world.

The South's unique history, traditions, and laws institutionalized employment, education, housing, and other forms of discrimination. A plethora of civil rights legislation was enacted to remedy inequities of Jim Crow laws and de facto segregation. Beginning in the 1970s the region was transformed into an economic "mecca."[24] Industrial growth was hailed as a panacea for the decades of neglect and second-class status accorded the region.

Even with the economic transformation, many of the region's old problems that were related to underdevelopment (e.g., poor education, large concentrations of unskilled labor, low wages, high unemployment, etc.) went unabated. New environmental problems were created with the influx of polluting industries. For example, in the 1970s four of the five states that led the nation in attracting polluting industries such as paper, chemical, and waste disposal firms were located in the South. These four states, Texas, South Carolina, North Carolina, and Florida, are not known for having strong environmental programs.[25]

Many industrial firms, especially waste disposal companies and industries that have a long history of pollution violations, came to view the black community as a "pushover lacking community organization, environmental consciousness, and with strong and blind pro-business politics."[26] These communities were ripe for exploitation. Residents of economically impoverished areas—intimidated by big corporations and deserted by local politicians—were slow to challenge private and governmental polluters of their neighborhoods. Moreover, the strong pro-jobs stance, a kind of "don't bite the hand that feeds you" sentiment, aided in institutionalizing risks at levels that are unacceptable in the larger society.[27]

According to nearly every quality-of-life indicator, black communities are worse off than their white counterparts. Environmental and land-use regulations are enforced on a less-than-routine basis in black communities. Because a large share of inhabitants in these communities are renters (many from low-income households) rather than home owners, it is difficult to organize and mobilize residents. This marginality also makes it hard for people to donate their time or money to fend off threats to the community. Logan and Molotch summarized this problem:

> Ghettos are organized less as attempts to defend the ongoing structure and institutional patterns of a specific neighborhood and more as assaults on the larger social order that denies basic resources to all deprived places and the people in them. It is organization around victimization. . . . [T]he special vulnerability of black neighborhoods to outside penetration and the difficulties of organizing around turf issues are caused by racist patterns of exploitation, exclusion, and stigma.[28]

There is, of course, a "direct historical connection between the exploitation of the land and the exploitation of people, especially black people."[29] Southern historian David R. Goldfield sees southern ecology as being tied to the race issue. Goldfield predicted that "as race relations continue to improve, so will Southern ecology."[30] Southern ecology has been shaped largely by excessive economic boosterism, a blind pro-business climate, lax enforcement of environmental regulations, and industrial strategies that had little regard for environmental cost.

Rapid and unrestrained development has ruined or threatened the region's unique habitat. A classic example of this ecological destruction is the transformation of the life-giving Mississippi River into a "deadly mixture of sewage, industrial waste, and insecticides below fire-belching Baton Rouge."[31] Public Data Access, Inc., in a study commissioned by the environmental group Greenpeace, discovered some startling facts on the pollution problem in the East Baton Rouge parish, as counties are called in Louisiana:

> East Baton Rouge parish had more violators of emissions permits, commercial toxic waste facilities, employees in petrochemicals, and toxic waste generation than any other county along the [Mississippi] River, and, in addition, ranked second or third for 6 toxic emissions measures, 3 toxic discharges measures, and for toxic waste landfills and incinerators.[32]

The entire Gulf Coast region, especially Mississippi, Alabama, Louisiana, and Texas, has been ravaged by "lax regulations and unbridled production."[33] Polluting industries exploit the pro-growth and pro-jobs sentiment exhibited among the poor, working-class, and minority communities.[34] Industries such as paper mills, waste disposal and treatment facilities, and chemical plants, searching for operation space, found these communities to be a logical choice for their expansion. Polluting smokestacks, to some individuals, were visible signs that plants were operating and employing people.

The smell of industrial operations was promoted as economic "progress." What civic-minded individual would advocate against economic progress? For example, a paper mill spewing its stench and poison in one of Alabama's poverty-ridden blackbelt counties led Governor George Wallace to declare: "Yeah, that's the smell of prosperity. Sho' does smell sweet, don't it."[35] Similar views have been reported of public officials in West Virginia's, Louisiana's, and Texas's "chemical corridor."[36]

Growing Black Militancy

Blacks did not launch a frontal assault on environmental problems affecting their communities until these issues were couched in a civil rights context beginning in the early 1980s. They began to treat their struggle for environmental equity as a struggle against institutionalized racism and an extension of the quest for social justice.[37] Just as black citizens fought for equal education, employment, and housing, they began to include the opportunity to live in a healthy environment as part of their basic rights. Moreover, they were convinced that disparate enforcement of environmental policies and regulations contributes to neighborhood decline much like housing discrimination, redlining practices, and residential segregation do.

Black resistance to environmental threats in the 1970s was confined to local issues and largely involved grassroots individuals. In the 1980s some changes occurred in the way black community groups and national advocacy groups dealt with the toxics issue. This new environmental activism among blacks did not materialize out of thin air nor was it an overnight phenomenon. It did, however, emerge out of the growing hostility to facility siting decisions that were seen as unfair, inequitable, and discriminatory toward poor people and people of color.

Toxic-waste disposal has generated demonstrations in many communities across the country.[38] The first national protest by blacks on the hazardous-waste issue occurred in 1982. Demonstrations and protests were triggered after Warren County, North Carolina, which is mostly black, was selected as the burial site for more than 32,000 cubic yards of soil contaminated with highly toxic PCBs (polychlorinated biphenyls). The soil had been illegally dumped along the roadways in fourteen North Carolina counties in 1978.

What was the source of the PCBs? The PCBs originated from the Raleigh-based Ward Transfer Company. A Jamestown, New York, trucking operation owned by Robert J. Burns obtained the PCB-laced oil from the Ward Transfer Company for resale. Faced with economic loss as a result of the Environmental Protection Agency (EPA) ban on resale of the toxic oil in 1979, the waste haulers chose the cheap way out by illegally dumping it along North Carolina's roadways. Burns and Ward were subsequently sent to jail for the criminal dumping of the tainted oil.[39]

This dumping was the largest PCB spill ever documented in the United States. More than 30,000 gallons of PCB-laced oil was left on 210 miles of roadway in the state for four years before the federal EPA and the state of North Carolina began clean-up activities. In 1982, after months of deliberations and a questionable site selection exercise, North Carolina Governor James B. Hunt in 1982 decided to bury the contaminated soil in the community of Afton located in Warren County. Local citizens later tagged the site "Hunt's Dump."

The Afton community is more than 84 percent black. Warren County has the highest percentage of blacks in the state and is one of the poorest counties in North Carolina. The county had a population of 16,232 in 1980. Blacks composed 63.7 percent of the county population and 24.2 percent of the state population. Per capita income for Warren County residents was $6,984 in 1982 compared with $9,283 for the state. The county ranked ninety-second out of 100 counties in median family income in 1980. The county unemployment rate was 13.3 percent in 1982 and 1983. More than 42 percent of the county's workforce commute out of the county for employment. Although the county lags far behind the rest of the state on a number of economic indicators, over three-fourths of Warren County residents own their homes. More than 78 percent of the whites and 64 percent of the blacks own their homes (nationally only 45 percent of blacks are home owners).[40]

Why was Warren County selected as the PCB landfill site? The decision made more political sense than environmental sense. In *Science for the People,* Ken Geiser and Gerry Waneck described the Warren County PCB siting decision:

The site at Afton was not even scientifically the most suitable. The water table of Afton, North Carolina, (site of the landfill) is only 5–10 feet below the surface, and the residents of the community derive all of their drinking water from local wells. Only the most optimistic could believe that the Afton landfill will not eventually leach into the groundwater. Unless a more permanent solution is found, it will only be a matter of time before the PCBs end up in these people's wells.[41]

Black civil rights activists, political leaders, and area residents marched and protested against the construction of the Warren County PCB landfill. Dr. Charles E. Cobb, who was director of the United Church of Christ's Commission for Racial Justice in 1982, voiced his strong opposition to the Warren County PCB landfill and other siting decisions that make blacks and the poor bear a heavier burden than other communities. His directive to blacks was clear:

> We must move in a swift and determined manner to stop yet another breach of civil rights. We cannot allow this national trend to continue. If it means that every jail in this country must be filled, then I say let it be. The depositing of toxic wastes within the black community is no less than attempted genocide.[42]

Local county residents did organize. They formed the Warren County Citizens Concerned About PCBs. This time local citizens were not standing alone. Grassroots groups were joined by national civil rights leaders, black elected officials, environmental activists, and labor leaders. For example, Reverend Leon White of the United Church of Christ's Commission for Racial Justice, Reverends Joseph Lowery and Ben Chavis and Fred Taylor of the Southern Christian Leadership Conference, District of Columbia Delegate Walter Fauntroy of the Congressional Black Caucus, and some 500 loyal supporters were able to focus the national limelight on the tiny black town of Afton.

The protests, however, did not stop the trucks from rolling in and dumping their loads. The state began hauling more than 6,000 truckloads of the PCB-contaminated soil to the landfills in mid-September of 1982. Just two weeks later, more than 414 protesters had been arrested. The protest demonstrations in

Warren County marked the first time anyone in the United States had been jailed trying to halt a toxic waste landfill.[43]

The Warren County protesters even got encouragement from the chief of EPA's hazardous waste implementation branch, William Sanjour. He urged the demonstrators to "keep doing what you are doing."[44] The EPA official questioned the disposal method selected over the alternatives (incineration and on-site neutralization). Sanjour's remarks at a rally at John Graham School in Warrenton reinforced what many of the protesters had suspected all along:

> Landfilling is cheap. It is cheaper than the alternative. The people who like to use landfills such as chemical industries are very powerful. No amount of science, truth, knowledge or facts goes into making this decision. It is a purely political decision. What they listen to is pressure.[45]

Residents of Warren County were searching for guarantees that the state was not creating a future "superfund" site that would threaten nearby residents. Of course, no guarantees could be given since there is no such thing as a 100-percent safe hazardous-waste landfill—one that will not eventually leak. The question is not *if* the facility will leak but *when* the facility will leak PCBs into the environment.

Waste Facility Siting Disparities

Although the demonstrations in North Carolina were not successful in halting the landfill construction, the protests brought a sharper focus to the convergence of civil rights and environmental rights and mobilized a nationally broad-based group to protest these inequities. The 1982 demonstrations prompted District of Columbia Delegate Walter E. Fauntroy, who had been active in the protest demonstrations, to initiate the 1983 U.S. General Accounting Office (GAO) study of hazardous-waste landfill siting in the region.[46]

The GAO study observed a strong relationship between the siting of offsite hazardous-waste landfills and race and socioeconomic status of surrounding communities. It identified four

TABLE 2.1 1980 Census Population, Income, and Poverty Data for Census Areas[a] Where EPA Region IV Hazardous-Waste Landfills Are Located (1983)

Landfill (State)	Population		Median family income ($)		Population below poverty level		
	Number	% Black	All Races	Blacks	Number	%	% Black
Chemical Waste Management (Alabama)	626	90	11,198	10,752	265	42	100
SCA Services (South Carolina)	849	38	16,371	6,781	260	31	100
Industrial Chemical Co. (South Carolina)	728	52	18,996	12,941	188	26	92
Warren County PCB Landfill (North Carolina)	804	66	10,367	9,285	256	32	90

[a]Areas represent subdivisions of political jurisdictions designated by the census for data gathering.

Source: U.S. General Accounting Office, *Siting of Hazardous Waste Landfills and Their Correlation with Racial and Economic Status of Surrounding Communities* (Washington, D.C.; General Accounting Office, 1983), p.4.

offsite hazardous-waste landfills in the eight states that compose EPA's Region IV (i.e., Alabama, Florida, Georgia, Kentucky, Mississippi, North Carolina, South Carolina, and Tennessee). The data in Table 2.1 detail the socio-demographic characteristics of the communities where the four hazardous-waste landfill sites are located.

The four hazardous-waste landfill sites included Chemical Waste Management (Sumter County, Alabama), SCA Services (Sumter County, South Carolina), Industrial Chemical Company (Chester County, South Carolina), and Warren County PCB landfill (Warren County, North Carolina). Blacks composed the majority in three of the four communities where the offsite hazardous-waste landfills are located. Blacks make up about one-fifth of the population in EPA's Region IV. The GAO study also revealed that more than one-fourth of the population in all four communities had incomes below the poverty level, and most of this population was black.[47] The facility siting controversy cannot be reduced solely to a class phenomenon because there is no shortage of poor white communities in the region. One only has to point to southern Appalachia to see widespread white poverty in America. Nevertheless, poor whites along with their more affluent counterparts have more options and leveraging mechanisms (formal and informal) at their disposal than blacks of equal status.

When the entire southern United States is studied, even more glaring siting disparities emerge. For example, there were twenty-seven hazardous-waste landfills operating in the forty-eight contiguous states with a total capacity of 127,897 acre-feet in 1987.[48] One-third of these hazardous-waste landfills were located in five southern states (i.e., Alabama, Louisiana, Oklahoma, South Carolina, and Texas). The total capacity of these nine landfills represented nearly 60 percent (76,226 acre-feet) of the nation's total hazardous-waste landfill capacity (see Table 2.2).

Four landfills in minority zip code areas represented 63 percent of the South's total hazardous-waste disposal capacity. Moreover, the landfills located in the mostly black zip code areas of Emelle (Alabama), Alsen (Louisiana), and Pinewood (South Carolina) in 1987 accounted for 58.6 percent of the region's hazardous-waste landfill capacity—although blacks make up only about 20 percent of the South's total population. These same three sites accounted for about 40 percent of the total estimated hazardous-waste landfill capacity in the entire United States.[49] Nationally, three of the five largest commercial hazardous-waste landfills are located in areas where blacks and Hispanics compose a majority of the population. These siting disparities expose minority citizens to greater risks than the general population.

It is not coincidental that the National Association for the Advancement of Colored People (NAACP) passed its first resolution on the hazardous-waste issue in 1983 after the national protest demonstration in Warren County, North Carolina. Subsequent protest actions were instrumental in getting the New York–based Commission for Racial Justice to sponsor its 1987 national study of toxic waste and race.[50] This national study,

TABLE 2.2 Operating Hazardous-Waste Landfills in the Southern United States and Ethnicity of Communities (1987)

Facility Name	Current Capacity in Acre-Feet[a]	Percent of Population in Zip code Area		
		Minority	Black	Hispanic
Chemical Waste Management (Emelle, AL)	30,000	79.5	78.9	0.0
CESCO International, Inc. (Livingston, LA)	22,400	23.8	21.6	1.8
Rollins Environmental Services (Scotlandville, LA)	14,440	94.7	93.0	1.5
Chemical Waste Management (Carlyss, LA)	5,656	6.8	4.6	1.7
Texas Ecologists, Inc. (Robstown, TX)	3,150	78.2	1.6	76.6
GSX Services of South Carolina (Pinewood, SC)	289	71.6	70.5	1.1
US Pollution Control, Inc. (Waynoka, OK)	118	37.3	23.2	12.3
Gulf Coast Waste Disposal Authority (Galveston, TX)	110	4.3	0.0	3.8
Rollins Environmental Services, Inc. (Deer Park, TX)	103	7.3	0.3	6.2

[a]Acre-feet is the volume of water needed to fill one acre to a depth of one foot. The total capacity of the nation's 27 hazardous-waste landfills was 127,989 in 1987.

Source: Commission for Racial Justice, *Hazardous Wastes and Race in the United States: A National Report on the Racial and Socio-Economic Characteristics of Communities with Hazardous Waste Sites* (New York: Commission for Racial Justice, 1987) Table B-10.

like the 1983 GAO report, found a strong association between race and the location of hazardous-waste facilities. Race was by far the most prominent factor in the location of commercial hazardous-waste landfills, more prominent than household income and home values. For example, the commission study found:

> Household incomes and home values were substantially lower when communities with hazardous-waste facilities were compared to communities in the surrounding county without such facilities. Mean household income was $2,745 less and mean value of owner-occupied homes was $17,301 less. The minority percentage of the population remained the most significant factor differentiating these groups of communities.[51]

Growing empirical evidence shows that toxic-waste dumps, municipal landfills, garbage incinerators, and similar noxious facilities are not randomly scattered across the American landscape. The siting process has resulted in minority neighborhoods (regardless of class) carrying a greater burden of localized costs than either affluent or poor white neighborhoods. Differential access to power and decision making found among black and white communities also institutionalizes siting disparities.

Toxic-waste facilities are often located in communities that have high percentages of poor, elderly, young, and minority residents.[52] An inordinate concentration of uncontrolled toxic-waste sites is found in black and Hispanic urban communities.[53] For example, when Atlanta's ninety-four uncontrolled toxic-waste sites are plotted by zip code areas, more than 82.8 percent of the city's black population compared with 60.2 percent of its white population were found living in waste site areas. Despite its image as the "capital of the New South," Atlanta is the most segregated big city in the region. More than 86 percent of the city's blacks live in mostly black neighborhoods. As is the case for other cities, residential segregation and housing discrimination limit mobility options available to black Atlantans.

Siting disparities also hold true for other minorities and in areas outside the southern United States. Los Angeles, the nation's second largest city, has a total of sixty uncontrolled toxic-waste sites. More than 60 percent of the city's Hispanic's live in waste-site areas compared with 35.3 percent of Los Angeles's white population. Although Hispanic's are less segregated than the black population, more than half of them live in mostly Hispanic neighborhoods. The city's Hispanic community is concentrated in the eastern half of the city where the bulk of the uncontrolled toxic-waste sites are found.

On the other hand, large commercial hazardous-waste landfills and disposal facilities are more likely to be found in rural communities in the southern blackbelt.[54] Many of these facilities that are located in black communities are invisible toxic time bombs waiting for a disaster to occur.

Finally, the burden, or negative side, of industrial development has not been equally distributed across all segments of the population. Living conditions in many communities have not improved very much with new growth. Black communities became the dumping grounds for various types of unpopular facilities, including toxic wastes, dangerous chemicals, paper mills, and other polluting industries.

The path out of this environmental quagmire is clearly one that involves more communities in activities designed to reclaim the basic right of all Americans—the right to live and work in a healthy environment. A political strategy is also needed that can draw from a wide cross-section of individuals and groups who share a common interest in preservation of environmental standards. In his keynote address to the 1983 Urban Environment Conference on toxics and minorities, Congressman John Conyers of Detroit pinpointed this strategy. The black congressman saw broad-based groups (e.g., similar to those attending the New Orleans meeting) as having an "opportunity to raise the fairness issue in all dimensions, including the toxic threat to the poor, minority and working class Americans."[55]

3

Analyzing long-enduring, self-organized, and self-governed CPRs

A direct attack on several of the key questions posed in this book can be launched by an examination of field settings in which (1) appropriators have devised, applied, and monitored their own rules to control the use of their CPRs and (2) the resource systems, as well as the institutions, have survived for long periods of time. The youngest set of institutions to be analyzed in this chapter is already more than 100 years old. The history of the oldest system to be examined exceeds 1,000 years. The institutions discussed in this chapter have survived droughts, floods, wars, pestilence, and major economic and political changes. We shall examine the organization of mountain grazing and forest CPRs in Switzerland and Japan and irrigation systems in Spain and the Philippine Islands.

By indicating that these CPR institutions have survived for long periods of time, I do not mean that their operational rules have remained fixed since they were first introduced. All of the environmental settings included in this chapter are complex and have varied over time. In such settings, it would be almost impossible to "get the operational rules right" on the first try, or even after several tries. These institutions are "robust" or in "institutional equilibrium" in the sense defined by Shepsle. Shepsle (1989b, p. 143) regards "an institution as 'essentially' in equilibrium if changes transpired according to an *ex ante* plan (and hence part of the original institution) for institutional change." In these cases, the appropriators designed basic operational rules, created organizations to undertake the operational management of their CPRs, and modified their rules over time in light of past experience according to their own collective-choice and constitutional-choice rules.

The cases in this chapter are particularly useful for gaining insight regarding how groups of self-organized principals solve two of the major puzzles discussed in Chapter 2: the problem of commitment and the problem of mutual monitoring. (The problem of supply of institutions is addressed in Chapter 4.) The continuing commitments of the appropriators to their institutions have been substantial in these cases. Restrictive rules have been established by the appropriators to constrain appropriation activities and mandate provisioning activities. Thousands of opportunities have arisen in which large benefits could have been reaped by breaking the rules, while the expected sanctions were comparatively low. Stealing water during a dry season in the Spanish *huertas* might on occasion save an entire season's crop from certain destruction. Avoiding spending day after day maintaining the Philippine irrigation systems might enable a farmer to earn needed income in other pursuits. Harvesting illegal timber in the Swiss or Japanese mountain commons would yield a valuable product. Given the temptations involved, the high levels of conformance to the rules in all these cases have been remarkable.

Sizable resources are invested in monitoring activities in these cases, but the "guards" are rarely "external" agents. Widely diverse monitoring arrangements are used. In all of them, the appropriators themselves play a major role in monitoring each other's activities. Even though mutual monitoring has aspects of being a second-order dilemma, the appropriators in these settings somehow solve this problem. Further, the fines assessed in these settings are surprisingly low. Rarely are they more than a small fraction of the monetary value that could be obtained by breaking the rules. In the conclusion to this chapter, I argue that commitment and monitoring are strategically linked and that monitoring produces private benefits for the monitor as well as joint benefits for others.

In explaining the robustness of these institutions and the resource systems themselves over time in environments characterized by high levels of uncertainty, one needs to search for the appropriate specificity of underlying commonalities that may explain this level of sustainability. Given the differences in environments and historical developments, one would hardly expect the particular rules used in these settings to be the same. And they are not. Given the length of time that they have had for trial-and-error learning about operational rules, the harshness of these environments as a stimulus toward improvement, and the low transformation costs in changing their own operational rules, one can, however, expect that these appropriators have "discovered" some underlying principles of good institutional design in a CPR environment. I do not claim that the institutions devised in these settings are in any sense "optimal." In fact, given the high levels of uncertainty involved and the difficulty of measuring benefits and costs, it would be extremely difficult to obtain a meaningful measure of optimality.[1]

On the other hand, I do not hesitate to call these CPR institutions successful. In all instances the individuals involved have had considerable autonomy to craft their own institutions. Given the salience of these CPRs to the appropriators using them, and their capacity to alter rules in light of past performance, these appropriators have had the incentives and the means to improve these institutions over time. The Swiss and Japanese mountain commons have been sustained, if not enhanced, over the centuries while being used intensively. Ecological sustainability in a fragile world of avalanches, unpredictable precipitation, and economic growth is quite an accomplishment for any group of appropriators working over many centuries. Keeping order and maintaining large-scale irrigation works in the difficult terrain of Spain or the Philippine Islands have been similarly remarkable achievements. That record has not been matched by most of the irrigation systems constructed around the world during the past 25 years. Consequently, I have attempted to identify a set of underlying design principles shared by successful CPR institutions and to determine how those design principles affect the incentives of appropriators so that the CPRs themselves and the CPR institutions can be sustained over time. When in Chapter 5 we discuss cases in which appropriators were not able to devise or sustain institutional arrangements to solve CPR problems, we shall consider to what extent the design principles used by appropriators in the "success" cases also characterize the "failure" cases.

The cases discussed in this chapter also help us to examine two other questions. First, the CPR institutions related to the use of precarious and delicately balanced mountain commons to provide fodder and forest products in Switzerland and Japan, in particular, help us to confront the question of the presumed superiority of private-property institutions for most allocational purposes, and specifically those related to the uses of land. Although many resource economists admit that technical difficulties prevent the creation of private property rights to fugitive resources, such as groundwater, oil, and fish, almost all share the presumption that the creation of private property rights to arable or grazing land is an obvious solution to the problem of degradation. Dasgupta and Heal (1979, p. 77), for example, assert that when private property rights are introduced in areas of arable or grazing land, "the resource ceases to be common property and the problem is solved at one stroke."

Many property-rights theorists presume that one of two undesirable outcomes is likely under communal ownership: (1) the commons will be destroyed because no one can be excluded, or (2) the costs of negotiating a set of allocation rules will be excessive, even if exclusion is achieved.[2] On the contrary, what one observes in these cases is the ongoing, side-by-side existence of private property and communal property in settings in which the individuals involved have exercised considerable control over institutional arrangements and property rights. Generations of Swiss and Japanese villagers have learned the relative benefits and costs of private-property and communal-property institutions related to various types of land and uses of land. The villagers in both settings have *chosen* to retain the institution of communal property as the foundation for land use and similar important aspects of village economies. The economic survival of these villagers has been dependent on the skill with which they have used their limited resources. One cannot view communal property in these settings as the primordial remains of earlier institutions evolved in a land of plenty. If the transactions costs involved in managing communal property had been excessive, compared with private-property institutions, the villagers would have had many opportunities to devise different land-tenure arrangements for the mountain commons.

Second, I have frequently been asked, when giving seminar presentations about the Swiss, Japanese, and Spanish institutions, if the same design principles are relevant for solving CPR problems in Third World settings. The last case discussed in this chapter – the *zanjera* institutions of the Philippines – provides a strong affirmative answer to this question. All of the design principles present in the Swiss, Japanese, and Spanish cases are also present in the Philippine case. An analysis of the underlying similarities of enduring CPR institutions, though based on a limited number of cases, may have broader applications.

COMMUNAL TENURE IN HIGH MOUNTAIN MEADOWS AND FORESTS[3]

Törbel, Switzerland

Our first case concerns Törbel, Switzerland, a village of about 600 people located in the Vispertal trench of the upper Valais canton, as described by Robert McC. Netting in a series of articles (1972, 1976) that were later incorporated into his book *Balancing on an Alp* (1981). Netting (1972, p. 133) identifies the most significant features of the general environment as "(1) the steepness of its slope and the wide range of microclimates demarcated by altitude, (2) the prevailing paucity of precipitation, and (3) the exposure to sunlight." For centuries, Törbel peasants have planted their privately owned plots with bread grains, garden vegetables, fruit trees, and hay for winter fodder. Cheese produced by a small group of herdsmen, who tend village cattle pastured on the communally owned alpine meadows during the summer months, has been an important part of the local economy.

Written legal documents dating back to 1224 provide information regarding the types of land tenure and transfers that have occurred in the village and the rules used by the villagers to regulate the five types of communally owned property: the alpine grazing meadows, the forests, the "waste" lands, the irrigation systems, and the paths and roads connecting privately and communally owned properties. On February 1, 1483, Törbel residents signed articles formally establishing an association to achieve a better level of regulation over the use of the alp, the forests, and the waste lands.

> The law specifically forbade a foreigner (*Fremde*) who bought or otherwise occupied land in Törbel from acquiring any right in the communal alp, common lands, or grazing places, or permission to fell timber. Ownership of a piece of land did *not* automatically confer any communal right (*genossenschaftliches Recht*). The inhabitants currently possessing land and water rights reserved the power to decide whether an outsider should be admitted to community membership.
>
> (Netting 1976, p. 139)

The boundaries of the communally owned lands were firmly established long ago, as indicated in a 1507 inventory document.

Access to well-defined common property was strictly limited to citizens, who were specifically extended communal rights.[4] As far as the summer grazing grounds were concerned, regulations written in 1517 stated that "no citizen could send more cows to the alp than he could feed during the winter" (Netting 1976, p. 139). That regulation, which Netting reports to be still enforced, imposed substantial fines for any attempt by villagers to appropriate a larger share of grazing rights. Adherence to this "wintering" rule was administered by a local official (*Gewalthaber*) who was authorized to levy fines on those who exceeded their quotas and to keep one-half of the fines for himself. The wintering rule is used by many other Swiss villages as a means for allocating appropriation rights (frequently referred to as "cow rights") to the commons. This and other forms of cow rights are relatively easy to monitor and enforce. The cows are all sent to the mountain to be cared for by the herdsmen. They must be counted immediately, as the number of cows each family sends is the basis for determining the amount of cheese the family will receive at the annual distribution.

The village statutes are voted on by all citizens and provide the general legal authority for an alp association to manage the alp. This association includes all local citizens owning cattle. The association has annual meetings to discuss general rules and policies and elect officials. The officials hire the alp staff, impose fines for misuse of the common property, arrange for distribution of manure on the summer pastures, and organize the annual maintenance work, such as building and maintaining roads and paths to and on the alp and rebuilding avalanche-damaged corrals or huts. Labor contributions or fees related to the use of the meadows usually are set in proportion to the number of cattle sent by each owner. Trees that will provide timber for construction and wood for heating are marked by village officials and assigned by lot to groups of households, whose members then are authorized to enter the forests and harvest the marked trees.

Private rights to land are well developed in Törbel and other Swiss villages. Most of the meadows, gardens, grainfields, and vineyards are owned by various individuals, and complex condominium-type agreements are devised for the fractional shares that siblings and other relatives may own in barns, granaries, and multistory housing units. The inheritance system in Törbel ensures that all legitimate offspring share equally in the division of the private holdings of their parents and consequently in access to the commons, but family property is not divided until surviving siblings are relatively mature (Netting 1972). Prior to a period of population growth in the nineteenth century, and hence severe population pressure on the limited land, the level of resource use was held in check by various population-control measures such as late marriages, high rates of celibacy, long birth spacing, and considerable emigration (Netting 1981).

Netting (1976, p. 140) dismisses the notion that communal ownership is simply an anachronistic holdover from the past by showing that for at least five centuries these Swiss villagers have been intimately familiar with the advantages and disadvantages of both private and communal tenure systems and have carefully matched particular types of land tenure to particular types of land use. He associates five attributes to land-use patterns with the differences between communal and individual land tenure. He argues that communal forms of land tenure are better suited to the problems that appropriators face when (1) the value of production per unit of land is low, (2) the frequency or dependability of use or yield is low, (3) the possibility of improvement or intensification is low, (4) a large territory is needed for effective use, and (5) relatively large groups are required for capital-investment activities. See Runge (1984a, 1986) and Gilles and Jamtgaard (1981) for similar arguments.

Communal tenure "promotes both general access to and optimum production from certain types of resources while enjoining on the entire community the conservation measures necessary to protect these resources from destruction" (Netting 1976, p. 145). Although yields are relatively low, the land in Törbel has maintained its productivity for many centuries. Overgrazing has been prevented by tight controls. The CPR not only has been protected but also has been enhanced by investments in weeding and manuring the summer grazing areas and by the construction and maintenance of roads.

Netting is clear that Törbel should not be considered the prototype for all Swiss alpine villages. A recent review of the extensive German literature on common-property regimes in Swiss alpine meadows reveals considerable diversity of legal forms for governing alpine meadows (Picht 1987). However, Netting's major findings are consistent with experience in many Swiss locations. Throughout the alpine region of Switzerland, farmers use *private* property for agricultural pursuits and a form of *common* property for the summer meadows, forests, and stony waste lands near their private holdings. Four-fifths of the alpine territory is owned by some form of common property: by local villages (*Gemeinden*), by corporations, or by cooperatives. The remaining alpine territory belongs either to the cantons or to private owners or groups of co-owners (Picht 1987, p. 4). Some villages own several alpine meadows and reallocate grazing rights to the use of a specific meadow every decade or so (Stevenson 1990).

In addition to defining who has access to the CPR, all local regulations specify authority rules to limit appropriation levels (Picht 1987). In most villages, some form of proportional-allocation rule is used. The proportion is based on (1) the number of animals that can be fed over the winter,[5] (2) the amount of meadowland owned by a farmer, (3) the actual amount of hay produced by a farmer, (4) the value of the land owned in the valley, or (5) the number of shares owned in a cooperative. A few villages allow all citizens to send equal numbers of animals to the summer alp (Picht 1987, p. 13). Overuse of alpine meadows is rarely reported.[6] Where overuse has occurred, the combined effects of entry rules and authority rules have not sufficiently limited grazing practices, or else several villages have owned and used a single alp without an overarching set of rules (Picht 1987, pp. 17–18; Rhodes and Thompson 1975; Stevenson 1990).[7]

All of the Swiss institutions used to govern commonly owned alpine meadows have one obvious similarity – the appropriators themselves make all major decisions about the use of the CPR.

The users/owners are the main decision making unit. They have to decide on all matters of importance and seem to have a considerable degree of autonomy. They can set up statutes and revise them, they can set limits for the use of the pastures and change them, they can adapt their organizational structure. . . . It can also be said that the user organizations are nested in a set of larger organizations (village, Kantone, Bund) in which they are perceived as legitimate.   (Picht 1987, p. 28)

Thus, residents of Törbel and other Swiss villages who own communal land spend time governing themselves. Many of the rules they use, however, keep their monitoring and other transactions costs relatively low and reduce the potential for conflict. The procedures used in regard to cutting trees for timber – a valuable resource unit that can be obtained from communal forests – illustrate this quite well. The first step is that the village forester marks the trees ready to be harvested. The second step is that the households eligible to receive timber form work teams and equally divide the work of cutting the trees, hauling the logs, and piling the logs into approximately equal stacks. A lottery is then used to assign particular stacks to the eligible households. No harvesting of trees is authorized at any other time of the year. This procedure nicely combines a careful assessment of the condition of the forest with methods for allocating work and the resulting products that are easy to monitor and are considered fair by all participants. Combining work days or days of reckoning (where the summer's cheese is distributed and assessments are made to cover the costs of the summer's work) with festivities is another method for reducing some of the costs associated with communal management.

In recent times, the value of labor has risen significantly, thus representing an exogenous change for many Swiss villages. Common-property institutions are also changing to reflect differences in relative factor inputs. Villages that rely on unanimity rules for changing their common-property institutions are not adjusting as rapidly as are those villages that rely on less inclusive rules for changing their procedures.

...

ZANJERA IRRIGATION COMMUNITIES
IN THE PHILIPPINES

The earliest recorded reference to the existing irrigation societies in the Ilocanos area of Ilocos Norte in the Philippines derives from Spanish priests writing in 1630 (H. Lewis 1980, p. 153). No serious effort has been made to determine if similar organizations were in existence before the Spanish colonial period, but it would not be unreasonable to assume that the modern *zanjeras* are derived from a mixture of traditions, including that of the Spanish. The most striking similarity between the *huerta* and *zanjera* systems is in the central role given to small-scale communities of irrigators who determine their own rules, choose their own officials, guard their own systems, and maintain their own canals. The internal organization of each *zanjera* has been tailored to its own history, and thus the specific rules in use vary substantially (Keesing 1962). In 1979 there were 686 communal irrigation systems in Ilocos Norte (Siy 1982, p. 25).[27]

Zanjeras have been established both by landowning farmers wanting to construct common irrigation works and by individuals organizing themselves so as to acquire land. The technologies used in *zanjera* systems are relatively crude and labor-intensive. The large number of operating systems and the amount of labor put into these by farmers – tenants as well as landowners – have meant that technological knowledge of how to construct dams and other works has been widely shared. With this knowledge, it has been possible for enterprising tenant farmers to band together to construct an irrigation system on previously nonirrigated land in return for the right to the produce from a defined portion of the newly irrigated land.

This type of contract – called a *biang ti daga* or a "sharing of the land" – allows the landowner to retain ownership. Use rights are extended to the *zanjera* dependent on continued maintenance of the irrigation system. At the time of forming an association, each original participant in the *zanjera* is issued one membership share or *atar*. The total number of *atars* is set at that point.[28] The share gives each member one vote and the right to farm a proportionate share of the land acquired by the *zanjera*, and it defines the obligation of the member for labor and material inputs. Each *atar*-holder is obligated to contribute one day's work during each work season declared by the *zanjera*, plus a share of the material required at construction time. The system was thus developed as a mode of acquiring long-term use rights to land and the water to irrigate it without prior accumulation of monetary assets.

Each *zanjera* is laid out differently, but all that were set up by a *biang ti daga* contract share an underlying pattern. The area is divided into three or more large sections. Each farmer is assigned a plot in each section. All members are thus in fundamentally symmetrical positions in relation to one another. Not only do they own rights to farm equal amounts of land, but they all farm some land in the most advantageous location near the head of the system, and some near the tail. In years when rainfall is not sufficient to irrigate all of the fields, a decision about sharing the burden of scarcity can be made rapidly and equitably by simply deciding not to irrigate the bottom section of land.

Several parcels are set aside for communal purposes. A few parcels, located at the tail end of the system, are assigned to officials of the association as payment for their services. This system not only provides a positive reward for services rendered but also enhances the incentives for those in leadership positions to try to get water to the tail end of the system. Other lands are retained to secure income for the *zanjera* itself. See the work of Coward (1979, 1985) for a detailed description of this system.

The members of each *zanjera* elect a *maestro* as their executive officer, a secretary, a treasurer, and a cook.[29] In the larger associations, they also select foremen and team leaders to supervise the construction activities. The *maestro* has the challenging job of motivating individuals to contribute many hours of physically exhausting labor in times of emergency, when control structures have been washed out, and for routine maintenance. Given the backbreaking efforts required during the monsoon season or during extremely hot weather, this motivational task is of substantial proportions. The *maestro* is, of course, not dependent simply on his persuasive powers. Many real inducements and sanctions are built into these systems by the rules that *zanjera* members have constructed for themselves.

To illustrate the task involved in governing these systems, we shall consider one of these systems – actually, a federation of nine *zanjeras* – in more detail, based on the work of Robert Siy (1982). The Bacarra-Vintar federation of *zanjeras* constructs and maintains a 100-meter-long brush dam that spans the Bacarra-Vintar River, located on the northwestern tip of Luzon Island approximately 500 kilometers north of Manila. The unpredictable and destructive Bacarra-Vintar River drains the northeastern parts of the provinces. During the rainy season each year, the river destroys the federation's dam, which is constructed of bamboo poles, banana leaves, sand, and rock. During some years the dam will be destroyed three or four times.

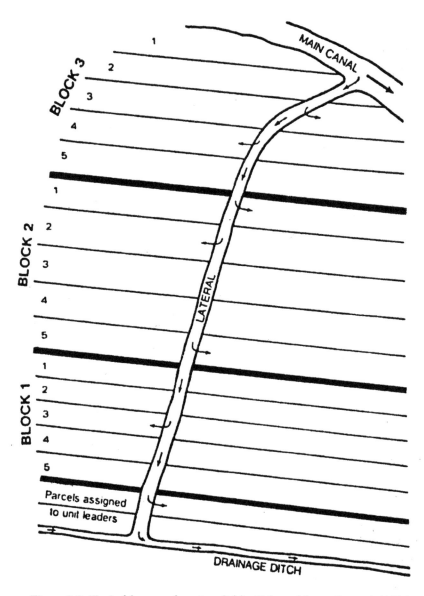

Figure 3.3. Typical layout of *zanjera* fields. (Adapted from Coward 1979.)

The histories of the nine component *zanjeras*, like that of the federation itself, have not been well preserved. What is known is that most of them were established independently and tried to construct and maintain their own diversion works from the river. The river has changed course several times in its history, and at various times some of the *zanjeras* have been cut off from their source of water by such changes. Two of the *zanjeras* were already associated during the nineteenth century and jointly constructed one dam and canal. A formal agreement dated 1906 was written when a third *zanjera* joined their federation. Other existing systems joined slowly through the 1950s. The last two *zanjeras* entered at the time of their formation (Siy 1982, pp. 67-8).

In 1978 the federation formally incorporated as a private corporation in response to the 1976 Philippine Water Code, which defined only individuals or "juridical persons" as eligible to obtain water rights. Given the history of litigation in the area (M. Cruz, Cornista, and Dayan 1987), members of the federation wanted secure water rights in the name of the federation itself, rather than in the name of individual *zanjeras*. The heads of all the component *zanjeras* form the board of directors, with the *maestro* of the Surgui *zanjera* – one of the founding *zanjeras* – named as the president and chairman of the board. In 1980 there were 431 individuals who owned shares, or parts of shares, in at least one *zanjera*. Many members were involved in more than one of the *zanjeras*. The smallest component *zanjera* had 20 members, and the largest had 73 members (Siy 1982, p. 85). Each *zanjera* is responsible for its own financial and internal affairs and owes no financial obligations to the federation.

The board of directors determines when the dam should be rebuilt or repaired. Rebuilding takes about a week – somewhat more when the weather is unfavorable – and involves several hundred persons. Each *zanjera* is responsible for bringing construction materials and providing work teams (and the cooks and food to feed them). After spending a day preparing banana and bamboo mats, work teams in heavy boats confront the swirling waters to begin pounding in the poles that form the foundation for the dam. Then the mats are woven around the poles and reinforced with sand and rock.

Each of the five *zanjeras* with the largest numbers of *atars* provides one work team. The four smaller associations form two work teams. As the dam is laid out, it is divided, by the use of a "flexible" rod, into seven sections that are roughly proportional to the sizes of the work teams and the difficulty of the terrain. This work assignment pattern allows each group to monitor the progress of other groups and engenders some spirited competition among them. The work of maintaining the main canal is also assigned in a similar manner. Work on distributory canals is organized by each *zanjera*, which has divided itself into smaller work teams called *gunglos*, composed of 5 to 10 members.

Siy computed the total obligations (including work as well as attendance at meetings and celebrations) of *zanjera* members to their own associations and to the federation for 1980. The owner of a full *atar* share of the Santo Rosario *zanjera* was obligated to contribute 86 days during 1980 (the largest obligation), whereas an owner of a full share in the Nibinib *zanjera* was obligated to contribute 32 days (the lowest). The average across the federation was 53 days (Siy 1982, p. 92). Given that some *atars* are held jointly by several farmers, the average number of days per working member is somewhat less – around 39 days for the year.

In terms of the contemporary schedule of 5 days per week, this amounts to two months of work supplied without direct monetary payment.[30]About 16,000 man-days were supplied by members to their own *zanjera* or federation during the year.[31] As Siy reflects, "there are definitely few rural organizations in the developing world which have been able to regularly mobilize voluntary [*sic*] labor to such extent" (Siy 1982, p. 95).[32] Given the rigorous and at times dangerous nature of the work, the level of attendance at these obligatory sessions is rather amazing.[33] On average, members were absent somewhat over 2 days out of their required 39, making the attendance rate about 94%. Fines assessed for nonattendance were fully paid in five of the *zanjeras*, and only one of the *zanjeras* had a substantial problem with the payment of fines (Siy 1982, p. 98).[34]

Over time, *zanjeras* face the problem of increased fragmentation of the original shares. A founding member with three sons, for example, may bequeath his plots to be distributed evenly among his sons, each of whom then assumes one-third of the obligations that their father had to fulfill (and having access to only one-third of the land). The individual *zanjeras* have responded to fragmentation in several ways. Some *zanjeras* appoint one person to be responsible for the fulfillment of *atar* responsibilities so that the associations do not have to monitor intra-*atar* work contributions or shirking. Some of the *zanjeras* now require prior approval before a share is sold or tenants are allowed to work *zanjera* land.

Prospective members are "screened," and made to understand the full extent of their obligations to the zanjera before the transaction or tenancy agreement is approved. In a few cases, new members have been required to sign an agreement affirming their recognition of the zanjera's by-laws. These by-laws usually stipulate that erring members may be suspended or expelled from the zanjera, and their lands, confiscated.   (Siy 1982, p. 101)

Given the great numbers of the landless population in the area, there is still fierce competition to gain access to land.

Water-allocation rules are not quite as restrictive in these systems as are work-contribution rules. In general, the supply of water to the irrigation system is more than adequate to meet the needs of the farmers, given the current cropping patterns and soil types involved. When water is abundant, water flows throughout the entire system, and anyone can irrigate at will. When water is scarce, rotation systems are established among the *zanjeras*, and within *zanjeras* among the various distributory canals. During extremely dry periods, downstream *zanjeras* are allowed the full flow of the system for several nights in a row. After notification and agreement, the downstream *zanjera* sends its *gunglos* upstream to set up checks and close turnouts. "Other members 'stand guard' to ensure that such temporary control devices remain in place. Other groups attend to the actual delivery of water to individual parcels" (Siy 1982, p. 122). Precedence is given to parcels with the greatest need, and then a regular rotation system is established.

Several of the downstream *zanjeras* harvest only one crop per year, but two crops are possible in the higher *zanjeras*. Siy presents clear evidence that it would be possible to reallocate water among the nine *zanjeras* so as to increase the productivity of the lower *zanjera* lands without a loss in productivity by the head-end *zanjeras* (Siy 1982, pp. 122–45). On the other hand, the distribution of water is roughly proportional to the contributions of labor and materials and to *atar* shares. Thus, the three *zanjeras* that contribute most of the labor and materials (48%) receive 55% of the water, the three *zanjeras* that contribute 30% of the labor and materials receive 25% of the water, and the three *zanjeras* that contribute 22% of the labor and materials receive 20% of the water.[35]

From the perspective of technical efficiency, the system is not as efficient in its water-allocation scheme as it could be. Siy is, however, extremely careful to point out that many costs besides those of output forgone are involved in designing and running such systems:

> The costs may be in the form of the time and energy expended in deciding on an acceptable arrangement or in adjusting to an externally-imposed procedure. . . . For example, a shift in the distribution of water may necessitate a shift in the distribution of obligations among zanjeras. A zanjera that ends up receiving more water may then be required to contribute a larger proportion of labor and materials for system maintenance in order to satisfy the demands for sharing obligations in proportion to the increased benefits received. However, there is always the danger that the individual zanjera involved may not possess the immediate capability to meet such requirements, and, as such, these new demands on their resources may actually undermine the stability or solidarity of the whole organization.   (Siy 1982, p. 146)

The major criterion used by irrigation engineers to evaluate the performance of an irrigation system is whether or not a system is technically efficient in the sense that water is allocated optimally to enhance crop production. The federation falls short in regard to this criterion, but it performs well in regard to mobilization of personnel for construction and maintenance activities. The members of the federation perceive the allocation of water to conform to legitimate formulas that they have themselves devised, rather than to formulas devised by external experts. As we shall see in Chapter 5, when external experts, working without the participation of the irrigators, have designed systems with the primary aim of achieving technical efficiency, they /frequently have failed to achieve either the hoped-for technical efficiency or the level of organized action required to allocate water in a regular fashion or to maintain the physical system itself.

Because many members of the lower *zanjeras* also participate in other *zanjeras*, many own lands that receive adequate or more than adequate quantities of water, thus offsetting those lands that are left dry part of the year. In a survey of *zanjera* members, respondents from the lower *zanjeras* were more likely than members of upstream *zanjeras* to report a lack of water during part of the year. But when asked what major irrigation problems they faced, none "had anything to say about the way water was allocated or about the fairness of water distribution" (Siy 1982, p. 141). The problem cited by 65% of the irrigators surveyed was the hardship associated with the annual damage to their dam.

SIMILARITIES AMONG ENDURING, SELF-GOVERNING CPR INSTITUTIONS

Despite all of the differences among the CPR settings described in this chapter – and substantial differences exist – all share fundamental similarities. One similarity is that all face uncertain and complex environments. In the mountain commons, the location and timing of rainfall cannot be predicted. In the irrigation systems, erratic rainfall is again a major source of uncertainty. Whereas the construction of physical works tends to reduce the level of uncertainty, it tends to increase the level of complexity in these systems. Irrigators must have practical engineering skills as well as farming skills.

In contrast to the uncertainty caused by these environments, the populations in these locations have remained stable over long periods of time. Individuals have shared a past and expect to share a future. It is important for individuals to maintain their reputations as reliable members of the community. These individuals live side by side and farm the same plots year after year. They expect their children and their grandchildren to inherit their land. In other words, their discount rates are low. If costly investments in provision are made at one point in time, the proprietors – or their families – are likely to reap the benefits.

Extensive norms have evolved in all of these settings that narrowly define "proper" behavior. Many of these norms make it feasible for individuals to live in close interdependence on many fronts without excessive conflict. Further, a reputation for keeping promises, honest dealings, and reliability in one arena is a valuable asset. Prudent, long-term self-interest reinforces the acceptance of the norms of proper behavior. None of these situations involves participants who vary greatly in regard to ownership of assets, skills, knowledge, ethnicity, race, or other variables that could strongly divide a group of individuals (R. Johnson and Libecap 1982).

The most notable similarity of all, of course, is the sheer perseverance manifested in these resource systems and institutions. The resource systems clearly meet the criterion of sustainability. The institutions meet Shepsle's (1989b) criterion of institutional robustness, in that the rules have been devised and modified over time according to a set of collective-choice and constitutional-choice rules. These cases were specifically selected because they have endured while others have failed. Now the task is to begin to explain their sustainability and robustness, given how difficult it must have been to achieve this record in such complex, uncertain, and interdependent environments in which individuals have continuously faced substantial incentives to behave opportunistically.

The specific operational rules in these cases differ markedly from one another. Thus, they cannot be the basis for an explanation across settings. In the Japanese mountain commons, for example, appropriation rights and provision duties are assigned to established family units in a village instead of to individuals. In the Swiss mountains, appropriation rights and provision duties are inherited by individual males who own private property in the village and remain citizens of the village. In eastern Spain, a farmer's right to irrigation water is based on the parcel of land inherited, purchased, or leased, not on a relationship to a village. In the Philippines, a complex contract among long-term usufructuary right-holders determines rights and provision duties. The rules defining when, where, and how an individual's allotted resource units can be harvested or how many labor days are required also vary considerably across cases.

Although the particular rules that are used within these various settings cannot provide the basis for an explanation of the institutional robustness and sustainability across these CPRs, part of the explanation that I offer is based on the fact that the particular rules differ. The differences in the particular rules take into account specific attributes of the related physical systems, cultural views of the world, and economic and political relationships that exist in the setting. Without different rules, appropriators could not take advantage of the positive features of a local CPR or avoid potential pitfalls that might be encountered in one setting but not others.

Instead of turning to the specific rules, I turn to a set of seven design principles that characterize all of these robust CPR institutions, plus an eighth principle used in the larger, more complex cases. These are listed in Table 3.1. By "design principle" I mean an essential element or condition that helps to account for the success of these institutions in sustaining the CPRs and gaining the compliance of generation after generation of appropriators to the rules in use. This list of design principles is still quite speculative. I am not yet willing to argue that these design principles are necessary conditions for achieving institutional robustness in CPR settings. Further theoretical and empirical work is needed before a strong assertion of necessity can be made. I am willing to speculate, however, that after

Table 3.1. *Design principles illustrated by long-enduring CPR institutions*

1. Clearly defined boundaries
   Individuals or households who have rights to withdraw resource units from the CPR must be clearly defined, as must the boundaries of the CPR itself.

2. Congruence between appropriation and provision rules and local conditions
   Appropriation rules restricting time, place, technology, and/or quantity of resource units are related to local conditions and to provision rules requiring labor, material, and/or money.

3. Collective-choice arrangements
   Most individuals affected by the operational rules can participate in modifying the operational rules.

4. Monitoring
   Monitors, who actively audit CPR conditions and appropriator behavior, are accountable to the appropriators or are the appropriators.

5. Graduated sanctions
   Appropriators who violate operational rules are likely to be assessed graduated sanctions (depending on the seriousness and context of the offense) by other appropriators, by officials accountable to these appropriators, or by both.

6. Conflict-resolution mechanisms
   Appropriators and their officials have rapid access to low-cost local arenas to resolve conflicts among appropriators or between appropriators and officials.

7. Minimal recognition of rights to organize
   The rights of appropriators to devise their own institutions are not challenged by external governmental authorities.

For CPRs that are parts of larger systems:
8. Nested enterprises
   Appropriation, provision, monitoring, enforcement, conflict resolution, and governance activities are organized in multiple layers of nested enterprises.

further scholarly work is completed, it will be possible to identify a set of necessary design principles and that such a set will contain the core of what has been identified here.[36]

For these design principles to constitute a credible explanation for the persistence of these CPRs and their related institutions, I need to show that they can affect incentives in such a way that appropriators will be willing to commit themselves to conform to operational rules devised in such systems, to monitor each other's conformance, and to replicate the CPR institutions across generational boundaries. I shall discuss each of the design principles in turn.

Clearly defined boundaries

1 Individuals or households who have rights to withdraw resource units from the CPR must be clearly defined, as must the boundaries of the CPR itself.

Defining the boundaries of the CPR and specifying those authorized to use it can be thought of as a first step in organizing for collective action. So long as the boundaries of the resource and/or the specification of individuals who can use the resource remain uncertain, no one knows what is being managed or for whom. Without defining the boundaries of the CPR and closing it to "outsiders," local appropriators face the risk that any benefits they produce by their efforts will be reaped by others who have not contributed to those efforts. At the least, those who invest in the CPR may not receive as high a return as they expected. At the worst, the actions of others could destroy the resource itself. Thus, for any appropriators to have a minimal interest in coordinating patterns of appropriation and provision, some set of appropriators must be able to exclude others from access and appropriation rights. If there are substantial numbers of potential appropriators and the demand for the resource units is high, the destructive potential should all be allowed to freely withdraw units from the CPR could push the discount rate used by appropriators toward 100%. The higher the discount rate, the closer the situation is to that of a one-shot dilemma in which the dominant strategy of all participants is to overuse the CPR.

Since the work of Ciriacy-Wantrup and Bishop (1975), the presence of boundaries concerning who is allowed to appropriate from the CPR has been used as the single defining characteristic of "common-property" institutions as contrasted to "open-access" institutions. The impression is sometimes given that this is all that is necessary to achieve successful regulation. Making this attribute one of seven, rather than a unique attribute, puts its importance in a more realistic perspective. Simply closing the boundaries is not enough. It is still possible for a limited number of appropriators to increase the quantity of resource units they harvest so that they either dissipate all potential rents or totally destroy the resource (Clark 1980). Consequently, in addition to closing the boundaries, some rules limiting appropriation and/or mandating provision are needed.

Congruence between appropriation and provision rules and local conditions

2 Appropriation rules restricting time, place, technology, and/or quantity of resource units are related to local conditions and to provision rules requiring labor, materials, and/or money.

Adding well-tailored appropriation and provision rules helps to account for the perseverance of these CPRs. In all these cases, the rules reflect the specific attributes of the particular resource. Among the four Spanish *huertas* that are located in fairly close proximity to one another, the specific rules for the various *huertas* differ rather substantially. It is only in the one system (Alicante) where there has been substantial storage available since the construction of Tibi Dam in 1594 that a water auction is held. At the time of the Sunday morning auction, substantial information about the level of water in the dam is made available to the Alicante irrigators. Consequently, they can know about how much water they will receive if they purchase an hour of water. In the systems without storage, water is strictly tied to the land, and some form of rotation is used. In Valencia, each farmer takes as much water as he can put to beneficial use in a defined order. Thus, each farmer has a high degree of certainty about the quantity of water to be received, and less certainty about the exact timing. In Murcia and Orihuela, where water is even more scarce, a tighter rotation system is used that rations the amount of time that irrigators can keep their gates open. Further, the rules attempt to solve the problem of getting water to a more diversified terrain than in Valencia. Subtly different rules are used in each system for assessing water fees used to pay for water guards and for maintenance activities, but in all instances those who receive the highest proportion of the water also pay the highest proportion of the fees. No single set of rules defined for all irrigation systems in the region could deal with the particular problems in managing each of these broadly similar, but distinctly different, systems.[37]

3 Most individuals affected by the operational rules can participate in modifying the operational rules.

CPR institutions that use this principle are better able to tailor their rules to local circumstances, because the individuals who directly interact with one another and with the physical world can modify the rules over time so as to better fit them to the specific characteristics of their setting. Appropriators who design CPR institutions that are characterized by these first three principles – clearly defined boundaries, good-fitting rules, and appropriator participation in collective choice – should be able to devise a good set of rules if they keep the costs of changing the rules relatively low.

The presence of good rules, however, does not ensure that appropriators will follow them. Nor is the fact that appropriators themselves designed and initially agreed to the operational rules in our case studies an adequate explanation for centuries of compliance by individuals who were not involved in the initial agreement. It is not even an adequate explanation for the continued commitment of those who were part of the initial agreement. Agreeing to follow rules ex ante is an easy commitment to make. Actually following rules ex post, when strong temptations arise, is the significant accomplishment.

The problem of gaining compliance to the rules – no matter what their origin – often is assumed away by analysts positing all-knowing and all-powerful *external* authorities who enforce agreements. In the cases described here, no external authority has had sufficient presence to play any role in the day-to-day enforcement of the rules in use.[38] Thus, external enforcement cannot be used to explain these high levels of compliance.

Some recent theoretical models of repeated siutations do predict that individuals will adopt contingent strategies to generate optimal equilibria without external enforcement, but with very specific information requirements rarely found in field settings (Axelrod 1981, 1984; Kreps et al. 1982; T. Lewis and Cowens 1983). In these models, participants adopt resolute strategies to cooperate so long as everyone else cooperates. If anyone deviates, the models posit that all others will deviate immediately and forever. Information about everyone's strategies in a previous round is assumed to be freely available. No monitoring activities are included in these models, because information is presumed to be already available.

It is obvious from our case studies, however, that even in repeated settings where reputation is important and where individuals share the norm of keeping agreements, reputation and shared norms are insufficient by themselves to produce stable cooperative behavior over the long run. If they had been sufficient, appropriators could have avoided investing resources in monitoring and sanctioning activities. In all of the long-enduring cases, however, active investments in monitoring and sanctioning activities are quite apparent. That leads us to consider the fourth and fifth design principles:

Monitoring

4 Monitors, who actively audit CPR conditions and appropriator behavior, are accountable to the appropriators or are the appropriators.

Graduated sanctions

5 Appropriators who violate operational rules are likely to be assessed graduated sanctions (depending on the seriousness and context of the offense) by other appropriators, by officials accountable to these appropriators, or by both.

Now we are at the crux of the problem – and with surprising results. In these robust institutions, monitoring and sanctioning are undertaken not by external authorities but by the participants themselves. The initial sanctions used in these systems are also surprisingly low. Even though it is frequently presumed that participants will not spend the time and effort to monitor and sanction each other's performances, substantial evidence has been presented that they do both in these settings. The appropriators in these CPRs somehow have overcome the presumed problem of the second-order dilemma.

To explain the investment in monitoring and sanctioning activities that occurs in these robust, self-governing CPR institutions, the term "quasi-voluntary compliance" can be useful, as applied by Margaret Levi (1988a, ch. 3) to describe the behavior of taxpayers in systems in which most taxpayers comply. Paying taxes is voluntary in the sense that individuals choose to comply in many situations in which they are not being directly coerced. On the other hand, it is "*quasi*-voluntary because the noncompliant are subject to coercion – if they are caught" (Levi 1988a, p. 52). Taxpayers, according to Levi, will adopt a strategy of quasi-voluntary compliance when they have

confidence that (1) rulers will keep their bargains and (2) the other constituents will keep theirs. Taxpayers are strategic actors who will cooperate only when they can expect others to cooperate as well. The compliance of each depends on the compliance of the others. No one perfers to be a "sucker." (Levi 1988a, p. 53)

Levi stresses the *contingent* nature of a commitment to comply with rules that is possible in a repeated setting. Strategic actors are willing to comply with a set of rules, Levi argues, when (1) they perceive that the collective objective is achieved, and (2) they perceive that others also comply. Levi is not the first to point to contingent behavior as a source of stable, long-term cooperative solutions. Prior work, however, had viewed contingent behavior as an *alternative* to coercion; see, for example, Axelrod (1981, 1984) and T. Lewis and Cowens (1983). Levi, on the other hand, views coercion as an *essential condition* to achieve quasi-voluntary compliance as a form of contingent behavior. In her theory, enforcement increases the confidence of individuals that they are not suckers. As long as they are confident that others are cooperating and the ruler provides joint benefits, they comply willingly to tax laws. In Levi's theory, enforcement is normally provided by an external ruler, although her theory does not preclude other enforcers.

To explain commitment in these cases, we cannot posit external enforcement. CPR appropriators create their own internal enforcement to (1) deter those who are tempted to break rules and thereby (2) assure quasi-voluntary compliers that others also comply.[39] As discussed in Chapter 2, however, the normal presumption has been that participants themselves will not undertake mutual monitoring and enforcement because such actions involve relatively high personal costs and produce public goods available to everyone. As Elster (1989, p. 41) states, "punishment almost invariably is costly to the punisher, while the benefits from punishment are diffusely distributed over the members." Given the evidence that individuals monitor, then the relative costs and benefits must have a different configuration than that posited in prior work. Either the costs of monitoring are lower or the benefits to an individual are higher, or both.

The costs of monitoring are low in many long-enduring CPRs as a result of the rules in use. Irrigation rotation systems, for example, usually place the two actors most concerned with cheating in direct contact with one another. The irrigator who nears the end of a rotation turn would like to extend the time of his turn (and thus the amount of water obtained). The next irrigator in the rotation system waits nearby for him to finish, and would even like to start early. The presence of the first irrigator deters the second from an early start, the presence of the second irrigator deters the first from a late ending. Neither has to invest additional resources in monitoring activities. Monitoring is a by-product of their own strong motivations to use their water rotation turns to the fullest extent. The fishing-site rotation system used in Alanya has the same characteristic that cheaters can be observed at low cost by those who most want to deter cheaters at that particular time and location.

Many of the ways that work teams are organized in the Swiss and Japanese mountain commons also have the result that monitoring is a natural by-product of using the commons. Institutional analysis that simply posits an external, zero-cost enforcer has not addressed the possibility that the rules devised by appropriators may themselves have a major effect on the costs, and therefore the efficiency, of monitoring by internal or external enforcers.

Similarly, it is apparent that personal rewards for doing a good job are given to appropriators who monitor. The individual who finds a rule-infractor gains status and prestige for being a good protector of the commons. The infractor loses status and prestige. Private benefits are allocated to those who monitor. When internal monitoring is accomplished as part of a specialized position accountable to the other appropriators, several mechanisms increase the rewards for doing a good job or exposing slackards to the risk of losing their positions. In the Spanish *huertas*, a portion of the fines is kept by the guards; the Japanese detectives also keep the saké they collect from infractors.[40] All of the formal guard positions are accountable to the appropriators, and thus the monitors can be fired easily if discovered slacking off. Because the appropriators tend to continue monitoring the guards, as well as each other, some redundancy is built into the monitoring and sanctioning system. Failure to deter rule-breaking by one mechanism does not trigger a cascading process of rule infractions, because other mechanisms are in place.

Consequently, the costs and benefits of monitoring a set of rules are not independent of the particular set of rules adopted. Nor are they uniform in all CPR settings. When appropriators design at least some of their own rules (design principle 3), they can learn from experience to craft enforceable rather than unenforceable rules. This means paying attention to the costs of monitoring and enforcing, as well as the benefits that accrue to those who monitor and enforce the rules.

In repeated settings in which appropriators face incomplete information, appropriators who undertake monitoring activities obtain valuable information for themselves that can improve the quality of the strategic decision they make. In most theoretical models, where contingent strategies are shown to lead to optimal and stable dynamic equilibria, actors are assumed to have complete information about past history. They know what others did in the last round of decisions and how those choices affected outcomes. No consideration is given to how this information is generated. In the settings we have examined in this chapter, however, obtaining information about behavior and outcomes is costly.

If the appropriators adopt contingent strategies – each agreeing to follow a set of rules, so long as most of the others follow the rules – each one needs to be sure that others comply and that their compliance produces the expected benefit. Thus, a previously unrecognized "private" benefit of monitoring in settings in which information is costly is that one obtains the information necessary to adopt a contingent strategy. If an appropriator who monitors finds someone who has violated a rule, the benefits of that discovery are shared by all who use the CPR, and the discoverer gains an indication of compliance rates. If the monitor does *not* find a violator, previously it has been presumed that private costs are involved without any benefit to the individual or the group. If information is not freely available about compliance rates, then an individual who monitors obtains valuable information from monitoring. The appropriator-monitor who watches how water is distributed to other appropriators not only provides a public good for all but also obtains information needed to make future strategic decisions.

By monitoring the behavior of others, the appropriator-monitor learns about the level of quasi-voluntary compliance in the CPR. If no one is discovered breaking the rules, the appropriator-monitor learns that others comply and that no one is being taken for a sucker. It is then safe for the appropriator-monitor to continue to follow a strategy of quasi-voluntary compliance. If the appropriator-monitor discovers a rule infraction, it is possible to learn about the particular circumstances surrounding the infraction, to participate in deciding the appropriate level of sanctioning, and then to decide whether or not to continue compliance. If an appropriator-monitor finds an offender who normally follows the rules but in one instance happens to face a severe problem, the experience confirms what everyone already knows: There will always be instances in which those who are basically committed to following the set of rules may succumb to strong temptations to break them.

The appropriator-monitor may want to impose only a modest sanction in this circumstance. A small penalty may be sufficient to remind the infractor of the importance of compliance. The appropriator-monitor might be in a similar situation in the future and would want some understanding at that time. Everyone will hear about the incident, and the violator's reputation for reliability will depend on complying with the rules in the future. If the appropriator-monitor presumes that the violator will follow the rules most of the time in the future, the appropriator-monitor can safely continue a strategy of compliance. The incident will also confirm for the appropriator-monitor the importance of monitoring even when most others basically are following the rules.

A real threat to the continuance of quasi-voluntary compliance can occur, however, if an appropriator-monitor discovers individuals who break the rules repeatedly. If this occurs, one can expect the appropriator-monitor to escalate the imposed sanctions in an effort to halt future rule-breaking by such offenders and any others who might start to follow suit. In any case, the appropriator-monitor has up-to-date information about compliance and sanctioning behavior on which to base future decisions about personal compliance.

Let us also look at the situation through the eyes of someone who breaks the rules and is discovered by a local guard (who will eventually tell everyone) or another appropriator (who also is likely to tell everyone). Being apprehended by a local monitor after having succumbed to the temptation to break the rules will have three results: (1) It will stop the infraction from continuing and may return contraband harvest to others. (2) It will convey information to the offender that someone else in a similar situation is likely to be caught, thus increasing confidence in the level of quasi-voluntary compliance. (3) A punishment in the form of a fine, plus loss of reputation for reliability, will be imposed. A large monetary fine may not be needed to return an occasional offender to the fold of those who are quasi-voluntary compliers with the rules. A large monetary fine imposed on a person facing an unusual problem may produce resentment and unwillingness to conform to the rules in the future. Graduated punishments ranging from insignificant fines all the way to banishment, applied in settings in which the sanctioners know a great deal about the personal circumstances of the other appropriators and the potential harm that could be created by excessive sanctions, may be far more effective than a major fine imposed on a first offender.

If quasi-voluntary compliance is contingent on the compliance rate of others, then the question is, What rate must be maintained to ensure that the commitment to comply will continue over time? Previous theoretical work has assumed that 100% is needed; but also see M. Taylor (1987, pp. 89–90), who posits less than 100%. It is assumed that any infraction (or error) will trigger a relentless process: Everyone will resolutely punish the offender (and themselves) by breaking their previous agreement. Although these trigger-strategy models have the attractive theoretical property of stable equilibria, they do not describe the behavior observed in our case studies (or any of the other cases I have read or observed in the field). Acceptable quasi-voluntary compliance rates that will lead appropriators to continue their own quasi-voluntary compliance will differ from one setting to another and will depend on economic or other circumstances within the CPR. Tolerance for rules infractions may be very high during a depression, so long as the higher rate appears temporary and not threatening to the survival of a CPR. This appears to have happened in one of the Japanese villages studied by McKean during the depression of the 1930s:

Almost all the villagers knew that almost all the other villagers were breaking the rules: sneaking around the commons at night, cutting trees that were larger than the allowed size, even using wood-cutting tools that were not permitted. This is precisely the behavior that could get a tragedy of the commons started, but it did not happen in Yamanaka. Instead of regarding the general breakdown of rules as an opportunity to become full-time free riders and cast caution to the winds, the violators themselves tried to exercise self-discipline out of deference to the preservation of the commons, and stole from the commons only out of desperation. Inspectors or other witnesses who saw violations maintained silence out of sympathy for the violators' desperation and out of confidence that the problem was temporary and could not really hurt the commons. (McKean 1986, pp. 565–6)

In other situations, the harm that a single infraction can inflict on others may be so substantial, and the potential for private gain so great, that 100% compliance is essential. McKean (1986, p. 565) describes a situation in the village of Shiwa when it suffered a severe drought. The temptation to break the dikes, in order to obtain water illegally, was so great for those serving as guards, as well as for the remaining farmers, that all adult males patrolled the dikes every night in mutual surveillance until the emergency was over.

The fourth and fifth design principles – monitoring and graduated sanctions – thus take their place as part of the configuration of design principles that can work together to enable appropriators to constitute and reconstitute robust CPR institutions. Let me summarize my argument to this point. When CPR appropriators design their own operational rules (design principle 3) to be enforced by individuals who are local appropriators or are accountable to them (design principle 4), using graduated sanctions (design principle 5) that define who has rights to withdraw units from the CPR (design principle 1) and that effectively restrict appropriation activities, given local conditions (design principle 2), the commitment and monitoring problem are solved in an interrelated manner. Individuals who think that a set of rules will be effective in producing higher joint benefits and that monitoring (including their own) will protect them against being suckered are willing to make a contingent self-commitment[41] of the following type:

I commit myself to follow the set of rules we have devised in all instances except dire emergencies if the rest of those affected make a similar commitment and act accordingly.

Once appropriators have made contingent self-commitments, they are then motivated to monitor other people's behaviors, at least from time to time, in order to assure themselves that others are following the rules most of the time. Contingent self-commitments and mutual monitoring reinforce one another, especially when appropriators have devised rules that tend to reduce monitoring costs. We are now ready to discuss the sixth design principle.

Conflict-resolution mechanisms

6 Appropriators and their officials have rapid access to low-cost local arenas to resolve conflicts among appropriators or between appropriators and officials.

In theoretical models of rule-governed behavior, the rules that structure the strategies available to participants are unambiguous and are enforced by external, all-knowing officials. In field settings, applying the rules is never unambiguous, even when the appropriators themselves are the monitors and sanctioners. Even such a simple rule as "each irrigator must send one individual for one day to help clean the irrigation canals before the rainy season begins" can be interpreted quite differently by different individuals. Who is or is not an "individual" according to this rule? Does sending a child below age 10 or an adult above age 70 to do heavy physical work meet this rule? Is working for four hours or six hours a "day" of work? Does cleaning the canal immediately next to one's own farm qualify for this community obligation? For individuals who are seeking ways to slide past or subvert rules, there are always various ways in which they can "interpret" a rule so that they can argue they have complied with the rule, but in effect subverting its intent. Even individuals who intend to follow the spirit of a rule can make errors. What happens if someone forgets about a labor day and does not show? Or what happens of the only able-bodied worker is sick, or unavoidably in another location?

If individuals are going to follow rules over a long period of time, there must be some mechanism for discussing and resolving what constitutes an infraction. If some individuals are allowed to free-ride by sending less able workers to a required labor day, others will consider themselves to be suckers if they send their strongest workers, who could be using that time to produce private goods rather than communal benefits. Should that continue over time, only children and old people would be sent to do work that would require strong adults, and the system would break down. If individuals who make honest mistakes or face personal problems that occasionally prevent them from following a rule do not have access to mechanisms that will allow them to make up for their lack of performance in an acceptable way, rules may come to be viewed as unfair, and conformance rates may decline.

Although the presence of conflict-resolution mechanisms does not guarantee that appropriators will be able to maintain enduring institutions, it is difficult to imagine how any complex system of rules could be maintained over time without such mechanisms. For those cases discussed earlier, such mechanisms sometimes are quite informal, and those who are selected as leaders are also the basic resolvers of conflict. In some cases – such as the Spanish *huertas* – the potential for conflict over a very scarce resource is so high that well-developed court mechanisms have been in place for centuries.

Minimal recognition of rights to organize

7 The rights of appropriators to devise their own institutions are not challenged by external governmental authorities.

Appropriators frequently devise their own rules without creating formal governmental jurisdictions for this purpose. In many inshore fisheries, for example, local fishers devise extensive rules defining who can use a fishing ground and what kind of equipment can be used. Provided the external governmental officials give at least minimal recognition to the legitimacy of such rules, the fishers themselves may be able to enforce the rules themselves. But if external governmental officials presume that only they have the authority to set the rules, then it will be very difficult for local appropriators to sustain a rule-governed CPR over the long run. In a situation in which one wishes to get around the rules created by the fishers, one may go to the external government and try to get local rules overturned. In Chapter 5 we shall examine several cases in which this design principle is not met.

Nested enterprises

8 Appropriation, provision, monitoring, enforcement, conflict resolution, and governance activities are organized in multiple layers of nested enterprises.

All of the more complex, enduring CPRs meet this last design principle. In the Spanish *huertas*, for example, irrigators are organized on the basis of three or four nested levels, all of which are then also nested in local, regional, and national governmental jurisdictions. There are two distinct levels in the Philippine federation of irrigation systems. The problems facing irrigators at the level of a tertiary canal are different from the problems facing a larger group sharing a secondary canal. Those, in turn, are different from the problems involved in the management of the main diversion works that affect the entire system. Establishing rules at one level, without rules at the other levels, will produce an incomplete system that may not endure over the long run.

In the last part of this chapter I have identified a set of design principles that characterize the long-enduring CPR institutions described in the first part. I have also attempted to examine why individuals utilizing institutional arrangements characterized by these design principles will be motivated to replicate the institutions over time and sustain the CPR to which they are related. We shall continue to discuss these design principles throughout the remainder of this study. In the next chapter we shall examine how individuals supply themselves with new institutions to solve CPR problems.

Sustainable Development: A Critical Review

SHARACHCHANDRA M. LÉLÉ*

Energy & Resources Group, University of California, Berkeley

Summary. — Over the past few years, "Sustainable Development" (SD) has emerged as the latest development catchphrase. A wide range of nongovernmental as well as governmental organizations have embraced it as the new paradigm of development. A review of the literature that has sprung up around the concept of SD indicates, however, a lack of consistency in its interpretation. More important, while the all-encompassing nature of the concept gives it political strength, its current formulation by the mainstream of SD thinking contains significant weaknesses. These include an incomplete perception of the problems of poverty and environmental degradation, and confusion about the role of economic growth and about the concepts of sustainability and participation. How these weaknesses can lead to inadequacies and contradictions in policy making is demonstrated in the context of international trade, agriculture, and forestry. It is suggested that if SD is to have a fundamental impact, politically expedient fuzziness will have to be given up in favor of intellectual clarity and rigor.

1. INTRODUCTION

The last few years have seen a dramatic transformation in the environment-development debate. The question being asked is no longer "Do development and environmental concerns contradict each other?" but "How can sustainable development be achieved?" All of a sudden the phrase Sustainable Development (SD) has become pervasive. SD has become the watchword for international aid agencies, the jargon of development planners, the theme of conferences and learned papers, and the slogan of developmental and environmental activists. It appears to have gained the broad-based support that earlier development concepts such as "ecodevelopment" lacked, and is poised to become the developmental paradigm of the 1990s.

But murmurs of disenchantment are also being heard. "What *is* SD?" is being asked increasingly frequently without, however, clear answers forthcoming. SD is in real danger of becoming a cliché like appropriate technology — a fashionable phrase that everyone pays homage to but nobody cares to define. Four years ago, Tolba lamented that SD had become "an article of faith, a shibboleth; often used, but little explained" (Tolba, 1984a); the situation has not improved since.

There are those who believe that one should not try to define SD too rigorously. To some extent, the value of the phrase does lie in its broad vagueness. It allows people with hitherto irreconcilable positions in the environment-development debate to search for common ground without appearing to compromise their positions.[1] If, however, this political meeting of minds and the concept of SD are both products of new insights into the relationship between social and environmental phenomena, then it should be advantageous to examine these insights and characterize the concept before it is misinterpreted, distorted, and even coopted.

Buttel and Gillespie (1988) contend that such cooptation has already taken place. Agencies such as the World Bank (Conable, 1986), the Asian Development Bank (Runnals, 1986) and the Organization for Economic Cooperation and Development (Environment Committee, 1985) have been quick to adopt the new rhetoric. The absence of a clear theoretical and analytical framework, however, makes it difficult to determine whether the new policies will indeed foster an environmentally sound and socially meaningful form of development. Further, the absence of semantic and conceptual clarity is hampering

*I am grateful to Richard Norgaard for pointing me towards significant references, and to him, Michael Maniates and Ken Conca for their extensive and invaluable comments on earlier drafts. Comments by John Harte, Arjun Makhijani, Paul Ekins, John Pezzey and an anonymous referee helped further refine the arguments, and are gratefully acknowledged. Special thanks to ExPro (Exploratory Project on the Conditions of Peace) for providing financial support.

a fruitful debate over what this form should actually be.

This paper is a critical review[2] of the literature on sustainable development with the foregoing in mind. The purpose is not to prove SD to be an intellectual oxymoron, nor to provide the "ultimate" definition of SD. The idea here is to clarify the semantics and to identify some critical weaknesses in concepts and reasoning — weaknesses that will have to be addressed if SD is to become a meaningful paradigm of development.[3]

I begin by examining the various interpretations of "sustainable development," contrasting the trivial or contradictory with the more meaningful ones. I then trace the evolution of the concept of SD, i.e., of its objectives and premises. I point out that the persuasive power of SD (and hence the political strength of the SD movement) stems from the underlying claim that new insights into physical and social phenomena force one to concur with the operational conclusions of the SD platform almost regardless of one's fundamental ethical persuasions and priorities. I argue that while these new insights are important, the argument is not inexorable, and that the issues are more complex than is made out to be. Hence (as is illustrated in Section 5), many of the policy prescriptions being suggested in the name of SD stem from subjective (rather than consensual) ideas about goals and means, and worse, are often inadequate and even counterproductive. I conclude with some thoughts about future research in SD.

2. INTERPRETING SUSTAINABLE DEVELOPMENT

The manner in which the phrase "sustainable development" is used and interpreted varies so much that while O'Riordan (1985) called SD a "contradiction in terms," Redclift suggests that it may be just "another development truism" (Redclift, 1987, p. 1). These interpretational problems, though ultimately conceptual, have some semantic roots. Most people use the phrase "sustainable development" interchangeably with "ecologically sustainable or environmentally sound development" (Tolba, 1984a). This interpretation is characterized by: (a) "sustainability" being understood as "ecological sustainability"; and (b) a conceptualization of SD as a process of change that has (ecological) sustainability added to its list of objectives.

In contrast, sustainable development is sometimes interpreted as "sustained growth," "sustained change," or simply "successful" development. Let us examine how these latter interpretations originate and why they are less useful than the former one, and try to define the terms for the rest of this discussion. Figure 1 is a "semantic map" that might help in this exercise.

(a) Contradictions and trivialities

Taken literally, sustainable development would simply mean "development that can be

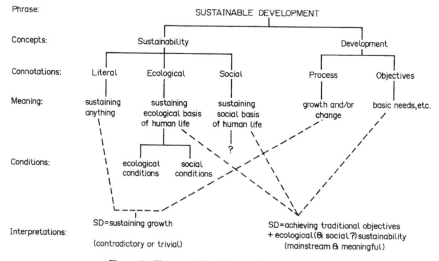

Figure 1. *The semantics of sustainable development.*

continued — either indefinitely or for the implicit time period of concern." But what is development? Theorists and practitioners have both been grappling with the word and the concept for at least the past four decades. (See Arndt, 1981, and Bartelmus, 1986, for semantic and conceptual histories of economic development.) Some equate development with GNP growth, others include any number of socially desirable phenomena in their conceptualization. The point to be noted is that development is *a process of directed change*. Definitions of development thus embody both (a) the objectives of this process, and (b) the means of achieving these objectives.

Unfortunately, a distinction between objectives and means is often not made in the development rhetoric. This has led to "sustainable development" frequently being interpreted as simply a process of change that can be continued forever (see Figure 1). Depending upon what characterization of the process is implicit, this interpretation is either impossible or trivial. When development is taken to be synonymous with growth in material consumption — which it often is even today — SD would be "sustaining the growth in material consumption" (presumably indefinitely). But such an idea contradicts the now general recognition that "*ultimate* limits [to usable resources] exist"[5] (WCED, p. 45, emphasis added). At best, it could be argued that growth in the per capita consumption of certain basic goods is necessary in certain regions of the world in the short term. To use "sustainable development" synonymously with "sustain[ing] growth performance" (Idachaba, 1987) or to cite the high rates of growth in agricultural production in South Asia as an example of SD (Hopper, 1987) is therefore a misleading usage, or at best a short-term and localized notion that goes against the long-term global perspective of SD.

One could finesse this contradiction by conceptualizing development as simply a process of socio-economic change. But one cannot carry on a meaningful discussion unless one states what the objectives of such change are and why one should worry about continuing the process of change indefinitely. Neoclassical economists define the objective of development as "increase in social welfare." They then proceed to measure social welfare in terms of economic output, and point out that "a growth in economic output does not necessarily mean growth in physical throughput of materials and energy" (Pezzey, 1989, p. 14), thus "proving" that there is no contradiction between sustainability and development. But this argument is at best circular (because to achieve continuous increases in social welfare one actual-

ly needs to know what constitutes social welfare, in which case one might as well explicitly state these constituents to be the objectives of development) and at worst fallacious (because there are serious theoretical problems with aggregating individual utility functions within and especially across generations, and serious practical problems with devising indicators for any social welfare function that are not as distorted as GNP). Again, it is not clear why the process of increasing welfare should continue indefinitely, or whether it can do so.

Sometimes, the adjective "sustainable" is simply used instead of "successful." For instance: "For economic development to be truly 'sustainable' requires 'tailoring the design and implementation of projects to the needs and capabilities of people who are supposed to benefit from them'" (Barbier, 1987). Since "beneficiary-oriented design" (or "grassroots participation") is a procedural imperative for any development program to be successful, such a statement tells us nothing about the overall goals of that developmental process. This usage is therefore not very useful; moreover, it is confusing, because sustainability has already acquired other specific connotations.

(b) *Sustainability*

What then are these specific connotations of "sustainability"? While a more conceptual discussion is reserved for later on, some basic terms and usages need to be clarified here. The concept of sustainability originated in the context of renewable resources such as forests or fisheries, and has subsequently been adopted as a broad slogan by the environmental movement (Lélé, 1988). Most proponents of sustainability therefore take it to mean "the existence of the ecological conditions necessary to support human life at a specified level of well-being through future generations," what I call *ecological sustainability* (see Figure 1).

Since ecological sustainability emphasizes the constraints and opportunities that nature presents to human activities, ecologists and physical scientists frequently dominate its discussion. But what they actually focus on are the ecological conditions for ecological sustainability — the biophysical "laws" or patterns that determine environmental responses to human activities and humans' ability to use the environment. The major contribution of the environment-development debate is, I believe, the realization that in addition to or in conjunction with these ecological conditions, there are social conditions that influence the ecological sustainability or unsustainability of the people-nature interaction. To give a stylized example, one could say that soil erosion undermining the agricultural basis for human society is a case of ecological (un)sustainability. It could be caused by farming on marginal lands without adequate soil conservation measures — the ecological cause. But the phenomenon of marginalization of peasants may have social roots, which would then be the social causes of ecological unsustainability.

Sometimes, however, sustainability is used with fundamentally social connotations. For instance, Barbier (1987) defines social sustainability as "the ability to maintain desired social values, traditions, institutions, cultures, or other social characteristics." This usage is not very common, and its needs to be carefully distinguished from the more common context in which social scientists talk about sustainability, viz., the social aspects of ecological sustainability. A war destroying human society would probably be an example of social (un)sustainability, and it in turn may have social or ecological causes. (Note that these categories are only conceptual devices for clarifying our thinking; real problems seldom fall neatly into one category or another.)

(c) *Sustainable development = development + sustainability?*

In the mainstream interpretation of SD, ecological sustainability is a desired attribute of any pattern of human activities that is the goal of the developmental process. In other words, SD is understood as "a form of societal change that, in addition to traditional developmental objectives, has the objective or constraint of ecological sustainability." Given an ever-changing world, the specific forms of and priorities among objectives, and the requirements for achieving sustainability, would evolve continuously. But sustainability — as it is understood at each stage — would remain a fundamental concern. Ecological sustainability is, of course, not independent of the other (traditional) objectives of development. Tradeoffs may sometimes have to be made between the extent to and rate at which ecological sustainability is achieved *vis-à-vis* other objectives. In other cases, however, ecological sustainability and traditional developmental objectives (such as satisfaction of basic needs) could be mutually reinforcing. This interpretation of SD dominates the SD debate; I shall therefore focus on it in the rest of this paper.

3. THE CONCEPT OF SUSTAINABLE DEVELOPMENT

What are the traditional objectives of development, and how have they been expanded or modified to include sustainability? If the pursuit of traditional development objectives has undermined ecological sustainability in the past, what new insights suggest that such undermining or contradiction can be avoided now and in the future? How does this help build a working consensus between different fundamental concerns? In this section, I examine how the SD debate has addressed these questions.

(a) *Evolution of objectives*

The term sustainable development came into prominence in 1980, when the International Union for the Conservation of Nature and Natural Resources (IUCN) presented the World Conservation Strategy (WCS) with "the overall aim of achieving sustainable development through the conservation of living resources" (IUCN, 1980). Critics acknowledged that "by identifying Sustainable Development as the basic goal of society, the WCS was able to make a profound contribution toward reconciling the interests of the development community with those of the environmental movement" (Khosla, 1987). They pointed out, however, that the strategy

> restricted itself to living resources, focussed primarily on the necessity of maintaining genetic diversity, habits and ecological processes. . . . It was also unable to deal adequately with sensitive or controversial issues — those relating to the international economic and political order, war and armament, population and urbanization (Khosla, 1987).

Moreover, the WCS was "essentially supply-sided, [in that] it assumed the level and structure of demand to be an independent and autonomous variable," and ignored the fact that "if a sustainable style of development is to be pursued, then both the level and particularly the structure of demand must be fundamentally changed" (Sunkel, 1987). In short, the WCS had really addressed only the issue of ecological sustainability, rather than sustainable development.

Many have responded to such criticisms during the eight years since the WCS. The United Nations Environment Program (UNEP) was at the forefront of the effort to articulate and popularize the concept. UNEP's concept of SD was said to encompass

(i) help for the very poor, because they are left with no options but to destroy their environment;

(ii) the idea of self-reliant development, within natural resource constraints;

(iii) the idea of cost-effective development using nontraditional economic criteria;

(iv) the great issues of health control [*sic*], appropriate technology, food self-reliance, clean water and shelter for all; and

(v) the notion that people-centered initiatives are needed (Tolba, 1984a).

This statement epitomizes the mixing of goals and means, or more precisely, of fundamental objectives and operational ones, that has burdened much of the SD literature. While providing food, water, good health and shelter have traditionally been the fundamental objectives of most development models (including UNEP's), it is not clear whether self-reliance, cost-effectiveness, appropriateness of technology and people-centeredness are additional objectives or the operational requirements for achieving the traditional ones.

A similar proliferation of objectives was apparent at the IUCN-UNEP-World Wildlife Fund sponsored conference on Conservation and Development held in Ottawa in 1986. Summarizing the debate, the rapporteurs Jacobs, Gardner and Munro (1987) said that "Sustainable Development seeks . . . to respond to five broad requirements: (1) integration of conservation and development, (2) satisfaction of basic human needs, (3) achievement of equity and social justice, (4) provision of social self-determination and cultural diversity, and (5) maintenance of ecological integrity." The all-encompassing nature of the first requirement, and the repetitions and redundancies between some of the others were acknowledged by Jacobs, Gardner and Munro, but they did not suggest a better framework.

In contrast to the aforementioned, the currently popular definition of SD — the one adopted by the World Commission on Environment and Development (WCED) — is quite brief:

> Sustainable development is development that meets the needs of the present without compromising the ability of future generations to meet their own needs (WCED, 1987; p. 43).

The constraint of "not compromising the ability of future generations to meet their needs" is (presumably) considered by the Commission to be equivalent to the requirement of some level of ecological and social sustainability.[5]

While the WCED's statement of the fundamental objectives of SD is brief, the Commission is much more elaborate about (what are essentially) the operational objectives of SD. It states that "the critical objectives which follow from the concept of SD" are:

(1) reviving growth;

(2) changing the quality of growth;

(3) meeting essential needs for jobs, food, energy, water, and sanitation;

(4) ensuring a sustainable level of population;

(5) conserving and enhancing the resource base;

(6) reorienting technology and managing risk;

(7) merging environment and economics in decision making; and

(8) reorienting international economic relations (WCED, 1987, p. 49).

Most organizations and agencies actively promoting the concept of SD subscribe to some or all of these objectives with, however, the notable addition of a ninth operational goal, viz.,

(9) making development more participatory.[6]

This formulation can therefore be said to represent the mainstream of SD thinking. This "mainstream" includes international environmental agencies such as UNEP (e.g., Tolba, 1987), IUCN and the World Wildlife Fund (WWF), developmental agencies including the World Bank (e.g., Warford, 1986), the US Agency for International Development, the Canadian and Swedish international development agencies, research and dissemination organizations such as the World Resources Institute (Repetto, 1985, 1986a), the International Institute for Environment and Development, the Worldwatch Institute (1984–88), and activist organizations and groups such as the Global Tomorrow Coalition (GTC, 1988).

The logical connection between the brief definition of fundamental SD objectives and the list of operational ones is not completely obvious — mainly because many of the operational goals are not independent of the others. Nevertheless, it is possible to infer that "meeting the needs of the present generation" is operationally equivalent to WCED's first and third operational goals (reviving growth and meeting basic needs), while the need to maintain the ecological basis for the satisfaction of these objectives in perpetuity can be operationalized through the remaining goals (especially 2, 4, 5, 6 and 7).

(b) *The strength of the concept*

The strength of the concept of SD stems from

the choice of an apparently simple definition of fundamental objectives — meeting current needs and sustainability requirements — from which can be derived a range of operational objectives that cut across most previous intellectual and political boundaries. Repetto has tried to express this idea of SD as a powerful tool for consensus:

> [SD has] three bases . . . scientific realities, consensus on ethical principles, and considerations of long-term self-interest. There is a broad consensus that pursuing policies that imperil the welfare of future generations . . . is unfair. Most would agree that . . . consign[ing] a large share of the world's population to deprivation and poverty is also unfair. Pragmatic self-interest reinforces that belief. Poverty . . . underlies the deterioration of resources and the population growth in much of the world and affects everyone (Repetto, 1986a, p. 17).

"Pragmatic self interest," however, is as much an ethical value judgment as feelings of unfairness over poverty or over intergenerational inequity, while (presumably) scientific reality is not. Therefore, the above could be rephrased as:

> The current state of scientific knowledge (particularly insights obtained in the last few decades) about natural and social phenomena and their interactions leads inexorably to the conclusion that anyone driven by *either* long-term self interest, *or* concern for poverty, *or* concern for intergenerational equity should be willing to support the operational objectives of SD.

Assuming that concern for intergenerational equity coincides with broad environmental concerns, and adding concern for local participation to the list, this formulation of SD has, in theory, the potential for building a very broad and powerful consensus.

(c) The premises of SD

So what are these insights that appear to be pushing us toward such an operational consensus? Most people now admit that many human activities are currently reducing the long-term ability of the natural environment to provide goods and services, as well as adversely affecting current human health and well-being. Many would also accept that grinding poverty is devastating the lives of millions of individuals all over the world. But neither of these insights has been able to generate a consensus between those concerned about environmental issues and those focusing upon economic and developmental ones (or even within each of these groups).

The insights that have pushed us toward this consensus pertain to the feedback between social and environmental phenomena. There is now a growing consensus that "many environmental problems in developing countries originate from the *lack of development*, that is from the struggle to overcome extreme conditions of poverty" (Bartelmus, 1986, p. 18; emphasis added), that environmental degradation impoverishes those dependent directly on the natural environment for survival, and conversely, that development must be environmentally sound if it is to be permanent (Dampier, 1982). Thus, "environmental quality and economic development are interdependent and in the long term, mutually reinforcing" (Tolba, 1984b), and the question is no longer whether they contradict each other but how to achieve this (environmentally) sustainable (form of) development.

More precisely, the perception in mainstream SD thinking of the environment-society link is based upon the following premises:

(i) *Environmental degradation:*
— Environmental degradation is already affecting millions in the Third World, and is likely to severely reduce human well-being all across the globe within the next few generations.
— Environmental degradation is very often caused by poverty, because the poor have no option but to exploit resources for short-term survival.
— The interlinked nature of most environmental problems is such that environmental degradation ultimately affects everybody, although poorer individuals/nations may suffer more and sooner than richer ones.

(ii) *Traditional development objectives:*
— These are: providing basic needs and increasing the productivity of all resources (human, natural and economic) in developing countries, and maintaining the standard of living in the developed countries.
— These objectives do not necessarily conflict with the objective of ecological sustainability. In fact, achieving sustainable patterns of resource use is necessary for achieving these objectives permanently.
— It can be shown that, even for individual actors, environmentally sound methods are "profitable" in the long run, and often in the short run too.

(iii) *Process:*
— The process of development must be participatory to succeed (even in the short run).

Given these premises, the need for a process of development that achieves the traditional objectives, results in ecologically sustainable patterns of resource use, and is implemented in a participatory manner is obvious.

Most of the SD literature is devoted to showing that this process is also feasible and can be made attractive to the actors involved. SD has become a bundle of neat fixes: technological changes that make industrial production processes less polluting and less resource intensive and yet more productive and profitable, economic policy changes that incorporate environmental considerations and yet achieve greater economic growth, procedural changes that use local non-governmental organizations (NGOs) so as to ensure grassroots participation, agriculture that is less harmful, less resource intensive and yet more productive, and so on. In short, SD is a "metafix" that will unite everybody from the profit-minded industrialist and risk-minimizing subsistence farmer to the equity-seeking social worker, the pollution-concerned or wildlife-loving First Worlder, the growth-maximizing policy maker, the goal-oriented bureaucrat, and therefore, the vote-counting politician.

4. WEAKNESSES

The major impact of the SD movement is the rejection of the notion that environmental conservation necessarily constrains development or that development necessarily means environmental pollution — certainly not an insignificant gain. Where the SD movement has faltered is in its inability to develop a set of concepts, criteria and policies that are coherent or consistent — both externally (with physical and social reality) and internally (with each other). The mainstream formulation of SD suffers from significant weaknesses in:

(a) its characterization of the problems of poverty and environmental degradation;
(b) its conceptualization of the objectives of development, sustainability and participation; and
(c) the strategy it has adopted in the face of incomplete knowledge and uncertainty.

(a) Poverty and environmental degradation: An incomplete characterization

The fundamental premise of mainstream SD thinking is the two-way link between poverty and environmental degradation, shown schematically in Figure 2.

In fact, however, even a cursory examination of the vast amount of research that has been done on the links between social and environmental phenomena suggests that both poverty and environmental degradation have deep and complex causes. While substantive disagreements still exist regarding the primacy of these causes and the feasibility and efficacy of different remedies, the diagram in Figure 3 is probably a reasonable approximation of the general consensus on the nature of the causes and their links.[7]

To say that mainstream SD thinking has completely ignored these factors would be unfair. But it would be fair to say that it has focused on an eclectically chosen few. In particular, inadequate technical know-how and managerial capabilities, common property resource management, and pricing and subsidy policies (e.g., Repetto, 1986b; World Bank, 1987a) have been the major themes addressed, and the solutions suggested have been essentially techno-economic ones. This approach is reflected in the "principles" suggested for policy making, such as "designing for efficiency, proper resource pricing, managing common resources, attending to basics, and building management capability" (Repetto, 1986a, pp. 23–40). Deeper socio-political changes (such as land reform) or changes in cultural values (such as overconsumption in the North) are either ignored or paid lip-service.

This is not to say that problems of the global commons or of the lack of techno-managerial expertise are unimportant. But the intellectual discourse needs to begin with an acknowledgement that the big picture in Figure 3 (or something similar) essentially holds in all cases, and then proceed to developing analytical methods to help estimate the relative importance of each causal factor in specific cases and identify means of and scope for change.

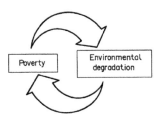

Figure 2. *The mainstream perception of the link between poverty and environmental degradation.*

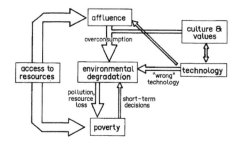

Figure 3. *A more realistic representation of the poverty-environmental degradation problem.*

(b) *Conceptual weaknesses*

Removal of poverty (the traditional developmental objective), sustainability and participation are really the three fundamental objectives of the SD paradigm. Unfortunately, the manner in which these objectives are conceptualized and operationalized leaves much to be desired. On the one hand, economic growth is being adopted as a major operational objective that is consistent with both removal of poverty and sustainability. On the other hand, the concepts of sustainability and participation are poorly articulated, making it difficult to determine whether a particular development project actually promotes a particular form of sustainability, or what kind of participation will lead to what kind of social (and consequently, environmental) outcome.

(i) *The role of economic growth*
By the mid-1970s, it had seemed that the economic growth and trickle-down theory of development had been firmly rejected, and the "basic needs approach" (Streeten, 1979) had taken root in development circles. Yet economic growth continues to feature in today's debate on SD. In fact, "reviving [economic] growth" heads WCED's list of operational objectives quoted earlier. Two arguments are implicit in this adoption of economic growth as an operational objective. The first, a somewhat defensive one, is that there is no fundamental contradiction between economic growth and sustainability, because growth in economic activity may occur simultaneously with either an improvement or a deterioration in environmental quality. Thus, "governments concerned with long-term sustainability need not seek to limit growth in economic output so long as they stabilize aggregate natural resource consumption" (Goodland and Ledec, 1987). But one could turn this argument around and suggest that, if economic growth is not correlated with environmental sustainability, there is no reason to have economic growth as an operational objective of SD.[8]

The second argument in favor of economic growth is more positive. The basic premise of SD is that poverty is largely responsible for environmental degradation. Therefore, removal of poverty (i.e., development) is necessary for environmental sustainability. This, it is argued, implies that economic growth is absolutely necessary for SD. The only thing that needs to be done is to "change the quality of [this] growth" (WCED, 1987, pp. 52–54) to ensure that it does not lead to environmental destruction. In drawing such an inference, however, there is the implicit belief that economic growth is necessary (if not sufficient) for the removal of poverty. But was it not the fact that economic growth *per se* could not ensure the removal of poverty that led to the adoption of the basic needs approach in the 1970s?

Thus, if economic growth by itself leads to neither environmental sustainability nor removal of poverty, it is clearly a "non-objective" for SD. The converse is a possibility worth exploring, viz., whether successful implementation of policies for poverty removal, long-term employment generation, environmental restoration and rural development will lead to growth in GNP, and, more important, to increases in investment, employment and income generation. This seems more than likely in developing countries, but not so certain in developed ones. In any case, economic growth may be the fallout of SD, but not its prime mover.

(ii) *Sustainability*
The World Conservation Strategy was probably the first attempt to carry the concept of sustainability beyond simple renewable resource systems. It suggested three ecological principles for ecological sustainability (see the nomenclature developed above), viz., "maintenance of essential ecological processes and life-support systems, the preservation of genetic diversity, and the sustainable utilization of species and resources" (IUCN, 1980). This definition, though a useful starting point, is clearly recursive as it invokes "sustainability" in resource use without defining it. Many subsequent attempts to discuss the notion are disturbingly muddled (e.g., Munn, 1988). There is a very real danger of the term becoming a meaningless cliché, unless a concerted effort is made to add precision and content to the discussion. While a detailed analysis of sustainability is given elsewhere (Lele, 1989), the following points may be made here.

Any discussion of sustainability must first

answer the questions "What is to be sustained? For whom? How long?" The value of the concept (like that of SD), however, lies in its ability to generate an operational consensus between groups with fundamentally different answers to these questions, i.e., those concerned either about the survival of future human generations, or about the survival of wildlife, or human health, or the satisfaction of immediate subsistence needs (food, fuel, fodder) with a low degree of risk. It is therefore vital to identify those aspects of sustainability that do actually cater to such diverse interests, and those that involve tradeoffs.

Differentiating between ecological and social sustainability could be a first step toward clarifying some of the discussion. Further, in the case of ecological sustainability, a distinction needs to be made between renewable resources, nonrenewable resources, and environmental processes that are crucial to human life, as well as to life at large. The few researchers who have begun to explore the idea of ecological sustainability emphasize its multidimensional and complex nature (e.g., Charoenwatana and Rambo, 1988).

In the context of sustainable use of renewable resources, it is necessary to go beyond the conventional simplistic notion of "harvesting the annual increment," and take into consideration the dynamic behavior of the resource, stochastic properties of and uncertainties about environmental conditions (e.g., climatic variations), the interactions between resources and activities (e.g., between forests, soils and agriculture), and between different uses or features of the "same" resource (e.g., tree foliage and stemwood).

In the rush to derive ecological principles of (ecological) sustainability, we cannot afford to lose sight of the social conditions that determine which of these principles are socially acceptable, and to what extent. Sociologists, eco-Marxists and political ecologists are pointing out the crucial role of socioeconomic structures and institutions in the pattern and extent of environmental degradation globally (also see discussion and notes in Section 4a). Neoclassical economists, whose theories have perhaps had the greatest influence in development policy making in the past and who therefore bear the responsibility for its social and environmental failures, however, have been very slow in modifying their theories and prescriptions. The SD movement will have to formulate a clear agenda for research in what is being called "ecological economics" (Ekins, 1986; Goodland and Ledec, 1987; Costanza, 1989) and press for its adoption by the mainstream of economics in order to ensure the possibility of real changes in policy making.

Social sustainability is a more nebulous concept than ecological sustainability. Brown et al. (1987), in a somewhat techno-economic vein, state that sustainability implies "the existence and operation of an infrastructure (transportation and communication), services (health, education, and culture), and government (agreements, laws, and enforcement)." Tisdell (1988) talks about "the sustainability of political and social structures" and Norgaard (1988) argues for cultural sustainability, which includes value and belief systems. Detailed analyses of the concept, however, seem to be nonexistent.[9] Perhaps achieving desired social situations is itself so difficult that discussing their maintainability is not very useful; perhaps goals are even more dynamic in a social context than in an ecological one, so that maintainability is not such an important attribute of social institutions/ structures. There is, however, no contradiction between the social and ecological sustainability; rather, they can complement and inform each other.

(iii) *Participation*

A notable feature of "ecodevelopment" — SD's predecessor — as well as some of the earlier SD literature was the emphasis placed on equity and social justice. For instance, in the IUCN-sponsored conference in Ottawa in 1986, "advancing equity and social justice [was perceived to be] so important" that the phrase used was "sustainable and equitable development" (Jacobs, Gardner and Munro, 1987). Subsequently, however, the mainstream appears to have quietly dropped these terms (suggesting at least a deemphasizing of these objectives), and has instead focused on "local participation."

There are, however, three problems with this shift. First, by using the terms equity, participation and decentralization interchangeably, it is being suggested that participation and decentralization are equivalent, and that they can somehow substitute for equity and social justice. This suggestion is at best a naive one. While all of these concepts are quite complex, it seems clear that some form of participation is necessary but not sufficient for achieving equity and social justice.

Second, the manner in which participation is being operationalized shows up the narrowminded, quick-fix and deceptive approach adopted by the mainstream promoters of SD. Cohen and Uphoff (1980) distinguished four types of participation — in decision making, implementation, benefit distribution and evaluation. Most of the SD literature does not make these distinctions at all. Mainstream SD literature blithely assumes and insists that "involvement of local NGOs" in project implementation will ensure project success (Maniates, 1990; he dubs this the "NGOization" of SD).

Third, there is an assumption that participation or at least equity and social justice will necessarily reinforce ecological sustainability. Attempts to test such assumptions rigorously have been rare. But preliminary results seem to suggest that equity in resource access may not lead to sustainable resource use unless new institutions for resource management are carefully built and nurtured. For instance, Jodha (1987) describes how land reform in Rajasthan (India) led to the neglect of village pastures that were well-maintained under the earlier feudal structure. Similarly, communal irrigation tanks in Tamil Nadu (India) fell into disrepair with the reduction in the feudal powers of the village landlords (von Oppen and Subba Rao, 1980). This should not be misconstrued as an argument against the need for equity, but rather as a word of caution against the tendency to believe that social equity automatically ensures environmental sustainability (or vice-versa).

5. POLICY PRESCRIPTIONS — INCONSISTENCIES AND INADEQUACIES

Given this confusion in terms, perceptions, and concepts, the policies being suggested by the mainstream of SD thinking cannot and do not conform to the basic idea of ecologically sound and socially equitable development. They are often seriously flawed, and reflect personal, organizational and political preferences. I use examples from three SD issues — international trade, agriculture, and tropical forests — to illustrate this argument.

(a) International trade and economic relations

Trade, multinational corporations, commercial lending and aid are the four dominant channels through which international economic relations manifest themselves today. That the manner in which these activities are currently pursued contradicts ecological sustainability and social well-being is being increasingly pointed out (Redclift, 1987; Rainforest Action Network, 1987). This is even being acknowledged in some cases by the parties involved (World Bank, 1987b).

Yet, the SD debate and policy prescriptions regarding international trade continue to be fundamentally flawed in two ways. First, the need to ensure a truly equitable basis for exchange by restructuring the international monetary system is completely ignored. Makhijani and Browne (1986) have described how the First World has in the past manipulated the system of exchange rates in order to maintain favorable terms of trade for itself. The International Monetary Fund (IMF) and the SD-promoting World Bank, however, continue to emphasize structural readjustment policies for debt-ridden Third World countries that include downward adjustment of exchange rates: in effect, maintaining the terms of trade of the colonial era.

Second, there seems to be a broad consensus cutting across almost all political and intellectual boundaries that free trade (interpreted as trade without any import or export barriers) is crucial to promoting SD. The Brandt Commission (1980 and 1983) argued that "the solution [to the dual problems facing the North and the South] is to make the South richer through increased trade." The WCED report cites protectionism as a major impediment to sustainable development (WCED, 1987, p. 83), and urges the removal of all such barriers. But this prescription is fundamentally flawed on the counts of development as well as sustainability.

In a succinct statement of the argument on the first count, Redclift (1987, p. 57) points out that although the neoclassical economics case is that "gains from [free] trade" outweigh the losses, (i) neoclassical theory itself acknowledges that the gains from trade may be very unevenly distributed between countries, (ii) in practice, there may be losers as well as gainers, and (iii) while freer trade will presumably stimulate economic growth, the assertion that economic growth is (socially) beneficial is questionable, to say the least.

Simultaneously, the contradictions between the neoclassical theory of international trade and ecological sustainability are being pointed out. Norgaard has described the impact of the modern globally integrated economy on biodiversity:

> [Comparative advantage, the efficiency of specialization, and the gains through exchange have] affected [biological] diversity in two ways. First, . . . development through capturing the gains of exchange has encouraged specialization and a reduction in crop and supporting species . . . [Second,] variation in aggregate economic welfare is reduced through increased variations in the activities of individual actors. This increased variation imposes stress on biological species that leads to extinction (Norgaard, 1987).

Further, he states, neoclassical trade theory assumes "that factors of production are mobile (that labor, capital and land can shift between lines of production, that labor can move to new locations). . . . *But environmental services which give land its value cannot freely shift from one product to another*" (Norgaard, 1987; emphasis added).

Examining the controversy over the Free Trade Agreement between the United States and Canada, McRobert (1988) makes the additional point that economics research on trade policy has studiously ignored the massive hidden environmental externalities (in the form of pollution and climate change) of the transportation implicit in international trade.

(b) *Sustainable agriculture: What? How?*

Agriculture is one of the foundations of human society and a major activity at the human-environment interface. Attempts to operationalize ecological sustainability have therefore focused significantly on agriculture. Unfortunately, while the literature on sustainable agriculture or SD in agriculture is proliferating, it is marked by the same confusion that afflicts the larger debate on sustainability. This confusion is obvious from the manner in which the terms "sustainable agriculture," "low-input agriculture" and "organic farming" are being used interchangeably when they actually differ significantly (Buttel and Gillespie, Jr., 1988). "Agroecology" is being proposed as the foundation for sustainable agriculture (Dover and Talbot, 1987), but it lacks a firm, consensual theoretical and practical framework. Moreover, the ability of a pattern of agriculture to simultaneously provide fair returns to the farmer and laborer, and to satisfy the needs of the nonagricultural population in an ecologically sound manner depends not only on ecological interactions but also on complex social conditions — conditions that are even less well understood today.

Struggling with these inadequacies, the SD movement has been slow in coming up with a clear definition of and agenda for sustainable agriculture. This has not only hampered efforts to redirect international agencies' policies (e.g., Committee on Agricultural Sustainability in Developing Countries, 1987) but, more important, it has also allowed the conventional Green Revolution experts to sell their old wine in the new bottle of "sustainable agriculture." At the World Bank workshops on SD in agriculture (Davis and Schirmer, 1987) most "experts" interpreted sustainability in agriculture as simply maintaining growth in agricultural production! Other research efforts (as in Parikh, 1987)

appear to be largely limited to designing and validating conventional crop production models, with simple environmental feedbacks added almost as an afterthought.

Not surprisingly, agricultural policy statements by the SD mainstream often give contradictory messages. For instance, the WCED report acknowledges that the increases in agricultural production in the Green Revolution have occurred through a nine-fold increase in fertilizer consumption, with reducing marginal gains, and at the cost of significant soil salinization and pollution. Nevertheless, it concludes that "many countries should increase yields by greater use of chemical fertilizers and pesticides, [although] countries can also improve yields by helping farmers use organic nutrients more efficiently" (WCED, 1987, p. 135).

(c) *Tropical Forests*

Tropical deforestation has been an item on the agenda of First World environmentalists for a long time (e.g., Myers, 1984). Rooted initially almost wholly in concern about wildlife and biological diversity, the movement to save the world's tropical forests broadened as the understanding of the phenomenon became more sophisticated in terms of the social context of forest use and the political economy of deforestation.

But when the World Resources Institute, the World Bank and the UNDP proposed their action plan for tropical forests (WRI, 1985), this plan — and a similar one outlined by the FAO (1985) — were both heavily criticized on exactly the same grounds of analytical incompleteness discussed earlier in Section 4. It suffices to quote Hildyard (1987):

> The WRI report is deeply flawed. . . . [First], it is based on the premise that poverty, over-population and ignorance are the prime cause of forest destruction. But making scapegoats of the poor and dispossessed not only obscures the reasons for their poverty but detracts from the real causes for deforestation, viz., the massive commercial development schemes being promoted in the Third World. [Second,] blaming the poor for deforestation also overlooks the fact that millions of peasant colonists have been actively encouraged to invade [forests] under government-sponsored colonization schemes. [Third,] blaming poverty also ignores the fact that best protected forests of the world are inhabited by those very tribal peoples who, by the standards of industrialized man, are among the world's poorest. [Fourth,] blaming the poor also serves to rationalise and hence legitimize the view that current development policies can (and should)

continue unabated. Indeed, the WRI plan goes further [and] interprets the problem in such a way as to justify socially and ecologically destructive [though politically and economically expedient] schemes.

Ross and Donovan (1986) present a similar, though milder, critique. Clearly, some serious introspection and rethinking is needed in the SD community on this issue.

6. CONCLUDING REMARKS: DILEMMAS AND AGENDAS

The proponents of SD are faced with a dilemma that affects any program of political action and social change: the dilemma between the urge to take strong stands on fundamental concerns and the need to gain wide political acceptance and support. Learning from the experience of ecodevelopment, which tended toward the former, SD is being packaged as the inevitable outcome of objective scientific analysis, virtually an historical necessity, that does not contradict the deep-rooted normative notion of development as economic growth. In other words, SD is an attempt to have one's cake and eat it too.

It may be argued that this is indeed possible, that the things that are wrong and need to be changed are quite obvious, and there are many ways of fixing them without significantly conflicting with either age-old power structures or the modern drive for a higher material standard of living. Therefore, it is high time that environmentalists and development activists put aside their differences and joined hands under the banner of sustainable development to tackle the myriad of problems facing us today. If, by using the politically correct jargon of economic growth and development and by packaging SD in the manner mentioned above, it were possible to achieve even 50% success in implementing this bundle of "conceptually imprecise" policies, the net reduction achieved in environmental degradation and poverty would be unprecedented.

I believe, however, that (analogous to the arguments in SD) in the long run there is no contradiction between better articulation of the terms, concepts, analytical methods and policymaking principles, and gaining political strength and broad social acceptance — especially at the grassroots. In fact, such clarification and articulation is necessary if SD is to avoid either being dismissed as another development fad or being coopted by forces opposed to changes in the status quo. More specifically, proponents and analysts of SD need to:

(a) clearly reject the attempts (and temptation) to focus on economic growth as a means to poverty removal and/or environmental sustainability;

(b) recognize the internal inconsistencies and inadequacies in the theory and practice of neoclassical economics, particularly as it relates to environmental and distributional issues; in economic analyses, move away from arcane mathematical models toward exploring empirical questions such as limits to the substitution of capital for resources, impacts of different sustainability policies on different economic systems, etc.;

(c) accept the existence of structural, technological and cultural causes of both poverty and environmental degradation; develop methodologies for estimating the relative importance of and interactions between these causes in specific situations; and explore political, institutional and educational solutions to them;

(d) understand the multiple dimensions of sustainability, and attempt to develop measures, criteria and principles for them; and

(e) explore what patterns and levels of resource demand and use would be compatible with different forms or levels of ecological and social sustainability, and with different notions of equity and social justice.

There are, fortunately, some signs that a debate on these lines has now begun (see, e.g., the December 1988 issue of *Futures*; also SGN, 1988, and Daly, 1991).

In a sense, if SD is to be really "sustained" as a development paradigm, two apparently divergent efforts are called for: making SD more precise in its conceptual underpinnings, while allowing more flexibility and diversity of approaches in developing strategies that might lead to a society living in harmony with the environment and with itself.

NOTES

1. For instance, the International Institute for Applied Systems Analysis published a collection of papers entitled "Sustainable Development of the Biosphere" (Clark and Munn, 1986). But nowhere in this large

volume was there any attempt to define development, sustainability, or sustainable development.

2. This is not the first review of the SD literature. Since the middle of 1987, reviews have appeared in at least three journal articles (Brown *et al.*, 1987; Barbier, 1987; Tisdell, 1988) and one book (Redclift, 1987) — in itself a striking indication of the proliferation of the SD literature. But while these authors — Redclift and Barbier in particular — have contributed to the discussion on SD, a comprehensive review of the SD literature that critically examines the semantic and conceptual issues is still lacking.

3. This indicates the necessarily subjective starting point of this analysis, viz., that I find at least some of the arguments being made by the SD movement (outlined later) more plausible than the arguments of those who would have us believe that no major environmental and social problems confront us, and/or if they do, that no major shifts in our thinking and in our individual and collective behavior and policy making will be called for to navigate through these problems.

4. More precisely, there are ultimate limits to the stocks of material resources, the flows of energy resources, and (in the event of these being circumvented by a major breakthrough in fission/fusion technologies) to the environment's ability to absorb waste energy and other stresses. The limits-to-growth debate, while not conclusive as to specifics, appears to have effectively shifted the burden of proof about the absence of such fundamental limits onto the diehard "technological optimists" who deny the existence of such limits.

5. Of course, "meeting the needs" is a rather ambiguous phrase that may mean anything in practice. Substituting this phrase with "optimizing economic and other societal benefits" (Goodland and Ledec, 1987) or "managing all assets, natural resources and human resources, as well as financial and physical assets for increasing long-term wealth and well-being (Repetto, 1986a, p. 15) does not define the objectives of development more precisely, although the importance attached to economic benefits or wealth is rather obvious.

6. It is tempting to conclude that this nine-point formulation of SD is identical with the concept of "ecodevelopment" — the original term coined by Maurice Strong of UNEP for environmentally sound development (see Sachs, 1977 and Riddell, 1981). Certainly the differences are less obvious than the similarities. Nevertheless, some changes are significant — such as the dropping of the emphasis on "local self-reliance" and the renewed emphasis on economic growth.

7. The diagram is necessarily unsatisfactory and incomplete, since the problem is basically not amenable to neat representation. Its only purpose is to illustrate the importance of access to or control over resources and technological and cultural factors in influencing (if not determining) *both* poverty and environmental degradation. Redclift (1987), Blaikie (1985), Blaikie and Brookfield (1987), and Little and Horowitz (1987) contain elaborations of this theme. Grossman (1984) and Hecht (1985) are examples of region-specific analyses. Eckholm (1976) typifies simpler analyses that focus on poverty and population growth.

8. Economists have responded by suggesting that currently used indicators of economic growth (GNP in particular) could be modified so as to somehow "build in" this correlation (e.g., Peskin, 1981). To what extent this is possible and whether it will serve more than a marginal purpose are, however, open questions (Norgaard, 1989).

9. Three other "social" usages of sustainability need to be clarified. Sustainable economy (Daly, 1980) and sustainable society (Brown, 1981) are two of these. The focus there, however, is on the patterns and levels of resource use that might be ecologically sustainable while providing the goods and services necessary to maintain human well-being, and the social reorganization that might be required to make this possible. The third usage is Chambers' definition of "sustainable livelihoods" as "a level of wealth and of stocks and flows of food and cash which provide for physical and social well-being and security against becoming poorer" (Chambers, 1986). This can be thought of as a sophisticated version of "basic needs", in that security or risk-minimization is added to the list of needs. It is therefore relevant to any paradigm of development, rather than to SD in particular.

REFERENCES

Arndt, H. W., "Economic development: a semantic history," *Economic Development and Cultural Change*, Vol. 29, No. 3 (1981), pp. 457–466.

Barbier, E. B., "The concept of sustainable economic development," *Environmental Conservation*, Vol. 14, No. 2 (1987), pp. 101–110.

Bartelmus, P., *Environment and Development* (London: Allen & Unwin, 1986).

Blaikie, P., *Political Economy of Soil Erosion in Developing Countries* (London: Longman, 1985).

Blaikie, P., and H. Brookfield (Eds.), *Land Degradation and Society* (New York: Methuen, 1987).

Brandt Commission, *Common Crisis* (London: Pan Books, 1983).

Brandt Commission, *North-South: A Programme for Survival* (London: Pan Books, 1980).

Brown, L. R., *Building a Sustainable Society* (New York: W. W. Norton, 1981).

Brown, B. J., M. Hanson, D. Liverman, and R. Merideth, Jr., "Global sustainability: Toward definition," *Environmental Management*, Vol. 11, No. 6 (1987), pp. 713–719.

Buttel, F. H., and G. W. Gillespie Jr., "Agricultural research and development and the appropriation of progressive symbols: Some observations on the politics of ecological agriculture," Bulletin no. 151 (Ithaca, NY: Department of Rural Sociology, Cornell University, 1988).

Chambers, R., *Sustainable Livelihoods: An opportunity for the World Commission on Environmental and Development* (Brighton, UK: Institute of Development Studies, University of Sussex, 1986).

Charoenwatana, T., and A. T. Rambo, "Preface," in T. Charoenwatana and A. T. Rambo (Eds.), *Sustainable Rural Development in Asia* (Khon Kaen, Thailand: KKU-USAID Farming Systems Research Project, Khon Kaen University and Southeast Asian Universities Agroecosystem Network, 1988), pp. vii–x.

Clark, W. C., and R. E. Munn (Eds.), *Sustainable Development of the Biosphere* (Cambridge: Cambridge University Press, 1986).

Cohen, J., and N. Uphoff, "Participation's place in rural development: Seeking to clarify through specificity," *World Development*, Vol. 8 (1980), pp. 213–235.

Committee on Agricultural Sustainability in Developing Countries, *The Transition to Sustainable Agriculture: An agenda for AID* (Washington, DC: International Institute for Environment and Development, 1987).

Conable, B., *Address to the Board of Governors of the World Bank and the International Finance Corporation, 30 September 1986* (Washington, DC: World Bank, 1986).

Costanza, R., "What is ecological economics?" *Ecological Economics*, Vol. 1, No. 1 (1989), pp. 1–8.

Daly, H., "Sustainable development: From concept and theory towards operational principles," in H. E. Daly, *Steady-state Economics: 2nd Edition with New Essays* (Washington, DC: Island Press, 1991).

Daly, H., *Economics, ecology, ethics: Essays toward a steady-state economy* (San Francisco: W. H. Freeman, 1980).

Dampier, W., "Ten years after Stockholm: A decade of environmental debate," *Ambio*, Vol. 11, No. 4 (1982), pp. 215–231.

Davis, T. J., and I. A. Schirmer (Eds.), *Sustainability Issues in Agricultural Development* (Washington, DC: World Bank, 1987).

Dover, M., and L. M. Talbot, *To Feed the Earth: Agro-Ecology for Sustainable Development* (Washington, DC: World Resources Institute, 1987).

Eckholm, E., *Losing Ground: Environmental Stress and World Food Prospects* (New York: W. W. Norton, 1976).

Ekins, P. (Ed.), *The Living Economy: A New Economics in the Making* (London: Routledge & Kegan Paul, 1986).

Environment Committee, *Environmental Assessment and Development Assistance: Final Report of the Ad-Hoc Group, 26 November 1985* (Paris: Organization for Economic Cooperation and Development, 1985).

FAO Committee on Forest Development in the Tropics, *Tropical Forestry Action Plan* (Rome: UN Food and Agriculture Organization, 1985).

GTC, *Making Common Cause: A Statement and Action Plan by U.S.-based Development, Environment, and Population NGOs* (New York: Public Education Working Group of the Global Tomorrow Coalition, 1988).

Goodland, R., and G. Ledec, "Neoclassical economics and principles of sustainable development," *Ecological Modelling*, Vol. 38 (1987), pp. 19–46.

Grossman, L. S., *Peasants, Subsistence Ecology, and Development in the Highlands of Papua New Guinea* (Princeton, NJ: Princeton University Press, 1984).

Hecht, S. B., "Environment, development, and politics: Capital accumulation and the livestock sector in Latin America," *World Development*, Vol. 13 (1985), pp. 663–684.

Hildyard, N., "Tropical forests: A plan for action," *The Ecologist*, Vol. 17, No. 4/5 (1987), pp. 129–133.

Hopper, W. D., "Sustainability, policies, natural resources and institutions," in T. J. Davis and I. A. Schirmer (Eds.), *Sustainability Issues in Agricultural Development* (Washington, DC: World Bank, 1987), pp. 5–16.

IUCN, *World Conservation Strategy: Living Resource Conservation for Sustainable Development* (Gland, Switzerland: International Union for Conservation of Nature and Natural Resources, United Nations Environment Program and World Wildlife Fund, 1980).

Idachaba, F. S., "Sustainability issues in agriculture development," in T. J. Davis and I. A. Schirmer (Eds.), *Sustainability Issues in Agricultural Development* (Washington, DC: World Bank, 1987), pp. 18–53.

Jacobs, P., J. Gardner, and D. Munro, "Sustainable and equitable development: An emerging paradigm," in P. Jacobs and D. A. Munro (Eds.), *Conservation with Equity: Strategies for Sustainable Development* (Cambridge: International Union for Conservation of Nature and Natural Resources, 1987), pp. 17–29.

Jodha, N. S., "A case study of the decline of common property resources in India," in P. Blaikie and H. Brookfield (Eds.), *Land Degradation and Society* (New York: Methuen, 1987), pp. 196–207.

Khosla, A., "Alternative strategies in achieving sustainable development," in P. Jacobs and D. A. Munro (Eds.), *Conservation with Equity: Strategies for Sustainable Development* (Cambridge: International Union for Conservation of Nature and Natural Resources, 1987), pp. 191–208.

Lélé, S., "A framework for sustainability and its application in visualizing a peace society," ExPro Working Paper (Chestnut Hill, Maryland: Exploratory Project on the Conditions of Peace, 1989).

Lélé, S., "The concept of sustainability," Paper presented at the Interdisciplinary Conference on Natural Resource Modelling and Analysis (Halifax, Canada: September 29–October 1, 1988).

Little, P. D. and M. M. Horowitz (Eds.), *Lands At Risk in the Third World: Local-Level Perspectives* (Boulder, CO: Westview Press, 1987).

Makhijani, A., and R. S. Browne, "Restructuring the international monetary system," *World Policy Journal*, Vol. 4, No. 1 (1986), pp. 61–80.

Maniates, M., "Organizing for rural energy development: Local organizations, improved cookstoves, and the state in Gujarat, India," PhD thesis, (Berkeley: Energy & Resources Group, University of California, 1990).

McRobert, D., "Questionable faith," *Probe Post*, Vol. 11, No. 1 (1988), pp. 24–29.

Munn, R. E., "Towards sustainable development: An environmental perspective," Paper presented at International Conference on Environment and Development (Milan, Italy: March 24–26, 1988).

Myers, N., *The Primary Source: Tropical Forests and Our Future* (New York: Norton, 1984).

Norgaard, R. B., "Three dilemmas of environmental accounting," *Ecological Economics*, Vol. 1, No. 4 (1989), pp. 303–314.

Norgaard, R. B., "Sustainable development: A coevolutionary view," *Futures*, Vol. 20, No. 6 (1988), pp. 606–620.

Norgaard, R. B., "Economics as mechanics and the demise of biological diversity," *Ecological Modelling*, Vol. 38 (1987), pp. 107–121.

O'Riordan, T., "Future directions in environmental policy," *Journal of Environment and Planning*, Vol. 17 (1985), pp. 1431–1446.

Parikh, J. K. (Ed.), *Sustainable Development in Agriculture* (Dordrecht, Netherlands: Martinus Nijhoff, 1988).

Peskin, H. M., "National Income Accounts and the Environment," *Natural Resources Journal*, Vol. 21 (1981), pp. 511–537.

Pezzey, J., "Economic analysis of sustainable growth and sustainable development," Environment Department Working Paper No. 15 (Washington, DC: World Bank, 1989).

Rainforest Action Network, *Financing Ecological Destruction: The World Bank and the International Monetary Fund* (San Francisco: The Rainforest Action Network, 1987).

Redclift, M., *Sustainable Development: Exploring the Contradictions* (New York: Methuen, 1987).

Repetto, R., *World Enough and Time* (New Haven, CT: Yale University Press, 1986a).

Repetto, R., *Economic Policy Reforms for Natural Resource Conservation* (Washington, DC: World Resources Institute, 1986b).

Repetto, R. (Ed.), *The Global Possible* (New Haven, CT: Yale University Press, 1985).

Riddell, R., *Ecodevelopment* (New York: St. Martin's Press, 1981).

Ross, M. S. and D. G. Donovan, "The world tropical forestry action plan: Can it save the tropical forests?" *Journal of World Forest Resource Management*, Vol. 2 (1986), pp. 119–136.

Runnalls, D., *Factors Influencing Environmental Policy in International Development Agencies* (Manila: Asian Development Bank, 1986).

Sachs, I., *Environment and Development — A New Rationale For Domestic Policy Formulation and International Cooperation Strategies* (Ottawa: Environment Canada and Canadian International Development Agency, 1977).

SGN, "Perspectives of sustainable development: Some critical issues related to the Brundtland report," Stockholm Studies on Natural Resources Management No. 1 (Stockholm: Stockholm Group for Studies on Natural Resources Management, 1988).

Streeten, P., "Basic needs: premises and promises," *Journal of Policy Modelling*, Vol. 1 (1979), pp. 136–146.

Sunkel, O., "Beyond the world conservation strategy: Integrating development and the environment in Latin America and the Caribbean," in P. Jacobs and D. A. Munro (Eds.), *Conservation with Equity: Strategies for Sustainable Development* (Cambridge: International Union for Conservation of Nature and Natural Resources, 1987), pp. 35–54.

Tisdell, C., "Sustainable development: Differing perspectives of ecologists and economists, and relevance to LDCs," *World Development*, Vol. 16, No. 3 (1988), pp. 373–384.

Tolba, M. K., *Sustainable Development: Constraints and opportunities* (London: Butterworths, 1987).

Tolba, M. K., *The premises for building a sustainable society — Address to the World Commission on Environment and Development, October 1984* (Nairobi: United Nations Environment Programme, 1984a).

Tolba, M. K., *Sustainable development in a developing economy — Address to the International Institute, Lagos, Nigeria, May 1984* (Nairobi: United Nations Environment Programme, 1984b).

von Oppen, M. and K. V. Subba Rao, *Tank Irrigation in Semi-Arid India* (Patancheru, Andhra Pradesh, India: International Crop Research Institute for the Semi-Arid Tropics, 1980).

WRI, *Tropical Forests: A Call for Action* (Washington, DC: World Resources Institute, 1985).

Warford, J., "Natural resource management and economic development," Projects Policy Department Working Paper (Washington, DC: World Bank, 1986).

World Bank, "Environment, growth and development," Development Committee Pamphlet No. 14 (Washington, DC: World Bank, 1987a).

World Bank, "Conable announces new steps to protect environment in developing countries," World Bank News Release no. 87/28 (Washington, DC: World Bank, 1987b).

World Commission on Environment and Development, *Our Common Future* (New York: Oxford University Press, 1987).

Worldwatch Institute, *State of the World* (New York: Norton, various years).

3
CHANGE AS A
COEVOLUTIONARY PROCESS

[Interesting philosophy] says things like "try thinking of it this way"—or more specifically, "try to ignore the apparently futile traditional questions by substituting the following new and possibly interesting questions."
(Richard Rorty 1989:9)

The real challenge of sustainability is to reframe the challenge. As conventionally understood, sustainable development contests our competence to predict the consequences of our interactions with nature and taxes our capability to control those interactions so that the old idea of development remains intact yet is sustainable. I trust, however, after elaborating on this framing of the challenge in the previous chapter, that it is clear that this challenge cannot be met. The world is far too complex for us to perceive and establish the conditions for sustainability. Those who—on realizing our limited ability to perceive and control—tout the use of markets fail to realize that the objectives that markets reach depend on the system of property rights underlying their performance. The design of the appropriate system of rights not only requires equivalent prescience but presumes a static world. Indeed, even the elaboration in the last chapter, with all of its complications, never went beyond assuming that societies and environments are static, complex systems.

This chapter initiates the framing of societies and environments as coevolving systems. By stressing complex processes instead of complex structures, the challenge of sustainability emerges anew. The new features and associations of this emerging challenge eventually lead to a revisioning of progress.

THE COEVOLUTION OF PESTS,
PESTICIDES, POLITICS, AND POLICY

The pesticide story in the United States during the twentieth century provides an excellent example of the coevolutionary process. Let's first recount key historical details of the interplay between pests, pesticides, politics, and policy with as little revolutionary language as possible, and then show how the coevolutionary framing puts it into perspective.

Prior to World War II, inorganic compounds such as arsenic, sulfur, and lead were used to control insects and other pests. The Pesticide Act of 1910 protected farmers from ineffective products. The Pure Food and Drug Act of 1906 protected consumers from contamination, and pesticide residues were specifically included in the Food, Drug, and Cosmetic Act of 1938. Regulation was seen as a matter of "truth in advertising," of seeing that farmers were getting useful chemicals and consumers were getting healthy food. The Federal Insecticide, Fungicide, and Rodenticide Act of 1947 expanded the range of products covered, but was still largely designed to facilitate the chemical industry and protect farmers from ineffective products.

The discovery of DDT in 1939, followed by other organochlorine insecticides soon after, and their expanding use after World War II changed the dynamics dramatically. By the early 1950s, the organic insecticides had driven inorganics nearly off the market because the organics were really effective initially. And because they were more effective, they were used on more crops and pests than the inorganics. They were so effective that they set in motion an interacting set of events that proved instrumental to our understanding of environmental problems.

The few insects that survived the application of DDT and other organochlorine pesticides were the individuals among the larger population who were the most resistant to the pesticide. When these surviving individuals reproduced, a high proportion of their offspring carried the genetic traits that favored resistance. Since most insects have many generations per season, the selective pressure of the insecticides on the evolution of resistance in insects was observable over a matter of years. I use evolutionary terminology here because evolution as biologists understand it was precisely the process.

In parallel with the problem of resistance were the problems of secondary pests and resurgence. Secondary pests are other species that can play a similar role, or fill a similar agroecological niche, as the initial pest. Reducing the population of the primary pest through the use of pesticides leaves an unfilled niche for secondary pests to fill. Secondary pests might come from neighboring fields or have been present in the sprayed field but in a phase of their life cycle that made them less susceptible at the time of the spraying. Resurgence also relates to the unfilled niche after spraying. Whatever primary pests are left after spraying have little competition after the demise of their cohorts and hence their populations rebuild, or resurge, rapidly. In the event that spraying has also reduced the predators of pests, or alternative prey of the predators, pest populations return even faster. Ecological language is appropriate here for the dynamics are characteristic of disturbed ecosystems.

The response of agricultural researchers and the chemical industry to the occurrence of greater pest problems after the initial success of organic pesticides was to recommend more frequent and heavier spraying. More pests demand more pesticides. This, of course, made sense for each individual farmer, but compounded the problems of resistance for farmers collectively. And, of course, heavier and more frequent spraying resulted in higher pest management costs, but now there was little choice. Many sensed they were on a "pesticide treadmill," but few could see it or how to get off it.

A few researchers were looking at the situation from an evolutionary and ecosystemic perspective and began to advocate more careful monitoring of pest populations before deciding whether to spray, the use of biological controls, and the collective selection of crop types, planting dates, and other practices to outwit insects. These integrated pest management programs included the use of chemicals, but at significantly lower levels. Some farmers adopted the integrated pest management philosophy of agroecosystem management, but the vast majority of farmers just wanted to kill pests.

The reduced effectiveness of the early organics opened up new opportunities for the chemical industry to introduce new insecticides and many were tried. Organophosphates and carbamates soon proved advantageous because of their higher acute toxicity and lower persistance in the environment. Higher acute toxicity was advantageous in that fewer individuals survived, slowing the evolution of resistance. Concern with the persistance of DDT was building with evidence that DDT, other organophosphates, and the derivative chemicals resulting from breakdown in the environment were affecting insect populations and hence bird populations beyond agriculture. Furthermore, these chemicals were accumulating in the food chain with unknown consequences. Rachel Carson and others stimulated a new environmental consciousness during the 1960s which eventually led to the Federal Environmental Pesticide Control Act (FEPCA) of 1972 with its provisions to protect the environment. Soon after DDT was banned and the elimination and regulation of other chemicals followed during the 1970s.

Organophosphates and carbamates, however, had new problems. Organophosphates such as parathion are deadly to people, carbamates to bees, and farmworkers and bees are essential to agriculture itself. Farm-workers and beekeepers joined environmentalists in the early 1970s in seeking protective regulatory decisions. By the later 1970s, communities in agricultural regions began to complain of illnesses, and agricultural chemicals' were found to be accumulating in groundwater used for drinking supplies. The increasing awareness of the effects of pesticides on other species, on workers' health, and on the health of adjacent communities intensified the interest in integrated pest management as an alternative.

The chemical industry's response was slowed by the difficulties of getting new chemicals registered and approved for use under the stricter requirements of FEPCA and the administrative procedures of the Environmental Protection Agency. More testing and registration delays also raised the costs of insecticides. The structure of the industry also changed as smaller chemical companies found they could not operate under the new restrictions, as moderate sized firms merged, and even as some larger ones left the field. Industry slowly responded, however, by providing insecticides that targeted narrower ranges of pests to reduce disrupting beneficial insects and which were much less toxic to people. These chemicals, largely synthetic pyrethroids modeled after some of nature's own insecticides, were far more expensive and their application frequently had to be more carefully timed. While farmers could afford these more expensive alternatives when farm prices were high in the mid-1970s, farm prices dropped significantly in the late 1970s and early 1980s and farmers became desperate. In spite of a long tradition of supporting chemical agriculture, in the 1980s, the U.S. Department of Agriculture initiated a program known as LISA (low input sustainable agriculture), which incorporated much of the philosophy and many of the techniques of integrated pest management.

The use of insecticides in U.S. agriculture is still very high, farmers are trying harder than ever to use them well, and other insect control techniques have been introduced as well. Nevertheless, crop losses to insects are about the same as they were before the use of modern insecticides. But we cannot simply stop using them because our agroecosystems and agroeconomy have been transformed by their use such that they must be used. All parties have suffered from unforeseen consequences of insecticide use which has led to a greater consensus on their environmental, social, and economic implications. If we had been able to foresee the diverse twists and turns of the pesticide story, and the foregoing only includes the most essential details, we would not have started using them in the first place.

We could not, however, have foreseen how the pesticide story unfolded because history is not deterministic; it is not like a missile on a predetermined course or even like a complex machine whose movements can be comprehended and thereafter forever predicted. Nevertheless, the changes that took place can be explained as a process of coevolution.

In biology, coevolution refers to the pattern of evolutionary change of two closely interacting species where the fitness of the genetic traits within each species is largely governed by the dominant genetic traits of the other. Coevolutionary explanations have been given for the shape of the beaks of hummingbirds and of the flowers they feed on, the behavior of bees and the distribution of flowering plants, the biochemical defenses of plants and the immunity of their insect prey, and the characteristics of other interactive species. Note that coevolutionary explanations invoke relationships between entities which affect the evolution of the entities. Entities and relationships are constantly changing, yet they constantly reflect each other, like the flowers and the hummingbirds' beaks. Everything is interlocked, yet everything is changing in accordance with the interlockedness.

Now pests and pesticides can also be thought of as interrelated and coevolving in response to the interrelatedness. Pests evolved resistance in response to pesticides, that is easy to see. But one can just as well argue that pesticides evolved new qualities in response to the evolution of resistance among pests. To be more precise, the distribution of individual insects by different traits shifted toward the trait of greater resistance in response to the application of organophosphates while the distribution of insecticides shifted away from organophosphates in response to the development of resistance. But the traits of pests and pesticides were also affected by pesticide legislation and regulatory decisions, which in turn were certainly affected by the characteristics of pest problems and the types of pesticides used. Pesticide legislation, however, did not evolve in response to pests and pesticides directly, but rather evolved in response to how political interests—environmentalists, laborers, beekeepers, and farmers—were affected by pests and pesticides. Similarly, the demand for integrated pest management evolved in response to all of these. In short, pests, pesticides, politics, policy, the pesticide industry, and integrated pest management evolved in response to changes in each other and in the relationships between them, or more simply, they coevolved. This process is illustrated in Figure 3.1 below.

The coevolutionary process will be elaborated in greater detail in later chapters, but a few things should be noted now. While Figure 3.1 could merely be the description of static relations, of a complex machine, it is

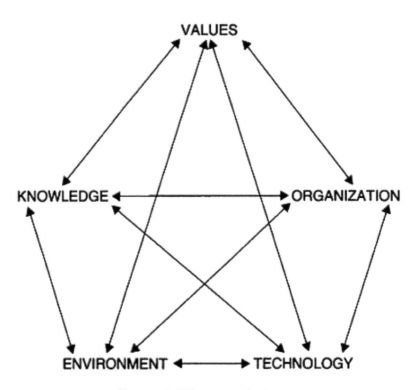

*Figure 3.1*The coevolutionary process

coevolutionary because random, or at least totally unforeseeable, introductions occur. Among biological species, the introduction of brand new genetic material through introductions from other ecosystems, genetic drift, and mutations occurs. Similar new chemicals are developed and introduced alongside existing chemicals. Whether these new introductions survive depends on their fitness. In coevolution, fitness is determined by the characteristics of other species or things with which an individual or thing interacts. Coevolutionary processes are not like the dynamics of a machine because the machine does not change. The parts of a machine and the relationships between the parts stay the same. By knowing the parts and relationships of a mechanical system, one can know how the system works. Coevolving systems, on the other hand, have parts and relations which change in unforeseeable ways. At any point in time, it can be described like an ecosystem, but over time it is as unpredictable as the evolution of life itself. Understanding change as coevolution helps us understand why the pesticide story could not have been foretold.

I have been asked many times why I am so entranced by a pattern of thinking that does not facilitate prediction and control. I hope to show how this relatively simple pattern of thinking contrasts with and thereby provides perspective on the pattern of thinking underlying modernity and how it acts as a catalyst for revisioning the future.

A PREVIEW OF THE OVERALL ARGUMENT

Critique is easy, reconstruction more difficult. While a consensus is emerging that modernity is in a shambles, most of the designs for its reconstruction rely on many of the same materials, the beliefs of modernism. This book pursues two parallel metaphysical and epistemological arguments in an effort to bring new material to the reconstruction of our future.

The first argument can be summarized as follows. Modernity coevolved with modern beliefs. The individual beliefs with which modernity coevolved are themselves highly complementary, they have been relied upon to order our social structure and environmental interactions, and hence any reconstruction around the same beliefs will quite likely have similar outcomes. The pursuit of appropriate technologies to make development work for people and not against the environment, for example, is certainly admirable, as is the more recent and broader pursuit of sustainable development. But if these pursuits build upon the same beliefs of modernism with respect to science, society, and the environment, they are doomed to result in unsatisfactory reconstructions because of inherent limitations in the materials. There are inherent limitations in how we can understand and interact with a complex and changing social and environmental reality. Western science consists of multiple ways of interpreting alternative slices of the complexity of reality. These multiple ways are not merging and cannot merge into one coherent whole. Recognition of the conceptual plurality of Western science is a necessary condition for a viable reconstruction. Acknowledging conceptual pluralism, furthermore, opens up possibilities for new materials with which to rebuild. This argument is presented within the dominant, deductive mode of Western thinking even while it exposes the limits of this mode.

The second line of reasoning draws upon, synthesizes, and argues from understandings which have emerged within Western science since Darwin. While evolutionary thought has a formal structure within Western science, the pattern of thinking more broadly interpreted explains and lends legitimacy to other ways of knowing. The traditional knowledge of peoples who have not been Westernized, and what is generally referred to as common sense among those who have, proved fit within the context in which they evolved. Much of the book addresses the meaning of knowledge from a coevolutionary perspective and explores its relations with values, social organization, technology, and the environment. This line of reasoning is inherently synthetic, contrasting with, indeed in contradiction with, the deductive mode of the first line of argument.

...

PART IV

SOCIO-ENVIRONMENTAL RESEARCH IN ECOLOGY

Patricia Balvanera

Ecology, which comes from the Greek word *oikos*, meaning household, home, or a place to live, is the study of the relationships between organisms and their environment. Ecological knowledge is grounded in natural history, the detailed observations of how plants, animals, fungi and bacteria interact with each other and with water, energy, and nutrients. Ecology provides unique insights about how human societies depend on resources from the environment to meet their needs as well as on the mechanisms that modulate such interactions.

Ecological perspectives on the interactions between people and nature are essential components of worldviews across cultures. The earliest explorations presented in this book are those by Humboldt **(Humboldt 1814, Part I)**, and Darwin **(Darwin 1859, Part I)**. Humboldt described the effect of human activity on forests, rivers, and other natural resources but also its impacts on the regulation of ecological processes. Darwin considered how people have shaped evolutionary processes through the domestication of plants and animals.

Ecology developed as a science at the end of the 1800s and in the first half of the 1900s (Real and Brown 1991). Stemming from a materialistic perspective, the natural world was largely conceived of as a machine made up of components. Each component could be studied through rigorous observations as well as through experiments that teased apart these components and interactions. Ecological principles could then be developed into general laws such as those developed in physics.

The impact of humans on ecological processes has been considered for decades, but more often in the background. For example, Forbes emphasized the crucial contributions of ecology to human welfare (Forbes 1922), conceiving it as a line of contact between the environment and societies. He called on ecologists to contribute to pressing societal issues, such as the regulation of insect populations, to serve human interests. Tansley (1935) coined the term *ecosystem* and identified *anthropogenic ecosystems*. Complementary approaches such as reductionism (focus on the pieces), and holism (focus on the whole assemblage), derive from early debates and clearly depict the different ways in which the complex, reciprocal relations between people and nature have been addressed.

G. Evelyn Hutchinson (1903–1991) was one of the most influential ecologists, providing the conceptual tools to explore the processes that determine the number of species that can be found in a particular place. Raised in Cambridge, UK, by a progressive feminist mother and a mineralogy professor father, his upbringing was enriched by many bright people visiting his home, including Sir George and Lady Darwin themselves. He worked in South Africa, then moved to Yale in 1928, just as post–World War I prosperity was rising in the US.

Hutchinson's *On Living in the Biosphere* (1948) is an exploration of the ecological, chemical, and geological interactions between humans and nature. Hutchinson analyzed the contributions of people to the global water, nitrogen, phosphorus, and carbon cycles. He wrote, "looking at man from a strictly geochemical standpoint, his most striking character is that he demands so much – not merely thirty or forty elements for physiological activity, but nearly all the others for cultural activity" (29). As a single dominant species, humans have profound effects on the whole biosphere. Increasing per capita demand beyond basic physiological needs is a major driver. The perils of overconsumption are portrayed as an "overheated kitchen, filled with too much electrical equipment and catering to overfed people" (29).

The work by Hutchinson set the stage for further explorations on the biogeochemical impacts of industrial society. It sets the stage for the contributions by Vitousek et al. (1997) on the quantitative assessment of the impacts of people on global biogeochemical cycles, land cover changes, and biodiversity loss. The conceptual framework of the Millennium Ecosystem Assessment (MEA 2003) builds on these early ideas about how people and nature interact. Hutchinson anticipated the most recent concerns on the disproportionate role of the most affluent households, those with highest incomes, often linked to greatest political power, in driving the current climate and biodiversity crisis (Wiedmann et al. 2020).

The 1950s and 1960s were marked by the post–World War II recovery, accelerating the exponential growth in gross domestic product and human population. Science and technology, especially the Green Revolution to increase agricultural yields, boosted such recovery. Institutions to fund science, such as the US National Science Foundation and research centers to explore global processes from land and earth[1], were created. The negative consequences of the accelerated growth of societal enterprise for the environment were already visible.

Rachel Carson (1907–1964) experienced herself the negative impacts of the use of pesticides. Born in rural Pennsylvania on a farm by the river, she began writing about nature at a young age. She abandoned her PhD studies in zoology and genetics and joined the US Fish and Wildlife Service in 1935. While visiting agricultural farms, she witnessed the dramatic impacts of aerial pesticide spraying. She corresponded extensively with a couple of women trained in an early form of organic agriculture during their three-year (1957–60) court action to stop the US Federal Government from spraying their property (Paull 2013). Carson herself died of cancer a few years later.

Carson provided one of the most compelling and influential narratives on the impacts of environmental degradation in her book *Silent Spring* (1962). Pollution, by toxic chemicals

---

[1] *Such as the National Center for Atmospheric Research, and the National Aeronautics and Space Administration.*

and nuclear radiation, was accumulating across the food chain, from insects to top predators and to people. It was happening across the whole landscape, from farms to water bodies to the oceans. Pesticides, so broad-scale as to be considered "biocides," were depleting complex ecosystems – that had developed over millions of years – of populations of insects, birds and fish within only a few years. This "war against nature" (30) was irreversible and non-selective. Carson calls for "prudent concern" (30) of those who decide on pesticide spraying, such as industries and government, for those who bear the resulting costs, the citizens.

Carson brought the term ecology into the public domain and impacted the minds and the hearts of many citizens, decision-makers, and scientists. She inspired the creation of environmental agencies, such as the United States Environmental Protection Agency, and environmental movements around the world, such as those around the fight against pesticides that led to the installment of Earth Day. Today, understanding and addressing dramatic shifts, such as those triggered by the sudden collapse of insects, birds and fish, is a central topic of socio-environmental research (Liu et al. 2007). Carson's call for prudence exemplified the precautionary principle, a critical guideline in environmental decision-making (Kriebel et al. 2001).

The 1960s and early 1970s saw the emergence of ecological tools (Real and Brown 1991) that have been incorporated into social–ecological research. These include mathematical models to assess the dynamics of interacting plant and animal populations, which inform today's models of future scenarios of interactions between societies and their environment. They also include descriptions of how social and ecological systems change across space and through time. Initiatives such as the International Biological Program and the United Nations' Man and Biosphere Initiative were launched to assess the effects of human-driven changes and protect biodiversity across the planet.

Eugene P. Odum (1913–2002) contributed key concepts, such as ecosystem services, far ahead of his colleagues. He was passionate about zoology from an early age and heavily influenced by his father, a professor at the University of Georgia, and by his older brother, the ecologist Howard T. Odum. *Fundamentals of Ecology*, first published by Howard Odum in 1953, and with a later edition co-written by the two Odum brothers, became a classic textbook, and the Odum School of Ecology at the University of Georgia is a testament to his lasting impact.

Odum's paper on "The Strategy of Ecosystem Development" (1969) set the stage for the concept of ecological sustainability and of the tools to measure it. He proposed that ecosystems could be broadly described through a predictable successional process, from younger to mature stages. A set of 24 ecosystem attributes, including community energetics, community structure, life history, nutrient cycling, selection pressures, and overall homeostasis, were identified. Such lists of indicators could guide the management of ecosystems and identify how management tends to emphasize younger successional stages to maximize production, at a cost.

Odum pioneered the identification of tradeoffs emerging from the management of ecosystems, "a basic conflict between the strategies of man and nature" (31). Agricultural fields, kept in early successional stages to maximize yields, do not allow nutrients to be restored. In contrast, mature forests, in later successional stages, provide important benefits such as "water-purification … and other protective functions of self maintaining ecosystems" (31), that "until recently mankind has more or less taken for granted" (31). The unavoidability of these tradeoffs has become a central tenet of socio-environmental research, as exposed by Liu et al. (2007), Liu actually being a student of Eugene's older brother, Howard Odum.

That same period also witnessed the contributions of a particularly innovative character, **Crawford S. Holling** (1930–2019). Raised in Ontario, Canada, Holling, who went by the name "Buzz," was trained in forestry and zoology. His thirst for flexibility, novelty, and adventure took him to several institutions across the US, Canada, and Europe (Holling 2017). He participated in the nascent Santa Fe Institute and Aspen Institute. He collaborated in 1973 with large interdisciplinary, international teams (>60 people), including an unusually large and diverse team at that time, at the International Institute for Applied Systems Analysis (IIASA) in Vienna.

Holling dramatically shifted contemporary visions on the stability of ecological systems in his contribution "Resilience and Stability of Ecological Systems" (1973). Departing from models of the dynamics of interacting species populations, he proposed the concept of domains of attraction. Rather than reaching a unique equilibrium point, systems oscillate around a certain range of possible conditions. Systems can shift from one domain of attraction to another. Lakes subjected to increasing nutrient availability (derived from sewage and agricultural runoff), can suddenly turn into a soup of algae devoid of oxygen and other species. Resilience was defined as "a measure of the ... ability of a system to absorb change and disturbance and still maintain the same relationships between state variables" (32).

Today, resilience is a key concept in social–ecological research (Folke 2006). Dramatic shifts between alternative states have been documented for a wide range of social–ecological systems. Yet, as societies change constantly, domains of attractions, such as poverty traps and social–ecological resilience, are accepted by some and questioned by others, especially in the social sciences. Holling sparked the creation of the Resilience Alliance, a research network, and the journal *Conservation Ecology*, now renamed *Ecology and Society*, to address these topics. Resilience has permeated the creation of research institutes and programs and science-policy initiatives, such as the Stockholm Resilience Center, the Programme for Ecosystem Change and Society, the MEA and Intergovernmental Platform on Biodiversity and Ecosystem Services (IPBES), and Future Earth.

A resilience perspective implies constant change rather than a predictable equilibrium, and the work by Holling led to the concept of adaptive management, which is embedded today in most management schemes and explorations of future alternative pathways (Allen and Garmestani 2015). "A resilience framework ... does not require precise capacity to predict the future, but only a qualitative capacity to devise systems that can absorb and accommodate future events in whatever unexpected form they may take" (Holling 1973, 32).

Multiple voices raised the alarm in the 1970s and early 1980s on the escalating impacts of human population growth **(Ehrlich and Holdren 1971, Part VI)** and the planetary limits to such an exponential trend **(Meadows et al. 1972, Part VI).** Even the most diverse and productive ecosystem on the planet, tropical rainforests, was shown to be non-renewable (Gómez-Pompa et al. 1972).

Policies and research agendas shifted towards sustainability for people and for nature, many of which were influenced by **Paul R. Ehrlich** (1932–) and **Harold A. Mooney** (1932–). Ehrlich, born in Philadelphia and trained in ecology, moved to Stanford University in 1959. While focusing on zoology, especially on butterflies, he also highlighted key global challenges associated with human population growth and mass extinctions. Mooney, born in California, moved to Stanford University in 1968. While pioneering work in physiological ecology, he played a central role in setting up large global research and science-policy initiatives, such as the International Geosphere Biosphere Program, the MEA, and IPBES.

"Extinction, Substitution, and Ecosystem Services" (1983) makes explicit the connection between species extinctions and human welfare. The paper puts forward the concept of ecosystem services, building upon early insights by Odum, and explores the societal consequences of species extinctions. The loss of species that play a key role in ecosystems would be expected to lead to large impacts on the supply of ecosystem services that could not be substituted by technological approaches. "Unfortunately humanity has been gambling with its future by saving fewer and fewer of its parts" as "the cost of substitution will almost certainly become unbearable" (33).

These insights were further expanded by their student Gretchen Daily and have set the stage for the prevalent importance of the concept of ecosystem services as a key connector between nature and society. Daily's 1997 book *Nature's Services* defines ecosystem services as the "conditions and processes through which natural ecosystems, and the species that make them up, sustain and fulfill human life" (p. 3). The book, together with a paper by Robert Costanza and collaborators (1997) in that same year, triggered an explosion of research, policy, and applications of the idea of ecosystem services (Costanza et al. 2017), including the creation of the journal *Ecosystem Services*. The MEA placed the ecosystem services concept at the center of interactions between society and environment, and since then it has been considered a key tool to guide policy design and decision-making (Chan and Satterfield 2020). The work by Ehrlich and Mooney reverberates in the recent IPBES assessments that have highlighted the dire consequences of the unsubstitutable loss of species, such as pollinators, for society.

As consumption-driven demands from nature grew and anthropogenic impacts escalated, multiple initiatives to manage terrestrial, aquatic, and marine ecosystems more sustainably were created. While global markets drove the depletion of resources from diverse areas in the Global South, development funds and researchers were sent from the Global North to address this degradation. Socio-environmental research gained from these colonial exchanges, expressed as explorations of ecosystems within the tropics by researchers from Europe and the United States.

James E. Ellis (1938–2002) and **David M. Swift** (1941–2020), both based at Colorado State University in the early 1970s, explored the East African pastoralist systems from an interdisciplinary perspective. Ellis was a systems thinker, able to conceptualize the complex interactions among the components of social–ecological systems and develop tools to understand their dynamics. Swift excelled at programming and data processing and was interested in a range of topics linking the nutrition and ecology of ruminants, the ecology of pastoral systems, human ecology, international development, and traditional ecological knowledge.

A complex systems perspective informed by multiple disciplines and field observations supports the contributions of "Stability of African Pastoral Systems: Alternate Paradigms and Implications for Development" (1988). The paper analyzed the interactions between pastoral nomads, livestock, vegetation, and climatic variability in northern Kenya. Ellis and Swift showed that rangeland vegetation dynamics were only minimally regulated by local consumption by livestock. Instead, seasonal and interannual variability were identified as major drivers, "external control mechanisms ... which are not subject to feedback control within the system" (34). Feedbacks occur when multi-year droughts lead to reductions in forage quantity and quality and pastoralists reduce the size of the herds by half, which reduces grazing pressure. Ellis and Swift also questioned how development programs were based on flawed assumptions by Western researchers on the irrationality of pastoral management and were detrimental to ecosystems and to people.

Ellis and Swift pioneered truly interdisciplinary, long-term, observational, and modeling work on the dynamics of complex systems, such as pastoral ones, providing important conceptual and methodological insights. The operation of reciprocal feedbacks between nature and people, of multiple drivers operating at several spatial and temporal scales, and the role of spatial and temporal heterogeneity are today central tenets of social–ecological research (Carpenter et al. 2009) and are embedded in the IPBES conceptual framework (Díaz et al. 2015).

The 1980s witnessed the creation of new disciplines at the interface between ecological systems and nature. As the Report of the Brundtland Commission was released and the United Nations Convention on Biological Diversity was negotiated, new hybrid disciplines such as conservation biology, landscape ecology, and ecological economics emerged to rise to the challenges.

One such nascent hybrid discipline was urban ecology, studied by **Mark J. McDonnell** (1953–) and **Steward T. A. Pickett** (1950–). McDonnell has contributed to bringing ecology into the design, building, and management of urban environments while based in New York, Singapore, and Melbourne. Pickett is the founding director of the Long-Term Ecological Research[2] site in Baltimore, one of the first such sites based in a city and the site with the longest history of social science research.

"Ecosystem Structure and Function along Urban–Rural Gradients: An Unexploited Opportunity for Ecology" (1990) delivers a framework for organizing the ecological study of urban systems. The authors contend that gradients of "environmental variation [are] ordered in space, and ... spatial environmental patterns govern the corresponding structure and function of ecosystems" (35). Rings of diminishing development from the dense center of urban areas are the basis for such gradients. Their paper provides a conceptual model of the effects of urbanization (e.g., their structural features) on ecological phenomena (e.g., physical and chemical characteristics) and on the dynamics of the whole city, conceived as an ecosystem. As an early roadmap for urban ecology, this paper sets key questions to be explored.

The contributions of McDonnell and Pickett still resonate today as cities are conceived as complex and highly interactive social–ecological systems (McPhearson et al. 2016). Current conceptual frameworks borrow from advances across disciplines and explore feedbacks among people, non-human species, and the functioning of urban systems, with the urban–rural gradient being one of the backbones guiding urban ecological research.

By the 1980s and 1990s, the interactions between societies and people within rural Indigenous communities were increasingly addressed by the expanding fields of

---

[2] *The Long-Term Ecological Research (LTER) network was established in 1980 by the US National Science Foundation to support long-term collection of ecological data. Currently, there are 40 LTER sites.*

ethnobotany, ethnoecology, and human ecology. Pioneering work by ecologists, many of them born or based in the diverse Global South, burgeoned.

Mahdav Gadgil (1942–), **Fikret Berkes** (1945–), and **Carl Folke** (1955–) provide a unique synthesis of such hybrid contributions. Gadgil was born in western India, where he studied biology and received a Master's degree in zoology. He graduated with a PhD from Harvard University and returned to India in 1971, where he founded two research centers and advised the Indian Government, deeply influencing conservation policies. Berkes was trained as an ecologist and joined the University of Manitoba, Canada, in 1991. He explores how societies interacting with their resources can avoid the "tragedy of the commons" through his contributions to community-based management, traditional ecological knowledge, and social–ecological resilience. Folke studied economics, business, and ecology and joined Stockholm University in 1999. He has incorporated resilience thinking into the management and stewardship of social–ecological systems at local to global scales. He has been instrumental in the development of the Resilience Alliance and the Stockholm Resilience Center.

Their article, "Indigenous Knowledge for Biodiversity Conservation" (reading 36), synthesizes ethnoecological explorations from Brazil, Indonesia, Australia, and the sub-Arctic. Gadgil and colleagues explore how Indigenous societies tend to be heavily reliant on the biotic resources found within their direct surroundings. Such dependence has fostered the development of strategies to enhance biodiversity at the landscape level, restore biodiversity in depleted landscapes, and restrain resource use to support their conservation. They highlight the paramount importance of the knowledge–practice–belief complex in informing sustainable management, in ways that differ from those of Western science.

The overview by Gadgil, Berkes, and Folke, as well as the contributions from many authors at the interface between ecology and Indigenous knowledge (e.g., Toledo 1990), has given rise to a blooming branch of socio-environmental research. Studies are being undertaken in local communities across the planet by researchers across disciplines, co-developed with Indigenous scholars and local communities. While several of these studies have made it into top international journals and science-policy platforms, many are found in specialized interdisciplinary journals, such as *Conservation and Society*, and many more are buried in books, reports, and theses not easily accessible through the Internet and written in a diversity of languages. Yet, they provide unique sustainability lessons that inform global sustainability (Nagendra 2018) and support Earth Stewardship initiatives grounded in local knowledge, values, and visions (Rozzi et al. 2015).

In 1989, the *Exxon Valdez* wreck produced one of the largest marine oil spills, exacerbating the increasing degradation of oceans and providing a reminder of how little these complex and vast ecosystems had been addressed by socio-environmental research. Visions of sustainable development were fleshed out at the United Nations' 2012 Rio conference, while voices from the Global South provided unique insights on the linkages between globalization, poverty, and the environment (e.g., **Lélé 1991, Part III**). The limits of science to fully embrace uncertainty and complexity and address the biodiversity crisis were debated (Funtowicz and Ravetz 1993). Interdisciplinary research was proliferating, and new tools and platforms to synthesize knowledge were being developed, such as the National Center for Ecological Analysis and Synthesis (NCEAS).

"Anecdotes and the Shifting Baseline Syndrome of Fisheries" (1995) highlights many of the above advances and challenges. **Daniel Pauly (1946–)**, born in Paris and raised in

Switzerland, studied marine biology in Germany. He started his career in the Philippines at the Center for Living Aquatic Resources Management, and then moved to the University of British Columbia in 1994. He has explored the impacts of fisheries on marine ecosystems, and developed databases and tools to track fishes and fisheries.

Pauly calls for new approaches to address the fundamental challenges that socio-environmental research faces. The majority of fish stocks were already collapsing, fostered by discarding bycatch and subsidies to unsustainable fleets. Models to guide sustainable management were flawed because of their lack of historical memory. The "shifting baseline syndrome ... has arisen because each generation of fisheries scientists accept as baseline the [conditions] that occurred at the beginning of their careers" (37). He shows how these models can be strengthened by integrating data from historical and anthropological literature, on the differential ecology of the huge diversity of fishes, and on the differential impacts of fishing methods. He probably was among the first to have an explicit gender position in fisheries science; he highlighted the need to account for the role of women and children in catching smaller reef organisms as well as the role of women scientists in making visible such unappreciated dimensions. Pauly's work calls for an in-depth exploration of the societal and ecological consequences of resource extraction from nature.

Today, social–ecological research of freshwater, coastal, and marine systems draws on a large diversity of disciplines. Integrated assessment of the dynamics and outcomes for nature and people of small-scale fisheries is underway (Kittinger et al. 2013). The United Nations Global Ocean Assessment (GRID-Arendal and UNEP 2016) provides a sobering overview of the state of fish and fisheries based on databases and tools that can monitor, for example, illegal vessels in the global oceans from space and that builds on conceptual frameworks and tools inspired by Pauly.

The legacy papers in this part highlight the contributions from ecology and ecologists to socio-environmental research. These include conceptual and analytical tools to assess the impacts of societies on the environment, the benefits that societies obtain from nature, and the complex feedbacks between them. These papers also encourage us to understand the complexity of socio-environmental systems, how they change across spatial and temporal scales, the tradeoffs that emerge from management or policy decisions, and the societal and ecological mechanisms that contribute to their long-term maintenance. The legacy papers have inspired a vast literature, relevant from local to global scales, and the creation of new journals, hybrid disciplines, and interdisciplinary research programs. The ecologists highlighted here, perhaps because they have been personally affected by environmental degradation, have been vocal promoters of environmental agencies, policies, and movements as well as of interdisciplinarity and global science–policy interfaces.

References

Allen, Craig R., and Ahjond S. Garmestani. "Adaptive Management." In *Adaptive Management of Social-Ecological Systems*, edited by Craig R. Allen and Ahjond S. Garmestani, 1–10. Dordrecht: Springer Netherlands, 2015.

Carpenter, Stephen R., Harold A. Mooney, John Agard, Doris Capistrano, Ruth S. DeFries, Sandra Díaz et al. "Science for Managing Ecosystem Services: Beyond the Millennium Ecosystem Assessment." *Proceedings of the National Academy of Sciences* 106, no. 5 (2009): 1305–1312.

Chan, Kai M. A., and Terre Satterfield. "The Maturation of Ecosystem Services: Social and Policy Research Expands, but Whither Biophysically Informed Valuation?" *People and Nature* 2, no. 4 (2020): 1021–1060.

Costanza, Robert, Ralph d'Arge, Rudolf De Groot, Stephen Farber, Monica Grasso, Bruce Hannon et al. "The Value of the World's Ecosystem Services and Natural Capital." *Nature* 387, no. 6630 (1997): 253–260.

Costanza, Robert, Rudolf De Groot, Leon Braat, Ida Kubiszewski, Lorenzo Fioramonti, Paul Sutton et al. "Twenty Years of Ecosystem Services: How Far Have We Come and How Far Do We Still Need To Go?" *Ecosystem Services* 28 (2017): 1–16.

Daily, Gretchen C. *Nature's Services*. Washington, DC: Island Press, 1997.

Díaz, Sandra, Sebsebe Demissew, Julia Carabias, Carlos Joly, Mark Lonsdale, Neville Ash et al. "The IPBES Conceptual Framework – Connecting Nature and People." *Current Opinion in Environmental Sustainability* 14 (2015): 1–16.

Folke, Carl. "Resilience: The Emergence of a Perspective for Social–Ecological Systems Analyses." *Global Environmental Change* 16, no. 3 (2006): 253–267.

Forbes, Stephen A. "The Humanizing of Ecology." *Ecology* 3, no. 2 (1922): 89–92.

Funtowicz, Silvio O., and Jerome R. Ravetz. "Science for the Post-Normal Age." *Futures* 25, no. 7 (1993): 739–755.

Gómez-Pompa, Arturo, Carlos Vazquez-Yanes, and Sergio Guevara. "The Tropical Rain Forest: A Nonrenewable Resource." *Science* 177, no. 4051 (1972): 762–765.

GRID-Arendal and United Nations Environmental Program. *World Ocean Assessment*. Arendal: GRID-Arendal, 2016.

Holling, C. S. *Bubbles and Spiral: The Memoirs of C. S. Buzz Holling,* 2nd ed. Stockholm: Stockholm Resilience Center, 2017.

Kittinger, John N., Elena M. Finkbeiner, Natalie C. Ban, Kenneth Broad, Mark H. Carr, Joshua E. Cinner et al. "Emerging Frontiers in Social-Ecological Systems Research for Sustainability of Small-Scale Fisheries." *Current Opinion in Environmental Sustainability* 5, no. 3–4 (2013): 352–357.

Kriebel, David, Joel Tickner, Paul Epstein, John Lemons, Richard Levins, Edward L. Loechler et al. "The Precautionary Principle in Environmental Science." *Environmental Health Perspectives* 109, no. 9 (2001): 871–876.

Liu, Jianguo, Thomas Dietz, Stephen R. Carpenter, Marina Alberti, Carl Folke, Emilio Moran et al. "Complexity of Coupled Human and Natural Systems." *Science* 317, no. 5844 (2007): 1513–1516.

McPhearson, Timon, Steward T. A. Pickett, Nancy B. Grimm, Jari Niemelä, Marina Alberti, Thomas Elmqvist et al. "Advancing Urban Ecology toward a Science of Cities." *BioScience* 66, no. 3 (2016): 198–212.

Millennium Ecosystem Assessment (MEA). "Summary." In *A Framework for Assessment*, 1–25. Washington, DC: Island Press, 2003.

Nagendra, Harini. "The Global South Is Rich in Sustainability Lessons That Students Deserve to Hear." *Nature* 577, no. 7706 (2018): 485–488.

Odum, Eugene, P. *Fundamentals of Ecology*. Philadelphia: W. B. Saunders Company, 1953.

Paull, John. "The Rachel Carson Letters and the Making of *Silent Spring*." *SAGE Open*, 3 (July–September), 2013.

Real, A. L., and Brown, J. H. (eds.). *Foundations of Ecology: Classic Papers with Commentaries*. Chicago: University of Chicago Press, 1991.

Rozzi, Ricardo, F. Stuart Chapin III, J. Baird Callicott, Steward T. A. Pickett, Mary E. Power, Juan J. Armesto, and Roy H. May Jr., eds. *Earth Stewardship: Linking Ecology and Ethics in Theory and Practice. Volume 2*. Dordrecht: Springer, 2015.

Tansley, Arthur G. "The Use and Abuse of Vegetational Concepts and Terms." *Ecology* 16, no. 3 (1935): 284–307.

Toledo, Víctor Manuel. "The Ecological Rationality of Peasant Production." In *Agroecology and Small Farm Development*, edited by Miguel A. Altieri and Susanna B. Hecht, 53–60. Boca Raton, FL: CRC Press, 1990.

Vitousek, Peter M., Harold A. Mooney, Jane Lubchenco, and Jerry M. Melillo. "Human Domination of Earth's ecosystems." *Science* 277, no. 5325 (1997): 494–499.

Wiedmann, Thomas, Manfred Lenzen, Lorenz T. Keyåer, and Julia K. Steinberger. "Scientists' Warning on Affluence." *Nature Communications* 11, no. 1 (2020): 1–10.

ON LIVING IN THE BIOSPHERE

G. E. HUTCHINSON

In the article below, which is from a paper presented in the symposium on "The World's Natural Resources" (AAAS Centennial, September 13–17), Professor Hutchinson claims originality only for some of the computations and part of the interpretation—most of the fundamental data are easily accessible in the scientific literature. Professor Hutchinson is on the staff of the Department of Zoology at Yale and consultant in biogeochemistry, American Museum of Natural History.

IN discussing the subject of "The World's Natural Resources," I want first to make a number of general observations that will provide an intellectual framework into which our developing knowledge, both academic and practical, may be fitted. We live in a rather restricted zone of our planet, at the base of its gaseous envelope and on the surface of its solid phase, with temporary excursions upwards, downwards, or sideways onto or into the oceans. These regions in which we can live and which we can explore are characterized by their temperature, which does not depart far from that at which water is a liquid, and by their closeness to regions on which solar radiation is being delivered. This zone of life is spoken of as the biosphere. Within it, certain natural products can be utilized in both biological and cultural life. It is customary to consider these resources as either material or energetic, but the two categories are not easily separable; contemporary solar radiation is an energetic resource, coal and oil are to be regarded as material resources valuable for their high energy content, which we may call, epigrammatically, fossil solar radiation. There is a third very important though inseparable aspect, namely, the pattern of distribution. Most fossil sunlight, or chemical energy of carbonaceous matter, is diffused through sedimentary rocks in such a way as to be useless to us. Schrödinger says that we feed on negative entropy, and I am almost tempted to regard pattern as being as fundamental a gift of nature as sunlight or the chemical elements.

The first major function of the sunlight falling on the earth's surface is as the energy of circulation of the oceans and atmosphere. The second is to increase the mobility of water molecules, to become latent heat of evaporation, and so to keep the water cycle operating. The third major function is photosynthesis. Apart from atomic energy, and a little volcanic heat which presumably is actually of radioactive origin, all industrial energy is solar and due to one or the other of these three processes.

The material requirements of life are extremely varied. Between thirty and forty chemical elements appear to be normally involved. Industrially, some use appears to be found for nearly all the natural elements, and some of the new synthetic ones also. Looking at man from a strictly geochemical standpoint, his most striking character is that he demands so much—not merely thirty or forty elements for physiological activity, but nearly all the others for cultural activity. What we may call the anthropogeochemistry of cultural life is worth examining. We find man scurrying about the planet looking for places where certain substances are abundant; then removing them elsewhere, often producing local artificial concentrations far greater than are known in nature. Such concentrations, whether a cube of sodium in a bottle in the laboratory, or the George Washington Bridge, have usually been brought into being by chemical changes, most frequently reductions, of such a kind that the product is unstable under the conditions in which accumulation takes place. Most artifacts are made to be used, and during use the strains to which they are submitted distort them, and they become worn-out or broken. This results in a very great quantity of the materials that are laboriously collected being lost again in city dumps and automobile cemeteries. The final fate of an object may depend on many factors, but it is probable that in most cases a very large quantity of any noncombustible, useful material is fated to be carried, either in solution or as sediment, into the sea. Modern man, then, is a very effective agent of zoogenous erosion, but the erosion is highly specific, affecting most powerfully arable soils, forests, accessible mineral deposits, and other parts of the biosphere which provide the things that *Homo sapiens* as a mammal and as an educatable social organism needs or thinks he needs. The process is continuously increasing in intensity, as popu-

From Hutchinson, G. Evlyn (1948). On living in the biosphere. The Scientific American 67: 393-397. Reprinted with permission from AAAS.

lations expand and as the most easily eroded loci have added their quotas to the air, the garbage can, the city dump, and the sea.

The most important general consideration to bear in mind in discussing the dynamics of the biosphere and its inhabitants is that some of the processes of significance are acyclical, and others, to a greater or less degree, cyclical. By an acyclical process will be meant one in which a permanent change in geochemical distribution is introduced into the system; usually a concentrated element tends to become dispersed. By a cyclical process will be meant one in which the changes involved introduce no permanent alteration in the large-scale geochemical pattern, concentration alternating with dispersion. The cyclical processes are not necessarily reversible in a thermodynamic sense; in fact, they are in general no more and no less reversible than the acyclical. Most of the cyclical processes operate because a continuous supply of solar energy is led into them, and sunlight will provide no problems for the conservationist for a very long time. It is important to realize that most of the acyclical processes are so slow that man appears as an active intruder into a passive pattern of distribution. They are safer to disturb because we know what the result of the disturbance will be. If we mine the copper in a given region sufficiently assiduously, we know that ultimately there will not be any more copper available there. Cyclical processes involve complex circular paths, regenerative circuits, feedback mechanisms, and the like. Small disturbances of such processes may merely result in small temporary changes, with a rapid return to the previous steady state. This has been beautifully demonstrated in the experiments of Einsele, who added single massive doses of phosphate to a lake, changing for a time, but only for a time, its entire chemistry and biology. This stability does not imply that if large disturbances strain the mechanism beyond certain critical limits very profound disruption will not follow; in fact, the very self-regulatory mechanisms that give the system stability against small disturbances are likely to accentuate the disruption when the critical limits are transcended. In disturbing cyclical processes we usually do not know what we are doing.

The most nearly perfect cyclical processes are those involving water and nitrogen. Some losses to the sediments of the deep oceanic basins must occur, but they are very small and are doubtless fully balanced or more than balanced by juvenile water and perhaps by molecular nitrogen and ammonia of volcanic origin.

The least cyclical processes are those in which material is removed from the continents and deposited in the permanent basins of the ocean. With one or two exceptions, the delivery to the deep-water sediments of the ocean is of little significance. Most of the mechanical and chemical sedimentation in the oceans takes place in relatively shallow water. The uplifting of shallow water sediments constitutes an important method of completing cycles. One, and perhaps two, exceedingly important elements are, however, sedimented less economically. Calcium, during the Paleozoic, was mainly precipitated in shallow water, but since the rise of the pelagic foraminifera in the Mesozoic, a great deal of calcium, along with an equivalent amount of carbon and oxygen, has been continually diverted to regions from which it is unlikely ever to be removed. At present the sedimentary rocks of the world are an adequate biological and commercial source of calcium, but, with progressive orogenic cycles, less and less of the element will be uplifted (Kuenen), and whatever organisms inherit the earth in that remote future will have to face the problem of the biosphere "going sour on them." For phosphorus, the case is less well established, but there is probably a slow loss in the form of sharks' teeth and the ear bones of whales (Conway), which are very resistant and which are known to be littered about on the floor of the abysses of the ocean.

It is desirable to consider two of the main geochemical cycles in order to gain an idea of the effect of man upon them. It must be admitted that we are ignorant of many matters of importance here. In the cycle of carbon we have a remarkable, possibly a unique, case in which man, the miner, increases the cyclicity of the geochemical process. It is generally admitted that our available store of carbon is ultimately of volcanic origin. A steady stream of carbon dioxide and lesser amounts of methane and carbon monoxide are entering the atmosphere from volcanic vents. Part, probably a major part, of this carbon dioxide is ultimately lost to the marine sediments as limestones; a very small part of it is then returned to the air, wherever lime kilns are in operation. Another part of the CO_2 entering the atmosphere is reduced in photosynthesis. A part of the organic matter so formed is fossilized, and a small part of this fossilized organic carbon is available as fuel, in the form of coal and oil. At the present time it appears that the combustion of coal and oil actually returns carbon to the atmosphere as CO_2 at a rate at least a hundred times greater than the rate of loss of all forms of carbon, oxidized and reduced, to the sediments (Goldschmidt). This particular process obviously cannot go on indefinitely. It concerns only the reduced carbon; to complete the cycle in the case of oxidized carbon, a great deal of energy would have to be supplied. It concerns only the reduced carbon which is aggregated. The poorest sources would be the poorest exploitable oil shales. Most of the reduced carbon is much more dispersed than this; in making an estimate of the total reduced carbon of the sediments, the commercially usable fuels constitute a negligible fraction that need not be considered.

Although the rate at which carbon dioxide is returned to the air by the human utilization of fossil fuels is so very much greater than is the primary production of carbon dioxide from volcanic sources, the rate is evidently a very small fraction—of the order of 1 percent—of the rate of photosynthetic fixation and subsequent respiratory liberation of CO_2 by the organisms of the earth. Since about 1890 a slight increase in CO_2 content in the air, at least at low altitudes over the land surfaces of the Northern Hemisphere, has been noted. This has been attributed to the accumulation of industrially produced CO_2, as the quantity of CO_2 that has appeared in the atmosphere is of the same order of magnitude as the total combustion of fuel (Callendar). In view of the small fraction of the total CO_2 production that industrial output represents, it seems very unlikely that merely adding an extra percent to the natural biological production should overload the cyclic process, so that it rejects quantitatively the additional load. It is known that the air at high altitudes still shows nineteenth-century values. The most reasonable explanation of the observed increase is that the photosynthetic machinery of the biosphere has been slightly impaired, probably by deforestation. It is clear that, in any intelligent long-term planning of the utilization of the biosphere, an extended study of atmospheric gases is desirable, even though for the moment it seems unlikely that the observed change is a particularly serious symptom.

The only other cycle that can be considered in any great detail is that of phosphorus. The chief event in the geochemical cycle of phosphorus is the leaching of the element from the rocks of the continents, and its transport by rivers to the sea. At the present time the rate of this transportaion is of the order of 20,000,000 tons of phosphorus per year for the entire earth. Part of this phosphorus, when it enters the sea, will ultimately be deposited in the sediments of the depths of the ocean. Such phosphorus will probably be largely lost to the geochemical cycle, as has just been indicated. The sedimentary rocks of the continents, therefore, will gradually lose phosphorus; there is some evidence that this has actually occurred (Conway). The main return path is by the uplifting of sediments formed in continental seas, which then undergo renewed chemical erosion. Of particular interest are methods by which concentrated phosphorus can be returned to the land surfaces. As far as is known, there are two such methods: The first is the formation of phosphatic nodules and other forms of phosphate rock in regions of upwelling in which water at a low pH, rich in phosphate, is brought up to the surface of the sea. The pH falls and an apatite-like phosphate is deposited. When the sea floor is later elevated, a commercial deposit may result. The second method is by the activity of sea birds, such as the guano birds of the Peruvian coast. There is little or no unequivocal evidence that guano deposits of great extent were formed prior to the late Pliocene or Pleistocene. Some of the well-known occurrences of rock phosphate, such as that of Quercy, have been explained in this way, but they are certainly not typical guano deposits. During the late Tertiary and Pleistocene, an extraordinary amount of phosphate was deposited on raised coral islands throughout the world, and bird colonies seem to provide the only reasonable agencies of deposition. The great deposits of Nauru, Ocean Island, Makatea, Angaur, the Daito Islands, Christmas Island south of Java, Curaçao, and some of the other West Indies all seem to have been formed in this way. This process is as characteristic of the time as is glaciation, though less grandiose. Its meaning is not clear, but it is probably connected with changes in vertical circulation of the ocean as glaciopluvial periods gave place to interpluvials. Today in certain regions massive amounts of guano are deposited, and it is probable that the oceanic birds of the world as a whole bring out from several tens to several hundreds of thousands of tons of phosphorus and deposit it on land. Only about 10,000 tons of the element are delivered in places where it is not washed away and where it can be carried by man to fertilize his fields.

The main processes that tend to reverse the phosphorus depletion of the continents are, therefore, the deposition of marine phosphorites on the continental shelves and subsequent elevation, and the formation of guano deposits. Both processes are evidently intermittent, and are quantitatively inadequate to arrest deflection of the element into the permanent ocean basins. Man contributes both to the loss and to the gain of phosphorus by land surfaces. He quarries phosphorite, makes superphosphate of it, and spreads it on his fields. Most of the phosphorus so laboriously acquired ultimately reaches the sea. At present the world's production of phosphate rock is about 10,000,000 tons per annum. This contains from 1,000,000 to 2,000,000 tons of elementary phosphorus. Human activity probably, therefore, accounts for from 5 to 10 percent of the loss of phosphorus from the land to the sea. Man also contributes to the processes bringing phosphorus from the sea to the land. This is done by fisheries. The total catch for the marine fisheries of the earth is of the order of 25–30.10^6 tons of fish, which corresponds to about 60,000 tons of elementary phosphorus. The human, no less than the nonhuman, processes tending to complete the cycle seem miserably inadequate. It is quite certain that ultimately man, if he is to avoid famine, will have to go about completing the phosphorus cycle on a large scale. It will be a harder task than that of solving the nitrogen problem, which would have loomed large in any symposium on "The World's Natural Resources" fifty years ago, but possibly an easier problem than some of the others that must be solved if we are to survive and really become the glory of the earth.

The population of the world is increasing, its available resources are dwindling. Apart from the ordinary biological processes involved in producing population saturation already known to Malthus, the current disharmony is accentuated by the effect of medical science, which has decreased death rates without altering birth rates, and by modern wars, which one may suspect put greater drains on resources than on populations. Terrible as these conclusions may appear, they have to be faced. The results of the interaction between population pressure and decline of resource potential are further partly expressed in such wars, which are pathological expressions of attempts to cope with these and other problems and which now invariably aggravate the situation. It is evident that the fundamental causes of war lie in those psychological properties of populations which make them attempt solutions in a warlike manner, and *not* in the existence of the problems themselves. The two problems of war and of resources are, however, at present very closely interrelated; it is probably impossible to find a solution for one without progress in the solution of the other. Any kind of reasonable use of the world's resources involves better international relations than now exist. It is otherwise impossible to operate on a planetary basis, or to avoid the fearful material and spiritual expense of living in a world divided into two armed camps. The lack of international trust is the first difficulty in achieving rational utilization of the resources of the world; the second difficulty for the United States is what may be called the problem of the transition from the pioneering to the old, settled community. For a pioneer, life may be hard, but in good country there is "plenty more where these came from," whether lumber or buffalo tongues or copper. This attitude is incorporated into popular thought very deeply; in a crude form, it is now completely destructive, but it seems possible that attitudes might be developed that could utilize such a point of view. There is at least one thing of which we have plenty more than we use, whether we call it Yankee ingenuity, American know-how, or the human intellect. In some industrial fields there has been a notable series of triumphs of the kind that really will give plenty more of a number of things, and give them in a cyclic manner. The rise of the magnesium industry, and the utilization of the magnesium of sea water as a source of the metal is an interesting example. The production of plastics, though it is probably at the moment not as geochemically economical as it should be, is another. These, along with the development of silicate building materials not involving any particularly uncommon elements, all point the way to the kind of material culture that permits a reasonably high living standard, at least in certain directions, without devastation of the earth.

The future outlook for the world, particularly in food resources, has been put before the public in several recent books, notably those of Fairfield Osborn and William Vogt. Anyone with any technical knowledge understands that the dangers described in these books are real enough, more real and more dangerous perhaps than the threat of an atomic world war. The problem of getting action to forestall such dangers in a culture that has developed under conditions of potentially unlimited abundance just around the corner, is obviously extremely difficult. I do not think it is impossible. The first requirement is a faith that the job can be done. The very difficulty of the task, its apparent impossibility, may here and now prove a challenge that brings the desired response. The difficult we do at once, the impossible takes a little longer. I doubt that a direct appeal to fear will produce any results except a disbelief in the prophets of doom. Cassandra seems even more unpopular in modern America than in ancient Ilium. There would seem to be forces operating in society which tend to reverse the destructive processes, or which could be made to do so. One of the most immediate needs is to find out what they are and do everything possible to strengthen them. I will give one example. A number of industrial concerns—particularly in the chemical and pharmaceutical field, but also some engineering and publishing firms—have used in their advertising a legitimate pride in the learning, skillful research, and development that have gone into the manufacture of their product. It is quite certain that there are many cases in which one particular product, the result of considerable research, and of taking risks in development, actually reduces the drain on the natural resources of the country and the world. I should like to see a small systematic experiment, on the part of some such concern, in advertising in which it is pointed out that by buying this product one is letting the industrial skill behind the product operate for the benefit of one's children. I am fully aware that if this point of view were worked up skillfully enough by a few responsible, public-spirited corporations and put into a form that would pay the corporation, as well as the country and the world, there would be other less responsible concerns who would use the method when their product is actually not one sparing natural resources. This, however, seems to be a lesser evil than the total neglect of the commercial advertising field, which is one of the most potent in determining the values of the public and which at present is largely disruptive. I do not doubt that a professional cultural anthropologist could pick out a great many other fields that could be used to promote the idea of an expanding economy based on an abundance of human ingenuity rather than on an excess of raw materials.

There is one further point that I should like to develop. Though the pursuit of happiness is embedded deeply in the constitutional foundations of the United States, we do not know much about it. It is fairly certain that no metric exists that can be applied directly to happiness, but intuitively we may proceed a little way by arguing as if such a metric could exist. It is obvious that only a very few people, with a genius for sanctity, can be happy if half-frozen and starving. If the temperature be raised, and the food supply and other amenities increased, the possibility of happiness is obviously at first also increased, but beyond a certain increase in the environmental resources available no further increase in happiness would be expected and we might begin to look for a decline. The image of an overheated kitchen, filled with too much electrical equipment and catering to overfed people, will, if adequately evoked, have a nightmarish quality. In more formal language, if we could find a function of the environmental resources that expressed the relationship of happiness to those resources, it is reasonably certain that the function would not be monotonic. For every resource there seems likely to be an optimum level of consumption, but we do not know if this optimum level is, in any particular case, widely exceeded, so that gross overutilization is actually producing avoidable distress.

The problem is not by any means as simple as that of determining discrete optima. In any given society all the cultural values are probably interrelated to form a coherent system, so that the existence of one set of values may greatly modify the others. If, as seems possible, our attitude toward food leads a considerable section of our population to be definitely overweight, it is legitimate to inquire to what extent, by changes in the upbringing of our children, the psychological needs filled by food can be satisfied in other ways so that the psychologically optimum intake falls to a level nearer the psychological optimum for the individual and the moral optimum for the world. We might ask, for instance, how we can substitute the delights of ballet and Mozartian opera, which are geochemically very cheap, for part of those provided by hot dogs or apple pie and ice cream, which may in the long run prove too expensive to use except as a source of energy and essential nutrients. This example is chosen with a view to indicating that the kind of substitutions that might be considered need not be in the least puritanical. It may appear overintellectualized, but that at least is a guarantee that it is not inhuman. Indeed it raises the very interesting problem that those of us in the educational world have to face, namely, why we are raising a generation in the belief that the majority of constructive, complicated, difficult activities are boring duties, when the same generation shows us that in certain specific fields, mainly concerned with electronic amplifiers and with the explosive combustion of hydrocarbons, complicated activity can be very entertaining. What we have to do is to show by example that a very large number of diversified, complicated, and often extremely difficult constructive activities are capable of giving enormous pleasure. This is, in fact, the reason why it is essential that the teachers in our colleges and universities should be enthusiastic investigators in the fields of scholarship or practitioners and critics in their arts. It ought to be possible to show that it is as much fun to repair the biosphere and the human societies within it as it is to mend the radio or the family car.

2. The Obligation to Endure

THE HISTORY OF LIFE on earth has been a history of interaction between living things and their surroundings. To a large extent, the physical form and the habits of the earth's vegetation and its animal life have been molded by the environment. Considering the whole span of earthly time, the opposite effect, in which life actually modifies its surroundings, has been relatively slight. Only within the moment of time represented by the present century has one species — man — acquired significant power to alter the nature of his world.

During the past quarter century this power has not only increased to one of disturbing magnitude but it has changed in character. The most alarming of all man's assaults upon the environment is the contamination of air, earth, rivers, and sea with dangerous and even lethal materials. This pollution is for the most part irrecoverable; the chain of evil it initiates not only in the world that must support life but in living tissues is for the most part irreversible. In this now universal contamination of the environment, chemicals are the sinister and little-recognized partners of radiation in changing the very nature of the world — the very nature of its life. Strontium 90, released through nuclear explosions into the air, comes to earth in rain or drifts down as fallout, lodges in soil, enters into the grass or corn or wheat grown there, and in time takes up its abode in the bones of a human being, there to remain until his death. Similarly, chemicals sprayed on croplands or forests or gardens lie long in soil, entering into living organisms, passing from one to another in a chain of poisoning and death. Or they pass mysteriously by underground streams until they emerge and, through the alchemy of air and sunlight, combine into new forms that kill vegetation, sicken cattle, and work unknown harm on those who drink from once pure wells. As Albert Schweitzer has said, "Man can hardly even recognize the devils of his own creation."

It took hundreds of millions of years to produce the life that now inhabits the earth — eons of time in which that developing and evolving and diversifying life reached a state of adjustment and balance with its surroundings. The environment, rigorously shaping and directing the life it supported, contained elements that were hostile as well as supporting. Certain rocks gave out dangerous radiation; even within the light of the sun, from which all life draws its energy, there were short-wave radiations with power to injure. Given time — time not in years but in millennia — life adjusts, and a balance has been reached. For time is the essential ingredient; but in the modern world there is no time.

The rapidity of change and the speed with which new situations are created follow the impetuous and heedless pace of man rather than the deliberate pace of nature. Radiation is no longer merely the background radiation of rocks, the bombardment of cosmic rays, the ultraviolet of the sun that have existed before there was any life on earth; radiation is now the unnatural creation of man's tampering with the atom. The chemicals to which life is asked to make its adjustment are no longer merely the calcium and silica and copper and all the rest of the minerals washed out of the rocks and carried in rivers to the sea; they are the synthetic creations of man's inventive mind, brewed in his laboratories, and having no counterparts in nature.

To adjust to these chemicals would require time on the scale that is nature's; it would require not merely the years of a man's life but the life of generations. And even this, were it by some miracle possible, would be futile, for the new chemicals come from our laboratories in an endless stream; almost five hundred annually find their way into actual use in the United States alone. The figure is staggering and its implications are not easily grasped — 500 new chemicals to which the bodies of men and animals are required somehow to adapt each year, chemicals totally outside the limits of biologic experience.

Among them are many that are used in man's war against nature. Since the mid-1940's over 200 basic chemicals have been created for use in killing insects, weeds, rodents, and other organisms described in the modern vernacular as "pests"; and they are sold under several thousand different brand names.

These sprays, dusts, and aerosols are now applied almost universally to farms, gardens, forests, and homes — nonselective chemicals that have the power to kill every insect, the "good" and the "bad," to still the song of birds and the leaping of fish in the streams, to coat the leaves with a deadly film, and to linger on in soil — all this though the intended target may be only a few weeds or insects. Can anyone believe it is possible to lay down such a barrage of poisons on the surface of the earth without making it unfit for all life? They should not be called "insecticides," but "biocides."

The whole process of spraying seems caught up in an endless spiral. Since DDT was released for civilian use, a process of escalation has been going on in which ever more toxic materials must be found. This has happened because insects, in a triumphant vindication of Darwin's principle of the survival of the fittest, have evolved super races immune to the particular insecticide used, hence a deadlier one has always to be developed — and then a deadlier one than that. It has happened also because, for reasons to be described later, destructive insects often undergo a "flareback," or resurgence, after spraying, in numbers greater than before. Thus the chemical war is never won, and all life is caught in its violent crossfire.

Along with the possibility of the extinction of mankind by nuclear war, the central problem of our age has therefore become the contamination of man's total environment with such substances of incredible potential for harm — substances that accumulate in the tissues of plants and animals and even penetrate the germ cells to shatter or alter the very material of heredity upon which the shape of the future depends.

Some would-be architects of our future look toward a time when it will be possible to alter the human germ plasm by design. But we may easily be doing so now by inadvertence, for many chemicals, like radiation, bring about gene mutations. It is ironic to think that man might determine his own future by something so seemingly trivial as the choice of an insect spray.

All this has been risked — for what? Future historians may well be amazed by our distorted sense of proportion. How could intelligent beings seek to control a few unwanted species by a method that contaminated the entire environment and brought the threat of disease and death even to their own kind?

Yet this is precisely what we have done. We have done it, moreover, for reasons that collapse the moment we examine them. We are told that the enormous and expanding use of pesticides is necessary to maintain farm production. Yet is our real problem not one of *overproduction?* Our farms, despite measures to remove acreages from production and to pay farmers *not* to produce, have yielded such a staggering excess of crops that the American taxpayer in 1962 is paying out more than one billion dollars a year as the total carrying cost of the surplus-food storage program. And is the situation helped when one branch of the Agriculture Department tries to reduce production while another states, as it did in 1958, "It is believed generally that reduction of crop acreages under provisions of the Soil Bank will stimulate interest in use of chemicals to obtain maximum production on the land retained in crops."

All this is not to say there is no insect problem and no need of control. I am saying, rather, that control must be geared to realities, not to mythical situations, and that the methods employed must be such that they do not destroy us along with the insects.

The problem whose attempted solution has brought such a train of disaster in its wake is an accompaniment of our modern way of life. Long before the age of man, insects inhabited the earth — a group of extraordinarily varied and adaptable beings. Over the course of time since man's advent, a small percentage of the more than half a million species of insects have come into conflict with human welfare in two principal ways: as competitors for the food supply and as carriers of human disease.

Disease-carrying insects become important where human beings are crowded together, especially under conditions where sanitation is poor, as in time of natural disaster or war or in situations of extreme poverty and deprivation. Then control of some sort becomes necessary. It is a sobering fact, however, as we shall presently see, that the method of massive chemical control has had only limited success, and also threatens to worsen the very conditions it is intended to curb.

Under primitive agricultural conditions the farmer had few insect problems. These arose with the intensification of agriculture — the devotion of immense acreages to a single crop. Such a system set the stage for explosive increases in specific insect populations. Single-crop farming does not take advantage of the principles by which nature works; it is agriculture as an engineer might conceive it to be. Nature has introduced great variety into the landscape, but man has displayed a passion for simplifying it. Thus he undoes the built-in checks and balances by which nature holds the species within bounds. One important natural check is a limit on the amount of suitable habitat for each species. Obviously then, an insect that lives on wheat can build up its population to much higher levels on a farm devoted to wheat than on one in which wheat is intermingled with other crops to which the insect is not adapted.

The same thing happens in other situations. A generation or more ago, the towns of large areas of the United States lined their streets with the noble elm tree. Now the beauty they hopefully created is threatened with complete destruction as disease sweeps through the elms, carried by a beetle that would have only limited chance to build up large populations and to spread from tree to tree if the elms were only occasional trees in a richly diversified planting.

Another factor in the modern insect problem is one that must be viewed against a background of geologic and human history: the spreading of thousands of different kinds of organisms from their native homes to invade new territories. This worldwide migration has been studied and graphically described by the British ecologist Charles Elton in his recent book *The Ecology of Invasions*. During the Cretaceous Period, some hundred million years ago, flooding seas cut many land bridges between continents and living things found themselves confined in what Elton calls "colossal separate nature reserves." There, isolated from others of their kind, they developed many new species. When some of the land masses were joined again, about 15 million years ago, these species began to move out into new territories — a movement that is not only still in progress but is now receiving considerable assistance from man.

The importation of plants is the primary agent in the modern spread of species, for animals have almost invariably gone along with the plants, quarantine being a comparatively recent and not completely effective innovation. The United States Office of Plant Introduction alone has introduced almost 200,000 species and varieties of plants from all over the world. Nearly half of the 180 or so major insect enemies of plants in the United States are accidental imports from abroad, and most of them have come as hitchhikers on plants.

In new territory, out of reach of the restraining hand of the natural enemies that kept down its numbers in its native land, an invading plant or animal is able to become enormously abundant. Thus it is no accident that our most troublesome insects are introduced species.

These invasions, both the naturally occurring and those dependent on human assistance, are likely to continue indefinitely. Quarantine and massive chemical campaigns are only extremely expensive ways of buying time. We are faced, according to Dr. Elton, "with a life-and-death need not just to find new technological means of suppressing this plant or that animal"; instead we need the basic knowledge of animal populations and their relations to their surroundings that will "promote an even balance and damp down the explosive power of outbreaks and new invasions."

Much of the necessary knowledge is now available but we do not use it. We train ecologists in our universities and even employ them in our governmental agencies but we seldom take their advice. We allow the chemical death rain to fall as though there were no alternative, whereas in fact there are many, and our ingenuity could soon discover many more if given opportunity.

Have we fallen into a mesmerized state that makes us accept as inevitable that which is inferior or detrimental, as though having lost the will or the vision to demand that which is good? Such thinking, in the words of the ecologist Paul Shepard, "idealizes life with only its head out of water, inches above the limits of toleration of the corruption of its own environment . . . Why should we tolerate a diet of weak poisons, a home in insipid surroundings, a circle of acquaintances who are not quite our enemies, the noise of motors with just enough relief to prevent insanity? Who would want to live in a world which is just not quite fatal?"

Yet such a world is pressed upon us. The crusade to create a chemically sterile, insect-free world seems to have engendered a fanatic zeal on the part of many specialists and most of the so-called control agencies. On every hand there is evidence that those engaged in spraying operations exercise a ruthless power. "The regulatory entomologists . . . function as prosecutor, judge and jury, tax assessor and collector and sheriff to enforce their own orders," said Connecticut entomologist Neely Turner. The most flagrant abuses go unchecked in both state and federal agencies.

It is not my contention that chemical insecticides must never be used. I do contend that we have put poisonous and biologically potent chemicals indiscriminately into the hands of persons largely or wholly ignorant of their potentials for harm. We have subjected enormous numbers of people to contact with these poisons, without their consent and often without their knowledge. If the Bill of Rights contains no guarantee that a citizen shall be secure against lethal poisons distributed either by private individuals or by public officials, it is surely only because our forefathers, despite their considerable wisdom and foresight, could conceive of no such problem.

I contend, furthermore, that we have allowed these chemicals to be used with little or no advance investigation of their effect on soil, water, wildlife, and man himself. Future generations are unlikely to condone our lack of prudent concern for the integrity of the natural world that supports all life.

There is still very limited awareness of the nature of the threat. This is an era of specialists, each of whom sees his own problem and is unaware of or intolerant of the larger frame into which it fits. It is also an era dominated by industry, in which the right to make a dollar at whatever cost is seldom challenged. When the public protests, confronted with some obvious evidence of damaging results of pesticide applications, it is fed little tranquilizing pills of half truth. We urgently need an end to these false assurances, to the sugar coating of unpalatable facts. It is the public that is being asked to assume the risks that the insect controllers calculate. The public must decide whether it wishes to continue on the present road, and it can do so only when in full possession of the facts. In the words of Jean Rostand, "The obligation to endure gives us the right to know."

The Strategy of Ecosystem Development

An understanding of ecological succession provides
a basis for resolving man's conflict with nature.

Eugene P. Odum

The principles of ecological succession bear importantly on the relationships between man and nature. The framework of successional theory needs to be examined as a basis for resolving man's present environmental crisis. Most ideas pertaining to the development of ecological systems are based on descriptive data obtained by observing changes in biotic communities over long periods, or on highly theoretical assumptions; very few of the generally accepted hypotheses have been tested experimentally. Some of the confusion, vagueness, and lack of experimental work in this area stems from the tendency of ecolo-

The author is director of the Institute of Ecology, and Alumni Foundation Professor, at the University of Georgia, Athens. This article is based on a presidential address presented before the annual meeting of the Ecological Society of America at the University of Maryland, August 1966.

gists to regard "succession" as a single straightforward idea; in actual fact, it entails an interacting complex of processes, some of which counteract one another.

As viewed here, ecological succession involves the development of ecosystems; it has many parallels in the developmental biology of organisms, and also in the development of human society. The ecosystem, or ecological system, is considered to be a unit of biological organization made up of all of the organisms in a given area (that is, "community") interacting with the physical environment so that a flow of energy leads to characteristic trophic structure and material cycles within the system. It is the purpose of this article to summarize, in the form of a tabular model, components and stages of development

at the ecosystem level as a means of emphasizing those aspects of ecological succession that can be accepted on the basis of present knowledge, those that require more study, and those that have special relevance to human ecology.

Definition of Succession

Ecological succession may be defined in terms of the following three parameters (1). (i) It is an orderly process of community development that is reasonably directional and, therefore, predictable. (ii) It results from modification of the physical environment by the community; that is, succession is community-controlled even though the physical environment determines the pattern, the rate of change, and often sets limits as to how far development can go. (iii) It culminates in a stabilized ecosystem in which maximum biomass (or high information content) and symbiotic function between organisms are maintained per unit of available energy flow. In a word, the "strategy" of succession as a short-term process is basically the same as the "strategy" of long-term evolutionary development of the biosphere—namely, increased control of, or homeostasis with, the physical environment in the sense of achieving maximum protection from its perturbations. As I illustrate below, the strategy of "maximum protection" (that is, trying to achieve maximum support of complex biomass structure) often conflicts with man's goal of "maximum production" (trying to obtain the highest possible yield). Recognition of the ecological basis for this conflict is, I believe, a first step in establishing rational land-use policies.

The earlier descriptive studies of succession on sand dunes, grasslands, forests, marine shores, or other sites, and more recent functional considerations, have led to the basic theory contained in the definition given above. H. T. Odum and Pinkerton (2), building on Lotka's (3) "law of maximum energy in biological systems," were the first to point out that succession involves a fundamental shift in energy flows as increasing energy is relegated to maintenance. Margalef (4) has recently documented this bioenergetic basis for succession and has extended the concept.

Changes that occur in major structural and functional characteristics of a developing ecosystem are listed in Table 1. Twenty-four attributes of ecological systems are grouped, for convenience of discussion, under six headings. Trends are emphasized by contrasting the situation in early and late development. The degree of absolute change, the rate of change, and the time required to reach a steady state may vary not only with different climatic and physiographic situations but also with different ecosystem attributes in the same physical environment. Where good data are available, rate-of-change curves are usually convex, with changes occurring most rapidly at the beginning, but bimodal or cyclic patterns may also occur.

Bioenergetics of Ecosystem Development

Attributes 1 through 5 in Table 1 represent the bioenergetics of the ecosystem. In the early stages of ecological succession, or in "young nature," so to speak, the rate of primary production or total (gross) photosynthesis (P) exceeds the rate of community respiration (R), so that the P/R ratio is greater than 1. In the special case of organic pollution, the P/R ratio is typically less than 1. In both cases, however, the theory is that P/R approaches 1 as succession occurs. In other words, energy fixed tends to be balanced by the energy cost of maintenance (that is, total community respiration) in the mature or "climax" ecosystem. The P/R ratio, therefore, should be an excellent functional index of the relative maturity of the system.

So long as P exceeds R, organic matter and biomass (B) will accumulate in the system (Table 1, item 6), with the result that ratio P/B will tend to decrease or, conversely, the B/P, B/R, or B/E ratios (where E = P + R) will increase (Table 1, items 2 and 3). Theoretically, then, the amount of standing-crop biomass supported by the available energy flow (E) increases to a maximum in the mature or climax stages (Table 1, item 3). As a consequence, the net community production, or yield, in an annual cycle is large in young nature and small or zero in mature nature (Table 1, item 4).

Comparison of Succession in a Laboratory Microcosm and a Forest

One can readily observe bioenergetic changes by initiating succession in experimental laboratory microecosystems. Aquatic microecosystems, derived from various types of outdoor systems, such as ponds, have been cultured by Beyers (5), and certain of these mixed cultures are easily replicated and maintain themselves in the climax state indefinitely on defined media in a flask with only light input (6). If samples from the climax system are inoculated into fresh media, succession occurs, the mature system developing in less than 100 days. In Fig. 1 the general pattern of a 100-day autotrophic succession in a microcosm based on data of Cooke (7) is compared with a hypothetical model of a 100-year forest succession as presented by Kira and Shidei (8).

During the first 40 to 60 days in a typical microcosm experiment, daytime net production (P) exceeds nighttime respiration (R), so that biomass (B) accumulates in the system (9). After an early "bloom" at about 30 days, both rates decline, and they become approximately equal at 60 to 80 days. the B/P ratio, in terms of grams of carbon supported per gram of daily carbon production, increases from less than 20 to more than 100 as the steady state is reached. Not only are autotrophic and heterotrophic metabolism balanced in the climax, but a large organic structure is supported by small daily production and respiratory rates.

While direct projection from the small laboratory microecosystem to open nature may not be entirely valid, there is evidence that the same basic trends that are seen in the laboratory are characteristic of succession on land and in large bodies of water. Seasonal successions also often follow the same pattern, an

early seasonal bloom characterized by rapid growth of a few dominant species being followed by the development later in the season of high B/P ratios, increased diversity, and a relatively steady, if temporary, state in terms of P and R (4). Open systems may not experience a decline, at maturity, in total or gross productivity, as the space-limited microcosms do, but the general pattern of bioenergetic change in the latter seems to mimic nature quite well.

These trends are not, as might at first seem to be the case, contrary to the classical limnological teaching which describes lakes as progressing in time from the less productive (oligotrophic) to the more productive (eutrophic) state. Table 1, as already emphasized, refers to changes which are brought about by biological processes within the ecosystem in question. Eutrophication, whether natural or cultural, results when nutrients are imported into the lake from outside the lake—that is, from the watershed. This is equivalent to adding nutrients to the laboratory microecosystem or fertilizing a field; the system is pushed back, in successional terms, to a younger or "bloom" state. Recent studies on lake sediments (10), as well as theoretical considerations (11), have indicated that lakes can and do progress to a more oligotrophic condition when the nutrient input from the watershed slows or ceases. Thus, there is hope that the troublesome cultural eutrophication of our waters can be reversed if the inflow of nutrients from the watershed can be greatly reduced. Most of all, however, this situation emphasizes that it is the entire drainage or catchment basin, not just the lake or stream, that must be considered the ecosystem unit if we are to deal successfully with our water pollution problems. Ecosystematic study of entire landscape catchment units is a major goal of the American plan for the proposed International Biological Program. Despite the obvious logic of such a proposal, it is proving surprisingly difficult to get tradition-bound scientists and granting agencies to look beyond their specialties toward the support of functional studies of large units of the landscape.

Food Chains and Food Webs

As the ecosystem develops, subtle changes in the network pattern of food chains may be expected. The manner in which organisms are linked together through food tends to be relatively sim-

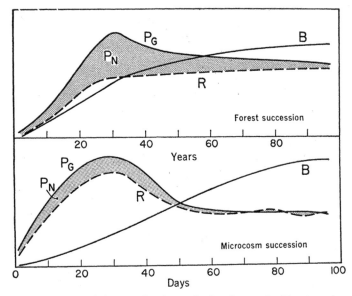

Fig. 1. Comparison of the energetics of succession in a forest and a laboratory microcosm. P_G, gross production; P_N, net production; R, total community respiration; B, total biomass.

ple and linear in the very early stages of succession, as a consequence of low diversity. Furthermore, heterotrophic utilization of net production occurs predominantly by way of grazing food chains—that is, plant-herbivore-carnivore sequences. In contrast, food chains become complex webs in mature stages, with the bulk of biological energy flow following detritus pathways (Table 1, item 5). In a mature forest, for example, less than 10 percent of annual net production is consumed (that is, grazed) in the living state (12); most is utilized as dead matter (detritus) through delayed and complex pathways involving as yet little understood animal-microorganism interactions. The time involved in an uninterrupted succession allows for increasingly intimate associations and reciprocal adaptations between plants and animals, which lead to the development of many mechanisms that reduce grazing—such as the development of indigestible supporting tissues (cellulose, lignin, and so on), feedback control between plants and herbivores (13), and increasing predatory pressure on herbivores (14). Such mechanisms enable the biological community to maintain the large and complex organic structure that mitigates perturbations of the physical environment. Severe stress or rapid changes brought about by outside forces can of course, rob the sys-

tem of these protective mechanisms and allow irruptive, cancerous growths of certain species to occur, as man too often finds to his sorrow. An example of a stress-induced pest irruption occurred at Brookhaven National Laboratory, where oaks became vulnerable to aphids when translocation of sugars and amino acids was impaired by continuing gamma irradiation (15).

Radionuclide tracers are providing a means of charting food chains in the intact outdoor ecosystem to a degree that will permit analysis within the concepts of network or matrix algebra. For example, we have recently been able to map, by use of a radiophosphorus tracer, the open, relatively linear food linkage between plants and insects in an early old-field successional stage (16).

Diversity and Succession

Perhaps the most controversial of the successional trends pertain to the complex and much discussed subject of diversity (17). It is important to distinguish between different kinds of diversity indices, since they may not follow parallel trends in the same gradient or developmental series. Four components of diversity are listed in Table 1, items 8 through 11.

The variety of species, expressed as

a species-number ratio or a species-area ratio, tends to increase during the early stages of community development. A second component of species diversity is what has been called equitability, or evenness (18), in the apportionment of individuals among the species. For example, two systems each containing 10 species and 100 individuals have the same diversity in terms of species-number ratio but could have widely different equitabilities depending on the apportionment of the 100 individuals among the 10 species—for example, 91-1-1-1-1-1-1-1-1 at one extreme or 10 individuals per species at the other. The Shannon formula,

$$- \Sigma \frac{ni}{N} \log_2 \frac{ni}{N}$$

where ni is the number of individuals in each species and N is the total number of individuals, is widely used as a diversity index because it combines the variety and equitability components in one approximation. But, like all such lumping parameters, Shannon's formula may obscure the behavior of these two rather different aspects of diversity. For example, in our most recent field experiments, an acute stress from insecticide reduced the number of species of insects relative to the number of individuals but increased the evenness in the relative abundances of the surviving species (19). Thus, in this case the "variety" and "evenness" components would tend to cancel each other in Shannon's formula.

While an increase in the variety of species together with reduced dominance by any one species or small group of species (that is, increased evenness) can be accepted as a general probability during succession (20), there are other community changes that may work against these trends. An increase in the size of organisms, an increase in the length and complexity of life histories, and an increase in interspecific competition that may result in competitive exclusion of species (Table 1, items 12–14) are trends that may reduce the number of species that can live in a given area. In the bloom stage of succession organisms tend to be small and to have simple life histories and rapid rates of reproduction. Changes in size appear to be a consequence of, or an adaptation to, a shift in nutrients from inorganic to organic (Table 1, item 7). In a mineral nutrient-rich environment, small size is of selective advantage, especially to autotrophs, because of the greater surface-to-volume ratio. As the ecosystem develops, however, inorganic nutrients tend to become more and more tied up in the biomass (that is, to become intrabiotic), so that the selective advantage shifts to larger organisms (either larger individuals of the same species or larger species, or both) which have greater storage capacities and more complex life histories, thus are adapted to exploiting seasonal or periodic releases of nutrients or other resources. The question of whether the seemingly direct relationship between organism size and stability is the result of positive feedback or is merely fortuitous remains unanswered (21).

Thus, whether or not species diversity continues to increase during succession will depend on whether the increase in potential niches resulting from increased biomass, stratification (Table 1, item 9), and other consequences of biological organization exceeds the countereffects of increasing size and competition. No one has yet been able to catalogue all the species in any sizable area, much less follow total species diversity in a successional series. Data are so far available only for segments of the community (trees, birds, and so on). Margalef (4) postulates that diversity will tend to peak during the early or middle stages of succession and then decline in the climax. In a study of bird populations along a successional gradient we found a bimodal pattern (22); the number of species increased during the early stages of old-field succession, declined during the early forest stages, and then increased again in the mature forest.

Species variety, equitability, and stratification are only three aspects of diversity which change during succession. Perhaps an even more important trend is an increase in the diversity of organic compounds, not only of those within the biomass but also of those excreted and secreted into the media (air, soil, water) as by-products of the increasing community metabolism. An increase in such "biochemical diversity" (Table 1, item 10) is illustrated by the increase in the variety of plant pigments along a successional gradient in aquatic situations, as described by Margalef (4, 23). Biochemical diversity within populations, or within systems as a whole, has not yet been systematically studied to the degree the subject of species diversity has been. Consequently, few generalizations can be made, except that it seems safe to say that, as succession progresses, organic extrametabolites probably serve increasingly important functions as regulators which stabi-

lize the growth and composition of the ecosystem. Such metabolites may, in fact, be extremely important in preventing populations from overshooting the equilibrial density, thus in reducing oscillations as the system develops stability.

The cause-and-effect relationship between diversity and stability is not clear and needs to be investigated from many angles. If it can be shown that biotic diversity does indeed enhance physical stability in the ecosystem, or is the result of it, then we would have an important guide for conservation practice. Preservation of hedgerows, woodlots, noneconomic species, noneutrophicated waters, and other biotic variety in man's landscape could then be justified on scientific as well as esthetic grounds, even though such preservation often must result in some reduction in the production of food or other immediate consumer needs. In other words, is variety only the spice of life, or is it a necessity for the long life of the total ecosystem comprising man and nature?

Nutrient Cycling

An important trend in successional development is the closing or "tightening" of the biogeochemical cycling of major nutrients, such as nitrogen, phosphorus, and calcium (Table 1, items 15–17). Mature systems, as compared to developing ones, have a greater capacity to entrap and hold nutrients for cycling within the system. For example, Bormann and Likens (24) have estimated that only 8 kilograms per hectare out of a total pool of exchangeable calcium of 365 kilograms per hectare is lost per year in stream outflow from a North Temperate watershed covered with a mature forest. Of this, about 3 kilograms per hectare is replaced by rainfall, leaving only 5 kilograms to be obtained from weathering of the underlying rocks in order for the system to maintain mineral balance. Reducing the volume of the vegetation, or otherwise setting the succession back to a younger state, results in increased water yield by way of stream outflow (25), but this

Table 1. A tabular model of ecological succession: trends to be expected in the development of ecosystems.

Ecosystem attributes	Developmental stages	Mature stages
Community energetics		
1. Gross production/community respiration (P/R ratio)	Greater or less than 1	Approaches 1
2. Gross production/standing crop biomass (P/B ratio)	High	Low
3. Biomass supported/unit energy flow (B/E ratio)	Low	High
4. Net community production (yield)	High	Low
5. Food chains	Linear, predominantly grazing	Weblike, predominantly detritus
Community structure		
6. Total organic matter	Small	Large
7. Inorganic nutrients	Extrabiotic	Intrabiotic
8. Species diversity—variety component	Low	High
9. Species diversity—equitability component	Low	High
10. Biochemical diversity	Low	High
11. Stratification and spatial heterogeneity (pattern diversity)	Poorly organized	Well-organized
Life history		
12. Niche specialization	Broad	Narrow
13. Size of organism	Small	Large
14. Life cycles	Short, simple	Long, complex
Nutrient cycling		
15. Mineral cycles	Open	Closed
16. Nutrient exchange rate, between organisms and environment	Rapid	Slow
17. Role of detritus in nutrient regeneration	Unimportant	Important
Selection pressure		
18. Growth form	For rapid growth ("r-selection")	For feedback control ("K-selection")
19. Production	Quantity	Quality
Overall homeostasis		
20. Internal symbiosis	Undeveloped	Developed
21. Nutrient conservation	Poor	Good
22. Stability (resistance to external perturbations)	Poor	Good
23. Entropy	High	Low
24. Information	Low	High

greater outflow is accompanied by greater losses of nutrients, which may also produce downstream eutrophication. Unless there is a compensating increase in the rate of weathering, the exchangeable pool of nutrients suffers gradual depletion (not to mention possible effects on soil structure resulting from erosion). High fertility in "young systems" which have open nutrient cycles cannot be maintained without compensating inputs of new nutrients; examples of such practice are the continuous-flow culture of algae, or intensive agriculture where large amounts of fertilizer are imported into the system each year.

Because rates of leaching increase in a latitudinal gradient from the poles to the equator, the role of the biotic community in nutrient retention is especially important in the high-rainfall areas of the subtropical and tropical latitudes, including not only land areas but also estuaries. Theoretically, as one goes equatorward, a larger percentage of the available nutrient pool is tied up in the biomass and a correspondingly lower percentage is in the soil or sediment. This theory, however, needs testing, since data to show such a geographical trend are incomplete. It is perhaps significant that conventional North Temperate row-type agriculture, which represents a very youthful type of ecosystem, is successful in the humid tropics only if carried out in a system of "shifting agriculture" in which the crops alternate with periods of natural vegetative redevelopment. Tree culture and the semiaquatic culture of rice provide much better nutrient retention and consequently have a longer life expectancy on a given site in these warmer latitudes.

Selection Pressure:

Quantity versus Quality

MacArthur and Wilson (26) have reviewed stages of colonization of islands which provide direct parallels with stages in ecological succession on continents. Species with high rates of reproduction and growth, they find, are more likely to survive in the early uncrowded stages of island colonization. In contrast, selection pressure favors species with lower growth potential but better capabilities for competitive survival under the equilibrium density of late stages. Using the terminology of growth equations, where r is the intrinsic rate of

increase and K is the upper asymptote or equilibrium population size, we may say that "r selection" predominates in early colonization, with "K selection" prevailing as more and more species and individuals attempt to colonize (Table 1, item 18). The same sort of thing is even seen within the species in certain "cyclic" northern insects in which "active" genetic strains found at low densities are replaced at high densities by "sluggish" strains that are adapted to crowding (27).

Genetic changes involving the whole biota may be presumed to accompany the successional gradient, since, as described above, quantity production characterizes the young ecosystem while quality production and feedback control are the trademarks of the mature system (Table 1, item 19). Selection at the ecosystem level may be primarily interspecific, since species replacement is a characteristic of successional series or seres. However, in most well-studied seres there seem to be a few early successional species that are able to persist through to late stages. Whether genetic changes contribute to adaptation in such species has not been determined, so far as I know, but studies on population genetics of *Drosophila* suggest that changes in genetic composition could be important in population regulation (28). Certainly, the human population, if it survives beyond its present rapid growth stage, is destined to be more and more affected by such selection pressures as adaptation to crowding becomes essential.

Overall Homeostasis

This brief review of ecosystem development emphasizes the complex nature of processes that interact. While one may well question whether all the trends described are characteristic of all types of ecosystems, there can be little doubt that the net result of community actions is symbiosis, nutrient conservation, stability, a decrease in entropy, and an increase in information (Table 1, items 20–24). The overall strategy is, as I stated at the beginning of this article, directed toward achieving as large and diverse an organic structure as is possible within the limits set by the available energy input and the prevailing physical conditions of existence (soil, water, climate, and so on). As studies of biotic communities become more functional and sophisticated,

one is impressed with the importance of mutualism, parasitism, predation, commensalism, and other forms of symbiosis. Partnership between unrelated species is often noteworthy (for example, that between coral coelenterates and algae, or between mycorrhizae and trees). In many cases, at least, biotic control of grazing, population density, and nutrient cycling provide the chief positive-feedback mechanisms that contribute to stability in the mature system by preventing overshoots and destructive oscillations. The intriguing question is, Do mature ecosystems age, as organisms do? In other words, after a long period of relative stability or "adulthood," do ecosystems again develop unbalanced metabolism and become more vulnerable to diseases and other perturbations?

Relevance of Ecosystem Development Theory to Human Ecology

Figure 1 depicts a basic conflict between the strategies of man and of nature. The "bloom-type" relationships, as exhibited by the 30-day microcosm or the 30-year forest, illustrate man's present idea of how nature should be directed. For example, the goal of agriculture or intensive forestry, as now generally practiced, is to achieve high rates of production of readily harvestable products with little standing crop left to accumulate on the landscape—in other words, a high P/B efficiency. Nature's strategy, on the other hand, as seen in the outcome of the successional process, is directed toward the reverse efficiency—a high B/P ratio, as is depicted by the relationship at the right in Fig. 1. Man has generally been preoccupied with obtaining as much "production" from the landscape as possible, by developing and maintaining early successional types of ecosystems, usually monocultures. But, of course, man does not live by food and fiber alone; he also needs a balanced CO_2–O_2 atmosphere, the climatic buffer provided by oceans and masses of vegetation, and clean (that is, unproductive) water for cultural and industrial uses. Many essential life-cycle resources, not to mention recreational and esthetic needs, are best provided man by the less "productive" landscapes. In other words, the landscape is not just a supply depot but is also the *oikos*—the home—in which we must live. Until recently mankind has more or less taken for granted the gas-exchange, water-purification, nutrient-cycling, and other protective functions of self-maintaining ecosystems, chiefly because neither his numbers nor his environmental manipulations have been great enough to affect regional and global balances. Now, of course, it is painfully evident that such balances are being affected, often detrimentally. The "one problem, one solution approach" is no longer adequate and must be replaced by some form of ecosystem analysis that considers man as a part of, not apart from, the environment.

The most pleasant and certainly the safest landscape to live in is one containing a variety of crops, forests, lakes, streams, roadsides, marshes, seashores, and "waste places"—in other words, a mixture of communities of different ecological ages. As individuals we more or less instinctively surround our houses with protective, nonedible cover (trees, shrubs, grass) at the same time that we strive to coax extra bushels from our cornfield. We all consider the cornfield a "good thing," of course, but most of us would not want to live there, and it would certainly be suicidal to cover the whole land area of the biosphere with cornfields, since the boom and bust oscillation in such a situation would be severe.

The basic problem facing organized society today boils down to determining in some objective manner when we are getting "too much of a good thing." This is a completely new challenge to mankind because, up until now, he has had to be concerned largely with too little rather than too much. Thus, concrete is a "good thing," but not if half the world is covered with it. Insecticides are "good things," but not when used, as they now are, in an indiscriminate and wholesale manner. Likewise, water impoundments have proved to be very useful man-made additions to the landscape, but obviously we don't want the whole country inundated! Vast man-made lakes solve some problems, at least temporarily, but yield comparative little food or fiber, and, because of high evaporative losses, they may not even be the best device for storing water; it might better be stored in the watershed, or underground in aquifers. Also, the cost of building large dams is a drain on already overtaxed revenues. Although as individuals we readily recognize that we can have too many dams or other large-scale environmental changes, governments are so fragmented and lacking in systems-analysis capabilities that there

Table 2. Contrasting characteristics of young and mature-type ecosystems.

Young	Mature
Production	Protection
Growth	Stability
Quantity	Quality

is no effective mechanism whereby negative feedback signals can be received and acted on before there has been a serious overshoot. Thus, today there are governmental agencies, spurred on by popular and political enthusiasm for dams, that are putting on the drawing boards plans for damming every river and stream in North America!

Society needs, and must find as quickly as possible, a way to deal with the landscape as a whole, so that manipulative skills (that is, technology) will not run too far ahead of our understanding of the impact of change. Recently a national ecological center outside of government and a coalition of governmental agencies have been proposed as two possible steps in the establishment of a political control mechanism for dealing with major environmental questions. The soil conservation movement in America is an excellent example of a program dedicated to the consideration of the whole farm or the whole watershed as an ecological unit. Soil conservation is well understood and supported by the public. However, soil conservation organizations have remained too exclusively farm-oriented, and have not yet risen to the challenge of the urban-rural landscape, where lie today's most serious problems. We do, then, have potential mechanisms in American society that could speak for the ecosystem as a whole, but none of them are really operational (29).

The general relevance of ecosystem development theory to landscape planning can, perhaps, be emphasized by the "mini-model" of Table 2, which contrasts the characteristics of young and mature-type ecosystems in more general terms than those provided by Table 1. It is mathematically impossible to obtain a maximum for more than one thing at a time, so one cannot have both extremes at the same time and place. Since all six characteristics listed in Table 2 are desirable in the aggregate, two possible solutions to the dilemma immediately suggest themselves. We can compromise so as to provide

moderate quality and moderate yield on all the landscape, or we can deliberately plan to compartmentalize the landscape so as to simultaneously maintain highly productive and predominantly protective types as separate units subject to different management strategies (strategies ranging, for example, from intensive cropping on the one hand to wilderness management on the other). If ecosystem development theory is valid and applicable to planning, then the so-called multiple-use strategy, about which we hear so much, will work only through one or both of these approaches, because, in most cases, the projected multiple uses conflict with one another. It is appropriate, then, to examine some examples of the compromise and the compartmental strategies.

Pulse Stability

A more or less regular but acute physical perturbation imposed from without can maintain an ecosystem at some intermediate point in the developmental sequence, resulting in, so to speak, a compromise between youth and maturity. What I would term "fluctuating water level ecosystems" are good examples. Estuaries, and intertidal zones in general, are maintained in an early, relatively fertile stage by the tides, which provide the energy for rapid nutrient cycling. Likewise, freshwater marshes, such as the Florida Everglades, are held at an early successional stage by the seasonal fluctuations in water levels. The dry-season drawdown speeds up aerobic decomposition of accumulated organic matter, releasing nutrients that, on reflooding, support a wet-season bloom in productivity. The life histories of many organisms are intimately coupled to this periodicity. The wood stork, for example, breeds when the water levels are falling and the small fish on which it feeds become concentrated and easy to catch in the drying pools. If the water level remains high during the usual dry season or fails to rise in the wet season, the stork will not nest (30). Stabilizing water levels in the Everglades by means of dikes, locks, and impoundments, as is now advocated by some, would, in my opinion, destroy rather than preserve the Everglades as we now know them just as surely as complete drainage would. Without periodic drawdowns and fires, the shallow basins would fill up with organic matter and succession would proceed from the present pond-and-prairie condition toward a scrub or swamp forest.

It is strange that man does not readily recognize the importance of recurrent changes in water level in a natural situation such as the Everglades when similar pulses are the basis for some of his most enduring food culture systems (*31*). Alternate filling and draining of ponds has been a standard procedure in fish culture for centuries in Europe and the Orient. The flooding, draining, and soil-aeration procedure in rice culture is another example. The rice paddy is thus the cultivated analogue of the natural marsh or the intertidal ecosystem.

Fire is another physical factor whose periodicity has been of vital importance to man and nature over the centuries. Whole biotas, such as those of the African grasslands and the California chaparral, have become adapted to periodic fires producing what ecologists often call "fire climaxes" (*32*). Man uses fire deliberately to maintain such climaxes or to set back succession to some desired point. In the southeastern coastal plain, for example, light fires of moderate frequency can maintain a pine forest against the encroachment of older successional stages which, at the present time at least, are considered economically less desirable. The fire-controlled forest yields less wood than a tree farm does (that is, young trees, all of about the same age, planted in rows and harvested on a short rotation schedule), but it provides a greater protective cover for the landscape, wood of higher quality, and a home for game birds (quail, wild turkey, and so on) which could not survive in a tree farm. The fire climax, then, is an example of a compromise between production simplicity and protection diversity.

It should be emphasized that pulse stability works only if there is a complete community (including not only plants but animals and microorganisms) adapted to the particular intensity and frequency of the perturbation. Adaptation—operation of the selection process—requires times measurable on the evolutionary scale. Most physical stresses introduced by man are too sudden, too violent, or too arrhythmic for adaptation to occur at the ecosystem level, so severe oscillation rather than stability results. In many cases, at least, modification of naturally adapted ecosystems for cultural purposes would seem preferable to complete redesign.

Prospects for a Detritus Agriculture

As indicated above, heterotrophic utilization of primary production in mature ecosystems involves largely a delayed consumption of detritus. There is no reason why man cannot make greater use of detritus and thus obtain food or other products from the more protective type of ecosystem. Again, this would represent a compromise, since the short-term yield could not be as great as the yield obtained by direct exploitation of the grazing food chain. A detritus agriculture, however, would have some compensating advantages. Present agricultural strategy is based on selection for rapid growth and edibility in food plants, which, of course, make them vulnerable to attack by insects and disease. Consequently, the more we select for succulence and growth, the more effort we must invest in the chemical control of pests; this effort, in turn, increases the likelihood of our poisoning useful organisms, not to mention ourselves. Why not also practice the reverse strategy—that is, select plants which are essentially unpalatable, or which produce their own systemic insecticides while they are growing, and then convert the net production into edible products by microbial and chemical enrichment in food factories? We could then devote our biochemical genius to the enrichment process instead of fouling up our living space with chemical poisons! The production of silage by fermentation of low-grade fodder is an example of such a procedure already in widespread use. The cultivation of detritus-eating fishes in the Orient is another example.

By tapping the detritus food chain man can also obtain an appreciable harvest from many natural systems without greatly modifying them or destroying their protective and esthetic value. Oyster culture in estuaries is a good example. In Japan, raft and long-line culture of oysters has proved to be a very practical way to harvest the natural microbial products of estuaries and shallow bays. Furukawa (*33*) reports that the yield of cultured oysters in the Hiroshima Prefecture has increased tenfold since 1950, and that the yield of oysters (some 240,000 tons of meat) from this one district alone in 1965 was ten times the yield of natural oysters from the entire country. Such oyster culture is feasible along the entire Atlantic and Gulf coasts of the United States. A large

investment in the culture of oysters and other seafoods would also provide the best possible deterrent against pollution, since the first threat of damage to the pollution-sensitive oyster industry would be immediately translated into political action!

The Compartment Model

Successful though they often are, compromise systems are not suitable nor desirable for the whole landscape. More emphasis needs to be placed on compartmentalization, so that growth-type, steady-state, and intermediate-type ecosystems can be linked with urban and industrial areas for mutual benefit. Knowing the transfer coefficients that define the flow of energy and the movement of materials and organisms (including man) between compartments, it should be possible to determine, through analog-computer manipulation, rational limits for the size and capacity of each compartment. We might start, for example, with a simplified model, shown in Fig. 2, consisting of four compartments of equal area, partitioned according to the basic biotic-function criterion—that is, according to whether the area is (i) productive, (ii) protective, (iii) a compromise between (i) and (ii) or (iv), urban-industrial. By continually refining the transfer coefficients on the basis of real world situations, and by increasing and decreasing the size and capacity of each compartment through computer simulation, it would be possible to determine objectively the limits that must eventually be imposed on each compartment in order to maintain regional and global balances in the exchange of vital energy and of materials. A systems-analysis procedure provides at least one approach to the solution of the basic dilemma posed by the question "How do we determine when we are getting too much of a good thing?" Also it provides a means of evaluating the energy drains imposed on ecosystems by pollution, radiation, harvest, and other stresses (*34*).

Implementing any kind of compartmentalization plan, of course, would require procedures for zoning the landscape and restricting the use of some land and water areas. While the principle of zoning in cities is universally accepted, the procedures now followed do not work very well because zoning restrictions are too easily overturned by short-term economic and population pressures. Zoning the landscape would require a whole new order of thinking. Greater use of legal measures providing for tax relief, restrictions on use, scenic easements, and public ownership will be required if appreciable land and water areas are to be held in the "protective" categories. Several states (for example, New Jersey and California), where pollution and population pressure are beginning to hurt, have made a start in this direction by enacting "open space" legislation designed to get as much unoccupied land as possible into a "protective" status so that future uses can be planned on a rational and scientific basis. The United States as a whole is fortunate in that large areas of the country are in national forests, parks, wildlife refuges, and so on. The fact that such areas, as well as the bordering oceans, are not quickly exploitable gives us time for the accelerated ecological study and programming needed to determine what proportions of different types of landscape provide a safe balance between man and nature. The open oceans, for example, should forever be allowed to remain protective rather than productive territory, if Alfred Redfield's (35) assumptions are correct. Redfield views the oceans, the major part of the hydrosphere, as the biosphere's governor, which slows down and controls the rate of decomposition and nutrient regeneration, thereby creating and maintaining the highly aerobic terrestrial environment to which the higher forms of life, such as man, are adapted. Eutrophication of the ocean in a last-ditch effort to feed the populations of the land could well have an adverse effect on the oxygen reservoir in the atmosphere.

Until we can determine more precisely how far we may safely go in expanding intensive agriculture and urban sprawl at the expense of the protective landscape, it will be good insurance to hold inviolate as much of the latter as possible. Thus, the preservation of natural areas is not a peripheral luxury for society but a capital investment from which we expect to draw interest. Also, it may well be that restrictions in the use of land and water are our only practical means of avoiding overpopulation or too great an exploitation of resources, or both. Interestingly enough, restriction of land use is the analogue of a natural behavioral control mechanism known as "territoriality" by which

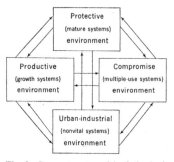

Fig. 2. Compartment model of the basic kinds of environment required by man, partitioned according to ecosystem development and life-cycle resource criteria.

many species of animals avoid crowding and social stress (36).

Since the legal and economic problems pertaining to zoning and compartmentalization are likely to be thorny, I urge law schools to establish departments, or institutes, of "landscape law" and to start training "landscape lawyers" who will be capable not only of clarifying existing procedures but also of drawing up new enabling legislation for consideration by state and national governing bodies. At present, society is concerned—and rightly so—with human rights, but environmental rights are equally vital. The "one man one vote" idea is important, but so also is a "one man one hectare" proposition.

Education, as always, must play a role in increasing man's awareness of his dependence on the natural environment. Perhaps we need to start teaching the principles of ecosystem in the third grade. A grammar school primer on man and his environment could logically consist of four chapters, one for each of the four essential kinds of environment, shown diagrammatically in Fig. 2.

Of the many books and articles that are being written these days about man's environmental crisis, I would like to cite two that go beyond "crying out in alarm" to suggestions for bringing about a reorientation of the goals of society. Garrett Hardin, in a recent article in *Science* (37), points out that, since the optimum population density is less than the maximum, there is no strictly technical solution to the problem of pollution caused by overpopulation; a solution, he suggests, can only be achieved through moral and legal means of "mutual coercion, mutually agreed upon by the majority of people."

Earl F. Murphy, in a book entitled *Governing Nature* (38), emphasizes that the regulatory approach alone is not enough to protect life-cycle resources, such as air and water, that cannot be allowed to deteriorate. He discusses permit systems, effluent charges, receptor levies, assessment, and cost-internalizing procedures as economic incentives for achieving Hardin's "mutually agreed upon coercion."

It goes without saying that the tabular model for ecosystem development which I have presented here has many parallels in the development of human society itself. In the pioneer society, as in the pioneer ecosystem, high birth rates, rapid growth, high economic profits, and exploitation of accessible and unused resources are advantageous, but, as the saturation level is approached, these drives must be shifted to considerations of symbiosis (that is, "civil rights," "law and order," "education," and "culture"), birth control, and the recycling of resources. A balance between youth and maturity in the socio-environmental system is, therefore, the really basic goal that must be achieved if man as a species is to successfully pass through the present rapid-growth stage, to which he is clearly well adapted, to the ultimate equilibrium-density stage, of which he as yet shows little understanding and to which he now shows little tendency to adapt.

References and Notes

1. E. P. Odum, *Ecology* (Holt, Rinehart & Winston, New York, 1963), chap. 6.
2. H. T. Odum and R. C. Pinkerton, *Amer. Scientist* **43**, 331 (1955).
3. A. J. Lotka, *Elements of Physical Biology* (Williams and Wilkins, Baltimore, 1925).
4. R. Margalef, *Advan. Frontiers Plant Sci.* **2**, 137 (1963); *Amer. Naturalist* **97**, 357 (1963).
5. R. J. Beyers, *Ecol. Monographs* **33**, 281 (1963).
6. The systems so far used to test ecological principles have been derived from sewage and farm ponds and are cultured in half-strength No. 36 Taub and Dollar medium [*Limnol. Oceanog.* **9**, 61 (1964)]. They are closed to organic input or output but are open to the atmosphere through the cotton plug in the neck of the flask. Typically, liter-sized microecosystems contain two or three species of nonflagellated algae and one to three species each of flagellated protozoans, ciliated protozoans, rotifers, nematodes, and ostracods; a system derived from a sewage pond contained at least three species of fungi and 13 bacterial isolates [R. Gordon, thesis, University of Georgia (1967)]. These cultures are thus a kind of minimum ecosystem containing those small species originally found in the ancestral pond that are able to function together as a self-contained unit under the restricted conditions of the laboratory flask and the controlled environment of a growth chamber [temperature, 65° to 75°F (18° to 24°C); photoperiod, 12 hours; illumination, 100 to 1000 footcandles].
7. G. D. Cooke, *BioScience* **17**, 717 (1967).
8. T. Kira and T. Shidei, *Japan. J. Ecol.* **17**, 70 (1967).
9. The metabolism of the microcosms was

monitored by measuring diurnal pH changes, and the biomass (in terms of total organic matter and total carbon) was determined by periodic harvesting of replicate systems.

10. F. J. H. Mackereth, *Proc. Roy. Soc. London Ser. B* **161**, 295 (1965); U. M. Cowgill and G. E. Hutchinson, *Proc. Intern. Limnol. Ass.* **15**, 644 (1964); A. D. Harrison, *Trans. Roy. Soc. S. Africa* **36**, 213 (1962).
11. R. Margalef, *Proc. Intern. Limnol. Ass.* **15**, 169 (1964).
12. J. R. Bray, *Oikos* **12**, 70 (1961).
13. D. Pimentel, *Amer. Naturalist* **95**, 65 (1961).
14. R. T. Paine, *ibid.* **100**, 65 (1966).
15. G. M. Woodwell, *Brookhaven Nat. Lab. Pub. 924(T-381)* (1965), pp. 1–15.
16. R. G. Wiegert, E. P. Odum, J. H. Schnell, *Ecology* **48**, 75 (1967).
17. For selected general discussions of patterns of species diversity, see E. H. Simpson, *Nature* **163**, 688 (1949); C. B. Williams, *J. Animal Ecol.* **22**, 14 (1953); G. E. Hutchinson, *Amer. Naturalist* **93**, 145 (1959); R. Margalef, *Gen. Systems* **3**, 36 (1958); R. MacArthur and J. MacArthur, *Ecology* **42**, 594 (1961); N. G. Hairston, *ibid.* **40**, 404 (1959); B. C. Patten, *J. Marine Res. (Sears Found. Marine Res.)* **20**, 57 (1960); E. G. Leigh, *Proc. Nat. Acad. Sci. U.S.* **55**, 777 (1965); E. R. Pianka, *Amer. Naturalist* **100**, 33 (1966); E. C. Pielou, *J. Theoret. Biol.* **13**, 131 (1966).
18. M. Lloyd and R. J. Ghelardi, *J. Animal Ecol.* **33**, 217 (1964); E. C. Pielou, *J. Theoret. Biol.* **10**, 370 (1966).
19. G. W. Barrett, *Ecology* **49**, 1019 (1969).
20. In our studies of natural succession following grain culture, both the species-to-numbers and the equitability indices increased for all trophic levels but especially for predators and parasites. Only 44 percent of the species in the natural ecosystem were phytophagous, as compared to 77 percent in the grain field.
21. J. T. Bonner, *Size and Cycle* (Princeton Univ. Press, Princeton, N.J., 1963); P. Frank, *Ecology* **49**, 355 (1968).
22. D. W. Johnston and E. P. Odum, *Ecology* **37**, 50 (1956).
23. R. Margalef, *Oceanog. Marine Biol. Annu. Rev.* **5**, 257 (1967).
24. F. H. Bormann and G. E. Likens, *Science* **155**, 424 (1967).
25. Increased water yield following reduction of vegetative cover has been frequently demonstrated in experimental watersheds throughout the world [see A. R. Hibbert, in *International Symposium on Forest Hydrology* (Pergamon Press, New York, 1967), pp. 527–543]. Data on the long-term hydrologic budget (rainfall input relative to stream outflow) are available at many of these sites, but mineral budgets have yet to be systematically studied. Again, this is a prime objective in the "ecosystem analysis" phase of the International Biological Program.
26. R. H. MacArthur and E. O. Wilson, *Theory of Island Biogeography* (Princeton Univ. Press, Princeton, N.J., 1967).
27. Examples are the tent caterpillar [see W. G. Wellington, *Can. J. Zool.* **35**, 293 (1957)] and the larch budworm [see W. Baltensweiler, *Can. Entomologist* **96**, 792 (1964)].
28. F. J. Ayala, *Science* **162**, 1453 (1968).
29. Ira Rubinoff, in discussing the proposed sea-level canal joining the Atlantic and Pacific oceans [*Science* **161**, 857 (1968)], calls for a "control commission for environmental manipulation" with "broad powers of approving, disapproving, or modifying all major alterations of the marine or terrestrial environments. . . ."
30. See M. P. Kahl, *Ecol. Monographs* **34**, 97 (1964).
31. The late Aldo Leopold remarked long ago [*Symposium on Hydrobiology* (Univ. of Wisconsin Press, Madison, 1941), p. 17] that man does not perceive organic behavior in systems unless he has built them himself. Let us hope it will not be necessary to rebuild the entire biosphere before we recognize the worth of natural systems!
32. See C. F. Cooper, *Sci. Amer.* **204**, 150 (April 1961).
33. See "Proceedings Oyster Culture Workshop, Marine Fisheries Division, Georgia Game and Fish Commission, Brunswick" (1968), pp. 49–61.
34. See H. T. Odum, in *Symposium on Primary Productivity and Mineral Cycling in Natural Ecosystems*, H. E. Young, Ed. (Univ. of Maine Press, Orono, 1967), p. 81; ———, in *Pollution and Marine Ecology* (Wiley, New York, 1967), p. 99; K. E. F. Watt, *Ecology and Resource Management* (McGraw-Hill, New York, 1968).
35. A. C. Redfield, *Amer. Scientist* **46**, 205 (1958).
36. R. Ardrey, *The Territorial Imperative* (Atheneum, New York, 1967).
37. G. Hardin, *Science* **162**, 1243 (1968).
38. E. F. Murphy, *Governing Nature* (Quadrangle Books, Chicago, 1967).

RESILIENCE AND STABILITY OF ECOLOGICAL SYSTEMS

C. S. Holling

Institute of Resource Ecology, University of British Columbia, Vancouver, Canada

INTRODUCTION

Individuals die, populations disappear, and species become extinct. That is one view of the world. But another view of the world concentrates not so much on presence or absence as upon the numbers of organisms and the degree of constancy of their numbers. These are two very different ways of viewing the behavior of systems and the usefulness of the view depends very much on the properties of the system concerned. If we are examining a particular device designed by the engineer to perform specific tasks under a rather narrow range of predictable external conditions, we are likely to be more concerned with consistent nonvariable performance in which slight departures from the performance goal are immediately counteracted. A quantitative view of the behavior of the system is, therefore, essential. With attention focused upon achieving constancy, the critical events seem to be the amplitude and frequency of oscillations. But if we are dealing with a system profoundly affected by changes external to it, and continually confronted by the unexpected, the constancy of its behavior becomes less important than the persistence of the relationships. Attention shifts, therefore, to the qualitative and to questions of existence or not.

Our traditions of analysis in theoretical and empirical ecology have been largely inherited from developments in classical physics and its applied variants. Inevitably, there has been a tendency to emphasize the quantitative rather than the qualitative, for it is important in this tradition to know not just that a quantity is larger than another quantity, but precisely how much larger. It is similarly important, if a quantity fluctuates, to know its amplitude and period of fluctuation. But this orientation may simply reflect an analytic approach developed in one area because it was useful and then transferred to another where it may not be.

Our traditional view of natural systems, therefore, might well be less a meaningful reality than a perceptual convenience. There can in some years be more owls and fewer mice and in others, the reverse. Fish populations wax and wane as a natural condition, and insect populations can range over extremes that only logarithmic transformations can easily illustrate. Moreover, over distinct areas, during long or short periods of time, species can completely disappear and then reappear. Different and useful insight might be obtained, therefore, by viewing the behavior of ecological systems in terms of the probability of extinction of their elements, and by shifting emphasis from the equilibrium states to the conditions for persistence.

An equilibrium centered view is essentially static and provides little insight into the transient behavior of systems that are not near the equilibrium. Natural, undisturbed systems are likely to be continually in a transient state; they will be equally so under the influence of man. As man's numbers and economic demands increase, his use of resources shifts equilibrium states and moves populations away from equilibria. The present concerns for pollution and endangered species are specific signals that the well-being of the world is not adequately described by concentrating on equilibria and conditions near them. Moreover, strategies based upon these two different views of the world might well be antagonistic. It is at least conceivable that the effective and responsible effort to provide a maximum sustained yield from a fish population or a nonfluctuating supply of water from a watershed (both equilibrium-centered views) might paradoxically increase the chance for extinctions.

The purpose of this review is to explore both ecological theory and the behavior of natural systems to see if different perspectives of their behavior can yield different insights useful for both theory and practice.

Some Theory

Let us first consider the behavior of two interacting populations: a predator and its prey, a herbivore and its resource, or two competitors. If the interrelations are at all regulated we might expect a disturbance of one or both populations in a constant environment to be followed by fluctuations that gradually decrease in amplitude. They might be represented as in Figure 1, where the fluctuations of each population over time are shown as the sides of a box. In this example the two populations in some sense are regulating each other, but the lags in the response generate a series of oscillations whose amplitude gradually reduces to a constant and sustained value for each population. But if we are also concerned with persistence we would like to know not just how the populations behave from one particular pair of starting values, but from all possible pairs since there might well be combinations of starting populations for which ultimately the fate of one or other of the populations is extinction. It becomes very difficult on time plots to show the full variety of responses possible, and it proves convenient to plot a trajectory in a phase plane. This is shown by the end of the box in Figure 1 where the two axes represent the density of the two populations.

The trajectory shown on that plane represents the sequential change of the two populations at constant time intervals. Each point represents the unique density of each population at a particular point in time and the arrows indicate the direction of change over time. If oscillations are damped, as in the case shown, then the trajectory is represented as a closed spiral that eventually reaches a stable equilibrium.

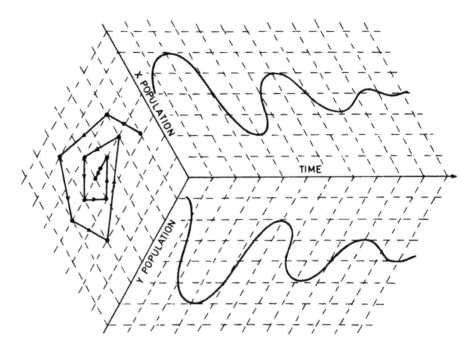

Figure 1 Derivation of a phase plane showing the changes in numbers of two populations over time.

We can imagine a number of different forms for trajectories in the phase plane (Figure 2). Figure 2a shows an open spiral which would represent situations where fluctuations gradually increase in amplitude. The small arrows are added to suggest that this condition holds no matter what combination of populations initiates the trajectory. In Figure 2b the trajectories are closed and given any starting point eventually return to that point. It is particularly significant that each starting point generates a unique cycle and there is no tendency for points to converge to a single cycle or point. This can be termed "neutral stability" and it is the kind of stability achieved by an imaginary frictionless pendulum.

Figure 2c represents a stable system similar to that of Figure 1, in which all possible trajectories in the phase plane spiral into an equilibrium. These three examples are relatively simple and, however relevant for classical stability analysis, may well be theoretical curiosities in ecology. Figures 2d–2f add some complexities. In a sense Figure 2d represents a combination of a and c, with a region in the center of the phase plane within which all possible trajectories spiral inwards to equilibrium. Those outside this region spiral outwards and lead eventually to extinction of one or the other populations. This is an example of local stability in contrast to the global stability of Figure 2c. I designate the region within which stability occurs as the domain of attraction, and the line that contains this domain as the boundary of the attraction domain.

The trajectories in Figure 2e behave in just the opposite way. There is an internal region within which the trajectories spiral out to a stable limit cycle and beyond

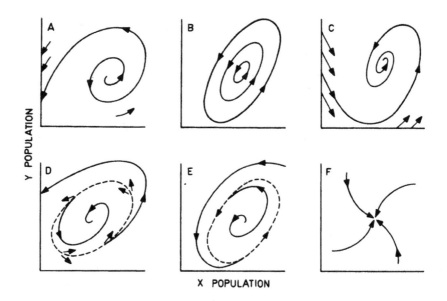

Figure 2 Examples of possible behaviors of systems in a phase plane; (a) unstable equilibrium, (b) neutrally stable cycles, (c) stable equilibrium, (d) domain of attraction, (e) stable limit cycle, (f) stable node.

which they spiral inwards to it. Finally, a stable node is shown in Figure 2f in which there are no oscillations and the trajectories approach the node monotonically. These six figures could be combined in an almost infinite variety of ways to produce several domains of attraction within which there could be a stable equilibrium, a stable limit cycle, a stable node, or even neutrally stable orbits. Although I have presumed a constant world throughout, in the presence of random fluctuations of parameters or of driving variables (Walters 39), any one trajectory could wander with only its general form approaching the shape of the trajectory shown. These added complications are explored later when we consider real systems. For the moment, however, let us review theoretical treatments in the light of the possibilities suggested in Figure 2.

The present status of ecological stability theory is very well summarized in a number of analyses of classical models, particularly May's (23–25) insightful analyses of the Lotka-Volterra model and its expansions, the graphical stability analyses of Rosenzweig (33, 34), and the methodological review of Lewontin (20).

May (24) reviews the large class of coupled differential equations expressing the rate of change of two populations as continuous functions of both. The behavior of these models results from the interplay between (*a*) stabilizing negative feedback or density-dependent responses to resources and predation, and (*b*) the destabilizing effects produced by the way individual predators attack and predator numbers respond to prey density [termed the functional and numerical responses, as in Holling (11)]. Various forms have been given to these terms; the familiar Lotka-Volterra model includes the simplest and least realistic, in which death of prey is caused only by predation, predation is a linear function of the product of prey and predator populations, and growth of the predator population is linearly proportional to the same product. This model generates neutral stability as in Figure 2b, but the assumptions are very unrealistic since very few components are included, there are no explicit lags or spatial elements, and thresholds, limits, and nonlinearities are missing.

These features have all been shown to be essential properties of the predation process (Holling 12, 13) and the effect of adding some of them has been analyzed by May (24). He points out that traditional ways of analyzing the stability properties of models using analytical or graphical means (Rosenzweig & MacArthur 33, Rosenzweig 34, 35) concentrate about the immediate neighborhood of the equilibrium. By doing this, linear techniques of analysis can be applied that are analytically tractable. Such analyses show that with certain defined sets of parameters stable equilibrium points or nodes exist (such as Figure 2c), while for other sets they do not, and in such cases the system is, by default, presumed to be unstable, as in Figure 2a. May (24), however, invokes a little-used theorem of Kolmogorov (Minorksy 26) to show that all these models have either a stable equilibrium point or a stable limit cycle (as in Figure 2e). Hence he concludes that the conditions presumed by linear analysis are unstable, and in fact must lead to stable limit cycles. In every instance, however, the models are globally rather than locally stable, limiting their behavior to that shown in either Figures 2c or 2e.

There is another tradition of models that recognizes the basically discontinuous features of ecological systems and incorporates explicit lags. Nicholson and Bailey initiated this tradition when they developed a model using the output of attacks and survivals within one generation as the input for the next (29). The introduction of this explicit lag generates oscillations that increase in amplitude until one or other of the species becomes extinct (Figure 2a). Their assumptions are as unrealistically simple as Lotka's and Volterra's; the instability results because the number of attacking predators at any moment is so much a consequence of events in the previous generation that there are "too many" when prey are declining and "too few" when prey are increasing. If a lag is introduced into the Lotka-Volterra formulation (Wangersky & Cunningham 40) the same instability results.

The sense one gains, then, of the behavior of the traditional models is that they are either globally unstable or globally stable, that neutral stability is very unlikely, and that when the models are stable a limit cycle is a likely consequence.

Many, but not all, of the simplifying assumptions have been relaxed in simulation models, and there is one example (Holling & Ewing 14) that joins the two traditions initiated by Lotka-Volterra and Nicholson and Bailey and, further, includes more realism in the operation of the stabilizing and destabilizing forces. These modifications are described in more detail later; the important features accounting for the difference in behavior result from the introduction of explicit lags, a functional response of predators that rises monotonically to a plateau, a nonrandom (or contagious) attack by predators, and a minimum prey density below which reproduction does not occur. With these changes a very different pattern emerges that conforms most closely to Figure 2d. That is, there exists a domain of attraction within which there is a stable equilibrium; beyond that domain the prey population becomes extinct. Unlike the Nicholson and Bailey model, the stability becomes possible, although in a limited region, because of contagious attack. [Contagious attack implies that for one reason or another some prey have a greater probability of being attacked than others, a condition that is common in nature (Griffiths & Holling 9).] The influence of contagious attack becomes significant whenever predators become abundant in relation to the prey, for then the susceptible prey receive the burden of attention, allowing more prey to escape than would be expected by random contact. This "inefficiency" of the predator allows the system to counteract the destabilizing effects of the lag.

If this were the only difference the system would be globally stable, much as Figure 2c. The inability of the prey to reproduce at low densities, however, allows some of the trajectories to cut this reproduction threshold, and the prey become extinct. This introduces a lower prey density boundary to the attraction domain and, at the same time, a higher prey density boundary above which the amplitudes of the oscillations inevitably carry the population below the reproduction threshold. The other modifications in the model, some of which have been touched on above, alter this picture in degree only. The essential point is that a more realistic representation of the behavior of interacting populations indicates the existence of at least one domain of attraction. It is quite possible, within this domain, to imagine stable equilibrium points, stable nodes, or stable limit cycles. Whatever the detailed configuration, the existence of discrete domains of attraction immediately suggests important consequences for the persistence of the system and the probability of its extinction.

Such models, however complex, are still so simple that they should not be viewed in a definitive and quantitative way. They are more powerfully used as a starting point to organize and guide understanding. It becomes valuable, therefore, to ask what the models leave out and whether such omissions make isolated domains of attraction more or less likely.

Theoretical models generally have not done well in simultaneously incorporating realistic behavior of the processes involved, randomness, spatial heterogeneity, and an adequate number of dimensions or state variables. This situation is changing very rapidly as theory and empirical studies develop a closer technical partnership. In what follows I refer to real world examples to determine how the four elements that tend to be left out might further affect the behavior of ecological systems.

SOME REAL WORLD EXAMPLES

Self-Contained Ecosystems

In the broadest sense, the closest approximation we could make of a real world example that did not grossly depart from the assumptions of the theoretical models would be a self-contained system that was fairly homogenous and in which climatic fluctuations were reasonably small. If such systems could be discovered they would reveal how the more realistic interaction of real world processes could modify the patterns of systems behavior described above. Very close approximations to any of these conditions are not likely to be found, but if any exist, they are apt to be fresh water aquatic ones. Fresh water lakes are reasonably contained systems, at least within their watersheds; the fish show considerable mobility throughout, and the properties of the water buffer the more extreme effects of climate. Moreover, there have been enough documented man-made disturbances to liken them to perturbed systems in which either the parameter values or the levels of the constituent populations are changed. In a crude way, then, the lake studies can be likened to a partial exploration of a phase space of the sorts shown in Figure 2. Two major classes of disturbances have occurred: first, the impact of nutrient enrichment from man's domestic and industrial wastes, and second, changes in fish populations by harvesting.

The paleolimnologists have been remarkably successful in tracing the impact of man's activities on lake systems over surprisingly long periods. For example, Hutchinson (17) has reconstructed the series of events occurring in a small crater lake in Italy from the last glacial period in the Alps (2000 to 1800 BC) to the present. Between the beginning of the record and Roman times the lake had established a trophic equilibrium with a low level of productivity which persisted in spite of dramatic changes in surroundings from *Artemesia* steppe, through grassland, to fir and mixed oak forest. Then suddenly the whole aquatic system altered. This alteration towards eutrophication seems to have been initiated by the construction of the Via Cassia about 171 BC, which caused a subtle change in the hydrographic regime. The whole sequence of environmental changes can be viewed as changes in parameters or driving variables, and the long persistence in the face of these major changes suggests that natural systems have a high capacity to absorb change without dramatically altering. But this resilient character has its limits, and when the limits are passed, as by the construction of the Roman highway, the system rapidly changes to another condition.

More recently the activities of man have accelerated and limnologists have recorded some of the responses to these changes. The most dramatic change consists of blooms of algae in surface waters, an extraordinary growth triggered, in most instances, by nutrient additions from agricultural and domestic sources.

While such instances of nutrient addition provide some of the few examples available of perturbation effects in nature, there are no controls and the perturbations are exceedingly difficult to document. Nevertheless, the qualitative pattern seems consistent, particularly in those lakes (Edmundson 4, Hasler 10) to which sewage has been added for a time and then diverted elsewhere. This pulse of disturbance characteristically triggers periodic algal blooms, low oxygen conditions, the sudden disappearance of some plankton species, and appearance of others. As only one example, the nutrient changes in Lake Michigan (Beeton 2) have been accompanied by the replacement of the cladoceran *Bosmina coregoni* by *B. Longirostris, Diaptomus oregonensis* has become an important copepod species, and a brackish water copepod *Eurytemora affinis* is a new addition to the zooplankton.

In Lake Erie, which has been particularly affected because of its shallowness and intensity of use, the mayfly *Hexagenia,* which originally dominated the benthic community, has been almost totally replaced by oligochetes. There have been blooms of the diatom *Melosira binderana,* which had never been reported from the

United States until 1961 but now comprises as much as 99% of the total phytoplankton around certain islands. In those cases where sewage has been subsequently diverted there is a gradual return to less extreme conditions, the slowness of the return related to the accumulation of nutrients in sediments.

The overall pattern emerging from these examples is the sudden appearance or disappearance of populations, a wide amplitude of fluctuations, and the establishment of new domains of attraction.

The history of the Great Lakes provides not only some particularly good information on responses to man-made enrichment, but also on responses of fish populations to fishing pressure. The eutrophication experience touched on above can be viewed as an example of systems changes in driving variables and parameters, whereas the fishing example is more an experiment in changing state variables. The fisheries of the Great Lakes have always selectively concentrated on abundant species that are in high demand. Prior to 1930, before eutrophication complicated the story, the lake sturgeon in all the Great Lakes, the lake herring in Lake Erie, and the lake whitefish in Lake Huron were intensively fished (Smith 37). In each case the pattern was similar: a period of intense exploitation during which there was a prolonged high level harvest, followed by a sudden and precipitous drop in populations. Most significantly, even though fishing pressure was then relaxed, none of these populations showed any sign of returning to their previous levels of abundance. This is not unexpected for sturgeon because of their slow growth and late maturity, but it is unexpected for herring and whitefish. The maintenance of these low populations in recent times might be attributed to the increasingly unfavorable chemical or biological environment, but in the case of the herring, at least, the declines took place in the early 1920s before the major deterioration in environment occurred. It is as if the population had been shifted by fishing pressure from a domain with a high equilibrium to one with a lower one. This is clearly not a condition of neutral stability as suggested in Figure 2b since once the populations were lowered to a certain point the decline continued even though fishing pressure was relaxed. It can be better interpreted as a variant of Figure 2d where populations have been moved from one domain of attraction to another.

Since 1940 there has been a series of similar catastrophic changes in the Great Lakes that has led to major changes in the fish stocks. Beeton (2) provides graphs summarizing the catch statistics in the lakes for many species since 1900. Lake trout, whitefish, herring, walleye, sauger, and blue pike have experienced precipitous declines of populations to very low values in all of the lakes. The changes generally conform to the same pattern. After sustained but fluctuating levels of harvest the catch dropped dramatically in a span of a very few years, covering a range of from one to four orders of magnitude. In a number of examples particularly high catches were obtained just before the drop. Although catch statistics inevitably exaggerate the step-like character of the pattern, populations must have generally behaved in the way described.

The explanations for these changes have been explored in part, and involve various combinations of intense fishing pressure, changes in the physical and chemical environment, and the appearance of a foreign predator (the sea lamprey) and foreign competitors (the alewife and carp). For our purpose the specific cause is of less interest than the inferences that can be drawn concerning the resilience of these systems and their stability behavior. The events in Lake Michigan provide a typical example of the pattern in other lakes (Smith 37). The catch of lake trout was high, but fluctuated at around six million pounds annually from 1898 to 1940. For four years catches increased noticeably and then suddenly collapsed to near extinction by the 1950s due to a complete failure of natural reproduction. Lake herring and whitefish followed a similar pattern (Beeton 2: Figure 7). Smith (37) argues that the trigger for the lake trout collapse was the appearance of the sea lamprey that had spread through the Great Lakes after the construction of the Welland Canal. Although lamprey populations were extremely small at the time of the collapse, Smith argues that even a small mortality, added to a commercial harvest that was probably at the maximum for sustained yield, was sufficient to cause the collapse. Moreover, Ricker (31) has shown that fishing pressure shifts the age structure of fish populations towards younger ages. He demonstrates that a point can come where only slight increases in mortality can trigger a collapse of the kind noted for lake trout. In addition, the lake trout was coupled in a network of competitive and predatory interconnections with other species, and pressures on these might have contributed as well.

Whatever the specific causes, it is clear that the precondition for the collapse was set by the harvesting of fish, even though during a long period there were no obvious signs of problems. The fishing activity, however, progressively reduced the resilience of the system so that when the inevitable unexpected event occurred, the populations collapsed. If it had not been the lamprey, it would have been something else: a change in climate as part of the normal pattern of fluctuation, a change in the chemical or physical environment, or a change in competitors or predators. These examples again suggest distinct domains of attraction in which the populations forced close to the boundary of the domain can then flip over it.

The above examples are not isolated ones. In 1939 an experimental fishery was started in Lake Windermere to improve stocks of salmonids by reducing the abundance of perch (a competitor) and pike (a predator). Perch populations were particularly affected by trapping and the populations fell drastically in the first three years. Most significantly, although no perch have been removed from the North Basin since 1947, populations have still not shown any tendency to return to their previous level (Le Cren et al 19).

The same patterns have even been suggested for terrestrial systems. Many of the arid cattle grazing lands of the western United States have gradually become invaded and dominated by shrubs and trees like mesquite and cholla. In some instances grazing and the reduced incidence of fire through fire prevention programs allowed invasion and establishment of shrubs and trees at the expense of grass. Nevertheless, Glendening (8) has demonstrated, from data collected in a 17-year experiment in which intensity of grazing was manipulated, that once the trees have gained sufficient size and density to completely utilize or materially reduce the moisture supply, elimination of grazing will not result in the grassland reestablishing itself. In short, there is a level of the state variable "trees" that, once achieved, moves the system from one domain of attraction to another. Return to the original domain can only be made by an explicit reduction of the trees and shrubs.

These examples point to one or more distinct domains of attraction in which the important point is not so much how stable they are within the domain, but how likely it is for the system to move from one domain into another and so persist in a changed configuration.

This sampling of examples is inevitably biased. There are few cases well documented over a long period of time, and certainly some systems that have been greatly disturbed have fully recovered their original state once the disturbance was removed. But the recovery in most instances is in open systems in which reinvasion is the key ingredient. These cases are discussed below in connection with the effects of spatial heterogeneity. For the moment I conclude that distinct domains of attraction are not uncommon within closed systems. If such is the case, then further confirmation should be found from empirical evidence of the way processes which link organisms operate, for it is these processes that are the cause of the behavior observed.

Process Analysis

One way to represent the combined effects of processes like fecundity, predation, and competition is by using Ricker's (30) reproduction curves. These simply represent the population in one generation as a function of the population in the previous generation, and examples are shown in Figures 3a, c, and e. In the simplest form, and the one most used in practical fisheries management (Figure 3a), the reproduction curve is dome-shaped. When it crosses a line with slope 1 (the straight line in the figures) an equilibrium condition is possible, for at such cross-overs the popula-

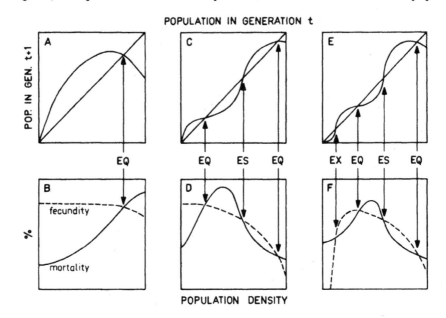

Figure 3 Examples of various reproduction curves (a, c, and e) and their derivation from the contributions of fecundity and mortality (b, d, and f).

tion in one generation will produce the same number in the next. It is extremely difficult to detect the precise form of such curves in nature, however; variability is high, typically data are only available for parts of any one curve, and the treatment really only applies to situations where there are no lags. It is possible to deduce various forms of reproduction curves, however, by disaggregating the contributions of fecundity and mortality. The three lower graphs in Figure 3b, 3d, and 3f represent this disaggregation of their counterpart reproduction curves. The simplest types of reproduction curve (Figure 3a) can arise from a mortality that regularly increases with density and either a constant fecundity or a declining one. With fecundity expressed as the percentage mortality necessary to just balance reproduction, the cross-over point of the curves represents the equilibrium condition. But we know that the effects of density on fecundity and mortality can be very much more complicated.

Mortality from predation, for example, has been shown to take a number of classic forms (Holling 11, 13). The individual attack by predators as a function of prey density (the functional response to prey density) can increase with a linear rise to a plateau (type 1), a concave or negatively accelerated rise to a plateau (type 2), or an S-shaped rise to a plateau (type 3). The resulting contribution to mortality from these responses can therefore show ranges of prey density in which there is direct density dependence (negative feedback from the positively accelerated portions of the type 3 response), density independence (the straight line rise of type 1), and inverse dependence (the positive feedback from the negatively accelerated and plateau portions of the curves). There are, in addition, various numerical responses generated by changes in the number of predators as the density of their prey increases. Even for those predators whose populations respond by increasing, there often will be a limit to the increase set by other conditions in the environment. When populations are increasing they tend to augment the negative feedback features (although with a delay), but when populations are constant, despite increasing prey density, the percent mortality will inevitably decline since individual attack eventually saturates at complete satiation (the plateaux of all three functional responses). In Figures 3d and 3f the mortality curves shown summarize a common type. The rising or direct density-dependent limb of the curve is induced by increasing predator populations and by the reduced intensity of attack at low densities, shown by the initial positively accelerated portion of the S-shaped type 3 response. Such a condition is common for predators with alternate prey, both vertebrates (Holling 14) and at least some invertebrates (Steele 38). The declining inverse density-dependent limb is induced by satiation of the predator and a numerical response that has been reduced or stopped.

Fecundity curves that decline regularly over a very wide range of increasing population densities (as in Figure 3d) are common and have been referred to as *Drosophila*-type curves (Fujita 6). This decline in fecundity is caused by increased competition for oviposition sites, interference with mating, and increased sterility. The interaction between a dome-shaped mortality curve and a monotonically decreasing fecundity curve can generate equilibrium conditions (Figure 3d). Two stable equilibria are possible, but between these two is a transient equilibrium designated as the escape threshold (ES in Figure 3). Effects of random changes on populations or parameters could readily shift densities from around the lower equilibrium to above this escape threshold, and in these circumstances populations would inevitably increase to the higher equilibrium.

The fecundity curves are likely to be more complex, however, since it seems inevitable that at some very low densities fecundity will decline because of difficulties in finding mates and the reduced effect of a variety of social facilitation behaviors. We might even logically conclude that for many species there is a minimum density below which fecundity is zero. A fecundity curve of this Allee-type (Fujita 6) has been empirically demonstrated for a number of insects (Watt 42) and is shown in Figure 3f. Its interaction with the dome-shaped mortality curve can add another transient equilibrium, the extinction threshold (EX in Figure 3f). With this addition there is a lower density such that if populations slip below it they will proceed inexorably to extinction. The extinction threshold is particularly likely since it has been shown mathematically that each of the three functional response curves will intersect with the ordinate of percent predation at a value above zero (Holling 13).

Empirical evidence, therefore, suggests that realistic forms to fecundity and mortality curves will generate sinuous reproduction curves like those in Figures 3c and 3e with the possibility of a number of equilibrium states, some transient and some stable. These are precisely the conditions that will generate domains of attraction, with each domain separated from others by the extinction and escape thresholds. This analysis of process hence adds support to the field observations discussed earlier.

The behavior of systems in phase space cannot be completely understood by the graphical representations presented above. These graphs are appropriate only when effects are immediate; in the face of the lags that generate cyclic behavior the reproduction curve should really produce two values for the population in generation $t + 1$ for each value of the population in generation t. The graphical treatment of Rosenzweig & MacArthur (33) to a degree can accommodate these lags and cyclic behavior. In their treatment they divide phase planes of the kind shown in Figure 2 into various regions of increasing and decreasing x and y populations. The regions are separated by two lines, one representing the collection of points at which the prey population does not change in density ($dx/dt = 0$, the prey isocline) and one in which the predator population does not so change ($dy/dt = 0$, the predator isocline). They deduce that the prey isocline will be dome-shaped for much the same reason as described for the fecundity curves of Figure 3f. The predator isocline, in the simplest condition, is presumed to be vertical, assuming that only one fixed level of prey is necessary to just maintain the predator population at a zero instantaneous rate of change.

Intersection of the two isoclines indicates a point where both populations are at equilibrium. Using traditional linear stability analysis one can infer whether these equilibrium states are stable (Figure 2c) or not (Figure 2a). Considerable importance is attached to whether the predator isocline intersects the rising or falling portion of the prey isocline. As mentioned earlier these techniques are only appropriate near equilibrium (May 24), and the presumed unstable conditions in fact generate stable limit cycles (Figure 2e). Moreover, it is unlikely that the predator isocline is a vertical one in the real world, since competition between predators at high predator densities would so interfere with the attack process that a larger number of prey would be required for stable predator populations. It is precisely this condition that was demonstrated by Griffiths & Holling (9) when they showed that a large number of species of parasites distribute their attacks contagiously. The result is a "squabbling predator behavior" (Rosenzweig 34, 35) that decreases the efficiency of predation at high predator/prey ratios. This converts an unstable system (Figure 2a) to a stable one (Figure 2c); it is likely that stability is the rule, rather than the exception, irrespective of where the two isoclines cross.

The empirical evidence described above shows that realistic fecundity and mortality (particularly predation) processes will generate forms that the theorists might tend to identify as special subsets of more general conditions. But it is just these special subsets that separate the real world from all possible ones, and these more realistic forms will modify the general conclusions of simpler theory. The ascending limb of the Allee-type fecundity curve will establish, through interaction with mortality, a minimum density below which prey will become extinct. This can at the same time establish an upper prey density above which prey will become extinct because the amplitude of prey fluctuations will eventually carry the population over the extinction threshold, as shown in the outer trajectory of Figure 2d. These conditions alone are sufficient to establish a domain of attraction, although the boundaries of this domain need not be closed. Within the domain the contagious attack by predators can produce a stable equilibrium or a stable node. Other behaviors of the mortality agents, however, could result in stable limit cycles.

More realistic forms of functional response change this pattern in degree only. For example, a negatively accelerated type of functional response would tend to make the domain of attraction somewhat smaller, and an S-shaped one larger. Limitations in the predator's numerical response and thresholds for reproduction of predators, similar to those for prey, could further change the form of the domain. Moreover, the behaviors that produce the sinuous reproduction curves of Figures 3c and 3e can add additional domains. The essential point, however, is that these systems are not globally stable but can have distinct domains of attraction. So long as the populations remain within one domain they have a consistent and regular form of behavior. If populations pass a boundary to the domain by chance or through intervention of man, then the behavior suddenly changes in much the way suggested from the field examples discussed earlier.

The Random World

To this point, I have argued as if the world were completely deterministic. In fact, the behavior of ecological systems is profoundly affected by random events. It is important, therefore, to add another level of realism at this point to determine how the above arguments may be modified. Again, it is applied ecology that tends to supply the best information from field studies since it is only in such situations that data have been collected in a sufficiently intensive and extensive manner. As one example, for 28 years there has been a major and intensive study of the spruce budworm and its interaction with the spruce-fir forests of eastern Canada (Morris

27). There have been six outbreaks of the spruce budworm since the early 1700s (Baskerville 1) and between these outbreaks the budworm has been an exceedingly rare species. When the outbreaks occur there is major destruction of balsam fir in all the mature forests, leaving only the less susceptible spruce, the nonsusceptible white birch, and a dense regeneration of fir and spruce. The more immature stands suffer less damage and more fir survives. Between outbreaks the young balsam grow, together with spruce and birch, to form dense stands in which the spruce and birch, in particular, suffer from crowding. This process evolves to produce stands of mature and overmature trees with fir a predominant feature.

This is a necessary, but not sufficient, condition for the appearance of an outbreak; outbreaks occur only when there is also a sequence of unusually dry years (Wellington 43). Until this sequence occurs, it is argued (Morris 27) that various natural enemies with limited numerical responses maintain the budworm populations around a low equilibrium. If a sequence of dry years occurs when there are mature stand of fir, the budworm populations rapidly increase and escape the control by predators and parasites. Their continued increase eventually causes enough tree mortality to force a collapse of the populations and the reinstatement of control around the lower equilibrium. The reproduction curves therefore would be similar to those in Figures 3c or 3e.

In brief, between outbreaks the fir tends to be favored in its competition with spruce and birch, whereas during an outbreak spruce and birch are favored because they are less susceptible to budworm attack. This interplay with the budworm thus maintains the spruce and birch which otherwise would be excluded through competition. The fir persists because of its regenerative powers and the interplay of forest growth rates and climatic conditions that determine the timing of budworm outbreaks.

This behavior could be viewed as a stable limit cycle with large amplitude, but it can be more accurately represented by a distinct domain of attraction determined by the interaction between budworm and its associated natural enemies, which is periodically exceeded through the chance consequence of climatic conditions. If we view the budworm only in relation to its associated predators and parasites we might argue that it is highly unstable in the sense that populations fluctuate widely. But these very fluctuations are essential features that maintain persistence of the budworm, together with its natural enemies and its host and associated trees. By so fluctuating, successive generations of forests are replaced, assuring a continued food supply for future generations of budworm and the persistence of the system.

Until now I have avoided formal identification of different kinds of behavior of ecological systems. The more realistic situations like budworm, however, make it necessary to begin to give more formal definition to their behavior. It is useful to distinguish two kinds of behavior. One can be termed stability, which represents the ability of a system to return to an equilibrium state after a temporary disturbance; the more rapidly it returns and the less it fluctuates, the more stable it would be. But there is another property, termed resilience, that is a measure of the persistence of systems and of their ability to absorb change and disturbance and still maintain the same relationships between populations or state variables. In this sense, the budworm forest community is highly unstable and it is because of this instability that it has an enormous resilience. I return to this view frequently throughout the remainder of this paper.

The influence of random events on systems with domains of attraction is found in aquatic systems as well. For example, pink salmon populations can become stabilized for several years at very different levels, the new levels being reached by sudden steps rather than by gradual transition (Neave 28). The explanation is very much the same as that proposed for the budworm, involving an interrelation between negative and positive feedback mortality of the kinds described in Figures 3d and 3f, and random effects unrelated to density. The same pattern has been described by Larkin (18) in his simulation model of the Adams River sockeye salmon. This particular run of salmon has been characterized by a regular four-year periodicity since 1922, with one large or dominant year, one small or subdominant, and two years with very small populations. The same explanation as described above has been proposed with the added reality of a lag. Essentially, during the dominant year limited numerical responses produce an inverse density-dependent response as in the descending limb of the mortality curves of Figure 3d and 3f. The abundance of the prey in that year is nevertheless sufficient to establish populations of predators that have a major impact on the three succeeding low years. Buffering of predation by the smolts of the dominant year accounts for the larger size of the subdominant. These effects have been simulated (Larkin 18), and when random influences are imposed in order to simulate climatic variations the system has a distinct probability of flipping into another stable configuration that is actually reproduced in nature by sockeye salmon runs in other rivers. When subdominant escapement reaches a critical level there is about an equal chance that it may become the same size as the dominant one or shrivel to a very small size.

Random events, of course, are not exclusively climatic. The impact of fires on terrestrial ecosystems is particularly illuminating (Cooper 3) and the periodic appearance of fires has played a decisive role in the persistence of grasslands as well as certain forest communities. As an example, the random perturbation caused by fires in Wisconsin forests (Loucks 21) has resulted in a sequence of transient changes that move forest communities from one domain of attraction to another. The apparent instability of this forest community is best viewed not as an unstable condition alone, but as one that produces a highly resilient system capable of repeating itself and persisting over time until a disturbance restarts the sequence.

In summary, these examples of the influence of random events upon natural systems further confirm the existence of domains of attraction. Most importantly they suggest that instability, in the sense of large fluctuations, may introduce a resilience and a capacity to persist. It points out the very different view of the world that can be obtained if we concentrate on the boundaries to the domain of attraction rather than on equilibrium states. Although the equilibrium-centered view is analytically more tractable, it does not always provide a realistic understanding of the systems' behavior. Moreover, if this perspective is used as the exclusive guide to the management activities of man, exactly the reverse behavior and result can be produced than is expected.

The Spatial Mosaic

To this point, I have proceeded in a series of steps to gradually add more and more reality. I started with self-contained closed systems, proceeded to a more detailed explanation of how ecological processes operate, and then considered the influence of random events, which introduced heterogeneity over time.

The final step is now to recognize that the natural world is not very homogeneous over space, as well, but consists of a mosaic of spatial elements with distinct biological, physical, and chemical characteristics that are linked by mechanisms of biological and physical transport. The role of spatial heterogeneity has not been well explored in ecology because of the enormous logistic difficulties. Its importance, however, was revealed in a classic experiment that involved the interaction between a predatory mite, its phytophagous mite prey, and the prey's food source (Huffaker et al 15). Briefly, in the relatively small enclosures used, when there was unimpeded movement throughout the experimental universe, the system was unstable and oscillations increased in amplitude. When barriers were introduced to impede dispersal between parts of the universe, however, the interaction persisted. Thus populations in one small locale that suffer chance extinctions could be reestablished by invasion from other populations having high numbers—a conclusion that is confirmed by Roff's mathematical analysis of spatial heterogeneity (32).

There is one study that has been largely neglected that is, in a sense, a much more realistic example of the effects of both temporal and spatial heterogeneity of a population in nature (Wellington 44, 45). There is a peninsula on Vancouver Island in which the topography and climate combine to make a mosaic of favorable locales for the tent caterpillar. From year to year the size of these locales enlarges or contracts depending on climate; Wellington was able to use the easily observed changes in cloud patterns in any year to define these areas. The tent caterpillar, to add a further element of realism, has identifiable behavioral types that are determined not by genetics but by the nutritional history of the parents. These types represent a range from sluggish to very active, and the proportion of types affects the shape of the easily visible web the tent caterpillars spin. By combining these defined differences of behavior with observations on changing numbers, shape of webs, and changing cloud patterns, an elegant story of systems behavior emerges. In a favorable year locales that previously could not support tent caterpillars now can, and populations are established through invasion by the vigorous dispersers from other locales. In these new areas they tend to produce another generation with a high proportion of vigorous behavioral types. Because of their high dispersal behavior and the small area of the locale in relation to its periphery, they then tend to leave in greater numbers than they arrive. The result is a gradual increase in the proportion of more sluggish types to the point where the local population collapses. But, although its fluctuations are considerable, even under the most unfavorable conditions there are always enclaves suitable for the insect. It is an example of a population with high fluctuations that can take advantage of transient periods of favorable conditions and that has, because of this variability, a high degree of resilience and capacity to persist.

A further embellishment has been added in a study of natural insect populations by Gilbert & Hughes (7). They combined an insightful field study of the interaction between aphids and their parasites with a simulation model, concentrating upon a specific locale and the events within it under different conditions of immigration from other locales. Again the important focus was upon persistence rather than degree of fluctuation. They found that specific features of the parasite-host interaction allowed the parasite to make full use of its aphid resources just short of driving the host to extinction. It is particularly intriguing that the parasite and its host were introduced into Australia from Europe and in the short period that the parasite has been present in Australia there have been dramatic changes in its developmental rate and fecundity. The other major difference between conditions in Europe and Australia is that the immigration rate of the host in England is considerably higher than in Australia. If the immigration rate in Australia increased to the English level, then, according to the model the parasite should increase its fecundity from the Australian level to the English to make the most of its opportunity short of extinction. This study provides, therefore, a remarkable example of a parasite and its host evolving together to permit persistence, and further confirms the importance of systems resilience as distinct from systems stability.

SYNTHESIS

Some Definitions

Traditionally, discussion and analyses of stability have essentially equated stability to systems behavior. In ecology, at least, this has caused confusion since, in mathematical analyses, stability has tended to assume definitions that relate to conditions very near equilibrium points. This is a simple convenience dictated by the enormous analytical difficulties of treating the behavior of nonlinear systems at some distance from equilibrium. On the other hand, more general treatments have touched on questions of persistence and the probability of extinction, defining these measures as aspects of stability as well. To avoid this confusion I propose that the behavior of ecological systems could well be defined by two distinct properties: resilience and stability.

Resilience determines the persistence of relationships within a system and is a measure of the ability of these systems to absorb changes of state variables, driving variables, and parameters, and still persist. In this definition resilience is the property of the system and persistence or probability of extinction is the result. Stability, on the other hand, is the ability of a system to return to an equilibrium state after a temporary disturbance. The more rapidly it returns, and with the least fluctuation, the more stable it is. In this definition stability is the property of the system and the degree of fluctuation around specific states the result.

Resilience versus Stability

With these definitions in mind a system can be very resilient and still fluctuate greatly, i.e. have low stability. I have touched above on examples like the spruce budworm forest community in which the very fact of low stability seems to introduce high resilience. Nor are such cases isolated ones, as Watt (41) has shown in his analysis of thirty years of data collected for every major forest insect throughout Canada by the Insect Survey program of the Canada Department of the Environment. This statistical analysis shows that in those areas subjected to extreme climatic conditions the populations fluctuate widely but have a high capability of absorbing periodic extremes of fluctuation. They are, therefore, unstable using the restricted definition above, but highly resilient. In more benign, less variable climatic regions the populations are much less able to absorb chance climatic extremes even though the populations tend to be more constant. These situations show a high degree of stability and a lower resilience. The balance between resilience and stability is clearly a product of the evolutionary history of these systems in the face of the range of random fluctuations they have experienced.

In Slobodkin's terms (36) evolution is like a game, but a distinctive one in which the only payoff is to stay in the game. Therefore, a major strategy selected is not one maximizing either efficiency or a particular reward, but one which allows persistence by maintaining flexibility above all else. A population responds to any environmental change by the initiation of a series of physiological, behavioral, ecological, and genetic changes that restore its ability to respond to subsequent unpredictable environmental changes. Variability over space and time results in variability in numbers, and with this variability the population can simultaneously retain genetic and behavioral types that can maintain their existence in low populations together with others that can capitalize on chance opportunities for dramatic increase. The more homogeneous the environment in space and time, the more likely is the system to have low fluctuations and low resilience. It is not surprising, therefore, that the commerical fishery systems of the Great Lakes have provided a vivid example of the sensitivity of ecological systems to disruption by man, for they represent climatically buffered, fairly homogeneous and self-contained systems with relatively low variability and hence high stability and low resilience. Moreover, the goal of producing a maximum sustained yield may result in a more stable system of reduced resilience.

Nor is it surprising that however readily fish stocks in lakes can be driven to extinction, it has been extremely difficult to do the same to insect pests of man's crops. Pest systems are highly variable in space and time; as open systems they are much affected by dispersal and therefore have a high resilience. Similarly, some Arctic ecosystems thought of as fragile may be highly resilient, although unstable. Certainly this is not true for some subsystems in the Arctic, such as Arctic frozen soil, self-contained Arctic lakes, and cohesive social populations like caribou, but these might be exceptions to a general rule.

The notion of an interplay between resilience and stabilty might also resolve the conflicting views of the role of diversity and stability of ecological communities. Elton (5) and MacArthur (22) have argued cogently from empirical and theoretical points of view that stability is roughly proportional to the number of links between species in a trophic web. In essence, if there are a variety of trophic links the same flow of energy and nutrients will be maintained through alternate links when a species becomes rare. However, May's (23) recent mathematical analyses of models of a large number of interacting populations shows that this relation between increased diversity and stability is not a mathematical truism. He shows that randomly assembled complex systems are in general less stable, and never more stable, than less complex ones. He points out that ecological systems are likely to have evolved to a very small subset of all possible sets and that MacArthur's conclusions, therefore, might still apply in the real world. The definition of stability used, however, is the equilibrium-centered one. What May has shown is that complex systems might fluctuate more than less complex ones. But if there is more than one domain of attraction, then the increased variability could simply move the system from one domain to another. Also, the more species there are, the more equilibria there may be and, although numbers may thereby fluctuate considerably, the overall persistence might be enhanced. It would be useful to explore the possibility that instability in numbers can result in more diversity of species and in spatial patchiness, and hence in increased resilience.

Measurement

If there is a worthwhile distinction between resilience and stability it is important that both be measurable. In a theoretical world such measurements could be developed from the behavior of model systems in phase space. Just as it was useful to disaggregate the reproduction curves into their constituent components of mortality and fecundity, so it is useful to disaggregate the information in a phase plane. There are two components that are important: one that concerns the cyclic behavior and its frequency and amplitude, and one that concerns the configuration of forces caused by the positive and negative feedback relations.

To separate the two we need to imagine first the appearance of a phase space in which there are no such forces operating. This would produce a referent trajectory containing only the cyclic properties of the system. If the forces were operating, departure from this referent trajectory would be a measure of the intensity of the forces. The referent trajectories that would seem to be most useful would be the neutrally stable orbits of Figure 2b, for we can arbitrarily imagine these trajectories as moving on a flat plane. At least for more realistic models parameter values can be discovered that do generate neutrally stable orbits. In the complex predator-prey model of Holling (14), if a range of parameters is chosen to explore the effects of different degrees of contagion of attack, the interaction is unstable when attack is random and stable when it is contagious. We have recently shown that there is a critical level of contagion between these extremes that generates neutrally stable orbits. These orbits, then, have a certain frequency and amplitude and the departure of more realistic trajectories from these referent ones should allow the computation of the vector of forces. If these were integrated a potential field would be represented with peaks and valleys. If the whole potential field were a shallow bowl the system would be globally stable and all trajectories would spiral to the bottom of the bowl, the equilibrium point. But if, at a minimum, there were a lower extinction threshold for prey then, in effect, the bowl would have a slice taken out of one side, as suggested in Figure 4. Trajectories that initiated far up on the side of the bowl would have amplitude that would carry the trajectory over the slice cut out of it. Only those

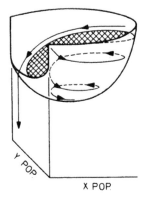

Figure 4 Diagramatic representation showing the feedback forces as a potential field upon which trajectories move. The shaded portion is the domain of attraction.

trajectories that just avoided the lowest point of the gap formed by the slice would spiral in to the bowl's bottom. If we termed the bowl the basin of attraction (Lewontin 20) then the domain of attraction would be determined by both the cyclic behavior and the configuration of forces. It would be confined to a smaller portion of the bottom of the bowl, and one edge would touch the bottom portion of the slice taken out of the basin.

This approach, then, suggests ways to measure relative amounts of resilience and stability. There are two resilience measures: Since resilience is concerned with probabilities of extinction, firstly, the overall area of the domain of attraction will in part determine whether chance shifts in state variables will move trajectories outside the domain. Secondly, the height of the lowest point of the basin of attraction (e.g. the bottom of the slice described above) above equilibrium will be a measure of how much the forces have to be changed before all trajectories move to extinction of one or more of the state variables.

The measures of stability would be designed in just the opposite way from those that measure resilience. They would be centered on the equilibrium rather than on the boundary of the domain, and could be represented by a frequency distribution of the slopes of the potential field and by the velocity of the neutral orbits around the equilibrium.

But such measures require an immeanse amount of knowledge of a system and it is unlikely that we will often have all that is necessary. Hughes & Gilbert (16), however, have suggested a promising approach to measuring probabilities of extinction and hence of resilience. They were able to show in a stochastic model that the distribution of surviving population sizes at any given time does not differ significantly from a negative binomial. This of course is just a description, but it does provide a way to estimate the very small probability of zero, i.e. of extinction, from the observed mean and variance. The configuration of the potential field and the cyclic behavior will determine the number and form of the domains of attraction, and these will in turn affect the parameter values of the negative binomial or of any other distribution function that seems appropriate. Changes in the zero class of the distribution, that is, in the probability of extinction, will be caused by these parameter values, which can then be viewed as the relative measures of resilience. It will be important to explore this technique first with a number of theoretical models so that the appropriate distributions and their behavior can be identified. It will then be quite feasible, in the field, to sample populations in defined areas, apply the appropriate distribution, and use the parameter values as measures of the degree of resilience.

APPLICATION

The resilience and stability viewpoints of the behavior of ecological systems can yield very different approaches to the management of resources. The stability view emphasizes the equilibrium, the maintenance of a predictable world, and the harvesting of nature's excess production with as little fluctuation as possible. The resilience view emphasizes domains of attraction and the need for persistence. But extinction is not purely a random event; it results from the interaction of random events with those deterministic forces that define the shape, size, and characteristics of the domain of attraction. The very approach, therefore, that assures a stable maximum sustained yield of a renewable resource might so change these deterministic conditions that the resilience is lost or reduced so that a chance and rare event that previously could be absorbed can trigger a sudden dramatic change and loss of structural integrity of the system.

A management approach based on resilience, on the other hand, would emphasize the need to keep options open, the need to view events in a regional rather than a local context, and the need to emphasize heterogeneity. Flowing from this would be not the presumption of sufficient knowledge, but the recognition of our ignorance; not the assumption that future events are expected, but that they will be unexpected. The resilience framework can accommodate this shift of perspective, for it does not require a precise capacity to predict the future, but only a qualitative capacity to devise systems that can absorb and accommodate future events in whatever unexpected form they may take.

Literature Cited

1. Baskerville, G. L. 1971. *The Fir-Spruce-Birch Forest and the Budworm.* Forestry Service, Canada Dept. Environ., Fredericton, N. B. Unpublished
2. Beeton, A. D. 1969. Changes in the environment and biota of the Great Lakes. *Eutrophication: Causes, Consequences, Correctives.* Washington DC: Nat. Acad. Sci.
3. Cooper, C. F. 1961. The ecology of fire. *Sci. Am.* 204:150–6, 158, 160
4. Edmondson, W. T. 1961. Changes in Lake Washington following increase in nutrient income. *Verh. Int. Ver. Limnol.* 14:167–75
5. Elton, C. S. 1958. *The Ecology of Invasions by Animals and Plants.* London: Methuen
6. Fujita, H. 1954. An interpretation of the changes in type of the population density effect upon the oviposition rate. *Ecology* 35:253–7
7. Gilbert, N., Hughes, R. D. 1971. A model of an aphid population—three adventures. *J. Anim. Ecol.* 40:525–34

8. Glendening, G. 1952. Some quantitative data on the increase of mesquite and cactus on a desert grassland range in southern Arizona. *Ecology* 33:319–28
9. Griffiths, K. J., Holling, C. S. 1969. A competition submodel for parasites and predators. *Can. Entomol.* 101:785–818
10. Hasler, A. D. 1947. Eutrophication of lakes by domestic sewage. *Ecology* 28:383–95
11. Holling, C. S. 1961. Principles of insect predation. *Ann. Rev. Entomol.* 6:163–82
12. Holling, C. S. 1966. The functional response of invertebrate predators to prey density. *Mem. Entomol. Soc. Can.* 48:1–86
13. Holling, C. S. 1965. The functional response of predators to prey density and its role in mimicry and population regulations. *Mem. Entomol. Soc. Can.* 45:1–60
14. Holling, C. S., Ewing, S. 1971. Blind man's buff: exploring the response space generated by realistic ecological simulation models. *Proc. Int. Symp. Statist. Ecol.* New Haven, Conn.: Yale Univ. Press 2:207–29
15. Huffaker, C. D., Shea, K. P., Herman, S. S. 1963. Experimental studies on predation. Complex dispersion and levels of food in an acarine predator-prey interaction. *Hilgardia* 34:305–30
16. Hughes, R. D., Gilbert, N. 1968. A model of an aphid population—a general statement. *J. Anim. Ecol.* 40:525–34
17. Hutchinson, G. E. 1970. Ianula: an account of the history and development of the Lago di Monterosi, Latium, Italy. *Trans. Am. Phil. Soc.* 60:1–178
18. Larkin, P. A. 1971. Simulation studies of the Adams River Sockeye Salmon *(Oncarhynchus nerka). J. Fish. Res. Bd. Can.* 28:1493–1502
19. Le Cren, E. D., Kipling, C., McCormack, J. C. 1972. Windermere: effects of exploitation and eutrophication on the salmonid community. *J. Fish. Res. Bd. Can.* 29:819–32
20. Lewontin, R. C. 1969. The meaning of stability. *Diversity and Stability of Ecological Systems, Brookhaven Symp. Biol.* 22:13–24
21. Loucks, O. L. 1970. Evolution of diversity, efficiency and community stability. *Am. Zool.* 10:17–25
22. MacArthur, R. 1955. Fluctuations of animal populations and a measure of community stability. *Ecology* 36:533–6
23. May, R. M. 1971. Stability in multi-species community models. *Math. Biosci.* 12:59–79
24. May, R. M. 1972. Limit cycles in predator-prey communities. *Science* 177:900–2
25. May, R. M. 1972. Will a large complex system be stable? *Nature* 238:413–14
26. Minorsky, N. 1962. *Nonlinear Oscillations.* Princeton, N.J.: Van Nostrand
27. Morris, R. F. 1963. The dynamics of epidemic spruce budworm populations. *Mem. Entomol. Soc. Can.* 31:1–332
28. Neave, F. 1953. Principles affecting the size of pink and chum salmon populations in British Columbia. *J. Fish. Res. Bd. Can.* 9:450–91
29. Nicholson, A. J., Bailey, V. A. 1935. The balance of animal populations—Part I. *Proc. Zool. Soc. London* 1935:551–98
30. Ricker, W. E. 1954. Stock and recruitment. *J. Fish. Res. Bd. Can.* 11:559–623
31. Ricker, W. E. 1963. Big effects from small causes: two examples from fish population dynamics. *J. Fish. Res. Bd. Can.* 20:257–84
32. Roff, D. A. 1973. Spatial heterogeneity and the persistence of populations. *J. Theor. Pop. Biol.* In press
33. Rosenzweig, M. L., MacArthur, R. H. 1963. Graphical representation and stability condition of predator-prey interactions. *Am. Natur.* 97:209–23
34. Rosenzweig, M. L. 1971. Paradox of enrichment: destabilization of exploitation ecosystems in ecological time. *Science* 171:385–7
35. Rosenzweig, M. L. 1972. Stability of enriched aquatic ecosystems. *Science* 175:564–5
36. Slobodkin, L. B. 1964. The strategy of evolution. *Am. Sci.* 52:342–57
37. Smith, S. H. 1968. Species succession and fishery exploitation in the Great Lakes. *J. Fish. Res. Bd. Can.* 25:667–93
38. Steele, J. H. 1971. Factors controlling marine ecosystems. *The Changing Chemistry of the Oceans,* ed. D. Dryssen, D. Jaquer, 209–21. Nobel Symposium 20, New York: Wiley
39. Walters, C. J. 1971. Systems ecology: the systems approach and mathematical models in ecology. *Fundamentals of Ecology,* ed. E. P. Odum. Philadelphia: Saunders. 3rd ed.
40. Wangersky, P. J., Cunningham, W. J. 1957. Time lag in prey-predator population models. *Ecology* 38:136–9
41. Watt, K. E. F. 1968. A computer approach to analysis of data on weather, population fluctuations, and disease. *Biometeorology, 1967 Biology Colloquium,* ed. W. P. Lowry. Corvallis, Oregon: Oregon State Univ. Press

42. Watt, K. E. F. 1960. The effect of population density on fecundity in insects. *Can. Entomol.* 92:674–95

43. Wellington, W. G. 1952. Air mass climatology of Ontario north of Lake Huron and Lake Superior before outbreaks of the spruce budworm and the forest tree caterpillar. *Can. J. Zool.* 30: 114–27

44. Wellington, W. G. 1964. Qualitative changes in populations in unstable environments. *Can. Entomol.* 96:436–51

45. Wellington, W. G. 1965. The use of cloud patterns to outline areas with different climates during population studies. *Can. Entomol.* 97:617–31

Extinction, Substitution, and Ecosystem Services

Paul R. Ehrlich and Harold A. Mooney

The loss of services to humanity following extinctions ranges from trivial to catastrophic, depending on the number of elements (populations, species, guilds) deleted and the degree of control each exerted in the system. Most attempts to substitute other organisms for those lost have been unsuccessful, to one degree or another, and prospects for increasing the success rate in the foreseeable future are not great. Attempts to supply the lost services by other means tend to be expensive failures in the long run. A conservative approach to the maintenance of services through minimizing anthropogenic extinctions is recommended. (*Accepted for publication 20 October 1982*)

At this moment humanity is forcing species and populations to extinction at what may well be an unprecedented rate (Council on Environmental Quality 1980, National Research Council 1980). This raises two critical questions. First, what can be predicted about the impact of extinctions of given elements (populations, species, guilds) in ecosystems upon the services delivered to humanity by those systems? Second, to what degree can those elements be replaced without impairing either short- or long-term ecosystem functioning? Economists have paid much attention to substitution for inorganic resources (e.g., Barnett and Morse 1963), but the potential for substituting for living resources has barely been explored.

We hope that this preliminary look at these questions and our conclusions may stimulate the research required to provide the data upon which rules for predicting the success of substitutions can be based. However, we agree with Ehrenfeld (1978) that there are compelling reasons for preserving the biotic diversity of Earth *regardless* of any present or future discoveries made about the benefits people may receive from other organisms.

Ehrlich and Mooney are with the Department of Biological Sciences, Stanford University. Stanford, CA 94305. © 1983 American Institute of Biological Sciences. All rights reserved.

ECOSYSTEM STRUCTURE AND SERVICES

We use the term *controllers* to refer to the organisms that determine the structure (e.g., species composition, trophic relationships) of the ecosystem and through which the principal flows of energy and materials pass. The term is used broadly, for we recognize that many controllers cannot be identified. One of the two most crucial sets of controllers in all major ecosystems is the one through which solar energy and carbon and other nutrients enter living systems (the producer trophic level). The other set consists of those that release the nutrients bound up in organisms by production for reuse (the decomposer trophic level). Loss of either of these sets brings the collapse of the entire system.

The number of controllers at any given level varies from ecosystem to ecosystem. For example, the trees in a redwood forest or tropical rainforest fix most of the solar energy captured by the system; they also accumulate and retain a large portion of the nutrients. The trees also exercise important control over water and air flows, microclimate, and, through their effect on the albedo and exchanges of CO_2 and water vapor with the atmosphere, macroclimate. In the redwood forest, a single species appears to be the primary controller; in the rainforest, a guild of many species shares control.

Internal configuration is largely controlled by the interactions among the species in the ecosystem. Population sizes and structures are influenced in ecological time by weather (itself influenced by other species), competitors, predators, pollinators, dispersers, seed bed providers, domicile providers, and other factors. In evolutionary time they are determined in large part by climate and coevolution (Schneider 1983), and the plant–herbivore interface has been proposed as the most important in the generation of structure (Ehrlich and Raven 1964). Here, there are usually major and minor controllers, determined by population sizes, physical size of individuals, degree of specialization, precise ecological role, and so on.

The degree of control exercised by a single species is often difficult to evaluate, even when its role is well understood—in part because controllers tend to function in series. For example, many trees depend on mycorrhizal fungi in the soil for nutrients (Wilde 1968). But how much credit the fungus should be given for the many crucial control functions of the tree is difficult to determine. Similarly, the role of the small moth *Cactoblastis cactorum* in Queensland is difficult to evaluate without knowing the history of its presence. One might decide it is a minor component. We know, however, if the moth were removed, millions of hectares would be choked by *Opuntia* cactus, transforming local ecosystems. That was their condition before *Cactoblastis* was imported from the South American home of *Opuntia* to control the cactus plague (DeBach 1974). These difficulties, however, do not mean that important conclusions cannot be drawn about the adverse effects of extinctions on ecosystems and the degree to which substitution can ameliorate them.

EXTINCTION AND ECOSYSTEM SERVICES

Extinctions occur continuously as part of evolutionary and ecological processes (Ehrlich et al. 1980), so ecosystems clearly can suffer some rate of extinction without serious degradation of their functional properties. Since the living components are known to be essential to those functions, however, there must also be some rate of extinction that a given ecosystem cannot absorb without undergoing reorganization leading to serious impairment of its functioning from the point of view of *Homo sapiens.*

The immediate impact of the deletion of a single element from an ecosystem obviously can be judged in theory from its quantitative involvement in the control of various functions. Judging the long-term impact is more difficult. For example, the deletion of one member of a controlling guild may or may not result in a significant change in ecosystem function in the long run, even if the species is quantitatively important. Abrupt decline of hemlock from a dominant role in the northeast in prehistoric times (~4800 B.P.) led to important ecosystem changes, notably enhancing lake productivity through greater leaf input (Whitehead and Crisman 1978). In contrast, the recent deletion of chestnut did not have a great impact on the ecosystem (Good 1968), although it may be too early to tell. Presumably other guild members, in the latter case, have expanded their population sizes, using the resource base left by the deletion. Of course, if all members of a guild are deleted, then there will be a significant and lasting impact on ecosystem function and hence on the services provided.

Deforestation, which is the permanent loss of the guild of dominant trees, and strip-mining, which deletes an entire biota, are classic examples of deletions causing massive disruption of ecosystem services. Strip-mining results in widespread disturbance of aquifers, water pollution, and loss of land from agriculture, support of wildlife, and other uses (Wali 1975). The newly exposed areas are prone to intense wind and water erosion (Hodder 1975).

Deforestation cancels numerous services, such as flood control (Revelle 1979), erosion prevention, filtration of atmospheric pollutants, and the continuous supplying of firewood and timber (Eckholm 1979). The loss of such services can be catastrophic for human populations dependent upon them. Even the

removal of arid-scrub vegetation by overgrazing can be disastrous. This contributed to the degradation of climate-ameliorating services in the Sahel, was a major factor in the large-scale famine there and made the return to moister conditions less likely (Charney et al. 1975). Other human impacts, such as dry deposition of air pollutants, acid rain, and assaults by other toxins may be more subtle but equally catastrophic (Harte 1983, Woodwell 1970, Likens et al. 1974).

Such massive assaults on ecosystems produce a number of well-known syndromes that are accompanied by severe degradation of the public service functions of the systems. By far the best documented is *desertification* (United Nations 1977). Related to desertification, and if anything more serious for *Homo sapiens,* is soil erosion. Recent estimates indicate that unless current rates of erosion are slowed, there will be a monotonic decline in agricultural productivity (Brown 1981). Equally ominous is the trend toward *toxification* of the entire planet (Charney et al. 1975). And, largely as a side-effect of other impacts on ecosystems and of human transport, we are witnessing the *weedification* of the planet. (Of course weeds can and often do play an important role in preserving some ecosystem services).

The ecosystemic effects of large-scale disruption can persist for very long periods. The extinction of the Pleistocene megafauna may have strongly influenced the distribution of the Central American plant species whose seeds were dispersed by the deleted guild of large browser (Janzen and Martin 1982). Whatever the impact of that change on ecosystem services, because of the long generation time of most of the plants (trees and large shrubs), the system has not yet adjusted evolutionarily.

Smaller interventions produce less dramatic disruptions. Consider the practice of using herbicides to kill shrubs following tree harvesting in the forests of the Pacific Northwest to reduce competition with regrowth of the desired trees (Newton 1981). Although tree growth is generally favored in the short term, community productivity may not be in the long term, since certain of the shrubs (e.g., *Ceanothus*) are nitrogen fixers. The setting back of this guild may eventually require a fertilizer subsidy. There are numerous other instances of the substantial consequences of guild reductions in both natural and artificial ecosystems caused by the use of biocides. The Can-

ête Valley disaster in Peru is a classic example. Early successes using insecticides against cotton pests led to larger and larger applications. Predatory insect populations were decimated, and several herbivorous species, freed from predation, were "promoted" to pest status. And, of course, the original pests became resistant. The result: a catastrophic drop in cotton yields (Barducci 1972).

At the other extreme from the massive disruptions, a service may be reduced only imperceptibly after an extinction. Deletion of one of an array of understory plants or of one insect herbivore attacking the trees might make no measurable difference in the delivery of services. Indeed, it might increase some of them. Similarly, to the degree that *Homo sapiens* raises extinction rates above the normal "background" rates, it is automatically disrupting the genetic library service. But in any given case the loss (no matter how serious in the long run) is small in proportion to the total library service and not easily detected. Losses in the genetic diversity of crops and their wild relatives clearly fall in this category (Ehrlich and Ehrlich 1981).

Thus ecosystem services can, in effect, be "killed outright," "nickled and dimed to death," or even enhanced by extinctions. The degree of alteration of services depends on the functional role(s) of the organisms that go extinct and on the pattern of extinctions (e.g., selective deletion from an ecosystem or destruction of most elements simultaneously).

ECOSYSTEM RESTORATION

Increasingly, assaults on ecosystems are of a nature and scale that basic physicochemical conditions are changed. This makes natural succession unlikely or impossible, or extends the time scale of restoration of the original ecosystem beyond the period of reasonable human interest. For example, when a large area of tropical moist forest ecosystem is cleared, climate change and soil and nutrient loss often dictate a pattern of succession that will not soon restore the forest. Confusion about the ease of recovery of such ecosystems is rife because of the often-cited return of moist forests on Krakatoa after that island was leveled by a volcanic eruption (Dubos 1980). But the destruction of the biota on that small island was not comparable to the leveling of large tracts of mainland rain forest. The local climate was not altered, sources of propagules were very nearby, and the soil was enriched with volcanic ash.

With enough human effort, deserts can be made to bloom and forests to grow in denuded areas of the tropics. But the key word is *enough*, and the key question is whether the replacement can adequately perform the functions of the natural system.

HOW FINE-TUNED ARE COMPONENTS?

Substitutability will, in part, be a function of how closely controllers are fitted to their ecosystemic roles. Some animals are very "fine-tuned." Many parasites are completely dependent on one other species for their existence and have their primary ecological impact on that species. Numerous attempted translocations of vertebrates have failed, in many cases apparently because genetic characteristics were inappropriate to the locality of introduction (Greig 1979).

Populations of many insect herbivore species have made delicate adjustments to both host plants and local conditions. For example, after all the British populations of the large copper butterfly were extinct, a series of attempts was made to introduce stock drawn from Dutch populations. Subtle differences in the ecological requirements of the genetically different populations resulted in failure of the attempted substitution (Duffey 1977). Similarly, checkerspot butterfly populations in the western United States show a high degree of local specialization. Stock from one area often will be unable to recolonize other areas where populations have been exterminated by transitory events (Ehrlich et al. 1975 and unpublished data).

Native Australian dung beetles are fine-tuned for handling the small fecal pellets of kangaroos, other marsupials, and emus. Few can satisfactorily process the moist, large droppings of grazing placental mammals—sheep and cattle—that have been introduced to the continent in huge numbers. This has led to an increase in populations of the pestiferous bushfly (which breeds in unburied droppings), lowered productivity of pasturelands because nutrients remain tied up in the feces, and the creation of a fecal pavement that retards plant growth. Australia has started to introduce African dung beetles in an attempt to solve these problems (Bornemissza 1960, Gillard 1967, Hughes et al. 1978).

Many plant populations are also fine-tuned to their physical environments. A large number of examples have been documented of the precise adaptive responses of plants to the conditions of their local habitat, including specific responses to photoperiod or soil type (e.g., Heslop-Harrison 1964). For example, it is recommended that seed sources for reforestation of douglas fir at a specific site should be collected from areas no further distant from the original site than 110 m difference in elevation, 1.6 degrees latitude and 2.7 degrees longitude (Rehfeldt 1979).

At the opposite end of the spectrum, there are generalists that are relatively insensitive to the disappearance of one or more resources or may have the phenotypic plasticity permitting them to thrive under a wide variety of environmental regimes, e.g., cabbage butterflies, black bears, and coyotes. The moth *Heliothis zea* feeds on an enormous variety of plants in many families. Some plants, such as the reed grass *Phragmites communis*, grow from the polar regions to the tropics (Haslam 1972). Replacing lost populations of such species with ones from elsewhere may be simpler than in more specialized species, unless they, too, prove to be subdivided into many ecotypes that cannot readily replace one another. Unfortunately the degree of ecotypic differentiation in most species remains uncertain.

SUBSTITUTION IN ECOSYSTEMS

The history of introductions gives some insights into the more general problem of substitutions. The most important of these is that a species that is not a controller in one system may turn out to be a major controller in another (Elton 1958). In a coevolved ecosystem, the resources available to a species are determined not only by its physiological capacity to gather these resources, but also on its relationships with competitors and predators. A species removed from its coevolved competitors and predators may reach unprecedented population sizes and have an ecosystem impact to match as, for example, the history of goats introduced onto islands attests.

Most successful weeds are preadapted to invade systems that have already been disrupted by human activities (Mooney and Godron 1983). Successful invasions into natural systems also occur, especially where the invaders represent guilds not previously present in the area. Disruption of island ecosystems by introduced species supplies numerous examples (e.g., Elton 1958, ch. 4).

Ease of substitution would be expected to follow a pattern parallel to that of the impact of loss. Dominant plants are usually long-lived, and even if a suitable substitute is available, restoration of functions dependent on such K-selected organisms is likely to be a long-term process. Furthermore, it has been estimated that every plant species that goes extinct takes with it an average of 10 to 30 other species of organisms (Raven 1976). In the tropical rain forests around Finca La Selva, Costa Rica, the fruit of the canopy tree *Casearia corymbosa* is a resource for 22 species of birds, several of which are entirely dependent on it during one part of the year. If *C. corymbosa* were to disappear, these birds would go also, with ramifications for other trees that depend on the birds for seed dispersal (Howe 1977)[1] and thus on the herbivores dependent on *those* trees, and so on. Species like the *Casearia*, whose loss would start a cascade of extinctions, have been termed "keystone mutualists" (Gilbert 1980). Substitution for a keystone mutualist would obviously be much more difficult than for a species that did not have a unique position in the trophic web.

At first glance, rapid substitution for decomposers might seem relatively easy, since most have short generation times. But, at least in terrestrial systems, most of their extinctions will probably be due to physical-chemical changes in soils. These may be very difficult to reverse, so substitutes will have to function in a different environment. Acid rain, for example, inhibits certain nitrogen-fixing organisms as well as decomposers (Cowling and Linthurst 1981). At low pH, fungi become predominant in the soil, and actinomycetes and other bacteria drop out (Alexander 1977).

Making predictions about substitutions for consumers is more difficult because of the complexity of plant–herbivore and predator–prey interactions in both ecological and coevolutionary time (Roughgarden 1979). Some predators do appear to occupy positions analogous to that of keystone mutualists, and their deletion from a system results in a rapid decay of diversity (Paine 1966). A review of experimental studies of predator removal from natural, predominantly marine, ecosystems, reveals that in most cases other species were lost from the system (Pimm 1980). Equally, removal of herbivores from an ecosystem can often have dramatic effects on plant spe-

---

[1] and personal communication, October, 1982.

cies composition and density (Bartholomew 1970). Indeed, the main generalization that can be made today about the
impact of the loss of a single consumer
population or species from an ecosystem
is that it is unpredictable, even though
theoreticians (Pimm 1980) are making
progress in developing rules for what to
expect from different patterns of consumer deletion. And inability to predict
the consequences of deletion makes
evaluation of the ease of substitution
much more difficult.

One standard of success in replacing
one population or species with another is
obviously the degree to which the previous delivery of ecosystem services is
maintained. Although this can be relatively easily measured in the case of
some services (e.g., flood control, genetic library), it, unfortunately, can be extremely difficult to measure and impossible to predict in others (weather
amelioration, pest control, soil generation). In addition, there is the difficult
question of stability of the reconstituted
system over time. We don't know how
long to wait before declaring that a substitution has succeeded or failed.

These are obviously questions that deserve further consideration and investigation by community ecologists. Here
we use, *faute de mieux*, a pragmatic
classification. A substitution may be
deemed successful if, after several generations, the ecosystem shows no obvious deterioration and no human intervention, especially continued inputs of
external energy, is required to maintain
its services.

UNSUCCESSFUL SUBSTITUTIONS

In most cases, substitutions appear to
be unsuccessful. The number of thoroughly studied attempts at substitution
is, however, small.

Trees for Trees

Considering the magnitude of the
losses due to deforestation, it is not
surprising that attempting to restore forest ecosystems is a major human activity. Frequently a single-species tree plantation is substituted for an original mixed
forest. The services provided by a forest
of one kind of trees, however, may not
be equivalent to those provided by another. For example, it has been estimated that shortly after the turn of the
century, there will be more than a million
hectares of forest plantations in Australia, planted primarily with the exotic

Monterey pine (Rymer 1981). It has been
shown that these trees cannot satisfactorily be substituted for native species and
that they lead to reduced energy flow
through the ecosystem, less cycling of
minerals, and a loss of soil nutrients, at
least under prevailing management practices (Feller 1978, Springett 1976). The
plantations are almost totally unable to
provide the genetic library service of the
forest ecosystem they replace.

Plantation monocultures are also less
likely to be self-sustaining in the long run
than natural forests. Not only is species
diversity likely to increase resistance to
insect outbreaks and epiphytotics, but so
is genetic diversity within species (Edmunds and Alstad 1978, Sturgeon 1979).
Polycultures generally do better. The
Kaingaroa Forest of New Zealand, a
multispecies stand of exotic conifers,
supports a wide diversity of native and
introduced consumers (Wilson and Willis 1975). The potential nutrient loss and
loss of the genetic library function in
temperate-zone exotic plantations, however, is trivial compared to those losses
when grasses are substituted for tropical
moist forest. Global climatic change that
might result from that substitution (Potter et al. 1975) could threaten Earth's last
bastion of large, dependable grain surpluses—the prairies of North America
(Poore 1976).

Crops for Prairies

It might be argued that the replacement of perennial prairie grasses by
wheat and corn represents a wholly successful substitution. It is true that the
energy capturing function of the ecosystem has been kept and modified in a way
that is of incalculable benefit to humanity. But annual crop species, which have
been derived from early successional
species, cannot be maintained without a
continual input of energy and resources.
Annuals allocate less carbon to underground production and thus do not contribute to the generation of soils to the
same degree as perennials. The loss of
the perennial grasses thus results in a
degradation of the site through a loss of
soil nutrient stores, which must then be
replaced by fertilization (Jackson 1980).

Fishes for Fishes

Since 1970 the total oceanic fisheries
yield has increased very little (Brown
1981), which means that per capita yields
have been rapidly declining. Behind this
widely recognized decline, however, is a

less well-known process of species substitutions. As stock after stock has been
driven to economic extinction, yields
have been maintained only by moving to
the exploitation of species previously
thought to be less desirable or less easily
harvested. This has already led to decreased catch per unit effort (Brown
1981). This process of substitution cannot go on forever. Although the overall
productivity of the sea may not be diminished by the removal of commercially
valuable fish stocks *seriatim*, there is no
evidence from history or theory to indicate that depletion of one such stock will
lead to the expansion of others that are
equally valuable (Clark 1976, Murphy
1966).

Cattle for African Grazers and Browsers

This substitution has produced disastrous consequences in much of semiarid
Africa. The cattle, unlike many antelopes, must trek daily to a water hole.
This both requires energy, reducing meat
yield, and accelerates the downhill spiral
toward desertification. Mixed herding of
antelopes utilizes the habitat much more
efficiently without degrading it (Hopcraft
1979).

Inorganic Substitutes for Extinct Organisms

F. H. Bormann (1976, p. 759) has summarized the problems of substituting
technologies for the loss of services following deforestation:

> We must find replacements for wood
> products, build erosion control works,
> enlarge reservoirs, upgrade air pollution
> control technology, install flood control
> works, improve water purification
> plants, increase air conditioning, and
> provide new recreational facilities.
> These substitutes represent an enormous
> tax burden, a drain on the world's supply
> of natural resources, and increased
> stress on the natural system that re-
> mains. Clearly, the diminution of solar-
> powered natural systems and the expan-
> sion of fossil-powered human systems
> are currently locked in a positive feed-
> back cycle. Increased consumption of
> fossil energy means increased stress on
> natural systems, which in turn means
> still more consumption of fossil energy to
> replace lost natural functions if the quali-
> ty of life is to be maintained.

A similar statement could be written
about substitution for natural pest control functions with synthetic organic pesticides, natural soil maintenance with

inorganic fertilizers, natural water purification with chlorine treatment, and so on. These substitutions are all unsuccessful because they require a continual energy subsidy; most also only partially substitute for the services originally supplied.

SUCCESSFUL SUBSTITUTION

Although there are numerous examples of unsuccessful substitutions, successful ones are hard to identify. In most cases, at least genetic library functions are degraded by large-scale substitutions, but these are usually not obvious and thus permit the substitution to be rated as largely "successful."

Honeybees for Natural Pollinators

Some 90 crops in the United States alone depend on insects to pollinate them, and 9 others benefit from insect pollination (USDA 1977). Some of these crops, however, can be adequately pollinated by honeybees and thus can be independent of that service from natural ecosystems.

Imported Plants for Native Plants

Many exotic plants seem to replace natives quite adequately in providing all but the genetic library service. In California, ice plants from South Africa function in erosion control, and exotic oleanders provide services on freeway verges where few plants of any kind can survive. Many specially selected street trees provide important services in a hostile environment, substituting for natives that could not adapt to such conditions. The grasslands of California, although now principally composed of annual weeds of Mediterranean origin, are self-maintaining. The weeds have replaced the native herbaceous ecosystems that, at least in coastal regions, were dominated by perennial grasses (Wester 1981). This conversion, evidently induced by overgrazing and drought, has, however, resulted in a substantial decrease in range productivity (Burcham 1957).

What Organisms Are Successful Substitutes?

Organisms that are tough generalists—often "weedy"—are most likely to make successful substitutes if they can supply the needed services. The species and populations that have gone extinct are,

naturally, unlikely to have been generalists. What evidence there is indicates that fine-tuned replacements are usually not available. But introduction of generalists to replace a lost service often presents grave risks of their uncontrolled spread resulting in the loss of other services, such as maintenance of the genetic library.

THE GENETIC LIBRARY

The degree to which genetic library functions can be replaced is especially difficult to evaluate. By looking to the past, one could easily conclude that substitution would be almost impossible. What if the wild progenitors of just wheat, maize, rice, horses, and cattle had been wiped out before domestication? For the last two it might have been a close call, considering the Pleistocene extinctions. And the array of other foods, medicines, industrial products, and esthetic treasures derived from other species is enormous (Myers 1979). Humanity has already withdrawn from the library the very basis of civilization, a priceless benefit.

If one looks to the present, library services seem equally impossible to replace. For example, the continuation of high-yield agriculture will become much more difficult if extinctions of close relatives of crops continue (Frankel and Soulé 1981). It is also clear that to function properly the library must have numerous editions of each "book"—series of genetically distinct populations that can provide the basis for future evolution under either natural or artificial selection. As Aldo Leopold (1953, p. 194) said, it is "the clear dictum of history that a species must be saved *in many places* if it is to be saved at all."

Looking to the future, one might imagine that substitution for the library function of natural ecosystems would become simpler. For instance, genetic engineering may hold some promise of eventually producing new crop varieties with highly desirable properties without reliance on the gene pools of wild plants. In the foreseeable future, however, genetic engineering will not be fabricating substitute organisms to replace those lost from ecosystems. This can be stated with confidence because so little is known of the subtleties of the ecological roles of even prominent organisms. In short, if geneticists could fashion an organism precisely "to order," ecologists would not know what to ask them to make. Indeed, a continuing epidemic of

extinctions will likely do much more to *reduce* the capability of the genetic engineering enterprise than geneticists can do to ameliorate other consequences of the extinctions (Eisner 1981). For now the engineers are confined to manipulating existing genetic material.

CONCLUSION

Aldo Leopold (1953, p. 190) also wrote: "If the biota, in the course of aeons, has built something we like but do not understand, then who but a fool would discard seemingly useless parts? To keep every cog and wheel is the first precaution of intelligent tinkering." Unfortunately, humanity has been gambling with its future by saving fewer and fewer of the parts. Therefore, if the tide of extinctions is not reversed, humanity obviously will be required increasingly to attempt substitutions to maintain ecosystem services. Satisfactory substitutes are unlikely to be found at anything like the rate that ecosystems are now being degraded. That situation may change somewhat in the future with advancing knowledge in community ecology (Wilson and Willis 1975). But there is also every reason to believe that the rate of destruction of ecosystems will also increase.

At some point the costs of substitution will almost certainly become unbearable. Therefore, it seems that a conservative approach, emphasizing the careful preservation of ecosystems and thus of the populations and species that function within them, is absolutely essential. The ways of accomplishing this are the subject of a growing literature in conservation biology (Soulé and Wilcox 1980) and even economics (Fisher 1981).

ACKNOWLEDGMENTS

F. Bazzaz, L. C. Birch, F. H. Bormann, Joel Cohen, A. H. Ehrlich, Thomas Eisner, A. C. Fisher, R. Goldstein, R. W. Holm, C. E. Holdren, J. P. Holdren, Nelson Johnson, N. Margaris, S. Gulmon, Jonathan Roughgarden, S. H. Schneider, R. H. Waring, M. K. Wali, B. A. Wilcox, and P. M. Vitousek were kind enough to read and comment on the manuscript. A. T. Harrison and D. Whitehead supplied important information, and an anonymous reviewer made helpful comments. This work was supported in part by a grant DEB 7922434 from the National Science Foundation and a grant from the Koret Foundation of San Francisco.

REFERENCES CITED

Alexander, M. 1977. *Introduction to Soil Microbiology.* John Wiley and Sons, New York.

Barducci, T. B. 1972. Ecological consequences of pesticides used for the control of cotton insects in the Cañete Valley, Peru. Pages 423–438 in M. T. Farver and J. P. Milton, eds. *The Careless Technology.* Natural History Press, Garden City, New York.

Barnett, H. O. and C. Morse. 1963. *Scarcity and Growth: The Economics of Natural Resource Availability.* Johns Hopkins University Press, Baltimore, MD.

Bartholomew, B. 1970. Bare zone between California shrub and grassland communities. *Science* 170: 1210–1212.

Bormann, F. H. 1976. An inseparable linkage: conservation of natural ecosystems and conservation of fossil energy. *BioScience* 26: 759.

Bornemissza, G. F. 1960. Could dung-eating insects improve our pastures? *J. Aust. Inst. Agric. Sci.* 26: 5–6.

Brown, L. R. 1981. *Building a Sustainable Society.* W. W. Norton, New York.

Burcham, L. T. 1957. *California Range Land.* State of California, Sacramento.

Charney, J., P. H. Stone, and W. J. Quirk. 1975. Drought in the Sahara: a biogeophysical feedback mechanism. *Science* 187: 434–435.

Clark, C. W. 1976. *Mathematical Bioeconomics.* John Wiley and Sons, New York.

Council of Environmental Quality. 1980. *Global 2000: Entering the 21st Century.* US Government Printing Office, Washington DC.

Cowling, E. B. and R. A. Linthurst. 1981. The acid precipitation phenomenon and its ecological consequences. *BioScience* 31: 649–654.

DeBach, P. 1974. *Biological Control by Natural Enemies.* Cambridge University Press, New York.

Dubos, R. 1980. *The Wooing of Earth.* Charles Scribner's Sons, New York.

Duffey, E. 1977. The reestablishment of the large copper butterfly, *Lycaena dispar batava* Obth. on Woodwalton Fen Nature Reserve, Cambridgeshire, England, 1969–73. *Biol. Conserv.* 12: 143–157.

Eckholm, E. 1979. *Planting for the Future: Forestry for Human Needs, Worldwatch Paper 26.* Worldwatch Institute, Washington, DC.

Edmunds, G. F. Jr. and D. N. Alstad. 1978. Coevolution in insect herbivores and conifers. *Science* 199: 941–945.

Ehrenfeld, D. 1978. *The Arrogance of Humanism.* Oxford University Press, New York.

Ehrlich, P. R. and A. H. Ehrlich. 1981. *Extinction: The Causes and Consequences of the Disappearance of Species.* Random House, New York.

Ehrlich, P. R. and P. H. Raven. 1964. Butterflies and plants: a study in coevolution. *Evolution* 18: 586–608.

Ehrlich, P. R., D. D. Murphy, M. Singer, C. B. Sherwood, R. R. White, and I. L. Brown. 1980. Extinction, reduction, stability and increase: the responses of checkerspot butterfly *(Euphydryas)* populations to the California drought. *Oecologia (Berl.)* 46: 101–115.

Ehrlich, P. R., R. R. White, M. C. Singer, S. W. McKechnie, and L. E. Gilbert. 1975. Checkerspot butterflies: a historical perspective. *Science* 188: 221–228.

Eisner, T. 1981. Testimony before US Senate *Committee on Environment and Public Works,* Hearings on *Endangered Species Act,* December 10.

Elton, C. S. 1958. *The Ecology of Invasions.* John Wiley and Sons, New York.

Feller, M. C. 1978. Nutrient movement in soils beneath eucalypt and exotic conifer forests in southern central Victoria. *Aust. J. Ecol.* 3: 357–372.

Fisher, A. C. 1981. Economic analysis and the extinction of species. *Energy and Resources Group Working Paper 81-4.* November.

Frankel, O. H. and M. E. Soulé. 1981. *Conservation and Evolution.* Cambridge University Press, Cambridge, England.

Gilbert, L. E. 1980. Food web organization and the conservation of neotropical diversity. Pages 11–33 in M. Soulé and B. Wilcox, eds. *Conservation Biology.* Sinauer, Sunderland, MA.

Gillard, P. 1967. Coprophagous beetles in pasture ecosystems. *J. Aust. Inst. Agric. Sci.* 33: 30–34.

Good, N. F. 1968. A study of natural replacement of chestnut in six stands in the Highlands of New Jersey. *Bull. Torrey Bot. Club* 95: 240–253.

Greig, J. C. 1979. Principles of genetic conservation in relation to wildlife management in southern Africa. *S. Afr. J. Wildl. Res.* 9: 57–78.

Harte, J. 1983. An investigation of acid precipitation in Qinghai Province, China. *Environmental Science and Technology,* in press.

Haslam, S. 1972. *Phragmites communis* Trin. *J. Ecol.* 60: 585–610.

Heslop-Harrison, J. 1964. Forty years of genecology. *Adv. Ecol. Res.* 2: 159–247.

Hodder, R. L. 1975. Montana reclamation problems and remedial techniques. Pages 90–106 in M. K. Wali, ed. *Practices and Problems of Land Reclamation in Western North America.* University of North Dakota Press, Grand Forks.

Hopcraft, D. 1981. Nature's technology. Pages 211–224 in J. Coomer, ed. *Quest for a Sustainable Society.* Pergamon Press, New York.

Howe, H. F. 1977. Bird activity and seed dispersal of a tropical wet forest tree. *Ecology* 58: 539–550.

Hughes, R. D., M. Tyndale-Biscoe, and J. Walker. 1978. Effects of introduced dung beetles (Coleoptera: Scarabaeidae) on the breeding and abundance of the Australian bushfly, *Musca vetustissima* Walker (Diptera: Muscidae). *Bull. Ent. Res.* 68: 361–372.

Jackson, W. 1980. *New Roots for Agriculture.* Friends of the Earth, San Francisco.

Janzen, D. H. and P. S. Martin. 1982. Neotropical anachronisms: the fruits the gomphotheres ate. *Science* 215: 19–27.

Leopold, Aldo. 1953. *Round River.* Oxford University Press, Oxford.

Likens, G. E. and F. H. Bormann. 1974. Acid rain: a serious regional environmental problem. *Science* 184: 1176–1179.

Mooney, H. and M. Godron, eds. 1983. *Disturbance and Ecosystems—Components of Response.* Springer-Verlag, Heidelberg, in press.

Murphy, G. I. 1966. Population biology of the Pacific sardine *(Sardinops caerulea). Proc. Calif. Acad. Sci.* 34: 1–84.

Myers, N. 1979. *The Sinking Ark.* Pergamon Press, New York.

National Research Council. 1980. *Research Priorities in Tropical Biology.* National Academy of Sciences, Washington, DC.

Newton, M. 1981. Chemical weed control in western forests. Pages 127–138 in H. A. Holt and B. C. Fisher, eds. *Weed Control in Forest Management.* Purdue University Press, West Lafayette.

Paine, R. T. 1966. Food web complexity and species diversity. *Am. Nat.* 100: 65–75.

Pimm, S. L. 1980. Properties of food webs. *Ecology* 61: 219–225.

Poore, D. 1976. Managment and utilization of the tropical moist forest. *Unasylva* 28: 1–148.

Potter, G., H. Ellsaesser, M. MacCracken, and F. Luther. 1975. Possible climatic impact of tropical deforestation. *Nature* 258: 697–698.

Raven, P. H. 1976. Ethics and attitudes. Pages 155–181 in J. Simmons, R. Beyer, P. Brandham, G. Lucas, and V. Parry, eds. *Conservation of Threatened Plants.* Plenum Publishing, New York.

Rehfeldt, G. E. 1979. Ecological adaptations in Douglas fir *(Pseudotsuga menziesii* var *glauca)* populations. 1. North Idaho and North-east Washington. *Heredity* 43: 383–397.

Revelle, R. 1979. Energy sources for rural development. *Energy* 4: 969–987.

Roughgarden, J. 1979. *Theory of Population Genetics and Evolutionary Ecology: An Introduction.* Macmillan, New York.

Rymer, L. 1981. Pine plantations in Australia as habitat for native animals. *Environ. Conserv.* 8: 95–96.

Schneider, S. H. 1983. *Weather Permitting: The Coevolution of Climate and Life.* Sierra Club, San Francisco, in press.

Soulé, M. E. and B. A. Wilcox, eds. *Conservation Biology.* Sinauer Associates, Sunderland, MA.

Springett, J. A. 1976. The effect of planting *Pinus pinaster* Ait. on populations of soil microarthropods and on litter decomposition at Gnangara, Western Australia. *Aust. J. Ecol.* 1: 83–87.

Sturgeon, K. B. 1979. Monoterpene variation in ponderosa pine xylem resin related to western pine beetle predation. *Evolution* 33: 803–814.

United Nations. 1977. *Desertifiction: Its Causes and Consequences*. Pergamon Press, Oxford.

United States Department of Agriculture. 1977. *Agricultural Statistics*. US Government Printing Office, Washington, DC.

Wali, M. K. 1975. The problem of land reclamation viewed in a systems context. Pages 1–17 in M. K. Wali, ed. *Practices and Problems of Land Reclamation in Western North America*. University of North Dakota Press, Grand Forks.

Wester, L. 1981. Composition of native grasslands in the San Joaquin Valley, California. *Madroño* 28: 231–241.

Whitehead, D. R. and T. L. Crisman. 1978. Paleolimnological studies of small New England (USA) ponds. Part I. Late-glacial and post-glacial trophic oscillations. *Pol. Arch. Hydrobiol.* 25: 471–481.

Wilde, S. A. 1968. Mycorrhizae and tree nutrition. *BioScience* 18: 482–484.

Wilson, E. O. and E. O. Willis. 1975. Applied biogeography. Pages 522–534 in M. L. Cody and J. M. Diamond, eds. *Ecology and Evolution of Communities*. Harvard University Press, Cambridge, MA.

Woodwell, G. M. 1970. Effects of pollution on the structure and physiology of ecosystems. *Science* 168: 429–433.

Invitational Synthesis Paper

The Editorial Board of the *Journal of Range Management* invited James E. Ellis and David M. Swift to prepare this synthesis paper in recognition of their many, valuable contributions to understanding of the ecology of grazing systems. Their work is characterized by a high level of imagination, by a steady commitment to thoroughness, and by distilled, clear thinking.

JIM ELLIS took undergraduate work in animal husbandry at the University of Missouri and also obtained his Master of Science degree there studying wildlife biology. In 1970, he received his Ph.D. in Zoology at the University of California at Davis, where he was a National Institute of Health trainee in systems ecology. Shortly thereafter, he held a National Science Foundation post-doctoral fellowship at the University of Bristol working on systems analysis of mammalian social systems. He joined the Natural Resource Ecology Laboratory of Colorado State University as a Research Ecologist in 1971. He is currently the Associate Director of the Laboratory. Jim has enjoyed immense success in developing research programs; during the last decade he directed or played a major role in 12 successful proposals, collectively exceeding three million dollars in support. He has published extensively, with most of his work focussing on processes regulating grazing systems. Jim has served as a consultant to the U.S. Senate, as well as the government of Saudi Arabia and the Norwegian Agency for International Development. His most recent project uses a systems approach to understand the controls on stability and persistence of a pastoral ecosystem in East Africa.

DAVE SWIFT studied forest botany as an undergraduate at the the New York State College of Forestry and received a master's degree in watershed resources at Colorado State University. In 1985 he earned his Ph.D. in Animal Science at Colorado State University. Since 1970 he has been a member of the staff of the Natural Resource Ecology Laboratory where he now serves as a

Jim Ellis and Dave Swift in Turkana District, Kenya.

Research Scientist. Dave has deep expertise in simulation of ecological systems. He has worked on models investigating energy and nitrogen balance in ruminants, control of diapause in grasshoppers, carrying capacity of elk winter range, effects of sulfur dioxide on grassland ecosystems, influences of grazing on primary production, and the nutritional benefits of herding. He frequently contributes to the applied as well as the theoretical ecological literature. Dave teaches courses in ruminant nutrition in the Range Science and Wildlife Biology departments at Colorado State University and has supervised the research of many masters and doctoral students. He is widely admired by those who have had the good fortune to study with him. His recent work has taken him to East Africa, Pakistan, and the People's Republic of China.

Stability of African pastoral ecosystems: Alternate paradigms and implications for development

JAMES E. ELLIS AND DAVID M. SWIFT

Abstract

African pastoral ecosystems have been studied with the assumptions that these ecosystems are potentially stable (equilibrial) systems which become destabilized by overstocking and overgrazing. Development policy in these regions has focused on internal alterations of system structure, with the goals of restoring equilibrium and increasing productivity. Nine years of ecosystem-level research in northern Kenya presents a view of pastoral ecosystems that are non-equilibrial but persistent, with system dynamics affected more by abiotic than biotic controls. Development practices that fail to recognize these dynamics may result in increased deprivation and failure. Pastoral ecosystems may be better supported by development policies that build on and facilitate the traditional pastoral strategies rather than constrain them.

Key Words: pastoral ecosystems, equilibrial ecosystems, non-equilibrial ecosystems, pastoral development

Authors are Associate Director and Senior Research Scientist, Natural Resource Ecology Laboratory, Colorado State University, Fort Collins, Colorado 80523, (303) 491-1643; Senior Research Scientist, Natural Resource Ecology Laboratory, and Associate Professor, Department of Range Science, Colorado State University, Fort Collins, Colorado 80523, (303) 491-1981.

Perceptions of Pastoral Ecosystems

Our view of the world, or our perception of any system, has a great deal of influence on how we go about dealing with that system. Believing that illness is caused by spirits leads to very different sorts of treatments than the perception that viruses or bacteria may be the cause. Likewise, our perception of how particular ecological systems operate determines the approaches that we advocate in attempting to modify or manipulate those ecosystems. For example, one school of thought held that unusually high concentrations of elephants were responsible for the wide-spread destruction of dry woodlands in many areas of east and central Africa (Beuchner and Dawkins 1961, Glover 1963, Laws 1970). An alternative hypothesis proposed that elephants and trees were locked in a stable limit-cycle in which elephant populations expanded at the expense of tree populations. Ultimately the loss of woodlands would cause reductions in elephant populations while woodlands would begin to recover, initiating a new cycle (Caughley 1976). The general perception of destructive elephant-tree interactions led to some large-scale programs of elephant "control" in an effort to preserve woodlands. However, later work suggested that climate change, varying water tables, or combinations of herbivores, fire, and climate changes were responsible for the decline of woodlands. These results showed that elephant control programs were, in some cases, a needless slaughter of a scarce species (Western 1973, Sinclair and Norton-Griffiths 1979, Pellew 1983, Dublin 1987).

In scientific parlance the elephant-tree interaction was perceived in terms of a particular paradigm; that paradigm influenced the structure of scientific discourse, the types of analyses and models used, and the kinds of management solutions applied to related problems. Alterations in management practices occurred when the dominant paradigm was questioned and another proposed. Such changes in scientific paradigms often promote the use of new models, different analytical approaches, and new solution regimes (Kuhn 1970).

We believe that the time is ripe to examine the paradigms which govern our thinking about African pastoral ecosystems. Discussions of these pastoral ecosystems usually revolve around problems of low productivity, overstocking, overgrazing, drought, range deterioration, dying livestock, starving people, and so on. In essence, the current paradigm focuses on pastoralism as a maladaptive and destructive system of exploitation. However, years of intervention into pastoral systems, by governments and development agencies, have failed to relieve the perceived problems; in fact, development has sometimes rendered the target population and its ecosystem somewhat worse off than before the intervention (Helland 1980, Swift 1977, Sanford 1983, Swift and Maliki, 1984). This lack of success has been so pervasive that some research organizations and major funding agencies have essentially given up on pastoralists and arid lands in Africa and are investing their resources elsewhere (ILCA 1987). It is possible that some intervention failures could be the result of technical incompetence on the part of development experts or intransigence on the part of pastoralists. However, it seems unlikely that experts are always technically incompetent or pastoralists always intransigent; the near universal failure of pastoral development suggests that something more fundamental is amiss. Universal failure would be expected if invalid paradigms underlie the development interventions.

Two sorts of inquiries are needed to assess the validity of the dominant pastoral paradigm and the models and interventions which follow therefrom. First, it is necessary to identify the dominant paradigm, the assumptions on which it is based, and their implications. Secondly, the paradigm needs to be tested against extant observations and empirical data. Do the observed dynamics fit the dominant paradigm? If not, the paradigm obviously needs to be altered to provide a more realistic model of pastoral ecosystems from which better intervention and management procedures would hopefully derive.

In this paper we propose that the dominant paradigm of pastoral systems is based on assumptions (1) that African pastoral ecosystems are potentially stable (equilibrial) systems; (2) that these potentially stable systems are frequently destabilized by improper use on the part of pastoralists; and (3) that alterations of system structure (reducing livestock numbers, changing land-tenure patterns, etc.) are needed to return these systems to an equilibrial and more productive state. We then provide evidence from our own work (1) that stable equilibria are not achievable in many pastoral ecosystems, although long-term persistence is; (2) that interventions aimed at achieving stability in non-equilibrial systems are likely to be irrelevant at best or disruptive and destructive at worst; and (3) that successful interventions will be designed to accommodate system dynamic variation rather than aimed at maintaining equilibrial conditions.

The Dominant Paradigm: Pastoral Degradation of Equilibrial Ecosystems

There are at least 3 separate but interrelated perceptions of pastoralists and their ecosystems in the literature. The work of early ethnographers tended to picture traditional pastoralists in a rather romantic light, i.e., self-reliant nomads, moving freely, living off the land, while defying governments and settled agriculturalists (Evans-Pritchard 1940, Spencer 1973, Jacobs 1965). This perception seems also to have incorporated the idea, at least implicitly, that pastoralists lived in some sort of balance with their natural environment if not with their neighbors; pastoral systems were not identified with environmental degradation.

The idea that pastoralists do not achieve a balance with their environment but routinely overstock and overgraze, is an old one (Stebbings 1935), but was stated most forcefully and coherently by Brown (1971), the Chief Agriculturalist of colonial Kenya, and more recently by Lamprey (1983). Brown's viewpoint arose no doubt from his own experiences, but also incorporated the concept of the irrationality of pastoral management practices first posed by Herskovits (1926), whose "cattle complex" hypothesis spawned the view that accumulation of vast numbers of livestock was an irrational pastoral tradition, ingrained in the social system; a practice clearly incompatible with good environmental management. Brown identified a different sort of irrationality; he asserted that pastoralists were engaged in dairy operations under environmental conditions much more suitable for meat production and he suggested that these dairy operations tended to be very inefficient in terms of milk produced per unit of forage available. He then presented a numerical analysis of pastoral subsistence needs based on inefficient milk production and used this to substantiate the case for obligatory overstocking, overgrazing, and environmental destruction by pastoralists.

Lamprey, in contrast, pointed out that pastoral management is adaptive and rational from the perspective of human survival, but does not incorporate environmental conservation as a management objective (1983). Further, Lamprey's incisive ecological perspective brought some important assumptions into the open which are often held implicitly, but seldom stated. There is, in this view, the potential for density-dependent equilibria between herbivores and vegetation in the regions of East Africa occupied by pastoralists and such equilibria are frequently achieved in natural ecosystems, not inhabited by man. Lamprey reasoned that overstocking by pastoralists causes departures from these equilibrial conditions and rangeland degradation. Clearly, continuing degradation of the environment would eventually lead to extinction of pastoralists, but this has not occurred over the thousands of years of pastoral occupation of Africa because, "Either they could move on from the degraded lands into new territories, or they could adapt their pastoral practices to increasingly marginal conditions (for instance, by herding camels instead of cattle) and remain where they were" (Lamprey 1983, p. 664).

Thus the pastoral paradigm materializes: pastoral systems are potentially stable systems but departures from equilibrial conditions necessarily follow from the inefficiency of pastoral resource exploitation and the resultant need to overstock the ecosystem. This environmentally destructive pattern has not led to extinction, however, because new regions were exploited while the degraded ones presumably recovered (this might now be considered a patch-dynamic equilibrium), or because alternate equilibrial states were possible (when a system was degraded by cattle it could still be utilized by camels or goats); i.e., these systems contain multiple stable points (Noy-Meir 1975).

The final element of the dominant paradigm proposes that in many cases development has exacerbated degradation in pastoral ecosystems. Veterinary care and reductions in tribal raiding are said to have released livestock populations from former density-dependent constraints, while curtailment of nomadism, losses of grazing lands to agriculture, security problems, and the settlement of some pastoralists have combined to reduce the area of rangeland available while herds are on the increase (Lusigi 1981, Lamprey and Yussuf 1981, Lamprey 1983). Overstocking and overgrazing have intensified while the reductions in rangeland area have removed the possiblity of achieving patch-dynamic equilibria. Resulting acceleration of degradation is hypothesized to be causing wide-spread shifts from cattle to camels in northern East Africa (Stiles 1983); while vegetation removal by livestock in the Sahel is believed to have increased soil surface albedo, leading to reductions in rainfall and rapid desertification (Charney et al. 1974, Sinclair and Fryxell 1985). To summarize, the dominant paradigm of pastoral ecosystem dynamics is based on perceptions that:

• Pastoralism is basically an inefficient and therefore environmentally destructive resource exploitation strategy.

• The ecosystems occupied by pastoralists have the capacity to support stable (equilibrial) populations of herbivores, but pastoral strategies necessarily lead to overstocking and tend to move the system away from the potential equilibrium conditions.

• Pastoralists have avoided large-scale extinctions by moving to new areas after degrading previously occupied environments or by changing strategies to accommodate the new but somewhat degraded environmental state.

• Modernization has made things worse by reducing the potential for large-scale movement and by removing density-dependent constraints on livestock populations (Fig. 1).

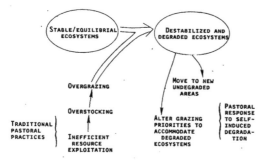

Fig. 1. The equilibrial paradigm of pastoral ecosystem dynamics.

Implications of the Equilibrial Paradigm

A similar perception of pastoral systems has been termed the "mainstream view" because it has dominated attitudes toward pastoralists and shaped pastoral development policy (Sandford 1983). Some of the major consequences which Sandford attributed to the mainstream view include the assumption that whatever pastoralists are doing is inappropriate and therefore range management programs aimed at reducing presumed degradation should be applied universally and rapidly. Likewise changes in land-tenure systems and existing institutions are assumed to be desirable and are therefore undertaken without consideration of what may be useful or valuable in the existing systems.

We think that the key assumption in the equilibrial paradigm has an even more fundamental impact on development policy than the mainstream view. The paradigm assumes that the ecosystems occupied by pastoralists generally function as equilibrial systems which are regulated by density-dependent feed-back controls; however, pastoralists override these feed-back controls to the detriment of themselves and their ecosystems. If this assumption is accepted, it is logical to reason that internal alterations in system structure can correct the imbalances and restore the system to equilibrial conditions. The most obvious adjustments to make are those involving the number of livestock per unit area. Hence two types of development procedures follow: reduction of stocking rates and alteration of land-tenure systems. Destocking is a very direct means of altering system structure, but it is hard to sell to pastoralists. Alteration of land-tenure has the advantage of less

immediate and direct impacts on pastoralists and it also may have a certain appeal to some local planners or officials. The assumption is that some form of privatization will alleviate the imbalances supposedly induced by communal grazing. This is the viewpoint expressed in Hardin's 'tragedy of the commons' concept (Hardin 1968) as applied to pastoral systems. Group ranches, grazing blocks, grazing cooperatives, etc., are all schemes aimed at attaching groups of pastoralists to specific tracts of land which will presumably induce better management practices (Helland 1980, Oxby 1982). Implicit in the strategy is the premise that the tracts of land have the capacity to support stable, balanced populations of livestock and people if only good management prevails, i.e., equilibrial conditions are attainable (Fig. 2). The results of interventions

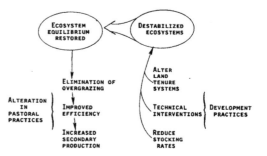

Fig. 2. Development practices and expected results based on the equilibrial paradigm.

and schemes based on these assumptions are not encouraging. They include disruption of pastoral societies and ecosystems and the retreat of development agencies from the arid and semi-arid lands of Africa (Helland 1980, Swift and Maliki 1984, ILCA 1987).

Some Theoretical Considerations

The idea that there is a balance of nature has always had great appeal to scientists, naturalists, and others. When the concept is stated in more analytical terms, it generally is accepted to mean that natural systems exist in some sort of homeostatic state, and that this state is maintained by interactions among the components of the system. The concept thus implies that the dominant type of system interaction is negative feed-back control of rate processes which maintain the system in a dynamic equilibrial state or within some limited domain of attraction. The antithetical view of natural system dynamics emphasizes the role of external control mechanisms, i.e., drivers, which are not subject to feed-back control from within the system. When system dynamics are dominated by external forces, the opportunity for the development of feed-back control is much reduced and the persistence of the system depends upon the development of other sorts of stabilizing mechanisms. These varying perceptions of natural system dynamics have given rise to famous scientific debates such as the density-dependent versus density-independent population control controversy (Nicholson 1958, Nicholson and Bailey 1935).

While these contrasting views of system dynamics are both now recognized as legitimate models of biological reality (Noy-Meir 1979/80, Wiens 1984), the concept that systems operate in a homeostatic fashion, based on equilibrial dynamics, has had by far the more pervasive role in ecological thought. The ideas that communities are structured by competitive interactions and that evolution proceeds through a continuous process of adaptive "fine tuning", are based on an equilibrial view of system dynamics. In addition, the development of the concept of ecological stability and the analysis of the stability properties of ecological models proceeded from an assumption of equilibrial conditions (Lotka 1925, Volterra 1928, Rosenzweig and MacArthur 1963). The equilibrial assumption, its implications for the competitive structuring of communities, and its validity in mathematical analyses, have been questioned (Wiens 1977, 1984, Caswell 1978, Connell and Sousa 1983); however, it remains a powerful concept in both field and mathematical assessments of population, community, and ecosystem dynamics. Clearly then, we should not be terribly surprised to find that equilibrial assumptions dominate our perceptions of pastoral ecosystems.

A major challenge to the equilibrial assumption was posed by Wiens, who after several years of field research on avian communities in arid and semi-arid environments, concluded that equilibrial conditions occurred only occasionally in these environments. He asserted that abiotic controls regulated bird populations in arid ecosystems and that since equilibrial conditions (those required for competitive interactions to develop) were seldom attained, the role of competition in shaping these communities was minimal (Wiens 1977). He extended these conclusions to propose that ecosystems exist along a gradient from equilibrial conditions where biotic interactions structure communities to non-equilibrial conditions where abiotic controls determine system structure and dynamics (Wiens 1984). This scheme has been further elaborated by DeAngelis and Waterhouse (1987), based on their review of model analyses of ecological stability. They propose that the spectrum of potential system behaviors centers on the concept of stable equilibrial systems, but that systems frequently fail to demonstrate equilibrial behavior due either to the disruptive effects of stochastic elements (such as abiotic controls) or due to instabilities induced by strong internal feed-backs. Strong feed-back or overconnectedness is exemplified where herbivores or predators over-exploit their food resources, resulting in departures from equilibrium points as proposed for the case of overgrazing by pastoralists. When strong feed-backs are accompanied by time-lags, stable-limit cycles may result. This is the dynamic suggested to explain the interdependent interactions between elephants and trees (Caughley 1976).

The question posed by DeAngelis and Waterhouse (1987) (and many theoretical modelers before them), focuses on how natural systems manage to persist in the face of pervasive destabilizing forces. Their resolution of the question is based on the operation of a number of different mechanisms incorporated into theoretical models to stabilize model performance. The crucial point is that different sorts of stabilizing mechanisms are required to ameliorate the effects of different destabilizing forces. Where destabilization results from strong stochastic forces, such as abiotic perturbations, persistence may be maintained by the introduction of compensatory mechanisms or by increasing the spatial scale of the model ecosystem. Where instability is caused by strong biotic interactions, stability may be regained by the introduction of disturbances which influence either resource or consumer density, or by incorporating mechanisms which reduce the strength of system interactions, e.g., interference, inefficiency, etc. Thus, the specific mechanisms employed in attempting to stabilize model ecosystem performance depend upon what one perceives as the critical destabilizing force. Although this analysis focuses on model ecosystems, it provides some important insights into the different ways in which real ecosystems may persist under the influence of potentially destabilizing controls and interactions (Fig. 3).

The implications for the persistence of African pastoral ecosystems are clear. If these are potentially equilibrial systems, controlled mainly by strong biotic interactions, then destabilizations and degradation may, in fact, be caused by overstocking and strong biotic coupling and feed-back from livestock to plants. This is a basic presumption of the dominant paradigm. Appropriate stabilizing procedures under these conditions would indeed involve fine-tuning the interaction between plants and livestock through such means as destocking and land privatization, provided these procedures actually reduced consumer density relative to vegetation resources. In other words, the currently applied development practices may be well-suited to correcting internally induced

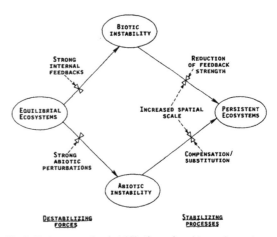

Fig. 3. Factors promoting destabilization and persistence of ecosystems. (Modified from DeAngelis and Waterhouse 1987).

instabilities in pastoral ecosystems. Why then is there such a poor record of success in the development of these ecosystems and why do degradation and famine continue in Africa? It is, in our view, because pastoral ecosystems are often strongly controlled by external forces rather than, or in addition to, internal biotic factors. Therefore the nominally applied development procedures are often irrelevant to the systems of interest, or worse, they comprise additional perturbations which act to further destabilize system dynamics, rather than damping them.

How does one determine whether a pastoral ecosystem is dominated by internal biotic controls or by external perturbations? Theoretical considerations and model analyses of grazing systems suggest that internal control mechanisms will come to dominate system dynamics when plant growing conditions are relatively invariant over time, whereas major variations in growing conditions will prevent the development of strong biotic interactions and feed-back controls on system dynamics (Noy-Meir 1975, Wiens 1977, 1984). A number of other structural characteristics and overall patterns of system dynamics may also be used to differentiate between equilibrial and non-equilibrial grazing systems (Table 1).

Non-Equilibrial Dynamics in a Pastoral Ecosystem

Nine years of research among pastoral nomads in northern Kenya has given us a very different view of the dynamical behavior of pastoral ecosystems than that portrayed in the equilibrial paradigm. It has also caused us to question the appropriateness of development procedures which are based on the paradigm's assumptions. The results of our work in Ngisonyoka Turkana, one of 15 sub-tribal territories in Turkana District, reveal anything but an equilibrial ecosystem. Here in the arid northwest corner of Kenya, pastoralists are locked in a constant battle against the vagaries of nature and the depredations of neighboring tribesmen. Droughts and raids are ongoing stresses; drought-induced livestock mortality frequently diminishes herds by 50 percent or more (McCabe 1985, in press, Ellis et al. 1987). Infant mortality rates are high, human nutritional status is quite dynamic, and emigration from the pastoral sector is common (Brainard 1986, Galvin 1985, McCabe 1985). However, despite the dynamic nature of the ecosystem there is little evidence of degradation or of imminent system failure (Ellis et al. 1985, Coughenour et al. 1985, McCabe and Ellis 1987). Instead, this ecosystem and its pastoral inhabitants are relatively stable in response to the major stresses on the system, e.g., frequent and severe droughts (Ellis et al. 1987). In theoretical terms this is a non-equilibrial but persistent ecosystem (Holling

Table 1. Characteristics of equilibrial and non-equilibrial grazing systems.

	Equilibrial Grazing Systems	Non-equilibrial Grazing Systems
Abiotic Patterns	Abiotic conditions relatively constant	Stochastic/variable conditions
	Plant growing conditions relatively invariant	Variable plant growing conditions
Plant-Herbivore Interactions	Tight coupling of interactions	Weak coupling of interactions
	Feedback control	Abiotic control
	Herbivore control of plant biomass	Plant biomass abiotically controlled
Population Patterns	Density dependence	Density independence
	Populations track carrying capacity	Carrying capacity too dynamic for close population tracking
	Limit cycles	Abiotically driven cycles
Community/ Ecosystem Characteristics	Competitive structuring of communities	Competition not expressed
	Limited spatial extent	Spatially extensive
	Self-controlled systems	Externalities critical to system dynamics

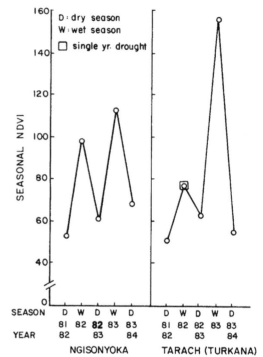

Fig. 4. Seasonal and annual NDVI dynamics.

1973, DeAngelis and Waterhouse 1987). Our results do not support the dominant paradigm, and they demonstrate that at least for some pastoral ecosystems, the assumptions of equilibrial dynamics, and the intervention practices which follow, are inappropriate.

Climatic Variation and Plant Production in Ngisonyoka, Turkana

A critical assumption about equilibrial grazing systems is that plant growing conditions are relatively invariant over time. Noy-Meir (1975) studied the stability properties of potentially equilibrial model grazing systems. His assumptions about the characteristics leading to equilibrium included time invariant growth during the growing season and no use of the forage resource outside of the growing season. These are very restrictive assumptions indeed and would not be expected to be met in many grazing ecosystems. It is probable, however, that such restrictive assumptions are not necessary for equilibrium dynamics to prevail. Equilibrium is probably possible or at least approachable in systems which demonstrate intraseasonal and intra-annual variation in plant growth provided the level of interannual variation is not too great. If the timing and magnitude of primary production is more or less predictable on an annual basis, a more or less constant carrying capacity develops and density dependent population processes permit the herbivore community to track gradual changes in the forage resource.

Pastoralists generally occupy arid or semi-arid environments where climatic variability causes distinct pulses of plant production followed by long periods of plant dormancy, but in which the pulses of production are not predictable in terms of time or magnitude. Rainfall averages between 200 mm and 600 mm over most of Ngisonyoka territory. The growing season ranges from 60 to 90 days (April–June) during normal years, leaving a 9–10 month dry season with little or no plant growth. Occasional good years occur when a short rainy period interrupts the dry season in October or November, causing another 20–30 days of plant growth. Satellite derived normalized vegetation index values (NDVI's) suggest that in the most productive regions of Turkana District, green plant biomass increases by 2–2.5 times during the growing season, relative to the long dry season (Fig. 4). Intra-annual growing conditions are highly variable even during the best of years and are not

indicative of equilibrial dynamics.

The major perturbations on the Turkana ecosystem are droughts of a year's duration or longer. In central Turkana, rainfall has dropped 33% or more below the long-term averages 13 times in the last 50 years, i.e., once every 3–4 years, and at least 4 of these have been severe multi-year droughts (Ellis et al. 1987). NDVI values recorded during a single-year drought show growing season values that are only 1–1.5 times those of the preceding dry season, indicating that plant biomass production was one-half or less of that achieved in a normal year (Fig. 4). This value may be even lower in the second year of a multi-year drought; production levels are known to drop to 1/3 or 1/4 of average values during droughts in some arid areas (Paulsen and Ares 1961). The dramatic response of plant biomass to dry seasons and drought demonstrates the pervasive role of climate in the vegetation dynamics of this ecosystem. The high degree of seasonality plus the great interannual variability lead to pulsed and undependable plant growth, rather than the constant or at least predictable growing conditions necessary for the development of equilibrial grazing interactions.

Herbivory and Plant Production

Another fundamental assumption about the operation of equilibrial grazing systems is that herbivores play a major role in controlling plant biomass through consumption and offtake (Noy-Meir 1975, McNaughton 1979). Using ecosystem-wide estimates of forage production and livestock consumption, we calculated that total livestock offtake is on the order of 10–12% of forage production during a good year in Ngisonyoka, Turkana. Likewise, Ngisonyoka stocking rates are less than one-fourth the theoretical maximum carrying capacity in a good year, allowing for 50% forage consumption (Coughenour et al. 1985, Ellis et al. 1987). Given these stocking levels and offtake rates it seems unlikely that livestock exert a major control on plant biomass.

Field and laboratory assessments of livestock grazing effects on herbaceous and dwarf shrub vegetation have revealed no significant grazing effects on plant production (Bamberg 1986, Coughenour et al. in prep). On the other hand, livestock do influence the morphology of important dwarf shrubs (Bamberg 1986, Mugambi in progress) and increase germination and establishment of important trees (Ellis et al. in prep, Reid in progress). Thus, while livestock may, in the long-run, alter the structure and composition of the plant community, they appear to have no role in regulating yearly plant production, only a minor role in regulating biomass levels and consequently little or no role in regulating the amount of forage available. The strong force exerted by climate on forage production and the minimal influence of livestock on forage availability means that there is little opportunity for the development of strong feedbacks from livestock to plants. The plant-herbivore interaction is therefore only loosely coupled in this respect, and probably operates as a density-independent relationship most of the time. This form of interaction is indicative of non-equilibrial systems, and is inconsistent with the characteristics of equilibrial grazing systems (DeAngelis and Waterhouse 1987).

Livestock Dynamics

Despite the fact that Ngisonyoka livestock consume only a small proportion of the total forage produced in a good year, livestock nutritional status and production rates closely track the seasonal dynamics of plant production. This is because nutritional condition and secondary production are limited by forage quality (protein content and digestibility) during the long dry seasons even though forage quantity is usually adequate. For most livestock species, diet quality drops to maintenance levels by the mid-dry season (Fig. 5), and loss of condition and reduction of production continue for several months, until the following rainy season. Though camels are able to do better than the other species, maintaining a diet of adequate nitrogen content, the digestibility of their diet declines to the point that their nutritional state is compromised also (Coppock et al. 1982, 1985, 1986a,b,c).

In drought years forage quantity as well as quality becomes limiting. While the quantity of forage removed by livestock is only moderate, that removal plus consumption by termites and losses due to decomposition and weathering eventually deplete the forage supply. During droughts, livestock are on starvation rations; nutritional condition and production spiral downward. Only the length of the drought determines how serious the effects on livestock condition and subsequently human nutrition, will be (Coppock et al. 1986b, c; Ellis et al. 1987). If droughts last two years or more, the decrements in livestock condition also begin to influence livestock population size through effects on reproduction and mortality.

We do not propose that there is no connection between livestock density and the degree of nutritional stress experienced during the annual dry period or during more extended droughts. Any time a forage resource varies in quality, there is an opportunity for density dependence to operate through competition for small quantities of the best available forage (Hobbs and Swift 1985). Thus, there is the potential for density dependent condition change during the dry season. In the case of longer droughts, it could be argued that a lower stocking rate prior to the drought results in a greater quantity of cured forage to carry the animals through the period of stress. This is correct to a point but termites, microbes, and abiotic factors deplete this resource regardless of the density of grazing animals. In the case of either seasonal dry periods or longer droughts there is simply no forage available that will maintain animals in positive nitrogen and energy balance (Coppock et al. 1986c). Once the forages cure, the animals have begun the process of starvation; and the length of time that the stress must be endured is more critical to the outcome than the number of animals enduring the stress. A similar conclusion was reached by Wallmo et al. (1977) regarding the carrying capacity of mule deer winter range. While density has a role, it is small compared to the role of environmental uncertainty.

Equilibrial grazing systems exhibit strong density-dependent

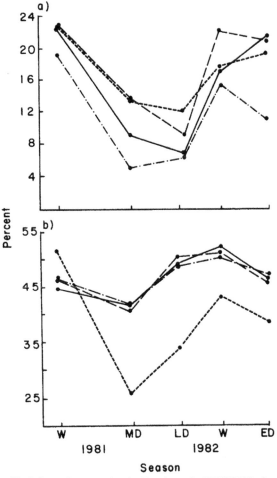

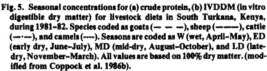

Fig. 5. Seasonal concentrations for (a) crude protein, (b) IVDDM (in vitro digestible dry matter) for livestock diets in South Turkana, Kenya, during 1981–82. Species coded as goats (— — —), sheep (————), cattle (—·—·—), and camels (----). Seasons are coded as W (wet, April–May), ED (early dry, June–July), MD (mid-dry, August–October), and LD (late-dry, November–March). All values are based on 100% dry matter. (modified from Coppock et al. 1986b).

population interactions in which livestock numbers are controlled by forage availability (Noy-Meir 1975). But in Turkana, drought perturbations appear to regulate livestock populations and this control operates independent of livestock density under most circumstances. Single-year droughts undermine livestock condition and may reduce reproductive rates but do not induce livestock mortality at low to moderate stocking rates. However, even at low stocking rates, multi-year droughts cause drastic increases in livestock mortality and reductions in reproduction. A two-year drought in 1979–80 caused losses of 50–70% of the livestock population in parts of Turkana District. But recovery from these losses was, in many cases, relatively rapid; some herds had returned to pre-drought levels four years after the drought ended. This rapid rate of recovery was largely the result of two mechanisms: rapid reproductive rates of small stock (goats and sheep) and immigration of cattle into the District after the drought ceased. Much of this immigration involved the return of animals which had been taken out of the region during the drought, while other recruitment was likely the result of raids on neighboring tribes (McCabe 1985,

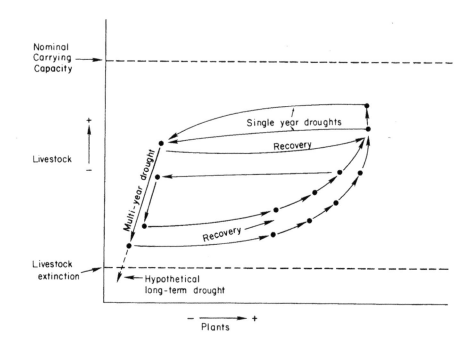

Fig. 6. Turkana plant-livestock interactions under the influence of frequent drought perturbations.

in press, Ecosystems Ltd. 1984, Ellis et al. 1987).

Livestock population dynamics in most portions of Turkana probably operate in a density-independent fashion most of the time. However, where stocking rates are relatively high, there is evidence of density-related mortality. In one small region of the District, Ngiboceros, stocking rates were about three times as high as in Ngisonyoka territory where stocking rates were only about 25% of the maximum "safe" level, based on consumption of 50% of forage. Ngiboceros stocking rates were about 75% of this maximum "safe" level. While there was no evidence of elevated mortality rates during dry seasons or single-year droughts in Ngisonyoka territory (or most of the rest of the District), Ngiboceros livestock did experience elevated mortality rates during dry seasons in non-drought years and during a single-year drought in 1984 (Ellis et al. 1987). Thus, density-related mortality occurs in Turkana, but appears to be exceptional because stocking rates seldom reach the levels required for this type of interaction to take place.

These observations suggest that Turkana livestock populations and vegetation interact as illustrated in Figure 6. During sequences of good years livestock populations expand at moderate rates. Single-year droughts have only limited effects on this pattern of increase; so based on the climatic patterns of the last 50 years, we might expect populations to continue to build between severe multi-year droughts, which have occurred about once per decade. These multi-year droughts cause population plunges, the magnitude of which may depend only weakly upon the levels of density reached prior to the drought. Between droughts, livestock populations tend toward but never reach the "nominal" carrying capacity of the vegetation, because of the relatively high frequency of multi-year drought perturbations. As a result, strong negative feedback relationships between plants and herbivores never develop. These negative feedbacks are necessary to the development of equilibrial systems and are the basis of traditional stocking rate determinations. Likewise, livestock populations approach but never reach extinction, because droughts do not last long enough to completely eliminate the livestock. The system is thus persistent in that it

operates within a limited range of variation, although populations are unstable over time. This view of plant/herbivore interactions in unpredictable, arid systems is consistent with the "autecological hypothesis" of system regulation proposed by Noy-Meir (1979/80), which suggests that arid systems may be structured more by the responses of individual species to stress than by interactions among species.

Human-Livestock Interactions

Although herbivore stocking rates in Ngisonyoka are low, human density is relatively high (Ecosystems Ltd. 1984), resulting in only about 2.5 tropical livestock units per person (one TLU=250 kg). This indicates that the human population may be near the carrying capacity of the livestock population and that the human-livestock interaction is much more closely coupled than the plant-livestock interaction. There are significant feedback controls from humans to livestock recruitment, since pastoralists compete with calves, kids, lambs etc. for milk (Donaldson 1986). It is usually assumed that pastoralists consume about 50% of the milk produced since they milk one side of the udder and leave the other for the calf. Empirical evidence suggests that human offtake may actually be considerably higher, leaving the offspring with a much reduced quantity of milk (Donaldson 1986). Lower human offtake would almost certainly result in greater survivability and better growth rates for immature livestock. Nevertheless, Ngisonyoka husbandry allows for moderate herd growth during good years. The primary control of livestock populations still appears to be the drought perturbations which periodically reduce herds by half or more. Because the Ngisonyoka are so dependent on their livestock, such losses comprise a severe stress to the pastoralists; they must resort to extreme measures to cope with these perturbations.

Pastoral Persistence

There is clear evidence that, like the plant-livestock system, the Ngisonyoka human population is persistent. The Turkana have occupied their present region since they descended into the Rift Valley from Uganda 150 to 200 years ago (Lamphere 1976), demonstrating long-term persistence in a difficult environment. Over the past decade, which included a serious drought in 1979–80 and a single-year drought in 1984, there has been no famine in Ngisonyoka territory and no overt evidence of environmental degradation (Ellis et al. 1984, 1987). Short-term persistence was maintained despite large-scale livestock losses in a situation in which 80% of pastoral food intake is derived directly from livestock products (Galvin 1985, Galvin and Waweru 1987).

Pastoral persistence in this ecosystem is achieved through a series of stabilizing strategies which vary with the strength of the environmental stress. Dry season decrements in forage availability and livestock production are mild stresses; the Ngisonyoka compensate for dry season diminution in livestock products by switching foods and by reducing their own activity levels and energy requirements. Nevertheless, dry seasons regularly result in small declines in human nutritional status and weight, averaging 3–4% of total body weight (Galvin 1985, Galvin and Waweru 1987). Likewise, there are indications of seasonality in human reproductive patterns with the majority of children born in the wet season, indicating that most conceptions occur when women reach their peak nutritional status in June and July (Leslie and Fry 1988).

Single-year droughts are only slightly more stressful than normal dry seasons for the Ngisonyoka; that is, the strategies employed to cope with a normal dry season are usually adequate for dealing with those relatively harsh dry seasons which follow a poor rainy season. Multi-year droughts provide a much more formidable stress and require more drastic responses. These generally take one of two forms: (1) expansion of the spatial scale of exploitation, or (2) compensation for losses of livestock products by wholesale food substitution or reduction in human demand.

Expansion of the Spatial Scale of Exploitation

Ngisonyoka territory covers about 9,000 km². In normal years only a portion of this territory is utilized by the pastoralists and their livestock (Ecosystems Ltd. 1984, McCabe 1985); however, during severe droughts the spatial scale of exploitation expands. This is accomplished through a variety of tactics, each of which entails increasing levels of risk.

As environmental stress increases, pastoralists divide their livestock herds and and accompanying human groups into smaller and smaller units. These small units tend also to move more frequently than do the larger units which exist during wetter periods (McCabe 1985, in press). Dispersion of the livestock and human population is facilitated by a widely distributed water supply; few large portions of the ecosystem are unusable because of lack of dry-season water (Dyson-Hudson and McCabe 1983). Division into smaller, more dispersed groups is referred to as "risk spreading" in the literature (den Boer 1968); however, smaller groups of pastoralists are actually more vulnerable to raids by bandits or other tribes, therefore incurring somewhat more risk. The real value of this dispersion is to spread grazing pressure more broadly and evenly over the region.

The fact that the Ngisonyoka have unused space to move into during periods of drought stress suggests that the system is stocked well below its average ecological carrying capacity. This is true and it is a critically important feature of the system; but not for the usual density-related reason that stocking too close to carrying capacity creates the risk of overgrazing and subsequent range degradation. In this case, the low level of stocking buys, not protection for the plant resource, but time for the pastoralist in the form of an ungrazed reserve. A single year drought is thus survivable. Large scale herd losses are avoided if the rains return before a second year of drought is encountered.

In some cases it is possible to move livestock completely out of the territory, into regions occupied by other tribes or into other subsections of Turkana, if the other pastoralists are agreeable. This tactic is especially useful if the other regions have been less affected by the impinging drought. If access to external grazing lands is not permitted, it may be necessary to move into boundary areas or "no man's lands" where the risk of intertribal raiding is especially high. Alternately, access to other tribes' resources is sometimes obtained by force, usually a difficult and bloody task (McCabe 1985, in press, Dyson-Hudson and McCabe 1983, McCabe and Ellis 1987, Ellis et al. 1987). Thus, a major strategy for coping with severe drought is to utilize fully the extensive scale of the ecosystem, or to obtain resources which are actually outside that system.

Compensation for Reductions in Livestock Production

The other major strategy for dealing with severe stress is compensation for reduction in livestock products, accomplished either by substituting other products which are available, or by reducing human demand. Alternative food products may include foods gathered from the bush; purchased foods, usually grain, obtained by selling livestock; meat from livestock which have died from starvation or disease: and relief foods, if these are available (Galvin 1985, Galvin and Waweru 1987, Ellis et al. 1987).

Reduction in human demand entails people actually leaving the pastoral system. During the multi-year drought of 1979–80, many Turkana pastoralists left the system temporarily by moving to villages or famine camps within Turkana or to highland areas where they stayed with friends or relatives or became laborers on farms, in abbatoirs, etc. (McCabe 1985). As much as 20% of the Turkana pastoral population may have emigrated from the pastoral sector in the early 1980's although many had returned by 1984 (Ecosystems Ltd. 1984). Individuals undertaking such temporary emigrations are usually of lower social status and not essential for the maintenance of pastoral herds, e.g., unmarried or widowed female relatives, young men who are not herders, etc. Their emigration does not substantially inhibit the normal functions of the system.

This combination of stabilizing strategies—food substitution, demand reduction through emigration, and expansion of the spatial scale of exploitation—allow the Ngisonyoka to persist through periods of severe stress without famine, without degrading their ecosystem, and without permanently reducing their own population. Each of these strategies depends to some extent on utilizing resources which are not exploited during non-stress periods and which lie outside the spatial scale of routine exploitation. It can be concluded that this ecosystem cannot and does not support the extant populations of humans and livestock during periods of severe environmental stress which occur about once per decade (10%–20% of the time). If external/peripheral resources were not available, the human and livestock populations would have to be maintained at considerably lower levels the remaining 80–90% of the time to avoid excessive livestock losses and human famine, during droughts. The ecosystem is not balanced, does not operate in an equilibrial fashion, and cannot be treated as if it did.

Implications for Development

The dominant paradigm assumes that pastoral ecosystems are potentially equilibrial grazing systems and that destabilization of these systems is due to overstocking and overgrazing by pastoralists. Conventional development practices are based on these assumptions; the goals include restoration of equilibrial conditions and increases in productivity. Conventional development procedures involve the establishment of group ranches, grazing blocks, or grazing associations where pastoralists are confined to particular tracts of land. Resources are developed and technical interventions are provided within those tracts to raise productivity and to better regulate the interaction between animals and plants. In theory these interventions have the potential to achieve the desired effects if, in fact, the problem is properly diagnosed; that is if they are applied in ecosystems which have the potential to operate as equilibrial systems, and which have truly been destabilized by overgrazing.

We have attempted to show that in Ngisonyoka Turkana and most probably in many other arid or semi-arid pastoral ecosystems, equilibrial conditions are not attainable. Rather, ecosystem dynamics are dominated by the stochastic perturbations of multi-year droughts. Under these conditions large-scale destocking would result in immediate deprivation for pastoralists even during mild stress periods. Likewise, confining pastoralists to grazing blocks or ranches would reduce the spatial scale of exploitation and result in disaster during serious droughts. The obvious conclusion is that conventional development procedures are destabilizing influences in ecosystems which are dominated by stochastic abiotic perturbations and which operate essentially as non-equilibrial ecosystems.

Can non-equilibrial pastoral systems be improved by development interventions? Or should pastoralists living under these conditions be left alone? In our view the latter is not an option. Pastoralists are coming under the influence of external forces regardless of their remoteness or of the relative success or failure of their traditional exploitation systems. Therefore we must explore the appropriateness of different development interventions for different ecosystems and design interventions to fit the dynamics of specific target systems. A cautious approach to pastoral development is to ask if intervention strategies can be formulated which will build upon the best aspects of traditional systems, rather than imposing wholesale alterations upon them.

The Ngisonyoka and several other Turkana groups seem adept at resisting the affects of severe droughts and at maintaining a sustainable and persistent, albeit unstable and dynamic, exploitation system. Their main strategies for maintaining this system in the face of perturbations are:

• expanding the spatial scale of exploitation during stress periods;
and

• compensating for productivity losses by product substitution and adjusting for loss by demand reduction.

Since these practices are key to pastoral success and persistence in non-equilibrial systems, interventions which facilitate rather than constrain these strategies should be considered. In the first case, it is clear that an extensive spatial scale is a prerequisite for a successful pastoral system where droughts are frequent. Reductions in scale or confining pastoralists to ranches is an invitation to disaster. There is no inherent reason why the maintenance of spatial scale could not be included as a development objective. However, development schemes often emphasize the delivery of "technical packages" and the improvement of resources within "tractable" areas and consider such issues as spatial scale beyond their purview.

In regard to the second strategy, food substitution and demand reduction can best be facilitated by maintaining an open economy in which there is a free flow of goods both in and out. Developing strong interactions with the national economy through viable marketing systems seems the best way to assure this flow. Improved economic flows could eventually reduce the necessary spatial scale, if the commodity flows included supplemental livestock feeds. However, these flows depend upon well-developed market and credit systems at local, regional, and national levels.

These development procedures, maintaining the spatial scale of pastoral ecosystems and facilitating the flow of goods into and out of the ecosystem through market development, are policy-oriented, rather than technical solutions. This is not to say that technical procedures are not useful; it simply suggests that they must be imbedded in a progressive pastoral development policy in order to be effective. Policy implementation is usually a national function, therefore, successful pastoral development probably can not be divorced from on-going national development. This suggests that we may need more emphasis and research on pastoral development policy and how best to use technical interventions to augment those policies, rather than emphasizing technical interventions alone, and ignoring policy-level concerns.

Should range ecologists and managers be involved in policy analysis and application as well as providing expertise on more technical issues? The main message which we wish to convey here is that appropriate policies and technical interventions can be applied only if the fundamental dynamics of the target systems are clearly understood. No one is better qualified than range ecologists to analyze and describe the dynamics of pastoral ecosystems and through this activity to provide the basic understanding necessary for enlightened development policy and intervention. Unless pastoral ecosystem dynamics are considered and used as guidelines for development policies, interventions are likely to be random activities which comprise development by trial and error, a practice with unfortunate implications for the ecosystems and people on which these "development experiments" are performed.

Acknowledgments

The research upon which this paper is based was supported by grants from the US National Science Foundation, Ecosystems Studies and Anthropology programs. Additional funding was provided by the Norwegian Aid Agency for International Development (NORAD). Our thanks to Jesse Logan, John Wiens, and Mike Coughenour for comments on the paper and to Sharren Sund for help with the manuscript.

Literature Cited

Bamberg, I.A. 1986. Effects of clipping and watering frequency on production of the African dwarf shrub *Indigofera spinosa*. M.S. Thesis, Colorado State University, Fort Collins.

Beuchner, H.K., and H.C. Dawkins. 1961. Vegetation change induced by elephants and fire in Murchison Falls National Park, Uganda. J. Wildl. Manage. 27:36-53.

Brainard, J. 1986. Differential mortality in Turkana agriculturalists and pastoralists. Amer. J. Physical Anthro. 70(4):525-536.

Brown, L.H. 1971. The biology of pastoral man as a factor in conservation. Biol. Conserv. 3(2):93-100.

Caswell, H. 1978. Predator mediated coexistence: a non-equilibrium model. Amer. Nat. 112:127-154.

Caughley, G. 1976. The elephant problem—an alternative hypothesis. E. Afr. Wildl. J. 14:265-283.

Charney, J.G., P.H. Stone, and W.J. Quirk. 1975. Drought in the Sahara: a biophysical feedback mechanism. Science 187:434-435.

Connell, J.H., and W.P. Sousa. 1983. On the evidence needed to judge ecological stability or persistence. Amer. Nat. 121:789-824.

Coppock, D.L., J.E. Ellis, J. Wienpahl, J.T. McCabe, D.M. Swift, and K.A. Galvin. 1982. A review of livestock studies of the South Turkana Ecosystem Project. pp. 168-172. *In:* Proceedings, Small Ruminant CRSP Workshop, Kenya. SR-CRSP, Nairobi.

Coppock, D.L., J.T. McCabe, J.E. Ellis, K.A. Galvin, and D.M. Swift. 1985. Traditional tactics of resource exploitation and allocation among nomads in an arid African environment. *In:* Proceedings of the International Rangeland Development Symposium, Society for Range Management, Salt Lake City, Utah.

Coppock, D.L., J.E. Ellis, and D.M. Swift. 1986a. Livestock feeding ecology and resource utilization in a nomadic pastoral ecosystem. J. Appl. Ecol. 23(2):573-583.

Coppock, D.L., D.M. Swift, and J.E. Ellis. 1986b. Seasonal nutritional characteristics of livestock diets in a nomadic pastoral ecosystem. J. Appl. Ecol. 23(2):585-595.

Coppock, D.L., D.M. Swift, J.E. Ellis, and K. Galvin. 1986c. Seasonal patterns of energy allocation and production for livestock in a nomadic pastoral ecosystem. J. Agr. Sci., Camb. 107:347-365.

Coughenour, M.B., J.E. Ellis, D.M. Swift, D.L. Coppock, K. Galvin, J.T. McCabe, and T.C. Hart. 1985. Energy extraction and use in a nomadic pastoral ecosystem. Science 20:619-625.

Coughenour, M.B., D.L. Coppock, J.E. Ellis, and M. Rowland. Herbaceous biomass and productivity in an arid pastoral ecosystem in Northwest Kenya (in prep.).

Coughenour, M.B., D.L. Coppock, M. Rowland, and J.E. Ellis. Dwarf shrub (*Indigofera* spp.) ecology in South Turkana, Kenya. (in prep.).

DeAngelis, D.L., and J.C. Waterhouse. 1987. Equilibrium and nonequilibrium concepts in ecological models. Ecol. Monogr. 57(1):1-21.

denBoer, P.J. 1968. Spreading the risk and stabilization of animal numbers. Acta Biotheoretica (Leiden) 18:165-194.

Dublin, H. 1987. Decline of the Mara woodlands: the role of fire and elephants. Ph.D. Thesis, University of British Columbia.

Dyson-Hudson, R., and J.T. McCabe. 1983. Final Report to the Norwegian Aid Agency on South Turkana Water Resources and Livestock Management.

Donaldson, T.J. 1986. Pastoralism and drought: a case study of the Borana of southern Ethiopia. M. Phil Thesis, University of Reading.

Ecosystems, Ltd. 1984. Turkana District resource survey. Turkana Rehabilitation Project, Draft Final Report, July 1984. Ministry of Energy and Regional Development, Government of Kenya, Nairobi.

Ellis, J.E., D.L. Coppock, J.T. McCabe, K. Galvin, and J. Wienpahl. 1984. Aspects of energy consumption in a pastoral ecosystem: wood use by the southern Turkana. *In:* C. Barnes, J. Ensminger and P. O'Keefe, (eds.) Wood, Energy and Households, Perspectives on Rural Kenya. Energy, Environment and Development in Africa (6). The Beijer Institute and the Scandinavian Institute of African Studies, Sweden.

Ellis, J.E., D.S. Schimel, M.B. Coughenour, T.C. Hart, J.G. Wyant, and S. Lewis. Enhancement of tree establishment by pastoral nomads in an arid tropical ecosystem. (in prep.).

Ellis, J.E., K. Galvin, J.T. McCabe, D.M. Swift. 1987. Pastoralism and drought in Turkana District, Kenya. A report to NORAD, Nairobi. 204pp.

Evans-Pritchard, E.E. 1973. The Nuer. Oxford University Press, London.

Galvin, K.A. 1985. Food procurement, diet and nutrition of Turkana pastoralists in the ecological and social context. Ph.D. Diss. State University of New York, Binghamton.

Galvin, K., and S.K. Waweru. 1987. Variation in the energy and protein content of milk consumed by nomadic pastoralists of northwest Kenya. *In:* A.A.J. Hansen, H.T. Horelli and V.J. Quinn (eds.) Food and Nutrition in Kenya. A Historical Review, UNICEF, Nairobi.

Glover, J. 1963. The elephant problem at Tsavo. E. Afr. Wildl. J. 1:30-39.

Hardin, G. 1968. The tragedy of the commons. Science 162:1243-48.

Helland, J. 1980. Five Essays on the Study of Pastoralists and the Development of Pastoralism. Occasional paper No. 20. Sosialanthropologisk Institut, Universitetet i Bergen, Bergen, Norway.

Herskovits, M.J. 1926. The cattle complex in East Africa. American Anthropologist Vol. 28.

Hobbs, N.T., and D.M. Swift. 1985. Estimates of habitat carrying capacity incorporating explicit nutritional constraints. J. Wildl. Manage. 49:814-822.

Holling, C.S. 1973. Resilience and stability of ecological systems. Ann. Rev. Ecol. Syst. 4:1-23.

ILCA (International Livestock Centre for Africa). 1987. ILCA's Strategy and Long-term Plan: A Summary. Addis Ababa, Ethiopia.

Jacobs, A.H. 1965. The traditional political organization of the pastoral Maasai. Unpublished Ph.D. Thesis, Oxford University. 427 pp.

Kuhn, T.S. 1970. The Structure of Scientific Revolutions, 2nd Ed. University of Chicago Press.

Lamphere, J. 1976. The Traditional History of the Jie. Oxford University Press, London.

Lamprey, H.F. 1983. Pastoralism yesterday and today: the overgrazing problem. pp. 643-666 *In:* F. Bourliere (ed.), Tropical Savannas: Ecosystems of the World, Vol. 13. Elsevier, Amsterdam.

Lamprey, H.F., and H. Yussuf. 1981. Pastoralism and desert encroachment in northern Kenya. Ambio 10:131-147.

Laws, R.M. 1970. Elephants as agents of habitat and landscape change in East Africa. Oikos 21:1-15.

Leslie, P.W., and P.H. Fry. Extreme seasonality of births among South Turkana pastoralists. (submitted).

Lusigi, W.J. 1981. Combatting Desertification and Rehabilitating Degraded Population Systems in Northern Kenya. Tech. Rep. A-4, UNESCO-UNEP Integrated Project in Arid Lands, Nairobi.

McCabe, J.T. 1985. Livestock management among the Turkana: a social and ecological analysis of herding in an East African pastoral population. Ph.D. Diss. State University of New York-Binghamton.

McCabe, J.T. Drought and recovery: Livestock dynamics among the Ngisonyoka Turkana of Kenya. Human Ecology (in press).

McCabe, J.T., and J.E. Ellis. 1987. Beating the odds in arid Africa. Nat. Hist. 96(1):32-41.

Nicholson, A.J. 1958. The self-adjustment of populations to change. Cold Spring Harbor Symp. Quant. Biol. 22:153-173.

Nicholson, A.J., and V.A. Bailey. 1935. The balance of animal populations. J. Animal Ecol. 2:132-178.

Norton-Griffiths, M. 1979. The influence of grazing, browsing and fire on the vegetation dynamics of the Serengeti. *In:* A.R.E. Sinclair and M. Norton-Griffiths (eds.), Serengeti: Dynamics of and Ecosystem. University of Chicago Press, Chicago.

Noy-Meir, I. 1975. Stability of grazing systems: an application of predator-prey graphs. J. Ecol. 63:459-481.

Noy-Meir, I. 1979/80. Structure and function of desert ecosystems. Isr. J. Bot. 28:1-19.

Oxby, C. 1982. Group ranches in Africa. Overseas Development Institute Review. 2:2-13.

Paulsen, H.A., and F.N. Ares. 1961. Trends in carrying capacity and vegetation on arid southwestern range. J. Range Manage. 14:78-83.

Pellew, R.A. 1983. The impacts of elephant, giraffe and fire upon the *Acacia tortilis* woodlands of the Serengeti. Afr. J. Ecol. 21:41-74.

Rosenzweig, M.L., and R.H. MacArthur. 1963. Graphical representation and stability condition of prey-predator interactions. Amer. Nat. 97:209-223.

Sandford, S. 1983. Management of Pastoral Development in the Third World. John Wiley and Sons, New York.

Sinclair, A.R.E., and J.M. Fryxell. 1985. The Sahael o Africa: ecology of a disaster. Can. J. Zool. 63:987-994.

Spencer, P. 1973. Nomads in Alliance. Symbiosis and Growth Among the Rendille and Samburu of Kenya. Oxford University Press, London.

Stebbings, E.P. 1935. The encroaching Sahara. Geo. J. 86:510.

Stiles, D.N. 1983. Camel pastoralism and desertification in Northern Kenya. Desert. Contr. 8:2-8.

Swift, J. 1977. Pastoral development in somalia: herding cooperatives as a strategy against desertification and famine. *In:* M.H. Glantz (ed.), Desertification: Environmental Degradation in and Around Arid Lands, Boulder Westview Press.

Swift, J., and A. Maliki. 1984. A cooperative development experiment among nomadic herders in Niger. Overseas Development Institute, paper 18c.

Volterra, V. 1928. Variations and fluctuations of the number of individuals in animal species living together. Translated in R. N. Chapman 1931. Animal Ecology. McGraw-Hill, New York.

Wallmo, O.C., L.H. Carpenter, W.L. Regelin, R.B. Gill, and D.L. Baker. 1977. Evaluation of deer habitat on a nutritional basis. J. Range Manage. 30:122-127.

Wiens, J.A. 1977. On competition and variable environments. Amer. Sci. 65:590-597.

Wiens, J.A. 1984. On understanding a non-equilibrium world: myth and reality in community patterns and processes. *In:* D.R. Strong, Jr., D. Simberloff, L.G. Abele, A.B. Thistle (eds.), Ecological Communities: Conceptual Issues and the Evidence, Princeton University Press, Princeton, New Jersey.

Western, D. 1973. Structure, dynamics and changes of the Amboseli Basin ecosystem. Ph.D. Diss. Nairobi University.

ECOSYSTEM STRUCTURE AND FUNCTION ALONG URBAN–RURAL GRADIENTS: AN UNEXPLOITED OPPORTUNITY FOR ECOLOGY[1]

M. J. McDonnell and S. T. A. Pickett
Institute of Ecosystem Studies, The New York Botanical Garden,
Mary Flagler Cary Arboretum, Box AB,
Millbrook, New York 12545 USA

INTRODUCTION

Urbanization is a massive, unplanned experiment that already affects large acreages and is spreading in many areas of the United States (Alig and Healy 1987). Urban areas are conservatively defined as those with human populations denser than 620 individuals/km² (United States Bureau of Census 1980, Bourne and Simmons 1982). In 1989, 74% of the United States population (203 million people) resided in urban areas and that number is expected to increase to >80% by the year 2025 (Fox 1987, Haub and Kent 1989). The increase in urban population throughout the country has resulted in the conversion of cropland, pastures, and forests into urban and suburban environments (Ehrenfeld 1970). Between 1960 and 1970 urban land in the United States increased by 9 million acres, and between 1970 and 1980 it increased by 13 million acres (Frey 1984).

Urbanization can be characterized as an increase in human habitation, coupled with increased per capita energy consumption and extensive modification of the landscape, creating a system that does not depend principally on local natural resources to persist. We can use the term "urbanization" as a convenient shorthand for the ecological forcing functions created by the growth of cities and associated human activities. However, the individual components (e.g., structures, physical and chemical environments, populations, communities, ecosystems, and human culture) must be quantified, and correlations among them assessed, to discover the ecologically important impacts of urban development and change.

The structure of metropolitan areas and their fringes consists of a variety of components, ranging from totally built environments to "natural" or seminatural areas (Mumford 1956, Dickinson 1966, Stearns and Montag 1974; Table 1). Natural areas in an urban context are those not intensively managed by people (e.g.,

wooded areas in city parks, lakes, ponds, streams, etc.; Andrews and Cranmer-Byng 1981, McDonnell 1988), and often include a high proportion of intentionally and accidentally introduced organisms as well as native species (Gill and Bonnett 1973, Noyes and Proqulske 1974, Numata 1977, Whitney and Adams 1980, Bornkamm et al. 1982, Kunick 1982, Dorney et al. 1984). The ecological study of the effects of urbanization can focus either on metropolitan areas as wholes or on natural areas within metropolitan areas. The study of the metropolis as an ecosystem, including its human inhabitants and institutions, would be a radical expansion of ecology. The study of natural areas along urban–rural gradients is an application of an existing ecological research strategy to a new situation.

This paper indicates how urbanization can be exploited as a research subject in ecology. We indicate how the effects of urbanization can provide a context for answering ecological questions of general importance and applicability, as well as questions that are specific and unique to urbanization. We also introduce a conceptual framework for the ecological study of urbanization.

URBANIZATION AS A COMPLEX ENVIRONMENTAL GRADIENT

The established and successful "gradient paradigm" (Whittaker 1967, Austin 1987, Stevens 1989) provides a useful basis for ecological studies of the spatially varying effects of urbanization (Ter Braak and Prentice 1988). The gradient paradigm can be summarized as the view that environmental variation is ordered in space, and that spatial environmental patterns govern the corresponding structure and function of ecological systems, be they populations, communities, or ecosystems. The degree of the environmental change in space determines, in part, the steepness of the gradient in system structure and function. Of course, interactions within the ecological systems, and between the environmental gradient and the ecological systems will affect the distribution and behavior of systems along the gradient (Terborgh 1971, Roberts 1987).

[1] For reprints of this Special Feature, see footnote 1, page 1231.

TABLE 1. Features of urbanization.

Structural features of urbanization
Dwellings
Factories
Office buildings
Warehouses
Roads
Pipelines
Power lines
Railroads
Channelized waterways
Reservoirs
Sewage disposal facilities
Dumps
Gardens
Parks
Cemeteries
Airports
Biota of urban areas
Crops
Ornamentals
Domestic pets
Pests
Disease organisms
Socio-economic factors

Because urban areas appear so often as a dense, highly developed core, surrounded by irregular rings of diminishing development (Dickinson 1966), the gradient paradigm is a powerful organizing tool for ecological research on urban influences on ecosystems. Like natural environmental gradients, urbanization should present ecologists with a rich spatial array to use in explaining or predicting environmental and ecological effects. Urban–rural gradients, moreover, provide an opportunity to explicitly examine the role of humans.

WHY STUDY ECOLOGICAL SYSTEMS ALONG URBAN–RURAL GRADIENTS?

From an ecologist's perspective, urbanization produces a variety of unprecedented and intense "experimental manipulations." Examples include changes in: (1) disturbance regimes, (2) biota, (3) landscape structure, (4) physiological stresses (e.g., air pollution), and (5) cultural, economic, and political factors. In most cases, both the spatial extent and magnitude of the manipulations are greater than those that ecologists are typically able to produce.

The coarse-scale, anthropogenic manipulations of ecological systems along urban–rural gradients provide an opportunity to address basic questions at various spatial scales. For example, questions related to hierarchy theory could be addressed. The central problem in this theory is to determine at what scale ecological processes and patterns uniquely appear (O'Neill et al.

1986). The relative influences of urban and natural environmental factors on ecosystem patterning, and the extent to which ecosystem processes are also influenced, could be examined most easily along urban–rural gradients, where human influences can be directly quantified.

Likewise, a number of questions that fall within the framework of disturbance theory could be examined. In disturbance theory, manipulations of disturbance regimes are used to determine the significance of different disturbance types, intensities, and frequencies in communities and ecosystems (Pickett and White 1985). The study of the interactions between urbanization and disturbance regimes and their effects on ecological properties provide an excellent opportunity to advance understanding in this general area.

One specific question that could be addressed using changes in disturbance regimes along urban–rural gradients is the balance between autogenesis and allogenesis (Kolasa and Pickett 1990). If various disturbance and stress factors can be attributed to forces either within or outside the community, then the balance of internal and external control of system organization can be contrasted along the gradient.

One additional area of ecological research that could benefit from studies along urban–rural gradients is that of species control on ecosystem fluxes. The simplification of community composition and the introduction of new species in urban areas provides an opportunity to address questions concerning the mechanistic role of species in ecological processes on higher levels of organization.

Finally, the intimate involvement of humans with the urban–rural gradient suggests that it would be an unparalleled situation in which to integrate humans as subjects for ecological study. Human ecology is the discipline that inquires into the patterns and process of interaction of humans with their environments (Boyden 1977, Boyden and Millar 1978, Vayda 1983). Human values, wealth, life-styles, resource use, and waste, etc. must affect and be affected by the physical and biotic environments along urban–rural gradients. The nature of these interactions is a legitimate ecological research topic and one of increasing importance.

Clearly, the interactions among various anthropogenic factors and between anthropogenic and natural variables make urban–rural gradients potentially complex. These interactions must be assessed before analyses such as those suggested above can be carried out. Furthermore, it is certain that urban–rural gradients are not appropriate for all ecological questions. Nevertheless, we believe that such gradients do provide new and sometimes unique opportunites for ecologists to test assumptions and predictions of many ecological theories.

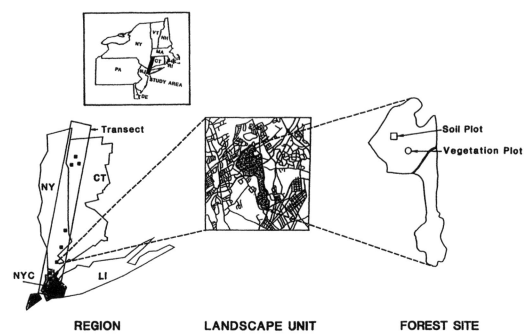

REGION   LANDSCAPE UNIT   FOREST SITE

Fig. 1.   The various scales involved in the study of ecological systems along urban–rural gradients illustrated by the New York City Metropolitan area. Each scale, ranging from the metropolitan region, through the landscape unit, to the site, can be the focus of ecological study.

Specific Questions Along the Urban–Rural Gradient: An Illustration

We have indicated how the study of ecological systems along urban–rural gradients could be used to address topics of general ecological interest. In this section we will discuss several specific uses of an urban–rural gradient, drawing in part on our study of ecosystems in the New York City Metropolitan area. The region includes a readily measurable gradient of land use radiating from the New York City urban core to suburban and rural areas at increasing distances from the city (Fig. 1). Remnant forest patches still exist in the city as well as elsewhere on the land-use gradient, providing an excellent opportunity to investigate long-term human impacts on forest ecosystems.

Soil resources

A preliminary study of the physical and chemical properties of forest soils along a land-use gradient from New York City to rural Dutchess County revealed that the soils at the urban end of the gradient were more hydrophobic than rural sites (White and McDonnell 1988, *unpublished manuscript*). This novel pattern provides a stimulus to address several ecological questions, including the following: Is the formation of the hydrophobic soil the result of natural processes (Adams et al. 1970, DeBano 1971, Reeder and Jurgensen 1979,

McGhie and Posner 1981) or new anthropogenically derived sources? How does the pattern of hydrophobicity vary in time and space? Does it limit resource availability to plants (e.g., by reducing N mineralization)? How does it affect litter decomposition rates and belowground processes? Does it affect gas fluxes from soil to the atmosphere? Is it amplified by other stresses and disturbances? Although these questions are generated by the pattern of contrasting hydrophobicity along the urban–rural gradient, they are relevant to larger concerns of biogeochemical fluxes in a broad range of ecosystems.

Community organization

We know already that forests in highly urban environments differ in both subtle and patent ways from those in the surrounding countryside (Bagnall 1979, Airola and Buchholz 1984, Hobbs 1988, Rudnicky and McDonnell 1989). For instance, urban and suburban forests have a conspicuous proportion of exotic and naturalized species (Bagnall 1979, Airola and Buchholz 1984, Hobbs 1988, Rudnicky and McDonnell 1989), and frequently a lower representation of certain native species. However, little is known about the functional importance of the differences in composition. Likewise, the structure of urban and rural forests differ (Rudnicky and McDonnell 1989). The canopy height

ANTHROPOGENIC CAUSES AND ECOLOGICAL EFFECTS
ALONG URBAN - RURAL GRADIENTS

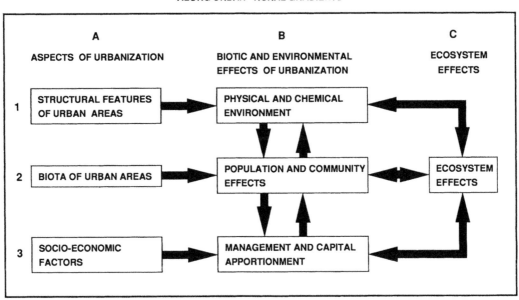

FIG. 2. A composite model of the effects of urbanization on ecological phenomena. The three columns are abstracted from the details of Tables 1 and 2, and the arrows indicate causal linkages between the features of urban areas (column A), as inputs, and the ecological phenomena (columns B, C) as results. The focus of the research program would be on the phenomena represented by rows 1 and 2, although the results would be helpful in decisions concerning societal phenomena represented by row 3. Feedbacks from columns B and C to A would receive attention in research building on that under the scope of this program.

is more uneven and the number and connectedness of treefall gaps differs from rural forests in the region. Furthermore, the fauna and its interactions (e.g., predation, Churcher and Lawton 1987) differ in urbanized areas. These conditions indicate that organization and dynamics of the forests should differ along the gradient.

Once the composition, size structure, and stand architecture are adequately documented, attention can turn to questions such as these: Is the regeneration of current or prior dominants limited in forests at the urban end of the gradient? What are the compositional or architectural correlates of regeneration success or failure? What are the abiotic, biotic, and anthropogenic causes of regeneration patterns? Do changes in the animal community have direct or indirect effects on the plant community dynamics by changing herbivory, predation, or dispersal relations of the plants? How do human perceptions, values, and behaviors affect the dynamics and persistence of forest communities along the gradient?

Landscape ecology

Landscape ecology is empirically a young discipline. Therefore, the questions that may be asked of it in the context of urban–rural gradients are quite exploratory.

Indeed, such questions may have a stimulatory effect on landscape ecology as a whole. In the spirit of an exploratory analysis, assume that the elements of a landscape are more distinct toward the urban end of the gradient (Godron and Forman 1983). Note, however, that the pattern of distinctness can be determined by direct measurement of aerial photographs and maps, so that whether distinctness of patches increases or decreases monotonically, or is humped in some middle distance on the gradient, is a simple empirical matter. Illustrative questions that may follow upon the pattern analysis include the following: Is dispersal of plants, over equivalent distances, more limited in urban than rural ranges of the gradient? Is any such change a direct result of the activities of people (including vehicles, pets, and pests) or some indirect effect of the size, shape, and arrangement of landscape elements? How does size, shape, distance, and arrangement of patches relate to their species composition and to ecosystem processes within them?

A CONCEPTUAL MODEL FOR THE ECOLOGY OF
URBAN–RURAL GRADIENTS

Because the study of ecology along urban–rural gradients is new, a framework to help guide the design

TABLE 2. An elaboration of the environmentally and biotically relevant effects of the features of urbanization (Table 1).

Physical and chemical environment
Local climate
Air pollution
Hydrologic changes
Water pollution
Soil changes and earth movement
Population and community characteristics
Altered disturbance regimes
Introduced species
Increase in morbidity
Altered assimilation
Altered reproductive status
Changes in growth rates
Phenological changes
Reduced longevity
Social and behavioral changes
Genetic drift and selection
Population size and structure
Altered successions
Reduced richness
Landscape fragmentation
Ecosystem structure and function
Debris dams
Forest floor
Sediment loading
Patchiness
Layering of vegetation
Productivity
Nutrient flux
Decomposition
Nutrient retention
Loss of redundant pathways
Loss of compartments
Alteration of equilibria
Management and capital apportionment

and integration of studies is required. The framework must account for: (A) the factors that constitute urbanization, (B) the effects of urbanization on the biota and physical environment, and (C) the resultant effects on ecosystems (Fig. 2). The constituent factors of urbanization and their biotic and environmental effects are each divisible into three realms. In order of increasing complexity, these realms are (1) physical structure, (2) biotic components, and (3) human culture and institutions (Fig. 2). The physical and chemical environment and the dynamics of natural and seminatural populations and communities can be affected by urbanization in many ways (Table 2). The conceptual model serves as a framework that will be filled in as the study of ecology in this new context matures. Ultimately, specific mechanistic, predictive, or explanatory models will be constructed to quantitatively describe the interactions and components of the conceptual framework (Fig. 2).

Ecologists usually only study two parts of the three-part model (Fig. 2b, c) and do so most often in nonurban systems. Explicit study of the aspects and features of urbanization (Fig. 2a) as well as their effects (Fig. 2b, c) is an underutilized area for ecological research, but one of increasing importance given the extension and magnitude of anthropogenic effects today.

CONCLUSION

The growth of metropolitan areas in North America and indeed worldwide indicates that knowledge of ecosystems under the influence of urbanization can only become increasingly important. The magnitude and nature of the change in the physical, chemical, and biotic environments that are associated with urbanization provide an unprecedented suite of "experimental manipulations" that ecologists can utilize. We propose a framework to guide the design and integration of ecological studies along urban–rural gradients and indicate its utility for addressing basic ecological questions. Finally, we suggest that the study of urban-rural gradients provides a new context in which to integrate humans as critical components of ecological systems. The results of these studies will not only contribute to our understanding of basic ecological principles, but are critical to the ecologically sound management of human-dominated ecosystems.

ACKNOWLEDGMENTS

We would like to thank J. Brown, G. Guntenspergen, J. Koch, P. Matson, P. Vitousek, and W. Zipperer for their comments and suggestions on the manuscript. G. Likens, G. Lovett, and R. Pouyat made helpful suggestions on earlier versions of the manuscript. Contribution to the program of the Institute of Ecosystem Studies, The New York Botanical Garden.

LITERATURE CITED

Adams, S., B. R. Strain, and M. S. Adams. 1970. Water-repellent soils, fire and annual plant cover in a desert shrub community in southeastern California. Ecology 51:696–700.

Airola, T. M., and K. Buchholz. 1984. Species structure and soil characteristics of five urban sites along the New Jersey Palisades. Urban Ecology 8:149–164.

Alig, R. J., and R. G. Healy. 1987. Urban and built-up land area changes in the United States: an empirical investigation of determinants. Land Economics 63:215–226.

Andrews, W. A., and J. L. Cranmer-Byng. 1981. Urban natural areas: ecology and preservation. Environmental Monograph Number 2. Institute for Environmental Studies, University of Toronto, Toronto, Canada.

Austin, M. P. 1987. Models for the analysis of species' response to environmental gradients. Vegetatio 69:35–45.

Bagnell, R. G. 1979. A study of human impact on an urban forest remnant: Redwood Bush, Tawa, near Wellington, New Zealand. New Zealand Journal of Botany 17:117–126.

Bornkamm, R., J. A. Lee, and M. R. D. Seaward. 1982. Urban ecology. Blackwell Scientific, Oxford, England.

Bourne, L. S., and J. W. Simmons. 1982. Defining the area of interest: definition of the city, metropolitan areas and extended urban regions. Pages 57–72 in L. S. Bourne, editor. Internal structure of the city. Oxford University Press, New York, New York, USA.

Boyden, S. V. 1977. Integrated ecological studies of human settlements. Impact of Science on Society 27:159–169.

Boyden, S. V., and S. Millar. 1978. Human ecology and the quality of life. Urban Ecology 3:263–287.

Churcher, P. B., and J. H. Lawton. 1987. Predation by domestic cats in an English village. Journal of Zoology 212: 439–455.

Debano, L. F. 1971. The effects of hydrophobic substances on water movement in soil during infiltration. Soil Science Society of America Proceedings 35:340–343.

Dickinson, R. E. 1966. The process of urbanization. Pages 463–478 in F. F. Darling and J. P. Milton, editors. Future environments of North America. Natural History Press, Garden City, New York, USA.

Dorney, J. R., G. R. Guntenspergen, J. R. Keough, and F. Stearns. 1984. Composition and structure of an urban woody plant community. Urban Ecology 8:69–90.

Ehrenfeld, D. W. 1970. Biological conservation. Holt, Rinehart, & Winston, New York, New York, USA.

Fox, R. 1987. Population images. United Nations Fund for Population Activities.

Frey, H. T. 1984. Expansion of urban area in the United States: 1960–1980. United States Department of Agriculture Economic Research Service Staff Report Number AGES830615.

Gill, D., and P. Bonnett. 1973. Nature in the urban landscape: a study of city ecosystems. York, Baltimore, Maryland, USA.

Godron, M., and R. T. T. Forman. 1983. Landscape modification and changing ecological characteristics. Pages 12–28 in H. A. Mooney and M. Godron, editors. Disturbance and ecosystems. Springer-Verlag, New York, New York, USA.

Haub, C., and M. M. Kent. 1989. 1989 world population data sheet. Population Reference Bureau, Washington, D.C., USA.

Hobbs, E. 1988. Using ordination to analyze the composition and structure of urban forest islands. Forest Ecology and Management 23:139–158.

Kolasa, J., and S. T. A. Pickett. 1990. Ecological systems and the concept of biological organization. Proceedings of the National Academy of Sciences (USA) 86:8837–8841.

Kunick, W. 1982. Comparison of the flora of some cities of the central European lowlands. Pages 13–22 in R. Bornkamm, J. A. Lee, and M. R. D. Seaward, editors. Urban ecology. Blackwell Scientific, Oxford, England.

McDonnell, M. J. 1988. A forest for New York. Public Garden 3:28–31.

McGhie, D. A., and A. M. Posner. 1981. The effect of plant top material on the water repellence of fired sands and water repellent soils. Australian Journal of Agricultural Research 32:609–520.

Mumford, L. 1956. The history of urbanization. Pages 382–400 in W. Thomas, editor. Man's role in changing the face of the earth. University of Chicago Press, Chicago, Illinois, USA.

Noyes, J., and D. Proqulske. 1974. Wildlife in an urbanizing environment. Cooperative Extension Service, University of Massachusetts, Amherst, Massachusetts, USA.

Numata, M. 1977. The impact of urbanization on vegetation in Japan. Pages 161–171 in A. Miyawaki and R. Tuxen, editors. Vegetation science and environmental protection. Proceedings of the International Symposium in Tokyo on Protection of the Environment and Excursion on Vegetation Science through Japan. 5–7 June. Maruzen, Tokyo, Japan. Chiba University, Chiba, Japan.

O'Neill, R. V., D. L. DeAngelis, J. B. Waide, and T. A. Allen. 1986. A hierarchical concept of ecosystems. Princeton University Press, Princeton, New Jersey, USA.

Pickett, S. T. A., and P. S. White, editors. 1985. The ecology of natural disturbance and patch dynamics. Academic Press, New York, New York, USA.

Reeder, C. J., and M. F. Jurgensen. 1979. Fire induced water repellency in forest soils of upper Michigan. Canadian Journal of Forest Research 9:369–373.

Roberts, D. W. 1987. A dynamical systems perspective on vegetation theory. Vegetatio 69:27–33.

Rudnicky, J. L., and M. J. McDonnell. 1989. Forty-eight years of canopy change in a hardwood-hemlock forest in New York City. Bulletin of the Torrey Botanical Club 116: 52–64.

Stearns, F., and T. Montag. 1974. The urban ecosystem: a holistic approach. Dowden, Hutchinson, and Ross, Stroudsburg, Pennsylvania, USA.

Stevens, G. C. 1989. The latitudinal gradient in species range: How so many species coexist in the tropics? American Naturalist 133:240–256.

Terborgh, J. 1971. Distribution on environmental gradients: theory and a preliminary interpretation of distributional patterns in the avifauna of the Cordillero Vilcabamba, Peru. Ecology 52:23–40.

Ter Braak, C. J. F., and I. C. Prentice. 1988. A theory of gradient analysis. Advances in Ecological Research 18:272–327.

United States Bureau of Census. 1980. Census user's guide. United States Department of Commerce, United States Government Printing Office, Washington, D.C., USA.

Vayda, A. P. 1983. Progressive contextualization methods for research in human ecology. Human Ecology 11:265–282.

White, C. S., and M. J. McDonnell. 1988. Nitrogen cycling processes and soil characteristics in an urban versus rural forest. Biogeochemistry 5:243–262.

Whitney, G. G., and S. D. Adams. 1980. Man as a maker of new plant communities. Journal of Applied Ecology 17: 341–448.

Whittaker, R. H. 1967. Gradient analysis of vegetation. Biological Reviews 49:207–264.

Madhav Gadgil, Fikret Berkes and Carl Folke

Indigenous Knowledge for Biodiversity Conservation

Indigenous peoples with a historical continuity of resource-use practices often possess a broad knowledge base of the behavior of complex ecological systems in their own localities. This knowledge has accumulated through a long series of observations transmitted from generation to generation. Such "diachronic" observations can be of great value and complement the "synchronic" observations on which western science is based. Where indigenous peoples have depended, for long periods of time, on local environments for the provision of a variety of resources, they have developed a stake in conserving, and in some cases, enhancing, biodiversity. They are aware that biological diversity is a crucial factor in generating the ecological services and natural resources on which they depend. Some indigenous groups manipulate the local landscape to augment its heterogeneity, and some have been found to be motivated to restore biodiversity in degraded landscapes. Their practices for the conservation of biodiversity were grounded in a series of rules of thumb which are apparently arrived at through a trial and error process over a long historical time period. This implies that their knowledge base is indefinite and their implementation involves an intimate relationship with the belief system. Such knowledge is difficult for western science to understand. It is vital, however, that the value of the knowledge-practice-belief complex of indigenous peoples relating to conservation of biodiversity is fully recognized if ecosystems and biodiversity are to be managed sustainably. Conserving this knowledge would be most appropriately accomplished through promoting the community-based resource-management systems of indigenous peoples.

INDIGENOUS AND MODERN KNOWLEDGE

Knowledge is an outcome of model-making about the functioning of the natural world. All societies, pre-scientific and scientific strive to make sense of how the natural world behaves and to apply this knowledge to guide practices of manipulating the environment. Before the elaboration of the modern hypothetico-deductive method of systematically accumulating understanding of the functioning of the natural world, pre-scientific societies accumulated knowledge at a rather slow pace. Much of this knowledge was qualitative and based on observations on a rather restricted geographical scale. Within these bounds, reference to habitat preferences, life histories and behavior patterns of prey species such as birds, could be amazingly detailed (1, 2). Models of how the natural world functions as well as prescriptions on how to manipulate it are inevitably linked to any society's world view. However, in pre-scientific societies such models and prescriptions are much more closely integrated with moral and religous belief systems, so that knowledge, practice and beliefs co-evolve.

Modern scientific knowledge, with its accompanying world view of humans as being apart from and above the natural world has been extraordinarily successful in furthering human understanding and manipulation of simpler systems. However, neither this world view nor scientific knowledge have been particularly successful when confronted with complex ecological systems. These complex systems vary greatly on spatial and temporal scales rendering the generalizations that positivistic science has come up with of little value in furnishing practical prescriptions for sustainable resource use (3, 4). Science-based societies have tended to overuse and simplify such complex ecological systems, resulting in a whole series of problems of resource exhaustion and environmental degradation.

It is in this context that the knowledge of indigenous societies accumulated over historical time, is of significance. The view of humans as a part of the natural world and a belief system stressing respect for the rest of the natural world is of value for evolving sustainable relations with the natural-resource base (5).

Not all pre-scientific societies have necessarily lived harmoniously with the natural world, and not all indigenous peoples, outside industrial societies do so today (6). For example, some nomadic hunter-gatherers, who are not tied to any specific resource base and without well-defined territories, may gain little from prudent resource use. The same is true for agriculturists colonizing new territories with options of moving on to new localities as resources are exhausted. Rather, it is the more sedentary fishing, horticultural or subsistence agricultural societies with considerable dependence on hunting and gathering in their immediate neighborhoods that are most likely to have accumulated long series of historical observations of relevance to sustainable resource use and conservation of biodiversity. Self-regulatory mechanisms tend to evolve in such societies when they are faced with resource limitations. Among these mechanisms is a recognition and accumulation of knowledge about the important role that species play in generating ecological services and natural resources. As several major studies point out, *indigenous knowledge*, or *traditional ecological knowledge*, is of significance from a conservation perspective and an attribute of societies with continuity in resource use practices (7, 8).

Indigenous knowledge is herein defined as a cumulative body of knowledge and beliefs handed down through generations by cultural transmission about the relationship of living beings, (including humans) with one another and with their environment (9).

Many indigenous societies depended on a rather limited resource catchment of a few hundred square kilometers to provide them with a wide diversity of resources. This is not to say that they were isolated societies; many had ongoing trade and social relationships with more complex societies. However, the extent to which indigenous societies transformed local resources through manufacturing was limited, as was their ability to supplement locally available resources with imports. Thus, there were strong incentives for indigenous people to nurture and sustain diversity in their immediate environs (10). They may therefore be expected not only to conserve locally present natural biodiversity, but also to augment it by manipulating the landscape. Such manipulations could increase landscape patchiness, for instance by introducing various successional stages, thereby enhancing diversity in local resource catchments.

This paper examines local biodiversity conservation and enhancement activities of indigenous peoples, the knowledge base

A sacred *Mimusops elengi* tree believed to harbor a nature spirit, Yakshi, near Siddapur in Karnataka state in the mid Western Ghats of south India. Myriads of such sacred trees still dot the Indian countryside. Photo: M. D. Subash Chandran.

underlying it, as well as linkages to practices dealing with ecosystems and related belief systems. This accumulation of experience, knowledge and beliefs in dealing with the natural environment may be viewed as a kind of "capital". The paper also considers how this stock of "cultural capital" (11) may be retained and put to use in broader efforts to conserve biodiversity.

ENHANCING BIODIVERSITY

There is ample evidence of indigenous knowledge and practices involved in enhancing biodiversity at the landscape level.

Recent work in the Amazon basin has concentrated on longer-term changes in the forest structure, and found practices that result in the creation, for example, of forest islands, *apete*, by the Kayapo Indians of Brazil who live at the southern limit of the rainforest (12). *Apete* begin as small mounds of vegetation about 2 m in diameter (*apete-nu*). As planted crop and tree seedlings grow and the planted area expands, the taller vegetation in the center of the mound are cut to allow light. A full-grown apete has an architecture that creates zones that vary in shade and moisture (Fig. 1). The species mix includes medicinal species, palms, and vines that produce drinking water. Of a total of 120 species found in ten apete, Posey estimated that 75% may have been planted.

New apete fields peak in crop production in 2 to 3 yrs, but some species continue to be productive for a longer period: sweet potatoes for 4 to 5 yrs, yams and taro for 5 to 6 yrs, papaya and banana for 5 or more years. Contrary to common belief, old fields (*ape-ti*) are not abandoned when the primary crop species disap-

pear; they keep producing a range of useful products. They become forest patches in the savannah-like open cerrado, managed for fruit and nut trees, and "game farms" that attract wildlife. This behavior actively promotes patchiness and heterogeneity in the landscape through a number of devices. Posey first became aware that these isolated forest patches were human-made in the seventh year of his field research (13).

Working in the Ecuador portion of the Amazon forest, Irvine (13) has also reported that Runa Indian swiddens resemble agroforestry systems rather than the slash-and-burn that merely results in temporary clearings in the forest canopy. Compared to unmanaged fallows, Irvine (14) found that management actually increased species diversity in 5-year-old fallows. Between 14 and 35% of this enhanced species diversity was attributed to direct planting and protection of secondary species. Irvine (14) characterized Runa agroforestry as a low-intensity succession management system which, nevertheless, alters forest composition and structure in the long run.

There is species diversity management also in some traditional aquaculture systems. Unlike many contemporary single-species, high-input, high-throughput aquaculture systems (15, 16), many of the ancient fish-rearing systems of China, Hawaii, Indonesia and elsewhere make use of a mix of species, taking advantage of the ecological characteristics of each, and making full use of waste recycled to provide food. Some Chinese systems combine, for example, grass carp, silver carp, bighead carp, common carp with three other aquaculture species, taking advantage of their complementary feeding specializations (17). The Chinese have also developed integrated agriculture-aquaculture systems in which agricultural waste feeds the fish, and fish waste fertilizes crops. An example is the fish-farming system integrated with mulberry-silkworm, vegetable and sugarcane production in the Zhujiang delta (18).

In Indonesia, traditional systems combined rice and fish culture, and wastes from this system often flowed downstream into brackish water aquaculture systems (*tambak*). The tambaks themselves were polyculture ponds, often combining fish, vegetables and tree crops (19). East Asia and Oceania had, and to some extent still has, a wealth of these systems. In ancient Hawaii, both freshwater and seawater fish ponds were integrated with agriculture, and river valleys were managed as integrated systems, from the upland forest (left uncut by taboo) all the way to the reef (20) (Fig. 2).

Costa-Pierce (20) provides a list of cultured species, some 10 to 20 in each of the four kinds of fish ponds. As with the Chinese carp culture systems, the number of cultivated species must have been fewer than those available in the natural environment. Nevertheless, the point is that both agriculture and aquaculture systems in these traditional societies were more diverse than contemporary food production systems. They were more "in

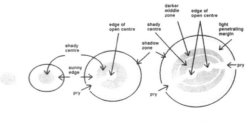

Figure 1. Enhancing biodiversity through the creation of forest islands, *apete*, by the Kayapo Indians of Brazil. Through a number of devices the behavior promotes patchiness and heterogeneity in the landscape in time and space (modified from 12).

tune" with natural ecosystem processes, and in fact one of them, the Hawaii case, may be considered an example of sustainable ecosystem management.

Habitat management by traditional agriculturalists and agroforesters is relatively well known. More controversial is habitat management by hunter-gatherers. Relatively recent work on fire management among Australian aborigines and northern Canadian Amerindians has shed new light on the controversy. Lewis and Ferguson (21) reported that Amerindians of northern Alberta, Canada, regularly used fire until the late 1940s to open up clearings (meadows and swales), corridors (trails, traplines, ridges, grass fringes of streams and lakes), and windfall forests. These clearings provided improved habitat for ungulates and waterfowl, thus, increasing hunting success, and the corridors and windfall areas improved accessibility.

Lewis' (22) work in Australia showed that aborigines possessed detailed technical knowledge of fire, and used it effectively to improve feeding habitat for game and to assist in the hunt itself. Lewis and Ferguson (21) theorized that cross-cultural comparisons of Amerindians from northwestern Canada, western Washington State and northwest California, and Australian aborigines from Tasmania, New South Wales, Western Australia and the Northern Territory indicate functionally parallel strategies in the ways that hunter-gatherers used fire.

Another widely used traditional practice, that of rotation of harvesting pressure, would similarly contribute to landscape heterogeneity. The principle of rotation in agriculture is well known: land is periodically fallowed or "rested", and often

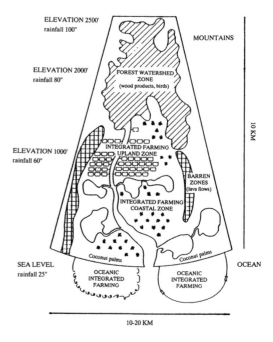

PLACEMENT OF SETTLEMENT

Figure 2. In the *ahupua'a* system of Hawaii, both freshwater and seawater fish ponds were integrated with agriculture, and river valleys were managed as integrated systems, from the upland forest (left uncut by taboo) all the way down to the reef (modified from 20).

Amerindians at a market in the Peruvian Andes. Photo: C. Folke.

planted with species that help restore soil fertility. Less well known is the use of rotation for grazing lands and for hunting and fishing grounds. In semiarid regions such as the fringe of Sahel, plant productivity is seasonal and follows the rains. Many of the larger herbivores have adapted to this pattern by migrating seasonally, and the migrations of traditional herding peoples also follow the same adaptation. Much of the problem of the Sahel is traceable to the disruption of this adaptation by the settlement of herding peoples (23). The yearly cycle of nomads and their cattle is a rotation, providing a chance for the recovery of heavily-grazed rangelands. Throughout arid and semiarid Africa, traditional herders followed migratory cycles, rotating grazing land seasonally and, in some cases, also rotating adjacent grazing areas in the same season (24).

Rotation of hunting and fishing lands of native Amerindian people have been described from eastern subarctic Canada. Feit (25) has reported on Waswanipi Cree Amerindian hunting-trapping territories in the James Bay area. The community lands of Cree groups are divided into a number of traditional hunting areas, each with a "stewart" in charge. In the ideal case, the steward divides his area into four segments, and concentrates on one segment per year, with a four-year rotation. Feit (25) found that the beaver harvest in areas rested for two or more years was significantly higher than in areas not rested. For moose, the trend was similar but statistically not significant. Berkes (26) found that Chisasibi Cree goose hunters rotated hunting areas on a 7-day cycle. In this case, the function of the rotation was to reduce disturbance to feeding and resting geese, and to harvest for subsistence needs with a minimum disruption of the large population that passes through the area.

Fish harvesting also followed rotations–rotations of different periodicities depending on the area. Remote lakes were fished on a 4 or 5 year cycle (27), a traditional pattern which has been incorporated into government fisheries management policy for remote northern lakes in Ontario and Manitoba as well. Chisasibi Cree fishermen used shorter cycles in coastal fishing near the community, fishing each cove hard until the catch per unit of effort fell below a threshold level and then moving to a different cove (26). The sustainability of the Chisasibi Cree fishery was tested and confirmed by comparing the age-frequency data for the two most important stocks, whitefish *Coregonus clupeaformis* and *C. artedii*, sampled 50 years apart (28). In each of these rotation cases, biodiversity conservation is an indirect effect of maintaining the general productivity of the habitat, whether it is grazing land or fishing area.

RESTORING BIODIVERSITY

With their interest in the availability of a wide diversity of resources within their resource catchments one expects indigenous people to contribute to restoration of biodiversity in the depleted landscapes as well. Where a stake in local resources has been created for them, indeed they do so, their detailed knowledge of succession and habitat preferences of the different species greatly contributing to such a process.

This, for example, has been the experience of village forest-protection committees in the Indian state of West Bengal (29). Over the last 10 years this state has pioneered involvement of local, mostly tribal communities of Midnapore and Purulia districts in protecting local forests to encourage natural regeneration. Once empowered and assured of a share in the produce of regenerated forest stocks, the local people have very enthusiastically nurtured regeneration of the local species-diverse forest dominated by the dipterocarp *Shorea robusta*. They have been especially motivated in this by their own immediately improved access to a wide variety of non-wood-forest-produce.

The National Wasteland Development Board (NWDB), an agency of the Government of India's Ministry of Environment and Forests, is now planning to seriously encourage such local initiatives for restoring productivity and biodiversity of degraded lands. NWDB's (30) experiment in microplanning for integrated development of wastelands now calls for preparation of a detailed land and water-management plan at a decentralized level through the agency of local village populations. It further recommends that all new plantations should include a nucleus of totally protected area, perhaps to the extent of 5% of the total, that would attempt to recreate a diverse plant community stocked by indigenous species. Such nuclei would be analogous to the traditional sacred groves or safety forests of Mizoram discussed in the next section, while the plantations would serve as supply forests. While this experiment is just being launched, the reaction of local people to the recreation of such species diverse patches appears to be entirely favorable.

CONSERVING BIODIVERSITY

That indigenous people are aware of a large variety of uses of local biodiversity including medicinal uses which have been incorporated in the modern pharmacopoeia is well known, as is their knowledge of habitat preference, life history, and behavior relevant to efficient foraging for such resources. Such knowledge is explicit and socially transmitted from one individual to another within and across generations in the same manner as scientific knowledge. The indigenous knowledge base pertaining to conservation is not as explicit. Conservation calls for restraints in resource use. The exact nature of such restraints is difficult to ascertain. Witness for example the bewildering array of fisheries regulations on mesh size, closed seasons, and quotas that may still prove insufficient to prevent the collapse of fisheries. Furthermore, resource management regulations may be difficult to implement because short-term individual interests may contradict long-term societal interests. Arriving at an appropriate set of restraints and implementing them is therefore not a simple matter of information transmission. Rather, implementation seems to be based on a complex set of "rules of thumb" arrived at through accumulated historical experience. Compliance is often facilitated through religious belief, ritual, and social conventions.

Four kinds of indigenous conservation practices are of particular relevance. They include:

(a) Total protection to many individual biological communities including pools along river courses, ponds, meadows and forests. Thus, sacred groves were once widely protected throughout the old world. They continue to be so protected even after conversion to Christianity in the tribal state of Mizoram in northeastern India, now being called "safety forest", while the village woodlot from which regulated harvests are made is called the "supply forest" (31). Ecological theory suggests that providing such absolute protection in "refugia" can be an effective way of ensuring persistence of biological populations (32).

(b) All individuals of certain species of plants and animals may be afforded total protection. Trees of all species of genus *Ficus* are protected in many parts of the old world. It is notable that *Ficus* is now considered a keystone resource significant to the conservation of overall biodiversity (33, 35). Local people seem to be aware of the importance of *Ficus* as affording food and shelter for a wide range of birds, bats and primates, and it is not difficult to imagine that such understanding was converted into widespread protection of the *Ficus* tree at some point in the distant past. It is more difficult to visualize the ecological significance of protection on a local scale, a large number of different plants and animals as being totemic.

(c) Certain particularly vulnerable stages in the life history of an organism may be given special protection. Thus, in south India fruit bats may be hunted when away foraging, but not at daytime roosts on trees that may be in the midst of villages. Many waders are hunted outside the breeding season, but not at heronaries, which may again be on trees lining village streets. The danger of overharvest and depletion of population is clearly far greater if these vulnerable stages are hunted and the protection afforded to them seems a clear case of ecological prudence (34, 35).

(d) Major events of resource harvest are often carried out as a group effort. Many tribal groups engage once a year in a large-scale communal hunt. Such a group exercise may have served the purpose of group-level assessment of the status of prey populations, and their habitats. This in turn may have helped in continually adjusting resource harvest practices so as to sustain yields and conserve diversity.

KNOWLEDGE, PRACTICE AND BELIEF

Joshi and Gadgil (32) provide a possible model of how such a complex of prescriptions and beliefs may evolve. They model an indigenous society attempting to maximize harvests from some biological population while keeping the risk of its extinction at a minimum. This is most effectively arrived at by using a decision rule involving total protection to a part of the population in a series of dispersed refugia. The optimum extent of such refugia may be arrived at by enhancing the total area under refugia whenever: (a) an increase in the area under refugia results in higher yields, or (b) a decrease in the area under refugia results in lower yields.

By the same token the total area under refugia should be reduced whenever: (a) an increase in the area under refugia leads to decreased yields, or (b) a decrease in the area under refugia leads to increased yields.

This of course holds when the harvesting effort outside refugia is constant. If the harvesting effort varies, decision rules can be appropriately amended. Joshi and Gadgil (32) found that such a system of decision rules incorporating refugia is far more likely to permit the social groups to arrive and stabilize at sustained yield levels, than a system where the level of harvesting effort is adjusted while harvesting from the entire population. Furthermore, a harvesting regime involving absolutely no harvests from refugia, while permitting harvests elsewhere is easier to implement than harvest quotas. It is also far easier to perceive and adjust the extent of refugia than levels of harvesting efforts.

As noted above, systems of refugia such as sacred ponds and groves are widely prevalent elements of indigenous resource-management systems. These systems might quite plausibly have been established through a process–involving decision rules of the kind sketched above. If so, the social group as a whole may be only vaguely aware of the rationale behind the system being implemented through religious beliefs. Such a process would inevitably result in a commingling of knowledge, practice and belief which seems to characterize conservation practices of indigenous people.

CONCLUSIONS

Resilience, the ability of the ecosystem to recover from surprises and shocks and to continue to function and provide ecological services, is probably the most critical ecosystem property to sustain (36). Generally, only a small number of species have a major role in ecosystem processes. But, when the ecosystem is stressed or perturbed, a larger number of other species perform a buffering role, in the sense that they contribute to the resilience of the system. Without these understudy species resilience is reduced, or perhaps even lost. Therefore, many ecologists argue that ecosystem resilience is promoted by biodiversity conservation. If resilience is the key property to be maintained to assure the generation of essential ecological services to human societies, then long-term historical experience of particular ecosystems is obviously of vital importance.

Indigenous people with their diachronic data, rather than western science with its reliance on synchronic data, may therefore have far more valuable knowledge relevant to biological conservation. But as indigenous knowledge is intricately linked to practice and belief, it is difficult to interpret this knowledge in the framework of western science. On the other hand, it is easier to abstract the knowledge directly related to utilization, for instance medical properties or fruiting times. Indeed, current efforts are mostly focussed on using indigenous knowledge for this purpose. Indigenous knowledge of conserving and enhancing biodiversity may be best put to use as an integrated system of knowledge, practice and beliefs. If this is to happen, then indigenous cultures must be conserved through the recovery of the right to decide their own destiny and the patterns of resource use they wish to pursue. This in turn calls for empowering communities of indigenous people to manage their own resource base.

Common property theory provides some general guidelines and policy prescriptions for the success of such indigenous peoples-based conservation (37, 38).

Eliminate open-access conditions

If access to the area to be conserved is open to any user, the commons will be degraded sooner or later. To eliminate open-access, property rights need to be defined; these can be private property or communal property rights. In areas with traditional peoples, private property is often not feasible as these peoples have social traditions of joint ownership. Controlling the access of others makes it possible for the local group to appropriate any benefits of the conservation effort. This establishes economic incentives to conserve.

Balance rights and responsibilities

Instituting resource-use rights for the local population or recognizing existing rights is only half of the solution. These rights should be balanced against responsibilities. This can be done by protecting and nurturing any existing communal property management systems. A characteristic of any well functioning communal property regime is the ability of users to limit access to the resource to members of the group, and the ability to make and enforce rules among themselves.

Legalize rights

The only way to provide long-term protection for local rights, and indirectly for local conservation, is to legalize communal resource-use rights. This may entail harmonization of national laws with local regulations, as for example in the case of appointing local leaders in charge of enforcing traditional rules, as game wardens.

Delegation of rule-making authority to the community assumes that the government is willing to share responsibilities, but remains ultimately responsible for conservation. A number of factors affect the ability of the local organization to function effectively. Internal factors include group size, social/cultural heterogeneity, and leadership; external factors include population pressure, commercialization of the resource and technology change.

Variety of consequences of restricting harvests of biological populations

Presumed consequences	Possible examples	
	Traditional societies	Modern societies
Save effort / avoid risk when returns are inadequate	Ban on sea fishing during the monsoon off west coast of India	Closure of whaling, as whale populations declined, by the American fleet
Enhance yields of same species in the long run	Protection to birds breeding in heronaries, or fruit bats in day time, South India	Closed seasons for hunting / fishing mesh-size regulations
Enhance yields of other species in the long run	Protection to keystone species like trees of genus *Ficus* in parts of Asia and Africa	
No discernible material benefits	Protection to all species of monkeys in many parts of India	Ban on hunting of cranes

Sharing conservation responsibility and benefits requires cooperative management (or co-management) arrangements between the local organization and the government, and the rights and benefits may be spelled out in a management plan. The biosphere reserve is one kind of conservation area that theoretically allows for joint management and compatible human activity in part of the protected area. New developments in Integrated Conservation Development Projects (ICDPs) and community-based conservation may be signalling a changing philosophy of conservation to allow for more biosphere reserve-style joint management (39). The general issue of the role which indigenous peoples should play in the management of protected areas, has come up for discussion only in recent years. There is a growing literature on this subject, but concerns have also been expressed that conservationists may be expecting too much too soon from ICBs in the absence of demonstrated success cases (40).

Furthermore, as traditional peoples are integrated into the global economy, and come under trade, acculturation and population pressures, they lose their attachment to their own restricted resource catchments. This may lead to a loss of motivation in sustainable uses of a diversity of local resources, along with the pertinent indigenous knowledge. The feedbacks between ecological carrying capacity of the local environment and social self-regulatory mechanisms, the shared values of local institutions, are cut. Thus, decoupling indigenous groups from their local resource base is likely to reduce the resilience of their social/cultural systems as well as the local ecosystems, making both more fragile.

In pace with the growing awareness that the natural capital is increasingly replacing human-made capital as the limiting factor for the development of human societies (41), there will be a growing demand for increased knowledge of the functioning of the natural environment. Much is yet to be learned about the contribution of particular species, or groups of species for the structure, functioning, resilience, and integrity of self-organizing ecosystems (42, 43). Indigenous knowledge does indeed hold valuable information on the role that species play in ecologically sustainable systems. Such knowledge is of great value for an improved use of natural resources and ecological services, and could provide invaluable insights and clues for how to redirect the behavior of the industrial world towards a path in synergy with the life-support environment on which it depends. Just as important as it is to conserve biodiversity for sustainability, it is as urgent to conserve the diversity of local cultures and the indigenous knowledge that they hold.

References and Notes

1. Diamond, J. 1989. This-fellow frog, name Belong-Him Dawko. *Nat. Hist. 89,* 16–20.
2. Diamond, J. 1989. The ethnobiologists's dilemma. *Nat. Hist. 89,* 26–30.
3. Ehrlich, P. 1987. Population biology, conservation biology, and the future of humanity. *BioScience 37,* 757–763.
4. Slobodkin, L.B. 1988. Intellectual problems of applied ecology. *BioScience 38,* 337–342.
5. Oldfield, M. L. and Alcorn, J. B. (eds.) 1991. *Biodiversity: Culture, Conservation and Ecodevelopment.* Westview Press.
6. Diamond, J. 1992. *The Rize and Fall of the Third Chimpanzee.* Vintage, London.
7. Lasserre, P. and Ruddle, K. 1983. *Traditional Knowledge and Management of Marine Coastal Systems.* Report of the ad hoc Steering Group. Unesco, Paris.
8. Ruddle, K. and Johannes, R.E. (eds.) 1989. *Traditional Marine Resource Management in the Pacific Basin: An Anthology.* Unesco/ROSTSEA, Jakarta.
9. Berkes, F. 1993. Traditional ecological knowledge in perspective. In: *Traditional Ecological Knowledge.* Unesco Canada/MAB, Ottawa. (In press).
10. Gadgil, M. 1987. Diversity: cultural and biological. *Trends Ecol. Evolut. 2,* 369–373.
11. Berkes, F. and Folke, C. 1992. A systems perspective on the interrelations between natural, human-made and cultural capital. *Ecol. Econ. 5,* 1–8.
12. Posey, D.A. 1985. Indigenous management of tropical forest ecosystems: The case of the Kayapo Indians of the Brazilian Amazon. *Agrofor. Syst. 3,* 139–158.
13. Taylor, K.I. 1988. Deforestation and Indians in the Brazilian Amazonia. In: *Biodiversity.* Wilson, E. O. (ed.). National Academy Press, Washington, DC, p. 138–144.
14. Irvine, D. 1989. Succession management and resource distribution in an Amazonian rain forest. *Adv. Econ. Bot. 7,* 223–237.
15. Folke, C. and Kautsky, N. 1989. The role of ecosystems for a sustainable development of aquaculture. *Ambio 18,* 234–243.
16. Folke, C. and Kautsky, N. 1992. Aquaculture with its environment: prospects for sustainability. *Ocean Coast. Mgmt 17,* 5-24.
17. Yan, J. and Yao, H. 1989. Integrated fish culture management in China. In: *Ecological Engineering.* Mitsch, W. J. and Jorgensen, S. E. (eds.) Wiley, N.Y. p. 375–408.
18. Ruddle, K. and Zhong, G. 1988. *Integrated Agriculture-Aquaculture in South China.* Cambridge University Press, Cambridge, UK.
19. Costa-Pierce, B.A. 1988. Traditional fisheries and dualism in Indonesia. *Naga 11,* 3–4.
20. Costa-Pierce, B.A. 1987. Aquaculture in ancient Hawaii. *BioScience 37,* 320–330.
21. Lewis, H.T. and T.A. Ferguson 1988. Yards, corridors and mosaics: How to burn a boreal forest. *Hum. Ecol. 16,* 57–77.
22. Lewis, H.T. 1989. Ecological and technical knowledge of fire: Aborigines versus park managers in Northern Australia. *Am. Anthropol. 91,* 940–961.
23. Sinclair, A.R.E. and Fryxell, J.M. 1985. The Sahel of Africa: Ecology of a disaster. *Can. J. Zool. 63,* 987–994.
24. Niamir, M. 1990. Herders' decision-making in natural resources management in arid and semi-arid Africa. *Community Forestry Note No. 4.* FAO, Rome.
25. Feit, H.A. 1986. James Bay Cree Indian management and moral consideration of fur-bearers. In: *Native People and Renewable Resource Management.* Alberta Society of Professional Biologists, Edmonton.
26. Berkes, F. 1982. Waterfowl management and northern native peoples with reference to Cree hunters of James Bay. *Musk-Ox 30,* 23–35.
27. Berkes, F. 1977. Fishery resource use in a subarctic Indian community. *Hum. Ecol. 5,* 289–307.
28. Berkes, F. 1979. An investigation of Cree Indian domestic fisheries in northern Quebec. *Arctic 32,* 46–70.
29. Malhotra, K.C. and Poffenberger, M. (ed.) 1989. Forest regeneration through community protection. *Proceedings of the Working Group Meeting on Forest Protection Committees, Calcutta, June 21–22, 1989.* West Bengal Forest Department, p. 47.
30. National Wastelands Development Board, Ministry of Environment and Forests, Government of India, 1991. *Guidelines for Microplanning.* NWDB, N. Dehli, p. 51.
31. Malhotra, K.C. 1990. Village supply and safety forest in Mizoram: a traditional practice of protecting ecosystem. In: *Abstracts of V International Congress of Ecology,* p. 439.
32. Joshi, N.V. and Gadgil, M. 1991. On the role of refugia in promoting prudent use of biological resources. *Theor. Popul. Biol. 40,* 211–229.
33. Terborgh, J. 1986. Keystone plant resources in the tropical forest. In: *Conservation Biology– The Science of Scarcity and Diversity.* Soule, M. E. (ed.). Sinauer Associates Inc. Sunderland, Massachusetts.
34. Slobodkin, L.B. 1968. How to be a predator. *Am. Zool. 8,* 43–51.
35. Gadgil, M. and Guha, R. 1992. *This Fissured Land: An Ecological History of India.* Oxford University Press, New Delhi, and University of California Press, Berkeley.
36. Holling, C.S. 1986. The resilience of terrestrial ecosystems: local surprise and global change. In: *Sustainable Development of the Biosphere.* Clarke, W. C. and Munn, R. E. (eds). Cambridge University Press, Cambridge, UK. p 292–317.
37. Berkes, F. (ed.). 1989. *Common Property Resources: Ecology and Community-based Sustainable Development.* Belhaven Press, London.
38. Ostrom, E. 1990. *Governing the Commons: The Evolution of Institutions for Collective Actions.* Cambridge University Press, Cambridge, UK.
39. Wells, M.P. 1992. Biodiversity conservation, affluence and poverty: mismatched costs and benefits and efforts to remedy them. *Ambio 21,* 237–243.
40. Wells, M.P. and Brandon, K. 1993. Conceptual and practical issues arising from the integration of biodiversity conservation with social and economic development outside park boundaries. *Ambio 22,* 157–162.
41. Daly, H.E. 1990. Towards some operational principles of sustainable development. *Ecol. Econ. 2,* 1–6.
42. Solbrig, O.T. (ed.). 1991. *From Genes to Ecosystems: A Research Agenda for Biodiversity.* The International Union of Biological Sciences, Paris.
43. Walker, B.H. 1992. Biodiversity and ecological redundancy. *Conserv. Biol. 6,* 18–23.
44. Acknowledgements. We thank Jeffrey McNeely and Shelton Davies for valuable comments on the manuscript. Madhav Gadgil's research is supported by the Ministry of Environment and Forests, Government of India, Fikret Berkes' work was supported by the Social Sciences and Humanities Research Council of Canada and Carl Folke's partly by grants from the Swedish Council for Forestry and Agricultural Research.
45. First submitted 3 December, 1992, accepted for publication after revision, 2 February, 1993.

Madhav Gadgil holds a PhD in biology from Harvard University and has served as a lecturer at Harvard University and a visiting professor at Stanford University. For the past 18 years he has been on the faculty of the Indian Institute of Science where he currently holds the Astra Professorship in Biological Sciences. His research interests encompass mathematical modelling as well as field studies in the areas of population biology, conservation biology and human ecology. He is also active in policy studies having served for 4 years on the Scientific Advisory Council to the Prime Minister of India. His address: Centre for Ecological Sciences, Indian Institute of Science, Bangalore 560012, India. Fikret Berkes, PhD, specializes in common property resources and community-based resource management. He has investigated, in a variety of geographical and cultural settings, the conditions under which the "tragedy of the commons" may be avoided. His address: Natural Resources Institute, University of Manitoba, Winnipeg, Manitoba R3T 2N2 Canada. Carl Folke, PhD, is deputy director of the Beijer International Institute of Ecological Economics, Royal Swedish Academy of Sciences and researcher at the Department of Systems Ecology, Stockholm University. His address: Beijer International Institute of Ecological Economics, Royal Swedish Academy of Sciences, Box 50005, S-104 05 Stockholm, Sweden.

Anecdotes and the shifting baseline syndrome of fisheries

Fisheries have recently become a topic for media with global audiences – but then again, fisheries are a global disaster: one of the few that affect, in very similar fashion, developed countries with well-established administrative and scientific infrastructure, newly industrialized countries, and developing countries.

This is quickly summarized:

• Heavily subsidized fleets, exceeding by a factor of 2 or 3 the numbers required to harvest nominal annual catches of about 90 million tonnes.
• Staggering levels of discarded bycatch, representing about one third of the nominal catch, a large unrecorded catch that perhaps raises the true global catch to about 150 million tonnes per year, well past most previous estimates of global potential.
• The collapse, depletion or recovery from previous depletion of the overwhelming majority of the over 260 fish stocks that are monitored by the Food and Agriculture Organization of the United Nations.

Fisheries science has responded as well as it could to the challenge this poses by developing methods for estimating targets for management – earlier the fabled Maximum Sustainable Yield (MSY)[1], now annual total allowable catch (TAC) or individual transferable quotas (ITQ). If these methods are to remain effective, fisheries scientists need to follow closely the behavior of fishers and fleets, but this has tended increasingly to separate us from the biologists studying marine or freshwater organisms and/or communities, and to factor out ecological and evolutionary considerations from our models. There are obviously exceptions to this, but I believe the rule generally applies, and it can be illustrated by our lack of an explicit model accounting for what may be called the 'shifting baseline syndrome'. Essentially, this syndrome has arisen because each generation of fisheries scientists accepts as a baseline the stock size and species composition that occurred at the beginning of their careers, and uses this to evaluate changes. When the next generation starts its career, the stocks have further declined, but it is the stocks at that time that serve as a new baseline. The result obviously is a gradual shift of the baseline, a gradual accommodation of the creeping disappearance of resource species, and inappropriate reference points for evaluating economic losses resulting from overfishing, or for identifying targets for rehabilitation measures.

These are strong claims that I can illustrate best by using analogies. For example, astronomy has a framework that uses ancient observations (including Sumerian and Chinese records that are thousands of years old) of sunspots, comets, supernovae or other phenomena that were recorded by ancient cultures, and this has made possible the testing of pertinent hypotheses. Similarly, oceanography has had, since the days of Commodore F. Maury, protocols for consolidating scattered observations on currents and winds, and later on sea surface temperatures; the latter have enabled the extending of the Comprehensive Ocean and Atmospheric Data Set (COADS) back to 1870, and infer that, indeed, global warming is occurring.

In contrast, fisheries science does not have formal approaches for dealing with early accounts of 'large catches' of presently extirpated resources, which are viewed as anecdotes . Yet the grandfather of my colleague Villy Christensen did report being annoyed by the bluefin tuna that entangled themselves in the mackerel nets he was setting in the waters of the Kattegat in the 1920s, and for which no market then existed. This observation is as factual as a temperature record, and one that should be of relevance to those dealing with bluefin tuna, whose range now excludes much, if not all, of the North Sea.

I could list hundreds of such observations – drawn from the historical or anthropological literature and elsewhere – but here it may be more useful to highlight two small fisheries-related studies that have attempted to consolidate them, and which have led, I believe, to important new insights. In the first, a (female) scientist[2] compiled scattered observations of (male) anthropologists reporting on fishing in the South Pacific, and concluded that, despite cultural emphasis on the catching of large fish by men, the gleaning of smaller reef organisms by women and children often accounted for as much catch as the more spectacular activities of the men (even though it does not enter official catch statistics). This fact, now widely confirmed by field studies, should lead to a re-evaluation of the fisheries potential of coral reefs.

The authors of the second study[3] used the anecdotes in Farley Mowat's *Sea of Slaughter*[4] to infer that the biomass of fish and other exploitable organisms along the North Atlantic coast of Canada now represents less than 10% of that two centuries ago. Some colleagues will find it difficult to accept that the early fishing methods should have had such impact, given their relative inefficiency when compared to our factory ships. However, it must be remembered that the large animals of low fecundity at the top of earlier food webs must have been less resilient to fishing than the survivors that are exploited today. That is, the big changes happened way back, but all that we have to recall them are anecdotes.

Developing frameworks for incorporation of earlier knowledge – which is what the anecdotes are – into the present models of fisheries scientists would not only have the effect of adding history to a discipline that has suffered from lack of historical reflection[1], but also of bringing into biodiversity debates an extremely speciose group of vertebrates: the fishes, whose ecology and evolution are as strongly impacted by human activities as the denizens of the tropical and other rain forests that presently occupy center stage in such debates. Frameworks that maximize the use of fisheries history would help us to understand and to overcome – in part at least – the shifting baselines syndrome, and hence to evaluate the true social and ecological costs of fisheries.

Daniel Pauly

*ICLARM, MC PO Box 2631,
0718 Makati City, Philippines, and Fisheries Centre,
UBC, 2204 Main Mall, Vancouver, B.C.,
Canada V6T 1Z4*

References
1 Smith, T.D. (1994) *Scaling Fisheries*, Cambridge University Press
2 Chapman, M.D. (1987) *Hum. Ecol.* 15, 267–288
3 MacIntyre, F., Estep, K.W. and Noji, T.T. (1995) NAGA (the ICLARM Quarterly) 18(3), 7–8
4 Mowat, F. (1984) *Sea of Slaughter*, Atlantic Monthly Press

PART V

ETHICAL, RELIGIOUS, AND HISTORICAL APPROACHES TO SOCIO-ENVIRONMENTAL RESEARCH

J. Baird Callicott

Ethical considerations that have influenced society–environment relationships are embedded in the worldviews of many ancient cultures (Callicott 1994). In Greek mythology, the world is not created; it is given birth to by Gaia (Earth), sired by Ouranos (Sky). Plato derives a law from this myth that requires the land to be tended as carefully as children care for their mother. In her legacy paper, **Vandana Shiva (1988)** indicates how ancient Hinduism ethically links society and environment. Ancient Chinese Daoism recommends harnessing – rather than forcefully bucking – natural tendencies and processes to achieve social goals. In American Indian plains cultures, Sky and Earth and the Winds of the four cardinal directions are Powers, united in the Great Spirit, and all living beings are regarded as relatives to be treated with respect. American Indian woodland cultures, along with the Navaho, as **Gregory Cajete (1994)** explains, represented other animals as having the same inner life as humans. Each species formed its own society, with which humans could unite through bonds of marriage and kinship to form a kind of spiritual analog of the ecological concept of a biotic community, a key concept in **Aldo Leopold's (1949)** land ethic. In the Dreamtime of Australian Aboriginal peoples, the landscape was at first featureless, until ancestral animal beings traveled across it creating hills, ravines, waterholes, and other topographical features and gave laws to each social group; and the species of each society's ancestor must be ceremonially renewed and their lands cleaned and rejuvenated with fire. As civilizations matured and spread, these various forms of ancient and Indigenous environmental ethics became attenuated. And with the Industrial Revolution they were replaced by human exceptionalism and overwhelmed by an ever-increasing reliance on technology and the substitution of artificial for natural capital.

The division of knowledge into a multitude of disciplines did not occur until the early twentieth century (McElreavy et al. 2016). The word "scientist" was coined only in

1833 by William Whewell; nor did it immediately catch on, as all researchers were called either natural or moral philosophers until around the turn of the century (Ross 1962). Some such natural philosophers worked informally in socio-environmental synthesis research – **Humboldt**, **Darwin**, and **Marsh (Part I)**, prominent among them. The influential "Frontier Thesis" of Turner (1894) – as **William Cronon (1992)** explains – was an early expression of socio-environmental history. **Lynn White, Jr. (1967)** and Nash (1967) were engaged in socio-environmental history in the mid-twentieth century, but the academic field only coalesced a decade later with the first issue of *Environmental History Review* in 1976. Several decades before the establishment of socio-environmental philosophy, both Thoreau (1862) and Muir (1916) adumbrated a socio-environmental ethic, more fully articulated by **Leopold (1949)**. A few academic philosophers began to reflect on the interaction between human societies and the natural environment in the early 1970s, and by the end of the decade, the journal *Environmental Ethics* enabled a new field to coalesce in philosophy. **White's (1967)** notorious allegation that "Christianity bears a huge burden of guilt" (40) for the environmental crisis provoked a speedy reaction in theology, also in the early 1970s, taking the form of a more environmentally friendly interpretation of scripture and claims that adopting other world religions could be our environmental salvation (e.g., Santmire 1970; Smith 1972). Among the humanities, literary criticism did not take an environmental turn – known as "ecocriticism" or, more formally, the study of literature and the environment – until the 1990s (Buell 1995). The genus, environmental humanities, was not identified as an academic taxon until the early twenty-first century; and the first journal, *Environmental Humanities*, was launched in 2012.

Research methods in the humanities differ from those in the natural and social sciences (Holm et al. 2017). In philosophy, religion, history, and literary criticism, scholars begin with a thesis instead of a hypothesis. A good example in this part is **White's (1967)** thesis that the historical roots of our environmental crisis are traceable to the biblical worldview set out in the first chapters of Genesis. And instead of testing hypotheses by performing experiments or collecting data, humanities scholars defend their theses by argumentation. Historians also frequently collect data, often in the form of archival documents, thus defying facile classification of history as among either the humanities or the social sciences (Reed 2018). As an academic taxon, the digital humanities are hybrids spawned by the union of one or another of the traditional humanities with computer science (Berry and Fagerjord 2017). Progress in the humanities consists in critical engagement by subsequent scholars building upon, revising, or rejecting the theses of their predecessors, often in response to developments in the sciences, such as evolutionary biology and ecology, or to larger social forces, such as revolutionary politics and economics. An example of the former from this part is **Leopold's (1949)** "The Land Ethic" and of the latter is **Leonardo Boff's (1995)** liberation theology.

The methodological differences between research in the humanities and the sciences render interdisciplinary cooperation and even communication between these two broad types of inquiry difficult (Rose et al. 2012). As a result, ethics, especially, has been marginalized in socio-environmental research, and is often misunderstood as consisting merely of subjective preferences in contrast to the putatively objective and verifiable claims of science. Presently, however, ethical quandaries – such as justice and equity for under-represented communities, treatment of animals in laboratory and field research, and editing the genomes of plants, animals, and humans – are becoming ever more insistent in socio-environmental research. Socio-environmental synthesis, therefore, requires natural and social scientists to become conversant with the humanities and for humanists actively to engage, in accessible terms, the conceptual and ethical concerns arising in the sciences (Paisson et al. 2013). And although ethics is considered to be a field of study in philosophy, it is obviously endemic in religion and implicit in history and literature, as the legacy papers in this part demonstrate.

The boyhood passion of **Aldo Leopold** (1887–1948) for the outdoors led to a mature study of nature informed by the emerging science of ecology. A graduate of the Yale Forest School (1909) and first employed by the US Forest Service, Leopold joined the University of Wisconsin in 1933, where he founded the discipline of wildlife management. The more recently emerged sciences of conservation biology and restoration ecology also claim him as a forebear, as do environmental philosophers.

"The Land Ethic" (Leopold 1949)[1] – composed shortly after the end of World War II and just as the ecosystem paradigm in ecology was emerging – is an outstanding example of the synthesis of science and the humanities. Leopold builds his land ethic on the account of the origin and social evolution of ethics in Darwin's *Descent of Man*. According to Leopold, an ethic is "a limitation on freedom of action in the struggle for existence," which provides "a differentiation of social from antisocial conduct" (38). Without ethics, Darwin (1874) argued, social cohesion and cooperation is impossible. Societies and their ethics originate when the emotional bond between parents and offspring extends to other relatives. As these small societies compete with one another, there is selective pressure for them to merge into larger and more diversified communities. This process continued right up to the emergence of nation states. Darwin (1874) even presciently envisioned the emergence of a single global society with a corresponding global ethic, which was expressed in the United Nations Declaration of Human Rights in 1948. (A twenty-first-century example of a global socio-environmental ethic is the Earth Charter, adopted by the United Nations in 2005.) Leopold then observes that ecology represents plants and animals as members of a biotic community. Because ethics have always been the bond that holds communities together and because, from an evolutionary point of view, humans are "plain member[s] and citizen[s]" of the biotic community, we need a land ethic to "preserve [its] integrity, stability, and beauty" (38), upon which we depend no less than we depend upon our social memberships. In contrast to what Leopold thought was the prevailing toxic mix of biblical human exceptionalism and consumerism, the key to realizing the land ethic is the universal adoption of an evolutionary–ecological worldview, which he develops in the foregoing sections of *A Sand County Almanac*.

Ping-ti Ho (1917–2012) was born in China and educated there and in the United States, earning a PhD from Columbia University in 1954. He was a member of the history department at the University of Chicago. In concert with **Malthus (1798, Part I)** and **Hong (1793, Part I)**, Ho treats *Homo sapiens* as a resource-limited species. The human population expanded in China in step with the expansion of rice cultivation, long the staple of the Chinese diet. And that expansion was made possible by the selection of varieties that matured more rapidly, thus enabling a harvest of two annual crops on one piece of land (and eventually even three, in places with favorable environmental conditions). Those

---

[1] *Published posthumously.*

varieties also required less water, enabling rice cultivation to expand uphill from well-watered river valleys and from clay soils to lighter soils. Ho demonstrates that the social aspect of the socio-environmental dynamic of the history of Chinese agriculture is in a dialectical relationship with the environmental aspect. For while the expansion of agriculture in China drove population increase, population increase drove the adaptation of agriculture to more challenging environments. Ho's argument though is not in concert with **Boserup's (1965, Part VI)** focus on improvements in technology but emphasizes the introduction of new domesticated plants, breeding new varieties, and intensifying labor. Complementing Leopold's description of a biotic ecosystem, Ho describes an anthropogenic agro-ecosystem adapting ever more creatively to China's diverse soils, topography, and climate.

Lynn White, Jr. (1907–1987) earned a PhD in medieval history from Stanford University and was professor of history at Princeton, Stanford, and University of California, Los Angeles. In the course of the argument of his legacy paper, White (1967) touches on the history of European agriculture in a way that offers a sharp contrast to Ho's (1959) history of Chinese agriculture. For the expansion of European agriculture was driven by advances in technology – the mold-board plow, wind and water mills – and less by plant breeding and human-labor intensification. And that social aspect of the socio-environmental dynamic of the history of European agriculture is again in a dialectical relationship with the environmental aspect. In the European case, it was "the wet climate and sticky soils of northern Europe" (reading 40) that necessitated technological innovation. White traces the historical roots of the environmental crisis back to the biblical ideas that "man" was created in the image of God, given dominion over nature, and commanded to subdue it. More generally, White argues, the environmental crisis is at bottom a crisis in how we collectively think of our human relationship with the environment. It requires worldview remediation. This paper led directly not only to the environmental turn in religions, but also in philosophy, because the main task of philosophy is critical thinking, which White directed toward the relationship between society and the environment. The influence of this article is vast and extends to twenty-first-century work in socio-environmental synthesis research (e.g., Stern 2000; Stibbe 2015; Berkes 2017).

Vandana Shiva (1952–) was born in India and educated in Catholic schools. She earned a BS in physics from Panjab University and an MA and PhD in philosophy of science at the University of Guelph and University of Western Ontario, respectively. She is an activist, writer, community organizer, and consultant. Her legacy reading (from Shiva 1988) represents one important branch of environmental philosophy called ecofeminism. Shiva discusses the many reasons for water shortages in India – from the introduction and irrigation of water-intensive crops to deforestation. The "maldevelopment" of water through "masculinist" technologies – dams, electric pumps, canals – principally harms Indian women who are responsible for obtaining drinking water and growing food plants. But the ultimate cause of all the proximate causes lies in a "Baconian" worldview characterized by reduction (of wholes to the sum of their parts), linear resource-exploitation processes (from source to consumption to sink), and commodification (the privatization of communal goods). Its endpoint is artificial desalinization as the solution for India's water-scarcity problems, which were created by this very way of thinking. The living planet Gaia, however, is holistic (the whole is greater than the sum of its parts), cyclical (the water cycle being perhaps the epitome thereof), and integral (such that humans are a part of, not apart from, cyclical Earth systems, and the well-being of each depends on the well-being of all). Shiva points out the irony in desalinization schemes:

The water cycle itself is a planetary-scale desalinization system directly powered by the sun, which evaporates water from the saline ocean and delivers it fresh to the land by the movement of the atmosphere (also solar-powered). Shiva turns to Hindu mythology for an alternative to this maladaptive worldview. She indicates how pre-colonial India's irrigation system, embedded in Hindu thought, worked with, not against, the natural courses and flows of India's rivers.

Early environmental philosophy was preoccupied with a debate about a value dichotomy: Do nonhuman natural entities and nature as a whole have only instrumental (use) value to humans or also intrinsic value? This dichotomy has largely been superseded by the concept of relational value, emerging from ecofeminism (Pascual et al. 2017). The relationships among the myriads of things and processes in the body of Gaia are mutually defining. Deer, for example, are what they are because of their phylogenetic heritage and complex relationships with the foods they eat, the predators and parasites that feed on them, and the microbes on and within them. And individual persons are who they are because of their unique socio-environmental relationships. As Naess (1973) graphically put it, "organisms are knots in the biospheric field of internal relations" (p. 95). In accordance with relational value, none of us can be personally healthy and secure apart from the socio-environmental matrix in which we are embedded, with which we are entangled, and on which we utterly depend.

Elinor G. K. Melville (1940–2006) was born in New Guinea and raised in Australia and New Zealand. She earned an MA and PhD from the University of Michigan and was a member of the history department of York University. Melville's (1990) essay is a model of socio-environmental synthesis. She mastered a number of very different areas of expertise – historiography of the most specialized kind (working with sixteenth-century Spanish-language primary sources), anthropology, climatology, ecology, geology, pedology – and brings them to bear on the environmental history of a well-defined region in Mexico. Melville artfully revises two distinct historical narratives: the first is place-specific, while the second has more general application. First, she demonstrates that the pre-conquest Indian population was large and was supported by forms of agriculture well adapted to the climate, soils, and slopes of the Valle del Mezquital. Second, she demonstrates that the introduction of cattle and sheep on small holds at first coexisted with nearby Indian farmlands, eventually encroached on those farmlands, and, along with mining and deforestation, radically altered the vegetation and consequently eroded away the soil over the entire landscape. As a result, only pastoral economies of scale were profitable, thus inaugurating the latifundia land-consolidation system a good century after the arrival of the conquistadores. Melville shows how synthesizing disciplines brings a new historical narrative to life.

Melville's story differs in one surprising way from Shiva's, which overflows with moral outrage at environmental injustice. Melville records the fact that the Spanish overlords used the unpaid labor of enslaved Africans and the Indigenous inhabitants with nary a word of reproach. And she records, without compassion, the fact that the Indian population was reduced by more than 90 percent by imported pathogens. Also, unlike Shiva's story, we learn nothing from Melville about the worldview in which the agricultural practices of the Indigenous people were embedded. Nor in Melville's essay do we find any plea for worldview remediation.

William Cronon (1954–) earned an MA and PhD from Yale University in history. He was a member of the history department of Yale prior to joining Wisconsin-Madison.

Cronon (1992) begins with the observation that storytelling is a human universal. Recognizing that science tells an ongoing and self-correcting story, the Millennium Ecosystem Assessment (Alcamo et al. 2003) and Intergovernmental Panel on Biodiversity and Ecosystem Services (Pascual et al. 2017) emphasize communication, outreach, and scenario building. Cronon's paper struggles with how to identify the most tenable among diverging and often contradictory historical narratives. As an example, he summarizes differing accounts of the history of the Great Plains, each insidiously reflecting the larger intellectual commitments and biases of its author. Thus, "we force our stories on a world that doesn't fit them. The historian's project of recovering past realities and representing them 'truly' or even 'fairly' is thus a delusion" (43). Cronon rejects such epistemic nihilism and also relativism, the view that everyone is entitled to choose the narrative that aligns with their preexisting beliefs and values. In such instances, each has their own truth; and all cannot agree on a shared reality. Cronon offers some criteria for assessing the veracity of historical narratives. The way Melville corrects both the local and national narratives of the Spanish colonization of Mexico provides an example of Cronon's criteria largely met. The critical key to getting as close to the truth as possible is to integrate information derived from the natural and social sciences and the humanities – in short, the narrative generated by socio-environmental synthesis research. Largely met, but not fully met, because input from the humanities and the moral perspective they bring is, as noted, absent in Melville's story.

Arturo Gómez-Pompa (1934–) was born in Mexico City and earned a PhD from Universidad Nacional Autónoma de México. He is Professor Emeritus of Botany at the University of California, Riverside. **Andrea Kaus** (1959–) earned a PhD in anthropology at the University of California, Riverside, in 1992 and retired from the University of California Institute for Mexico and the United States in 2019. Gómez-Pompa and Kaus (1992) begin with a call for remediating the conservation worldview according to which originally nature was, and ideally should be, in a wilderness condition where, in the words of the US Wilderness Act of 1964, "man is a visitor who does not remain" (Callicott and Nelson 1999). The first published assault on the wilderness idea came from Guha (1989), an Indian sociologist, noting how, when densely populated countries like India imported this idea from the United States, the result was expulsion of Indigenous peoples from their homelands. Next came my critique of it, which explained that the wilderness idea perpetuated the false dichotomy between humans and nature and was a colonial tool for masking genocide on a continental scale (Callicott 1991). Cronon (1995) added his voice to the rising chorus of wilderness critiques by repackaging those foregoing; and his became the most frequently cited. The critique of Gómez-Pompa and Kaus rests on three important paradigm shifts. (1) Evidence of long-standing and ubiquitous human inhabitation and anthropogenic disturbance on all continents except Antarctica, including a wave of Pleistocene megafauna extinctions, correlated with the migration of *Homo sapiens* out of Africa (Martin and Klein 1984). (2) The discovery that communities and associated ecosystems are not in a steady state, unless anthropogenically disturbed, but subject to frequent natural disturbance and directionless change (**Holling 1973, Part IV**; Pickett et al. 1992). (3) The realization that Indigenous populations of the Western Hemisphere were much larger when demographers took account of the holocaust caused by zoonotic diseases brought by European colonists (Denevan 1992). So, the notion that North and South America were virtually in a wilderness condition in 1491 was abandoned. That forests in Central America and the Amazon Basin appear to be pristine is not due to the absence of people but to the knowledge and skill of their Indigenous inhabitants in managing them sustainably.

Gregory Cajete (1952–), a Tewa, earned a BA from New Mexico Highlands University, an MA from the University of New Mexico, and a PhD in science education from the International College of Los Angeles. He taught at the Institute of American Indian Arts and is Director of Native American Studies at the University of New Mexico. Anthropologist Philippe Descola (2013) notes that in the prevailing Western worldview the natural world is regarded as a robust and predictable steadfast reality, while consciousness varies in degree and kind from species to species and its content changes unpredictably. In the American Indian worldview consciousness is regarded as a robust and steadfast reality that varies not at all from species to species while the natural world is fluid and shifting. And we see that general characterization well expressed in the Navaho story Cajete (1994) relates. In "The Hunter of Good Heart" the consciousness of the deer is the same as human consciousness, and the deer people transform themselves into humans, plants, and even into rocks. They require respect, and the hunter of good heart must negotiate terms of the hunt with them, if he is to succeed. Stories like these vividly anticipate the way contemporary socio-environmental research in ethology recognizes robust animal agency as well as consciousness – and indeed affirms rather than rejects anthropomorphic description thereof (de Waal 2016). Cajete calls this a "spiritual ecology," and one might even call it a spiritual conservation biology, because spiritual preparation, proper decorum in the hunt, ritual after the kill, and care with disposing of the bones prevent wanton slaughter. Stories like this are not unique to the Navaho. On the contrary, they are the heritage of American Indian peoples from the Arctic to Tierra del Fuego and from the Atlantic to the Pacific (Callicott and Nelson 2004). They provide intergenerational continuity for Indigenous peoples. Fernández-Llamazares and Cabeza (2018) reinforce the power and importance of stories noted by Cronon. Cajete's book is cited by several socio-environmental researchers (Louis 2007; Tuck and McKenzie 2014; Datta 2018).

Leonardo Boff (1938–) was born in Brazil and earned a PhD in theology and philosophy from the University of Munich. He is a member of the Franciscan Order and Professor Emeritus of Ethics, Philosophy of Religion, and Ecology at the Rio de Janeiro State University. Gómez-Pompa and Kaus (1992) begin with an epigraph by Aldo Leopold, and Boff (1995) begins his legacy paper with an allusion to these words of Leopold (1953): "One of the penalties of an ecological education is that one lives alone in a world of wounds" (46). But for Boff there are two worlds of wounds, the other being that of poverty and its attendant human suffering. The obligation we have to free the poor from the vicissitudes of poverty is not to do for them but to make possible their own freedom and agency. Boff greatly expands the scale of the socio-environmental matrix to that of the entire universe, following his fellow Catholic theologian Thomas Berry, who crafted a new creation story in line with modern cosmology (Swimme and Berry 1992). Part of that story provides a new and unique role for humanity through whom the universe attains self-consciousness (the "noosphere") and becomes "co-pilot" of the frankly teleological cosmic evolution. The two wounds of ecological degradation and human poverty are linked together by common causes: capitalism (or the neoliberal economic order), consumerism, and unlimited accumulation that leads to the commodification and plundering of nature and to the exploitation of powerless people and ever-increasing inequality. The alternative worldview sketched by Boff remains decidedly Christian, but also decidedly unorthodox. Following White, who suggests an alternative Franciscan theology, Boff elaborates the intersectionality of socio-environmental synthesis, illuminating a path for socio-environmental synthesis research motivated by an ethic of care.

The common thread running through the legacy papers in this part is the importance of story or narrative, whether explicitly stated, as in those of White, Shiva, Cronon, Gómez-Pompa and Kause, Cajete, and Boff, or implicit, as in those of Leopold, Ho, and Melville. White and Boff argue that we need a new religious story and agree (suggested by the order to which Boff belongs) that it could be inspired by Saint Francis of Assisi, whom Pope John Paul II made the patron saint of ecologists in 1979. Leopold (in the foreword to *A Sand County Almanac*), Shiva, Gómez-Pompa and Kaus, and Cajete explicitly call for worldview remediation – changing the narrative – which, in one way or another, springs from the geosciences of evolutionary biology, ecology, and Earth system science. Shiva and Cajete link the contemporary geoscientific narrative with the ancient wisdoms indigenous to India and North America. Ho, Cronon, and Melville all focus more narrowly on human social history, but in relationship to the peculiar environments of China, the North American Great Plains, and central Mexico, respectively. We must not give up hope for the realization of another and better-adapted story: the advent of a new, universal understanding of the dynamic, intertwined relationship of society and environment underpinning a relational socio-environmental ethic, upon which the future flourishing of both society and environment utterly depend. The papers in this part can fuel our dreams of a better tomorrow and inspire socio-environmental research to help make it a waking reality.

References

Alcamo, Joseph, Neville J. Ash, Colin D. Butler, J. Baird Callicott, Doris Capistrano, Stephen R. Carpenter et al. *Ecosystems and Human Well-being: A Framework for Assessment*. Washington, DC: Island Press, 2003.

Berkes, Fikret. *Sacred Ecology*, 4th ed. New York: Routledge, 2017.

Berry, David M., and Anders Fagerjord. *The Digital Humanities: Knowledge and Critique in a Digital Age*. Hoboken, NJ: John Wiley and Sons, 2017.

Buell, Lawrence. *The Environmental Imagination: Thoreau, Nature Writing, and the Formation of American Culture*. Cambridge, MA: The Belknap Press of Harvard University, 1995.

Callicott, J. Baird. "The Wilderness Idea Revisited: The Sustainable Development Alternative." *Environmental Professional* 13, no. 3 (1991): 235–247.

Callicott, J. Baird. *Earth's Insights: A Multicultural Survey of Ecological Ethics from the Mediterranean Basin to the Australian Outback*. Berkeley: University of California Press, 1994.

Callicott, J. Baird, and Michael P. Nelson, eds. *The Great New Wilderness Debate*. Athens, GA: University of Georgia Press, 1999.

Callicott, J. Baird, and Michael P. Nelson. *American Indian Environmental Ethics: An Ojibwa Case Study*. Upper Saddle River, NJ: Prentice-Hall, 2004.

Cronon, Willian. "The Trouble with Wilderness, or Getting Back to the Wrong Nature." In *Uncommon Ground: Toward Reinventing Nature*, edited by William Cronon, 68–90. New York: W.W. Norton and Company, Inc., 1995.

Darwin, Charles. *The Descent of Man and Selection in Relation to Sex*, 2nd ed. London: John Murray, 1874.

Datta, Ranjan. "Decolonizing Both Researcher and Research and Its Effectiveness in Indigenous Research." *Research Ethics* 14, no. 2 (2018): 1–24.

Denevan, William M. "The Pristine Myth: The Landscape of the Americas in 1492." *Annals of the Association of American Geographers* 82, no. 3 (1992): 369–385.

Descola, Philippe. *Beyond Nature and Culture*, translated by Janet Lloyd. Chicago: University of Chicago Press, 2013.

de Waal, Frans. (2016). *Are We Smart Enough to Recognize How Smart Animals Are?* New York: W. W. Norton and Company, Inc., 2016.

Fernández-Llamazares, Álvaro, and Mar Cabeza. "Rediscovering the Potential of Indigenous Storytelling for Conservation Practice." *Conservation Letters* 11, no. 3 (2018): e12398.

Guha, Ramachandra. "Radical American Environmentalism and Wilderness Preservation: A Third World Critique." *Environmental Ethics* 11, no. 1 (1989): 71–83.

Holm, Paul, Arne Jarrick, and Dominic Scott. *Humanities World Report 2015*. London: Palgrave Macmillan, 2017.

Leopold, Aldo. *Round River: From the Journals of Aldo Leopold*, edited by Luna B. Leopold. New York: Oxford University Press, 1953.

Louis, Renee Pualani. "Can You Hear Us Now? Voices from the Margin: Using Indigenous Methodologies in Geographic Research." *Geographical Research* 45, no. 2 (2007): 130–139.

McElreavy, Christine, Victoria Tobin, Taylor Martin, Mickayla Bea Damon, Nicole Crate, Andrew Godinez, and Kayleigh Bennett. "The History of the Academy and the Disciplines." *Interdisciplinary Studies: A Connected Learning Approach* (2016). https://press.rebus.community/idsconnect/chapter/the-history-of-the-academy-and-the-disciplines.

Martin, Paul S., and Richard G. Klein, eds. *Quaternary Extinctions: A Prehistoric Revolution*. Tucson: University of Arizona Press, 1984.

Muir, John. *A Thousand Mile Walk to the Gulf*. Boston: Houghton Mifflin, 1916.

Naess, Arne. "The Shallow and the Deep, Long-Range Ecology Movement: A Summary." *Inquiry* 16, no. 1–4 (1973): 95–100.

Nash, Roderick Frazier. *Wilderness and the American Mind*. New Haven, CT: Yale University Press, 1967.

Paisson, Gisli, Bronislaw Szerszynski, Sverker Sörlin, John Marks, Bernard Avril, Carole Crumley et al. "Reconceptualizing the 'Anthropos' in the Anthropocene: Integrating the Social Sciences and Humanities in Global Environmental Change Research." *Environmental Science & Policy* 28 (2013): 3–13.

Pascual, Unai, Patricia Balvanera, Sandra Díaz, György Pataki, Eva Roth, Marie Stenseke et al. "Valuing Nature's Contributions to People: The IPBES Approach." *Current Opinion in Environmental Sustainability* 26 (2017): 7–16.

Pickett, Steward T. A., V. Thomas Parker, and Peggy L. Fiedler. "The New Paradigm in Ecology: Implications for Conservation Biology above the Species Level." In *Conservation Biology: The Theory and Practice of Nature Conservation, Preservation, and Management*, edited by Peggy L. Fielder and Subodh K. Jain, 65–88. Boston, MA: Springer, 1992.

Reed, Matt. "Where Should History Go?: Inside Higher Ed." *Confessions of a Community College Dean*, May 28, 2018. www.insidehighered.com/blogs/confessions-community-college-dean/where-should-history-go

Rose, Deborah Bird, Thom van Dooren, Matthew Chrulew, Stuart Cooke, Matthew Kearnes, and Emily O'Gorman. "Thinking through the Environment, Unsettling the Humanities." *Environmental Humanities* 1, no. 1 (2012): 1–5.

Ross, Sydney. "Scientist: The Story of a Word." *Annals of Science* 18, no. 2 (1962): 65–85.

Santmire, H. Paul. *Brother Earth: Nature, God, and Ecology in a Time of Crisis*. Nashville: Thomas Nelson Publishing Company, 1970.

Smith, Huston. "Tao Now: An Ecological Testament." In *Earth Might be Fair: Reflections on Ethics, Religion, and Ecology*, edited by Ian Barbour, 62–81. Englewood Cliffs, NJ: Prentice-Hall, Inc., 1972.

Stern, Paul C. "New Environmental Theories: Toward a Coherent Theory of Environmentally Significant Behavior." *Journal of Social Issues* 56, no. 3 (2000): 407–424.

Stibbe, Arran. *Ecolinguistics: Language, Ecology and the Stories We Live By*. New York: Routledge, 2015.

Swimme, Brian, and Thomas Berry. *The Universe Story: From the Primordial Flaring Forth to the Ecozoic Era, a Celebration of the Unfolding of the Cosmos*. New York: Harper Collins, 1992.

Thoreau, Henry David. "Walking." *Atlantic Monthly* 9 (1862): 657–674.

Tuck, Eve, and Marcia McKenzie. *Place in Research: Theory, Methodology, and Methods*. New York: Routledge, 2014.

Turner, Frederick Jackson. "The Significance of the Frontier in American History." In *Annual Report of the American Historical Association, 1893*, edited by Anonymous, 197–227. Washington, DC: Government Printing Office, 1894.

The Land Ethic

When god-like Odysseus returned from the wars in Troy, he hanged all on one rope a dozen slave-girls of his household whom he suspected of misbehavior during his absence.

This hanging involved no question of propriety. The girls were property. The disposal of property was then, as now, a matter of expediency, not of right and wrong.

Concepts of right and wrong were not lacking from Odysseus' Greece: witness the fidelity of his wife through the long years before at last his black-prowed galleys clove the wine-dark seas for home. The ethical structure of that day covered wives, but had not yet been extended to human chattels. During the three thousand years which have since elapsed, ethical criteria have been extended to many fields of conduct, with corresponding shrinkages in those judged by expediency only.

The Ethical Sequence

This extension of ethics, so far studied only by philosophers, is actually a process in ecological evolution. Its sequences may be described in ecological as well as in philosophical terms. An ethic, ecologically, is a limitation on freedom of action in the struggle for existence. An ethic, philosophically, is a differentiation of social from anti-social conduct. These are two definitions of one thing. The thing has its origin in the tendency of interdependent individuals or groups to evolve modes of co-operation. The ecologist calls these symbioses. Politics and economics are advanced symbioses in which the original free-for-all competition has been replaced, in part, by co-operative mechanisms with an ethical content.

The complexity of co-operative mechanisms has increased with population density, and with the efficiency of tools. It was simpler, for example, to define the anti-social uses of sticks and stones in the days of the mastodons than of bullets and billboards in the age of motors.

The first ethics dealt with the relation between individuals; the Mosaic Decalogue is an example. Later accretions dealt with the relation between the individual and society. The Golden Rule tries to integrate the individual to society; democracy to integrate social organization to the individual.

There is as yet no ethic dealing with man's relation to land and to the animals and plants which grow upon it. Land, like Odysseus' slave-girls, is still property. The land-relation is still strictly economic, entailing privileges but not obligations.

The extension of ethics to this third element in human environment is, if I read the evidence correctly, an evolutionary possibility and an ecological necessity. It is the third step in a sequence. The first two have already been taken. Individual thinkers since the days of Ezekiel and Isaiah have asserted that the despoliation of land is not only inexpedient but wrong. Society, however, has not yet affirmed their belief. I regard the present conservation movement as the embryo of such an affirmation.

An ethic may be regarded as a mode of guidance for meeting ecological situations so new or intricate, or involving such deferred reactions, that the path of social expediency is not discernible to the average individual. Animal instincts are modes of guidance for the individual in meeting such situations. Ethics are possibly a kind of community instinct in-the-making.

The Community Concept

All ethics so far evolved rest upon a single premise: that the individual is a member of a community of interdependent parts. His instincts prompt him to compete for his place in that community, but his ethics prompt him also to co-operate (perhaps in order that there may be a place to compete for).

The land ethic simply enlarges the boundaries of the community to include soils, waters, plants, and animals, or collectively: the land.

This sounds simple: do we not already sing our love for and obligation to the land of the free and the home of the brave? Yes, but just what and whom do we love? Certainly not the soil, which we are sending helter-skelter downriver. Certainly not the waters, which we assume have no function except to turn turbines, float barges, and carry off sewage. Certainly not the plants, of which we exterminate whole communities without batting an eye. Certainly not the animals, of which we have already extirpated many of the largest and most beautiful species. A land ethic of course cannot prevent the alteration, management, and use of these 'resources,' but it does affirm their right to continued existence, and, at least in spots, their continued existence in a natural state.

In short, a land ethic changes the role of *Homo sapiens* from conqueror of the land-community to plain member and citizen of it. It implies respect for his fellow-members, and also respect for the community as such.

In human history, we have learned (I hope) that the conqueror role is eventually self-defeating. Why? Because it is implicit in such a role that the conqueror knows, *ex cathedra*, just what makes the community clock tick, and just what and who is valuable, and what and who is worthless, in community life. It always turns out that he knows neither, and this is why his conquests eventually defeat themselves.

In the biotic community, a parallel situation exists. Abraham knew exactly what the land was for: it was to drip milk and honey into Abraham's mouth. At the present moment, the assurance with which we regard this assumption is inverse to the degree of our education.

The ordinary citizen today assumes that science knows what makes the community clock tick; the scientist is equally sure that he does not. He knows that the biotic mechanism is so complex that its workings may never be fully understood.

That man is, in fact, only a member of a biotic team is shown by an ecological interpretation of history. Many historical events, hitherto explained solely in terms of human enterprise, were actually biotic interactions between people and land. The characteristics of the land determined the facts quite as potently as the characteristics of the men who lived on it.

Consider, for example, the settlement of the Mississippi valley. In the years following the Revolution, three groups were contending for its control: the native Indian, the French and English traders, and the American settlers. Historians wonder what would have happened if the English at Detroit had thrown a little more weight into the Indian side of those tipsy scales which decided the outcome of the colonial migration into the cane-lands of Kentucky. It is time now to ponder the fact that the cane-lands, when subjected to the particular mixture of forces represented by the cow, plow, fire, and axe of the pioneer, became bluegrass. What if the plant succession inherent in this dark and bloody ground had, under the impact of these forces, given us some worthless sedge, shrub, or weed? Would Boone and Kenton have held out? Would there have been any overflow into Ohio, Indiana, Illinois, and Missouri? Any Louisiana Pur-

chase? Any transcontinental union of new states? Any Civil War?

Kentucky was one sentence in the drama of history. We are commonly told what the human actors in this drama tried to do, but we are seldom told that their success, or the lack of it, hung in large degree on the reaction of particular soils to the impact of the particular forces exerted by their occupancy. In the case of Kentucky, we do not even know where the bluegrass came from—whether it is a native species, or a stowaway from Europe.

Contrast the cane-lands with what hindsight tells us about the Southwest, where the pioneers were equally brave, resourceful, and persevering. The impact of occupancy here brought no bluegrass, or other plant fitted to withstand the bumps and buffetings of hard use. This region, when grazed by livestock, reverted through a series of more and more worthless grasses, shrubs, and weeds to a condition of unstable equilibrium. Each recession of plant types bred erosion; each increment to erosion bred a further recession of plants. The result today is a progressive and mutual deterioration, not only of plants and soils, but of the animal community subsisting thereon. The early settlers did not expect this: on the ciénegas of New Mexico some even cut ditches to hasten it. So subtle has been its progress that few residents of the region are aware of it. It is quite invisible to the tourist who finds this wrecked landscape colorful and charming (as indeed it is, but it bears scant resemblance to what it was in 1848).

This same landscape was 'developed' once before, but with quite different results. The Pueblo Indians settled the Southwest in pre-Columbian times, but they happened *not* to be equipped with range livestock. Their civilization expired, but not because their land expired.

In India, regions devoid of any sod-forming grass have been settled, apparently without wrecking the land, by the simple expedient of carrying the grass to the cow, rather than vice versa. (Was this the result of some deep wisdom, or was it just good luck? I do not know.)

In short, the plant succession steered the course of history; the pioneer simply demonstrated, for good or ill, what successions inhered in the land. Is history taught in this spirit? It will be, once the concept of land as a community really penetrates our intellectual life.

The Ecological Conscience

Conservation is a state of harmony between men and land. Despite nearly a century of propaganda, conservation still proceeds at a snail's pace; progress still consists largely of letterhead pieties and convention oratory. On the back forty we still slip two steps backward for each forward stride.

The usual answer to this dilemma is 'more conservation education.' No one will debate this, but is it certain that only the *volume* of education needs stepping up? Is something lacking in the *content* as well?

It is difficult to give a fair summary of its content in brief form, but, as I understand it, the content is substantially this: obey the law, vote right, join some organizations, and practice what conservation is profitable on your own land; the government will do the rest.

Is not this formula too easy to accomplish anything worth-while? It defines no right or wrong, assigns no obliga-

tion, calls for no sacrifice, implies no change in the current philosophy of values. In respect of land-use, it urges only enlightened self-interest. Just how far will such education take us? An example will perhaps yield a partial answer.

By 1930 it had become clear to all except the ecologically blind that southwestern Wisconsin's topsoil was slipping seaward. In 1933 the farmers were told that if they would adopt certain remedial practices for five years, the public would donate CCC labor to install them, plus the necessary machinery and materials. The offer was widely accepted, but the practices were widely forgotten when the five-year contract period was up. The farmers continued only those practices that yielded an immediate and visible economic gain for themselves.

This led to the idea that maybe farmers would learn more quickly if they themselves wrote the rules. Accordingly the Wisconsin Legislature in 1937 passed the Soil Conservation District Law. This said to farmers, in effect: *We, the public, will furnish you free technical service and loan you specialized machinery, if you will write your own rules for land-use. Each county may write its own rules, and these will have the force of law.* Nearly all the counties promptly organized to accept the proffered help, but after a decade of operation, *no county has yet written a single rule.* There has been visible progress in such practices as strip-cropping, pasture renovation, and soil liming, but none in fencing woodlots against grazing, and none in excluding plow and cow from steep slopes. The farmers, in short, have selected those remedial practices which were profitable anyhow, and ignored those which were profitable to the community, but not clearly profitable to themselves.

When one asks why no rules have been written, one is

told that the community is not yet ready to support them; education must precede rules. But the education actually in progress makes no mention of obligations to land over and above those dictated by self-interest. The net result is that we have more education but less soil, fewer healthy woods, and as many floods as in 1937.

The puzzling aspect of such situations is that the existence of obligations over and above self-interest is taken for granted in such rural community enterprises as the betterment of roads, schools, churches, and baseball teams. Their existence is not taken for granted, nor as yet seriously discussed, in bettering the behavior of the water that falls on the land, or in the preserving of the beauty or diversity of the farm landscape. Land-use ethics are still governed wholly by economic self-interest, just as social ethics were a century ago.

To sum up: we asked the farmer to do what he conveniently could to save his soil, and he has done just that, and only that. The farmer who clears the woods off a 75 per cent slope, turns his cows into the clearing, and dumps its rainfall, rocks, and soil into the community creek, is still (if otherwise decent) a respected member of society. If he puts lime on his fields and plants his crops on contour, he is still entitled to all the privileges and emoluments of his Soil Conservation District. The District is a beautiful piece of social machinery, but it is coughing along on two cylinders because we have been too timid, and too anxious for quick success, to tell the farmer the true magnitude of his obligations. Obligations have no meaning without conscience, and the problem we face is the extension of the social conscience from people to land.

No important change in ethics was ever accomplished without an internal change in our intellectual emphasis, loyalties, affections, and convictions. The proof that conservation has not yet touched these foundations of conduct lies in the fact that philosophy and religion have not yet heard of it. In our attempt to make conservation easy, we have made it trivial.

Substitutes for a Land Ethic

When the logic of history hungers for bread and we hand out a stone, we are at pains to explain how much the stone resembles bread. I now describe some of the stones which serve in lieu of a land ethic.

One basic weakness in a conservation system based wholly on economic motives is that most members of the land community have no economic value. Wildflowers and songbirds are examples. Of the 22,000 higher plants and animals native to Wisconsin, it is doubtful whether more than 5 per cent can be sold, fed, eaten, or otherwise put to economic use. Yet these creatures are members of the biotic community, and if (as I believe) its stability depends on its integrity, they are entitled to continuance.

When one of these non-economic categories is threatened, and if we happen to love it, we invent subterfuges to give it economic importance. At the beginning of the century songbirds were supposed to be disappearing. Ornithologists jumped to the rescue with some distinctly shaky evidence to the effect that insects would eat us up if birds failed to control them. The evidence had to be economic in order to be valid.

It is painful to read these circumlocutions today. We have

no land ethic yet, but we have at least drawn nearer the point of admitting that birds should continue as a matter of biotic right, regardless of the presence or absence of economic advantage to us.

A parallel situation exists in respect of predatory mammals, raptorial birds, and fish-eating birds. Time was when biologists somewhat overworked the evidence that these creatures preserve the health of game by killing weaklings, or that they control rodents for the farmer, or that they prey only on 'worthless' species. Here again, the evidence had to be economic in order to be valid. It is only in recent years that we hear the more honest argument that predators are members of the community, and that no special interest has the right to exterminate them for the sake of a benefit, real or fancied, to itself. Unfortunately this enlightened view is still in the talk stage. In the field the extermination of predators goes merrily on: witness the impending erasure of the timber wolf by fiat of Congress, the Conservation Bureaus, and many state legislatures.

Some species of trees have been 'read out of the party' by economics-minded foresters because they grow too slowly, or have too low a sale value to pay as timber crops: white cedar, tamarack, cypress, beech, and hemlock are examples. In Europe, where forestry is ecologically more advanced, the non-commercial tree species are recognized as members of the native forest community, to be preserved as such, within reason. Moreover some (like beech) have been found to have a valuable function in building up soil fertility. The interdependence of the forest and its constituent tree species, ground flora, and fauna is taken for granted.

Lack of economic value is sometimes a character not only of species or groups, but of entire biotic communities: marshes, bogs, dunes, and 'deserts' are examples. Our formula in such cases is to relegate their conservation to government as refuges, monuments, or parks. The difficulty is that these communities are usually interspersed with more valuable private lands; the government cannot possibly own or control such scattered parcels. The net effect is that we have relegated some of them to ultimate extinction over large areas. If the private owner were ecologically minded, he would be proud to be the custodian of a reasonable proportion of such areas, which add diversity and beauty to his farm and to his community.

In some instances, the assumed lack of profit in these 'waste' areas has proved to be wrong, but only after most of them had been done away with. The present scramble to reflood muskrat marshes is a case in point.

There is a clear tendency in American conservation to relegate to government all necessary jobs that private landowners fail to perform. Government ownership, operation, subsidy, or regulation is now widely prevalent in forestry, range management, soil and watershed management, park and wilderness conservation, fisheries management, and migratory bird management, with more to come. Most of this growth in governmental conservation is proper and logical, some of it is inevitable. That I imply no disapproval of it is implicit in the fact that I have spent most of my life working for it. Nevertheless the question arises: What is the ultimate magnitude of the enterprise? Will the tax base carry its eventual ramifications? At what point will governmental conservation, like the mastodon, become handicapped by its own dimensions? The answer, if there is any, seems to be in a land ethic, or some other force which assigns more obligation to the private landowner.

Industrial landowners and users, especially lumbermen and stockmen, are inclined to wail long and loudly about the extension of government ownership and regulation to land, but (with notable exceptions) they show little disposition to develop the only visible alternative: the voluntary practice of conservation on their own lands.

When the private landowner is asked to perform some unprofitable act for the good of the community, he today assents only with outstretched palm. If the act costs him cash this is fair and proper, but when it costs only forethought, open-mindedness, or time, the issue is at least debatable. The overwhelming growth of land-use subsidies in recent years must be ascribed, in large part, to the government's own agencies for conservation education: the land bureaus, the agricultural colleges, and the extension services. As far as I can detect, no ethical obligation toward land is taught in these institutions.

To sum up: a system of conservation based solely on economic self-interest is hopelessly lopsided. It tends to ignore, and thus eventually to eliminate, many elements in the land community that lack commercial value, but that are (as far as we know) essential to its healthy functioning. It assumes, falsely, I think, that the economic parts of the biotic clock will function without the uneconomic parts. It tends to relegate to government many functions eventually too large, too complex, or too widely dispersed to be performed by government.

An ethical obligation on the part of the private owner is the only visible remedy for these situations.

The Land Pyramid

An ethic to supplement and guide the economic relation to land presupposes the existence of some mental image of land as a biotic mechanism. We can be ethical only in relation to something we can see, feel, understand, love, or otherwise have faith in.

The image commonly employed in conservation education is 'the balance of nature.' For reasons too lengthy to detail here, this figure of speech fails to describe accurately what little we know about the land mechanism. A much truer image is the one employed in ecology: the biotic pyramid. I shall first sketch the pyramid as a symbol of land, and later develop some of its implications in terms of land-use.

Plants absorb energy from the sun. This energy flows through a circuit called the biota, which may be represented by a pyramid consisting of layers. The bottom layer is the soil. A plant layer rests on the soil, an insect layer on the plants, a bird and rodent layer on the insects, and so on up through various animal groups to the apex layer, which consists of the larger carnivores.

The species of a layer are alike not in where they came from, or in what they look like, but rather in what they eat. Each successive layer depends on those below it for food and often for other services, and each in turn furnishes food and services to those above. Proceeding upward, each successive layer decreases in numerical abundance. Thus, for every carnivore there are hundreds of his prey, thousands of their prey, millions of insects, uncountable plants. The pyramidal form of the system reflects this numerical progression from apex to base. Man shares an intermediate layer with the bears, raccoons, and squirrels which eat both meat and vegetables.

The lines of dependency for food and other services are called food chains. Thus soil-oak-deer-Indian is a chain that has now been largely converted to soil-corn-cow-farmer. Each species, including ourselves, is a link in many chains. The deer eats a hundred plants other than oak, and the cow a hundred plants other than corn. Both, then, are links in a hundred chains. The pyramid is a tangle of chains so complex as to seem disorderly, yet the stability of the system proves it to be a highly organized structure. Its functioning depends on the co-operation and competition of its diverse parts.

In the beginning, the pyramid of life was low and squat; the food chains short and simple. Evolution has added layer after layer, link after link. Man is one of thousands of accretions to the height and complexity of the pyramid. Science has given us many doubts, but it has given us at least one certainty: the trend of evolution is to elaborate and diversify the biota.

Land, then, is not merely soil; it is a fountain of energy flowing through a circuit of soils, plants, and animals. Food chains are the living channels which conduct energy upward; death and decay return it to the soil. The circuit is not closed; some energy is dissipated in decay, some is added by absorption from the air, some is stored in soils, peats, and long-lived forests; but it is a sustained circuit, like a slowly augmented revolving fund of life. There is always a net loss by downhill wash, but this is normally small and offset by the decay of rocks. It is deposited in the ocean and, in the course of geological time, raised to form new lands and new pyramids.

The velocity and character of the upward flow of energy depend on the complex structure of the plant and animal community, much as the upward flow of sap in a tree depends on its complex cellular organization. Without this complexity, normal circulation would presumably not occur. Structure means the characteristic numbers, as well as the characteristic kinds and functions, of the component species. This interdependence between the complex structure of the land and its smooth functioning as an energy unit is one of its basic attributes.

When a change occurs in one part of the circuit, many other parts must adjust themselves to it. Change does not necessarily obstruct or divert the flow of energy; evolution is a long series of self-induced changes, the net result of which has been to elaborate the flow mechanism and to lengthen the circuit. Evolutionary changes, however, are usually slow and local. Man's invention of tools has enabled him to make changes of unprecedented violence, rapidity, and scope.

One change is in the composition of floras and faunas. The larger predators are lopped off the apex of the pyramid; food chains, for the first time in history, become shorter rather than longer. Domesticated species from other lands are substituted for wild ones, and wild ones are moved to new habitats. In this world-wide pooling of faunas and floras, some species get out of bounds as pests and diseases, others are extinguished. Such effects are seldom intended or foreseen; they represent unpredicted and often untraceable readjustments in the structure. Agricultural science is largely a race between the emergence of new pests and the emergence of new techniques for their control.

Another change touches the flow of energy through plants and animals and its return to the soil. Fertility is the ability of soil to receive, store, and release energy. Agriculture, by overdrafts on the soil, or by too radical a substitution of domestic for native species in the superstructure, may derange the channels of flow or deplete storage. Soils depleted of their storage, or of the organic matter which anchors it, wash away faster than they form. This is erosion.

Waters, like soil, are part of the energy circuit. Industry, by polluting waters or obstructing them with dams, may exclude the plants and animals necessary to keep energy in circulation.

Transportation brings about another basic change: the plants or animals grown in one region are now consumed and returned to the soil in another. Transportation taps the energy stored in rocks, and in the air, and uses it elsewhere; thus we fertilize the garden with nitrogen gleaned by the guano birds from the fishes of seas on the other side of the Equator. Thus the formerly localized and self-contained circuits are pooled on a world-wide scale.

The process of altering the pyramid for human occupation releases stored energy, and this often gives rise, during the pioneering period, to a deceptive exuberance of plant and animal life, both wild and tame. These releases of biotic capital tend to becloud or postpone the penalties of violence.

＊　　　　　＊　　　　　＊

This thumbnail sketch of land as an energy circuit conveys three basic ideas:

(1) That land is not merely soil.

(2) That the native plants and animals kept the energy circuit open; others may or may not.

(3) That man-made changes are of a different order than evolutionary changes, and have effects more comprehensive than is intended or foreseen.

These ideas, collectively, raise two basic issues: Can the land adjust itself to the new order? Can the desired alterations be accomplished with less violence?

Biotas seem to differ in their capacity to sustain violent conversion. Western Europe, for example, carries a far different pyramid than Caesar found there. Some large animals are lost; swampy forests have become meadows or plowland; many new plants and animals are introduced, some of which escape as pests; the remaining natives are greatly changed in distribution and abundance. Yet the soil is still there and, with the help of imported nutrients, still fertile; the waters flow normally; the new structure seems to function and to persist. There is no visible stoppage or derangement of the circuit.

Western Europe, then, has a resistant biota. Its inner processes are tough, elastic, resistant to strain. No matter how violent the alterations, the pyramid, so far, has developed some new *modus vivendi* which preserves its habitability for man, and for most of the other natives.

Japan seems to present another instance of radical conversion without disorganization.

Most other civilized regions, and some as yet barely touched by civilization, display various stages of disorganization, varying from initial symptoms to advanced wastage. In Asia Minor and North Africa diagnosis is confused by climatic changes, which may have been either the cause or the effect of advanced wastage. In the United States the degree of disorganization varies locally; it is worst in the Southwest, the Ozarks, and parts of the South, and least in New England and the Northwest. Better land-uses may still arrest it in the less advanced regions. In parts of Mexico, South America, South Africa, and Australia a violent and accelerating wastage is in progress, but I cannot assess the prospects.

This almost world-wide display of disorganization in the land seems to be similar to disease in an animal, except that it never culminates in complete disorganization or death. The land recovers, but at some reduced level of complexity, and with a reduced carrying capacity for people, plants, and animals. Many biotas currently regarded as 'lands of opportunity' are in fact already subsisting on exploitative agriculture, i.e. they have already exceeded their sustained carrying capacity. Most of South America is overpopulated in this sense.

In arid regions we attempt to offset the process of wastage by reclamation, but it is only too evident that the prospective longevity of reclamation projects is often short. In our own West, the best of them may not last a century.

The combined evidence of history and ecology seems to support one general deduction: the less violent the manmade changes, the greater the probability of successful readjustment in the pyramid. Violence, in turn, varies with human population density; a dense population requires a more violent conversion. In this respect, North America has a better chance for permanence than Europe, if she can contrive to limit her density.

This deduction runs counter to our current philosophy, which assumes that because a small increase in density enriched human life, that an indefinite increase will enrich it indefinitely. Ecology knows of no density relationship that holds for indefinitely wide limits. All gains from density are subject to a law of diminishing returns.

Whatever may be the equation for men and land, it is improbable that we as yet know all its terms. Recent discoveries in mineral and vitamin nutrition reveal unsuspected dependencies in the up-circuit: incredibly minute quantities of certain substances determine the value of soils to plants, of plants to animals. What of the down-circuit? What of the vanishing species, the preservation of which we now regard as an esthetic luxury? They helped build the soil; in what unsuspected ways may they be essential to its maintenance? Professor Weaver proposes that we use prairie flowers to reflocculate the wasting soils of the dust bowl; who knows for what purpose cranes and condors, otters and grizzlies may some day be used?

Land Health and the A-B Cleavage

A land ethic, then, reflects the existence of an ecological conscience, and this in turn reflects a conviction of individual responsibility for the health of the land. Health is the capacity of the land for self-renewal. Conservation is our effort to understand and preserve this capacity.

Conservationists are notorious for their dissensions. Superficially these seem to add up to mere confusion, but a more careful scrutiny reveals a single plane of cleavage common to many specialized fields. In each field one group (A) regards the land as soil, and its function as commodity-production; another group (B) regards the land as a biota, and its function as something broader. How much broader is admittedly in a state of doubt and confusion.

In my own field, forestry, group A is quite content to grow trees like cabbages, with cellulose as the basic forest commodity. It feels no inhibition against violence; its ideology is agronomic. Group B, on the other hand, sees forestry as fundamentally different from agronomy because it employs natural species, and manages a natural environment rather than creating an artificial one. Group B prefers natural reproduction on principle. It worries on biotic as well as economic grounds about the loss of species like chestnut, and the threatened loss of the white pines. It worries about a whole series of secondary forest functions: wildlife, recreation, watersheds, wilderness areas. To my mind, Group B feels the stirrings of an ecological conscience.

In the wildlife field, a parallel cleavage exists. For Group A the basic commodities are sport and meat; the yardsticks of production are ciphers of take in pheasants and trout.

Artificial propagation is acceptable as a permanent as well as a temporary recourse—if its unit costs permit. Group B, on the other hand, worries about a whole series of biotic side-issues. What is the cost in predators of producing a game crop? Should we have further recourse to exotics? How can management restore the shrinking species, like prairie grouse, already hopeless as shootable game? How can management restore the threatened rarities, like trumpeter swan and whooping crane? Can management principles be extended to wildflowers? Here again it is clear to me that we have the same A-B cleavage as in forestry.

In the larger field of agriculture I am less competent to speak, but there seem to be somewhat parallel cleavages. Scientific agriculture was actively developing before ecology was born, hence a slower penetration of ecological concepts might be expected. Moreover the farmer, by the very nature of his techniques, must modify the biota more radically than the forester or the wildlife manager. Nevertheless, there are many discontents in agriculture which seem to add up to a new vision of 'biotic farming.'

Perhaps the most important of these is the new evidence that poundage or tonnage is no measure of the food-value of farm crops; the products of fertile soil may be qualitatively as well as quantitatively superior. We can bolster poundage from depleted soils by pouring on imported fertility, but we are not necessarily bolstering food-value. The possible ultimate ramifications of this idea are so immense that I must leave their exposition to abler pens.

The discontent that labels itself 'organic farming,' while bearing some of the earmarks of a cult, is nevertheless biotic in its direction, particularly in its insistence on the importance of soil flora and fauna.

The ecological fundamentals of agriculture are just as poorly known to the public as in other fields of land-use. For example, few educated people realize that the marvelous advances in technique made during recent decades are improvements in the pump, rather than the well. Acre for acre, they have barely sufficed to offset the sinking level of fertility.

In all of these cleavages, we see repeated the same basic paradoxes: man the conqueror *versus* man the biotic citizen; science the sharpener of his sword *versus* science the searchlight on his universe; land the slave and servant *versus* land the collective organism. Robinson's injunction to Tristram may well be applied, at this juncture, to *Homo sapiens* as a species in geological time:

> Whether you will or not
> You are a King, Tristram, for you are one
> Of the time-tested few that leave the world,
> When they are gone, not the same place it was.
> Mark what you leave.

The Outlook

It is inconceivable to me that an ethical relation to land can exist without love, respect, and admiration for land, and a high regard for its value. By value, I of course mean something far broader than mere economic value; I mean value in the philosophical sense.

Perhaps the most serious obstacle impeding the evolution of a land ethic is the fact that our educational and economic system is headed away from, rather than toward, an intense consciousness of land. Your true modern is separated from the land by many middlemen, and by innumerable physical gadgets. He has no vital relation to it; to him it is the space between cities on which crops grow. Turn him loose for a day on the land, and if the spot does not happen to be a golf links or a 'scenic' area, he is bored stiff. If crops could be raised by hydroponics instead of farming, it would suit him very well. Synthetic substitutes for wood, leather, wool, and other natural land products suit him better than the originals. In short, land is something he has 'outgrown.'

Almost equally serious as an obstacle to a land ethic is the attitude of the farmer for whom the land is still an adversary, or a taskmaster that keeps him in slavery. Theoretically, the mechanization of farming ought to cut the farmer's chains, but whether it really does is debatable.

One of the requisites for an ecological comprehension of land is an understanding of ecology, and this is by no means co-extensive with 'education'; in fact, much higher education seems deliberately to avoid ecological concepts. An understanding of ecology does not necessarily originate in courses bearing ecological labels; it is quite as likely to be labeled geography, botany, agronomy, history, or economics. This is as it should be, but whatever the label, ecological training is scarce.

The case for a land ethic would appear hopeless but for the minority which is in obvious revolt against these 'modern' trends.

The 'key-log' which must be moved to release the evolutionary process for an ethic is simply this: quit thinking about decent land-use as solely an economic problem. Examine each question in terms of what is ethically and esthetically right, as well as what is economically expedient. A thing is right when it tends to preserve the integrity, stability, and beauty of the biotic community. It is wrong when it tends otherwise.

It of course goes without saying that economic feasibility limits the tether of what can or cannot be done for land. It always has and it always will. The fallacy the economic determinists have tied around our collective neck, and which we now need to cast off, is the belief that economics determines *all* land-use. This is simply not true. An innumerable host of actions and attitudes, comprising perhaps the bulk of all land relations, is determined by the land-users' tastes and predilections, rather than by his purse. The bulk of all land relations hinges on investments of time, forethought, skill, and faith rather than on investments of cash. As a land-user thinketh, so is he.

I have purposely presented the land ethic as a product of social evolution because nothing so important as an ethic is ever 'written.' Only the most superficial student of history supposes that Moses 'wrote' the Decalogue; it evolved in the minds of a thinking community, and Moses wrote a tentative summary of it for a 'seminar.' I say tentative because evolution never stops.

The evolution of a land ethic is an intellectual as well as emotional process. Conservation is paved with good intentions which prove to be futile, or even dangerous, because they are devoid of critical understanding either of the land, or of economic land-use. I think it is a truism that as the ethical frontier advances from the individual to the community, its intellectual content increases.

The mechanism of operation is the same for any ethic: social approbation for right actions: social disapproval for wrong actions.

By and large, our present problem is one of attitudes and implements. We are remodeling the Alhambra with a steam-shovel, and we are proud of our yardage. We shall hardly relinquish the shovel, which after all has many good points, but we are in need of gentler and more objective criteria for its successful use.

CHAPTER VIII

Land Utilization and Food Production

One factor beneficial to population growth has been the continual improvement in land utilization in Ming, Ch'ing, and modern times. The acreage in cultivated land has been expanding and land has been yielding more, partly as a result of increasingly labor-intensive farming and even more of the steady improvement in the cropping system. It is partially true that for centuries there has been no technological revolution in Chinese agriculture, evidenced by the fact that the same kind of agricultural implements have been used by Chinese peasants of certain areas for centuries.[1] Yet such a generalization requires qualification. During the Ming period there were significant improvements in agricultural implements, particularly in various kinds of water-pumps. The increase of draft animals, especially oxen and water buffalo, and the construction of irrigation works in the southeast both deserve further study.[2] In any event, not every important change in the technique of production is necessarily associated with the introduction of new implements and machinery. The first major revolution in the iron and steel industry of the modern West, to name an outstanding example, was caused by the use of a new fuel rather than by machinery. The core of Chinese agriculture has been its cropping system. In the absence of major technological inventions the nature of the crops has done more than anything else to push the agricultural frontier further away from the low plains, basins, and valleys to the more arid hilly and mountainous regions and has accounted for an enormous increase in national food production.

I

The first long-range revolution in land utilization and food production in China during the past millennium was brought about by the early-ripening rice which began to be disseminated extensively at the opening of the eleventh century. Agricultural and etymological evidence indicates that although ancient and medieval China had a limited number of indigenous early-ripening rice varieties, they were far less important than the heavy yielding late- and medium-ripening varieties which usually require well-watered lowlands and rich clayey soil. *Ch'i-min yao-shu*, the earliest systematic Chinese agricultural treatise extant, compiled in the first half of the sixth century, states as a general rule that the rice plant required 150 days to mature after it was transplanted from the nursery bed to the paddy field.[3] Since rice seeds are grown in nursery bed for four to six weeks, the total time required by ancient and medieval rice was evidently at least 180 days. Such a long period made it difficult for the same piece of land to be sown to a second crop after the harvest. Furthermore, it is common geographic knowledge that the well-watered lowlands in China suitable for the cultivation of late- and medium-ripening rice are of rather limited extent. For these reasons the area of rice culture up to the end of the first millennium of the Christian era had been practically confined to the deltas, basins, and valleys of the Yangtze region.[4]

The increase in the population of the rice area throughout the T'ang (618–907) and Five Dynasties (908–960) brought about a shift of the economic and demographic center of gravity from the northwest to the southeast and made it necessary for the early Sung emperors to expand the frontier of rice culture. The Sung Emperor Chen-tsung (998–1022), who was deeply concerned with the food supply, introduced to Fukien an early-ripening and relatively drought-resistant rice from Champa, a state in central Indochina. In 1012, when the lower Yangtze and Huai River regions suffered from drought, 30,000 bushels of Champa seeds were shipped from Fukien to be distributed in the drought areas. To familiarize peasants of the lower Yangtze and Huai areas with this valuable strain the government circulated pamphlets in which the methods of cultivating it were explained. The publicity given to the Champa rice in 1012 not only accounted for its cultivation in the southeastern provinces but called the attention of Chinese rice farmers to the importance of developing further early-ripening varieties with which to use land of poorer quality.

Sung local histories of Chekiang and southern Kiangsu recorded that the original Champa rice matured one hundred days after transplantation. It was suited to relatively well-watered hills on which indigenous late- and medium-ripening varieties could not be successfully grown. By the twelfth century ingenious Chinese farmers had developed new strains which matured in but sixty days after transplantation. The effect of the introduction of Champa rice and the further development of early-ripening varieties was manifold. First, these early-ripening varieties helped greatly to ensure the success of the double-cropping system, for which Chinese agriculture, particularly that of the rice area, is world famous. Since rice in the Yangtze area is usually a summer crop, a shorter period of growth made it possible for the same piece of land to be sown to wheat, rape, or other winter crops after the rice was harvested. The average late-ripening rice of ancient and medieval China would have made such a rotation of crops precarious if not entirely impossible in many places. Secondly, for centuries peasants had worried about the precarious food supply during the long interval between the harvest of the winter crop and the harvest of rice. Early-ripening rice now became a valuable fill-in crop, much superior in taste and food value to other auxiliary crops. Thirdly, the early-ripening rice, which requires much less water than the other varieties, made possible the cultivation of higher land and hilly slopes that could be fed by spring or rain water. The first long-range revolution in land utilization and food production, based on the conquest of relatively well-watered hills, thus got under way.

The dissemination of the early-ripening rice, like the propagation of other food crops, was necessarily a slow process. As far as can be ascertained, up to the end of the Southern Sung (1127–1270) the early-ripening varieties seem to have been disseminated to a significant extent only in Chekiang, southern Kiangsu, Fukien, and Kiangsi. The lowlands of southern Anhwei and large sections of Hupei and Hunan, which are the main rice baskets of modern China, were markedly lacking in early-ripening varieties and therefore agriculturally underdeveloped. The Huai River plain, which serves as an area of demarcation between the rice-producing south and the wheat-producing north and into which Champa rice was introduced by the imperial order of 1012, suffered a long-term economic decline after the downfall of the Northern Sung dynasty in 1126. After that it became a major battleground between the Chinese and the Juchens, who had conquered north China, and throughout the twelfth and thirteenth centuries remained a vast desolate area. Any agricultural gains that this area may have made during the eleventh century must have been nullified by these wars and extensive southward migrations by the Chinese. Hupei suffered for similar reasons.

Although the dissemination of Champa rice during the Sung was confined mainly to three or four provinces, its effect on food production and population growth seems to have been already felt within one century of its introduction, for in 1102 and 1110 the number of officially registered households for the entire country exceeded 20,000,000. Since it has been shown by some leading Japanese historians that the official Sung household returns, in sharp contrast to the notoriously under-registered "total" population, were comparatively accurate,[5] it seems reasonable to believe that by the opening of the twelfth century China's actual population exceeded, for the first time in her history, the 100,000,000 mark.

The dissemination of Champa varieties must have continued throughout the subsequent periods. From available Ming records and local histories it appears that the sixty-day variety, which had been confined essentially to the lower Yangtze area during the Sung, had reached as far as Kwangtung and become fairly common throughout the rice area. The "white-Champa," "red-Champa," and other early-ripening varieties had struck permanent root in Kwangsi before the compilation of the 1531 edition of its provincial history. The 1574 edition of the history of Yunnan province places early-ripening rice as the foremost crop of Meng-hua prefecture. While giving no detailed listing of rice varieties, the 1621 edition of the history of Ch'eng-tu prefecture, in the heart of the Szechwan basin, particularly points out that the "white-Champa" had been instrumental in conquering dry fields. Even in Honan during the sixteenth century the early-ripening rice varieties were very common.

Most of the vulgar names for early-ripening varieties are too fanciful to be of real use for our purpose, but the continual dissemination of early-ripening rice may be conveniently indicated by the names of those varieties which reveal their geographic origins. In the lowland districts of southern Anhwei, an area that had been agriculturally underdeveloped during the Sung, there were in the fifteenth-century varieties such as the "white sixty-day" and "red sixty-day," and also varieties bearing names like the "Kiangsi early,"

"Hu-kuang (Hupei and Hunan) early," and "Kwangtung early."
These lowland districts around Wu-hu have since become the leading producing centers of early-ripening rice. By the early seventeenth century, possibly much earlier, in as southerly a place as
Ch'üan-chou in southern coastal Fukien, there were early-ripening
varieties called the "Su-chou red" and "Honan early." In the
eighteenth century another valuable northern variety was added
which bore the name "Shantung seed." Hupei and Hunan, two
provinces that during the Ming forged ahead as leading rice-export-
ing areas, must have introduced and developed many new rice
varieties in the course of time. Unfortunately, neither the handful
of extant Ming local histories of these provinces nor the 1684 edition
of the combined history of Hupei and Hunan yield any systematic
information. However, when Hunan compiled its first independent
provincial history in 1820, there were not only certain extremely
early-ripening varieties called "fifty-day Champa" and "forty-day
Champa" but over a dozen varieties that, judging by their names,
had originated in Su-chou, Nanking, Liang-shan in southwestern
Szechwan, Kuei-yang and Ssu-nan in Kweichow, Kiangsi, Kwangtung, Yunnan, and Indochina.

For centuries after the introduction of Champa rice the quickest
ripening variety required 60 days after transplantation. The availability of a fairly large number of quickly ripening varieties ranging
from 120 to 60 days had unquestionably helped to solve the problem
of crop rotation in the rice area. But there were marginal rice areas
which called for varieties that would ripen even more rapidly or for
special strains that could survive unusually menacing natural conditions. The marshy flats of Kiangsu north of the Yangtze, visited
by the annual midsummer flood and consequently submerged for a
greater part of the year, made up one such blighted area. For this
reason Kao-yu and T'ai-chou, both in the heart of the Kiangsu flats,
became virtually the experimental farm for extremely early-ripening varieties. To beat the annual midsummer flood, peasants of Kao-
yu developed the fifty-day in the sixteenth century. Some inland
districts in southwestern Chekiang and lakeside Kiangsi may also
have independently developed this fifty-day variety. It was the
fifty-day that saved Kao-yu's peasants from total crop failure
during the terrible flood of 1720–21. In the eighteenth century the
forty-day was developed, probably independently, in Kao-yu and

in Heng-chou in southern Hunan. When Kiangsu was particularly hard hit by the flood of 1834–35, the thirty-day variety, which was said to have been developed in Hupei, was rushed in and distributed to peasants of the Kiangsu flats. The development and dissemination of these extremely early-ripening varieties by the peasants of marginal rice areas indicate that rice culture in China Proper was about to reach its saturation point. The thirty-day variety is probably the quickest ripening rice ever recorded in world history.

There was also a need in certain localities for a rice that could be planted late and would also ripen quickly. This need was created by the limited local supply of good land and the local rotation system. By Ming times at the latest certain Champa varieties had been developed that could withstand cold so well that they were planted in midsummer after the field was entirely cleared of spring crops or early rice. Since they were harvested in late fall or early winter, they were called "cold Champa" or "winter Champa," names that are recorded in a number of early modern local histories of Chekiang, Fukien, Kiangsi, and Hunan. In addition to facilitating a multiple cropping system, they gave peasants of lakeside Kiangsi a weapon with which to combat the annual flood. A particular reddish rice was usually planted in the lake districts of Kiangsi in the sixth lunar month, when the flood began to recede. It became so famous that when in 1608 a northern Chekiang county suffered from an unusually severe flood and was on the point of losing all its rice crop, this reddish variety was introduced by the magistrate to be planted after the retreat of the flood waters. In spite of its late planting, it ripened quickly enough to ensure the peasants a fair harvest. T'ai-chou in the Kiangsu flats was equally famous for this late-planting but quick-ripening reddish rice. Since then the reddish rice of Chiu-chiang, the main port of Kiangsi, and T'ai-chou has been recorded in a number of local histories. By the early eighteenth century the practice of growing quick-ripening and late-planting rice after the retreat of flood waters had become increasingly common in the rice area.

It is true that in the history of so slowly moving an industry as agriculture, let alone the self-sustaining agriculture of China, few events can really be called revolutionary. Yet the introduction of Champa rice and the subsequent dissemination of all early-ripening varieties, native and imported, ultimately produced results beyond the fondest dream of the early Sung emperors. Within two centuries of the introduction of Champa rice the landscape of the eastern half of China's rice area had already been substantially changed. By the thirteenth century much of the hilly land of the lower Yangtze region and Fukien, where water resources, climatic and soil conditions were sufficiently suitable for the cultivation of early-ripening rice, had been turned into terraced paddies. The early-ripening rice not only ensured the success of the double-crop system but also prolonged the economic hegemony of the Yangtze area. One certain indication of such hegemony was that the population of the rice area was increasing much more rapidly than the population of north China throughout the Sung, Yüan and Ming periods. During the Yüan and Ming the cultivation of early-ripening rice became common in the southwestern provinces and in Hupei and Hunan, the two provinces that have since become China's rice bowl. By the time of Matteo Ricci (1553–1610) double, and sometimes triple, rice cropping was common.[6]

In the 1830's, when early-ripening rice culture was approaching its saturation in the main rice area, it was estimated by Lin Tse-hsü (of Opium War fame and an authority on early-ripening rice) that the total national output of early-ripening rice approximately equaled that of the late- and medium-ripening rice.[7] There can be little doubt that, with the extension of irrigated areas and some not insignificant improvements in water pumps and other agricultural implements during the early modern centuries, the national production of late- and medium-ripening rice must have increased somewhat since the eleventh century; hence it seems reasonable to conclude that China's total rice output had greatly increased, perhaps even doubled between 1000 and 1850. As the per-acre yield of early-ripening rice is considerably less than that of late- and medium-ripening rice, it seems possible that the total area under early-ripening rice may have exceeded the area under the latter. In other words, China's rice area, too, may have more than doubled itself in the course of some eight and a half centuries. Since there is a good deal of qualitative evidence showing that the Chinese agricultural system was becoming increasingly labor-intensive, owing to an almost continual growth of population during Ming and early Ch'ing times, and since in most places the early-ripening rice indirectly enhanced local food production by promoting a busier and better rotation system, the effect of early-ripening rice on the overall food supply of early-modern China must have been even greater than Lin's statement indicates.

In retrospect, it may be suggested that during the greater parts of the past millennium the food situation was in all probability better in China, where the early-ripening rice brought about a major long-range revolution in land utilization and food production, than in Europe, where important changes in agriculture did not occur until the eighteenth century. If this is correct, it seems worth further notice by historians of world population because, in terms of food supply, China's population could probably increase, and apparently did begin to increase substantially, from 1000 onward. If it did not assume a continuous or more or less linear growth until after the founding of the Ming dynasty, this was due to political and institutional factors rather than to the most basic of economic restraints imposed by a shortage of food.

...

The Historical Roots of Our Ecologic Crisis

Lynn White, Jr.

A conversation with Aldous Huxley not infrequently put one at the receiving end of an unforgettable monologue. About a year before his lamented death he was discoursing on a favorite topic: Man's unnatural treatment of nature and its sad results. To illustrate his point he told how, during the previous summer, he had returned to a little valley in England where he had spent many happy months as a child. Once it had been composed of delightful grassy glades; now it was becoming overgrown with unsightly brush because the rabbits that formerly kept such growth under control had largely succumbed to a disease, myxomatosis, that was deliberately introduced by the local farmers to reduce the rabbits' destruction of crops. Being something of a Philistine, I could be silent no longer, even in the interests of great rhetoric. I interrupted to point out that the rabbit itself had been brought as a domestic animal to England in 1176, presumably to improve the protein diet of the peasantry.

All forms of life modify their contexts. The most spectacular and benign instance is doubtless the coral polyp. By serving its own ends, it has created a vast undersea world favorable to thousands of other kinds of animals and plants. Ever since man became a numerous species he has affected his environment notably. The hypothesis that his fire-drive method of hunting created the world's great grasslands and

The author is professor of history at the University of California, Los Angeles. This is the text of a lecture delivered 26 December 1966 at the Washington meeting of the AAAS.

helped to exterminate the monster mammals of the Pleistocene from much of the globe is plausible, if not proved. For 6 millennia at least, the banks of the lower Nile have been a human artifact rather than the swampy African jungle which nature, apart from man, would have made it. The Aswan Dam, flooding 5000 square miles, is only the latest stage in a long process. In many regions terracing or irrigation, overgrazing, the cutting of forests by Romans to build ships to fight Carthaginians or by Crusaders to solve the logistics problems of their expeditions, have profoundly changed some ecologies. Observation that the French landscape falls into two basic types, the open fields of the north and the *bocage* of the south and west, inspired Marc Bloch to undertake his classic study of medieval agricultural methods. Quite unintentionally, changes in human ways often affect nonhuman nature. It has been noted, for example, that the advent of the automobile eliminated huge flocks of sparrows that once fed on the horse manure littering every street.

The history of ecologic change is still so rudimentary that we know little about what really happened, or what the results were. The extinction of the European aurochs as late as 1627 would seem to have been a simple case of overenthusiastic hunting. On more intricate matters it often is impossible to find solid information. For a thousand years or more the Frisians and Hollanders have been pushing back the North Sea, and the process is culmi-

nating in our own time in the reclamation of the Zuider Zee. What, if any, species of animals, birds, fish, shore life, or plants have died out in the process? In their epic combat with Neptune have the Netherlanders overlooked ecological values in such a way that the quality of human life in the Netherlands has suffered? I cannot discover that the questions have ever been asked, much less answered.

People, then, have often been a dynamic element in their own environment, but in the present state of historical scholarship we usually do not know exactly when, where, or with what effects man-induced changes came. As we enter the last third of the 20th century, however, concern for the problem of ecologic backlash is mounting feverishly. Natural science, conceived as the effort to understand the nature of things, had flourished in several eras and among several peoples. Similarly there had been an age-old accumulation of technological skills, sometimes growing rapidly, sometimes slowly. But it was not until about four generations ago that Western Europe and North America arranged a marriage between science and technology, a union of the theoretical and the empirical approaches to our natural environment. The emergence in widespread practice of the Baconian creed that scientific knowledge means technological power over nature can scarcely be dated before about 1850, save in the chemical industries, where it is anticipated in the 18th century. Its acceptance as a normal pattern of action may mark the greatest event in human history since the invention of agriculture, and perhaps in nonhuman terrestrial history as well.

Almost at once the new situation forced the crystallization of the novel concept of ecology; indeed, the word *ecology* first appeared in the English language in 1873. Today, less than a century later, the impact of our race upon the environment has so increased in force that it has changed in essence. When the first cannons were fired, in the early 14th century, they affected ecology by sending workers scrambling to the forests and moun-

From White, Lynn (1967). The historical roots of our ecologic crisis. Science 155: 1203–1207. Reprinted with permission from AAAS.

tains for more potash, sulfur, iron ore, and charcoal, with some resulting erosion and deforestation. Hydrogen bombs are of a different order: a war fought with them might alter the genetics of all life on this planet. By 1285 London had a smog problem arising from the burning of soft coal, but our present combustion of fossil fuels threatens to change the chemistry of the globe's atmosphere as a whole, with consequences which we are only beginning to guess. With the population explosion, the carcinoma of planless urbanism, the now geological deposits of sewage and garbage, surely no creature other than man has ever managed to foul its nest in such short order.

There are many calls to action, but specific proposals, however worthy as individual items, seem too partial, palliative, negative: ban the bomb, tear down the billboards, give the Hindus contraceptives and tell them to eat their sacred cows. The simplest solution to any suspect change is, of course, to stop it, or, better yet, to revert to a romanticized past: make those ugly gasoline stations look like Anne Hathaway's cottage or (in the Far West) like ghost-town saloons. The "wilderness area" mentality invariably advocates deep-freezing an ecology, whether San Gimignano or the High Sierra, as it was before the first Kleenex was dropped. But neither atavism nor prettification will cope with the ecologic crisis of our time.

What shall we do? No one yet knows. Unless we think about fundamentals, our specific measures may produce new backlashes more serious than those they are designed to remedy.

As a beginning we should try to clarify our thinking by looking, in some historical depth, at the presuppositions that underlie modern technology and science. Science was traditionally aristocratic, speculative, intellectual in intent; technology was lowerclass, empirical, action-oriented. The quite sudden fusion of these two, towards the middle of the 19th century, is surely related to the slightly prior and contemporary democratic revolutions which, by reducing social barriers, tended to assert a functional unity of brain and hand. Our ecologic crisis is the product of an emerging, entirely novel, democratic culture. The issue is whether a democratized world can survive its own implications. Presumably we cannot unless we rethink our axioms.

The Western Traditions of Technology and Science

One thing is so certain that it seems stupid to verbalize it: both modern technology and modern science are distinctively *Occidental*. Our technology has absorbed elements from all over the world, notably from China; yet everywhere today, whether in Japan or in Nigeria, successful technology is Western. Our science is the heir to all the sciences of the past, especially perhaps to the work of the great Islamic scientists of the Middle Ages, who so often outdid the ancient Greeks in skill and perspicacity: al-Rāzi in medicine, for example; or ibn-al-Haytham in optics; or Omar Khayyám in mathematics. Indeed, not a few works of such geniuses seem to have vanished in the original Arabic and to survive only in medieval Latin translations that helped to lay the foundations for later Western developments. Today, around the globe, all significant science is Western in style and method, whatever the pigmentation or language of the scientists.

A second pair of facts is less well recognized because they result from quite recent historical scholarship. The leadership of the West, both in technology and in science, is far older than the so-called Scientific Revolution of the 17th century or the so-called Industrial Revolution of the 18th century. These terms are in fact outmoded and obscure the true nature of what they try to describe—significant stages in two long and separate developments. By A.D. 1000 at the latest —and perhaps, feebly, as much as 200 years earlier—the West began to apply water power to industrial processes other than milling grain. This was followed in the late 12th century by the harnessing of wind power. From simple beginnings, but with remarkable consistency of style, the West rapidly expanded its skills in the development of power machinery, labor-saving devices, and automation. Those who doubt should contemplate that most monumental achievement in the history of automation: the weight-driven mechanical clock, which appeared in two forms in the early 14th century. Not in craftsmanship but in basic technological capacity, the Latin West of the later Middle Ages far outstripped its elaborate, sophisticated, and esthetically magnificent sister cultures, Byzantium and Islam. In 1444 a great Greek ecclesiastic, Bessarion, who had

gone to Italy, wrote a letter to a prince in Greece. He is amazed by the superiority of Western ships, arms, textiles, glass. But above all he is astonished by the spectacle of waterwheels sawing timbers and pumping the bellows of blast furnaces. Clearly, he had seen nothing of the sort in the Near East.

By the end of the 15th century the technological superiority of Europe was such that its small, mutually hostile nations could spill out over all the rest of the world, conquering, looting, and colonizing. The symbol of this technological superiority is the fact that Portugal, one of the weakest states of the Occident, was able to become, and to remain for a century, mistress of the East Indies. And we must remember that the technology of Vasco da Gama and Albuquerque was built by pure empiricism, drawing remarkably little support or inspiration from science.

In the present-day vernacular understanding, modern science is supposed to have begun in 1543, when both Copernicus and Vesalius published their great works. It is no derogation of their accomplishments, however, to point out that such structures as the *Fabrica* and the *De revolutionibus* do not appear overnight. The distinctive Western tradition of science, in fact, began in the late 11th century with a massive movement of translation of Arabic and Greek scientific works into Latin. A few notable books—Theophrastus, for example—escaped the West's avid new appetite for science, but within less than 200 years effectively the entire corpus of Greek and Muslim science was available in Latin, and was being eagerly read and criticized in the new European universities. Out of criticism arose new observation, speculation, and increasing distrust of ancient authorities. By the late 13th century Europe had seized global scientific leadership from the faltering hands of Islam. It would be as absurd to deny the profound originality of Newton, Galileo, or Copernicus as to deny that of the 14th century scholastic scientists like Buridan or Oresme on whose work they built. Before the 11th century, science scarcely existed in the Latin West, even in Roman times. From the 11th century onward, the scientific sector of Occidental culture has increased in a steady crescendo.

Since both our technological and our scientific movements got their start, acquired their character, and achieved world dominance in the Middle Ages, it would seem that we cannot understand their nature or their present impact upon ecology without examining fundamental medieval assumptions and developments.

Medieval View of Man and Nature

Until recently, agriculture has been the chief occupation even in "advanced" societies; hence, any change in methods of tillage has much importance. Early plows, drawn by two oxen, did not normally turn the sod but merely scratched it. Thus, cross-plowing was needed and fields tended to be squarish. In the fairly light soils and semiarid climates of the Near East and Mediterranean, this worked well. But such a plow was inappropriate to the wet climate and often sticky soils of northern Europe. By the latter part of the 7th century after Christ, however, following obscure beginnings, certain northern peasants were using an entirely new kind of plow, equipped with a vertical knife to cut the line of the furrow, a horizontal share to slice under the sod, and a moldboard to turn it over. The friction of this plow with the soil was so great that it normally required not two but eight oxen. It attacked the land with such violence that cross-plowing was not needed, and fields tended to be shaped in long strips.

In the days of the scratch-plow, fields were distributed generally in units capable of supporting a single family. Subsistence farming was the presupposition. But no peasant owned eight oxen: to use the new and more efficient plow, peasants pooled their oxen to form large plow-teams, originally receiving (it would appear) plowed strips in proportion to their contribution. Thus, distribution of land was based no longer on the needs of a family but, rather, on the capacity of a power machine to till the earth. Man's relation to the soil was profoundly changed. Formerly man had been part of nature; now he was the exploiter of nature. Nowhere else in the world did farmers develop any analogous agricultural implement. Is it coincidence that modern technology, with its ruthlessness toward nature, has so largely been produced by descendants of these peasants of northern Europe?

This same exploitive attitude appears slightly before A.D. 830 in Western illustrated calendars. In older calendars the months were shown as passive personifications. The new Frankish calendars, which set the style for the Middle Ages, are very different: they show men coercing the world around them—plowing, harvesting, chopping trees, butchering pigs. Man and nature are two things, and man is master.

These novelties seem to be in harmony with larger intellectual patterns. What people do about their ecology depends on what they think about themselves in relation to things around them. Human ecology is deeply conditioned by beliefs about our nature and destiny—that is, by religion. To Western eyes this is very evident in, say, India or Ceylon. It is equally true of ourselves and of our medieval ancestors.

The victory of Christianity over paganism was the greatest psychic revolution in the history of our culture. It has become fashionable today to say that, for better or worse, we live in "the post-Christian age." Certainly the forms of our thinking and language have largely ceased to be Christian, but to my eye the substance often remains amazingly akin to that of the past. Our daily habits of action, for example, are dominated by an implicit faith in perpetual progress which was unknown either to Greco-Roman antiquity or to the Orient. It is rooted in, and is indefensible apart from, Judeo-Christian teleology. The fact that Communists share it merely helps to show what can be demonstrated on many other grounds: that Marxism, like Islam, is a Judeo-Christian heresy. We continue today to live, as we have lived for about 1700 years, very largely in a context of Christian axioms.

What did Christianity tell people about their relations with the environment?

While many of the world's mythologies provide stories of creation, Greco-Roman mythology was singularly incoherent in this respect. Like Aristotle, the intellectuals of the ancient West denied that the visible world had had a beginning. Indeed, the idea of a beginning was impossible in the framework of their cyclical notion of time. In sharp contrast, Christianity inherited from Judaism not only a concept of time as nonrepetitive and linear but also a striking story of creation. By gradual stages a loving and all-powerful God had created light and darkness, the heavenly bodies, the earth and all its plants, animals, birds, and fishes. Finally, God had created Adam and, as an afterthought, Eve to keep man from being lonely. Man named all the animals, thus establishing his dominance over them. God planned all of this explicitly for man's benefit and rule: no item in the physical creation had any purpose save to serve man's purposes. And, although man's body is made of clay, he is not simply part of nature: he is made in God's image.

Especially in its Western form, Christianity is the most anthropocentric religion the world has seen. As early as the 2nd century both Tertullian and Saint Irenaeus of Lyons were insisting that when God shaped Adam he was foreshadowing the image of the incarnate Christ, the Second Adam. Man shares, in great measure, God's transcendence of nature. Christianity, in absolute contrast to ancient paganism and Asia's religions (except, perhaps, Zoroastrianism), not only established a dualism of man and nature but also insisted that it is God's will that man exploit nature for his proper ends.

At the level of the common people this worked out in an interesting way. In Antiquity every tree, every spring, every stream, every hill had its own *genius loci*, its guardian spirit. These spirits were accessible to men, but were very unlike men; centaurs, fauns, and mermaids show their ambivalence. Before one cut a tree, mined a mountain, or dammed a brook, it was important to placate the spirit in charge of that particular situation, and to keep it placated. By destroying pagan animism, Christianity made it possible to exploit nature in a mood of indifference to the feelings of natural objects.

It is often said that for animism the Church substituted the cult of saints. True; but the cult of saints is functionally quite different from animism. The saint is not *in* natural objects; he may have special shrines, but his citizenship is in heaven. Moreover, a saint is entirely a man; he can be approached in human terms. In addition to saints, Christianity of course also had angels and demons inherited from Judaism and perhaps, at one remove, from Zoroastrianism. But these were all as mobile as the saints themselves. The spirits *in* natural objects, which formerly had protected nature from man, evaporated. Man's effective monopoly on spirit in this world was confirmed, and the old inhibitions to the exploitation of nature crumbled.

When one speaks in such sweeping terms, a note of caution is in order.

Christianity is a complex faith, and its consequences differ in differing contexts. What I have said may well apply to the medieval West, where in fact technology made spectacular advances. But the Greek East, a highly civilized realm of equal Christian devotion, seems to have produced no marked technological innovation after the late 7th century, when Greek fire was invented. The key to the contrast may perhaps be found in a difference in the tonality of piety and thought which students of comparative theology find between the Greek and the Latin Churches. The Greeks believed that sin was intellectual blindness, and that salvation was found in illumination, orthodoxy—that is, clear thinking. The Latins, on the other hand, felt that sin was moral evil, and that salvation was to be found in right conduct. Eastern theology has been intellectualist. Western theology has been voluntarist. The Greek saint contemplates; the Western saint acts. The implications of Christianity for the conquest of nature would emerge more easily in the Western atmosphere.

The Christian dogma of creation, which is found in the first clause of all the Creeds, has another meaning for our comprehension of today's ecologic crisis. By revelation, God had given man the Bible, the Book of Scripture. But since God had made nature, nature also must reveal the divine mentality. The religious study of nature for the better understanding of God was known as natural theology. In the early Church, and always in the Greek East, nature was conceived primarily as a symbolic system through which God speaks to men: the ant is a sermon to sluggards; rising flames are the symbol of the soul's aspiration. This view of nature was essentially artistic rather than scientific. While Byzantium preserved and copied great numbers of ancient Greek scientific texts, science as we conceive it could scarcely flourish in such an ambience.

However, in the Latin West by the early 13th century natural theology was following a very different bent. It was ceasing to be the decoding of the physical symbols of God's communication with man and was becoming the effort to understand God's mind by discovering how his creation operates. The rainbow was no longer simply a symbol of hope first sent to Noah after the Deluge: Robert Grosseteste, Friar Roger Bacon, and Theodoric of Frei-

berg produced startlingly sophisticated work on the optics of the rainbow, but they did it as a venture in religious understanding. From the 13th century onward, up to and including Leibnitz and Newton, every major scientist, in effect, explained his motivations in religious terms. Indeed, if Galileo had not been so expert an amateur theologian he would have got into far less trouble: the professionals resented his intrusion. And Newton seems to have regarded himself more as a theologian than as a scientist. It was not until the late 18th century that the hypothesis of God became unnecessary to many scientists.

It is often hard for the historian to judge, when men explain why they are doing what they want to do, whether they are offering real reasons or merely culturally acceptable reasons. The consistency with which scientists during the long formative centuries of Western science said that the task and the reward of the scientist was "to think God's thoughts after him" leads one to believe that this was their real motivation. If so, then modern Western science was cast in a matrix of Christian theology. The dynamism of religious devotion, shaped by the Judeo-Christian dogma of creation, gave it impetus.

An Alternative Christian View

We would seem to be headed toward conclusions unpalatable to many Christians. Since both *science* and *technology* are blessed words in our contemporary vocabulary, some may be happy at the notions, first, that, viewed historically, modern science is an extrapolation of natural theology and, second, that modern technology is at least partly to be explained as an Occidental, voluntarist realization of the Christian dogma of man's transcendence of, and rightful mastery over, nature. But, as we now recognize, somewhat over a century ago science and technology—hitherto quite separate activities—joined to give mankind powers which, to judge by many of the ecologic effects, are out of control. If so, Christianity bears a huge burden of guilt.

I personally doubt that disastrous ecologic backlash can be avoided simply by applying to our problems more science and more technology. Our science and technology have grown out of Christian attitudes toward man's

relation to nature which are almost universally held not only by Christians and neo-Christians but also by those who fondly regard themselves as post-Christians. Despite Copernicus, all the cosmos rotates around our little globe. Despite Darwin, we are *not*, in our hearts, part of the natural process. We are superior to nature, contemptuous of it, willing to use it for our slightest whim. The newly elected Governor of California, like myself a churchman but less troubled than I, spoke for the Christian tradition when he said (as is alleged), "when you've seen one redwood tree, you've seen them all." To a Christian a tree can be no more than a physical fact. The whole concept of the sacred grove is alien to Christianity and to the ethos of the West. For nearly 2 millennia Christian missionaries have been chopping down sacred groves, which are idolatrous because they assume spirit in nature.

What we do about ecology depends on our ideas of the man-nature relationship. More science and more technology are not going to get us out of the present ecologic crisis until we find a new religion, or rethink our old one. The beatniks, who are the basic revolutionaries of our time, show a sound instinct in their affinity for Zen Buddhism, which conceives of the man-nature relationship as very nearly the mirror image of the Christian view. Zen, however, is as deeply conditioned by Asian history as Christianity is by the experience of the West, and I am dubious of its viability among us.

Possibly we should ponder the greatest radical in Christian history since Christ: Saint Francis of Assisi. The prime miracle of Saint Francis is the fact that he did not end at the stake, as many of his left-wing followers did. He was so clearly heretical that a General of the Franciscan Order, Saint Bonaventura, a great and perceptive Christian, tried to suppress the early accounts of Franciscanism. The key to an understanding of Francis is his belief in the virtue of humility—not merely for the individual but for man as a species. Francis tried to depose man from his monarchy over creation and set up a democracy of all God's creatures. With him the ant is no longer simply a homily for the lazy, flames a sign of the thrust of the soul toward union with God; now they are Brother Ant and Sister Fire, praising the Creator in their own ways as Brother Man does in his.

Later commentators have said that Francis preached to the birds as a rebuke to men who would not listen. The records do not read so: he urged the little birds to praise God, and in spiritual ecstasy they flapped their wings and chirped rejoicing. Legends of saints, especially the Irish saints, had long told of their dealings with animals but always, I believe, to show their human dominance over creatures. With Francis it is different. The land around Gubbio in the Apennines was being ravaged by a fierce wolf. Saint Francis, says the legend, talked to the wolf and persuaded him of the error of his ways. The wolf repented, died in the odor of sanctity, and was buried in consecrated ground.

What Sir Steven Ruciman calls "the Franciscan doctrine of the animal soul" was quickly stamped out. Quite possibly it was in part inspired, consciously or unconsciously, by the belief in reincarnation held by the Cathar heretics who at that time teemed in Italy and southern France, and who presumably had got it originally from India. It is significant that at just the same moment, about 1200, traces of metempsychosis are found also in western Judaism, in the Provençal *Cabbala*. But Francis held neither to transmigration of souls nor to pantheism. His view of nature and of man rested on a unique sort of pan-psychism of all things animate and inanimate, designed for the glorification of their transcendent Creator, who, in the ultimate gesture of cosmic humility, assumed flesh, lay helpless in a manger, and hung dying on a scaffold.

I am not suggesting that many contemporary Americans who are concerned about our ecologic crisis will be either able or willing to counsel with wolves or exhort birds. However, the present increasing disruption of the global environment is the product of a dynamic technology and science which were originating in the Western medieval world against which Saint Francis was rebelling in so original a way. Their growth cannot be understood historically apart from distinctive attitudes toward nature which are deeply grounded in Christian dogma. The fact that most people do not think of these attitudes as Christian is irrelevant. No new set of basic values has been accepted in our society to displace those of Christianity. Hence we shall continue to have a worsening ecologic crisis until we reject the Christian axiom that nature has no reason for existence save to serve man.

The greatest spiritual revolutionary in Western history, Saint Francis, proposed what he thought was an alternative Christian view of nature and man's relation to it: he tried to substitute the idea of the equality of all creatures, including man, for the idea of man's limitless rule of creation. He failed. Both our present science and our present technology are so tinctured with orthodox Christian arrogance toward nature that no solution for our ecologic crisis can be expected from them alone. Since the roots of our trouble are so largely religious, the remedy must also be essentially religious, whether we call it that or not. We must rethink and refeel our nature and destiny. The profoundly religious, but heretical, sense of the primitive Franciscans for the spiritual autonomy of all parts of nature may point a direction. I propose Francis as a patron saint for ecologists.

6. Women and the Vanishing Waters

The disappearing source

The drying up of India, like that of Africa, is a man-made rather than a natural disaster. The issue of water, and water scarcity, has been the most dominant one in the '80s as far as struggles. for survival in the subcontinent are concerned. The manufacture of drought and desertification is an outcome of reductionist knowledge and modes of development which violate cycles of life in rivers, in the soil, in mountains. Rivers are drying up because their catchments have been mined, deforested or over-cultivated to generate revenue and profits. Groundwater is drying up because it has been over-exploited to feed cash crops. Village after village is being robbed of its lifeline, its sources of drinking water, and the number of villages facing water famine is in direct proportion to the number of 'schemes' implemented by government agencies to 'develop' water. Since women are the water providers, disappearing water sources have meant new burdens and new drudgery for them. Each river and spring and well drying up means longer walks for women for collecting water, and implies more work and less survival options. In Uttar Pradesh, Rajasthan, Gujarat, Madhya Pradesh, Maharashtra, Karnataka, Andhra Pradesh and Tamil Nadu, most villages are facing new water scarcities created by maldevelopment and a reductionist science.

In Uttar Pradesh, as many as 43 out of 57 districts were reeling under an acute drinking water famine in 1983. The crisis is clearly man-made: in the '60s, the number of villages with drinking water problems was 17,000 — in 1972 it went up to 35,000. New schemes were implemented to bring water to 34,144 of the villages, which should have left only 856 villages with water problems. But 1985 saw 25,000 new villages facing acute scarcity; the schemes failed because water sources had dried up.[1]

The worst hit regions of U.P. are Banda, Hamirpur, Jhansi, Allahabad, Mirzapur, Varanasi, Ballia, Jaunpur and the hill districts. Sources of potable water are drying up everywhere, and because of this, handpumps and piped water supply schemes are becoming useless. In Banda, trains have been used to provide water; in Hamirpur, bullock-carts are being used, while women now have to walk for 15-20 miles to fetch water.[2]

In the hill districts of U.P. 2,300 out of 2,700 projects which were implemented for drinking water supply have failed because the sources have dried up.[3] How this translates into a burden for women is evident from the fact that no woman is willing to marry a man from Dharchula because of the water scarcity in Dharchula district.[4] The Chipko message that forests produce water, is becoming a truism as continuing deforestation is leading to increased scarcity of water in the hills. Madhya Pradesh, the forested heartland of India, was famous for water at every step. It lost 18 lakh ha. of forests from 1975 to 1982. Whenever afforestation has been undertaken it has made the situation worse because species like eucalyptus further deplete water resources. Today Madhya Pradesh is trapped in an irreversible depletion of water resources: most of its rivers, ponds, wells and springs have dried up. In 1985, an official memorandum to the Central government stated that all 45 districts were in the grip of an unprecedented crisis: 'If adequate steps to provide drinking water facilities are not taken immediately, it can be said without exaggeration that a large population will have no water to drink at all.' In towns, the water scarcity is leading to violence. In May 1985 hundreds of people, including policemen, were hurt in clashes over water in Jabalpur. Sagar is without water, because the Debus river which provides it with drinking water dried up for the first time in 1985. Water was sold for Rs. 10 a drum, and people are keeping their drinking water, supplied under police protection, under lock and key. As the Superintendent of Police stated, 'We had to post policemen with each water tanker and lorry because of frequent cases of quarrels,

[1] 'Water Crisis Hits Most U.P. Areas', *Hindustan Times*, June 13, 1983.
[2] 'Acute Water Crisis Grips Uttar Pradesh', *Indian Express*, May 19, 1984.
[3] 'Serious Water Crisis in U.P. Hill Districts', *Indian Express*, June 15, 1984.
[4] 'No Water, No Wife, *Indian Express*, July 6, 1984.

assaults on drivers and attempts to snatch water.'[5] The Malwa region, once known for its abundance of water, is today dry, both above and below ground. While, earlier, water was normally struck at 80 ft., it is now difficult to find even after drilling 300 ft. below ground. As a result of the over-exploitation of groundwater, the number of villages whose water sources have dried up increases every year. Even those villages where the problem is supposed to have been tackled by new schemes are experiencing a recurrence of drinking water shortages. In 1980, out of 70,000 villages of the region, 36,420 reported water shortage; in 1982, this number rose to 50,000 and in 1985 it was 64,565. In other words, nearly all the villages of the state suffer from a water crisis. Commercial exploitation of forests, over-exploitation of ground water for commercial agriculture and inappropriate afforestation are the major reasons identified for the water crisis.[6] Meteorologically, neither Madhya Pradesh nor neighbouring Orissa are arid zones. Desertification and dessication in these regions has been manufactured by maldevelopment. Kalahandi in Orissa is a glaring example: 30 years ago it was an unending stretch of lush green forests, rich in teak and sal which provided a livelihood to the tribal population. Today, 830 out of its 2,842 villages are desertified. One hundred and ninety villages have been deserted, with some people migrating to cities, others into forests where edible roots and fruits help them survive. Nowapura subdivision which was, until recently, densely forested is today a stretch of parched land. A systematic exploitation of its forest resources has left the region barren and dry. Each year Kalahandi faces more acute water scarcity which in turn leads to scarcity of food, employment and means of livelihood. The Adivasis, Harijans and other poor people who were supported by forest resources have started fleeing their parched houses. According to one estimate 40,000 people have left the district over the last few years to escape starvation. Those who stay behind are largely women and children, and they are the worst victims of scarcity conditions. In the summer of 1985, four children and two women died of starvation in Kamna block. Panasi Punji, a 35 year old shepherd woman in Amrapali village in Kalahandi, is an example of how women are special victims of desertification. Panasi's husband left her to search for work; initially, she supported her children and her 14 year old sister-in-law, Vanita, by working on people's farms. With increasing water scarcity, agricultural employment also came to an end. Finally, Panasi survived for a little longer by selling Vanita to a rich farmer who paid her Rs. 50.[7]

Gujarat's biggest problem today is drinking water. For the first time in the history of the state, the shortage of drinking water has assumed alarming proportions because most of the wells, ponds and dams have gone dry. The number of villages declared as 'no source' villages has been increasing with each passing year, inspite of an expenditure of Rs. 400 crores on drinking water supply schemes. At the end of the Fifth Plan, 3,844 villages had drinking water problems. But surprisingly, in the first year of the Seventh Plan, the number of no-source villages shot up to more than 6,000. In 1985, the figure went up to 8,000 and in 1986, 12,250 villages out of a total of 18,000 were without water. In 1985-86, potable water was being supplied to Gujarat by special trains, tankers, camels and bullock-carts. The government's crash programme in 1985-86 to provide drinking water, estimated to have cost nearly Rs. 86 crore, has left the problem as acute as ever. New sources have dried up, and the 4,000-tubewells dug have run dry. The government is now ready to spend another Rs. 93 crores on long distance transfer and on more tubewells. Gujarat also has a World Bank aided water supply project of Rs. 136 crores, but both technology inputs and financial inputs are failing in providing water in the face of the depletion of water sources themselves.[8]

The cause of the water crisis and the failure of solutions both arise from reductionist science and maldevelopment working against the logic of the water cycle, and hence violating the integrity of water flows which allows rivers, streams and wells to regenerate themselves. The arrogance of these anti-nature and anti-women development programmes lies in their belief that they

---

[5] 'Sagar Crying Out For Water,' *Indian Express,* June 16, 1986 and 'Drought in M.P. Leaves Trail of Misery', *Indian Express,* June 19, 1985.

[6] 'Alarming Fall in M.P. Water Resources', *Indian Express,* June 23, 1985.

[7] 'A Drought-hit People,' *Times of India,* July 26, 1986; 'Stage-show and Survival Struggle,' *Indian Express,* June 26, 1985; 'Severe Scarcity Conditions in Orissa', *Times of India,* July 3, 1986; 'Plight of Women', *Indian Express,* July 28, 1986; 'Stir in Orissa over Water Shortage', *Indian Express,.* April 21, 1985.

[8] 'Gujarat in for Acute Water Famine', *Times of India,* December 20, 1986; 'Solutions that Hold no Water', *Times of India,* December 8, 1986.

create water and have the power to 'augment' it. They fail to recognise that humans, like all living things, are participants in the water cycle and can survive sustainably only through that participation. Working against it, assuming one is controlling and augmenting water while over-exploiting or disrupting it, amounts at one point to a breakdown of the cycle of life. That is why in water management, it is imperative to think and act ecologically, to 'think like a river' and to flow with the nature of water.[9] All attempts that have violated the logic of the water's natural flow in renewing itself have ended up worsening the problem of water scarcity. Water circulates from seas to clouds, to land and rivers, to lakes and to underground streams, and ultimately returns to the oceans, generating life wherever it goes. It is a renewable resource by virtue of this endless cyclic flow between sea, air and land. Despite what engineers like to think, water cannot be 'augmented' or 'built'. It can be diverted and redistributed and it can be wasted, but the availability of water on earth is united and limited by the water cycle. Since it is volatile, and since most of its flow is invisible, in and below the soil, it is rarely seen as being the element that places the strictest limits on sustainable use. Used within these limits, water can be available forever in all its forms and abundance; stretched beyond these limits, it disappears and dries up. Over-exploitation for a few decades or even a few years can destroy sources that have supported life over centuries. Violence to the water cycle is probably the worst but most invisible form of violence because it simultaneously threatens the survival of all.

Dominant approaches to water utilisation and management are reductionist and fail to perceive the cyclical nature of water flows. They linearise and commoditise thinking about water as a resource and create an illusion of producing abundance while manufacturing scarcity. The submersion of catchments and the diversion of surface water by large dams; the depletion of groundwater caused by diverting river flows as well as by over-exploitation made possible by energised pumping and tubewells; and the overuse of water by surface cultivation of water intensive crops and trees are some major causes for the drying up of water systems. Yet the crisis mind proposes an extension of the disease as the cure —its solution to

desertification is more dams, more tubewells, more water intensive cultivation on the one hand, and more technology intensive solutions to the drinking water crisis on the other. Nature's natural flow is further violated, destroying the feminine principle and sustaining power of water, and destroying women's knowledge and productivity in providing sustenance.

Dams as violence to the river

India is a riparian civilisation. The temples of ancient India have often been temples dedicated to rivers and their sources, and one of the best descriptions of the vital ecoprocesses of the water cycle is the story of the mighty river Ganga, roaring down the Himalayan slopes with no one to hold the Earth together in the face of her might. Brahma, the creator of the universe according to Indian mythology, was deeply concerned about the ecological problem of the descent of Ganga from the heavens to the Earth. He said,

> *Ganga, whose waves in swarga flow*
> *Is daughter of the Lord of Snow*
> *Win Shiva, that his aid be lent*
> *To hold her in her mid-descent*
> *For earth alone will never bear*
> *These torrents travelled from the upper air.*[10]

The above metaphor is a description of the hydrological problem associated with the descent of mighty rivers like the Ganga, which are fed by seasonal and powerful monsoonic rains. Reiger, the eminent Himalayan ecologist, describes the material rationality of the myth in the following words:

> In the scriptures a realisation is there that if all the waters which descend upon the mountain were to beat down upon the naked earth, then earth would never bear the torrents. . . In Shiva's hair we have a very well known physical device which breaks the force of the water coming down . . . the vegetation of the mountains.[11]

A Chipko song by Ghanshyam 'Shailani', inspired by a Garhwali

---

[9] D. Worster, 'Thinking Like a River', in W. Jackson, *et al.* (eds.), *Meeting the Expectations of the Land*, San Francisco: Northpoint Press, 1984, p 57.

[10] H.C. Reiger, 'Whose Himalaya? A Study in Geopiety', in T. Singh, (ed.), *Studies in Himalayan Ecology and Development Strategies*, New Delhi: The English Book Store, 1980, p.2.

[11] *Ibid.*

woman, talks of the natural broadleaved forests of oak-rhododendron on mountain tops, inviting the rain and yielding water from their roots. Rivers have thus been perceived and used in the total integration of their relationship with rainfall, mountains, forests, land and sea. Natural forests in catchments have been viewed as the best mechanism for water control and flood control in Indian thought. Catchment forests of rivers and streams have therefore always been treated as sacred.

Rapidly, however, the temples of ancient India, dedicated to the river goddesses, were substituted by dams, the temples of modern India, dedicated to capitalist farmers and industrialists, built and managed by engineers trained in patriarchal, western paradigms of water management. Water management has been transformed from the management of an integrated water cycle by those who participate in it, particularly women, into the exploitation of water with dams, reservoirs and canals by experts and technocrats in remote places, with masculinist minds. These engineering and technological feats are part of the Baconian vision of substituting sacred rivers with inert, passive water resources which can be managed and exploited by scientific man in the service of profit. The desacralisation of rivers and their sources has removed all constraints from the overuse and abuse of water. Projects of controlling the rivers, of damming and diverting them against their logic and flow to increase water availability and provide 'dependable' water supplies have proved to be self-defeating. The illusion of abundance created by dams has been created by ignoring the abundance provided by nature. The role of the river in recharging water sources throughout its course, and in its distributive role in taking water from high-rainfall catchments through diverse ecosystems has been ignored. When dams are built by submerging large areas of forested catchments, and river waters are diverted from the river course into canals, four types of violence are perpetrated on the river's water cycle:

1. Deforestation in the catchment reduces rainfall and hence reduces river discharges and turns perennial flows into seasonal flows.
2. Diversion of water from its natural course and natural irrigation zones to engineered 'command' areas leads to problems of water-logging and salinity.

3. Diversion of water from its natural course prevents the river from recharging groundwater sources downstream.
4. Reduced inflows of fresh water into the sea disturb the fresh water-sea water balance and lead to salinity ingress and sea erosion.

Violence is not intrinsic to the use of river waters for human needs. It is a particular characteristic of gigantic river valley projects which work *against*, and not *with*, the logic of the river. These projects are based on reductionist assumptions which relate water use not to nature's processes but to the processes of revenue and profit generation.

Impounding rivers and streams for irrigation is not in itself an example of modern western technology. The ancient anicuts on the Kaveri and Krishna rivers in South India are examples of how riparian societies in India used river water to increase benefits to man without violence to the river. In the indigenous system, water storage and distribution were based on nature's logic, and worked in harmony with nature's cycles. Among these non-violent irrigation systems was the major tank system of Mysore. Major Sankey, one of the first British engineers who came to Mysore observed that 'to such an extent has the principle of storage been followed that it would require some ingenuity to discover a site within this great area suitable for a new tank. While restorations are of course feasible, any absolutely new work of this description would, within this area, almost certainly be found to cut off the supply of another, lower down the same basin.'[12] These tank systems constructed over centuries also endured over centuries. Their management was based on local participation with women and men desilting the tank-beds and repairing the breaches during February, March and April. On Bhim Ekadashi day, villagers imitated the epic hero, Bhim, by collective desilting of field channels. Though observed as a religious festival, it had the effect of preventing water-logging.[13] Small tanks in the village were replenished by women who carried water from the river.

The sophisticated engineering sense, built on an ecological

---

[12] B.V. Krishna Murthy, *Eco-development in Southern Mysore*, New Delhi: Dept. of Environment, p.30, 1983, p. 7.

[13] K.M. Munshi, *Land Management in India*, New Delhi: Ministry of Agriculture, 1952.

sense that provided the foundation for irrigation in India, has been commented on again and again by famous British engineers who learnt water management from indigenous techniques. Major Arthur Cotton, credited as the 'founder' of modern irrigation programmes, wrote in 1874:

> There are multitudes of old native works in various parts of India. . . . These are noble works, and show both boldness and engineering talent. They have stood for hundreds of years. . . . When I first arrived in India, the contempt with which the natives justifiably spoke of us on account of this neglect of material improvements was very striking; they used to say we were a kind of civilized savages, wonderfully expert about fighting, but so inferior to their great men that we would not even keep in repair the works they had constructed, much less even imitate them in extending the system.[14]

The East India Company which took control of the Kaveri delta in 1799 was unable to check the rising river bed. Company officials struggled for a quarter century; finally, using indigenous technology, Cotton was able to solve the problem by renovating the Grand Anicut. As he wrote later:

> It was from them (the native Indians) we learnt how to secure a foundation in loose sand of unmeasured depth. . . . The Madras river irrigations executed by our engineers have been from the first the greatest financial success of any engineering works in the world, solely because we learnt from them. . . . With this lesson about foundations, we built bridges, weirs, aqueducts and every kind of hydraulic work. . . . We are thus deeply indebted to the native engineers.[15]

Throughout the country, irrigation works, big and small, protected agriculture in the dry season. Persian wheels and counterpoise lifts, rope and bucket lifts and water ladders used renewable human and animal energy and kept water use within the limits of renewability. So adequate were these diverse irrigation systems that when the agriculture policy was being formulated in independent India, the only task considered for irrigation was restoration and repair of old works.[16] With independence, the project to build a modern India got a new impetus. Dam-building took the form of an epidemic, with large structures being built for flood control, irrigation and power generation.

River valley projects are considered the usual solution to meeting the water needs of agriculture, for controlling floods or mitigating drought. More than 1,554 large dams have been built in India during the past three decades. It is estimated that about 79 mha metre of water can be used annually from the surface in Indian rivers, but less than 25 mha metres is actually utilised. The obvious answer so far has been to provide storage capacity in large reservoirs behind huge and costly dams. Between 1951 and 1980, India has spent Rs. 75,100 million on major or medium irrigation dams. Yet the return from this large investment has been far less than anticipated. In fact, where irrigated lands should yield at least five tonnes of grain per ha, in India yield has remained at 1.7 tonnes per ha. The annual losses from irrigation projects caused by unexpectedly low water availability, heavy siltation, reducing storage capacity, water-logging, etc., now amount to Rs. 4,270 million. The Kabini project in Karnataka is a good example of how water development projects can themselves become the cause of disruption of the hydrological cycle and destruction of water resources in the basin. It has a submersion area of 6,000 acres, but it entailed the clear-felling of 30,000 acres of primeval forest in the catchments to rehabilitate displaced villages. As a consequence, the local rainfall fell from 60 inches to 45 inches, and high siltation rates have already drastically reduced the life of the project. In the command area, large areas of well developed coconut gardens and paddy fields have been laid waste through water-logging and salinity within two years of irrigation from the project. The story of the Kabini project is a classic case of how the water crisis is being created by the very projects aimed at increasing water availability or stabilising water flows.[17]

The damming of two of India's most sacred rivers, the Ganga and the Narmada, have been seriously resisted by women, peasants and tribals whose sacred sites will be destroyed and whose life-support

[14] N. Sengupta, 'Irrigation: Traditional vs. Modern,' Madras: Institute of Development Studies, 1985, p. 17.

[15] Ibid., p. 18.

[16] K.M. Munshi, op.cit., p. 9.

[17] B. Prabhakar, 'Social Forestry Dissertation', Dehradun: Forest Research Institute, 1983.

systems are being disrupted. But the people of Narmada Valley, resisting dislocation and displacement from the Sardar Sarovar and Narmada Sagar dams,[18] or the people of Tehri, resisting the Tehri dam,[19] do not merely struggle to preserve their homeland. Their resistance is against the destruction of entire civilizations and ways of life in the very process of dam building which involves the large scale dislocation of peoples and river systems. As the women of Tehri state on the site where they have been protesting daily for nearly two decades, 'Tehri Dam is a symbol of total destruction.' (Tehri dam *sampurna vinash ka pratik hai*).

The reductionist mind which sees 'environment' as passive and fragmented has viewed the 'recovery' of ecological balance merely as a matter of creating plantations in the command. However the destruction of forests in the catchments cannot be restored by planting trees elsewhere because catchments are where rain falls most bountifully, and catchment forests contribute to the overall precipitation and its conservation. Research by the United Nations University has established that 75 per cent of rainfall in rainforest regions is contributed by the rainforest itself. Moist forests in the tropics create rain and conserve it for perennial discharge. Destroying the rainforest implies decreasing the available rainfall. Plantations somewhere else cannot recover these biospheric functions because they are not ecologically equivalent to the catchment forest — for one they are man-made plantations and not forests, for another, they are in the command and not in the catchments.

Most rivers in India have been used for irrigation, over centuries. Irrigation systems were created like the 'round river', taking off from the river to nurture agriculture, and going back to the river to recharge it. Modern irrigation, overpowered by the masculinist trend of the large and spectacular and by the principle of overpowering the river, has created systems that work against nature's own drainage. On the one hand, this leads to a destruction of irrigated agriculture in the river valley and turns skilled farmers into unskilled 'refugees'. The Soliga, displaced from Kabini, were originally irrigated-paddy cultivators; today they are ignorant dryland farmers. The Soliga women complain about how they are now

captives of pesticide firms and banks which come to give them new 'expertise' for cash crop cultivation. The peasants uprooted by the Srisailam dam lost irrigated land along the Krishna and are today living in abject poverty.[20] Probably they too, like the Santhal in Bengal, created songs about dams that caused their destruction:

> *Which company came to my land to open a karkhana?*
> *It awakened its name in the rivers and the ponds*
> *calling itself the DVC*?*
> *It throws earth, dug by a machine, into the river.*
> *It has cut the mountain and made a bridge.*
> *The water runs beneath.*
> *Roads are coming, they are giving us electricity,*
> *having opened the karkhana.*
> *The praja all question them.*
> *Then ask what this name belongs to.*
> *When evening falls they give paper notes as pay.*
> *Where will I keep these paper notes?*
> *They dissolve in the water.*
> *In every house there is a well which gives water*
> *for brinjal and cabbage.*
> *Every house is bounded by walls which make it look*
> *like a palace*
> *This Santhal tongue of ours has been destroyed in the district.*
> *You came and made this a bloody burning ghat,*
> *calling yourself the DVC.*[21]

Every major new dam in modern India has displaced people from fertile river valleys, both upstream and downstream of the dam, and has left fertile alluvial soils submerged or barren. This destruction of irrigation potential is never accounted for in new irrigation projects.

The new command areas created have topographies, soils, climates which were never intended to manage large water inputs. Water-logging and salinity are therefore the result. The water cycle can be destabilised by adding more water to an ecosystem than the natural drainage potential of that system. This leads to desertifica-

---

[18] 'The Narmada Project,' *Kalpavriksha*, New Delhi, 1988; Medha Patkar, 'Development or Destruction? A Case of Sardar Sarovar Project on the Narmada River,' paper, 1987.

[19] *The Tehri Dam: A Prescription for Disaster*, New Delhi: INTACH, 1987.

[20] *Lokayan Bulletin*, Reports on Displacement by Srisailam Dam.

[21] Quoted in Shiv Viswanathan, From the Annals of a Lab. State', *Lokayan Bulletin*, Vol. 3, No: 4/5, p. 39.

* Damodar Valley Corporation. (A song sung in the Purulia District of West Bengal.)

tion through water-logging and salinisation of land. Desertification of this kind is also a form of water abuse rather than water use. It is associated with large irrigation projects and water intensive cultivation patterns. About 25 per cent of the irrigated land in the U.S. suffers from salinisation and water-logging. In India 10 mha of canal-irrigated land have become waterlogged and another 25 mha are threatened with salinity. Land gets waterlogged when the water table is within 1.5 to 2.1 metres below the ground surface. The water table goes up if water is added to a basin faster than it can drain out. Certain types of soils and topography are most vulnerable to waterlogging. The rich alluvial plains of Punjab and Haryana suffer seriously from desertification induced by the introduction of excessive irrigation water to make green revolution farming possible. Heavy water-logging and salinity threaten three southern districts of Punjab, viz., Faridkot, Ferozepur and Bhatinda. In Haryana, in nearly 6,80,000 hectares of land the water table is within a depth of three metres and in another 3,00,000 hectares it is approaching this level.[22] A 10-year Rs. 800 crore phased programme aided by the world financial institutions to save the heartland of the state from the scourge of rising saline groundwater has been planned by the Haryana Minor Irrigation and Tubewell Corporation.[23] When this cost is added to the cost of supplying irrigation water, water intensive cropping patterns will not emerge as more productive than rainfed ones. Just as, in the case of desertification due to water depletion, the cause is mistakenly identified as drought, in the case of desertification due to water-logging, the cause is mistakenly identified as absence of adequate surface and subsurface drainage. The engineering solution offered by the reductionist mind is capital intensive, artificial drainage works — some including trenching machines which have to be imported. A simpler ecological solution, which recovers the productivity of soils and women as food producers, is a shift in cropping patterns, away from thirsty cash crops to water prudent staple foods so that less water is introduced into the system and the threat of water-logging is immediately removed. Intensive irrigation which requires intensive drainage works is a counterproductive strategy and results in the abuse of water resources.

Black cotton soils are extra prone to water-logging, while they are highly productive in a sustainable manner under rainfed conditions.[24] Such soils have a natural advantage for being rainfed because they have a high water-holding capacity and are very retentive of moisture. They are considered the most fertile and are suited for dry cultivation. Cotton, *jowar, bajra* and wheat grow principally with underground moisture alone. The retentive nature of the soil, especially when it has much depth in addition makes possible the dry cultivation of several crops which are ordinarily grown only under irrigation in other soils. It is because the natural productivity of black cotton soils is being destroyed through irrigation and consequent water-logging that farmers in these regions have been resisting the government's irrigation policy. The Mitti Bachao Abhiyan (Save the Soil Movement) in Tawa and the Ryot Sangha resistance in Malaprabha Ghatprabha (Karnataka) are signals of how productive rainfed land has been laid waste by irrigation. Visveshwara, the 19th century scientific genius of Mysore, had categorically ruled out irrigation schemes for black cotton soil regions while building large dams for Mysore State. Yet the reductionist mind continues to build these and large canal networks, threatening ecological stability everywhere. It is predicted that the massive irrigation project in Rajasthan, the Indira Gandhi Canal, will render more than 30 per cent of the 15 lakh hectares of command into waterlogged and saline wastelands.[25]

Taking water in large canals to arid regions to 'make the desert bloom' has been a particularly favourite masculinist project. In regions of scarce rainfall, the earth contains a large amount of unleached salts; pouring excessive water into these canals brings the salts to the surface, and also leaches them to other water sources. As the irrigation water evaporates, it leaves a whitish residue of salt behind. Finally, more water is used to flush away salt left by earlier irrigation. The cure to the problem of water scarcity as created by the crisis mind demands more water and more energy — a cure that at some point becomes even worse than the illness.

The reductionist mind-set treats the river as a linear, not a circular flow, and is indifferent to the diversity of soils and topography. Its engineering feats continue to be ecological failures because it

[22] 'The High Cost of Irrigation', *Indian Express*, Nov. 4, 1986.
[23] 'Rising Saline Groundwater in Haryana', *Economic Times*, October 13, 1984.
[24] J.S. Kanwar, *Rainwater Management*, Hyderabad: ICRISAT, 1983.
[25] 'Indira Gandhi Canal to Create More Problems', *Times of India*, January 16, 1987.

thinks against the logic of the river. This violence to the river is a sporadic act, ill-considered and destructive. As Worster points out, 'The natural river has been regarded by a succession of planners as an unruly dangerous beast that must be tamed and disciplined by modern science and its commodities.'[26] It is this mentality that is tapped in an advertisement for cement which says, 'The river is furious, but the dam will hold. The cement is Vikram.' Yet we know that dams do not always hold. The Koyna and Morvi disasters are witness to the vulnerability of the 'invulnerable' projects of modern man, whose reductionist mind, in tearing nature apart, reduces her capacity to renew herself and support life.

This engineering logic, by taking water away to where it does not belong, creates wet and salt-laden deserts. In addition, dams also divert water away from where it belongs in nature's logic, and leave entire regions with dry river beds and wells. The perennial river is not merely a surface flow, it also renews water below the ground. The diversion of rivers results in the depletion and drying up of groundwater. Nowhere have I seen this more clearly than in Maharashtra where the damming up of the Yarala river led to a drying up of the river downstream as well as the drying up of all the wells the river used to recharge. It was an old woman who quietly said to me, 'They do not see the huge water reservoir nature provides below the ground. They do not see nature's work and our work in distributing water. All they can see is the structures they build.'

The masculinist mind, by wanting to tame and control every river in ignorance of nature's ways, is in fact sowing the seeds of large scale desertification and famine. The Ethiopian famine which has killed nearly one million people and affected eight million, is not merely related to the failure of rainfall. It is more closely linked with the damming of the Awash river. Before the construction of dams, more than 1,50,000 people were supported by agriculture in the Awash Valley. The building of a series of dams on the Awash with World Bank funds to provide water to the sugarcane, cotton and banana plantations of rich Ethiopians and Dutch, Italian, Israeli and British firms dried up the lands downstream and flooded the lands upstream, uprooting more than 20,000 people. The Afar, the traditional pastoralists of Awash Valley, were pushed up the fragile slopes, which their herds turned bare in the struggle for survival. The 1972 drought killed 30 per cent of the Afar tribe.

How many other rivers must have run dry, how many millions of acres of land in other regions been turned into desert because the reductionist vision failed to see the invisible flows of water when it dammed and diverted the rivers? How many peasants must be left with parched fields because engineers and planners take their water away to produce cash crops and commodities? The links between new capital and technology intensive irrigation works and cash crop farming have already been discussed. The example of emerging femicide trends in Tamil Nadu (discussed earlier in the section on Food) shows that the devaluation of the work of the river is associated with the devaluation of the work of women, and both arise from the commoditisation of the economy which forces violence on nature and women. Rivers, instead of being seen as sources of life, become sources of cash. In Worster's words the river ends up becoming an assembly line, rolling increasingly toward the goal of unlimited production. The irrigated factory drinks the region dry. The premium on visibility and dramatic impact, and ecological blindness towards the water cycle have facilitated the commercialisation of land and water use. 'Engineers enjoy the challenge of designing irrigation schemes, particularly when they are on a large scale, and therefore speak of water "wasted" when it runs into the sea; if it runs into the sea through a good dam site or a desert they become almost uncontrollable'.[27] But the water flowing into the sea is not waste: it is a crucial link in the water cycle. With the link broken, the ecological balance of land and oceans, fresh water and sea water, also gets disrupted. Saline water starts intruding inwards, sea water starts swallowing the beaches and eroding the coast. Marine life is depleted, deprived of nutrients that rivers bring. In the lower Indus, fishing as a livelihood has come to an end because all the water in the lean period is extracted for irrigation. In the Nile Basin, the building of the Aswan High Dam has led to a disruption of fisheries, caused by the loss of 18,000 metric tons of Nile nutrients per year.[28] Rivers imprisoned in dams and wasted by giant hydraulic systems are

---

[26] D. Worster, op.cit., p. 34.

[27] Carruthers Clark, *The Economics of Irrigation*, Liverpool: English Language Book Society, p. 184.

[28] 'The Seven Deadly Sins of Egypt's Aswan High Dam,' in E. Goldsmith & N. Hildyard, *The Social and Environmental Effects of Large Dams*, Cornwall: Wadebridge Ecological Centre, 1986, Vol. II p. 181.

prevented from performing the multi-dimensional functions of maintaining the diversity of life throughout the basin. Dams create dead rivers, and dead rivers cannot support life. A song by Daya Pawar, sung by Dalit women in Maharashtra captures the anti-life force of the dammed river which irrigates commodity crops like sugarcane, while women and children thirst for drinking water.

> *As I build this dam*
> *I bury my life.*
> *The dawn breaks*
> *There is no flour in the grinding stone.*
>
> *I collect yesterday's husk for today's meal*
> *The sun rises*
> *And my spirit sinks.*
> *Hiding my baby under a basket*
> *And hiding my tears*
> *I go to build the dam*
>
> *The dam is ready*
> *It feeds their sugarcane fields*
> *Making the crop lush and juicy.*
> *But I walk miles through forests*
> *In search of a drop of drinking water*
> *I water the vegetation with drops of my sweat*
> *As dry leaves fall and fill my parched yard.*

Environmental and Social Change in the Valle del Mezquital, Mexico, 1521–1600

ELINOR G. K. MELVILLE

York University

The often disastrous consequences of the introduction of exotic animals into a New World environment are very clearly demonstrated by the sixteenth-century history of the Valle del Mezquital, Mexico. A rapid and profound process of environmental degradation, caused by overstocking and indiscriminate grazing of sheep in the post-conquest era, leads us to ask whether the Spanish always acted in their own long-term interests in the New World.[1]

Recent analyses of the growth and development of colonial societies accord a crucial role to the natural resources: It is argued that the conquerors were forced to adapt their expectations, and their institutions, to New World realities. Differences in local resources are used to explain evidence of marked regional differentiation in social and economic growth; the strengths and deficiencies of the various regions at the time of the conquest, as defined by their relation to the growing world market, are thought to have shaped future development (or underdevelopment). A basic premise of these studies is that, given the available resources and the technology of the early modern era, the Spanish developed the best possible means to exploit the riches of the Americas.[2]

A preliminary version of this paper was read at the annual meeting of the American Historical Association, December 30, 1985. I would like to acknowledge the help of all the people who helped with earlier drafts (even though I did not always take heed of their advice): Christon Archer, Robert Claxton, Bruce Drewitt, Bonnie Kettel, Barbara Kilbourn, Herman Konrad, Michael Levin, Rebecca Scott, Gavin Smith, Jacque Solway, Shelia Van Wyck, Lorna Woods, and Eric Van Young. I am very grateful to the Social Sciences and Humanities Research Council of Canada for support during research and writing.

[1] The ecological consequences of the European diaspora have been clearly and comprehensively set out in Alfred Crosby's works, see especially *The Columbian Exchange. Biological and Cultural Consequences of 1492* (Westport, 1972) and *Ecological Imperialism. The Biological Expansion of Europe, 900–1900* (London, 1986).

[2] Charles Gibson's classic study, *The Aztecs under Spanish Rule* (Stanford, 1967), which was aimed at uncovering the details of Indian adaptation to Spanish rule, also demonstrated how Spanish institutions were modified and adapted to colonial realities and is the departure point for the modern histories. As others followed Gibson's lead and carried out detailed regional studies and investigated the internal structure and external relations of the hacienda, his ideas about the entrepreneurial spirit of the landowners in the colonial era were confirmed (*Aztecs*, 326 *ff.*), but striking differences between the regions were uncovered. See Eric Van Young's review article for

0010-4175/90/2215-5384 $5.00 © 1990 Society for Comparative Study of Society and History

The idea that the settlers were sufficiently in control of events to implement desired changes in land use, let alone develop the best systems of production, is called into question by the case of the Valle del Mezquital. In this paper I will demonstrate that the system of land use which dominated regional production in the Valle del Mezquital after the conquest was not only not the best choice—it was not even planned. Moreover, the system of production which characterised the mature colony was neither based on the initial status of the natural resources, nor was it the result of conscious planning. My research indicates that extensively exploited latifundia were formed during the last two decades of the sixteenth century, when small-scale intensive pastoralism collapsed as a result of the deterioration of the range from overgrazing.[3]

By demonstrating a close relationship between environmental change and economic growth and development, this study challenges the idea that the natural resource base acted simply as a passive constraint on development and that external developments, such as the expansion of the international market, are sufficient to explain growth in the colonial economies.[4] Change in the natural resource base, not the initial resources, constrained regional development and made the Valle del Mezquital into one of the most economically depressed regions of Mexico; and local events assumed critical importance in shaping local developments.

SIXTEENTH-CENTURY CHANGES IN LAND USE

The Valle del Mezquital is the archetype of the barren, eroded regions of Mexico.[5] The high treeless hills, which separate the wide flat valleys and plains typical of the neo-volcanic plateau, are bare rock; the lower slopes are eroded down to a rock-like layer of hard pan (caliche) in which only scattered cacti and thorn scrub grow. Strip mining for lime has chewed away whole hills, and cement factories pollute the land with dust. Since the 1940s, ever-expanding

---

a discussion of the current understanding of reasons for regional differences in Mexico: "Mexican Rural History Since Chevalier: The Historiography of the Colonial Hacienda," *Latin American Research Review*, 18:3 (1983), 5–61. See also Magnus Mörner, "The Spanish American Hacienda: A Survey of Recent Research and Debate," *Hispanic American Historical Review*, 53:2 (1973), 183–216.

[3] This paper is based on research carried out for my doctoral dissertation: "The Pastoral Economy and Environmental Degradation in Highland Central Mexico, 1530–1600" (Department of Anthropology, University of Michigan, 1983).

[4] The most explicit statement of the importance of the international market for local development in Latin America was made by André Gunder Frank, *Capitalism and Underdevelopment in Latin America. Historical Studies of Chile and Brazil* (New York, 1967). Immanuel Wallerstein, *The Modern World-System*, 2 vols. (New York, 1974 and 1980), has followed with a detailed analysis of world history according to this thesis.

[5] I have followed Miguel Orthón de Mendizabal's definition (in *Obras Completas*, 6 vols. [Mexico, 1946–47], VI) of the region: *south*, the northern end of the Valley of Mexico and the high mountains separating the region from the Toluca Valley; *east*, up to but not including the Sierra de Pachuca; *north*, the southern slopes of the Sierra de Juárez; and *west*, the San Francisco and Moctezuma Rivers.

irrigation systems using the waste water from Mexico City have transformed the plains and valleys of much of the region, and permanent crops are grown here for the market in Mexico City; but the rest remains the dry and barren wasteland traditionally associated with the Valle del Mezquital.

This landscape is in striking contrast with the picture obtained from the early post-conquest documents.[6] Up to the middle of the sixteenth century, the Valle del Mezquital was a densely populated and complex agricultural mosaic of planted and fallow fields, villages, quarries, lakes, dams, woods, and native grasslands. Maize production predominated in the south and west, and the maguey and nopal cactus in the drier northeast.[7] Hills and mountains were forested; willows lined the banks of rivers and streams; and few villages were without groves of trees such as cedar, native cherries, and mesquite.[8] Limestone was extracted from quarries in the hills of the southeastern quarter, which had ample woodlands to supply the lime kilns.[9] Springs, rivers, streams, dams, and *jagueyes* (depressions which catch runoff) supplied extensive irrigation systems present in most of the region; and humid bottom lands and swamps provided rich moist soils, even in areas lacking surface water for irrigation.[10]

The vegetation, both wild and domesticated, was far more varied than at the present and desert species more limited in extent. As one moved from south to north across the region, the species comprising the savannahs, woodlands, and edge habitats shifted with decreasing rainfall and increasing temperatures from grasses, oaks, pines, native cherries, willows, and the occasional mesquite (as isolated trees, not chaparral) to semidesert species of wild maguey, wild cactus, yucca, mesquite, and thorns.[11] The proportions of

[6] Documentation for this study was taken from the *Archivo General de Indias* (specific sections: Audiencia de México, Contaduría, Escribanía de cámara, Indiferente Justicia, and Patronato) in Seville, Spain, and the *Archivo General de la Nación* (specific sections: Civil, General de Parte, Historia, Indios, Libro de Congregaciones, Mercedes, and Tierras) in Mexico City [hereafter referred to as *AGI* and *AGN* respectively]. The major published source was *Papeles de Nueva Espanña*, Francisco del Paso y Troncoso, ed. (Madrid and Mexico, 1905–48), Vols. I, III and VI [hereafter cited as *PNE* I, III or VI]. Where the documentary citations are extensive, reference will be made to my doctoral dissertation.

[7] For a discussion of the production of maize in the southeast quarter of the region, see Sherburne F. Cook, *The Historical Demography and Ecology of the Teotlalpan* (Berkeley, 1949), 33–41. For evidence of the importance of the maguey and nopal in the north, see PNE I, 219, 220; and *PNE* VI, 22.

[8] Melville, ''The Pastoral Economy,'' ch. 2, n.s 25–30. *AGN*, ''Mercedes,'' supplied the bulk of the information about forests and woodlands; *AGN*, ''Tierras;'' and *PNE* I also provided a great quantity of information; and occasional references were found in *PNE* III and VI; *AGN*, ''Indios;'' and *AGI*, ''Escribanía de cámara,'' ''Indiferente,'' and ''Justicia.''

[9] Melville, ''The Pastoral Economy,'' ch. 2, n. 31.

[10] Melville, ''The Pastoral Economy,'' ch. 2, n. 24. The major source for information about the irrigation systems was *PNE* I; *AGN*, ''Tierras'' and ''Mercedes'' were next in importance; followed by *AGN*, ''Historia'' and ''Indios;'' and *AGI*, ''Justicia'' and ''México.'' References were mostly made to irrigation (*rregadio* or *irrigación*) but occasionally irrigation ditches or canals were mentioned (*acequias* or *canales*).

[11] Melville, ''The Pastoral Economy,'' ch. 2, n.s 25–30, 37. The names given in the docu-

cultigens such as maize, beans, squash, chia, chilies, cotton, and the domesticated maguey and nopal cactus also shifted in a roughly south to north direction with the availability of water for irrigation and increasing average temperatures.[12]

The Spaniards who surveyed this region twenty-seven years after the conquest (*circa* 1548) noted that excellent soils promised good wheat crops, good but limited native grasslands, the possibility of an expanded lime industry, and extensive woodlands necessary for mining ventures. There were, however, several obstacles to Spanish expansion at this date: the still dense indigenous populations (which took up space and used the available water resources to irrigate their fields); the undependable rainfall; and the frequent frosts.[13]

By the time of the conquest in 1521, this region had been exploited by agriculturalists for centuries. They had modified the landscape by clearing land and growing crops, building dams and irrigation systems, terraces, roads, and villages; and the population in the southeastern quarter was so dense that houses were practically contiguous.[14] There is evidence, however, that the human population was in reasonable ecological balance with its physical environment. For example, forests were still extensive; and the extent of the irrigation systems (only eight out of a total of thirty-seven head towns were without irrigation) not only indicates the essential aridity of the region and the need to use irrigation to secure crops,[15] but also the good health of the catchment area and the fact that sufficient water was internally generated for these systems. Finally, there is an almost total lack of evidence of environmental degradation before the last three decades of the sixteenth century.[16]

---

ments for these plants were: *pasto, zacate* (grasses); *rrobles, encinos* (oaks, live oaks); *pinos, oyamel* (pines); *capulies* (native cherries); *sauces* (willows); *ahuehuetes* (cedars); *mesquite* (mesquite); *lechuguilla* (wild maguey); *tunal, nopal* (wild cactus); *palmas sylvestres* (yuccas); *espinos* (thorns). The records of the land grants found in *AGN,* "Mercedes," were the most fruitful source for this type of information and were used together with *AGN,* "Tierras;" *PNE* I; and *AGI,* "Justicia."

[12] *PNE* I, 60, 125, 159–60, 193, 217–8, 220. *PNE* III, 69, 72. *AGN,* "Mercedes," vol. 2, fols. 48, 95–6; vol. 5, fol. 260; vol. 6, fol. 515; vol. 7, fol. 349; vol. 9, fols. 132–3; vol. 12, fol. 485. *AGN,* "Tierras," vol. 3, exp. 1; vol. 1529, exp. 1. *AGI,* "Justicia," leg. 124 #1, fol. 19.

[13] *PNE* I, 2–3, 21, 22, 57, 59, 60, 110, 125, 159–60, 166, 193–4, 207, 208, 209–10, 217–20, 223–4, 289, 292, 310.

[14] Archaeological surveys of the Tula River basin and headwaters demonstrate very dense conquest population, see A. B. Mastache de E. and Ana Maria Crespo O., "La Ocupación Prehispánica en el Area de Tula, Hgo.," in *Proyecto Tula,* Eduardo Matos Moctezuma, ed. (Mexico, 1974–76), no. 33; and W. T. Sanders, J. R. Parsons, and R. Santley, *The Basin of Mexico* (New York, 1979), 179, 213–6. See also Peter A. Gerhard, *A Guide to the Historical Geography of New Spain* (Cambridge, 1972), 295.

[15] *AGN,* "Tierras," vol. 3, exp. 1; vol. 64, exp. 1; vol. 79, exp. 6; vol. 1486, exp. 8; vol. 1487, exp. 1; *AGN,* "Mercedes," vol. 5, fol. 122; *AGI,* "Justicia," vol. 207:2, ramo 3. *AGI,* "Escribanía de cámara," leg. 161–C, fol. 250.

[16] I found evidence of only two instances of environmental degradation prior to 1560, both pertaining to the Jilotepec sub-area, *AGI,* México, leg. 96, ramo 1, and *AGN,* "Tierras," vol. 1872, exp. 10.

By the end of the sixteenth century the picture had changed. The Valle del Mezquital had been transformed into a sparsely settled mesquite desert considered to be fit only for sheep, and the decimated Indian communities were congregated in villages separated by an homogenous vista of mesquite-dominated desert scrub. Hills and mountainsides were deforested, and Spaniards and Indians fought over the trees now confined to *quebradas* (gorges) in areas which had been forested only thirty years before.[17] The heavy summer rains washed the stones off the now denuded hills onto formerly fertile slopes; sheet erosion removed the fertile topsoils and exposed the *tepetate* (hard pan); and gullies cut down through all the layers. Springs all over the region were failing, and both Indian and Spanish landowners had problems with their water supply. Instead of the intensive irrigation agriculture practised by the Indians at mid-century, regional production was now dominated by grazing on extensively exploited latifundia.

Three major processes underlay the transformation of the landscape of the Valle del Mezquital: the conversion of land use to grazing; the demographic collapse of the indigenous populations; and the ecological changes that accompanied the expansion of intensive sheep grazing. I argue that these processes also transformed the early colonial systems of production associated with intensive sheep grazing and initiated latifundia development.

Before discussing these processes, I will comment on the documents used for this study, especially those used to document the process of environmental change, because the evidence is scattered through a wide variety of sources which all pose some problems in their use. The best known and most obvious sources for sixteenth-century Mexican ecological history are the *Relaciones Geográficas*—regional descriptions that were drawn up circa 1548, 1569–70, and 1579–81 to supply the crown and the church with information about the Spanish possessions in the New World. While these documents are a rich source of information about populations and settlement patterns, the subsistance base and modes of land use of the Indian communities, as well as the natural resource base, they are clearly subject to observer bias.[18] The interests and capabilities of the local officials (clerics and representatives of the crown) obviously influenced the completeness and accuracy of these descriptions. Some officials were clearly fascinated by the history and the culture of the indigenous peoples of their regions and have left invaluable records, while others filled out the questionares with the minimum of information. While these reports were ostensibly made to record the Indians' use of the natural

[17] A more complete discussion of ecological changes is given below. References relating to deforestation and fighting over trees is to be found in: *AGN*, ''Mercedes,'' vol. 6, fol. 456; vol. 7, fol. 87, *AGN*, ''Tierras,'' vol. 2697, exp. 11; *PNE* I, 217–8. *PNE* VI, 33.

[18] For a discussion of the distorting effect of the ''preconceptions and expectations'' of settlers in descriptions of early New England, see William Cronon, *Changes in the Land. Indians, Colonists and the Ecology of New England* (New York, 1983), 20. See also ch. 1, in which Cronon discusses the types of sources available for the study of ecological changes in New England.

resources, they were actually made to assess the potential for Spanish expansion and this requirement shaped the perception as to which natural resources were worth mentioning. Carefully used, however, these sources not only supply general descriptions of the landscape at specific points in time but also—and perhaps more importantly for the purposes of this study—indicate broad changes over time.

The most fruitful sources for the details on ecological change are those surveys carried out for land grants, land suits, and *congregaciones* (concentration of Indian populations in selected centers). Unfortunately the most common type of survey, the boundary survey, is subject to statistical error because the information is limited to a strip of land around the edge of a property and can be regarded at best as a random sample. Very often notes were recorded, however, about the condition of the vegetation and soils of a hillside or a valley, or the productivity and history of use of the land; and these give us a somewhat broader picture. The surveys of village resources made in preparation for *congregaciones,* while providing a more complete idea of resources, are subject to the same type of observer bias noted in geographic relations since the Spanish had an obvious interest in the lands vacated by the Indian communities (although the takeover of these pueblo lands was prohibited).[19]

The problem of nomenclature was encountered in all sources: Generic terms were most often used to describe the vegetation (oaks, pines, thorns), and there is the possibility that a plant was assigned to an incorrect genera because it looked like an Old-World species. It proved possible to correct for this to some extent by collecting a lot of data from different sources and comparing the results with evidence of known plant associations for this and other semiarid regions of Mexico. In spite of these problems, surprisingly detailed evidence of the character of the landscape at specific points in time, as well as the processes by which the landscape changed, was obtained by using several different types of sources together.

The Spanish introduced cattle, horses, sheep, and pigs into the Valle del Mezquital probably as early as the 1520s but definitely in the 1530s. They also accelerated the production of lime in the southeastern quarter of the region for the rebuilding of Mexico–Tenochtitlan and founded silver and lead mines in Ixmiquilpan in the 1540s and in the Sierra de Pachuca in 1552. European crops and fruit trees were grown in a limited way as tribute crops for *encomenderos* (recipient of a grant of Indian tribute and labor) and the crown.

The initial expansion of pastoralism developed according to an alien perception of the natural resource base and rights to its exploitation in a process that combined legal resource exploitation, illegal land grabbing, and force. In

---

[19] See Charles Gibson, *Aztecs,* 281, for a discussion of Spanish invasion of vacated pueblo lands.

the early years of the expansion of pastoralism, livestock were introduced directly into the densely populated agricultural lands of the Indians in which they grazed on native grasslands and fallow fields, and browsed in edge habitats and crops. Spanish custom, developed to promote the needs of small pastoralists, recognised grass as a common resource wherever it grew; and crop lands were open to grazing after the harvest.[20] This meant that the expansion of grazing animals into the Indian crop lands could be viewed as an entirely legal and necessary use of an unexploited natural resource—especially given the Spaniards' consumption of animal products. After being a habitat for the wild animals that formed part of the subsistence base of the Indians, grass became a common resource of the Spanish pastoralists. Land for stations (corrals and shepherds' huts), on the other hand, was acquired illegally: The Spaniards simply took village land for the stations, often placing them next to the houses of the Indians. The practise was so extensive that when Viceroy Mendoza began to give out land grants for station sites in the 1530s, most of the early grants legalised the squatters' rights.[21]

The pastoralists maintained access to the pasture by force. Extremely violent relationships quickly developed between the herdsmen (mostly black slaves) and the villagers, and were, I believe, an important element in the process by which access to grazing was maintained in an era of common grazing in a predominantly agricultural region. For example, in 1556 the Indians of Jilotepec accused the local Spanish livestock owners of inciting their slaves to steal the women, food, and belongings of the Indians. The slaves, they said, came in large bands by order of their owners to rob the Indians, who were beaten with staffs when they tried to defend themselves. On one occasion an Indian who tried to rescue his wife from the station in which she was being held was tied to the tail of a horse and dragged until he died—naturally, said the Indians, the slave owners pretended this had not happened.[22] Undoubtedly such stories were improved in the telling to support the Indians' claims of bad treatment, but they do indicate unequal power relations; and it is clear that slaves were one of the instruments used by the Spanish to implement their rule.

At first the herds were small. Although the animals were at a relatively low density in the immediate post-conquest decades, the fact that they were maintained within lands used for agricultural production meant that they had

[20] See François Chevalier, *La formación de los latifundios en México* (Mexico, 1975), 12, 119–20; and also David E. Vassberg, *Land and Society in Golden Age Castile* (Cambridge, 1984), especially 13–18 for a discussion of stubble grazing and 19–56 for a discussion of municipal property.

[21] Simpson, Gibson, and Chevalier also noted that grants were very often issued for lands in the possession of the grantees; see Lesley B. Simpson, *Exploitation of Land in Central Mexico in the Sixteenth Century* (Berkeley, 1952), 6; Gibson, *Aztecs*, 275; Chevalier, *La Formación*, 131.

[22] *AGN*, "Mercedes," vol. 4, fols. 330–2. Herman W. Konrad has come to the same conclusion regarding the slaves' relations with the villagers and thinks that perhaps the station owners ignored orders to control their slaves' actions (verbal communication, 1985).

caused more damage than the small numbers imply.[23] As part of his efforts to relieve the destruction of the Indian crop lands by Spanish livestock, Viceroy Velasco expelled cattle from the Valle dei Mezquital and other densely populated central regions in the mid-1550s—sheep, being less of an obvious menace, were allowed to stay.[24]

During the 1550s the flocks of sheep quadrupled in size from approximately 1,000 to 3,900 head. This sudden increase in the sheep population in a region that was still densely populated by Indian agriculturalists[25] was due to a combination of factors: The animals were by this time acclimatised to the highlands; the pressure on grazing space had been lessened by the expulsion of cattle; and the domesticated grazing animals were moving into underused ecological niches, such as grasslands, edge habitats, and forests.

The expansion of the grazing animals into the nondomesticated niches helps to explain both the underdevelopment of Spanish agriculture and the incredible speed with which pastoralism expanded. The slow development of Spanish agriculture can be partly explained by fact that the Indian agriculturalists supplied the Spaniards' needs for more than fifty years after the conquest. Spaniards here, as elsewhere in New Spain, really only began to take a serious interest in agriculture after the severe 1576–81 epidemic of unknown origin (the Great *Cocoliztle*) when Indian production declined abruptly. The slow development of Spanish agriculture does not in itself explain the rapid expansion of the grazing animals since, as noted above, the region was still densely populated by Indian agriculturalists when the sheep population began to expand in the 1550s. It also does not explain why the region did not develop into an agricultural region later in the century, when the early descriptions reported excellent natural resources for grain production; irrigation systems were already in place; and there were growing markets for agricultural products in the nearby mines, Mexico City, and the mule trains going north to Zacatecas. I suggest that the explanation lies in the contrasting ways by which the Old World agricultural species and grazing animals were introduced into this region (and indeed all of New Spain).

European agricultural species were introduced by way of the *encomienda*-village system of production. Following the conquest, the *conquistadores* received grants of Indian tribute and labor (the *encomienda*) as booty. The encomenderos could specify the type of tribute they required and use their tributaries for a wide range of entrepreneurial activities, but they had no legal

[23] See Melville, ''The Pastoral Economy,'' ch. 2, n.s 35 and 40, for references to problems with animals in the Valle del Mezquital. The problems caused in Indian lands by grazing animals is well documented, see for example Gibson, *Aztecs,* 280, and Simpson, *Exploitation,* 4–6.

[24] Chevalier, *La Formación,* 133–5. Evidently the order to remove cattle was enforced. For example, Rodrigo de Castañeda was compelled to remove his cattle from Jilotepec in 1557, see *AGI,* ''México,'' leg. 1841, fols. 1r–8r.

[25] An investigation carried out in 1564 of the towns to be granted in encomienda to a son of Moctzuma shows that there was still little room between the towns in the Tula sub-area at this date. See *AGI,* ''Justicia,'' leg. 207, #2, ramo 3.

rights to the land within their encomiendas. They thus had direct control over *what* was produced but were ultimately dependent on Indian land and labor: The encomienda was embedded in the village systems of production.

Pastoralism, by contrast, was not dependent on Indian lands and labor. As noted above, grass was a common resource wherever it grew, making the ownership of land not strictly necessary; and, at least in the early years, black slaves were used as herdsmen, thus bypassing the need for Indian labor. Grazing developed outside the constraints of the encomienda–village system of production as a parallel system of resource exploitation. Thus the grazing animals and their owners had all the advantages of untrammeled exploitation of new ecological and social niches; and this, combined with the use or threat of force, meant that the grazing animals were able to expand extremely rapidly in spite of the density of the Indian agricultural populations.

By the time Spaniards began to take an interest in agriculture in the 1580s, sheep grazing not only dominated regional production but had so degraded the environment of large areas of the Valle del Mezquital that the land was fit only for sheep. Opportunities were lost in the early post-conquest period, and the degradation of the fragile ecology of this semiarid region prevented the exploitation of the growing markets for agricultural products later in the century.

During the 1560s and 1570s, the flocks continued their spectacular increase. Although the legal stocking rate (number of head per holding) was 2,000 head of *ganado menor* for a station of 7.8 square kilometers, 10,000 to 15,000 head was considered normal, even desirable during these decades.[26] By the 1570s regional production had shifted from intensive irrigation agriculture and was dominated by intensive sheep grazing.

It became clear in the course of the analysis of documentation of land use and land tenure that the domination of regional production by intensive grazing involved the displacement of the Indian agriculturalists. Once the sheep had reached a critical mass in their population increase, they simply took off and increased so rapidly and to such densities that they converted extensive areas of land to pastoralism by use—before they were transferred to the Spanish system of land tenure.

The Indians were displaced not only from their crop lands but also from such resource areas as forests, and traditional systems of land use were modified as the natural resource base of the Indian communities was diminished and in some cases destroyed. For example, the Indians complained bitterly that sheep trampled the stands of the shrub *tlacotl*, which was used as a wood substitute in the production of lime; ate the leaves of edible roots so that they

[26] See footnote 39. In sixteenth-century Mexico, the term *ganado menor* included sheep, goats and pigs. In the Valle del Mezquital, sheep appear to have constituted the bulk of the flocks, although by the end of the century the proportion of goats in the flocks increased.

could not be harvested; and left the soils bare.[27] Those Indian villages which had managed to retain some of the best agricultural lands often found the usefulness of these lands reduced by lack of water and heavy secondary growth.[28]

In a classic study of land exploitation in sixteenth-century Mexico, Lesley Byrd Simpson proposed that the post-conquest demographic collapse of the indigenous populations freed land for the expansion of the domestic grazing animals introduced by the Spanish. Simpson inversely correlated the human population decline with a steady increase in the animal population up to circa 1620, when the Indian population began to stabilise.[29]

In the case of the Valle del Mezquital, however, the growth of the sheep population was not related to the human population decline in a straightforward way. The human population of the Valle del Mezquital declined from an estimated tributary population at the conquest of between 452,623 and 226,311, to a documented tributary population in 1600 of 20,447.5 (that is, an overall decline of between 90.9 percent and 95.4 percent).[30] The sheep population began to increase rapidly in the 1550s, and it simply took off— after reaching a critical point sometime in the 1560s at which it outstripped the human population decline—and reached an estimated total of nearly four million head by 1579. In the late 1570s and early 1580s, there was, however, a sudden drop in flock size of up to 50 percent with a slowing up of the rate of increase; and by 1599 the estimated sheep population in the region had dropped to 72 percent of the total in the 1570s. The demographic collapse of the Indian population undoubtedly facilitated the formal takeover of land, but the sheep population increased and then decreased independently of the human population decline.

Even if the Indians of the Valle del Mezquital did not wholeheartedly

---

[27] *AGN,* "Tierras," vol. 1525, exp. 1; vol. 2697, exp. 11; *AGN,* "Mercedes," vol. 4, fols. 330–2; vol. 7, fol. 87.

[28] *AGN,* "Tierras," vol. 64, exp. 1; vol. 83, exp. 10; vol. 2717, exp. 10; vol. 2766, exp. 3.

[29] Simpson, *Exploitation,* ii.

[30] See Melville, "The Pastoral Economy," Appendix A, for sources used in estimating tributary population totals. Evidence for the tributary population was based almost entirely on tribute records from the Contaduría of the *AGI.* Ecclesiastical records were used to check doubtful totals or for the few cases in which totals were not available in the Contaduría. My totals agree fairly well with those estimated for this region by Sherburne F. Cook and Woodrow Borah, *The Indian Population of Central Mexico, 1531–1610* (Berkeley, 1960) but have the advantage of being based almost entirely on a single source.

According to contemporary witnesses the Indian population of New Spain had declined by as much as two-thirds and possibly five-sixths by the mid-1560s (Gibson, *The Aztecs,* 138). The tributary population of the Valle del Mezquital in 1570 was 76,946. If a 66 percent decline is applied to the 1570 total, we get an estimated tributary population in the Valle del Mezquital of 226,311 at the time of the conquest; while an 83 percent decline gives 452,623. By 1600 the tributary population had declined to 20,447.5. The total population decline for the Valle del Mezquital over the period 1519–1600 is therefore somewhere between 90.9 percent and 95.4 percent.

TABLE 1

"Spanish" Landowning: Sheep Stations Taken up Each Decade

Decade Ending	Grantees[a]		Squatters[b]		
	Spaniards	Others[c]	Spaniards	Others[c]	Totals
1539	5	—	29	—	34
1549	5	—	36	—	41
1559	13	—	16	4	33
1565	24	90	49	5	168
1569	6	5	7	—	18
1579	18	26.5	54	7	105.5
1589	55	15	100	14	184
1599	49.5	95	95	39	278.5
Totals:	175.5	231.5	386	69	862

[a]*Grantees* were recipients of land grants.
[b]*Squatters* were undocumented landowners.
[c]*Others* refers to non-Spanish landowners: Indians; Mestizos (two grantees in the 1580s and one in the 1590s); and one Mulatto squatter (1590s).
SOURCES: Melville "The Pastoral Economy," Appendix E, 272–95. See also n.31 of this article.

accept sheep herding as a way of life, they recognised the profitability of pastoralism and the need to retain control of the land. Indian nobles, and some communities, asked for and received 56 percent of all the grants for sheep stations, and acquired 34 percent of all holdings in the Valle del Mezquital in the sixteenth century[31] (see Table 1). However, while they retained formal control of the land within the Spanish system of land tenure, it was lost for all intents and purposes to the traditional Indian systems of land use.

By the end of the 1570s, large numbers of pastoralists who based their operations in small holdings dominated regional production and exploitation of the natural resources. Mining and lime manufacture were, in terms of overall regional production, limited in importance and extent. Even as the region came to be dominated by grazing, however, the pasture failed; the size of the flocks dropped markedly; the animals became underweight and re-

[31] Sources of land holding are listed by *cabecera* and in chronological order. The two primary sources for land use and land tenure were: *AGN*, "Tierras," which contains documentation of land suits concerning ownership, boundaries, land use and inheritance; and *AGN*, "Mercedes," the records of land grants, grazing rights and licences of various sorts. Other sources in order of importance were: *AGI*, "México," "Justicia," "Escribanía de cámara;" *AGN*, "General de parte," "Indios"; and *PNE* I.

Of the 862 sheep stations documented for the Valle del Mezquital for the sixteenth century, only 407 appeared in the records as formal land grants. I have designated the remaining 455 stations as "squatters' holdings" because I found no evidence of formal title for these stations; they were identified by reference to holdings which had either formed the boundaries of new grants, been mentioned as being located in their vicinity, or been the subject of court cases, wills, *informes, diligencias,* or *relaciones.*

produced more slowly; and the production of wool, meat, and tallow began to decline. As the carrying capacity of the range declined, the intensive pastoralism of the 1560s and 1570s was no longer possible and was replaced by extensive pastoralism by the end of of the 1580s.[32]

OVERGRAZING AND ECOLOGICAL CHANGE

The argument that the rapid expansion of intensive sheep raising caused environmental degradation in the Valle del Mezquital brings us to the issue of the relationship between grazing and ecological change.

The ecological changes documented for the Valle del Mezquital during the last half of the sixteenth century have been reported for many other semiarid regions in which domestic grazing animals have been introduced in large numbers. The sequence of events in all these regions is similar: reduction of the vegetative cover, diminution of the number of grass species, and invasion by unpalatable arid-zone species; followed by gullying, sheet erosion, flooding, and a drop in the water table. The relationship between grazing and environmental change, especially vegetative change, is not a straightforward one; and several variables have been proposed, singly or in combination, to account for the ecological changes. The more frequently discussed for all semiarid regions include: overgrazing and trampling by large numbers of introduced grazing animals; climatic change; and the suppression by Europeans of regular burns in woodlands and grasslands. There is, however, a general consensus of opinion that removal of the ground cover results in the reduction of the infiltration rate of rainfall (with a consequent drop in the water table and loss of spring flow) and increased overland flow of surface water, which results in increased flooding, erosion, and gullying.[33]

Most historical studies which investigate the relationship of domestic livestock to environmental degradation, especially the increase in density of woody species such as the mesquite, stress the increase in numbers and density of grazing animals. Nevertheless, while I agree that the reduction and weakening of the native grasses under high densities of grazing animals produced conditions favourable to the invasion by arid-zone species, I argue that the abrupt *drop* in numbers and density of grazing animals in the late 1570s and 1580s, together with changes in the fire regime, is the key to understanding the rapid increase of mesquite-dominated desert vegetation in the Valle del Mezquital during the last quarter of the sixteenth century.

[32] Carl L. Johanessen writes that the "carrying capacity of the range may be described as the number of animals it can support in health, during the period when grass is palatable and nutritious, without reducing forage production in subsequent years. . . . Overgrazing occurs when the number of stock exceeds the carrying capacity of the range" (*Savannas of Interior Honduras* [Berkeley, 1963], 106).

[33] For a discussion of the variables and the supporting arguments, see James R. Hastings and Raymond M. Turner, *The Changing Mile. An Ecological Study of Vegetation Change With Time in the Lower Mile of an Arid and Semi Arid Region* (Tucson, 1965), 3–6, 275–83.

Very briefly, the timing and sequence of changes in the density of the sheep population and in the environment of the Valle del Mezquital during the last half of the sixteenth century were as follows. Large numbers of sheep grazed the native grasses, herbs, and shrubs during the 1560s and 1570s until the ground was left bare. Arid-zone species began to invade these grazed areas in the late 1560s but were kept in check by the density of the grazing animals and the regular burns carried out by herdsmen to stimulate pasture growth. In the 1570s the grasses began to fail—probably as a result of heavy grazing and repeated burns. As the pasture failed, the flocks declined in size and quality. It is most likely that fires no longer spread for lack of fine fuel (dry grasses). At the same time, arid-zone species began to expand in extent and density. I argue that the lessening of the controlling factors (the high density of grazing animals, frequent burns, and competing grasses) allowed a rapid increase in the spread of the arid-zone species in the last quarter of the sixteenth century. The formation of a stony pavement and exposure of the impervious hard-pan layer produces a micro-environment hostile to grasses. Such soil conditions, which were reported for the Valle del Mezquital during the 1580s, would have reinforced the shift to arid-zone species.[34]

Climatic change cannot be ruled out as a complicating factor in the process of environmental change in the Valle del Mezquital, especially since the second half of the sixteenth century falls within the peak years of the Little Ice Age (1550–1700); and a shift to drier climatic conditions, such as occurred in some areas during the Little Ice Age, would favour arid-zone species over grasses. However, the meagre amount of climatic data for this period makes it difficult to see a trend in any direction, and the climate appears to have been consistently unpredictable throughout the sixteenth century. The major argument against climatic change as the primary variable in the process of environmental degradation in the Valle del Mezquital is that environmental deterioration did not occur in all parts of the region to the same degree. As I will demonstrate below, there is a clear correlation between high densities of sheep and environmental degradation in certain sub-areas, which indicates that overgrazing was the primary variable in the process of ecological change in this region during the sixteenth century.

Regular burns in grasslands and woodlands were carried out by the indigenous populations in many parts of the New World to promote optimum conditions for game animals. The suppression of these burns by Europeans allowed for the spread of a dense undergrowth of woody species; however, I have no evidence that the indigenous populations of the Valle del Mezquital used fire—either as a means to clear agricultural land or to maintain the forest floor clear of undergrowth. Burning the grasslands was evidently instituted soon after the conquest, and permission to burn-off had to be obtained from

[34] See Johanessen, *Savannas*, 78, for a similar interpretation of the relationship between vegetative changes and livestock densities.

the *guarda mayor* of the *Mesta*.[35] Fire is therefore a new element in the ecological equation, but one about which there is surprisingly little data.

In order to examine the relationship between grazing and environmental change in the Valle del Mezquital, I grouped the sub-areas[36] of the region into five types according to differences in (1) the extent of land formally converted to pastoralism each decade[37] and (2) the grazing rates.[38] This typology presents the changing levels of exploitation of the sub-areas between 1530–1600 in schematic form (see Table 2). It can be seen that exploitation varied from almost complete conversion of lands to pastoralism and high grazing rates (Type 1) to limited conversion and relatively low grazing rates (Type 5); and from intensive exploitation at an early date to extensive exploitation throughout the century.

The number of head present in each sub-area forms the basis for estimates of grazing rates and is dependent on the stocking rate (the average number of head per holding), a figure where was determined for the region as a whole. The stocking rate, in turn, depended on the state of the pasture, human population densities, markets, and changed throughout the century according to shifts in these variables. In the early decades the flocks were restricted by the extent of the Indian agricultural lands, but they increased from 1,000 to 3,900 head per station when the amount of available grazing land expanded in in the 1550s. Although the legal stocking rate for a sheep station of 7.8 square kilometers was 2,000 head, there are enough examples in the 1550s to show

[35] *AGN*, "Mercedes," vol. 3, fols. 95–96, 113. Although the regulations did not allow burning in the forests, accidents sometimes did happen, and forests were burnt; see *AGN*, "Mercedes," vol. 3, exp. 249, fols. 95–96. Jeffrey R. Parsons has not found archaeological evidence of regular burns in the Valley of Mexico for the late pre-conquest period (verbal communication, 1983).

[36] The division of the region into ten sub-areas was based on geopolitical criteria. The Valle del Mezquital consists of eight wide flat plains and valleys, an area of low rolling hills forming the headwaters of the Tula River, and the high mountain valleys of the northern end of the Sierra de las Cruces. The broad division of the region is based on these ten geographic areas, and the final boundaries have been taken to be coterminous with the land under the jurisdiction of the *cabeceras* (head towns) located within their borders.

[37] The area of land converted to pastoralism was obtained by multiplying the number of stations present in each sub-area at the end of each decade by 7.8 (the area in square kilometers of a grant for a sheep station) and was expressed as a percentage of the total sub-area in order to compare sub-areas of widely different surface extension.

Land use was specified in the land grants, and these documents have been used in this study as the major source of evidence for land use practises. Although compliance with these orders is somewhat problematic, I have found that only in the 1550s, when cattle were expelled from this region, was the order disobeyed and in this case the owner was forced to comply (see n. 24). In his study Simpson asked: "What assurance do we have that land was actually used for the purposes stipulated in the grants?" and replied that this question "may . . . be safely answered in the affirmative" (*Exploitation*, 20).

[38] The grazing rates for each sub-area were calculated by dividing the total number of head within a sub-area at the end of each decade by the square kilometers of grazing land: $G = Sn/a$, where G is the grazing rate in head per square kilometers, S is the stocking rate in head per station, n is the number of stations, and a is the surface area of the sub-area in square kilometers.

TABLE 2A

Land Converted to Pastoralism[a]

	Decade Ending							
	1539	*1549*	*1559*	*1565*	*1569*	*1579*	*1589*	*1599[b]*
TYPE 1:								
Tula	.6	2.5	9.5	44.0	45.9	61.2	72.0	93.6
Southern Plain	6.4	9.6	11.3	24.2	32.2	47.2	73.0	81.6
TYPE 2:								
Central Plain	2.5	3.8	3.8	29.7	34.9	45.5	54.6	80.3
North-South Plain	—	1.0	2.0	14.5	16.5	33.1	63.1	76.7
TYPE 3:								
Alfaxayuca	4.9	6.1	12.3	31.9	31.9	45.5	59.0	61.1
Jilotepec	9.4	18.9	22.1	28.7	30.0	34.1	43.9	68.4
Chiapa de Mota	—	7.8	10.1	25.8	25.8	26.9	33.7	63.3
TYPE 4:								
Huichiapan	—	—	1.3	11.0	11.0	21.2	45.7	66.1
TYPE 5:								
Northern Valley	—	.7	.7	1.5	2.3	5.3	16.8	18.4
Ixmiquilpan	—	1.5	3.0	7.5	7.5	7.5	9.8	9.8
REGION:	2.6	5.8	8.4	21.5	22.9	31.1	45.5	60.0

[a]Land converted to pastoralism as expressed as a percentage of the total surface area of each sub-area, by decade.
[b]Estimates for the 1590s are based on totals adjusted downwards by 30 percent to account for undocumented sales.
SOURCES: Melville "The Pastoral Economy," ch. 5, n.s 8–16; Appendix E, 272–95.

that the injunction to carry 2,000 head was taken to mean 2,000 ewes, with a total flock consisting of 3,900 ewes and lambs. Rates varied greatly in the first half of the 1560s, ranging from 4,000 to 10,000 head; and an average rate of 7,500 head has been used for the period 1560–65. During the period from the mid-1560s to the late 1570s stocking rates were very high: The highest stocking rate recorded for the Valle del Mezquital was for the 20,000 sheep corralled on one station in 1576; however, the rather conservative figure of 10,000 head has been used to calculate the grazing rates in the period 1566–79.[39]

[39] See Melville, "The Pastoral Economy," ch. 5, n.s 8–16 for references to changes in the stocking rates up to 1579. Evidence for the stocking rates was taken from documentation of court cases, wills, complaints lodged by Indians, and censuses. The information contained in these sources has obvious biases. For instance, in order to make their case Indians undoubtedly reported larger numbers of animals than in fact existed on Spanish lands. The numbers admitted to by the Spanish pastoralists depended on the case they were arguing. For example, if they needed more Indian laborers assigned to them, they inflated the numbers of stock; but if they were responding to the complaints of the Indians about overstocking, they played down the numbers.

TABLE 2B

Grazing Rates on Common Pasture

	Decade ending							
	1539	*1549*	*1559*	*1565*	*1569*	*1579*	*1589*	*1599*
TYPE 1:								
Tula	.8	3	48	423	589	785	712	440
Southern Plain	8	12	56	233	414	605	703	384
TYPE 2:								
Central Plain	3	5	19	286	448	584	525	378
North-South Plain	—	1	10	139	212	425	607	361
TYPE 3:								
Alfaxayuca	6	8	61	307	410	583	567	288
Jilotepec	12	24	111	276	385	437	423	322
Chiapa de Mota	—	10	50	248	331	346	324	298
TYPE 4:								
Huichiapan	—	—	7	107	142	273	440	310
TYPE 5:								
Northern Valley	—	.9	4	15	29	69	162	86
Ixmiquilpan	—	2	15	73	97	97	95	47
REGION:	3	7	42	207	294	399	438	288

SOURCES: Melville "The Pastoral Economy," ch. 5, n.s 8–16; Appendix E, 272–95.

The average rate of 7,500 head per station used to calculate the grazing rates of the 1580s reflects a decline in the stocking rates reported for the region as whole and is based on a 1588 census of Spanish holdings in the Alfaxayuca and Huichiapan sub-areas.[40] During the 1590s the flocks further declined in numbers and in quality. Ewes were so underweight that they failed to reproduce before four years of age, whereas they had begun bearing lambs at fourteen months to two years in the era of expansion; and the animals slaughtered for meat and tallow were smaller and weighed less. The stocking rate for the 1590s is estimated at 3,700.[41]

Although there was no official grazing rate, the legal stocking rate of 2,000 head on 7.8 square kilometers may indicate an official estimate of the carrying

---

Censuses carried out by royal officials are probably the most reliable source, although officials can always be bribed. The best way to deal with these difficulties is to collect many samples from different sources, so that the estimation of the average stocking rate is predicated on as broad a sample as possible.

[40] *AGI*, México, leg. 111, ramo 2, doc. 12.

[41] The principal source for the stocking rate of the 1590s is de Mendizábal, *Obras Completas*, 114–7.

capacity of the pastures of New Spain at conquest, of 256 per square kilometer. It is likely, however, that if estimates were made of grazing rates, they were predicated on areas far larger than the legal area of a land grant, since common grazing was the custom and land grants were viewed as operational bases in the immediate post-conquest decades. The problem of estimating the amount of land actually used for grazing complicates the calculation of grazing rates for the sixteenth century. Given the customs of *agostadero* (grazing harvest stubble) and grazing in common, together with the fact that Indians could graze flocks within village lands (which were prohibited to Spaniards after the 1550s),[42] it can be argued that all lands provided grazing at some point in the year. For this reason the simple estimate of grazing rates predicted on the total area of each sub-area was accepted as providing an adequate indication of the changing densities of animals. Even when the total surface area of a sub-area is used to estimate grazing rates, grazing rates were excessively high by twentieth-century standards.[43]

Having constructed the typology, I then compared it to the changes in the environment documented for each sub-area. The results of this comparison indicated that those areas subjected to high densities of sheep and heavy grazing between 1560–80 exhibited environmental deterioration by the end of the 1560s and environmental degradation by the 1580s. In areas subjected to lower densities of sheep and lighter grazing, the deterioration of the environment was correspondingly less and later in appearance.

TYPE 1: TULA, THE SOUTHERN PLAIN

These southern sub-areas, which had the best initial resources for both pastoralism and agriculture, were the first to be intensively grazed and the most markedly affected by grazing. As can be seen from the typology (Table 1), both of these sub-areas were under very heavy grazing after the 1560s; and by 1580 they were over-exploited. By 1600 between 81.6 percent and 93.6 percent of the total surface area of both sub-areas was formally converted to pastoralism. The increased rate in land takeover during the last two decades of the century occurred even though there was a sharp drop in the carrying capacity, as indicated by the grazing rates: from 711.9 head per square kilo-

[42] Chevalier, *La Formación,* 121–2, 131, 141 *ff.,* Gibson, *Aztecs,* 276–7, 280. Simpson, *Exploitation,* 21.

[43] Grazing rates reported for other regions far exceeded those estimated for the Valle del Mezquital in the sixteenth century. For example, Simpson reports a grazing rate of 2,857 per square kilometer in Tlaxcala in 1542 (*Exploitation,* 13). In the Bajío in 1582, 200,000 sheep, along with 100,000 cows and 10,000 horses, grazed an area of nine leagues square (1,417.5 square kilometers): Richard J. Morrisey, "Colonial Agriculture in New Spain," *Agricultural History* 31 (July, 1957), 24–29. Using sixteenth-century estimates of adequate stocking rates for sheep (2000 head/7.8 square kilometers equal 256 per square kilometer) and cattle and horses (500 head divided by 17.5 square kilometers equals 28.5 per square kilometer), it can be seen that cattle and horses needed nine times as much grazing land; if we convert cattle and horses into sheep, we get a total of 1,190,000 head and a grazing rate of 839 per square kilometer.

meter in the 1580s to 440 head per square kilometer in the 1590s in Tula; and from 703 per square kilometer to 384 per square kilometer in the Southern Plain. The accelerated exploitation of limestone after the conquest in the northern hills of the Southern Plain and in the hills separating the two sub-areas represents another significant drain on the natural resources of these two sub-areas during the sixteenth century.[44]

Environmental deterioration was apparent by the 1560s and consisted of deforestation, denudation of the soils, conversion of abandoned crop lands and hillsides to grasslands, and the subsequent invasion of these grasslands by a secondary growth of arid-zone species: wild maguey (*lechuguilla*: Agave lechuguilla), yucca (*palmas sylvestres*: Yucca species), cacti (*nopal, tunal sylvestre*: Opuntia species), thorn bushes (*espinos*: possibly ocotillo: Fouqueria species), mesquite (*mesquite*: Prosopis species), and a type of thistle or wild artichoke (*cardon*: possibly the introduced Cynara carnunculus). Environmental degradation was evident by the 1580s: Hills were eroded and gullied, and the remaining soils in many areas were thin and stony. In 1595 deterioration of the catchment value of the Southern Plain was evident in the failure of the springs that fed the north-flowing streams. By 1601 sheet erosion of slopes had exposed extensive areas of *tepetate* (hard-pan). The deep soils in the center of the valleys were still very fertile; they had not been eroded to tepetate and were not covered by stony slope-wash debris, but because they were covered with desert scrub and lacked water for irrigation they were not considered to be suitable for agriculture.[45]

A few examples taken from different parts of these sub-areas will serve to illustrate the changes in the landscape and land use in the last quarter of the sixteenth century. The lands around Hueypustla in the Southern Plain were noted for their good soils and suitability for growing wheat circa 1548, but there were increasing references to stony soils during the last two decades of the century; and by 1606 these lands were characterised by thin, ruined soils and slopes eroded down to *tepetate*.[46]

A Spaniard who asked for permission in 1590 to convert arable land on the north bank of the Tepexi River (a tributary of the Tula River) to sheep grazing said that the soils were no longer suitable for cultivation because they were stony and eroded tepetate—they were suitable only for sheep. Sheep stations had been granted in the hills surrounding this holding in the 1560s, but by 1590 they were described as being gullied and eroded tepetate.[47] It was said

[44] For Spanish lime manufacture, see *AGN*, "Mercedes," vol. 6, fols. 455–6; vol. 7, fol. 87; vol. 8, fols. 227–8; vol. 13, fols. 71, 176; vol. 14, fol. 292; vol. 16, fols. 201–2. *AGN*, "Indios," vol. 6-2, exp. 998. *AGN*, "Tierras," vol. 2697, exp. 10; exp. 11.

[45] Melville, "The Pastoral Economy," ch. 5, n.s 27–68.

[46] "Son tierras ruinas y lomas en tepetate y tierras delgadas," *AGN*, "Tierras," vol. 2812, exp.12.

[47] "Por ser tierra pedregosa tepetate barrancas no es por sembrarlas . . . esta rrodeada de cerros y barrancas y todo pedregal calichal y tepetate . . . no servir sino para traer en ellas ganado menor," *AGN*, "Tierras," vol. 2735, 2ª pte., exp. 9.

that only sheep were to be found in Sayula, a town in the Tula Basin, in 1580. By the early 1600s the banks of the Tula River were covered with mesquite, and sheep grazing predominated over agriculture—even though there was water for irrigation, and the soils had not been affected by erosion.[48] Just to the south of Sayula, a village was surveyed in 1561 as part of the encomienda assigned to Pedro de Moctezuma. At this time there was no mention of erosion, but in a report made in 1603 this village was described as being unsuitable as a center for congregation because it was situated on lands that had been eroded down to tepetate and was surrounded by mesquite. It was, however, "suitable for sheep."[49]

TYPE 2: NORTH–SOUTH PLAIN, CENTRAL VALLEY

Less land was converted to pastoralism at the height of the era of high stocking rates (1570s) and the densities of animals were therefore somewhat lower in these sub-areas than in Type 1, but by the end of the century there was almost complete conversion of land to pastoralism with 76.7–80.3 percent of the land formally converted. Although grazing rates dropped from very high in the 1580s to high in the 1590s (from 525 per square kilometer in the Central Valley and 607 per square kilometer in the North–South Plain, to 377 per square kilometer and 361 per square kilometer) they still represented a continuous pressure on the land.

The sequence of environmental change recorded for these sub-areas is similar to that reported for Type 1, although the appearance of actual degradation is somewhat later. By 1580 the forests in the foothills of the Sierra de Pachuca had been heavily cut by the Indian communities for sale to the miners in Pachuca, and deforestation was noticeable in the limestone hills of the southwestern Central Valley and the western North–South Plain. The process of destruction of the ground cover was completed by pastoralists who cleared and burned the grasslands, thus providing the soil–climate suitable for desert species.[50]

The formerly densely populated flat lands of the southern half of the North–South Plain were said to have "always been grazing lands" in 1580. The Central Valley and the northern half of the North–South Plain, areas that had been considered unsuitable for grazing for lack pasture and surface water in the 1540s, were also reported to have good grazing at this date.[51] Since the primary native vegetation in the northern end of the North–South Plain consisted of desert species, which were invading the southern half and the Central Valley, it would appear that not only had grazing replaced agriculture but also

[48] "No hay ni se halla en este pueblo mas de ovexas y desto hay buen multiplico," *PNE* VI, 181.
[49] "Que esta en un llano sobre tepetate y entre unos mesquitales no comodo por congregacion . . . comodo el sitio para ganado menor" (*AGN,* Historia, vol. 410, exp. 5).
[50] Melville, "The Pastoral Economy," ch. 5, n.s 89–92.
[51] *Ibid.,* ch. 5, n.s 69–73.

that desert species had replaced grasses as the norm for adequate fodder by 1580.

By the 1590s the North–South Plain was covered by mesquite-dominated scrub so dense that it was impossible to cut a way through it, and incipient forest regeneration in the southern half of the North–South Plain was displaced in the early 1590s by desert species. The soils underlying this heavy secondary growth were thin, stony, and infertile. Erosion to *tepetate* on the lower slopes of the eastern hills of the Central Valley was apparent by 1599, and the failure of springs and loss of stream flow was reported here by that same date. In 1607 the encomendero of Tlacotlapilco in the northern end of the North–South Plain commented that the person who took up and populated a station in this area merited a prize, because that was the only way to clear out the sheep stealers and escaped slaves who found refuge in these badlands.[52]

During the last two decades of the sixteenth century, formal transference of land from the Indian to the Spanish systems of land tenure in the four sub-areas comprising Types 1 and 2 outstripped the rate of Indian population decline. Grants were made of the leftovers (*sobras y demasias*) between earlier grants; some grants were a fraction of the legal size—800 *pasos* (paces) square (1.24 square kilometers) instead of 2,000 pasos square (7.8 square kilometers); and Indian communal lands were being granted as *mercedes* (land grants).[53] The majority of these holdings were for sheep grazing; only a very small percentage of the total surface area of these sub-areas was granted for agriculture during the sixteenth century (see Table 3).

By 1600 Indian communities were left with less than the legal square league of communal lands. The Tula sub-area, in which twelve *cabeceras* (head towns) required 17 percent (210.7 square kilometers) of the total surface area, only .8 percent (9.7 square kilometers) remained unconverted to the Spanish system (see Table 3). In the Southern Plain six cabeceras theoretically had title to community lands comprising at least 105.3 square kilometers or 22 percent of the total surface area, but only 7.6 percent of the surface area (36.7 square kilometers) remained unconverted. In the North–South Plain 17.2 percent (129.5 square kilometers) remained unconverted by 1600, but the nine cabeceras in this sub-area needed 21 percent (158 square kilometers) of the surface area for communal lands. In the Central Valley 18.4 percent (110.9 square kilometers) remained—barely sufficient for the six cabeceras that should have had 104.5 square kilometers, or 17.4 percent of the surface area.[54] It must be remembered, however, that Indian *caciques*

[52] *Ibid.*, ch. 5, n.s 74–88, 93–102.

[53] *Ibid.*, ch. 5, n.s 24–26.

[54] One quote from an application for a *merced* in the Tula sub-area in 1594 makes the position very clear: "Y declara y declaro no aver lugar de dar sitio de estancia en ella a otra persona y asi lo mande poner por auto" (*AGN*, "Mercedes," vol. 18, fol. 156).

TABLE 3

Land Converted to the Spanish System of Land Tenure[a]

Sub-Area	Percent of Surface Converted to Pastoralism	Percent of Surface Converted to Agriculture	Percent of Surface Converted to Spanish Tenure
Tula	93.6	5.6	99.2
Southern Plain	81.6	10.8	92.4
Central Valley	80.3	1.3	81.6
North-South Plain	76.7	6.1	82.8
Alfaxayuca	61.1	1.3	62.4
Jilotepec	68.4	3.1	71.5
Chiapa de Mota	63.3	5.8	69.1
Huichiapan	66.1	7.3	73.4
Northern Valley	18.4	.5	18.9
Ixmiquilpan	9.8	.4	10.2

[a]Land converted to the Spanish system of land tenure, as expressed in a percentage of the surface area of each sub-area.
SOURCES: Melville "The Pastoral Economy," Appendix E, 272–95.

(headmen) and communities held over one-third of the sheep stations in the Valle del Mezquital by the end of the century. So that while the traditional Indian resource base and systems of land use had been drastically modified and the total area of land at the disposal of the communities reduced, the Indian caciques, and the villages to a lesser extent, benefited by a system that legally allowed them to hold land within the Spanish land tenure system.

TYPE 3: JILOTEPEC, ALFAXAYUCA, AND CHIAPA DE MOTA

This type does not exhibit the very high densities of animals nor the complete conversion of land found in the first two types; however, the sub-areas making up this type were quite intensively exploited from an early date and demonstrate environmental degradation by the end of the century.

As early as 1551 the Indian communities of Jilotepec complained that the animals grazed by the Spaniards in the savannas (30,000 sheep and many mares and cows) had laid the soils bare. After the 1550s the pressure on the densely populated eastern half of the Jilotepec plateau was relieved by the measures enforced by Velasco. Cattle were removed from the Valle del Mezquital; sheep were to be grazed 3,000 pasos from Indian villages; and grants were made in lightly populated areas of the western half of the plateau and in the mountains to the south. Following thirty to forty years of relatively light grazing, scattered deterioration of the environment—deforestation of oak woodlands, and the appearance of the yucca—was reported in the western half of Jilotepec.[55]

After forty years of grazing (1530–70), the flat lands of the northern half of the Alfaxayuca Valley were covered in a dense cover of mesquite-dominated scrub. Young live oaks reported as growing in the valley in the 1560s appear to have been superceded by desert scrub later in the century, and by 1611 the formerly fertile Indian agricultural lands in the north were covered by debris eroded off the surrounding high hills.[56] In Chiapa de Mota environmental degradation in the form of erosion first appears in the records for the 1590s. There are no reports of deterioration prior to this date, and it is probable that the heavy forests masked changes on the steep mountain slopes.[57]

TYPE 4: HUICHIAPAN

Environmental deterioration was apparent by the 1580s in areas which had been grazed from the 1550s and 1560s; environmental degradation, evident in the failure of springs, was reported in the 1590s.[58] Spanish settlement of the Huichiapan Plateau was slowed by the threat of invasion by the nomadic Indians (Chichimecs), who moved further south with the opening of the route to the northern mines and made this sub-area so unsafe for both Indian villagers and Spanish settlers that it was called *tierra de guerra* (war zone). Early grants for grazing were concentrated along the southeastern and eastern borders with Jilotepec and Alfaxayuca. Because the estimated grazing rates are predicated on total surface area, the early rates indicate much lower densities than probably obtained in areas actually grazed.

The history of Spanish agriculture in the Huichiapan sub-area provides a clear instance of the problems facing agriculturalists in semiarid regions in which the lands forming the catchment areas are intensively grazed. It was not until the 1570s that grants for both grazing and agriculture were made for all parts of the plateau, even though the Chichimecs still posed a threat and Tecozautla in the northwest sector of this sub-area was a walled town in which all men (including Indians) had to bear arms.[59] The number of Spanish-owned agricultural holdings in this sub-area in 1588 (eighty-two in all) demonstrates very clearly the interest shown by the Spanish in intensive agriculture after the 1576–81 epidemics. Of the total number of Spanish holdings at this date, 50 percent were sheep stations and the other 50 percent were *labores* (agricultural holdings approximately 85–255 hectares in size).[60] There were more

[55] Melville, ''The Pastoral Economy,'' ch. 5, n.s 103–12.
[56] *Ibid.*, ch. 5, n.s 113–7.
[57] *Ibid.*, ch. 5, n. 118.
[58] Melville, ''The Pastoral Economy,'' ch. 5, n.s 119–36.
[59] *AGN,* ''Tierras,'' vol. 79, exp. 6.
[60] *Labores* consisted of two to six *caballerias de tierra* (agricultural land grants) worked as a unit.

of these small intensively worked agricultural holdings in Huichiapan by the 1590s than in any other sub-area of the Valle del Mezquital. By contrast, Alfaxayuca and other sub-areas in which pastoralism had been introduced at an early date and dominated production had few agricultural holdings that were not ancillary to a station.[61]

Intensive methods of cultivation and irrigation were used to grow wheat, barley, grape vines, and fruit trees on the labores, and they were very productive and profitable until the end of the century when many were abandoned for lack of water. By 1600 springs were failing in Tecozautla in which 500 fig trees were lost for lack of irrigation, and there was insufficient water to irrigate 4,000 grape vines belonging to Catalina Méndez and to supply the village of San José Atlán to the south of the cabecera, Huichiapan. When an agreement between Méndez and the village was first completed in 1590 to share the water, there was enough for all; but a new agreement in 1600 had to be drawn up according to which Méndez had sole use of the water from Saturday night to Monday morning, restricting the village to the rest of the week. Indian witnesses in the Méndez case made it very clear that the problem did not arise from increased use by the failure of the springs: "When the agreement was made there was twice as much water in the springs, since then some springs have dried up and others don't give as much water as they did, and if it were necessary to irrigate the hacienda and the Indian lands today according to the agreement between them, there would not be sufficient water."[62]

Although one-half of the Spanish holdings in the 1580s were for agriculture, sheep grazed an area at least five times as large as the agricultural lands—and they grazed the hills forming the catchment area. Deforestation and removal of the ground cover in the high eastern hills meant that the ground water was not replenished and spring flow was reduced along the lower western edge of the plateau, where the rich agricultural lands were situated.

TYPE 5: IXMIQUILPAN, NORTHERN VALLEY

These two sub-areas did not attract many pastoralists in the sixteenth century, and the number of stock which grazed there was never very large. Less than 20 percent of the surface area was converted to pastoralism, and grazing rates remained low to very low throughout the century. Little environmental deterioration was recorded for these arid sub-areas during the sixteenth century, and it is possible that they were considered to be so poor in natural resources that

[61] Melville, "The Pastoral Economy," ch. 5, n. 135. Huichiapan had forty-one labores; the Southern Plain, two, Tula, two, Jilotepec, one.

[62] The quote reads: "Quando se hizo el concierto avia doble mas agua en los fuentes y despues aca se an secado algunos ojos y no mana tanta cantidad como solia y si huviesen de rregar el dia de hoy la dicha hazienda y los dichos yndios por el orden y concierto que an tenido no ay bastante agua," AGN, "Tierras," vol. 3, exp. 1.

changes went unnoticed and unrecorded. It is very difficult to distinguish long-term trends in the environment from those associated with sixteenth-century changes in land use. For example, the appearance of a dense thorn scrub on the flat lands of the Northern Valley by the late 1570s obscured earlier specialisation in the cultivation of the maguey and nopal. Since this type of vegetation is native to the area and because the invasion coincided with the epidemic known as the Great *Cocolistle* (1576–81), it could simply have been the regrowth of primary vegetation on fallow lands. However, gully formation was accelerated in the Northern Valley in the 1590s and followed a sharp increase in the number of stations and the density of animals when grazing rates rose from 68.8 per square kilometer to 162.2 per square kilometer in the 1580s.[63]

The mines that were founded in Ixmiquilpan in the 1540s caused a heavy drain on the stands of mesquite because these trees were favoured by the miners for wheels and lanterns.[64] By the beginning of the seventeenth century there was a dense growth of thorns, yucca, and thistles on former agricultural lands in Ixmiquilpan; and while these plants represent the normal vegetation for this area, the presence of the thistle indicates disturbance by grazing animals.[65]

The sudden increase in interest in these arid regions during the last two decades of the century, especially the Northern Valley, reflected a lack of room for continued expansion in the rest of the region; it also indicated an acceptance of desert species as adequate fodder for the goats, which made up an increasing proportion of the flocks at the end of the century, and sheep.

The intensity of exploitation by grazing in the Valle del Mezquital in the sixteenth century and the severity of environmental degradation from over-grazing depended on the initial status of the natural resources. Those areas with excellent resources (water, pasture, and soils) were the first to be intensively grazed, were over-exploited as early as the 1560s, and exhibited the most advanced state of degradation by the end of the sixteenth century. As pastoralism spread, however, the original diversity of the vegetative cover was lost everywhere; and the region was considered to be fit only for sheep. Continued grazing (albeit at reduced levels of intensity) maintained pressure on the environment and the possibility of recovery was reduced. Indeed the region continued to deteriorate under grazing until the late 1940s, when irrigation systems using the waste waters of Mexico City were initiated.

This process of environmental deterioration is not the first to afflict the Valle del Mezquital. In his study of the southeastern quarter of the Valle del Mezquital (the Teotlalpan), Sherburne F. Cook demonstrated that three cycles of erosion and deposition have occurred here during the past one thousand

---

[63] Melville, ''The Pastoral Economy,'' ch. 5, n.s 137–46.

[64] *PNE* VI, 4.

[65] Melville, ''The Pastoral Economy,'' ch. 5, n.s 147–9.

TABLE 4

Timetable of Erosion Cycles in the Teotlalpan

Cycle 1 (1000–1200 A.D.): Removal of the A horizon; Toltec Era.
Cycle 2 (1400–1521 A.D.): Sheet erosion and deposition of slope wash; Aztec Era.
Cycle 3 (1600–1800 A.D.): Gullying and barranca formation; Post–Conquest.

SOURCE: Cook, *The Historical Demography*, 41–59.

years.[66] He developed a timetable linking these erosion events to increases in population densities and such destructive land use techniques as hillside agriculture and concluded that the periods of maximum deterioration "had coincided with the era of high cultures and great populations before the Conquest"[67] (see Table 4).

While he conceded that the Spanish vastly accelerated the destructive processes by increased deforestation for lime manufacture and mining and the introduction of the plough and grazing animals, Cook thought that by 1521 the processes of environmental degradation "had gone a long way towards completion, and that had the Spanish not arrived when they did the Teotlalpan, in any case, *was ecologically and demographically doomed to destruction*" (emphasis in original).[68] Evidence of accelerated environmental degradation in a period of rapid population decline (circa 1548–80) did seem, however, to cause some difficulties for Cook, who attempted to resolve these problems by postulating that: "There can be no alternative to the conclusion that hillside agriculture, as introduced into and practised extensively in the Teotlalpan in the fifteenth century, was accompanied by severe wastage of the mature top soil. This wastage was caused primarily by sheet erosion, rather than gully erosion."[69] He concluded that "the Spanish accounts . . . lead to the supposition that agriculture in the Teotlalpan had undergone a sharp decline in the sixteenth century. If it is assumed that such a decline had already been initiated under the last Aztec rulers, *or at least was on the point of becoming manifest*, then the influence of the Spaniards was simply to induce a rapid acceleration of the process" (emphasis added).[70]

I found no documentary evidence for the extensive sheet erosion and deforestation that Cook proposed as being present at conquest, rather, I found that the region was ecologically quite stable up to the middle of the sixteenth century. Like Cook, I did find evidence that the situation deteriorated between 1548 and 1580; and recent archaeological surveys of the Valley of Mexico

[66] Cook, *The Historical Demography*, 41–59.
[67] Sherburne F. Cook, *Soil Erosion and Population in Central Mexico* (Berkeley, 1949), 84.
[68] Cook, *Historical Demography*, 54.
[69] *Ibid.*, 52.
[70] *Ibid.*, 54.

and the Tula River Basin, which include parts of the Teotlalpan in their survey areas, not only confirm Cook's timetable of preconquest population cycles and the move to hillside agriculture in the immediate preconquest era (Late Horizon 1350–1519 A.D.) but also demonstrate the extensive use of erosion control techniques such as terracing.[71]

It is instructive, at this point, to compare the post-conquest history of the Valle del Mezquital with the conclusions reached in a study of the relationship of the human populations of southern Greece to their environment over the last 50,000 years.[72] Tjeerd H. van Andel and Curtis Runnels concluded that two modes of land stabilization (terracing; or total abandonment) and two modes of destabilization (overly intensive use, including clearing very marginal lands; or a shift to pastoralism) alternated to produce the cycles of erosion and recovery evident in the southern Argolid. They note that while terraces require unceasing maintenance, total abandonment of terraced land does not necessarily lead to erosion since rapid growth of grasses and bushes binds the soils and walls and prevents slippage. By contrast, partial abandonment and a shift to pastoralism can lead to catastrophic erosion. When care is not taken to maintain the terrace walls by replacing stones dislodged by grazing animals or by protecting them with thorny branches, they are destroyed by the movement of animals across them; and the contained earth then slides down into the valleys. These authors correlate the process of partial abandonment and expansion of grazing onto terraced lands with economic downturns, when agriculturalists use their good bottom lands for grains and rent their terraces to shepherds.[73]

If this model of the replacement of hillside agriculture by grazing and the resulting erosion is kept in mind, it is possible to accomodate the results of my study with Cook's model of human population maximums and ecological deterioration by modifying the erosion timetable to take into account the lack of evidence of environmental degradation in the immediate post-conquest era and the documented consequences of intensive sheep grazing in the last half of the sixteenth century.

I suggest that truncation of the A horizons (Cycle 1) under the intensive indigenous systems of agriculture continued up to the late 1570s, when the 1576–81 epidemic caused a sharp reduction in Indian population and agriculture. Deforestation, the rapid expansion of grazing, and removal of the vegetative cover from overgrazing during the 1560s and 1570s, accelerated this process and led to the catastrophic erosion events of the 1580s and 1590s that left bare stretches of tepetate on the piedmont and deep deposits of slope wash on the flats (Cycle 2). I principally agree with Cook's dating of the last

[71] Mastache de E. and Crespo O., "La Ocupación Prehispánica en el Area de Tula, Hgo.," 76–77; and W. T. Sanders *et al. The Basin of Mexico*, 179, 213–6.

[72] Tjeerd van Andel and Curtis Runnels, *Beyond the Acropolis: A Rural Greek Past* (Stanford, 1987).

[73] *Ibid.*, ch. 8.

TABLE 5

Revised Timetable Events

Cycle 1 (1000–1580 A.D.):	Removal of the A horizon; indigenous agricultural systems.
Cycle 2 (1580–1600 A.D.):	Sheet erosion and slope-wash deposition; deforestation and overgrazing.
Cycle 3 (1580–Present):	Gullying; continued grazing or overgrazing.

erosion cycle, gullying and barranca formation (Cycle 3), but in some parts of the region this process started as early as the 1580s and has continued up to the present (see Table 5).

THE FORMATION OF LATIFUNDA

During the last two decades of the sixteenth century, more land was transferred to the Spanish system of land tenure for grazing than in the previous five decades.[74] The reason for the intensified interest in sheep raising in a region characterised by declining production is to be found in the market for pastoral products at the end of the century. Prices for all foodstuffs rose sharply in the 1580s and 1590s, in part because of the general inflation and dropping supplies of the late sixteenth century. It is generally agreed that the supply of agricultural products dropped because of the demographic collapse of the Indian population, especially after the 1576–81 epidemic, since the Indians up to that time were the primary producers of agricultural products. The drop in the supply of pastoral products was based on declining production (a decrease in the herds which was reported for all parts of New Spain during the last quarter of the sixteenth century). Even though the number of consumers dropped with the dwindling population of Indians, who were not replaced in equal numbers by the growing Spanish and Mestizo (people of Spanish and Indian descent) populations, the demand for meat exceeded the supply.[75] Pastoralists were assured of a market and ample profits if production could be sustained—the problem was to maintain or increase the number of stock.

The pastoralists of the Valle del Mezquital responded to the decline in the carrying capacity of the region and high prices for their products by attempting to restrict access to the pasture on their lands. The rich and powerful *latifundistas* of this early period, who had the capital needed to buy out the small land owners, acquired title to large numbers of stations, increasing their

---

[74] The number of stations (cumulative totals) in the region by the end of each decade: 1539 (34), 1549 (75), 1559 (108), 1565 (276), 1569 (294), 1579 (399.5), 1589 (583.5), 1599 (862).

[75] Chevalier, *La Formación*, 139–140. Simpson, *Exploitation*, 22 *ff.*, André Gunder Frank, *Mexican Agriculture, 1521–1630*, (Cambridge, 1979), ch. 6.

flocks at the expense of the smaller individual pastoralists and thus monopolised access to the land. For example, in 1603 when Jerónimo López entailed
three estates totalling 500 square kilometers in the eastern half of the region,
the annual proceeds from his urban and rural holdings totalled 51,720 *pesos
de oro comun*.[76]

The most striking aspect of the haciendas that developed in the Valle del
Mezquital at the end of the sixteenth century is their size. While they do not
compare with the vast northern estates, which measured up to thousands of
square kilometers in size, they were large in comparison with the haciendas in
the Valley of Mexico. They represent a marked monopolization of land and
indicate the extent to which the ownership of the means of production had
shifted from many small land owners to a small number of large landowners.[77]

Contrary to current thought on the development of extensively exploited
latifundia in Mexico, the emergence of a small powerful elite that monopolised land and production in the Valle del Mezquital by the early seventeenth
century was neither based on inherently poor natural resources and a small
initial population, nor did it emerge as a direct result of the displacement of
Indian agriculturalists. Rather, it was based on the decline of the productive
potential of the region. The very rapid process of environmental deterioration
that accompanied the over-exploitation of the region by grazing, with mining
and limeworking to a lesser extent, produced an environment in which (given
the technology of the age) only large holdings were economically feasible.

The formation of haciendas in late sixteenth-century New Spain is generally associated with the monopolization of land; and both François Chevalier
and André Gunder Frank note that the hacienda developed out of small holdings during a period of waning supply and high prices for both agricultural
and pastoral products.[78] Although recent research has demonstrated an immense amount of variability in the size of colonial haciendas, as well as their
internal structure and relation to regional and interregional trade networks, it
does appear that especially large holdings were formed in pastoral regions—

---

[76] Mendizábal, *Obras,* vol. 6, 112. The *peso* of common gold was a low-grade gold coin of
approximately 14 carats, equal in market value to eight silver *reales.* See Wilbur Meeks, *The
Exchange Media of Colonial Mexico* (New York, 1948), 34–38.

[77] Examples of some haciendas in the Valle del Mezquital in the early seventeenth century by
approximate size: 500 square kilometers (Mendizábal, *Obras,* VI, 112); 420 square kilometers
(*AGN,* "Tierras," vol. 2711, exp. 10); 293 square kilometers (*AGN,* "Tierras," vol. 1520, exp.
5); 180 square kilometers (*AGN,* "Tierras," vol. 2692, exp. 6); 124 square kilometers (*AGN,*
"Tierras," vol. 2813, exp. 13). Ixmiquilpan was monopolised by ten haciendas, each averaging
102.8 square kilometers (*AGN,* "Civil," vol. 77, exp. 11, fol. 80v). The huge Santa Lucia
Hacienda of the Jesuits, which extended over the eastern half of the Northern Valley and
Ixmiquilpan sub-areas by the early eighteenth century, already extended well into the Northern
Valley by the early seventeenth century (Herman W. Konrad, *A Jesuit Hacienda in Colonial
Mexico. Santa Lucia, 1576–1767,* (Stanford, 1980), ch. 3).

[78] Chevalier, *La Formación,* 144. Frank, *Mexican Agriculture,* ch. 6.

even in central regions such as the Toluca and Puebla Valleys—as well as in the Valle del Mezquital.[79]

I argue that the carrying capacity of many parts of New Spain deteriorated under intensive grazing in the sixteenth century and that environmental degradation, by making large areas of grazing land necessary to maintain profits, was an important variable in the development of large-scale haciendas in many pastoral regions. Similar ecological changes to those documented for the Valle del Mezquital are indicated for other pastoral regions by the generalised decline in flock size.[80]

Many early writers commented on the spectacular increase of the grazing animals introduced shortly after the conquest, and New Spain was known for cheap and plentiful meat,[81] but in the last decades of the century herds were decreasing; animals were underweight; and production of pastoral products declined. Several reasons were given by contemporaries for the decline in the herds: the extremely wasteful practise of killing huge numbers of animals for their tallow or hides alone; the high consumption of meat by the Indians; cattle rustling by nomadic Indians; and the depredations of the bands of wild dogs that roamed the land.[82] Viceroy Martín Enríquez wrote in 1574 that "livestock does not multiply as it used to: cows used to bear calves before two years of age—because the land was not trampled and pasture was fertile and extensive; but now that [the pasture] has failed the cows do not bear before three or four years of age."[83]

Most modern writers would agree with Viceroy Enríquez that the primary cause of the decline in the herds was deterioration of the natural resource base from overstocking and overgrazing. Simpson saw a direct correlation between overgrazing by sheep and goats and the decline in flock size because sheep and goats destroy their own subsistence base, but he thought that perhaps the shortage of cattle was confined to the Spanish cities and simply a factor of growing demand. However, Chevalier wrote that diminution of the herds of cattle was reported in the north as well as the center of Mexico and quite clearly resulted from deterioration of the natural resource base.[84]

Measures were taken to correct the drop in supply of pastoral products

[79] Jack A. Licate, *Creation of a Mexican Landscape. Territorial Organization and Settlement in the Eastern Puebla Basin, 1520–1605*, (Department of Geography, University of Chicago, Research Paper no. 201, 1981), 124. James Lockhart, "Españonles entre indios: Toluca a fines del siglo XVI," in *Estudios sobre la ciudad iberoamericana*, Francisco de Solano, coordinator (Madrid, 1975), 448.

[80] Several writers have correlated the spread of pastoralism and overgrazing in Mexico with deforestation and erosion. For example, see Gibson, *Aztecs*, 5–6, 305; Simpson, *Exploitation*, 23; Wolfgang Trautmann, *Las transformaciones en el paisaje cultural de Tlaxcala durante la época colonial* (Weisbaden, 1981), 177.

[81] See Simpson, *Exploitation*, 2 *ff.*, for contemporary witness to the increase of grazing animals.

[82] Chevalier, *La Formación*, 137 *ff.*

[83] Quoted in Chevalier, *La Formación*, 138.

[84] Simpson, *Exploitation*, 23; Chevalier, *La Formación*, 139 *ff.*

during the last decades of the sixteenth century. For example, Indian meat consumption was restricted, and laws were passed that forbade the killing of female stock and limited the numbers of animals butchered. Viceroy Luis de Velasco the Younger wrote to his successor in 1595 that although he had taken measures to correct the situation, he did not ''see that [he had] been able to make good the damage or to restore the herds to their former size.''[85]

I suggest that a process of environmental degradation associated with the expansion of intensive pastoralism underlay a structural shift in the pastoral sector of the political economy of New Spain during the last quarter of the sixteenth century. At mid-century, supply kept pace with and even exceeded demand; prices were low, although rising with the general inflation of the sixteenth century; and production was in the hands of a large number of small and medium landowners. By the end of the sixteenth century, the pastoral sector was, however, characterised by an absolute decline in productive potential as well as production levels; increasing demand (a factor of both declining supply and growing numbers of Spaniards and mestizos); high prices; and control of the means of production by a small group of large landowners. Although the pastoral sector of the emerging Mexican economy at the beginning of the seventeenth century was monopolised by the owners of the great estates—in confirmation of the classic picture of Mexican society—this situation was neither an inevitable outcome of either the conquest of Mexico by the Spaniards nor the development of mercantile capitalism; it grew out of local processes and history.

[85] Quoted in Simpson, *Exploitation*, 22.

A Place for Stories:
Nature, History, and Narrative

William Cronon

> Children, only animals live entirely in the Here and Now. Only nature knows neither memory nor history. But man — let me offer you a definition — is the story-telling animal. Wherever he goes he wants to leave behind not a chaotic wake, not an empty space, but the comforting marker-buoys and trail-signs of stories. He has to go on telling stories. He has to keep on making them up. As long as there's a story, it's all right. Even in his last moments, it's said, in the split second of a fatal fall — or when he's about to drown — he sees, passing rapidly before him, the story of his whole life.
>
> — Graham Swift, *Waterland*

In the beginning was the story. Or rather: many stories, of many places, in many voices, pointing toward many ends.

In 1979, two books were published about the long drought that struck the Great Plains during the 1930s. The two had nearly identical titles: one, by Paul Bonnifield, was called *The Dust Bowl;* the other, by Donald Worster, *Dust Bowl.*[1] The two authors dealt with virtually the same subject, had researched many of the same documents, and agreed on most of their facts, and yet their conclusions could hardly have been more different.

Bonnifield's closing argument runs like this:

William Cronon is professor of history at Yale University.

I would like to thank the many friends and colleagues who have read and criticized various versions of this essay. David Laurence was responsible for convincing me, rather against my will, that the perspective I've adopted here could be neither ignored nor evaded, and he offered generous guidance as I tried to acquire the critical vocabulary that would allow me to tackle these problems. As always, David Scobey has been my most faithful guide in helping me find my way through the dense thickets of literary theory. Comments and suggestions from Thomas Bender, Elise Broach, Robert Burt, Michael P. Cohen, James Davidson, David Brion Davis, Kai Erikson, Ann Fabian, Peter Gay, Amy Green, Michael Goldberg, Ramachandra Guha, Reeve Huston, Susan Johnson, Howard Lamar, Jonathan Lear, Patricia Limerick, Arch McCallum, George Miles, Katherine Morrissey, Jim O'Brien, Robert Shulman, Thompson Smith, Alan Taylor, Paul Taylor, Sylvia Tesh, Thompson Webb III, Timothy Weiskel, Richard White, Bryan Wolf, Donald Worster, and two anonymous readers likewise helped shape my thoughts on this subject. Finally, I owe a special debt to David Thelen and Steven Stowe for their persistence in encouraging me to return to an essay I had all but abandoned. I am grateful to all.

The opening quotation is from Graham Swift, *Waterland* (New York, 1983), 53–54.

[1] Paul Bonnifield, *The Dust Bowl: Men, Dirt, and Depression* (Albuquerque, 1979); Donald Worster, *Dust Bowl: The Southern Plains in the 1930s* (New York, 1979). On Dust Bowl historiography in general, see the collection of essays in *Great Plains Quarterly*, 6 (Spring 1986).

In the final analysis, the story of the dust bowl was the story of people, people with ability and talent, people with resourcefulness, fortitude, and courage. . . . The people of the dust bowl were not defeated, poverty-ridden people without hope. They were builders for tomorrow. During those hard years they continued to build their churches, their businesses, their schools, their colleges, their communities. They grew closer to God and fonder of the land. Hard years were common in their past, but the future belonged to those who were ready to seize the moment. . . . Because they stayed during those hard years and worked the land and tapped her natural resources, millions of people have eaten better, worked in healthier places, and enjoyed warmer homes. Because those determined people did not flee the stricken area during a crisis, the nation today enjoys a better standard of living.[2]

Worster, on the other hand, paints a bleaker picture:

The Dust Bowl was the darkest moment in the twentieth-century life of the southern plains. The name suggests a place — a region whose borders are as inexact and shifting as a sand dune. But it was also an event of national, even planetary significance. A widely respected authority on world food problems, George Borgstrom, has ranked the creation of the Dust Bowl as one of the three worst ecological blunders in history. . . . It cannot be blamed on illiteracy or overpopulation or social disorder. It came about because the culture was operating in precisely the way it was supposed to. . . . The Dust Bowl . . . was the inevitable outcome of a culture that deliberately, self-consciously, set itself [the] task of dominating and exploiting the land for all it was worth.[3]

For Bonnifield, the dust storms of the 1930s were mainly a natural disaster; when the rains gave out, people had to struggle for their farms, their homes, their very survival. Their success in that struggle was a triumph of individual and community spirit: nature made a mess, and human beings cleaned it up. Worster's version differs dramatically. Although the rains did fail during the 1930s, their disappearance expressed the cyclical climate of a semiarid environment. The story of the Dust Bowl is less about the failures of nature than about the failures of human beings to accommodate themselves to nature. A long series of willful human misunderstandings and assaults led finally to a collapse whose origins were mainly cultural.

Whichever of these interpretations we are inclined to follow, they pose a dilemma for scholars who study past environmental change — indeed, a dilemma for all historians. As often happens in history, they make us wonder how two competent authors looking at identical materials drawn from the same past can reach such divergent conclusions. But it is not merely their *conclusions* that differ. Although both narrate the same broad series of events with an essentially similar cast of characters, they tell two entirely different *stories*. In both texts, the story is inextricably bound to its conclusion, and the historical analysis derives much of its force from the upward or downward sweep of the plot. So we must eventually ask a more basic question: where did these stories come from?

[2] Bonnifield, *The Dust Bowl*, 202.
[3] Worster, *Dust Bowl*, 4.

The question is trickier than it seems, for it transports us into the much contested terrain between traditional social science and postmodernist critical theory. As an environmental historian who tries to blend the analytical traditions of history with those of ecology, economics, anthropology, and other fields, I cannot help feeling uneasy about the shifting theoretical ground we all now seem to occupy. On the one hand, a fundamental premise of my field is that human acts occur within a network of relationships, processes, and systems that are as ecological as they are cultural. To such basic historical categories as gender, class, and race, environmental historians would add a theoretical vocabulary in which plants, animals, soils, climates, and other nonhuman entities become the coactors and codeterminants of a history not just of people but of the earth itself. For scholars who share my perspective, the importance of the natural world, its objective effects on people, and the concrete ways people affect it in turn are not at issue; they are the very heart of our intellectual project. We therefore ally our historical work with that of our colleagues in the sciences, whose models, however imperfectly, try to approximate the mechanisms of nature.[4]

And yet scholars of environmental history also maintain a powerful commitment to narrative form. When we describe human activities within an ecosystem, we seem always to tell *stories* about them.[5] Like all historians, we configure the events of the past into causal sequences—stories—that order and simplify those events to give them new meanings. We do so because narrative is the chief literary form that tries to find meaning in an overwhelmingly crowded and disordered chronological reality. When we choose a plot to order our environmental histories, we give them a unity that neither nature nor the past possesses so clearly. In so doing, we move well beyond nature into the intensely human realm of value. There, we cannot avoid encountering the postmodernist assault on narrative, which calls into question not just the stories we tell but the deeper purpose that motivated us in the first place: trying to make sense of nature's place in the human past.

By writing stories about environmental change, we divide the causal relationships of an ecosystem with a rhetorical razor that defines included and excluded, relevant and irrelevant, empowered and disempowered. In the act of separating story from non-story, we wield the most powerful yet dangerous tool of the narrative form. It is a commonplace of modern literary theory that the very authority with which narrative presents its vision of reality is achieved by obscuring large portions of that reality. Narrative succeeds to the extent that it hides the discontinuities, ellipses,

---

[4] For a wide-ranging discussion that explores the emerging intellectual agendas of environmental history, see "A Round Table: Environmental History," *Journal of American History*, 76 (March 1990), 1087–1147.

[5] Throughout this essay, I will use "story" and "narrative" interchangeably, despite a technical distinction that can be made between them. For some literary critics and philosophers of history, "story" is a limited genre, whereas narrative (or *narratio*) is the much more encompassing part of classical rhetoric that organizes all representations of time into a configured sequence of completed actions. I intend the broader meaning for both words, since "storytelling" in its most fundamental sense is the activity I wish to criticize and defend. I hope it is emphatically clear at the outset that I am *not* urging a return to "traditional" narrative history that revolves around the biographies of "great" individuals (usually elite white male politicians and intellectuals); rather, I am urging historians to acknowledge storytelling as the necessary core even of *longue durée* histories that pay little attention to individual people. Environmental history is but one example of these, and most of my arguments apply just as readily to the others.

and contradictory experiences that would undermine the intended meaning of its story. Whatever its overt purpose, it cannot avoid a covert exercise of power: it inevitably sanctions some voices while silencing others. A powerful narrative reconstructs common sense to make the contingent seem determined and the artificial seem natural. If this is true, then narrative poses particularly difficult problems for environmental historians, for whom the boundary between the artificial and the natural is the very thing we most wish to study. The differences between Bonnifield's and Worster's versions of the Dust Bowl clearly have something to do with that boundary, as does my own uneasiness about the theoretical underpinnings of my historical craft.[6]

The disease of literary theory is to write too much in abstractions, so that even the simplest meanings become difficult if not downright opaque. Lest this essay wander off into litcrit fog, let me ground it on more familiar terrain. I propose to examine the role of narrative in environmental history by returning to the Great Plains to survey the ways historians have told that region's past. What I offer here will *not* be a comprehensive historiography, since my choice of texts is eclectic and I will ignore many major works. Rather, I will use a handful of Great Plains histories to explore the much vexed problems that narrative poses for all historians. On the one hand, I hope to acknowledge the deep challenges that postmodernism poses for those who applaud "the revival of narrative"; on the other, I wish to record my own conviction—chastened but still strong—that narrative remains essential to our understanding of history and the human place in nature.

If we consider the Plains in the half millennium since Christopher Columbus crossed the Atlantic, certain events seem likely to stand out in any long-term history of the region. If I were to try to write these not as a *story* but as a simple *list*—I will not entirely succeed in so doing, since the task of *not* telling stories about the past turns out to be much more difficult than it may seem—the resulting chronicle might run something like this.

Five centuries ago, people traveled west across the Atlantic Ocean. So did some plants and animals. One of these—the horse—appeared on the Plains. Native peoples used horses to hunt bison. Human migrants from across the Atlantic even-

---

[6] Much of the reading that lies behind this essay cannot easily be attached to a single argument or footnote. Among the works that helped shape my views on the importance and problems of narrative are the following: William H. Dray, *Philosophy of History* (Englewood Cliffs, 1964); Robert Scholes and Robert Kellogg, *The Nature of Narrative* (New York, 1966); Frank Kermode, *The Sense of an Ending: Studies in the Theory of Fiction* (New York, 1967); Hayden White, *Metahistory: The Historical Imagination in Nineteenth-Century Europe* (Baltimore, 1973); Hayden White, *Tropics of Discourse: Essays in Cultural Criticism* (Baltimore, 1978); Robert H. Canary and Henry Kozicki, eds., *The Writing of History: Literary Form and Historical Understanding* (Madison, 1978); W. J. T. Mitchell, ed., *On Narrative* (Chicago, 1981); Fredric Jameson, *The Political Unconscious: Narrative as a Socially Symbolic Act* (Ithaca, 1981); Jonathan Culler, *On Deconstruction: Theory and Criticism after Structuralism* (Ithaca, 1982); Terry Eagleton, *Literary Theory: An Introduction* (Minneapolis, 1983); Paul Ricoeur, *Time and Narrative* (3 vols., Chicago, 1984, 1985, 1988), trans. Kathleen Blamey and David Pellauer; Dominick LaCapra, *Rethinking Intellectual History: Texts, Contexts, Language* (Ithaca, 1983); Arthur C. Danto, *Narration and Knowledge: Including the Integral Text of Analytical Philosophy of History* (New York, 1985); James Clifford and George E. Marcus, eds., *Writing Culture: The Poetics and Politics of Ethnography* (Berkeley, 1986); Wallace Martin, *Recent Theories of Narrative* (Ithaca, 1986); Louis O. Mink, *Historical Understanding* (Ithaca, 1987); Hayden White, *The Content of the Form: Narrative Discourse and Historical Representation* (Baltimore, 1987); and Kai Erikson, "Obituary for Big Daddy: A Parable," unpublished manuscript (in William Cronon's possession).

tually appeared on the Plains as well. People fought a lot. The bison herds disappeared. Native peoples moved to reservations. The new immigrants built homes for themselves. Herds of cattle increased. Settlers plowed the prairie grasses, raising corn, wheat, and other grains. Railroads moved people and other things into and out of the region. Crops sometimes failed for lack of rain. Some people abandoned their farms and moved elsewhere; other people stayed. During the 1930s, there was a particularly bad drought, with many dust storms. Then the drought ended. A lot of people began to pump water out of the ground for use on their fields and in their towns. Today, Plains farmers continue to raise crops and herds of animals. Some have trouble making ends meet. Many Indians live on reservations. It will be interesting to see what happens next.

I trust that this list seems pretty peculiar to anyone who reads it, as if a child were trying to tell a story without quite knowing how. I've tried to remove as much sense of *connection* among these details as I can. I've presented them not as a narrative but as a *chronicle*, a simple chronological listing of events as they occurred in sequence.[7] This was not a pure chronicle, since I presented only what I declared to be the "most important" events of Plains history. By the very act of separating important from unimportant events, I actually smuggled a number of not-so-hidden stories into my list, so that such things as the migration of the horse or the conquest of the Plains tribes began to form little narrative swirls in the midst of my ostensibly story-less account. A pure chronicle would have included every event that ever occurred on the Great Plains, no matter how large or small, so that a colorful sunset in September 1623 or a morning milking of cows on a farm near Leavenworth in 1897 would occupy just as prominent a place as the destruction of the bison herds or the 1930s dust storms.

Such a text is impossible even to imagine, let alone construct, for reasons that help explain historians' affection for narrative.[8] When we encounter the past in the form of a chronicle, it becomes much less recognizable to us. We have trouble sorting out why things happened when and how they did, and it becomes hard to evaluate the relative significance of events. Things seem less *connected* to each other, and it becomes unclear how all this stuff relates to us. Most important, in a chronicle we easily lose the thread of what was going on at any particular moment. Without some plot to organize the flow of events, everything becomes much harder—even impossible—to understand.

How do we discover a story that will turn the facts of Great Plains history into something more easily recognized and understood? The repertoire of historical plots

---

[7] This distinction between chronicle and narrative is more fully analyzed in White, *Metahistory*, 5–7; White, *Tropics of Discourse*, 109–11; Louis O. Mink, "Narrative Form as a Cognitive Instrument," in *Writing of History*, ed. Canary and Kozicki, 141–44; David Carr, *Time, Narrative, and History* (Bloomington, 1986), 59; Danto, *Narration and Knowledge*; and Paul A. Roth, "Narrative Explanations: The Case of History," *History and Theory*, 27 (no. 1, 1988), 1–13.

[8] There are deeper epistemological problems here that I will not discuss, such as how we recognize what constitutes an "event" and how we draw boundaries around it. It should eventually become clear that "events" are themselves defined and delimited by the stories with which we configure them and are probably impossible to imagine apart from their narrative context.

we might apply to the events I've just chronicled is endless and could be drawn not just from history but from all of literature and myth. To simplify the range of choices, let me start by offering two large groups of possible plots. On the one hand, we can narrate Plains history as a story of improvement, in which the plot line gradually ascends toward an ending that is somehow more positive — happier, richer, freer, better — than the beginning. On the other hand, we can tell stories in which the plot line eventually falls toward an ending that is more negative — sadder, poorer, less free, worse — than the place where the story began. The one group of plots might be called "progressive," given their historical dependence on eighteenth-century Enlightenment notions of progress; the other might be called "tragic" or "declensionist," tracing their historical roots to romantic and antimodernist reactions against progress.

If we look at the ways historians have actually written about the changing environment of the Great Plains, the upward and downward lines of progress and declension are everywhere apparent. The very ease with which we recognize them constitutes a warning about the terrain we are entering. However compelling these stories may be as depictions of environmental change, their narrative form has less to do with nature than with human discourse. Their plots are cultural constructions so deeply embedded in our language that they resonate far beyond the Great Plains. Historians did not invent them, and their very familiarity encourages us to shape our storytelling to fit their patterns. Placed in a particular historical or ideological context, neither group of plots is innocent: both have hidden agendas that influence what the narrative includes and excludes. So powerful are these agendas that not even the historian as author entirely controls them.

Take, for instance, the historians who narrate Great Plains history as a tale of frontier progress. The most famous of those who embraced this basic plot was of course Frederick Jackson Turner, for whom the story of the nation recapitulated the ascending stages of European civilization to produce a uniquely democratic and egalitarian community. Turner saw the transformation of the American landscape from wilderness to trading post to farm to boomtown as the central saga of the nation.[9] If ever there was a narrative that achieved its end by erasing its true subject, Turner's frontier was it: the heroic encounter between pioneers and "free land" could only become plausible by obscuring the conquest that traded one people's freedom for another's. By making Indians the foil for its story of progress, the frontier plot made their conquest seem natural, commonsensical, inevitable. But to say this is only to affirm the narrative's power. In countless versions both before and after it acquired its classic Turnerian form, this story of frontier struggle and progress remains among the oldest and most familiar narratives of American history. In its ability to turn ordinary people into heroes and to present a conflict-ridden invasion as an epic march toward enlightened democratic nationhood, it perfectly fulfilled the ideological needs of its late-nineteenth-century moment.[10]

[9] Frederick Jackson Turner, *The Frontier in American History* (New York, 1920), 12.
[10] I have written about the rhetorical structure of Turner's work in two essays: William Cronon, "Revisiting the Vanishing Frontier: The Legacy of Frederick Jackson Turner," *Western Historical Quarterly,* 18 (April 1987), 157–76;

The Great Plains would eventually prove less tractable to frontier progress than many other parts of the nation. Turner himself would say of the region that it constituted the American farmer's "first defeat," but that didn't stop the settlers themselves from narrating their past with the frontier story.[11] One of Dakota Territory's leading missionaries, Bishop William Robert Hare, prophesied in the 1880s that the plot of Dakota settlement would follow an upward line of migration, struggle, and triumph:

> You may stand ankle deep in the short burnt grass of an uninhabited wilderness—next month a mixed train will glide over the waste and stop at some point where the railroad has decided to locate a town. Men, women and children will jump out of the cars, and their chattel will be tumbled out after them. From that moment the building begins. The courage and faith of these pioneers are something extraordinary. Their spirit seems to rise above all obstacles.[12]

For Hare, this vision of progress was ongoing and prospective, a prophecy of future growth, but the same pattern could just as easily be applied to retrospective visions. An early historian of Oklahoma, Luther Hill, could look back in 1909 at the 1890s, a decade that had "wrought a great change in Oklahoma territory": in a mere ten years, settlers had transformed the "stagnant pool" of unused Indian lands into the "waving grain fields, the herds of cattle, and the broad prospect of agricultural prosperity [which] cause delight and even surprise in the beholder who sees the results of civilization in producing such marvels of wealth."[13] Ordinary people saw such descriptions as the fulfillment of a grand story that had unfolded during the course of their own lifetimes. As one Kansas townswoman, Josephine Middlekauf, concluded,

> After sixty years of pioneering in Hays, I could write volumes telling of its growth and progress. . . . I have been singularly privileged to have seen it develop from the raw materials into the almost finished product in comfortable homes, churches, schools, paved streets, trees, fruits and flowers.[14]

Consider these small narratives more abstractly. They tell a story of more or less linear progress, in which people struggle to transform a relatively responsive environment. There may be moderate setbacks along the way, but their narrative role is to play foil to the heroes who overcome them. Communities rapidly succeed in becoming ever more civilized and comfortable. The time frame of the stories is brief, limited to the lifespan of a single generation, and is located historically in the moment just after invading settlers first occupied Indian lands. Our attention

and William Cronon, "Turner's First Stand: The Significance of Significance in American History," in *Writing Western History: Classic Essays on Classic Western Historians*, ed. Richard Etulain (Albuquerque, 1991), 73–101. See also Ronald H. Carpenter, *The Eloquence of Frederick Jackson Turner* (San Marino, 1983).

[11] Turner, *Frontier in American History*, 147.

[12] William Robert Hare, ca. 1887, as quoted in Howard R. Lamar, "Public Values and Private Dreams: South Dakota's Search for Identity, 1850–1900," *South Dakota History*, 8 (Spring 1978), 129.

[13] Luther B. Hill, *A History of the State of Oklahoma* (Chicago, 1909), 382, 386, 385.

[14] Josephine Middlekauf, as quoted in Joanna L. Stratton, *Pioneer Women: Voices from the Kansas Frontier* (New York, 1981), 204.

as readers is focused on local events, those affecting individuals, families, townships, and other small communities. All of these framing devices, which are as literary as they are historical, compel us toward the conclusion that this is basically a happy story. It is tempered only by a hint of nostalgia for the world that is being lost, a quiet undercurrent of elderly regret for youthful passions and energies now fading.

If the story these narrators tell is about the drama of settlement and the courage of pioneers, it is just as much about the changing stage on which the drama plays itself out. The transformation of a Kansas town is revealed not just by its new buildings but by its shade trees, apple orchards, and gardens; the triumphant prosperity of Oklahoma resides in its wheat fields, cattle pastures, and oil derricks. As the literary critic Kenneth Burke long ago suggested, the scene of a story is as fundamental to what happens in it as the actions that comprise its more visible plot. Indeed, Burke argues that a story's actions are almost invariably consistent with its scene: "there is implicit in the quality of a scene," he writes, "the quality of the action that is to take place within it."[15]

If the way a narrator constructs a scene is directly related to the story that narrator tells, then this has deep implications for environmental history, which after all takes scenes of past nature as its primary object of study. If the history of the Great Plains is a progressive story about how grasslands were turned into ranches, farms, and gardens, then the end of the story requires a particular kind of scene for the ascending plot line to reach its necessary fulfillment. Just as important, the closing scene has to be different from the opening one. If the story ends in a wheatfield that is the happy conclusion of a struggle to transform the landscape, then the most basic requirement of the story is that the earlier form of that landscape must either be neutral or negative in value. It must *deserve* to be transformed.

It is thus no accident that these storytellers begin their narratives in the midst of landscapes that have few redeeming features. Bishop Hare's Dakota Territory begins as "an uninhabited wilderness," and his railroad carries future settlers across a "waste." Just so does narrative revalue nature by turning it into scenery and pushing to its margins such characters as Indians who play no role in the story—or rather, whose roles the story is designed to obscure. When Luther Hill's Oklahoma was still controlled by Indians, it remained "a stagnant pool," while Josephine Middle-kauf perceived the unplowed Kansas grasslands chiefly as "raw materials." Even so seemingly neutral a phrase as this last one—"raw materials"—is freighted with narrative meaning. Indeed, it contains buried within it the entire story of progressive development in which the environment is transformed from "raw materials" to "finished product." In just this way, story and scene become entangled—with each other, and with the politics of invasion and civilized progress—as we try to understand the Plains environment and its history.

Now in fact, these optimistic stories about Great Plains settlement are by no means typical of historical writing in the twentieth century. The problems of settling a semiarid environment were simply too great for the frontier story to proceed

---

[15] Kenneth Burke, *A Grammar of Motives* (Berkeley, 1969), 6–7.

without multiple setbacks and crises. Even narrators who prefer an ascending plot line in their stories of regional environmental change must therefore tell a more complicated tale of failure, struggle, and accommodation in the face of a resistant if not hostile landscape.

Among the most important writers who adopt this narrative strategy are Walter Prescott Webb and James Malin, the two most influential historians of the Great Plains to write during the first half of the twentieth century. Webb's classic work, *The Great Plains,* was published over half a century ago and has remained in print to this day.[16] It tells a story that significantly revises the Turnerian frontier. For Webb, the Plains were radically different from the more benign environments that Anglo-American settlers had encountered in the East. Having no trees and little water, the region posed an almost insurmountable obstacle to the westward march of civilization. After describing the scene in this way, Webb sets his story in motion with a revealing passage:

> In the new region—level, timberless, and semi-arid—[settlers] were thrown by Mother Necessity into the clutch of new circumstances. Their plight has been stated in this way: east of the Mississippi civilization stood on three legs—land, water, and timber; west of the Mississippi not one but two of these legs were withdrawn,—water and timber,—and civilization was left on one leg—land. It is small wonder that it toppled over in temporary failure.[17]

It is easy to anticipate the narrative that will flow from this beginning: Webb will tell us how civilization fell over, then built itself new legs and regained its footing to continue its triumphant ascent. The central agency that solves these problems and drives the story forward is human invention. Unlike the simpler frontier narratives, Webb's history traces a dialectic between a resistant landscape and the technological innovations that will finally succeed in transforming it. Although his book is over five hundred pages long and is marvelously intricate in its arguments, certain great inventions mark the turning points of Webb's plot. Because water was so scarce, settlers had to obtain it from the only reliable source, underground aquifers, so they invented the humble but revolutionary windmill. Because so little wood was available to build fences that would keep cattle out of cornfields, barbed wire was invented in 1874 and rapidly spread throughout the grasslands. These and other inventions—railroads, irrigation, new legal systems for allocating water rights, even six-shooter revolvers—eventually destroyed the bison herds, created a vast cattle kingdom, and broke the prairie sod for farming.

Webb closes his story by characterizing the Plains as "a land of survival where nature has most stubbornly resisted the efforts of man. Nature's very stubbornness has driven man to the innovations which he has made."[18] Given the scenic requirements of Webb's narrative, his Plains landscape must look rather different from that of earlier frontier narrators. For Webb, the semiarid environment is neither a wilder-

[16] Walter Prescott Webb, *The Great Plains* (New York, 1931).
[17] *Ibid.*, 9.
[18] *Ibid.*, 508.

ness nor a waste, but itself a worthy antagonist of civilization. It is a landscape the very resistance of which is the necessary spur urging human ingenuity to new levels of achievement. Webb thus spends much more time than earlier storytellers describing the climate, terrain, and ecology of the Great Plains so as to extol the features that made the region unique in American experience. Although his book ends with the same glowing image of a transformed landscape that we find in earlier frontier narratives, he in no way devalues the "uncivilized" landscape that preceded it. Quite the contrary: the more formidable it is as a rival, the more heroic become its human antagonists. In the struggle to make homes for themselves in this difficult land, the people of the Plains not only proved their inventiveness but built a regional culture beautifully adapted to the challenges of their regional environment.

Webb's story of struggle against a resistant environment has formed the core of most subsequent environmental histories of the Plains. We have already encountered one version of it in Paul Bonnifield's *The Dust Bowl*. It can also be discovered in the more ecologically sophisticated studies of James C. Malin, in which the evolution of "forest man" to "grass man" becomes the central plot of Great Plains history.[19] Malin's prose is far less story-like in outward appearance than Webb's, but it nonetheless narrates an encounter between a resistant environment and human ingenuity. Malin's human agents begin as struggling immigrants who have no conception of how to live in a treeless landscape; by the end, they have become "grass men" who have brought their culture "into conformity with the requirements of maintaining rather than disrupting environmental equilibrium." So completely have they succeeded in adapting themselves that they can even "point the finger of scorn at the deficiencies of the forest land; grassless, wet, with an acid, leached, infertile soil."[20] Human inhabitants have become one with an environment that only a few decades before had almost destroyed them.

The beauty of these plots is that they present the harshness of the regional environment in such a way as to make the human struggle against it appear even more positive and heroic than the continuous ascent portrayed in earlier frontier narratives. The focus of our attention is still relatively small-scale, though both the geographical and the chronological context of the plot have expanded. The story is now much more a regional one, so that the histories of one family or town, or even of Kansas or Oklahoma, become less important than the broader history of the grass-land environment as a whole. The time frame too has advanced, so that the history of technological progress on the Plains moves well into the twentieth century. Because the plot still commences at the moment that Euroamerican settlers began to

[19] These terms appear, for instance, in Malin's magnum opus, James C. Malin, *The Grassland of North America: Prolegomena to Its History* (Gloucester, Mass., 1967), but this basic notion informs virtually all of his work on the grasslands. See also James C. Malin, *Grassland Historical Studies: Natural Resources Utilization in a Background of Science and Technology* (Lawrence, Kan., 1950); and the collection of essays, James C. Malin, *History and Ecology: Studies of the Grassland*, ed. Robert P. Swierenga (Lincoln, 1984).

[20] Malin, *Grassland of North America*, 154.

occupy the grasslands, though, there is no explicit *backward* extension of the time frame. The precontact history of the Indians is not part of this story.

Most interestingly, the human subject of these stories has become significantly broader than the earlier state and local frontier histories. Rather than focus primarily on individual pioneers and their communities, these new regional studies center their story on "civilization" or "man." The inventions that allowed people to adapt to life on the Great Plains are thus absorbed into the broader story of "man" and "his" long conquest of nature. No narrative centered on so singular a central character could be politically innocent. More erasures are at work here: Indians, yes, but also women, ethnic groups, underclasses, and any other communities that have been set apart from the collectivity represented by Man or Civilization. The narrative leaves little room for them, and even less for a natural realm that might appropriately be spared the conquests of technology. These are stories about a progress that, however hard-earned, is fated; its conquests are only what common sense and nature would expect. For Webb and Malin, the Great Plains gain significance from their ties to a world-historical plot, Darwinian in shape, that encompasses the entire sweep of human history. The ascending plot line we detect in these stories is in fact connected to a much longer plot line with the same rising characteristics. Whether that longer plot is expressed as the Making of the American Nation, the Rise of Western Civilization, or the Ascent of Man, it still lends its grand scale to Great Plains histories that outwardly appear much more limited in form. This may explain how we can find ourselves so entranced by a book whose principal subject for five hundred pages is the invention of windmills and barbed wire.

But there is another way to tell this history, one in which the plot ultimately falls rather than rises. The first examples of what we might call a "declensionist" or "tragic" Great Plains history began to appear during the Dust Bowl calamity of the 1930s. The dominant New Deal interpretation of what had gone wrong on the Plains was that settlers had been fooled by a climate that was sometimes perfectly adequate for farming and at other times disastrously inadequate. Settlement had expanded during "good" years when rainfall was abundant, and the perennial optimism of the frontier had prevented farmers from acknowledging that drought was a permanent fact of life on the Plains. In this version, Great Plains history becomes a tale of self-deluding hubris and refusal to accept reality. Only strong government action, planned by enlightened scientific experts to encourage cooperation among Plains farmers, could prevent future agricultural expansion and a return of the dust storms.

The classic early statement of this narrative is that of the committee that Franklin D. Roosevelt appointed to investigate the causes of the Dust Bowl, in its 1936 report on *The Future of the Great Plains*. Its version of the region's history up until the 1930s runs as follows:

> The steady progress which we have come to look for in American communities was beginning to reverse itself. Instead of becoming more productive, the Great Plains

"The Great Plains of the Past," an illustration for *The Future of the Great Plains* (1936). This and its companion illustrations on the following pages illustrate the New Deal story of the Great Plains. The caption for this one reads, in part, "As the first white settlers drove their covered wagons slowly westward . . . they found the Red Man living in rude but productive harmony with Nature."

were becoming less so. Instead of giving their population a better standard of living, they were tending to give them a poorer one. The people were energetic and courageous, and they loved their land. Yet they were increasingly less secure in it.[21]

One did not have to look far to locate the reason for this unexpected reversal of the American success story. Plains settlers had failed in precisely the agricultural adaptations that Webb and Malin claimed for them. Radical steps would have to be taken if the Dust Bowl disaster were not to repeat itself. "It became clear," said the

---

[21] *The Future of the Great Plains: Report of the Great Plains Committee* (Washington, 1936), 1. On this report, see Gilbert F. White, "*The Future of the Great Plains* Re-Visited," *Great Plains Quarterly,* 6 (Spring 1986), 84–93.

"The Great Plains of the Present."
The original caption was, in part, "The White Man . . . came as a conqueror first of the
Indian, then of Nature. . . . The plough ignores Nature's 'Keep Off' signs;
communities, for all the courage of their people, fall into decay."

planners, describing their own controversial conclusions with the settled authority of the past tense, "that unless there was a permanent change in the agricultural pattern of the Plains, relief always would have to be extended whenever the available rainfall was deficient."[22]

Whatever the scientific or political merits of this description, consider its narrative implications. The New Deal planners in effect argued that the rising plot line of our earlier storytellers not only was false but was itself the principal cause of the environmental disaster that unfolded during the 1930s. The Dust Bowl had occurred because people had been telling themselves the wrong story and had tried

[22] *Future of the Great Plains*, 1.

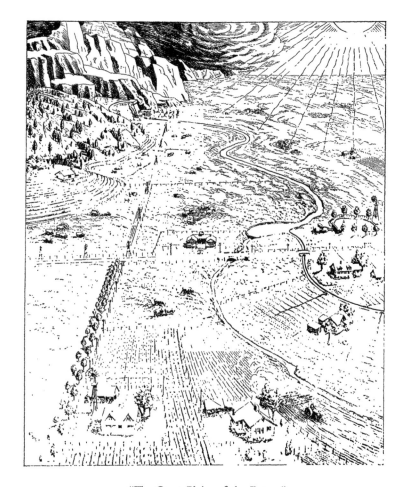

"The Great Plains of the Future."
The original caption was, in part, "The land may bloom again if man once more makes his
peace with Nature. Careful planting will give him back the foothill trees;
. . . fewer and larger farms on scientifically selected sites may yield . . .
a comfortable living. . . . This is no Utopian dream. It is a
promise, to be realized if we will."

to inscribe that story—the frontier—on a landscape incapable of supporting it.[23]
The environmental rhythms of the Plains ecosystem were cyclical, with good years
and bad years following each other like waves on a beach. The problem of human
settlement in the region was that people insisted on imposing their linear notions
of progress on this cyclical pattern. Their perennial optimism led them always to
accept as "normal" the most favorable part of the precipitation cycle, and so they
created a type and scale of agriculture that could not possibly be sustained through

[23] This image of colonial invaders seeking to "inscribe" their ideology on an alien landscape is one of the central
notions of a fascinating monograph: Tzvetan Todorov, *The Conquest of America* (New York, 1984).

the dry years. In effect, bad storytelling had wreaked havoc with the balance of nature.

By this interpretation, the "plot" of Great Plains history rises as Euroamerican settlement begins, but the upward motion becomes problematic as farmers exceed the natural limits of the ecosystem. From that moment forward, the story moves toward a climax in which the tragic flaws of a self-deluding people finally yield crisis and decline. Although the geographical and chronological frame of this narrative are much the same as in the earlier progressive plots, the *scene* has shifted dramatically. For Webb and Malin, the Plains environment was resistant but changeable, so that struggle and ingenuity would finally make it conform to the human will. In this early New Deal incarnation of a pessimistic Great Plains history, the environment was not only resistant but in some fundamental ways unchangeable. Its most important characteristics—cyclical drought and aridity—could not be altered by human technology; they could only be accommodated. If the story was still about human beings learning to live in the grasslands, its ultimate message was about gaining the wisdom to recognize and accept natural limits rather than strive to overcome them. Although the close of the New Deal committee's story still lay in the future when its report was released in 1936, its authors clearly intended readers to conclude that the only appropriate ending was for Americans to reject optimistic stories such as Webb's and Malin's in favor of environmental restraint and sound management.

The political subtext of this story is not hard to find. Whereas the heroes of earlier Great Plains narratives had been the courageous and inventive poeple who settled the region, the New Dealers constructed their stories so as to place themselves on center stage. Plainspeople, for all their energy, courage, and love of the land, were incapable of solving their own problems without help. They had made such a mess of their environment that only disinterested outsiders, offering the enlightened perspective of scientific management, could save them from their own folly. In this sense, the New Deal narrative is only partially tragic, for in fact the planners still intended a happy ending. Like Webb and Malin, they saw the human story on the Plains as a tale of adaptation, but their vision of progressive modernization ended in regional coordination and centralized state planning. Federal planners would aid local communities in developing new cooperative institutions and a more sustainable relation to the land. This was the conclusion of Pare Lorentz's famous New Deal propaganda film, *The Plow that Broke the Plains* (1936), in which a seemingly inevitable environmental collapse is finally reversed by government intervention. Technology, education, cooperation, and state power—not individualism—would bring Plains society back into organic balance with Plains nature and thereby avert tragedy to produce a happy ending.

Seen in this light, James Malin's storytelling takes on new meaning. Malin wrote in the wake of the New Deal and was a staunch conservative opponent of everything it represented. His narratives of regional adaptation expressed his own horror of collectivism by resisting the New Deal story at virtually every turn. The planners, he said, had exaggerated the severity of the Dust Bowl to serve their own statist ends and had ignored the fact that dust storms had been a natural part of the Plains environment as far back as anyone remembered. Their scientistic faith in ecology had grave political dangers, for the ecologists had themselves gone astray in viewing the Plains environment as a stable, self-equilibrating organism in which human action inevitably disturbed the balance of nature.[24] Ecosystems were dynamic, and so was the human story of technological progress: to assert that nature set insurmountable limits to human ingenuity was to deny the whole upward sweep of civilized history. The New Dealers' affection for stories in which nature and society were metaphorically cast as organisms only revealed their own hostility to individualism and their flirtation with communist notions of the state. "Scientism," Malin declared, "along with statism, have become major social myths that threaten freedom."[25]

If the New Dealers' Great Plains was a constrained environment forcing inhabitants to accept its natural limits, Malin's was a landscape of multiple possibilities, a stage for human freedom. The story of the one began in balance, moved into chaos, and then returned to the wiser balance of a scientifically planned society. The story of the other had no such prophetic return to an organic whole but expressed instead a constant process of readaptation that continued the long march of human improvement that was the core plot of Malin's history. In both cases, the shape of the landscape conformed to the human narratives that were set within it and so became the terrain upon which their different politics contested each other. Malin's commitment to individualist freedom led him to probe more deeply into grassland ecology than any historian before him, but always in an effort to find human possibilities rather than natural limits. The scene he constructed for his story was an environment that responded well to human needs unless misguided bureaucrats interfered with people's efforts to adapt themselves to the land.

It is James Malin's anti–New Deal narrative that informs Paul Bonnifield's *The Dust Bowl*. Writing in the late 1970s, at a time when conservative critiques of the welfare state were becoming a dominant feature of American political discourse, Bonnifield argues less urgently and polemically than Malin, but he tells essentially the same story. For him, the Great Plains did pose special problems to the people who settled there, but no one grappled with those problems more successfully than they. When the Dust Bowl hit, it was the people who lived there, not government scientists, who invented new land-use practices that solved earlier problems. New Deal planners understood little about the region and were so caught up in their own ideology that they compounded its problems by trying to impose their vision of a planned society.

Rather than allow residents to come up with their own solutions, Bonnifield argues, the planners used every means possible to drive farmers from their land. They did this not to address the environmental problems of the Plains, but to solve their own problem of reducing the national overproduction of wheat. To justify this deceit, they caricatured Plains inhabitants as "defeated, poverty-ridden people without hope" in such propaganda as *The Plow that Broke the Plains* and the Farm

---

[24] On the role of the Dust Bowl in reshaping the science of ecology itself, see Ronald C. Tobey, *Saving the Prairies: The Life Cycle of the Founding School of American Plant Ecology, 1895–1955* (Berkeley, 1981).

[25] Malin, *Grassland of North America*, 168.

Security Administration photographs, with their mini-narratives of environmental destruction and social despair.[26] In fact, Bonnifield argues, the Plains contained some of the best farming soil in the world. The landscape was difficult but ultimately benign for people who could learn to thrive upon it. Their chief problem was less a hostile nature than a hostile government. The narrative echoes Malin's scenic landscape but gains a different kind of ideological force when placed at the historical moment of its narration — in the waning years of the Carter administration just prior to Ronald Reagan's triumphant election as president. Bonnifield's is a tale of ordinary folk needing nothing so much as to get government off their backs.

If Bonnifield elaborates the optimistic Dust Bowl narrative of a conservative critic of the New Deal, Donald Worster returns to the New Deal plot and deepens its tragic possibilities. Worster, who is with Webb the most powerful narrator among these writers, accepts the basic framework of Roosevelt's planners — the refusal of linear-minded Americans to recognize and accept cyclical environmental constraints — but he shears away its statist bias and considerably expands its cultural boundaries. One consequence of the New Deal tale was to remove the history of the Plains from its role in the long-term ascent of civilization; instead, the region became merely an unfortunate anomaly that imposed unusual constraints on the "steady progress" that was otherwise typical of American life. Worster rejects this reading of Plains history and argues instead that the Plains were actually a paradigmatic case in a larger story that might be called "the rise and fall of capitalism."

For Worster, the refusal to recognize natural limits is one of the defining characteristics of a capitalist ethos and economy. He is therefore drawn to a narrative in which the same facts that betokened progress for Webb and Malin become signs of declension and of the compounding contradictions of capitalist expansion. The scene of the story is world historical, only this time the plot leads toward catastrophe:

> That the thirties were a time of great crisis in American, indeed, in world, capitalism has long been an obvious fact. The Dust Bowl, I believe, was part of that same crisis. It came about because the expansionary energy of the United States had finally encountered a volatile, marginal land, destroying the delicate ecological balance that had evolved there. We speak of farmers and plows on the plains and the damage they did, but the language is inadequate. What brought them to the region was a social system, a set of values, an economic order. There is no word that so fully sums up those elements as "capitalism." . . . Capitalism, it is my contention, has been the decisive factor in this nation's use of nature.[27]

By this reading, the chief agent of the story is not "the pioneers" or "civilization" or "man"; it is capitalism. The plot leads from the origins of that economic system, through a series of crises, toward the future environmental cataclysm when the system will finally collapse. The tale of Worster's Dust Bowl thus concerns an intermediate crisis that foreshadows other crises yet to come; in this, it proclaims an apocalyptic prophecy that inverts the prophecy of progress found in earlier frontier narratives. Worster's inversion of the frontier story is deeply ironic, for it implies that

[26] Bonnifield, *The Dust Bowl*, 202.
[27] Worster, *Dust Bowl*, 5.

the increasing technological "control" represented by Webb's and Malin's human ingenuity leads only toward an escalating spiral of disasters. He also breaks rank with the New Dealers at this point, for in his view their efforts at solving the problems of the Dust Bowl did nothing to address the basic contradictions of capitalism itself. For Worster, the planners "propped up an agricultural economy that had proved itself to be socially and ecologically erosive."[28]

Given how much his basic plot differs from Webb's and Malin's, the scene Worster constructs for his narrative must differ just as dramatically. Since Worster's story concerns the destruction of an entire ecosystem, it must end where the frontier story began: in a wasteland. His plot must move downward toward an ecological disaster called the Dust Bowl. Whereas the frontier narratives begin in a negatively valued landscape and end in a positive one, Worster begins his tale in a place whose narrative value is entirely good. His grasslands are "an old and unique ecological complex" that nature had struggled for millions of years to achieve, "determining by trial and error what would flourish best in this dry corner of the good earth."[29] Delicate and beautiful, the Plains were an ecosystem living always on the edge of drought, and their survival depended on an intricate web of plants and animals that capitalism was incapable of valuing by any standard other than that of the marketplace. From this beginning, the story moves down a slope that ends in the dust storms whose narrative role is to stand as the most vivid possible symbol of human alienation from nature.

The very different scenes that progressive and declensionist narrators choose as the settings for their Great Plains histories bring us to another key observation about narrative itself: where one chooses to begin and end a story profoundly alters its shape and meaning. Worster's is not, after all, the only possible plot that can organize Great Plains history into a tale of crisis and decline. Because his metanarrative has to do with the past and future of capitalism, his time frame, like that of the frontier storytellers, remains tied to the start of white settlement—the moment when the American plot of progress or decline begins its upward or downward sweep. Although he acknowledges the prior presence of Indians in the region, he devotes only a few pages to them. They are clearly peripheral to his narrative. This is true of *all* the stories we have examined thus far, for reasons that have as much to do with narrative rhetoric as with historical analysis. In their efforts to meet the narrative requirements that define a well-told tale—organic unity, a clear focus, and only the "relevant" details—these historians have little to say about the region's earlier human inhabitants. They therefore ignore the entire first half of my original chronicle of "key events" in Great Plains history. If we shift time frames to encompass the Indian past, we suddenly encounter a new set of narratives, equally tragic in their sense of crisis and declension, but strikingly different in plot and scene. As such, they offer further proof of the narrative power to reframe the past so as to include certain events and people, exclude others, and redefine the meaning of landscape accordingly.

[28] *Ibid.*, 163.
[29] *Ibid.*, 66.

One can detect this process of inclusion and exclusion in the passing references that progressive frontier narrators make to the prior, less happy stories of Indians. Sometimes, the tone of such references is elegiac and melancholy, as in the classic image of a "vanishing race"; sometimes the tone is simply dismissive. As Webb put it, "The Plains Indians were survivals of savagery," and "when there was nowhere else to push them they were permitted to settle down on the reservations."[30] If progressive change was inevitable, then so too was the eventual death or removal of the Indians. Their marginalization is thus a necessary requirement of the narrative. The feature of the environment that served as the best scenic indicator of this inevitability was the American bison, whose destruction was among the most crucial steps in undermining Indian subsistence. Even if one did not feel favorably disposed toward Indians, one could still mourn the bison. Webb again: "The Great Plains afforded the last virgin hunting grounds in America, and it was there that the most characteristic American animal made its last stand against the advance of the white man's civilization."[31]

These passing references to Indian "pre-history" are essentially framing devices, the purpose of which is to set the stage for the more important drama that is soon to follow. Historians who focus more centrally on Indians in their narratives almost inevitably construct very different plots from the ones I have described thus far. Among such scholars, one of the most sophisticated is Richard White.[32] Although his work too can be seen as a metaplot about the expansion of capitalism, the landscape he constructs is defined by Indian stories. White's narrative of Pawnee history, for instance, begins with a people living in the mixed grasslands on the eastern margins of the Plains, dividing their activities in a seasonally shifting cycle of farming, gathering, and bison hunting. As one would expect of a declensionist plot, the initial scene is basically a benign and fruitful landscape, despite occasionally severe droughts. At the moment that the Pawnees began their encounter with Euroamerican culture—first with the arrival of the horse, then with the fur trade—the Plains environment was furnishing them a comfortable subsistence. In narrative terms, its meaning was that of a much-loved home.

The downward line of White's narrative records the steady erosion of the Pawnees' landscape. European disease wiped out much of their population. The expanding Sioux tribes made it harder for them to hunt bison and raise crops. As hunting became more difficult, the material and spiritual underpinnings of Pawnee subsistence began to disintegrate. Pawnee life was increasingly in crisis, and by the 1870s—when the great herds were finally destroyed—the tribe was forced to abandon its traditional homeland and remove to Indian Territory. The story ends as a classic tragedy of exodus and despair: "When the Pawnees decided to leave the

[30] Webb, *Great Plains*, 508.

[31] *Ibid.*, 509. For a similar use of the bison story as the symbol of an earlier Indian world that in some sense "vanished" during the last third of the nineteenth century, see William Cronon, *Nature's Metropolis: Chicago and the Great West* (New York, 1991), 213–18.

[32] Richard White, *The Roots of Dependency: Subsistence, Environment, and Social Change among the Choctaws, Pawnees, and Navajos* (Lincoln, 1983), 147–211.

Loup Valley, it was in the hope that to the south in Indian territory lay a land where they could hunt the buffalo, grow corn, and let the old life of the earthlodges flower beyond the reach of the Sioux and American settlers."[33] Unfortunately, this hoped-for ending to the Pawnee story would never be achieved, because the scene it required no longer existed. As White says, "Such a land had disappeared forever."[34]

The frame of this story differs from anything we have seen thus far. It ends at the moment most of the other plots begin. It starts much further back in time, as European animals and trade goods begin to change the Plains landscape, offering opportunities and improvements in Pawnee life. Eventually a downward spiral begins, and the tragedy of the narrative becomes unrelenting as the Pawnees lose control of their familiar world. As for the scene of this plot, we have already encountered it in a different guise. The "wilderness" in which the progressive frontier narrators begin their stories is nothing less than the destroyed remnant of the Pawnees' home. It is less a wasteland than a land that has been wasted.

Narratives of this sort are by no means limited to white historians. Plenty Coups, a Crow Indian chief, tells in his 1930 autobiography of a boyhood vision sent him by his animal Helper, the Chickadee. In the dream, a great storm blown by the Four Winds destroyed a vast forest, leaving standing only the single tree in which the Chickadee—smallest but shrewdest of animals—made its lodge. The tribal elders interpreted this to mean that white settlers would eventually destroy not only the buffalo but also all tribes who resisted the American onslaught. On the basis of this prophetic dream, the Crows decided to ally themselves with the United States, and so they managed to preserve a portion of their homelands. Saving their land did not spare them from the destruction of the bison herds, however, and so they shared with other Plains tribes the loss of subsistence and spiritual communion that had previously been integral to the hunt. As Plenty Coups remarks at the end of his story, "when the buffalo went away the hearts of my people fell to the ground, and they could not lift them up again. After this nothing happened."[35]

Few remarks more powerfully capture the importance of narrative to history than this last of Plenty Coups: "After this nothing happened." For the Crows as for other Plains tribes, the universe revolved around the bison herds, and life made sense only so long as the hunt continued. When the scene shifted—when the bison herds "went away"—that universe collapsed and history ended. Although the Crows continued to live on their reservation and although their identity as a people has never ceased, for Plenty Coups their subsequent life is all part of a different story.[36] The story he loved best ended with the buffalo. Everything that has happened since is part of some other plot, and there is neither sense nor joy in telling it.

---

[33] White, *Roots of Dependency*, 211.

[34] *Ibid*.

[35] Frank Linderman, *Plenty-coups: Chief of the Crows* (1930; reprint, Lincoln, 1962), 311.

[36] The danger in the way Plenty Coups ends his story, and in Richard White's ending as well, is that the close of these tragic narratives can all too easily be taken as the end of their protagonists' cultural history. The notion that Indian histories come to an end is among the classic imperialist myths of the frontier, wherein a "vanishing race" "melts away" before the advancing forces of "civilization." Plenty Coups's declaration that "after this nothing happened" conveys with great power the tragedy of an older Indian generation but says nothing about the generations of Indians who still live within the shadow of that narrative punctuation mark.

The nothingness at the end of Plenty Coups's story suggests just how completely a narrative can redefine the events of the past and the landscapes of nature to fit the needs of its plot. After this nothing happened: not frontier progress, not the challenge of adaptation to an arid land, not the Dust Bowl. Just the nothingness that follows the end of a story. It is this nothingness that carries me back to the place where I began, to my own awareness of a paradox at the heart of my intellectual practice as an historian. On the one hand, most environmental historians would be quite comfortable in asserting the importance of the nonhuman world to any understanding of the human past. Most would argue that nature is larger than humanity, that it is not completely an invention of human culture, that it impinges on our lives in ways we cannot completely control, that it is "real," and that our task as historians is to understand the way it affects us and vice versa. Black clouds bringing dust and darkness from the Kansas sky, overturned sod offering itself as a seedbed for alien grains sprouting amid the torn roots of dying prairie grasses, dry winds filled with the stench of rotting bison flesh as wolves and vultures linger over their feasts: these are more than just stories.

And yet — they are stories too. As such, they are human inventions despite all our efforts to preserve their "naturalness." They belong as much to rhetoric and human discourse as to ecology and nature. It is for this reason that we cannot escape confronting the challenge of multiple competing narratives in our efforts to understand both nature and the human past. As I hope my reading of Great Plains history suggests, the narrative theorists have much to teach us. Quite apart from the environmental historian's analytical premise that nature and culture have become inextricably entangled in their process of mutual reshaping, the rhetorical practice of environmental history commits us to narrative ways of talking about nature that are anything but "natural." If we fail to reflect on the plots and scenes and tropes that undergird our histories, we run the risk of missing the human artifice that lies at the heart of even the most "natural" of narratives.

And just what *is* a narrative? As the evidence of my Great Plains chronicle would imply, it is not merely a sequence of events. To shift from chronicle to narrative, a tale of environmental change must be structured so that, as Aristotle said, it "has beginning, middle, and end."[37] What distinguishes stories from other forms of discourse is that they describe an action that begins, continues over a well-defined period of time, and finally draws to a definite close, with consequences that become meaningful because of their placement within the narrative. Completed action gives a story its unity and allows us to evaluate and judge an act by its results. The moral of a story is defined by its ending: as Aristotle remarked, "the end is everywhere the chief thing."[38]

Narrative is a peculiarly human way of organizing reality, and this has important implications for the way we approach the history of environmental change. Some

[37] Aristotle, *Poetics,* in *The Complete Works of Aristotle: The Revised Oxford Translation,* ed. Jonathan Barnes (2 vols., Princeton, 1984), II, 2321.
[38] *Ibid.* On the importance of a story's ending in determining its configured unity, see Kermode, *Sense of an Ending;* this can be usefully combined with Edward W. Said, *Beginnings: Intention and Method* (New York, 1975).

nonhuman events can be said to have properties that conform to the Aristotelian beginning-middle-end requirement of storytelling, as when an individual organism (or a species or a mountain range or even the universe itself) is born, persists, and dies. One *can* tell stories about such things — geologists and evolutionary biologists often do — but they lack the compelling drama that comes from having a judgeable protagonist. Things in nature usually "just happen," without raising questions of moral choice. Many natural events lack even this much linear structure. Some are cyclical: the motions of the planets, the seasons, or the rhythms of biological fertility and reproduction. Others are random: climate shifts, earthquakes, genetic mutations, and other events the causes of which remain hidden from us. One does not automatically describe such things with narrative plots, and yet environmental histories, which purport to set the human past in its natural context, all have plots. Nature and the universe do not tell stories; we do. Why is this?

Two possible answers to this question emerge from the work that philosophers and post-structuralist literary critics have done on the relationship between narrative and history. One group, which includes Hayden White and the late Louis Mink as well as many of the deconstructionists, argues that narrative is so basic to our cultural beliefs that we automatically impose it on a reality that bears little or no relation to the plots we use in organizing our experience.[39] Mink summarizes this position nicely by asserting that "the past is not an untold story." The same could presumably be said about nature: we force our stories on a world that doesn't fit them.[40] The historian's project of recovering past realities and representing them "truly" or even "fairly" is thus a delusion. Trapped within our narrative discourse, we could not do justice either to nature or to the past no matter how hard we tried — presuming, of course, that "nature" or "the past" even exist at all.

An alternative position, most recently defended by David Carr but originally developed by Martin Heidegger, is that although narrative may not be intrinsic to events in the physical universe, it is fundamental to the way we humans organize our experience. Whatever may be the perspective of the universe on the things going on around us, our human perspective is that we inhabit an endlessly storied world. We narrate the triumphs and failures of our pasts. We tell stories to explore the alternative choices that might lead to feared or hoped-for futures. Our very habit of partitioning the flow of time into "events," with their implied beginnings, middles, and ends, suggests how deeply the narrative structure inheres in our experience of the world. As Carr puts it, "Narrative is not merely a possibly successful way of describing events; its structure inheres in the events themselves. Far from being a

[39] See White, *Tropics of Discourse;* White, *Metahistory;* Mink, "Narrative Form as Cognitive Instrument"; a less extreme position that ultimately leads toward a similar conclusion can be found in Ricoeur, *Time and Narrative,* I. For a useful, if biased, explication of these debates, see Hayden White, "The Question of Narrative in Contemporary Historical Theory," *History and Theory,* 23 (no. 1, 1984), 1–33. A valuable survey can be found in Martin, *Recent Theories of Narrative.*

[40] Mink, "Narrative Form as Cognitive Instrument," 148. See also Richard T. Vann, "Louis Mink's Linguistic Turn," *History and Theory,* 26 (no. 1, 1987), 14.

formal distortion of the events it relates, a narrative account is an extension of one of their primary features."[41]

Carr's position will undoubtedly be attractive to most historians, since it argues that, far from being arbitrary, our narratives reflect one of the most fundamental properties of human consciousness. It also gives us a way of absorbing the lessons of narrative theory without feeling we have abandoned all ties to an external reality. Insofar as people project their wills into the future, organizing their lives to make acts in the present yield predictable future results—to just that extent, they live their lives as if they were telling a story. It is undoubtedly true that we all constantly tell ourselves stories to remind ourselves who we are, how we got to be that person, and what we want to become. The same is true not just of individuals but of communities and societies: we use our histories to remember ourselves, just as we use our prophecies as tools for exploring what we do or do not wish to become.[42] As Plenty Coups's story implies, to recover the narratives people tell themselves about the meanings of their lives is to learn a great deal about their past actions and about the way they *understand* those actions. Stripped of the story, we lose track of understanding itself.

The storied reality of human experience suggests why environmental histories so consistently find plots in nature and also why those plots almost always center on people. Environmental history sets itself the task of including within its boundaries far more of the nonhuman world than most other histories, and yet human agents continue to be the main anchors of its narratives. Dust storms have been occurring on the Plains for millennia, and yet the ones we really care about—those we now narrate under the title "Dust Bowl"—are the ones we can most easily transform into stories in which people become the heroes or victims or villains of the piece. In this, historians consistently differ from ecologists, who more often than not treat people as exogenous variables that fit awkwardly if at all into the theoretical models of the discipline. The historian's tendency is quite opposite. The chief protagonists and antagonists of our stories are almost always human, for reasons that go to the very heart of our narrative impulse.

Our histories of the Great Plains environment remain fixed on people because what we most care about in nature is its meaning for human beings. We care about the dust storms because they stand as a symbol of human endurance in the face of natural adversity—or as a symbol of human irresponsibility in the face of natural fragility. Human interests and conflicts create *values* in nature that in turn provide the moral center for our stories. We want to know whether environmental change

[41] David Carr, "Narrative and the Real World: An Argument for Continuity," *History and Theory*, 25 (no. 2, 1986), 117.

[42] See Robert Cover, "Nomos and Narrative," *Harvard Law Review*, 97 (Nov. 1983), 3–68. Carr's argument that all human experience is narrated does not address a deeper relativist claim, that there is no necessary correlation between the stories people tell in their own lives and the stories historians tell in reconstructing those lives. On this issue, see Noel Carroll, review of *Time, Narrative, and History* by David Carr, *History and Theory*, 27 (no. 3, 1988), 297–306.

is good or bad, and that question can only be answered by referring to our own sense of right and wrong. Nature remains mute about such matters. However passionately we may care about the nonhuman world, however much we may believe in its innate worth, our historical narratives, even those about the nonhuman world, remain focused on a human struggle over values. If these values are in effect the meanings we attach to judgeable human actions—nonhuman actions being generally unjudgeable by us—then the center of our stories will remain focused on human thoughts, human acts, and human values.

It is because we care about the consequences of actions that narratives—unlike most natural processes—have beginnings, middles, and ends. Stories are intrinsically teleological forms, in which an event is explained by the prior events or causes that lead up to it. This accounts for one feature that all these Great Plains histories have in common: all are designed so that the plot and its changing scene—its environment—flow toward the ultimate end of the story. In the most extreme cases, if the tale is of progress, then the closing landscape is a garden; if the tale is of crisis and decline, the closing landscape (whether located in the past or the future) is a wasteland. As an obvious but very important consequence of this narrative requirement, opening landscapes must be different from closing ones to make the plot work. A trackless waste must become a grassland civilization. Or: a fragile ecosystem must become a Dust Bowl. The difference between beginning and end gives us our chance to extract a moral from the rhetorical landscape. Our narratives take changes in the land and situate them in stories whose endings become the lessons we wish to draw from those changes.

However serious the epistemological problems it creates, this commitment to teleology and narrative gives environmental history—all history—its moral center. Because stories concern the consequences of actions that are potentially valued in quite different ways, whether by agent, narrator, or audience, we can achieve no neutral objectivity in writing them. Historians may strive to be as fair as they can, but as these Plains examples demonstrate, it remains possible to narrate the same evidence in radically different ways. Within the field of our narratives we too—as narrators—are moral agents and political actors. As storytellers we commit ourselves to the task of judging the consequences of human actions, trying to understand the choices that confronted the people whose lives we narrate so as to capture the full tumult of their world. In the dilemmas they faced we discover our own, and at the intersection of the two we locate the moral of the story. If our goal is to tell tales that make the past meaningful, then we cannot escape struggling over the values that define what meaning is.

This vision of history as an endless struggle among competing narratives and values may not seem very reassuring. How, for instance, are we to choose among the infinite stories that our different values seem capable of generating? This is the question that lurks so threateningly at the intersections of the different Great Plains histories we have encountered. Are nature and the past infinitely malleable in the face of our ability to tell stories about them? The uneasiness that many historians feel in confronting the postmodernist challenge comes down to this basic concern, which potentially seems to shake the very foundations of our enterprise. If our choice of narratives reflects only our power to impose our preferred version of reality on a past that cannot resist us, then what is left of history?[43]

Most practicing historians, of course, do not believe that all stories about the past are equally good, even if we are not very articulate in explaining why one is better or worse than another. Usually we just declare that we recognize good history when we see it. If pressed, we may perhaps offer a few rules of thumb to help define what we are looking for. Some might argue for depth, saying that the narrative that explains more, that is richer in its suggestions about past causes, meanings, and ambiguities, is the better history. Others might seek breadth, preferring the historical narrative that accommodates the largest number of relevant details without contradicting any relevant facts.[44] Then again, less may be more: A simple story well told may reveal far more about a past world than a complicated text that never finds its own center. Inclusiveness is another virtue: a history is better, surely, when it incorporates many different voices and events to reflect the diversity of past human experiences. But maybe coherence is more important: we might demand of good history that its components be tightly enough linked that it contains no unnecessary parts or extraneous details, lest we call it antiquarian. We might ask that a good history reflect the full historiographical tradition that lies behind it while simultaneously pushing the boundaries of that tradition. We of course want it to offer a subtle and original reading of primary sources. It should surprise us with new perspectives and interpretations. We would prefer that it be lucid, engaging, a good read. And so the list goes on.

All of these are plausible criteria, and most of us would agree that they play a part in helping us recognize good history when we see it. The trouble, obviously, is that they themselves can all too easily become objects of disagreement and struggle. Indeed, many of them reflect the same sorts of aesthetic judgments that we make when encountering any narrative, historical or nonhistorical, fictional or nonfictional. It is not at all clear that they would help us very much in deciding whether Webb or Worster or Bonnifield or Plenty Coups is the better narrator of Great Plains history. If the criteria we use in deciding the relative merits of historical narratives are open to the same sorts of value judgments as the narratives themselves, then we have hardly escaped the dilemma that postmodernist theory has posed for us. We seem still to be rudderless in an endless sea of stories.

Before going any further, I should probably confess my own uncertainty about how to navigate from here to a safe harbor, wherever it might be. I first wrote this essay nearly five years ago in an effort to acknowledge the rich insights that postmodernism has given us into the complexities of narrative discourse. I assembled

[43] This question, in a somewhat different form, is the chief topic of Peter Novick, *That Noble Dream: The "Objectivity Question" and the American Historical Profession* (Cambridge, Eng., 1988).

[44] As with most of these criteria, there are deep problems here. To say that historical narratives must include all relevant details and contradict no relevant facts begs the most important question, for the tool we use to define relevance is narrative itself. Does this particular fact belong to this particular story? Only the story can tell us. To test a narrative by its ability to include facts—the relevance of which is defined by the narrative's own plot—is to slide rapidly into tautology.

a small collection of stories about the Great Plains to see what narrative theory might tell me about the way those stories shape our sense of a landscape and the people who live upon it. The exercise persuaded me that plot and scene and character, beginnings and middles and ends, the rhetoric of storytelling, the different agendas of narrators and readers, all permeate our activities as historians. To deny the richness of this insight would be an evasion of self-knowledge, a willful refusal to recognize the power and the paradoxes that flow from our narrative discourse.

And yet despite what I have learned in writing this essay, it has also been a frustrating struggle, because I, like most practicing historians, am only willing to follow the postmodernists so far. The essay has gone through four radically different versions, each with a different title, each trying to make a different kind of peace with the dilemmas these Great Plains histories pose. My goal throughout has been to acknowledge the immense power of narrative while still defending the past (and nature) as real things to which our storytelling must somehow conform lest it cease being history altogether. Alas, I shared each new version of the essay with a different group of readers and critics, and each time they persuaded me that my efforts to find safe harbor had failed. Each new version of the essay, and each letter and conversation that critiqued it, returned me to where I began: each became a different story about the meaning of stories, a different argument about how narrative does and does not ground itself in nature and the past. The essay, in other words, recapitulated the very problems it set out to solve.

But perhaps there lies hidden in this seemingly frustrating fact a partial solution to the narrative dilemma. (Watch: I try one more tack to seek some shelter in this rhetorical storm.) The same process of criticism that shaped the different versions of this essay typifies the production and consumption of all historical texts. The stories we tell about the past do not exist in a vacuum, and our storytelling practice is bounded in at least three ways that limit its power. First, our stories cannot contravene known facts about the past. This is so much a truism of traditional historical method that we rarely bother even to state it, but it is crucial if we wish to deny that all narratives do an equally good job of representing the past. At the most basic level, we judge a work bad history if it contradicts evidence we know to be accurate and true. Good history does not knowingly lie. A history of the Great Plains that narrated a story of continuous progress without once mentioning the Dust Bowl would instantly be suspect, as would a history of the Nazi treatment of Jews that failed to mention the concentration camps. Historical narratives are bounded at every turn by the evidence they can and cannot muster in their own support.

Environmental historians embrace a second set of narrative constraints: given our faith that the natural world ultimately transcends our narrative power, our stories must make ecological sense. You can't put dust in the air—or tell stories about putting dust in the air—if the dust isn't there.[45] Even though environmental histories transform ecosystems into the scenes of human narratives, the biological and geological processes of the earth set fundamental limits to what constitutes a plausible nar-

---

[45] I borrow this lovely epigram from a remark of Patricia Limerick's.

rative. The dust storms of the 1930s are not just historical facts but natural ones: they reflect the complex response of an entire ecosystem—its soils, its vegetation, its animals, its climate—to human actions. Insofar as we can know them, to exclude or obscure these natural "facts" would be another kind of false silence, another kind of lying.

In choosing to assign narrative meaning to "natural" events of this sort, we face a special problem, for nature does not tell us whether a dust storm is a good or bad thing; only we can do that. Nature is unlike most other historical subjects in lacking a clear voice of its own. The very fact that Great Plains historians can ascribe to the same landscape such different meanings is one consequence of this lack of voice. Still, nature is hardly silent. No matter what people do, their actions have real consequences in nature, just as natural events have real consequences for people. In narrating those consequences, we inevitably interpret their meaning according to human values—but the consequences themselves are as much nature's choice as our own. To just that extent, nature coauthors our stories. A Bonnifield and a Worster may draw radically different lessons from the Dust Bowl, but neither can deny the great storms themselves. The power of narrative does not extend nearly so far.

Finally, historical narratives are constrained in a third important way as well. Historians do not tell stories by themselves. We write as members of communities, and we cannot help but take those communities into account as we do our work. Being American, being male, being white, being an upper-middle-class academic, being an environmentalist, I write in particular ways that are not all of my own choosing, and my biases are reflected in my work. But being a scholar, I write also for a community of other scholars—some very different from me in their backgrounds and biases—who know nearly as much about my subject as I do. They are in a position instantly to remind me of the excluded facts and wrong-headed interpretations that my own bias, self-delusion, and lack of diligence have kept me from acknowledging.

The stories we write, in other words, are judged not just as narratives, but as nonfictions. We construct them knowing that scholars will evaluate their accuracy, and knowing too that many other people and communities—those who have a present stake in the way the past is described—will also judge the fairness and truth of what we say. Because our readers have the skill to know what is *not* in a text as well as what is in it, we cannot afford to be arbitrary in deciding whether a fact does or does not belong in our stories. Someone among our readers—a bemused colleague, an angry partisan, a wounded victim—will eventually inform us of our failings. Nature, of course, will not bother to construct such a critique, but plenty of others will step forward to speak on its behalf as we ourselves have done. We therefore struggle to anticipate criticisms, to absorb contradictory accounts, and to fit our narratives to what we already know about our subject. Criticism can sometimes do more harm than good—sapping the life from a story, burying strong arguments beneath nitpicking caveats, reinforcing conventional wisdom at the expense of new or radical insights, and murdering passion—but it can also keep us honest by forcing us to confront contradictory evidence and counternarratives. We tell stories *with* each other and *against* each other in order to speak *to* each other. Our readers, in short, play crucial roles in shaping the stories we tell. Just so has this essay gone through four separate incarnations to reach its present form, each of them responding in different ways to the critical communities that in a very real sense helped author them. No matter how frustrating this process of revision may be, the resulting text is in this case unquestionably better as a result.[46]

And what of my own story here? What kind of tale have *I* been telling about Great Plains history? My most visible narrative has of course been a story about storytellers who express their own times and political visions. Each told tales that embodied the values of a particular community. Each tried to be true to the "facts" as they then appeared. Each looked back to earlier storytellers, accommodating them when possible and trying to demonstrate their inadequacy when this was necessary to the success of the newer story. The result was a sequence of contesting stories, from tales of frontier progress to the New Deal tragedies, to Malin's and Bonnifield's stories of local resistance in the face of a hostile environment and bureaucracy, to Worster's tragedy of environmental crisis and capitalist self-destruction.

But the meaning of my story about stories also reflects that other, more personal, narrative, the one about my struggle to accommodate the lessons of critical theory without giving in to relativism. That story began with a question. If postmodernism is correct in arguing that narrative devices are deeply present even in such a field as environmental history, which takes for its subject the least human and least storied of worlds—nature—must we then accept that the past is infinitely malleable, thereby apparently undermining the entire historical project? Given my biases, the answer to this question has got to be no, and so my story has worked its way toward an ending about the ultimate justification of history in community, past reality, and nature itself. For me, there is something profoundly unsatisfying and ultimately self-deluding about an endless postmodernist deconstruction of texts that fails to ground itself in history, in community, in politics, and finally in the moral problem of living on earth. Against it, I would assert the virtues of narrative as our best and most compelling tool for searching out meaning in a conflicted and contradictory world.

The danger of postmodernism, despite all the rich insights it offers into the contested terrain of narrative discourse, is that it threatens to lose track of the very thing that makes narrative so compelling a part of history and human consciousness both. After all, the principal difference between a chronicle and a narrative is that a good story makes us *care* about its subject in a way that a chronicle does not.[47] My list of "significant Great Plains events" surely had no effect on anyone's emotions or moral vision, whereas I doubt anyone can read Donald Worster's *Dust Bowl* without being moved in one way or another. More powerfully still, the nothingness at the end of Plenty Coups's story suggests that even silence—the ability of narrative to

[46] I owe this argument about the role of criticism in limiting historical narratives to Richard White's comments on an earlier version of this essay. His help, and the way it has reshaped the text you now read, precisely illustrates my point about the critical praxis of scholarly communities.

[47] Jim O'Brien pointed me toward the importance of this insight.

rupture the flow of time in the service of its meaning—can touch us deeply with its eloquence. When a narrator honestly makes an audience care about what happens in a story, the story expresses the ties between past and present in a way that lends deeper meaning to both. This process, like everything else in history, is open to criticism, since the rhetorical devices for making an audience care can become all too manipulative and sentimental. At its best, however, historical storytelling helps keep us morally engaged with the world by showing us how to care about it and its origins in ways we had not done before.

If this is true, then the special task of environmental history is to assert that stories about the past are better, all other things being equal, if they increase our attention to nature and the place of people within it. They succeed when they make us look at the grasslands and their peoples in a new way. This is different from saying that our histories should turn their readers into environmentalists or convince everyone of a particular political point of view. Good histories rarely do this. But if environmental history is successful in its project, the story of how different peoples have lived in and used the natural world will become one of the most basic and fundamental narratives in all of history, without which no understanding of the past could be complete. Despite the tensions that inevitably exist between nature and our narrative discourse, we cannot help but embrace storytelling if we hope to persuade readers of the importance of our subject. As Aristotle reminded us so long ago, narrative is among our most powerful ways of encountering the world, judging our actions within it, and learning to care about its many meanings.

Because I care so much about nature and storytelling both, I would urge upon environmental historians the task of telling not just stories about nature, but stories about stories about nature.[48] I do so because narratives remain our chief moral compass in the world. Because we use them to motivate and explain our actions, the stories we tell change the way we act in the world. They are not just passive accounts: in a very literal sense, the frontier stories helped *cause* the Dust Bowl, just as the New Deal stories helped cause the government response to that disaster. We find in such stories our histories and prophecies both, which means they remain our best path to an engaged moral life. In organizing ecological change into beginnings, middles, and ends—which from the point of view of the universe are fictions, pure and simple—we place human agents at the center of events that they themselves may not fully understand but that they constantly affect with their actions. The end of these human stories creates their unity, the telos against which we judge the efficacy, wisdom, and morality of human actions.

Historians and prophets share a common commitment to finding the meaning of endings. However much we understand that an ecosystem transcends mere humanity, we cannot escape the valuing process that defines our relationship to it. To see how much this is so, one has only to consider the various labels Americans have attached to the Great Plains since 1800: the Land of the Buffalo; the Great American

---

[48] An extraordinary example of such stories about stories, set within the boundaries of a single Kansas county on the eastern Plains, is William Least Heat-Moon, *PrairyErth (a deep map)* (Boston, 1991).

Desert; the Great Plains; the Wheat Belt; the Dust Bowl; the Breadbasket of the World; the Land Where the Sky Begins.[49] These are not simply names or descriptive phrases. Each implies a different possible *narrative* for environmental histories of the region, and different possible endings for each of those stories. Narrative is thus inescapably bound to the very names we give the world. Rather than evade it — which is in any event impossible — we must learn to use it consciously, responsibly, self-critically. To try to escape the value judgments that accompany storytelling is to miss the point of history itself, for the stories we tell, like the questions we ask, are all finally about value. So it is with questions that I will end:

What do people care most about in the world they inhabit?

How do they use and assign meaning to that world?

How does the earth respond to their actions and desires?

What sort of communities do people, plants, and animals create together?

How do people struggle with each other for control of the earth, its creatures, and its meanings?

And on the grandest scale: what is the mutual fate of humanity and the earth?

Good questions all, and starting points for many a story . . . .

[49] The shifting meanings of the Plains as "Great American Desert" are explored in Martyn J. Bowden, "The Great American Desert in the American Mind: The Historiography of a Geographical Notion," in *Geographies of the Mind: Essays in Historical Geography*, ed. David Lowenthal and Martyn J. Bowden (New York, 1976), 119–47. See also William E. Riebsame, "The Dust Bowl: Historical Image, Psychological Anchor, and Ecological Taboo," *Great Plains Quarterly*, 6 (Spring 1986), 127–36.

Taming the Wilderness Myth

Environmental policy and education are currently based on Western beliefs about nature rather than on reality

Arturo Gómez-Pompa and Andrea Kaus

Despite nearly a century of propaganda, conservation still proceeds at a snail's pace; progress still consists largely of letterhead pieties and convention oratory. On the back forty we still slip two steps backward for each forward stride. The usual answer to this dilemma is "more conservation education." No one will debate this, but is it certain that only the volume of education needs stepping up? Is something lacking in the content as well?
  Aldo Leopold (1966, p. 222–223)

N ever before has the western world been so concerned with issues relating to humankind's relationship with the environment. As concerned members of this industrialized civilization, we have recognized that humanity is an integral part of the biosphere, at once the transformer and the self-appointed protector of the world. We assume that we have the answers. We assume that our perceptions of environmental problems and their solutions are the correct ones, based as they are on Western rational thought and scientific analysis. And we often present the preservation of wilderness as part of the solution toward a better planet under the presumption that we

Arturo Gómez-Pompa is a professor in the Department of Botany and Plant Science and director of the University of California Consortium on Mexico and the United States and Andrea Kaus is a graduate student in the Department of Anthropology at the University of California, Riverside, CA 92521. © 1992 American Institute of Biological Sciences.

The perspectives of rural populations are missing in our concept of conservation

know what is to be preserved and how it is to be managed.
  However, we need to evaluate carefully our own views of the environment and our own self-interests for its future use. Until now, a key component of the environmental solution has been left out of our conservation policies and education. The perspectives of the rural populations are missing in our concept of conservation. Many environmental education programs are strongly biased by elitist urban perceptions of the environment and issues of the urban world. This approach is incomplete and insufficient to deal with the complex context of conservation efforts at home and abroad. It neglects the perceptions and experience of the rural population, the people most closely linked to the land, who have a firsthand understanding of their surrounding natural environment as a teacher and provider. It neglects those who are most directly affected by the current policy decisions that are made in urban settings regarding natural resource use. It neglects those who feed us.
  Environmental policies and education reflect a collective perception of nature, the consolidation of what is held to be true about the natural

world and what is necessary to pass on to future generations. And this perception underlies and shapes the visions of alternative actions and appropriate actions formed by individuals and conservation groups. How accurate and sound is this vision? Our perceptions and knowledge of the environment are based on common beliefs, basic experience, and scientific research. Through time and generations, certain patterns of thought and behavior have been accepted and developed into what can be termed a Western tradition of environmental thought and conservation.

Western concepts of wilderness

Traditional conservationist beliefs have generally held that there is an inverse relationship between human actions and the well-being of the natural environment. The natural environment and the urban world are viewed as a dichotomy and the concern is usually focused on those human actions that negatively affect the quality of life by urban standards. Mountains, deserts, forests, and wildlife all make up that which is conceived as "wilderness," an area enhanced and maintained in the absence of people. According to the 1964 US Wilderness Act, wilderness is a place "where man himself is a visitor who does not remain." These areas are seen as pristine environments similar to those that existed before human interference, delicately balanced ecosystems that need to be preserved for our enjoyment and use and that of

The ancient Maya site of Yaxchilán (Chiapas, Mexico) shows the presence of past human civilizations and perturbations in the green wilderness of the Mexican tropics.

future generations. The wilderness is valued for its intrinsic worth, as places of reverence for nature, as sacred places for the preservation of the wilderness image (Nash 1988).

Wild lands are also seen as areas useful to modern civilization. They are presented to the public as natural-resource banks of biodiversity that merit protection from human actions and as outdoor laboratories that deserve unhindered exploration by the scientific community. And they are seen as a vital part of the environmental machinery that must be maintained to provide an acceptable quality of life in developed areas, as exemplified by current concerns about air pollution, global climate change, and deforestation. All of these concepts fall under the general term *conservation*, yet they represent mostly urban beliefs and aspirations. All too often they do not correspond with scientific findings or first-hand experience of how the world works.

In addition, the validity of widely accepted environmental truths needs to be challenged, from our belief in the virgin nature of the tropical forests to our newly developing thoughts on global warming. Scientific findings are often accepted as if they are the gospel word. But a scientific truth is really a conclusion drawn from a limited data set. It is an explanation of what scientists know to date about a topic, based on their training and interpretation of the information available. It may be replaced by another truth in light of new information that does not fit the old paradigm.

Concepts of climax communities and ecological equilibrium, for example, have been used for most of this century as a basis for scientific research, resource management, and conservation teaching. As long-term studies are analyzed and their findings tested against the old truths, the previous paradigms are being challenged (see Botkin 1990). Today, few ecologists defend the equilibrium and climax concepts. Nonequilibrium models now influence ecological theory, and nature is increasingly perceived as being in a state of continuous change. Some changes are in part random and independent of each other, whereas others are human-induced.

Other accepted truths relating to the environment are myths about nature that come from nonscientific sources. For instance, the concept of wilderness as an area without people has influenced thought and policy throughout the development of the western world (Manning 1989, Nash 1967, Stankey 1989, Whyte 1967). People see in the wilderness a window to the past, to the remote beginnings of humankind long before the comforts of modern life. We wish to set aside and preserve that which both reminds us of our place in evolutionary time and contrasts with our beliefs of human nature. Yet, recent research indicates that much wilderness has long been influenced by human activities (see Gómez-Pompa and Kaus 1990).

> The ongoing public conversation about the environment is grounded in the ancient dichotomy of man versus nature. So far we have sought to resolve the argument through a series of truces—either sequestering large tracts of wilderness in a state of imagined innocence, say, or limiting the ways in which man can domesticate nature's imagined savagery (Pollan 1990, p. 24).

The Western world has also seen the wilderness as a challenge, a frontier to be tamed and managed. Agricultural landscapes are often admired for their intrinsic beauty, as living masterpieces created by human hands from the wild. They are confirmation of an underlying belief in human technological superiority over primal forces. It confirms our faith in our ability to manage the environment, a legacy from the Industrial Revolution rooted in the concept of progress and a biblical notion of human dominion over nature. In Genesis (1:28), God says to Adam and Eve, "Be fruitful, and multiply, and fill the earth, and subdue it."

The danger is that this theoretical delineation between the realms of civilized and wild, of the intrinsic value of each realm separately, and of human mastery over natural forces, has only too tangible consequences. Emerging from Western history and experience in temperate zones, a belief in an untouched and untouchable wilderness has permeated global policies and politics in resource management from the tropics to the deserts, causing serious environmental problems.

We must begin to challenge some of our most fundamental and contradictory beliefs regarding the natural environment: the scientific capacity and knowledge available to harness and manage nature the way we see fit, and the perceived pristine state of uninhabited areas. Both beliefs, combined with the concept of the balance of nature, have led to unrealistic and contradictory tenets in our natural-resource management policies. On the utilitarian side, these policies are permeated with an acceptance of destructive practices, generated from a belief that mitigating measures can halt or reverse environmental depletion and degradation. Yet, on the preservationist side, conventional resource-management policy also includes practices based on the belief that setting aside so-called pristine tracts of land will automatically preserve their biological integrity. Neither belief takes into consideration the possibilities for natural-resource management that might arise from the integration of alternative environmental perceptions and current scientific information.

Alternative perceptions and conservation practices

The concept of wilderness as the untouched or untamed land is mostly an urban perception, the view of people who are far removed from the natural environment they depend on for raw resources. The inhabitants of rural areas have different views of the areas that urbanites designate as wilderness, and they base their land-use and resource management practices on these alternative visions. Indigenous groups in the tropics, for example, do not consider the tropical forest environment to be wild; it is their home. To them, the urban setting might be perceived as a wilderness.

> As a city dweller never looks at bricks, so the Indian never looks at a tree. There are saplings for making bows, and jatoba for making canoes, and certain branches where animals like to sit, but there are never trees noticeable for self-conscious reasons—beauty, terror, wonder (Cowell 1990, p. 25).

Many agriculturalists enter into a personal relationship with the environment. Nature is no longer an object, an *it,* but a world of complexity whose living components are often personified and deified in local myths. Some of these myths are based on generations of experience, and their representations of ecological relationships may be closer to reality than scientific knowledge. Conservation may not be part of their vocabulary, but it is part of their lifeway and their perceptions of the human relationship to the natural world.

Throughout the world, communally held resources have been managed and conserved by diverse human societies via cultural mechanisms that attach symbolic and social significance to land and resources beyond their immediate extractive value (see Feeny et al. 1990, McCay and Acheson 1990). In the Brazilian Amazon, the Kayapó belief system and ecological management, as described by Posey (1983), revolves around maintaining an energy balance between the natural and spiritual world by regulating animal and plant use via ritual and custom. Indigenous fishermen of Northern California used to place a ritual moratorium on fishing during the first few days of the salmon runs, thereby protecting the perpetuation of the salmon resource and maintaining intergroup relations along the river (Swezey and Heizer 1982).

External economic and political demands for natural resources have placed conflicting demands on land and resources maintained by local inhabitants. Often backed by powerful government or corporate business interests, conflicting perceptions of how the land and resources should be used have led to the replacement or collapse of previous resource management systems and subsequent unrestricted or uneducated use of the region. In Chiapas, Mexico, for example, the Lacandon Maya perceived the forest as the provider of subsistence. Forests were converted temporarily to agricultural lands for corn, beans, and squash within a shifting cultivation system, or the forest fallows managed to attract wildlife (Nations and Nigh 1980). Before the entry of outsider groups with other objectives and other interests, Maya people lived within the tropical ecosystem of southern Mexico and Guatemala for centuries in ways that allowed continuous forest regeneration. Yet, the majority of Maya groups that inhabited the Lacandon forest were never consulted in the government policy decisions of land use that ultimately led to its destruction.

These same lands have been, and still are, viewed from the outside as lands to be conquered, colonized, grazed, or preserved. The forests contain hardwoods valuable on the international market. The cleared forest then provides land for the landless and pastures for the cattle industry. Deforestation is not perceived as a problem by the representatives of these interests but rather as a mechanism by which to gain land tenure rights. Traditional conservationists, on the other hand, see the aesthetic, biological, and ecological value of the same land but do not necessarily see the people. They often fail to see the effects of past or current human actions, to differentiate among types of human use, or to recognize the economic value of sustainable use.

The well-known cycle of initial lumber or mineral extraction followed by colonization, land acquisition, and subsequent conversion to pasture lands has been a common denominator in most of the American tropics (Myers 1981). Though we tend to focus on the actions of the local people, on what is immediately observable, such actions are often the result of higher-level policies, such as government concessions to extractive industries (Parsons 1976, Partridge 1984, Reppetto 1990).

Even with the documentation of this cycle, even with the evidence that it is our own outside interests that are ultimately responsible for the greater part of tropical deforestation, we continue to place the blame on poverty and on the land practices of the rural sector when they are only the visible symptoms of much deeper underlying problems. More important, our beliefs and assumptions blind us to the fact that, in many cases, the traditional land-use practices of the rural sector are responsible for maintaining and protecting the biodiversity of our wilderness and have often provided the genetic diversity that strengthens the world's major food crop varieties (Altieri and Merrick 1987, Brush 1986, Nabhan 1985, Oldfield and Alcorn 1987, Reganold et. al. 1990).

Footprints in the wilderness

Scientific findings indicate that virtually every part of the globe, from the boreal forests to the humid tropics, has been inhabited, modified, or managed throughout our human past (Gómez-Pompa 1987, Kunstadter 1978, Lundell 1937, Parsons 1975,

Sauer 1958). Although they may appear untouched, many of the last refuges of wilderness our society wishes to protect are inhabited and have been so for millenia. In any current dialogue regarding tropical forests, for instance, the Amazon Basin is usually mentioned as a vital area to be left untouched and protected. Yet, archaeological, historical, and ecological evidence increasingly shows not only a high density of human populations in the past and sites of continuous human occupation over many centuries but an intensively managed and constantly changing environment as well (Anderson and Posey 1989, Balée 1989, Denevan 1976, Hartshorn 1980, Hecht and Cockburn 1990, Roosevelt 1989).

The Amazon is still the homeland for many indigenous groups who have inhabited the area since long before the arrival of Europeans, and it contains the resources on which they and other nonindigenous people depend for their livelihood. The Kayapó of Central Brazil currently occupy a two-million hectare Indian reserve, but at one time they practiced their nomadic agriculture in an area approximately the size of France (Hecht and Cockburn 1990, Posey 1983). In addition, new evidence from the Maya region suggests that the seemingly natural forests we are trying to protect from our version of civilization supported high densities of human populations and were managed by past civilizations.

Present-day parks, reserves, and refuges in the region are filled with archaeological sites. According to Turner (1976), the Maya population of southeastern Mexico may have ranged from 150 to 500 people per km^2 in the Late Classic Period, contrasting sharply with current population densities of 4.5 to 28.1 people per km^2 in the same region (Pick et. al. 1989). These past civilizations apparently managed the forests for food, fiber, wood, fuel, resins, and medicines (Gómez-Pompa 1987). Many of the tree species now dominant in the mature vegetation of tropical areas were and still are the same species protected, spared, or planted in the land cleared for crops as part of the practice of shifting agriculture (Gómez-Pompa and Kaus 1990).

It is only relatively recently that research on shifting agriculture and other tropical production systems has started to change from its previous focus on the initially cleared field to an examination of the management of the fallow after abandonment of the area for annual crops. The current composition of mature vegetation may well be the legacy of past civilizations, the heritage of cultivated fields and managed forests abandoned hundreds of years ago. Our late realization of this possibility stems from the long-held belief that only cleared and planted areas are managed, as in the ploughed fields of our experience, and that mature vegetation represents a climax community, a stable endpoint reflecting the order of nature given no human interference. Until we understand and teach that the tropical forests are "both artifact and habitat" (Hecht 1990), we will be advocating policies for a mythical pristine environment that exists only in our imagination.

As our knowledge and understanding of the anthropogenic influence on the composition of mature vegetation increases, it is necessary to redefine and qualify what is meant by undisturbed habitat. The issue is not simply the presence or density of humans, but the tools, technologies, techniques, knowledge, and experience that accompany a given society's production system. The ancient societies discussed previously, for example, were more closely bound to the local environment and more dependent on regional resources for basic subsistence. Increased productivity would have come from principally internal modifications and increased human labor for more intensive ecosystem management. Production systems that were viable remained; those that failed disappeared.

In contrast, modern production systems have advanced technologies, from chemical fertilizers to hydroelectric dams, that are external to the local environment. These technologies have the potential to impose irreversible transformations on the environment that cannot be predicted by traditional knowledge (i.e., cumulative knowledge specific to the local environment). What is recognized by the environmental and conservation movements is an ability to destroy the environment at a much greater scale

than ever before in human history. When we speak of protecting undisturbed habitat or wilderness, then, it is important to clarify that the word *undisturbed* refers to the absence of disturbance by modern technologies.

However, not all modern societies use destructive technologies, and the benefits of human interference in ecological processes are not restricted to tropical zones or past times. Present-day farmers in remote areas all over the world have managed, conserved, and even created some of the biodiversity we value so highly (Alcorn 1990, Felger and Nabhan 1978, Gliessman et. al. 1981). In the Sonoran Desert, a study of two oases on either side of the Mexico–United States border indicates that the customary land-use practices of Papago farmers on the Mexican side of the border contributed to the biodiversity of the oasis. In turn, the protection from land use of an oasis 54 km to the northwest, within the US Organ Pipe Cactus National Monument, resulted in a decline in the species diversity over a 25-year period (Nabhan et. al. 1982).

In addition, many rare varieties and species related to our major food crops can be found maintained within or bordering agricultural fields in cultivated regions. In the Sierra de Manantlan (Jalisco, Mexico), the discovery of a new perennial corn, *Zea diploperennis*, led to the establishment of a biosphere reserve to protect this species and the ecosystem in which it survives (Iltis 1988). (Biosphere reserves are part of an international reserve system established by the UNESCO Man and the Biosphere Programme, which contain zones of human use within the overall management agenda; theoretically, a biosphere reserve integrates the goals and strategies of conservation, development, reserch, and education.) The difficulty is that *Z. diploperennis* is a secondary species that grows in abandoned cornfields. To protect the species, the slash-and-burn techniques of this form of traditional agriculture have to be continued to provide the habitat that it requires. Without all the human cultural practices that go with the habitat, the species will be lost forever. Yet, this dimension of conservation has been neglected in our own tradition of natural-resource management.

Anthropogenic fires in natural-resource management

It is of utmost importance to understand both the beneficial and destructive ecological consequences of anthropogenic perturbations and to incorporate this knowledge in research and education programs. Future scientists, leaders, farmers, fishermen, and ranchers need to be exposed to alternative views and taught to see natural environmental issues within their historical, social, and cultural contexts. The view of the white ashes of forest trees that have been felled and burned for an agricultural plot may appear to an urbanite outsider to be a desecration of the wilderness, but a farmer may see it as an essential stage of renewal. One could argue that the trees felled are the representatives of rare and endangered species, and, in selected sites, this argument might be reasonable. However, most often, many of the cut or burned trunks resprout, providing the bulk of the new forest.

Slash-and-burn agriculture has been an integral part of the tropical forests ecosystems for millenia. This ancient form of agriculture is not to be confused with the widespread destructive fires set by recent colonists or squatters who have little local experience or land-tenure security. Fire is today used to open new forest land, often on the edge of new timber or mining roads, or, even worse, used as a mechanism to vent anger at the impotence of poverty and inadequate government programs. Though such rapid clearing of the forest by landless peasants is also improperly referred to as slash-and-burn or shifting agriculture, in reality the planted areas are not fallowed, but successively replanted and eventually abandoned. This sequence is very different from the continual process of clearing, planting, and fallowing that is typical of more long-standing forms of shifting agriculture, which creates a mosaic of different ages of forest growth, including large patches of mature vegetation.

To give a concrete example, when a major forest fire in 1989 burned 120,000 hectares near Cancún, Mexico, the news media conveyed an image of ecocide, covering the fire's daily progress with statements about the extinction of species and the loss of invaluable forests. Environmentalists, conservationists, and most nongovernmental organizations connected to environmental issues protested the lack of any fire management plan to prevent, stop, or control forest fires. Yet, no attempt was made to understand why a fire of this magnitude had occurred in the first place.

The Cancún fire began in several different places at the same time, and its cause remains unclear. It is unlikely that it was the result of an agricultural fire that escaped from an area cleared for crops. In all of the tropical Maya region, no official form of fire control has ever existed beyond that of the individual farmers. Yet, fires have seldom been as large or extensive as this one. Agricultural fires are carefully controlled by the farmers. One of the most critical decisions they must make is when to burn the slash: when the conditions are finally dry enough, but before the first seasonal rains. The farmers know the winds, the annual climatic shifts, and past fire histories. They know how to control the size and intensity of their fires to protect the neighboring forests from burning.

The patchy mosaic of forests, forest fallows, and agricultural plots is an ideal landscape for controlling forest fires. The view from a helicopter flying over the burned area around Cancún revealed that the line of fire had stopped in areas of slash-and-burn agriculture (Figure 1). Local residents and forest authorities say that the forest burned most dramatically in areas where the valuable woods had been mined out and been subsequently devastated by Hurricane Gilberto.[1] The actual commercial and biological value of the forest was low. Biological surveys indicate that the burned zone was in fact not rich in endemic organisms (López Portillo et. al. 1990).

Although Mexico contains a multitude of unique sites in which rare and endangered species are truly threatened, these sites have not received the same visibility in the public conciousness as did the Cancún fire. But neither are they situated so close to a major international tourist location. The concern over the Cancún fire was due to a desire to have an attractive forest landscape for the increasing tourist trade in the area. This concern is not necessarily invalid, but the entire fire coverage was presented out of context and based on unfounded arguments (López Portillo et. al. 1990). The underlying problem was the general lack of understanding of the ecological processes that shape forests and landscapes. We confuse all too

---

[1] A. Gómez-Pompa, 1989, interviews.

Figure 1. Photograph taken after the 1989 forest fire in the Cancún region of the Yucatan Peninsula shows where the line of fire stopped at the edge of a slash-and-burn agricultural site.

easily the great need to protect rare and endangered species with the protection of wilderness, and we confuse our admiration of forests with the conservation of nature.

Cancún is not an isolated example. Ongoing research on the chapparral environments on both sides of the Mexico–United States California border has shown the role of fire in combatting fire (Minnich 1983, 1989). These studies indicate that the mosaic vegetation pattern in Baja California, the result of repeated small burns, has prevented the large catastrophic burns so characteristic of the equivalent ecological zone in Southern California. The composition and structure of so-called virgin forests and wilderness areas are in part artifacts of previous burns, both natural and anthropogenic (Komarek 1973, Savonen 1990, Thompson and Smith 1971). A policy of fire suppression in the United States has eliminated natural barriers to fires. Fire control in wilderness areas, from the big trees of California to the Northern and Northeastern forests, has also led to undesirable changes in their environments (Botkin 1990, Heinselman 1971, Kilgore 1973, Wells 1969, Wright and Bailey 1982).

Due to our limited knowledge of the role and experience of local human populations in managing fire, fire suppression remains the dominant policy in our management of natural resources and many national parks. We fear, and are trying to prevent, a repeat of the 1988 fire in Yellowstone National Park without fully understanding the underlying causes for its great extent, intensity, and damage. In addition, without the knowledge of the role of fire in a given ecosystem, we have developed areas so that they can no longer be subjected to prescribed burns without great risk. Yet, in so doing, these areas are also at risk from fires that cannot be controlled once they start.

The integration of alternative views of the environment

The differences between the perceived and actual environmental effects of forest fires, fire suppression, shifting agriculture, or preservationist policies are only a few examples of the contradictions and confusion that exist in connection with environmental issues and conservation. In the city or in the rural areas, inaccurate information is passed from our own education system to the rest of society and the next generations of environmental users, managers, and abusers. Research and education programs need to be redesigned to inform urban as well as rural populations (from children to adults) about appropriate and alternative resource management practices and policies. Most policy agendas and education curricula neglect rural perceptions of the environment or traditional systems of food production and resource management. They do not address the current difficulties confronted by these systems and lifeways or their contributions to conservation and our own survival. Beyond opening our eyes to the realities of the areas we call wilderness, we must learn how to listen to their caretakers (both good and bad ones) to include local needs, experience, and aspirations within our perspectives (Gómez-Pompa and Bainbridge in press).

To adjust our recommendations for better use of the environment to reflect reality rather than myths, we must learn how local inhabitants in rural areas understand their environment and must bring this vision into both the urban and rural classrooms. The first step is to recognize that conservation traditions exist in other cultural practices and beliefs that are separate from Western traditional conservation. The rural sector is not a homogeneous group, however, and research and education efforts also need to be directed toward the social or economic constraints and incentives that lead to destructive practices or conflict with institutional conservation policies.

Several priorities for research and education programs can be mentioned to improve the information and alternatives available for natural-resource management programs and future resource managers:

• Research on the influence of human activities on past and present environments to understand the influence of all forms of management, whether modern or traditional, intensive or extensive, on the shape and content of the environment.

• Long-term monitoring of environmental change that includes the social and economic variables affecting such change.

• Documentation of the views and perceptions of nature and conservation found in rural populations and integration of these beliefs and corresponding empirical realities in the general pool of collective knowledge. Knowledge of the beliefs, constraints, and aspirations of local residents in ecologically fragile lands will aid in coordinating conservation and rural development policy and practice.

• Continued emphasis on the co-ordination of research efforts in different scientific disciplines to present conservation and management alternatives with balanced representations of the different interests in conservation and rural development.

• Collaboration with interested individuals in the rural sector to establish demonstration and experimental sites for resource management alternatives and environmental restoration techniques.

• Development of environmental educational programs that integrate the knowledge and experience of scientists, educators, and local practitioners. This development should include programs that not only take scientists and educators out to rural communities, but also that encourage rural residents with successful land-use techniques to teach, whether in their own communities, other rural areas, or the urban setting.

• Development of graduate programs in conservation and natural-resource management that train a new generation of professionals, scientists, and decision makers with a view of conservation issues that includes the role of humans in both environmental deterioration and enrichment. These programs need to instill a sense of the tremendous responsibility current and future generations have to shape our own environment and of the risks of failing.

We have reached a time where the lines are not drawn between the known and unknown but between belief systems. This situation leads to an unfortunate set of circumstances in which we divide ourselves on issues where our opponents are not villains. They are often others who, like us, are working toward the protection of the environment. Yet, we line up behind the banners of preservation, conservation, development, or restoration and then subdivide on points of human involvement, responsibility, and equity in resource management. The only realities left between these polemics are the resources and the people who use them. This middle ground is where future research and education needs to be concentrated if we are to emerge from this seeming morass of controversial arguments held at a level far removed from the field.

As scientists or conservationists, we need to enter the field, literally. We speak of local participation and of developing a dialogue among the rural, research, and education communities. However, the presence of local rural residents in a classroom or conference hall does not necessarily engender participation. These locations and procedures are a standard part of our own traditional education process; they are unlikely to be familiar to the majority of indigenous or remote communities and unlikely to be conducive to an exchange of information among researchers and local people.

We sometimes forget that experience is often the best teacher and place great emphasis instead on the letters that proceed or follow a person's name, on the way that person talks, and on the material he or she writes. In so doing, we have created a barrier of formally structured education and language that is imposing to rural populations. One rancher from North Mexico once commented on the researchers with whom he had worked, "We tell them what it is like here, but they write about it differently."[2]

We know, in fact, very little about how environmental knowledge held by farmers, ranchers, fishermen, hunters, and gatherers from the deserts to the tropics is passed from one generation or society to another. This understanding requires learning the setting and language that people use to describe their environment and their relationship with the land. It implies understanding the underlying concepts of their words and the corresponding actions that are considered appropriate. Such environmental perceptions may not exactly match concepts of sustainable use or restricted access to limited or fragile resources, but overlaps of conservation concepts and practices do exist.

In an informal survey,[3] 15 people of a remote region of Durango, Mexico, were asked what the word *conservación* meant. None of them knew. "No," they replied, shaking their heads. "Qué será? (What would it mean?)" Earlier, one man from this group had pointed out ways in which he and his family were trying to protect the rangeland from the effects of drought and overgrazing and to protect the wildlife from poachers. When asked why, he turned in his saddle, viewing the range stretching away from him into the distance, and said, "Hay que cuidar, verdad? [You've got to care for it, don't you?]"

A shared perception of caring for the land can be emphasized in conservation policy and education. However, integrating this perception requires acknowledging the presence of humans in wilderness areas. Part of the problem in working with local people stems from our perception of wilderness as uninhabited. The attention automatically falls on the land first and the people second. We think of local people living in the buffer zone surrounding an uninhabited area and do not stop to consider that perhaps the buffer zone should be the principal area of conservation.

Botkin (1990) describes how resource management policies to both protect and control elephant populations in the Tsavo National Park of East Africa led to a severe deterioration of the land within the park boundaries. The inhabited area surrounding the park remained forested. The clear demarcation of the boundaries in the LANDSAT images and aerial photos appeared "as a photographic negative of one's expectation of a park. Rather than an island of green in a wasted landscape, Tsavo appeared as a wasted island amid a green land" (Botkin 1990, p. 16).

Perceptions of wilderness and protected areas as uninhabited means that local-level collaboration is often neglected or considered only as an afterthought and in terms of our own priorities. We talk easily of the role of local people in our conservation programs but often do not stop to think of the role we play in their lives. Local cooperation, participation, or collaboration are not free commodities. They influence lives and futures and both deserve and require negotiation. In the Chihuahuan Desert, for example, inhabitants in the region of the Mapimí Biosphere Reserve have included a policy of wildlife conservation and an ecological research program within their own lifeway. Their willingness to stop eating the endangered Bolsón tortoise, *Gopherus flavomarginatus*, and protect it from poaching has resulted in an increase in population of this endemic species within the reserve. In turn, the researchers have opened a window to another world outside that arid basin by providing a vision of the national importance and value of local resources and efforts.

However, the local-level efforts up to now have not been equal. Some of the local people say they have benefitted the reserve more than the reserve has benefitted them.[4] Why, then, did the local people accept the researchers in the first place? They say it was for *la convivencia*, the willingness of the initial researchers to live and work side by side with them, to accept their help and advice, and to include their concerns in the decision-making process. It was a matter of trust. The local people trust that their perceptions, their world, will be part of what is taught to others who have never set foot in the Bolsón de Mapimí, part of what is taken into consideration by those who wish to alter either local land use or the reserve management.

Environmental conservation responsibilities

Cooperative relationships with local residents in ecologically fragile areas

---

[2]A. Kaus, 1990, interviews.

[3]A. Kaus, 1989–1990, interviews.

[4]A. Kaus, 1989, interviews.

are of utmost importance to our understanding of the natural environment and the effects of resource use. Yet, we cannot neglect our responsibilities in such relationships or underestimate the effect (positive or negative) that we can have on a rural community. We need to contribute in turn and impart the information to which we have access. In this way, local people can come to understand their situation in a larger context and make informed decisions about their lives and land. But it also means orienting some of the research toward local benefits and including local-level perspectives in research design and dissemination. More important, it means including the local people in the same education process that we, ourselves, are undergoing to understand the natural environment and society's effects on it.

The benefits of local-level collaboration to our urban communities are perhaps even more than we can realistically offer in return. Perceptions, knowledge, and experience of the rural sector incorporated into the urban classroom can guide our global civilization to more informed decisions about what is termed *wilderness* and about what is meant by *conservation*. The wilderness we have envisioned up to this point is not the same when viewed from the field. In reality, the frontier does not exist between people and the wild, but between the known and unknown.

The point here is not to create a new myth or fall into the trap of the "ecologically noble savage" (Redford 1990). Not all farmers or ranchers are sages, folk scientists, or unrecognized conservationists. Yet, within the rural sector can be found individuals who directly depend on the land for their physical and cultural subsistence. And within that group of individuals exists a set of knowledge about that terrain, a knowledge of successes and failures that should be taken into account in our environmental assessments. Currently, we are discussing and designing policies for something about which we still know little. And those who do know more have rarely been included in the discussion. The fundamental challenge is not to conserve the wilderness, but to tame the myth with an understanding that humans are not apart from nature.

Acknowledgments

We would like to thank David Bainbridge and Denise Brown for their timely comments and suggestions and the external reviewers of *BioScience* for their excellent critical assessments and arguments. This article is based on a presentation given at the 19th Annual Conference of the North American Association for Environmental Education, 2–7 November 1990, in San Antonio, Texas. An abstract, in Spanish, of the presentation, "Desafáo Al Mito de la Virginidad de los Ecosistemas," was published in the proceedings of the conference's Latin American Symposium, *Estableciendo la Agenda de Educaión Ambiental para la Decada de los Noventa*, I. Castillo and A. Medina, editors.

References cited

Alcorn, J. B. 1990. Indigenous agroforestry systems in the Latin American tropics. Pages 203–213 in M. A. Altieri and S. Hecht, eds. *Agroecology and Small Farm Development*. CRC Press, Boca Raton, FL.

Altieri, M. A., and L. C. Merrick. 1987. In situ conservation of crop genetic resources through maintenance of traditional farming systems. *Econ. Bot.* 41: 86–96.

Anderson, A. B., and D. A. Posey. 1989. Management of a tropical scrub savanna by the Gorotire Kayapó of Brazil. Pages 159–173 in D. A. Posey and W. Balée, eds. *Resource Management in Amazonia: Indigenous and Folk Strategies*. New York Botanical Garden, Bronx.

Balée, W. C. 1989. The culture of Amazon forests. Pages 1–21 in D. A. Posey and W. Balée, eds. *Resource Management in Amazonia: Indigenous and Folk Strategies*. New York Botanical Garden, Bronx.

Botkin, D. B. 1990. *Discordant Harmonies: A New Ecology for the Twenty-First Century*. Oxford University Press, New York.

Brush, S. B. 1986. Genetic diversity and conservation in traditional farming systems. *J. Ethnobiol.* 6: 151–167.

Cowell, A. 1990. *The Decade of Destruction: The Crusade to Save the Amazon Rain Forest*. Henry Holt and Co., New York.

Denevan, W. M. 1976. *The Native Population of the Americas in 1492*. University of Wisconsin Press, Madison.

Feeny, D., F. Berkes, B. J. McCay, and J. M. Acheson. 1990. The tragedy of the commons: twenty-two years later. *Human Ecol.* 18: 1–19.

Felger, R. S., and G. P. Nabhan. 1978. Agroecosystem diversity: a model from the Sonoran Desert. Pages 129–149 in N. L. González, ed. *Social and Technological Management in Dry Lands: Past and Present, Indigenous and Imposed*. Westview Press, Boulder, CO.

Gliessman, S. R., R. Garcia E., and M. Amador A. 1981. The ecological basis for the appli-

cation of traditional agricultural technology in the management of tropical agro-ecosystems. *Agro-Ecosystems* 7: 173–185.

Gómez-Pompa, A. 1987. On Maya silviculture. *Mexican Studies* 3: 1–17.

Gómez-Pompa, A., and D. A. Bainbridge. In press. Tropical forestry as if people mattered. In A. E. Lugo and C. Lowe, eds. *A Half Century of Tropical Forest Research*. Springer-Verlag, New York.

Gómez-Pompa, A., and A. Kaus. 1990. Traditional management of tropical forests in Mexico. Pages 45–64 in A. B. Anderson, ed. *Alternatives to Deforestation: Steps Toward Sustainable Use of the Amazon Rain Forest*. Columbia University Press, New York.

Hartshorn, G. S. 1980. Neotropical forest dynamics. *Biotropica* 12(Suppl.): 23–30.

Hecht, S. 1990. Tropical deforestation in Latin America: myths, dilemmas, and reality. Paper presented at the Systemwide Workshop on Environment and Development Issues in Latin America, University of California, Berkeley, 16 October 1990.

Hecht, S., and A. Cockburn. 1990. *The Fate of the Forest: Developers, Destroyers and Defenders of the Amazon*. Harper Perennial, New York.

Heinselman, M. L. 1971. Restoring fire to the ecosystems of Boundary Waters Canoe Area, Minnesota, and to similar wilderness areas. Pages 9–21 in *Proceedings of the Annual Tall Timbers Fire Ecology Conference No. 10*. Tall Timbers Research Station, Tallahassee, FL.

Iltis, H. H. 1988. Serendipity in the exploration of biodiversity: what good are weedy tomatoes? Pages 98–105 in E. O. Wilson, ed. *Biodiversity*. National Academy Press, Washington, DC.

Kilgore, B. M. 1973. The impact of prescribed burning on a sequoia–mixed conifer forest. Pages 345–375 in *Proceedings of the Annual Tall Timbers Fire Ecology Conference No. 12*. Tall Timbers Research Station, Tallahassee, FL.

Komarek, E. V. 1973. Ancient fires. Pages 219–240 in *Proceedings of the Annual Tall Timbers Fire Ecology Conference No. 12*. Tall Timbers Research Station, Tallahassee, FL.

Kundstadter, P. 1978. Ecological modification and adaptation: an ethnobotanical view of Lua' swiddeners in northwestern Thailand. *Anthropological Papers* 67: 169–200.

Leopold, A. 1966. *A Sand County Almanac*. Oxford University Press, New York.

López-Portillo, J., M. R. Keyes, A. González, E. Cabrera C., and Odilón Sánchez. 1990. Los incendios de Quintana Roo: ¿catástrofe ecológica o evento periódico? *Ciencia y Desarrollo* 16(91): 43–57.

Lundell, C. L. 1937. *The Vegetation of Petén*. Carnegie Institute, Washington, DC.

Manning, R. E. 1989. The nature of America: visions and revisions of wilderness. *Nat. Res. J.* 29: 25–40.

McCay, B. J., and J. M. Acheson, eds. 1990. *The Question of the Commons: The Culture and Ecology of Communal Resources*. University of Arizona Press, Tucson.

Minnich, R. A. 1983. Fire mosaics in Southern California and Northern Baja California. *Science* 219: 1287–1294.

———. 1989. Chaparral fire history in San Diego County and adjacent Northern Baja California: an evaluation of natural fire regimes and the effects of suppression management. *Nat. Hist. Mus. Los Angel. Cty. Contrib. Sci.* 34: 37–47.

Myers, N. 1981. Deforestation in the tropics: who gains, who loses? Pages 1–21 in V. H. Sutlive, N. Altshuler, and M. Zamora, eds. *Where Have All the Flowers Gone? Deforestation in the Third World.* Studies in Third World Societies 13, Dept. of Anthropology, College of William and Mary, Williamsburg, VA.

Nabhan, G. P. 1985. Native crop diversity in Aridoamerica: conservation of regional gene pools. *Econ. Bot.* 39: 387–399.

Nabhan, G. P., A. M. Rea, K. L. Reichhardt, E. Mellink, and C. F. Hutchinson. 1982. Papago influences on habitat and biotic diversity: Quitovac oasis ethnoecology. *J. Ethnobiol.* 2: 124–143.

Nash, R. F. 1967. *Wilderness and the American Mind.* Yale University Press, New Haven, CT.

———. 1988. The United States—why wilderness? Pages 194–201 in V. Martin, ed. *For the Conservation of Earth.* Fulcrum, Golden, CO.

Nations, J. D., and R. B. Nigh. 1980. The evolutionary potential of Lacandon Maya sustained-yield tropical forest agriculture. *J. Anthropol. Res.* 36: 1–30.

Oldfield, M. L., and J. B. Alcorn. 1987. Conservation of traditional agroecosystems. *BioScience* 37: 199–208.

Parsons, J. J. 1975. The changing nature of New World tropical forests since European colonization. Pages 28–38 in *Proceedings of the International Meeting on the Use of Ecological Guidelines for Development in the American Humid Tropics.* IUCN Publications New Series 31, Morges, Switzerland.

———. 1976. Forest to pasture: development or destruction? *Rev. Biol. Trop.* 24 (Suppl. 1): 121–138.

Partridge, W. L. 1984. The humid tropics cattle ranching complex: cases from Panama reviewed. *Hum. Org.* 43: 76–80.

Pick, J. B., E. W. Butler, and E. L. Lanzer. 1989. *Atlas of Mexico.* Westview Press, Boulder, CO.

Pollan, M. 1990. Only man's presence can save nature. *J. For.* 88(7): 24–33.

Posey, D. A. 1983. Indigenous knowledge and development: an ideological bridge to the future. *Ciencia e Cultura* 35: 877–894.

Redford, K. H. 1990. The ecologically noble savage. *Orion* 9(3): 25–29.

Reganold, J. P., R. I. Papendick, and J. F. Parr. 1990. Sustainable agriculture. *Sci. Am.* 262(6): 112–120.

Repetto, R. 1990. Deforestation in the tropics. *Sci. Am.* 262(4): 36–42.

Roosevelt, A. 1989. Resource management in Amazonia before the conquest: beyond ethnographic projection. Pages 30–62 in D. A. Posey, and W. Balé, eds. *Resource Management in Amazonia: Indigenous and Folk Strategies.* New York Botanical Garden, Bronx.

Sauer, C. 1958. Man in the ecology of tropical America. *Proceedings of the Ninth Pacific Science Congress* 20: 105–110

Savonen, C. 1990. Ashes in the Amazon. *J. For.* 88(9): 20–25.

Stankey, G. H. 1989. Beyond the campfire's light: historical roots of the wilderness concept. *Natural Resources Journal* 29:9–24.

Swezey, S. L., and R. F. Heizer. 1982. Ritual management of salmonid fish resources in California. *Journal of California Anthropology* 4: 7–29.

Thompson, D. Q., and R. H. Smith. 1971. The forest primeval in the Northeast—a great myth? Pages 255–265 in *Proceedings of the Annual Tall Timbers Fire Ecology Conference No. 10.* Tall Timbers Research Station, Tallahassee, FL.

Turner, B. L. II. 1976. Population density in the Classic Maya lowlands: new evidence for old approaches. *Geographical Review* 66: 73–82.

Wells, G. S. 1969. *Garden in the West.* Dodd, Mead and Co., New York.

Whyte, L. 1967. The historical roots of our ecological crisis. *Science* 155: 1203–1207.

Wright, H. A., and A. W. Bailey. 1982. *Fire Ecology: United States and Southern Canada.* John Wiley & Sons, New York.

The Hunter of Good Heart

In North American hunting cultures, the hunter of good heart is a metaphoric ideal that reveals the nature of journeying toward completeness. This goal is reached through a direct relationship with animals and through hunting a path to our innermost spirit. The hunter is an archetypal form that resides in each of us, man or woman. The hunter is that part of us who searches for the completion that each of us, in our own way, strives to find. The Huichol Indians follow the tracks of the five pointed deer, their elder brother and the spirit of their sacred peyote. They say the act of hunting is about trying to find our life. Hunting from this perspective becomes a sacred act tied to our primal beginnings as hunter/gatherers. This is the perspective of becoming the animals we hunted, when humans and animals spoke to each other, and when humans began learning to be human, to be People.

Humans have always hunted to satisfy their hunger. It is one of the most basic of all human activities. It is the most basic reciprocal activity of all living things, that is, to eat and be eaten. Hunting, and all it involves, is a primal foundation of human learning and teaching. All the physical, emotional, and cognitive abilities of humans are applied in the act of hunting. Hunting is a complete and completing human act. It touches the heart of the drama of being born, living, and dying. Hunting characterizes our species to such an extent that, even in modern society, men and women far removed from hunting animals for sustenance still metaphorically "hunt" to be filled, to be satisfied, to be complete. Hunting involves coming to terms with elemental relationships at the physical, social, and spiritual levels. Hunting, in its most spiritual sense, is based on following the path of one's own spirit, manifested in the physical form of what you are hunting. In modern society, people are no longer hunting an animal, but something they have come to yearn for, the soul our ancestors were able to touch through their hunting and ritual; "that place Indians talk

Reprinted with permission of the author, G. Cajete.

about." So it is in our times that we continue to hunt for ourselves, in our families, in our communities, in our careers, in our schools, in our institutions, and our relationships.

Mimbres Hunter of Good Heart

The Hunter of Good Heart is an ancient metaphor of one path of human meaning. It is a path of spiritual ecology with various descriptions in all ancient cultures. Before there were warriors, there were hunters, and if war was waged in that world of the first people, it more than likely was waged for advantage in hunting. The Indigenous art of war has its roots firmly planted in the strategies and single-minded focus of hunting. They are both spiritual acts, matters of life, death, and learning about spirit.

The Hunter of Good Heart is a state of completeness reached through an educational process involving the hunter's whole being. The culmination is the hunter's initiation into a communion of his spirit with the animals and the world. This process gives life to the hunter and his people. The good hunt is a responsibility undertaken not only by the individual, but by the community as well. The hunter hunts to perpetuate the life of family, clan, and community. The hunter represents the community to the world of animals and spirits; therefore, the community as well as the hunter is judged through his behavior. Wife, children, parents, relatives, the whole community participates with the hunter in a cycle of life and spirit involving prayer, hunting technique, ritual, art, sacrifice, and celebration. It is, as Plains Indian people say, "a medicine circle."

The act of hunting in the context of a medicine circle is a mythical act based on the central foundation of life and the taking of life to have life. Life feeds on life, continually transforming the physical substance of life and spirit through the never-ending cycle of living, dying, and being reborn. Like the predator revealing its most primal characteristic during the moment of predation, so humans awaken when they hunt with full consciousness of their actions. This higher context of hunting allows full awareness, understanding, and reverence for the spiritual meaning of hunting. Hunters participate in the "great dance of life"– a dance of holy communion that has been joined since the dawn of human history. The hunter/shaman paintings of Tres Frieres, Altimira, and Lascaux, the Eland dance of the Bushmen, and other Indigenous art forms around the world attest to the hunter's special covenant with the life of animals and the life of a community. It is a communion merging inner and outer, with the hunter coming to know his own spirit through the spirit of that which he hunts. In this seemingly paradoxical union, a window appears into the reality of spirit where the most elusive prey resides – one's true and complete self.

Hunting is a path to a spiritual reality where each participant gains a set of universal understandings; deep relationship, abiding respect, fulfillment, and love for life meld into one. It becomes a spiritual foundation for teaching and learning ecological relationship. The foundational qualities that the hunter internalizes include: complete attentiveness to the moment, integration of action, being, and thinking, a sense for concentric rings of relationship, and humility.

Indigenous stories of Hunters reflect in the most eloquent way the nature of this path and way of spiritual learning. The following story from the Navajo Deer Hunting Way provides advice to all Hunters of Good Heart. It is presented here with deep respect for the "Diné' way of Indigenous being.

The Navajo Deer Hunting Way

The basis of Navajo ceremonialism is establishing and maintaining a harmonious balance with all that surrounds them. The Deer Hunting Way is one of the Pathways that form ceremonial structures. These ceremonies establish the harmony and balance the Navajo deemed essential for being prosperous. As is true of all tribes in whose territory deer lived, the Navajo venerated deer as a source of food and life. Deer were considered spirit beings who had to be respected and properly treated. Deer were teachers of a source of life and knowledge of spirit that originated with the First People, the Holy Ones. But they had to be hunted. For his own well-being and success, the hunter had to possess, not only the technical skills for hunting them, but the necessary spiritual preparations as well. The hunter had to know the origin of the deer. He needed to know how to behave and be successful in taking their life, for his own life and that of his family and clan. With respect and honor, the story goes this way:

"There was a hunter who waited in ambush. Wind had told him, 'This is where the tracks are. The deer will come marching through in single file.' The hunter had four arrows: one was made from sheet lightning, one from zigzag lightning, one of sunlight roots, and one of rainbow.

"Then the first deer, a large buck with many antlers, came. The hunter got ready to shoot the buck. His arrow was already in place. But just as he was ready to shoot, the deer transformed himself into a mountain mahogany bush, *tse esdaazii*. After a while, a mature man stood up from behind the bush. He stood up and said, 'Do not shoot! We are your neighbors. These are the things that will be in the future, when human beings come into existence. This is the way you will eat us.' And he told the hunter how to kill and eat the deer. So the hunter let the mature Deerman go for the price of his information. And the Deerman left.

"Then the large doe, a shy doe, appeared behind the one who had left. The hunter was ready again to shoot the doe in the heart. But the doe turned into a cliffrose bush, *aweets aal.* A while later a young woman stood up from the bush. The woman said, 'Do not shoot! We are your neighbors. In the future, when man has been created, men will live because of us. Men will use us to live on.' So then, for the price of her information the hunter let the Doewoman go. And she left.

"Then a young buck, a two pointer, came along. And the hunter got ready to shoot. But the deer transformed itself into a dead tree, *tsin bisga*. After a while, a young boy stood up from behind the dead tree and said, 'In the future, after man has been created, if you talk about us in the wrong way we will cause trouble for you when you urinate, and we will trouble your eyes. We will also trouble your ears if we do not approve of what you say about us.' And for the price of his information, the hunter let the young Deerman go.

"Then the little fawn appeared. The hunter was ready to shoot the fawn, but she turned into a lichen-spotted rock, *tse dlaad*. After a while, a young girl stood up from the rock and spoke: 'In the future all this will happen if we approve, and whatever we shall disapprove shall be up to me. I am in charge of the other Deer People. If you talk badly about us, and if we disapprove of what you say, I am the one who will respond with killing you with what I am. If you hear the cry of my voice, you will know that trouble is in store for you. If you do not make use of us properly, even in times when we are numerous, you will not see us anymore. We are the four deer who have transformed themselves into different kinds of things. Into these four kinds of things we can transform ourselves. Moreover, we can assume the forms of all different kinds of plants. Then when you look you will not see us. In the future only those of whom we approve shall eat the mighty deer. If when you hunt, you come across four deer, you will not kill all of them. You may kill three and leave one. But if you kill all of us, it is not good.

"'These are the things that will bring you happiness. When you kill a deer, you will lay him with the head toward your house. You will cover the earth with plants or with branches of trees lengthwise, with the growing tips of the plants pointing the direction of the deer's head, toward your house. Then you will take us to your house and eat of us. You will place our bones under any of the things whose form we can assume. At these places you may put our bones. You will sprinkle the place with yellow pollen. Once, Twice. Then you lay the bones. And then you sprinkle yellow pollen on top of the bones. This is for protection of the game animals. In this manner they will live on; their bones can live again and live a lasting life.'"[11]

In the story Fawn, speaking on behalf of the Deer people, gives advice to First Hunter about how to respect the spirit of the deer. Fawn also refers to Talking God as the holy being who decides when and how the Deer People will give their life to sustain humans. The proper relationship of hunters to Deer People is thus determined in the realm of spirit. It is then passed on through ritual practice and understanding rooted in the guiding myths of the Navajo. The story presents a foundation for developing a spiritual orientation to the act of hunting. The first step in this hunting way is the development of the hunter's heart in preparation for the hunter's journey. This is an education through the spirits and landscape of Nature. It is an education founded on a deep level of ecological understanding and responsibility. When it takes hold, this education transforms the hunter and firmly establishes his journey toward becoming complete. Thus, a Pathway to self is found through deeper understanding and responsibility.

Teaching spiritual ecology in Indigenous settings is really teaching personal power. The gaining of Personal Medicine, as Plains Indian traditions relate, is the real goal of a Pathway of learning. Gaining personal power begins with establishing a context of respect. Personal power and a deep sense of respect are what "empowerment" reality means!

The Indigenous hunter not only learned how to attract his prey, but in the process of development as a hunter, he also learned how to obtain luck by gaining the attention of the spirits of Nature. The power gained reflected and transformed the hunter at physical, psychological, social, and spiritual levels.

Indigenous complexes of hunting found throughout the world followed a pattern that, while finding a diversity of expressions, included the following basic component processes. The first stage of Indigenous hunting began with setting one's intention through prayerful asking; this was followed by a process of purification. These preparations were meant to set the tone of respect and proper relationship. The second stage included intense questing and application of skill and attractive behavior toward the goal of a successful hunt. The third stage included the community process of respectful treatment of the prey, celebration, and thanksgiving. In all these processes, comprising the circle of Indigenous hunting, the connecting thread is reverence for a giving Earth and the spirit relationships to the animals. In this traditional context of hunting, the Hunter and the Hunted joined in a kinship, a conscious and respectful sharing of the flow of life. The lesson for the hunter lies in gaining a deeper understanding of spiritual relationship and transformation. He and the community he represented reinforced their understanding of the deep ecological meaning of mutual reciprocity. The Hunter and his community entered a spiritual exchange, a creative process of learning and teaching that has formed the foundation of human meaning since the dawn of history.

While these relationships with animals have undergone significant transformations since earlier times, they are still there in spirit if not in reality. The crisis of our times beckons us to retrace the tracks of our ancestors and the animals they knew. We must understand the meaning of our lives in relationship with the natural world. The Hunter of Good Heart shows us spirituality and ourselves as spiritual beings. It is an aspect of our humanity that we must reclaim if we are to find our life. The compacts made by those first hunters were covenants of spirit and were for life. They were based on an abiding relationship of respect.

Modern perceptions regarding animals are infused with representations of attitudes about the other. Moderns no longer experience a daily and direct relationship with animals. They are informed by a media that skews the images of animals. This has led to a host of reactions ranging from fear to romanticization to outright inhumanity. Modern society's biased orientation to animals mirrors contemporary misunderstanding about other races, especially those who have become few in number, such as American Indians. To truly understand animals, is also to truly understand the other.

Liberation Theology and Ecology

Rivals or Partners?

Liberation theology and ecological discourse have something in common: they start from two bleeding wounds. The wound of poverty breaks the social fabric of millions and millions of poor people around the world. The other wound, systematic assault on the Earth, breaks down the balance of the planet, which is under threat from the plundering of development as practiced by contemporary global societies. Both lines of reflection and practice have as their starting point a cry: the cry of the poor for life, freedom, and beauty (cf. Ex 3:7), and the cry of the Earth groaning under oppression (cf. Rom 8:22-23). Both seek liberation, a liberation of the poor by themselves as active subjects who are organized, conscious, and networked to other allies who take on their cause and their struggle; and a liberation of the Earth through a new covenant between it and human beings, in a brotherly and sisterly relationship and with a kind of sustainable development that respects the different ecosystems and assures a good quality of life for future generations.[1]

Now is the time to bring these two discourses together, to see to what extent they are different, or in opposition, or complementary. We begin with ecological discourse because it represents a truly comprehensive standpoint. We are going to take up once more the perspectives presented thus far—with some risk of repetition—so as to bring these two major movements together.

THE ECOLOGICAL AGE

Ecology was first understood to be the subheading of biology that studies the inter(retro)relationships of living beings among themselves and with their environment. That was how it was understood by Ernst Haeckel,

---

who first formulated it in 1866. Soon, however, that understanding spread into three ecologies.[2] *Environmental* ecology is concerned with the environment and the relations that various societies have with it in history, whether they are easy or harsh on it, whether they integrate human beings into or distance them from nature. *Social* ecology is primarily concerned with social relations as belonging to ecological relations; that is, because human beings (who are personal and social) are part of the natural world their relationship with nature passes through the social relationship of exploitation, collaboration, or respect and reverence. Hence social justice—the right relationship between persons, roles, and institutions—implies some achievement of ecological justice, which is the right relationship with nature, easy access to its resources, and assurance of quality of life. Finally, *mental* ecology starts from the recognition that nature is not outside human beings but within them, in their minds, in the form of psychic energies, symbols, archetypes, and behavior patterns that embody attitudes of aggression or of respect and acceptance of nature.

In its early stages ecology was still a regional discourse, concerned with the preservation of certain endangered species (whales, panda bears, and so forth) or the creation of nature reserves to guarantee favorable conditions for the various ecosystems. In short, it was concerned with what is green on the planet, such as forests, and particularly tropical forests, which are the sites of the greatest biodiversity on Earth. However, as awareness of the unwanted effects of the industrial development process grew, ecology became increasingly a global discourse. It is not just species or ecosystems that are threatened; the Earth as a whole is ill and must be treated and healed. The alarm went out in 1972 with the well-known report of the Club of Rome, *The Limits to Growth.* The death machine is unrelenting. By 1990 ten species of living beings a day were disappearing. By the year 2000 the rate of disappearance will be one an hour, and by that time twenty percent of the life forms on the planet will have disappeared.[3]

A vigorous social critique has arisen out of ecology.[4] An arrogant anthropocentrism is at work, one which lies at the root of contemporary societies. Human beings understand ourselves as beings above other beings and lords of life and death over them. During the past three centuries, scientific and technical advances have provided us with the tools for dominating the world and systematically plundering its riches, which are reduced to "natural resources" with no regard for their relative autonomy.

The Earth sciences that have developed, especially since the 1950s with the deciphering of the genetic code and of various kinds of knowledge drawn from space projects, present us a new cosmology, that is, a unique image of the universe, a different perspective on Earth and on the role of the human being in the evolutionary process, which many now call *cosmogenesis.*[5]

To begin with, we have attained an absolutely new vision. For the first time in the history of humankind, we can see the Earth from outside, as the astronauts do.[6] "From the moon," said astronaut John Jung, "the Earth fits on the palm of my hand; on it there are no blacks and whites, Marxists and democrats; we must love this marvelous blue and white planet, because it is under threat."

Second, from a spaceship it is obvious that the Earth and humankind are a single entity, as Isaac Asimov noted in 1982 on the twenty-fifth anniversary of the launching of Sputnik, which began the space age.[7] Such is perhaps the most fundamental intuition of the ecological perspective: the discovery of Earth as a living superorganism called Gaia.[8] Rocks, waters, atmosphere, life, and consciousness are not just put together side by side, separated from one another, but have always been interconnected in an entire inclusion and reciprocity, constituting a single organic reality.

Third, human beings are not so much beings on Earth as beings of Earth. We are the most complex and unique expression of the Earth and the cosmos thus far. Man and woman are the Earth–thinking, hoping, loving, dreaming, and entering into the phase in which decision is no longer by instinct but conscious.[9] The noosphere (the specifically human sphere, that of the spirit) represents an emergence from the biosphere that in turn entails an emergence from the atmosphere, the hydrosphere, and the geosphere. All is related to all at all points and at all times. A radical interdependence of living systems and of apparently nonliving systems is at work. This is the basis for cosmic community and planetary community. Human beings must discover our place in this global community along with other species, not outside or above them. There is no justification for anthropocentrism, but that does not mean ceasing to regard the human being as unique, as that being of nature through whom nature itself achieves its own spacial curve, breaks out in reflex awareness, becomes capable of copiloting the evolutionary process, and emerges as an ethical being assuming responsibility for bringing the entire planet to a happy fate (the meaning of the anthropic principle). As the great American ecologist Thomas Berry says, "This is the ultimate daring venture for the Earth, this confiding its destiny to human decision, the bestowal upon the human community of the power of life and death over its basic life systems."[10] In other words, it is the Earth itself that through one of its expressions–the human species–takes on a conscious direction in this new phase of the evolutionary process.

Finally, all these perceptions create a new awareness, a new vision of the universe, and a redefinition of human beings in the cosmos and their practices toward it. We therefore face a new paradigm. The foundation is laid for a new age, the ecological age. After centuries of confronting nature and being isolated from the planetary community, human beings are finding their way back to their shared home, the great, good, and bountiful Earth. They wish to initiate a new covenant of respect and kinship with it.

HEARING THE CRY OF THE OPPRESSED

Where does liberation theology stand with regard to ecological concern? We must acknowledge that the initial setting within which liberation theology emerged was not that of ecological concern as we have sketched it above. The most salient and challenging fact was not the threat to Earth as a whole but to the sons and daughters of Earth exploited and condemned to die prematurely, the poor and oppressed.[11] That does not mean, however, that its basic intuitions have little to do with ecology. The relationship to ecology is direct, for the poor and the oppressed belong to nature and their situation is objectively an ecological aggression. But all of this was considered within a more restricted historical and social framework and in the context of classical cosmology.

Liberation theology was set in motion back in the 1960s by ethical indignation (true sacred wrath of the prophets) in the face of the dire poverty of the masses, especially in the Third World. This situation seemed—and still seems—unacceptable to the Christian conscience, which reads in the faces of the poor and the outcast the reembodiment of the passion of the Crucified One, who cries out and wants to arise for the sake of life and freedom.

The option for the poor, against their poverty and for their liberation, has constituted and continues to constitute the core of liberation theology. To opt for the poor entails a practice; it means assuming the place of the poor, their cause, their struggle, and at the limit, their often tragic fate.

Never in the history of Christian theologies have the poor become so central. To seek to build an entire theology starting from the perspective of the victims and so to denounce the mechanisms that have made them victims and to help overcome those mechanisms by drawing on the spiritual storehouse of Christianity, thereby collectively forging a society that offers greater opportunity for life, justice, and participation: this is the unique intuition of liberation theology.

That is why for this theology the poor occupy the epistemological locus; that is, the poor constitute the point from which one attempts to conceive of God, Christ, grace, history, the mission of the churches, the meaning of the economy, politics, and the future of societies and of the human being. From the standpoint of the poor, we realize to what extent current societies are exclusionary, to what extent democracies are imperfect, to what extent religions and churches are tied to the interests of the powerful.

From the beginning Christianity has cared for the poor (cf. Gal 2:10) but never have they been accorded so central a place in theology and for political transformation as they have been given by liberation theology.

The understanding of the poor in liberation theology has never been reduced to a single focus on them as poor. The poor are not simply beings made up of needs, but they are also beings of desire for unrestricted communication, beings hungering for beauty. Like all human beings, the poor—as the Cuban poet José Roberto Retamar puts it nicely—have two basic hungers, one for bread, which can be sated, and another for beauty, which is insatiable. Hence, liberation can never be restricted to the material, social, or merely spiritual realm. It is only true when it remains open to the full sweep of human demands. It has been the merit of liberation theology to have maintained its comprehensive scope since its origins; it did so because it was correctly interpreting what human liberation is about, not because of the demands of doctrinal authorities in the Vatican.

To be genuine, liberation must not only remain comprehensive in scope, but it must also and primarily be achieved by the poor themselves. Perhaps this is one of the unique features of liberation theology when compared with other practices of tradition, which have also shown concern for the poor. The poor are generally regarded as those who do not have food, shelter, clothes, work, culture. Those who have, so it is said, must help those who do not, so as to free them from the inhumanity of poverty. This strategy is full of good will and is well meaning; it is the basis for all assistance and paternalism in history. However, it is neither efficient nor sufficient. It does not liberate the poor, because it keeps them in a situation of dependence; worse yet, it does not even appreciate the liberating potential of the poor. The poor are not simply those who do not have; they *do* have. They have culture, ability to work, to work together, to get organized, and to struggle. Only when the poor trust in their potential, and when the poor opt for others who are poor, are conditions truly created for genuine liberation. As people sing in our Christian base communities: "I believe that the world will be better, when the child who suffers believes in the child." The poor become the agents of their own liberation; they become free persons, capable of their own self-determination with others who are different from them, so that together they may be free in a society that is more just, family spirited, and ecologically integrated.

Hence we should insist that it is not the churches that liberate the poor, nor the welfare state (whether socialist or social democrat), nor those classes that aid them. They may be allies of the poor, provided that they do not deprive them of their leading role and leadership. We can speak of liberation only when the poor themselves emerge as the primary agent of their journey, even when they have support from others.

Certainly one of the permanent merits of liberation theology flows from the method that it has brought into theological reflection.[12] Its starting point is not ready-made doctrines, not even what is revealed or Christian traditions. All of that is present on the horizon of the Christian and the theologian as background to illuminate convictions and as the starting point of reflection. It is what we call the horizon prior to any knowledge that becomes known thematically. But liberation theology's starting point is the anti-reality, the cry of the oppressed, the open wounds that have been bleeding for centuries.

Its first step is to honor reality in its more stark and problematic side. This is the moment of *seeing*, of feeling and suffering the impact of human passion, both personal and social. This is an overall experience of compassion, of protest, of mercy, and of a will to liberating action. This entails direct contact with the anti-reality, an experience of existential shock. Without this first step, it is difficult to set in motion any liberation process intended to change society.

The second moment is the analytical *judging* in a twofold sense, in the sense of critical knowledge (analytical meditation) and the sense of illumination on the basis of the contents of faith (hermeneutic mediation). It is important to decipher the causes that produce suffering, to seek their cultural roots, in the interplay of economic. political, and ideological power. Poverty is neither innocent nor natural; it is produced. That is why the poor are exploited and impoverished. It has been the merit of Marxist rationality to have shown that the poor are oppressed, people who have been dehumanized by an objective process of exploitation that is economic, political, ecological, and cultural in nature.

The data of revelation, the faith tradition, and Christian practice over the centuries condemn this situation of poverty as sin; that is, as something that also has to do with God, insofar as it denies the historic realization of God's design, which goes by way of the mediation of justice, of kindness toward the poor, and of participation and communion. From the standpoint of faith, the poor represent the suffering Savior and the supreme and eschatological judge. That is why they have an unsuspected theological intensity, especially in view of the degradation caused by dire poverty. Within the logic of faith, it is precisely this degradation that prompts God to intervene and that as a countermovement initiates a sacramental presence of God.

The third moment is transformative *action*, the most important moment, for that is where everything should culminate. Christian faith must make its contribution in the transformation of relationships of injustice into relationships that foster greater life and happiness due to living in participation and in a decent quality of life for all. Christian faith has no monopoly on the idea of transformation, but it joins other forces that also take up the cause and the struggle of the poor, contributing by means of its religious and symbolic uniqueness, its way of organizing the faith of the people, and its presence in society. Where faith and the church come into play is not in the economic realm or in what is directly political, but rather in the realm of the cultural and the symbolic. The church bears powerful messages that can create solidarity movements and project values for resistance, protest, and commitment to the specific liberation of the oppressed; it can organize celebrations and nourish their imagination, so that through that imagination they may be able to reject the present oppressive situation and dream of some new situation that is yet to be achieved by historic practice.

Finally, there is the moment of *celebrating*. This is a decisive dimension for faith, for it is here that the more gratuitous and symbolic side of liberation emerges. In celebration the Christian community recognizes that the specific advances achieved through commitment go further than the social, community, or political dimensions. Besides all that, those advances also signify the anticipatory signs of the goods of the Reign, the advent of divine redemption, embodied in actual liberations in society, the moment when the utopia of integral liberation is anticipated in frail signs, symbols, and rites. Faith identifies the spirit in action in liberation processes. It discerns the power of resurrection acting in the salvaging of a minimally worthy life. It sees the Reign taking place as process within the history of the oppressed. All this is uncovered in celebration and transformed into material for praising God.

Because of its liberating commitment, which is the basis for theological reflection, Christianity has shown that the idea of revolution (liberation, transformation) is not a monopoly of the "left" but can be a summons from Christianity's central message, which proclaims someone who was a political prisoner, was tortured, and was nailed onto the cross as a result of the way he led his life. If he rose, it was likewise to show—in addition to its strictly theological content—the truth of this practice and the utopian realization of the drives for life and freedom.

THE MOST THREATENED BEINGS IN CREATION: THE POOR

At this point the discourses of ecology and of liberation theology must be brought together for comparison. In analyzing the causes of the impoverishment that afflicts most of the world's population, liberation theology became aware that a perverse logic was at work. The very same logic of the prevailing system of accumulation and social organization that leads to the exploitation of workers also leads to the pillaging of whole nations and ultimately to the plundering of nature. It no longer suffices merely to adjust technologies or to reform society while keeping the same basic logic, although such things should always be done; the more important thing is to overcome such logic and the sense of being that human beings have held for at least the last three centuries. It will not be possible to deal with nature as our societies have tried to do, as though it were a supermarket or a self-service restaurant. It is our common wealth that is being mercilessly plundered, and that inheritance must be safeguarded.

Moreover, conditions for nature's further evolution must be assured for our own generation and for those to come, for the whole universe has been working for fifteen billion years so that we could come to the point at which we have arrived. Human beings must educate themselves so that far from being the Satan of Earth they may serve as its guardian angel, able to save Earth, their cosmic homeland and earthly mother.

The astronauts have accustomed us to see the Earth as a blue-and-white spaceship floating in space, bearing the common fate of all beings. Actually, on this spaceship Earth, one-fifth of the population is traveling in the passenger section. They consume 80 percent of the supplies for the journey. The other fourth-fifths are traveling in the cargo hold. They suffer cold, hunger, and all kinds of hardships. They are slowly becoming aware of the injustice of this distribution of goods and services. They plan rebellion: "Either we die passively of hunger or we make changes that will benefit everyone," they say. It is not a difficult argument: either we are all saved in a system of participatory common life in solidarity with and on spaceship Earth or in an explosion of wrath we could blow up the ship, sending us all falling into the abyss. Such an awareness is on the rise and can be terrifying.

The most recent arrangements of the world order led by capital and under the regime of globalization and neoliberalism have brought marvelous material progress. Leading-edge technologies produced by the third scientific revolution (computerization and communications) are being employed and are increasing production enormously. However, they dispense with human labor and hence the social effect is perverse: many workers and whole regions of the world are left out, since they are of little relevance for capital accumulation and are met by an attitude of the cruelest indifference.[13]

Recent data indicate that today globally integrated accumulation requires a Hiroshima and Nagasaki in human victims every two days.[14] There has been huge progress, but it is profoundly inhuman. The individual and peoples with their needs and preferences do not stand at its center, but rather the commodity and the market to which everything must submit. Hence, the most threatened creatures are not whales but the poor, who are condemned to die before their time. United Nations' statistics indicate that each year fifteen million children die of hunger or hunger-related diseases before they are five days old; 150 million children are under-nourished, and 800 million go hungry all the time.[15]

This human catastrophe is liberation theology's starting point for con-sidering ecology. In other words, its starting point is social ecology, the way human beings, the most complex beings in creation, relate to one another and how they organize their relationship with other beings in nature. At present it all takes place under a very exploitative and cruelly exclusionary system. We are faced with the cry of the oppressed and the excluded. A minimum of social justice is most urgently sought in order to assure life and the basic dignity that goes along with it. Once this basic level of social justice (social relationship between human beings) has been achieved, it will be possible to propose a possible ecological justice (rela-tionship of human beings with nature). Such justice entails more than so-cial justice. It entails a new covenant between human beings and other beings, a new gentleness toward what is created, and the fashioning of an ethic and mystique of kinship with the entire cosmic community. The Earth is also crying out under the predatory and lethal machinery of our model of society and development. To hear these two interconnected cries and to see the same root cause that produces them is to carry out integral liberation.

The social and political framework for this kind of integral liberation is an extended and enriched democracy. Such democracy will have to be biocracy, socio-cosmic democracy; in other words, a democracy that is centered on life, one whose starting point is the most downtrodden hu-man life. It must include elements of nature like mountains, plants, water, animals, the atmosphere, and landscapes as new citizens participating in human common life and human beings sharing in the cosmic life in com-mon. There will only be ecological and societal justice when peace is as-sured on planet Earth.

Liberation theology should draw from ecological discourse the emerg-ing world view or cosmology; that is, the vision that understands Earth as a living superorganism connected to the entire universe in cosmogenesis. It must comprehend the mission of human beings, man and woman, as expressions of Earth itself and as manifestations of the principle of intel-ligibility existing in the universe, that the human being—the noosophere—represents a more advanced stage of the cosmic evolutionary process at its conscious level, one of copiloting it, utilizing the guiding principles of the universe that have governed the whole process since the moment of ex-pansion-explosion fifteen billion years ago. Human beings were created for the universe—not vice versa—so as to attain a higher and more complex stage of universal evolution, namely, in order to be able to celebrate and glorify the Creator who wanted to have companions in love.

This background should first of all lead to a broader understanding of liberation. It is not only the poor and oppressed who must be liberated but all human beings, rich and poor, because all are oppressed by a paradigm—abuse of the Earth, consumerism, denial of otherness, and of the inherent value of each being—that enslaves us all. We all must seek a paradigm that will enable Gaia to live and all beings in creation, especially human beings, to exist in solidarity. We suggest the paradigm of the connectedness of all with all, which allows for the emergence of a religion, convergence in religious diversity, and will achieve peace between humans and Earth.

Second, the starting point must also be redefined, namely, the option for the poor, including the most threatened beings in creation. The first of these is planet Earth as a whole. It has not become widely enough accepted that the supreme value is the preservation of planet Earth and the maintenance of conditions for the fulfillment of the human species. Such a way of looking at things shifts the central focus of all issues. The basic question is not the future of Christianity or of Christ's church. Nor is it the fate of the West. Rather, the basic question is what kind of future there will be for planet Earth and the humankind that is its expression. To what extent does the West with its applied science and its culture and Christianity with its spiritual resources assure this collective future?

Third, it is urgent to reaffirm an option for the poor of the world, those huge masses of the human species who are exploited and slaughtered by a small minority of the same species. The challenge will be to bring human beings to realize that they are a large earthly family together with other species and to discover their way back to the community of the other living beings, the community of the planet and the cosmos.

The final question is how we are to assure the sustainability not of a particular type of development but of planet Earth, in the short, medium, and long term. This will happen only through a non-consumeristic type of cultural practice that is respectful of ecosystems, ushers in an economy of what is sufficient for all, and fosters the common good not only of humans but also of the other beings in creation.

LIBERATION AND ECOLOGY:
THE BRIDGE BETWEEN NORTH AND SOUTH

Two major issues will occupy the minds and hearts of humankind from now on: what is the destiny and future of planet Earth if the logic of pillage to which the present type of development and consumption have accustomed us continues? What hope is there for the poor two-thirds of humankind? There is a danger that "the culture of the satisfied" will become enclosed in its consumeristic selfishness and cynically ignore the devastation of the poor masses of the world. There is also a danger that the "new barbarians" will not accept their death sentence and will set out on a desperate struggle for survival, threatening everything and destroying everything. Humankind may find itself facing violence and destruction at levels never before seen on the face of the Earth unless we collectively decide to change the course of civilization and shift its thrust from the logic of means at the service of an exclusionary accumulation toward a logic of ends serving the shared well-being of planet Earth, of human beings, and of all beings in the exercise of freedom and cooperation among all peoples.

These two issues, with different accents, are shared concerns in the North and South of the planet. They are also the central content of liberation theology and of ecological thought. These two directions of thought make possible dialogue and convergence in diversity between the geographical and ideological poles of the world. They must be an indispensable mediation in the safeguarding of everything created and in rescuing the dignity of the poor majorities of the world. Hence liberation theology and ecological discourse need each other and are mutually complementary.

SONS AND DAUGHTERS OF THE RAINBOW

Theologically speaking, a truly ecumenical challenge is opening up: to inaugurate a new covenant with the Earth in such a way that it will signify the covenant that God established with Noah after the destruction wrought by the flood. There we read: "I set my bow in the clouds to serve as a sign of the covenant between me and the earth . . . everlasting covenant that I have established between God and all living beings—all mortal creatures that are on earth" (Gn 9:13-16). Human beings must feel that they are sons and daughters of the rainbow, those who translate this divine covenant with Gaia, the living superorganism, and with all the beings existing and living on it, with new relationships of kindness, compassion, cosmic solidarity, and deep reverence for the mystery that each one bears and reveals. Only then will there be integral liberation, of the human being and of Earth, and rather than the cry of the poor and the cry of the Earth there will be common celebration of the redeemed and the freed, human beings in our own house, on our good, great, and bountiful Mother Earth.

PART VI

TECHNOLOGY, ENERGY, MATERIALS, AND SOCIO-ENVIRONMENTAL RESEARCH

Marina Fischer-Kowalski

The Intellectual Context for a Systemic Perspective on Society–Nature Relations

This part assembles seven legacy readings on socio-environmental research, published across a timespan of four decades, from the 1960s to the 1990s. Intellectually, this marks a period in which gradually a systemic perspective emerged, with new methodological and technical means (i.e., increasingly powerful computers) to deal with the dynamic interaction of natural and social processes.

This new systems paradigm was, among others, introduced by Ludwig von Bertalanffy (1968). According to him, the quest for a general system theory offers the opportunity for a unification of sciences after a long phase of ever increasing specialization where disciplines became "encapsulated in their private universes" (p. 30). Independently of one another and each in its own way, several disciplines had started to transcend their preoccupation with the behavior of elements to consider questions of "wholeness," problems of organization of interacting parts, and theories of what von Bertalanffy terms "organised complexity." Authors from both the natural and the social sciences embraced this novel approach, with a similar focus on systems, feedbacks, and long-term processes.

The first paper in this part, chapter 3 from **Ester Boserup's** (1910–1999) pathbreaking 1965 book, *The Conditions of Agricultural Growth*,[1] allows readers to catch a glimpse of her heretical view of technological progress as a slow process that may also be reversed, as

---

[1] *This book originally was published in 1965, and later by five different publishing houses in 17 issues from 1965 to 2008, translated into French, Swedish, Japanese, and Estonian.*

it may increase the workload of humans (and not reduce it, as commonly assumed).[2] She thus contradicts the thesis put forward by the British Reverend and scholar Thomas Malthus in the eighteenth century (see **Malthus 1798, Part I** in this volume). Malthus had postulated that population growth would invariably lead to poverty because agricultural production would not keep up with the increasing demand for food. Boserup responded by documenting the ability of rural societies to innovate as they grew. But she not only suggested that it was possible for agricultural production to keep pace with a growing population, she also indicated that some of the innovation required in the process depended on a certain population density, i.e., was even driven by population growth, rather than being hindered by it. Boserup had thus positioned herself in opposition to the mainstream theories and policies of development of the 1960s.[3] Trained as an economist, she describes her research agenda as follows: "My own research focused on the interplay of economic and non-economic factors in the process of social change, both today and in the past, viewing human societies as dynamic relationships between natural, economic, cultural, and political structures, instead of trying to explain them within the framework of one or a few disciplines" (Boserup, 1999, p. 7). Boserup's approach represents a system dynamic variant of some of the historical and ecological anthropology approaches assembled in **Part II** of this volume (e.g., **Childe 1936**, **Rappaport 1967**, and in particular **Sahlins 1972**) that had focused on social metabolism as molding cultures and societal organization as well as the natural environment.

Robert U. Ayres (1932–) and **Allan V. Kneese** (1930–2001), like Boserup, offer a systemic critique of the economic theories of their time, but focus on fully industrial rather than industrializing economies. The only overarching theory in economics bridging the "great divide" between society and nature had been developed by Marx (see **Marx 1867, Part I**). He talked about the "metabolism" between society and nature, and thus about the interlinkage of economic development and, building upon Liebig's (1842) foundational work on agrochemistry, the thereby induced changes in the natural environment. This approach had been thoroughly discriminated against during the Cold War and the evolving realities in the Soviet Union and its allies, and it survived but in niches. Instead, in economics, Hayek (1948) turned towards an economic model focusing almost exclusively on monetary flows and individual market actions of producers and consumers, and he discouraged a systems perspective that would include biophysical variables and processes.[4] Also, standard economic models were better suited to predicting short-term dynamics, such as for election periods, rather than longer-term economic feedback processes between society and the natural environment. Ayres and Kneese (1969) address these deficiencies, with no reference to Marx, criticizing the inability of this dominant economic model to handle the "negative externalities," such as pollution and wastes, within the price system of the economy. Instead, it views these externalities as exceptional cases that "may distort the allocation

---

[2] *In a book published in honor of Boserup's 100th anniversary (Fischer-Kowalski et al., 2014b), this thesis was confirmed on a broader database also quantitatively. What Boserup did not anticipate, though, was the huge impact of fossil fuel use on lowering both the labor burden and population growth in agricultural modernization.*

[3] *This position is, to a certain degree, still defended in Ehrlich and Holdren (1971) in this part.*

[4] *The few economists that show up in this part are all deviants from their tribe, as we have seen above with Boserup. This also applies to Alan Kneese in coauthorship with R. Ayres, trained as a physicist. Ayres rightly claims himself also to be an economist – but apart from the article represented in this collection, he never found access to high-ranking economic journals (pers. comm.).*

of resources, but can be dealt with adequately through simple *ad hoc* arrangements" (reading 48). More fundamentally, they claim neoclassical economic theory looks upon the production and consumption processes "in a manner that is somewhat at variance with the fundamental law of conservation of mass" (reading 48). Standard economics conceptualizes "the 'final consumption' of goods as though material objects such as fuels, materials, and finished goods somehow disappeared into the void." Instead, Ayres and Kneese suggest viewing pollution as a "materials balance problem for the entire economy" (reading 48), and present a table on the materials production (including imports) for the USA in tons for the early 1960s (reading 48).[5] They then proceed with a formal model showing that the system of prices "in a spaceship economy" (with reference to Boulding 1966) would have to be radically different. Their idea of looking upon the economy as a metabolic system organizing an energetic and material exchange between human society and its natural environment proved to bear fruits in the further debate, and laid the ground both for a new field of research (Fischer-Kowalski et al. 2011) and novel worldwide statistics for societal resource use, regularly published by the United Nations Environment Program (UNEP), the Organization for Economic Co-operation and Development (OECD), and the European Union (European Communities 2003; OECD 2019).

Sociologists, in the face of the environmental damages caused by society, also raised questions about the further validity of fundamental assumptions of their discipline. **Ulrich Beck** (1944–2015), professor of sociology at Munich University, in 1986 wrote a bestselling book, *Risk Society*, in which he challenged the traditional social science understanding of inequality. "Advanced modernity," Beck wrote, was characterized not just by a social stratification of wealth, but simultaneously by inequalities socially produced through environmental risks. "Earth has become an ejector seat that no longer recognises any distinctions between rich and poor, Black and white, North and South nor East and West" (reading 51). Beck thus recognizes a reciprocal relationship between society and the environment, and the term environmental risk reverberates in the social science literature, but he does not leave the social science terrain to venture into interdisciplinarity nor, beyond his core feedback assumption, to elaborate on a systemic approach. It was left to Luhmann (1986), another German sociologist, to use a system paradigm and discuss the challenge for social systems of controlling the negative impacts of their systemic functioning on the environment. Luhmann conceptualized social systems as systems of communication, to which the biophysical world (even the human body) was clearly external, permitting no direct feedbacks. Similarly, the German sociologist and historian R. P. Sieferle (2001) analyzed the emergence of the new energy system in England from the sixteenth century onward, one based upon coal,[6] in an outstanding systemic analysis of society–nature interaction. Other sociologists, among them Fischer-Kowalski and Haberl (1993), in this part, later extended Luhmann's and Sieferle's perspectives, describing

---

[5] *Interestingly, in the same period there appeared an article by K. G. Gofman (Gofman et al., 1974) in a Russian journal, with about the same idea: 'The Economics of Nature Management: Objectives of a New Science'.*

[6] *His meticulous description of Britain's transition from wood to coal as fuel during the sixteenth century, and its consequences for society and the environment (long before the "industrial transformation" based upon the steam engine set in), was a stand-alone achievement following a thoroughly systemic paradigm. It received recognition among his co-historians only in the early 2000s when he initiated a broad interdisciplinary research program 'Europe's Special Course' (Sieferle, 2001).*

systemic biophysical exchange relations between society and nature across time, in a historical perspective.

The boldest example for a full-fledged realization of von Bertalanffy's modeling vision is represented in this part by **Donella H. Meadows**[7] (1941–2001), **Dennis L. Meadows** (1942–), **Jørgen Randers** (1945–), and **William W. Behrens III** (1949–), in Meadows et al. (1972), *The Limits to Growth*, their report for the Club of Rome.[8] With the help of the first powerful computer in civil use at the Massachusetts Institute of Technology and a fine sense for the mathematical form of timelines and relationships, they constructed a system dynamics model comprising essentially five variables: population, food, natural resources, pollution absorption capacity, and the world economy (termed "industry"). On the basis of scattered statistical data and smart assumptions about feedback mechanisms, they reconstructed the timelines of their variables from 1900 to 1970 to seed various model runs until 2100. The standard run shows a peak of per capita food and industrial output around 2020 that then declines sharply, followed about two decades later by a peak in pollution and then, around 2060, in population – a picture of "overshoot and collapse" (reading 50). This bold exercise landed as a bomb: it challenged common beliefs in eternal progress, and humans no longer appeared as masters of the planet who should just learn to behave a little better towards their natural environment. "Human exceptionalism" (see **Catton and Dunlap 1978, Part III**) was at once put fundamentally at stake. An army of critics immediately mobilized to rescue the crumbling edifice of eternal human progress (e.g., Simon 1980), and the study was thoroughly discredited.[9] Its coauthor Randers (2012) stated: "Now, at sixty-six, I see that I have been worrying in vain. Not because the future looks problem free and rosy. My worrying has been in vain because it hasn't had much impact on global evolution over the long generation since I started worrying" (p. 1).

Driven by widespread public protest, environmental issues had risen to become a core policy agenda in the course of the decades covered by our selection, both on civil society and governmental levels, starting in the early 1970s. Environmental scientists as actors on both levels gained in prominence. The next reading in this part was born in the heated environmental atmosphere of the early 1970s, and its authors were strongly involved in public debate. **Paul R. Ehrlich** (1932–) and **John P. Holdren's** (1944–) paper was originally published in *Science* and was thus visible to scientists worldwide. It blames population growth in the USA and elsewhere as the main cause of environmental deterioration. Population growth,[10] they claimed, has a disproportionate impact on the environment because of resource depletion according to the law of diminishing returns (the well-known Malthusian argument) instead of what is broadly believed as advantageous effects because of economies of scale. They argue the environmental per capita impact rises

[7] It should be noted that Donella (not Dennis) Meadows figures as first author.

[8] The Club of Rome was founded in 1968 by A. Peccei (an Italian top industrialist) and A. King (a British science manager in OECD) and a number of managers from industry as well as scientists and policy-makers. Today, it is composed of more than 100 members from 50 countries, meeting annually. The Limits to Growth was its first report. It sold 30 million copies in more than 30 languages, making it the bestselling environmental book in history (Simmons 2000).

[9] Further runs of the (partly modified) model with new data, by Meadows and colleagues (Meadows et al., 1992; Randers, 2012; Turner, 2014), more or less confirmed the fit of the curves for another 40 years. The interesting peaks, though, are yet to come!

[10] As the authors note in their piece, in the United States there has been an increase in population of 53 percent between 1940 and 1969 (p. 171). What they do not see is that this has largely been a result of immigration.

disproportionally with population size, instead of being invariant.[11] Unless population growth were addressed, they anticipate severe consequences, such as "awesome" (reading 49) effects of high-yield agriculture, climate problems as a consequence of CO_2 emissions, and the invasion of "deadly viruses." While this view was and remained scientifically and politically contested,[12] the relatively simple formula that evolved from the debate with Barry Commoner (1971), one of their critics, namely that environmental impact equals per capita impact times population size, underwent a number of extensions and modifications and is still widely in use as the "IPAT formula" (Impact = Population size × Affluence × Technology; see Chertow, 2000). Chertow tracks the various forms the IPAT equation has taken over 30 years as a means of examining an underlying shift among many environmentalists toward a more accepting view of the role technology can play in sustainable development. Although the IPAT equation was once used to determine which single variable was the most damaging to the environment, an industrial ecology view, as discussed below, reverses this usage, recognizing that increases in population and affluence can often be balanced by improvements to the environment offered by technological systems.

This last component of the formula, technology, is addressed by **Thomas E. Graedel** (1938–), **Braden R. Allenby** (1950–) and **Peter B. Linhart** (1926–) in 1993, i.e., 20 years after Ehrlich and Holdren's publication. It seeks to provide a systemic technological recipe for departing from the usual dissipative use of materials towards optimizing materials cycles from "virgin material, to finished material, to component, to product, to waste product and to ultimate disposal" (reading 52). What Ayres and Kneese (1969, this part) had laid the grounds for intellectually, and later elaborated on (e.g., Ayres and Ayres, 2003), Graedel et al. outline as a new strategy for "industrial ecology," using empirical examples of production processes. This shift in focus from publicly scandalizing the deteriorating impacts of technologies on the environment towards developing practical solutions also reflects a change in the political atmosphere of the time (1980s–1990s). Today, such efforts contribute toward work on "circular economy," strongly promoted and publicized by the MacArthur Foundation (MacArthur, 2013) and taken up by politicians in Japan ("reduce–reuse–recycle"), the European Union, and China.[13] One of the merits of Graedel and his team consisted in the patient and insistent development of "industrial ecology" into an internationally respected academic discipline around an international scientific society and a journal of the same name (founded in 1997).[14]

The final reading in this part, by **Marina Fischer-Kowalski** (1946–) and **Helmut Haberl** (1963–), pioneers by presenting a systemic perspective on biophysical society–nature exchanges from a social science perspective. We address societies as

[11] *The ensuing debate between Commoner and Ehrlich finally led to the still widely used IPAT formula, I=P*A*T, where I is environmental impact, P population, A affluence and T technology (Note: the additional factor "affluence" was never addressed by either Commoner or by Ehrlich and Holdren!). The interdependence of population numbers and per capita impact claimed by Ehrlich and Holdren was not confirmed by later research.*

[12] *Barry Commoner, their long-term adversary in the environmental debate in the United States, states that perhaps population growth may be challenged – but technology is "much more effective in doing damage" (Commoner 1971).*

[13] *Empirical analyses, though, disclose that the circularity of Western industrial economies has not matured very far, yet (Mayer et al., 2019)*

[14] *For the history of this discipline, see Erkman (1997).*

historically highly variable organizations that "are supposed to serve one ultimate purpose: sustaining the species mankind ... They have evolved by organizing a continuous flow of energy and materials from their natural environments, transforming them in various ways useful for human life, and discharging them again into natural environments" (reading 53). This process is covered by the term "social metabolism" and connects well to the contributions in this part by Boserup (1965), Ayres and Kneese (1969) and Graedel et al. (1993). Our approach reaches beyond the industrial social metabolism and discusses the modes of society–nature exchanges in earlier societies of hunter-gatherers and agrarian societies (still extant in many parts of the world).[15] In doing so, we flag a pattern of human domination of natural processes, one transcending input–output patterns: "By 'colonizing [sic]' parts of the environment, society intentionally changes some elements of the environment so that it is rendered more exploitable for social needs" (reading 53). The colonization of terrestrial ecosystems for agriculture as well as of plant and animal species by breeding are core technological inventions of the Neolithic revolution that fed back on social organization, promoting the evolution of sedentary, hierarchical, and urban human societies (see **Childe 1936, Part II**). Finally, the exploitation of fossil fuels has permitted human societies to operate at an energy level that far surpasses the historical level of energy that could be extracted from land and sea by human and animal labour before – but that also led these societies into a sustainability trap.

This "socioecological" approach,[16] with participation of the Vienna Institute of Social Ecology, helped to develop public information on systems of material and energy flow accounting that became standard in the European Union, the OECD, Japan, and the UN-Environment's International Resource Panel,[17] which on a global level matches the Intergovernmental Panel on Climate Change (IPCC) and the Intergovernmental Science-Policy Platform on Biodiversity and Ecosystem Services.

Concluding Remarks

The papers in this part, while they build upon many of the rich insights gathered throughout the decades of socio-environmental research represented in this book, laid the ground for an overarching, fundamentally systemic perspective on human–nature relations. This perspective underscores the challenge ahead: it has become conceivable that humanity changes the natural conditions on planet Earth in a way that may make it uninhabitable for future generations of humans, or at least for advanced civilizations. Climate science insists on the urgency of change; in order to avoid disastrous global warming, human societies must, within the next decades, abstain from burning fossil fuels (in order to become "CO_2

---

[15] Later further elaborated in the Anthropocene Review, see Fischer-Kowalski et al. (2014a).

[16] Fathered by Bookchin (Bookchin, 1982), and related to the so-called Chicago school of human ecology (e.g., Duncan, 1964), it is still very distinct in its concern for a historical dimension and its focus on developing national indicators to guide policies (see Fischer-Kowalski, 2015).

[17] See lately Oberle et al. (2019).

neutral"). Although, how to run – in the case of limited substitutability of fossil fuels – world society on a much lower level of primary energy,[18] in my opinion, still remains unresolved.

References

Ayres, Robert U., and Leslie W. Ayres. *Industrial Ecology: Towards Closing the Material Cycle*. Cheltenham: Edward Elgar, 2003.

Bookchin, M. *The Ecology of Freedom: The Emergence and Dissolution of Hierarchy.* Palo Alto, CA: Cheshire Books, 1982.

Boserup, Ester. *My Professional Life and Publications*. Copenhagen: Museum Tusculanum Press, 1999.

Boulding, Kenneth E. "The Economics of the Coming Spaceship Earth." In *Environmental Quality in a Growing Economy: Essays from the Sixth RFF Forum*, edited by H. Jarrett, 3–14. Baltimore: John Hopkins Press, 1966.

Chertow, Marian R. "The IPAT Equation and Its Variants." *Journal of Industrial Ecology* 4, no. 4 (2000): 13–29.

Commoner, Barry. *The Closing Circle: Nature, Man and Technology*. New York: Alfred A. Knopf, 1971.

Duncan, Otis D. "Social Organization and the Ecosystem." In *Handbook of Modern Sociology*, edited by R. E.. L. Faris, 36–85. Chicago: Rand McNally, 1964.

Erkman, Suren. "Industrial Ecology: An Historical View." *Journal of Cleaner Production* 5, no. 1–2 (1997): 1–10.

European Communities. *Material Use in the European Union 1980–2000: Indicators and Analysis*. Luxemburg: Office for Official Publications of the European Communities, 2003.

Fischer-Kowalski, Marina. "Social Ecology." In *International Encyclopedia of the Social & Behavioral Sciences*, 2nd ed., edited by J. D. Wright, 254–262. Oxford: Elsevier, 2015.

Fischer-Kowalski, Marina, Fridolin Krausmann, Stefan Giljum, Stephan Lutter, Andreas Mayer, Stefan Bringezu et al. "Methodology and Indicators of Economy Wide Material Flow Accounting: State of the Art and Reliability across Sources." *Journal of Industrial Ecology* 15, no. 6 (2011): 855–876.

Fischer-Kowalski, Marina, Fridolin Krausmann, and Irene Pallua. "A Sociometabolic Reading of the Anthropocene: Modes of Subsistence, Population Size and Human Impact on Earth." *Anthropocene Review* 1, no. 1 (2014a): 8–33.

---

[18] *For illustration: primary energy use (TPES) per capita in the highly developed industrial countries, according to the statistics of the Environmental Energy Agency (EEA), is now twice as high as it had been in 1970. Currently, more than 40 percent of the world population have less than 10 percent of this per capita energy level at their disposal.*

Fischer-Kowalski, Marina, Andreas Mayer, Anke Schaffartzik, and Anette Reenberg. *Ester Boserup's Legacy on Sustainability: Orientations for Contemporary Research.* Dordrecht: Springer Nature, 2014b.

Gofman, Konstantin, M. Lemeschew, and N. Reimers. "Die Ökonomie der Naturnutzung-Aufgaben einer Neuen Wissenschaft (orig. russ.)." *Nauka i shisn* 6 (1974): 12.

Hayek, Friedrich A. *Individualism and Economic Order.* London: Routledge, 1948.

Liebig, Justus von. *Animal Chemistry or Organic Chemistry in Its Application to Physiology and Pathology.* New York: Johnson Reprint, 1964 [1842].

Luhmann, Niklas. *Ökologische Kommunikation: kann die moderne Gesellschaft sich auf ökologische Gefährdungen einstellen?* Opladen: Westdeutscher Verlag, 1986.

MacArthur, Ellen. "Towards the Circular Economy." *Journal of Industrial Ecology* 2 (2013): 23–44.

Mayer, Andreas, Willi Haas, Dominik Wiedenhofer, Fridolin Krausmann, Philip Nuss, and Gian Andrea Blengini. "Measuring Progress towards a Circular Economy: A Monitoring Framework for Economy-Wide Material Loop Closing in the EU28." *Journal of Industrial Ecology* 23, no. 1 (2019): 62–76.

Meadows, Donella H., Dennis L. Meadows, and Jørgen Randers. *Beyond the Limits: Confronting Global Collapse, Envisioning a Sustainable Future.* Post Mills, VT.: Chelsea Green Pub. Co., 1992.

Oberle, Bruno, Stefan Bringezu, Steve Hatfield-Dodds, Stefanie Hellweg, Heinz Schandl, and Jessica Clement. *Global Resources Outlook 2019: Natural Resources for the Future We Want: Summary for Policymakers.* Zurich: ETH Zurich, 2019.

Organization for Economic Co-operation and Development (OECD). *Global Material Resources Outlook to 2060: Economic Drivers and Environmental Consequences.* Paris: OECD Publishing, 2019.

Randers, Jorgen. "2052: A Report to the Club of Rome Commemorating the 40th Anniversary of *The Limits to Growth.*" White River Junction, VT: Chelsea Green Publishing, 2012.

Sieferle, Rolf P. *The Subterranean Forest: Energy Systems and the Industrial Revolution,* translated by Michael P. Osman. Cambridge: The White Horse Press, 2001.

Simmons, Matthew R. "Revisiting *The Limits to Growth*: Could the Club of Rome Have Been Correct after All?" Club of Rome, 2000.

Simon, J. "Resources, Population, Environment: An Over-supply of False Bad News." *Science* 208 (1980): 1431–1437.

Turner, Graham. "Is Global Collapse Imminent? An Updated Comparison of the Limits to Growth with Historical Data." *MSSI Research Paper* 4 (2014).

von Bertalanffy, Ludwig. *General System Theory.* New York: G. Braziller, 1968.

THE COEXISTENCE OF CULTIVATION SYSTEMS

One of the main contentions of this book is that the growth of population is a major determinant of technological change in agriculture. If this is so, it follows that there must have been some similarity between the rates at which population was growing and the rates at which technological change was occurring. This chapter examines these problems in some more detail, and in the light of historical experience.

Until recently rates of population growth were low or very low in most pre-industrial communities and from time to time the size of the population would be reduced by wars, famines or epidemics. Thus, we should expect the rate of technological change in agriculture to have been slow and interrupted by periods of stagnation or even regression of techniques; before intensive systems of land use had time to become applied over the whole territory of a given village or region, a set-back to population growth might often have occurred and the process of change would be interrupted. Thus, the slowly penetrating new systems of land use and techniques would be expected to coexist for long periods with older systems within the same village or in different villages within the same region.

Is such a theory supported by the available information about past agricultural development and about the present agricultural landscape in underdeveloped countries? In my opinion the hypothesis set forth in this book fits the facts better than does the traditional theory of autonomous technical progress.

Consider, for instance, the spread of the use of the plough.[1] This was an exceedingly slow process. In Europe, the plough was in use several thousand years ago and even in the most remote corners of Europe some ploughs seem to have been in use at least 3,000 years

[1] G. Haudricourt and M. J. Delamarre, *L'homme et la charrue* (Paris, 1955), gives a systematic review of the use of the plough all over the world since antiquity.

ago. Stone carvings of ploughs dating from that period have been found as far north as Sweden. Nevertheless, most of Europe seems to have been cultivated by the system of fire, axe and forest fallow as late as the time of the Roman Empire and as late as the eighteenth century it was not uncommon in many villages even in France and Germany that plots in forest and bush land were cleared and cultivated for only a year.[1] Even in the twentieth century, cultivation by forest fallow is found in some mountain districts of central Europe,[2] and until recently, forest fallow was widely used also on flat land in the sparsely populated Scandinavian countries. It was not before the end of the First World War that the burning of forest plots for short-term cultivation ceased in Sweden.[3] Thus, in a country with a rather homogeneous culture like Sweden, where geographical variations in techniques can hardly be explained by a lack of contacts, the plough coexisted with axe and fire for at least 3,000 years.

In Asia, where the plough was used long before it penetrated to Europe, it has never completely replaced fire and axe. There is still some forest fallow in southern China, considerably more in India and very large areas in South East Asia, apart from Java, are under that system of cultivation. The plough has been in use in all these regions for several thousand years, and much cultivation under forest fallow is done by villagers who at the same time use more intensive cultivation systems on part of their village land.

ADAPTATION OF LAND USE TO NATURAL CONDITIONS

In order to understand the process of intensification of land use, it is necessary to take account of the differences in natural conditions for agriculture as between various localities and between different parts of a given village territory. When increasing population density makes it necessary to change the pattern of land use in a given territory the changes are likely to be made in a way which takes account of the differences in natural conditions.

This point may need some explanation. Suppose, for instance, that the population of a community living by the system of forest fallow is increasing and let it be assumed for the sake of simplicity, that the rate at which the population is growing is relatively low, but regular.

[1] Marc Bloch, op. cit., vol. I (Paris, 1931), pp. 26–30, and vol. II (Paris, 1956), pp. 23–6, 33–8 and 68–75.

[2] Statistics of the area under the cultivation system of forest fallow in central Europe around 1900 are available in the article 'Haubergwirtschaft', in *Handwörterbuch der Staatswissenschaften*, vol. IV (Jena, 1900), pp. 1123–4.

[3] H. C. Darby, 'The Clearing of the Woodland in Europe', in W. L. Thomas, op. cit., p. 210.

Our cultivators would begin by gradually spreading their cultivation —under the system of forest fallow—over the territory under their command. At a certain point, when they have become so numerous that it is getting difficult to find suitable forest, they would begin to shorten the period of fallow and cultivate first a little part of the land, but later more and more of it, with the system of bush fallow instead of that of forest fallow. If they are aware of plough techniques and have access to draught animals they are unlikely to proceed far with the intensification of bush fallow. Instead, they may prefer to clear a few plots more thoroughly, removing stones and roots, and begin to use them in a system of short fallow with ploughing while they continue with long fallow in most of the area. Further fields for ploughing would be prepared in step with the increase of population. If the rate of population growth were very low, the complete change from one cultivation system to another might take many centuries.

At the time when our community was still relatively sparsely populated and the plough was being slowly introduced in replacement of long fallow, a few arable fields would be prepared each year, or at longer intervals, and the number of temporary plots cleared annually in the forest would be reduced a little, so as to prevent the complete disappearance of the forest. Centuries later, when our community had become much more densely populated, and for instance was replacing short-fallow cultivation with annual cropping of irrigated land, a few fields would each year—or at longer intervals —be provided with irrigation facilities and some dry fields would be allowed to lapse into a state of permanent grazing, so as to provide fodder for the necessary increase in the number of domestic animals. Finally, when our community had become extremely densely populated a few fields would now and then be added to the part of that cultivated area, which could be sown and harvested twice a year instead of once, or three times instead of two. If the rate of population growth was sufficiently low, an observer at any given moment would not see many signs of change. He would be inclined to describe the community as technically stagnant.

At every stage of this process of transformation, the fields first chosen for more frequent cropping would be among those relatively well suited for the next step in the development towards more intensive patterns of land use. This would be flat fields at the stage when plough cultivation was to be introduced, it would be land near rivers or wells when annual cropping with irrigation was to appear, and at the point of transition to intensive manuring, fields close to centres of settlement would be chosen for the conversion.

Once a particular system of land use had become the dominant one, the fields still under the previous system would probably not be particularly well suited for the new system of land use. Some cultivators might therefore find it preferable to introduce a third, and still more intensive system, instead of continuing with the shift away from the oldest and most extensive system to the one of intermediate intensity. There would then be three systems of land use coexisting in the same village territory, each of them occupying land which seemed relatively well suited for that particular degree of intensity.[1]

A contemporary observer, who were to make a cross-section survey of the village territory at any given moment during this period of gradual change, would perhaps be inclined to interprete it in terms of a static geographic theory of land use. He would explain that different types of land use were adaptations to differences in natural conditions, because each part of the village territory seemed to be used in the way for which it was best suited. If our observer happened to be a staunch Malthusian he would point out with conviction that the differences in intensity of land use had developed because the population had grown up to the size which could be supported in a territory with these particular natural conditions; and he might go on to say that it would be impossible to apply more labour productively in that territory, and that additional population would have to face the choice between underemployment and starvation, or emigration.

THE CASES OF JAVA AND JAPAN

Dutch observers have many times taken such cross-section views of the island of Java and concluded from the dense population in the fertile volcanic districts that the island was overpopulated.[2] Such a statement was made already one and a half centuries ago[3] but today the number of inhabitants is more than ten times as large; and Java is nearly self-sufficient in food, even though hardly any chemical fertilizer is used for food production.

The striking expansion of agricultural production in Java is some-

[1] A. T. Grove has made the observation that where population densities in Northern Nigeria are below the range of 150–200 to the square mile, the pattern of land utilization is largely unrelated to variations in the inherent capabilities of land resources, but any particular patch may be under woodland at one time, rough grazing at another, cropped for a few years and then abandoned while in areas with higher population densities land use is much more closely correlated with variations of soil conditions. See A. T. Grove, 'Population Densities and Agriculture in Northern Nigeria', in Barbour and Prothero, op. cit., pp. 125–6.

[2] For a recent example of this see F. A. van Baaren, 'Soils in Relation to Population in Tropical Regions', in *Tijdschrift voor Economische en Sociale Geografie* (Rotterdam, September 1960).

[3] In 1816, a former director of the province of Java's North-East Coast remarked that in his time the rice fields were cultivated on rotation because the 'population far exceeded the cultivation'. (Quoted from J. M. van der Kroef, *Indonesia in the Modern World*, vol. II (Bandung, 1956), p. 67.)

times explained as the result of the agricultural education and the capital investment undertaken by the Dutch. But these activities can account for only a small share of the additional food production in the island. The main explanation of the high elasticity of production is in the gradual spread over the island of methods applied in parts of its territory for centuries before the arrival of the Dutch.

The cultural pattern of Java is rather homogeneous, and it is impossible to explain geographical differences in technical methods within the island in terms of insufficient cultural contacts. The plough was in use in the tenth century and it was probably introduced much earlier. When the Dutch arrived, most of the island was still under the systems of forest and bush fallow, but long fallow has gradually been eliminated as a result of the rapid growth of population in the Dutch period. Systematic comparison of reports from various periods of the colonial era has revealed successive changes in some areas from long fallow to short fallow and from short fallow to annual and multiple cropping.

The changes that took place in Java are well summarized by H. Bartlett: 'During the last century typical *ladang* cultivation (i.e. long-fallow cultivation) as it persists elsewhere in Indonesia fell more and more into disuse on account of the great pressure on land that arose from excessive increase in the population and finally practically disappeared. *Ladang* land of the sort that required felling and burning of forest and a long recovery period of ten or twenty years after from one to three seasons of satisfactory agricultural productivity, ceased to exist, being replaced by the permanent cultivation of dry fields ... when *ladang* cultivation evolved through necessity into a higher type of land use. Before *ladang* cultivation with a forest fallow was abandoned it sometimes went into a limited phase of long grass-land fallow, which could only exist where excessive expenditure of labour was made necessary by dearth of suitable land.'[1] Commenting on the Government report of 1911 for Middle Java he says: 'This entire report is interesting as indicating what happens, when population pressure finally becomes too great after a long period of unplanned land use, or abuse. Forest is swept into grass land. By excessive grazing useless bushy and thorny wooded vegetation is enabled to take over the grass land. Permanent agriculture on non-irrigated land supersedes the primitive shifting agriculture through various intervening stages in which the labour demand grows, the fallow stages grow short and finally, by unwilling adoption of deep cultivation and manuring, the stage of annual cropping is reached, but by a wasteful and devastating series of steps.'[2]

[1] H. H. Bartlett, 'Fire in Relation to Primitive Agriculture', op. cit., p. 554.
[2] Ibid., p. 807. See also pp. 158, 651, 721 and 801–9.

Today long fallow has nearly vanished from Java and multiple cropping has become widespread in the most densely populated parts of the island. In step with these changes, the Javanese peasant has been transformed from the careless idler of the old times, who had just to scratch the land to get food enough. In the most densely populated regions the peasants of today are working hard in tiny fields in order to keep them completely free of weeds and so level that no water runs to waste or damages the crop.

The agriculture of Japan in the period from about 1600 to about 1850 provides another example of rising population and gradual change to more intensive systems of land use.[1] At the beginning of the period dynastic change created internal peace after a period of turbulence, and population grew rapidly, particularly during the first half of the period. From a certain point, the increase of population seems to have caused a reduction of the average size of agricultural holdings and a thorough change of methods. Ploughing with the help of draught animals became more frequent, and double cropping was made possible by irrigation and by the use of purchased night soil and dried fish for fertilizer as a supplement to or substitution for the traditional method of trampling grass, leaves or ashes into the fields. A number of publications propagated improved methods of farming and the spreading of new crops and varieties.

Thomas C. Smith makes it clear, in his interesting analysis of Japanese agriculture in this period, that few of these changes were the result of contemporary inventions; most of them resulted from the spread of known techniques from localities where they were already well established to areas where they had been unknown or unused. He also emphasizes that for the majority of the holdings the effect of the changes was to raise not only yields per hectare, but also labour requirements per hectare. Nevertheless, he thinks that the cause of the changes 'remains obscure' and he explicitly rejects the idea that population increase was the driving force. His main argument against such an explanation is that the bulk of the population increase came before 1725, whereas both the technical changes and the trend towards smaller holdings were most marked after that time.[2] This argument, however, is not wholly convincing since there may have been considerable scope around 1600 for the accommodation of additional people and increased production of food without the recourse to subdivision of existing holdings with use of intensive methods. These became indispensable at a later stage, when the density of population had reached a certain critical level.

[1] Thomas C. Smith, *The Agrarian Origins of Modern Japan* (Stanford University Press, 1959). See especially the chapter on agricultural technology, pp. 87–107.
[2] Ibid., pp. 87 and 104–5.

REDUCTIONS OF POPULATION DENSITY

In both Java and Japan the size of populations was steadily growing for several centuries. This, however, is somewhat exceptional; agrarian history knows relatively few examples of such steady growth and many more of population growth interrupted by frequent set-backs, as already mentioned. According to Malthusian theory we would expect population quickly to regain its former size after such set-backs, but historical evidence does not confirm this expectation. It took a century to regain the losses after the Thirty Years' War in central Europe and there are many examples from all parts of the world of still longer periods of recuperation.

In cases where population density was reduced by wars or other catastrophes there often seems to have been a relapse into more extensive systems of cultivation. Many of the permanent fields which were abandoned after wars or epidemics during the early Middle Ages remained uncultivated for centuries after. The use of labour-intensive methods of fertilization, such as marling, were abandoned for several centuries in France and then reappeared in the same region, when population again became dense.[1] The disappearance and reappearance of the cultivation of leguminous crops have already been mentioned.

Latin America is the region which suffered most from population decline in recent centuries. In many regions, the population density of pre-Columbian times has never been regained and the Indian population has regressed in agricultural techniques. More revealingly, the same process of technical regression can be seen, when migrants from more densely populated regions with much higher technical levels become settlers in the sparsely populated regions of Latin America. Let me illustrate this by quoting the statement made at an international conference by a specialist on rural conditions in Latin America:

'It seems to me that two points should be made. First of all, as most of us realize, a large proportion of all of the farmed area and of the farm population of Latin America is in a pre-scientific stage. . . .

[1] During the period of population increase in England and some parts of France in the sixteenth century, marl and lime were put on the land for the first time since the Roman period and the thirteenth century (i.e. the two previous periods of high population densities). See B. H. Slicher van Bath, op. cit., p. 205. In regions of northern Germany, where the land before the Thirty Years' War had been fallowed only every fourth, fifth, sixth or eighth year, the population decline caused by the war led to a return to the three-course rotation, which survived in these regions until the eighteenth century. Ibid., p. 245. See also the enumeration of the eleven systems of land use which coexisted in western Europe in the seventeenth and eighteenth centuries. Ibid., p. 244.

This is because the people are isolated by distance, or by completely different cultural levels, or by cultural inertia—defined as the reluctance of a conservative rural society to adopt new ideas. . . . The other point I would like to make is that there are certain segments of the Latin American farm population which according to field investigators have been descending in the technological scale rather than ascending. They are going the wrong way. Observers such as Waibel and Lynn Smith, who have studied the relatively recent European colonization in South Brazil, for example, tell us that these colonists, who came (or whose ancestors came) from countries like Germany and Italy with relatively advanced techniques, have lost many of these techniques. This is true even of such fairly simple practices as the use of the plow and crop rotation and the inclusion of livestock and forage crops in the farm economy for the maintenance of soil fertility.'[1]

This Latin American experience of apparent technical regression, when population declines or when people move to less densely populated areas, is by no means unique.[2] Many observers report of apparent technical regression after migrations to less densely populated regions even in cases where the migrations took place at government initiative and were designed to promote the spread of intensive methods to the regions of immigration. In both Tanganyika, Vietnam, Ceylon and India extension service administrations have made the experience that cultivators who used intensive methods in their densely settled home districts give up these methods after they have been resettled in less densely populated districts and given more land per family. Many settlement areas, meant to serve as model farms for the local population, provide sad sights of poor yields obtained from unweeded and unwatered fields.

EFFECTS OF RAPID POPULATION GROWTH

The examples above referred partly to the case of a steady but relatively slow population increase accompanied by changes of agricultural techniques, and partly to the case of a decline of population or migration to less densely populated regions accompanied by a return

[1] See the statement made by Henry S. Sterling at the 1951 International Conference on Land Tenure and Related Problems in World Agriculture. Kenneth H. Parsons et al. (ed.), *Land Tenure* (Madison, 1956), pp. 349–50.

[2] Professor Sauvy is emphatic on this point: 'On ne peut trouver, dans l'histoire, aucune stagnation ou recul démographique heureux. Une telle accumulation de démentis montre que la théorie est défectueuse, pèche par quelque point.' A. Sauvy, *Théorie générale de la population*, vol. II (Paris, 1954), p. 20. P. Gourou gives examples of a return to extensive land use in 'Les pays tropicaux', op. cit., pp. 101, 109 and 142.

to extensive and apparently more primitive techniques. It now remains to consider the effects of very rapid population growth on agricultural methods.

With rapid population growth, the process of intensification would need to take place much more quickly than in the cases we have dealt with so far. Not just a few fields, but a large number of fields would each year have to be cleared or provided with irrigation facilities, perhaps with the result that two harvests could be taken annually rather than one. A large amount of land clearing, land improvement and drainage or investment in irrigation facilities would have to take place simultaneously. Contemporary observers would not fail to notice this increased activity and they might well describe the period of rapidly rising population as a period of agricultural revolution. The agricultural revolution in eighteenth-century western Europe seems to have been of this type, and the agricultural changes which are now occurring in many underdeveloped countries seem to provide us with another example of rapid spread of the techniques of intensive agriculture owing to population pressure. Future historians will perhaps describe the decades after 1950 as those of the Indian agrarian revolution.

When population growth becomes so rapid, in a pre-industrial economy, that its effects may appear as an agricultural revolution to contemporary observers or economic historians, new and difficult problems, unknown under slow or moderate growth, present themselves. The cultivators must be able to adapt themselves quickly to methods which are new to them, although they may have been used for millenia in other parts of the world, and—perhaps even more difficult—they must get accustomed, within a relatively short period, to regular, hard work instead of a more leisurely life with long periods of seasonal idleness. Moreover, the community must somehow be able to bear the burden of a high rate of investment and perhaps to undertake sweeping changes in land tenure. I shall revert to the problems of investment and land tenure in later chapters.

Production, Consumption, and Externalities

By Robert U. Ayres and Allen V. Kneese*

"For all that, welfare economics can no more reach conclusions applicable to the real world without some knowledge of the real world than can positive economics" [21].

Despite tremendous public and governmental concern with problems such as environmental pollution, there has been a tendency in the economics literature to view externalities as exceptional cases. They may distort the allocation of resources but can be dealt with adequately through simple *ad hoc* arrangements. To quote Pigou:

> When it was urged above, that in certain industries a wrong amount of resources is being invested because the value of the marginal social net product there differs from the value of the marginal private net product, it was tacitly assumed that in the main body of industries these two values are equal [22][1].

And Scitovsky, after having described his cases two and four which deal with technological externalities affecting consumers and producers respectively, says:

> The second case seems exceptional, because most instances of it can be and usually are eliminated by zoning ord-

nances and industrial regulations concerned with public health and safety. The fourth case seems unimportant, simply because examples of it seem to be few and exceptional [25].

We believe that at least one class of externalities—those associated with the disposal of residuals resulting from the consumption and production process—must be viewed quite differently.[2] They are a normal, indeed, inevitable part of these processes. Their economic significance tends to increase as economic development proceeds, and the ability of the ambient environment to receive and assimilate them is an important natural resource of increasing value.[3] We will argue below that

---

* The authors are respectively visiting scholar and director, Quality of the Environment Program, Resources for the Future, Inc. We are indebted to our colleagues Blair Bower, Orris Herfindahl, Charles Howe, John Krutilla, and Robert Steinberg for comments on an earlier draft. We have also benefited from comments by James Buchanan, Paul Davidson, Robert Dorfman, Otto Eckstein, Myrick Freeman, Mason Gaffney, Lester Lave, Herbert Mohring, and Gordon Tullock.

[1] Even Baumol who saw externalities as a rather pervasive feature of the economy tends to discuss external diseconomies like "smoke nuisance" entirely in terms of particular examples [3]. A perspective more like that of the present paper is found in Kapp [16].

[2] We by no means wish to imply that this is the only important class of externalities associated with production and consumption. Also, we do not wish to imply that there has been a lack of theoretical attention to the externalities problem. In fact, the past few years have seen the publication of several excellent articles which have gone far toward systematizing definitions and illuminating certain policy issues. Of special note are Coase [9], Davis and Whinston [12], Buchanan and Stubblebine [6], and Turvey [27]. However, all these contributions deal with externality as a comparatively minor aberration from Pareto optimality in competitive markets and focus upon externalities between two parties. Mishan, after a careful review of the literature, has commented on this as follows: "The form in which external effects have been presented in the literature is that of partial equilibrium analysis; a situation in which a single industry produces an equilibrium output, usually under conditions of perfect competition, some form of intervention being required in order to induce the industry to produce an "ideal" or "optimal" output. If the point is not made explicitly, it is tacitly understood that unless the rest of the economy remains organized in conformity with optimum conditions, one runs smack into Second Best problems" [21].

[3] That external diseconomies are integrally related to economic development and increasing congestion has been noted in passing in the literature. Mishan has commented: "The attention given to external effects in

the common failure to recognize these facts may result from viewing the production and consumption processes in a manner that is somewhat at variance with the fundamental law of conservation of mass.

Modern welfare economics concludes that if (1) preference orderings of consumers and production functions of producers are independent and their shapes appropriately constrained, (2) consumers maximize utility subject to given income and price parameters, and (3) producers maximize profits subject to the price parameters; a set of prices exists such that no individual can be made better off without making some other individual worse off. For a given distribution of income this is an efficient state. Given certain further assumptions concerning the structure of markets, this "Pareto optimum" can be achieved via a pricing mechanism and voluntary decentralized exchange.

If waste assimilative capacity of the environment is scarce, the decentralized voluntary exchange process cannot be free of uncompensated technological external diseconomies unless (1) all inputs are fully converted into outputs, with no unwanted material residuals along the way,[4] and all final outputs are utterly destroyed in the process of consumption, or (2) property rights are so arranged that all relevant environmental attributes are in private ownership and these rights are exchanged in competitive markets. Neither of these conditions can be expected to hold in an actual economy and they do not.

Nature does not permit the destruction of matter except by annihilation with anti-matter, and the means of disposal of unwanted residuals which maximizes the internal return of decentralized decision units is by discharge to the environment, principally, watercourses and the atmosphere. Water and air are traditionally examples of free goods in economics. But in reality, in developed economies they are common property resources of great and increasing value presenting society with important and difficult allocation problems which exchange in private markets cannot resolve. These problems loom larger as increased population and industrial production put more pressure on the environment's ability to dilute and chemically degrade waste products. Only the crudest estimates of present external costs associated with residuals discharge exist but it would not be surprising if these costs were in the tens of billions of dollars annually.[5] Moreover, as we shall emphasize again, technological means for processing or purifying one or another type of waste discharge do not destroy the residuals but only alter their form. Thus, given the level, patterns, and technology of production and consumption, recycle of materials into productive uses or discharge into an alternative medium are the only general options for protecting a particular environmental medium such as water. Residual problems must be seen in a broad regional or economy-wide context rather

the recent literature is, I think, fully justified by the unfortunate, albeit inescapable, fact that as societies grow in material wealth, the incidence of these effects grows rapidly ... " [21]; and Buchanan and Tullock have stated that as economic development proceeds, "congestion" tends to replace "co-operation" as the underlying motive force behind collective action, i.e., controlling external diseconomies tends to become more important than cooperation to realize external economies [7].

[4] Or any residuals which occur must be stored on the producer's premises.

[5] It is interesting to compare this with estimates of the cost of another well known misallocation of resources that has occupied a central place in economic theory and research. In 1954, Harberger published an estimate of the welfare cost of monopoly which indicated that it amounted to about .07 percent of GNP [15]. In a later study, Schwartzman calculated the allocative cost at only .01 percent of GNP [24]. Leibenstein generalized studies such as these to the statement that " ... in a great many instances the amount to be gained by increasing allocative efficiency is trivial ... " [19]. But Leibenstein did not consider the allocative costs associated with environmental pollution.

than as separate and isolated problems of disposal of gas, liquid, and solid wastes.

Frank Knight perhaps provides a key to why these elementary facts have played so small a role in economic theorizing and empirical research.

> The next heading to be mentioned ties up with the question of dimensions from another angle, and relates to the second main error mentioned earlier as connected with taking food and eating as the type of economic activity. The basic economic magnitude (value or utility) is service, not good. It is inherently a stream or flow in time . . . [18].[6]

Almost all of standard economic theory is in reality concerned with services. Material objects are merely the vehicles which carry some of these services, and they are exchanged because of consumer preferences for the services associated with their use or because they can help to add value in the manufacturing process. Yet we persist in referring to the "final consumption" of goods as though material objects such as fuels, materials, and finished goods somehow disappeared into the void—a practice which was comparatively harmless so long as air and water were almost literally free goods.[7] Of course, residuals from both the production and consumption processes remain and they usually render disservices (like killing fish, increasing the difficulty of water treatment, reducing public health, soiling and deteriorating buildings, etc.) rather than services. Control efforts are aimed at eliminating or reducing those disservices which flow to consumers and pro-

ducers whether they want them or not and which, except in unusual cases, they cannot control by engaging in individual exchanges.[8]

I. *The Flow of Materials*

To elaborate on these points, we find it useful initially to view environmental pollution and its control as a materials balance problem for the entire economy.[9] The inputs to the system are fuels, foods, and raw materials which are partly converted into final goods and partly become waste residuals. Except for increases in inventory, final goods also ultimately enter the waste stream. Thus goods which are "consumed" really only render certain services. Their material substance remains in existence and must either be reused or discharged to the ambient environment.

In an economy which is closed (no imports or exports) and where there is no net accumulation of stocks (plant, equipment, inventories, consumer durables, or residential buildings), the amount of residuals inserted into the natural environment must be approximately equal to the weight of basic fuels, food, and raw materials entering the processing and production system, plus oxygen taken from the atmosphere.[10] This result, while obvious

---

[6] The point was also clearly made by Fisher: "The only true method, in our view, is to regard uniformly as income the *service* of a dwelling to its owner (shelter or money rental), the *service* of a piano (music), and the *service* of food (nourishment) . . . " (emphasis in original) [14].

[7] We are tempted to suggest that the word consumption be dropped entirely from the economist's vocabulary as being basically deceptive. It is difficult to think of a suitable substitute, however. At least, the word consumption should not be used in connection with goods, but only with regard to services or flows of "utility."

[8] There is a substantial literature dealing with the question of under what conditions individual exchanges can optimally control technological external diseconomies. A discussion of this literature, as it relates to waterborne residuals, is found in Kneese and Bower [17].

[9] As far as we know, the idea of applying materials balance concepts to waste disposal problems was first expressed by Smith [26]. We also benefitted from an unpublished paper by Joseph Headley in which a pollution "matrix" is suggested. We have also found references by Boulding to a "spaceship economy" suggestive [4]. One of the authors has previously used a similar approach in ecological studies of nutrient interchange among plants and animals; see [1].

[10] To simplify our language, we will not repeat this essential qualification at each opportunity, but assume it applies throughout the following discussion. In addition, we must include residuals such as NO and NO_2 arising from reactions between components of the air itself but occurring as combustion by-products.

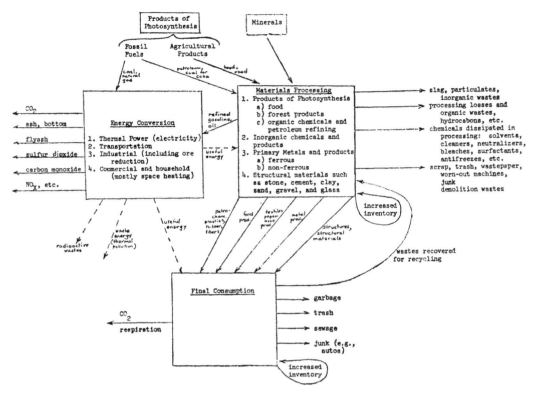

FIGURE 1.—MATERIALS FLOW

upon reflection, leads to the, at first rather surprising, corollary that residuals disposal involves a greater tonnage of materials than basic materials processing, although many of the residuals, being gaseous, require no physical "handling."

Figure 1 shows a materials flow of the type we have in mind in greater detail and relates it to a broad classification of economic sectors for convenience in our later discussion, and for general consistency with the Standard Industrial Classification. In an open (regional or national) economy, it would be necessary to add flows representing imports and exports. In an economy undergoing stock or capital accumulation, the production of residuals in any given year would be less by that amount than the basic inputs. In the entire U.S. economy, accumulation accounts for about 10–15 percent of basic annual inputs, mostly in the form of construction materials, and there is some net importation of raw and partially processed materials amounting to 4 or 5 percent of domestic production. Table 1 shows estimates of the weight of raw materials produced in the United States in several recent years, plus net imports of raw and partially processed materials.

Of the active inputs,[11] perhaps three-quarters of the overall weight is eventually discharged to the atmosphere as carbon (combined with atmospheric oxygen in the form of CO or CO_2) and hydrogen (combined with atmospheric oxygen as H_2O) under current conditions. This results from combustion of fossil fuels and from animal respiration. Discharge of carbon dioxide can be considered harmless in the short run. There are large "sinks" (in the form of vegetation and large water bodies,

[11] See footnote to Table 1.

TABLE 1—WEIGHT OF BASIC MATERIALS PRODUCTION
IN THE UNITED STATES PLUS NET IMPORTS,
1963 (10^6 tons)

	1963	1964	1965
Agricultural (incl. fishery and wildlife and forest) products			
Food { Crops (excl. livestock feed)	125	128	130
Livestock	100	103	102
Other products	5	6	6
Fishery	3	3	3
Forestry products (85 per cent dry wt. basis)			
Sawlogs	53	55	56
Pulpwood	107	116	120
Other	41	41	42
Total	434	452	459
Mineral fuels	1,337	1,399	1,448
Other minerals			
Iron ore	204	237	245
Other metal ores	161	171	191
Other nonmetals	125	133	149
Total	490	541	585
Grand total[a]	2,261	2,392	2,492

[a] Excluding construction materials, stone, sand, gravel, and other minerals used for structural purposes, ballast, fillers, insulation, etc. Gangue and mine tailings are also excluded from this total. These materials account for enormous tonnages but undergo essentially no chemical change. Hence, their use is more or less tantamount to physically moving them from one location to another. If this were to be included, there is no logical reason to exclude material shifted in highway cut and fill operations, harbor dredging, land-fill, plowing, and even silt moved by rivers. Since a line must be drawn somewhere, we chose to draw it as indicated above.

Source: R. U. Ayres and A. V. Kneese [2, p. 630].

mainly the oceans) which reabsorb this gas, although there is evidence of net accumulation of CO_2 in the atmosphere. Some experts believe that the latter is likely to show a large relative increase, as much as 50 per cent by the end of the century, possibly giving rise to significant —and probably, on balance, adverse— weather changes.[12] Thus continued combustion of fossil fuels at a high rate could produce externalities affecting the entire world. The effects associated with most residuals will normally be more confined, however, usually limited to regional air and water sheds.

The remaining residuals are either gases (like carbon monoxide, nitrogen dioxide, and sulfur dioxide—all potentially harmful even in the short run), dry solids (like rubbish and scrap), or wet solids (like garbage, sewage, and industrial wastes suspended or dissolved in water). In a sense, the dry solids are an irreducible, limiting form of waste. By the application of appropriate equipment and energy, most undesirable substances can, in principle, be removed from water and air streams[13]— but what is left must be disposed of in solid form, transformed, or reused. Looking at the matter in this way clearly reveals a primary interdependence between the various waste streams which casts into doubt the traditional classification of air, water, and land pollution as individual categories for purposes of planning and control policy.

Residuals do not necessarily have to be discharged to the environment. In many instances, it is possible to recycle them back into the productive system. The materials balance view underlines the fact that the throughput of new materials necessary to maintain a given level of production and consumption decreases as the technical efficiency of energy conversion and materials utilization increases. Similarly, other things being equal, the longer that cars, buildings, machinery, and other durables remain in service, the fewer new materials are required to compensate for loss, wear, and obsolescence— although the use of old or worn machinery (e.g., automobiles) tends to increase other residuals problems. Technically efficient combustion of (desulfurized) fossil fuels

[12] See [30]. There is strong evidence that discharge of residuals has already affected the climate of individual cities; see Lowry [20].

[13] Except CO_2, which may be harmful in the long run, as noted.

would leave only water, ash, and carbon dioxide as residuals, while nuclear energy conversion need leave only negligible quantities of material residuals (although thermal pollution and radiation hazards cannot be dismissed by any means).

Given the population, industrial production, and transport service in an economy (a regional rather than a national economy would normally be the relevant unit), it is possible to visualize combinations of social policy which could lead to quite different relative burdens placed on the various residuals-receiving environmental media; or, given the possibilities for recycle and less residual-generating production processes, the overall burden to be placed upon the environment as a whole. To take one extreme, a region which went in heavily for electric space heating and wet scrubbing of stack gases (from steam plants and industries), which ground up its garbage and delivered it to the sewers and then discharged the raw sewage to watercourses, would protect its air resources to an exceptional degree. But this would come at the sacrifice of placing a heavy residuals load upon water resources. On the other hand, a region which treated municipal and industrial waste water streams to a high level and relied heavily on the incineration of sludges and solid wastes would protect its water and land resources at the expense of discharging waste residuals predominantly to the air. Finally, a region which practiced high level recovery and recycle of waste materials and fostered low residual production processes to a far reaching extent in each of the economic sectors might discharge very little residual waste to any of the environmental media.

Further complexities are added by the fact that sometimes it is possible to modify an environmental medium through investment in control facilities so as to improve its assimilative capacity. The clearest, but far from only, example is with respect to

watercourses where reservoir storage can be used to augment low river flows that ordinarily are associated with critical pollution (high external cost situations).[14] Thus internalization of external costs associated with particular discharges, by means of taxes or other restrictions, even if done perfectly, cannot guarantee Pareto optimality. Investments involving public good aspects must enter into an optimal solution.[15]

To recapitulate our main points briefly: (1) Technological external diseconomies are not freakish anomalies in the processes of production and consumption but an inherent and normal part of them. (2) These external diseconomies are quantitatively negligible in a low-population or economically undeveloped setting, but they become progressively (nonlinearly) more important as the population rises and the level of output increases (i.e., as the natural reservoirs of dilution and assimilative capacity become exhausted).[16] (3) They cannot be properly dealt with by considering environmental media such as air and water in isolation. (4) Isolated and *ad hoc* taxes and other restrictions are not sufficient for their optimum control, although they are essential elements in a more systematic and coherent program of environmental quality management. (5) Public investment programs, particularly including transportation systems, sewage disposal, and river flow regulation, are intimately related to the amounts and

[14] Careful empirical work has shown that this technique can fit efficiently into water quality management systems. See Davis [11].

[15] A discussion of the theory of such public investments with respect to water quality management is found in Boyd [5].

[16] Externalities associated with residuals discharge may appear only at certain threshold values which are relevant only at some stage of economic development and industrial and population concentrations. This may account for their general treatment as "exceptional" cases in the economics literature. These threshold values truly were exceptional cases for less developed economies.

effects of residuals and must be planned in light of them.

It is important to develop not only improved measures of the external costs resulting from differing concentrations and duration of residuals in the environment but more systematic methods for forecasting emissions of external-cost-producing residuals, technical and economic trade-offs between them, and the effects of recycle on environmental quality.

In the hope of contributing to this effort and of revealing more clearly the types of information which would be needed to implement such a program, we set forth a more formal model of the materials balance approach in the following sections and relate it to some conventional economic models of production and consumption. The main objective is to make some progress toward defining a system in which flows of services and materials are simultaneously accounted for and related to welfare.

II. Basic Model

The take off point for our discussion is the Walras-Cassel general equilibrium model,[17] extended to include intermediate consumption, which involve the following quantities:

resources and services
$$\overbrace{r_1, \cdots\cdots\cdots, r_M}$$
products or commodities
$$\overbrace{X_1, \cdots\cdots\cdots, X_N}$$
resource prices
$$\overbrace{v_1, \cdots\cdots\cdots, v_M}$$
product or commodity prices
$$\overbrace{p_1, \cdots\cdots\cdots, p_N}$$
final demands
$$\overbrace{Y_1, \cdots\cdots\cdots, Y_N}$$

[17] The original references are Walras [28] and Cassel [8]. Our own treatment is largely based on Dorfman et al. [13].

The M basic resources are allocated among the N sectors as follows:

$$r_1 = a_{11}X_1 + a_{12}X_2 + \cdots + a_{1N}X_N$$
$$r_2 = a_{21}X_1 + a_{22}X_2 + \cdots + a_{2N}X_N$$
$$\vdots$$
$$r_M = a_{M1}X_1 + a_{M2}X_2 + \cdots + a_{MN}X_N$$

(1a) or

$$r_j = \sum_{k=1}^{N} a_{jk}X_k \qquad j = 1, \cdots, M$$

In (1a) we have implicitly assumed that there is no possibility of factor or process substitution and no joint production. These conditions will be discussed later. In matrix notation we can write:

(1b) $$[r_{j1}]_{M,1} = [a_{jk}]_{M,N} \cdot [X_{k1}]_{N,1}$$

where $[a]$ is an $M \times N$ matrix.

A similar set of equations describes the relations between commodity production and final demand:

(2a) $$X_k = \sum_{l=1}^{N} A_{kl}Y_l \qquad k = 1, \cdots, N$$

(2b) $$[X_{k1}]_{N,1} = [A_{kl}]_{N,N} \cdot [Y_{l1}]_{N,1}$$

and the matrix $[A]$ is given by

(3) $$[A] = [I - C]^{-1}$$

where $[I]$ is the unit diagonal matrix and the elements C_{ij} of the matrix $[C]$ are essentially the well known Leontief input coefficients. In principle these are functions of the existing technology and, therefore, are fixed for any given situation.

By combining (1) and (2), we obtain a set of equations relating resource inputs directly to final demand, viz.,

(4a)
$$r_j = \sum_{k=1}^{N} a_{jk} \sum_{l=1}^{N} A_{kl}Y_l = \sum_{k,l=1}^{N} a_{jk}A_{kl}Y_l$$
$$= \sum_{l=1}^{N} b_{jl}Y_l \qquad j = 1, \cdots, M$$

or, of course, in matrix notation (4b).

(4b)
$$[r_{j1}]_{M,1} = [a_{jk}]_{M,N} \cdot [A_{kl}]_{N,N} \cdot [Y_{l1}]_{N,1}$$
$$= [b_{j1}]_{M,N} \cdot [Y_{l1}]_{N,1}$$

We can also impute the prices of N intermediate goods and commodities to the prices of the M basic resources, as follows:

(5a) $\quad p_k = \sum_{j=1}^{M} v_j b_{jk} \qquad k = 1, \cdots, N$

(5b) $\quad [p_{1k}]_{1,N} = [v_{1j}]_{1,M} \cdot [b_{jk}]_{M,N}$

To complete the system, it may be supposed that demand and supply relationships are given, a priori, by Pareto-type preference functions:

(6) Demand: $Y_k = F_k(p_1, \cdots, p_N)$
$$k = 1, \cdots, N$$

(7) Supply: $r_k = G_k(v_1, \cdots, v_M)$
$$k = 1, \cdots, M$$

where, of course, the p_j are functions of the v_j as in (5b).

In order to interpret the X's as physical production, it is necessary for the sake of consistency to arrange that outputs and inputs always balance, which implies that the C_{ij} must comprise *all* materials exchanges including residuals. To complete the system so that there is no net gain or loss of physical substances, it is also convenient to introduce two additional sectors, viz., an "environmental" sector whose (physical) output is X_0 and a "final consumption" sector whose output is denoted X_f. The system is then easily balanced by explicitly including flows both to and from these sectors.

To implement this further modification of the Walras-Cassel model, it is convenient to subdivide and relabel the resource category into tangible raw materials $\{r^m\}$ and services $\{r^s\}$:

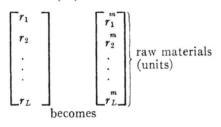

raw materials (units)

becomes

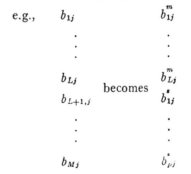

service (units)

becomes

where, of course,

(8) $L + P = M$

It is understood that services, while not counted in tons, can be measured in meaningful units, such as man-days, with well defined prices. Thus, we similarly relabel the price variables as follows:

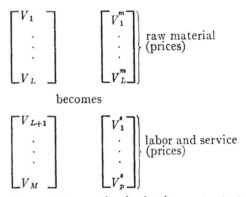

raw material (prices)

becomes

labor and service (prices)

The coefficients $\{a_{ij}\}$, $\{b_{ij}\}$ are similarly partitioned into two groups,

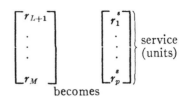

e.g., b_{1j} b_{1j}^m
        $\vdots$ $\vdots$
        b_{Lj} b_{Lj}^m
              becomes b_{1j}^s
        $b_{L+1,j}$
        $\vdots$ $\vdots$
        b_{Mj} b_{pj}^s

These notational changes have no effect whatever on the substance of the model, although the equations become somewhat more cumbersome. The partitioned matrix notation simplifies the restatement of the basic equations. Thus (1b) becomes (9), while (5b) becomes (10).

(9)
$$M \left\{ \begin{bmatrix} \vdots \\ r \\ \vdots \end{bmatrix} \equiv \begin{bmatrix} & r^m & \\ & \cdots & \\ & r^s & \end{bmatrix} \begin{matrix} \} L \\ \\ \} P \end{matrix} \right. = M \left\{ \begin{bmatrix} \} L & b^m \\ & \cdots \\ \} P & b^s \end{bmatrix} \begin{bmatrix} \\ Y \\ \end{bmatrix} \right\} N $$

(10)
$$[p_1, \cdots, p_N] = [v^m \vdots v^s] \underbrace{\begin{bmatrix} b^m \\ \cdots \\ b^s \end{bmatrix}}_{N} \Big\} M$$
$$ = [\cdots v^m \cdots] \begin{bmatrix} \cdot & & b^m & \cdot \\ \vdots & & & \vdots \\ \cdot & & \cdot & \cdot \end{bmatrix} + [\cdots v^s \cdots] \begin{bmatrix} \cdot & & b^s & \cdot \\ \vdots & & & \vdots \\ \cdot & & \cdot & \cdot \end{bmatrix}$$

The equivalent of (5a) is:

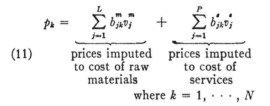

(11)
$$p_k = \underbrace{\sum_{j=1}^{L} b_{jk}^{m} v_j^{m}}_{\substack{\text{prices imputed} \\ \text{to cost of raw} \\ \text{materials}}} + \underbrace{\sum_{j=1}^{P} b_{jk}^{s} v_j^{s}}_{\substack{\text{prices imputed} \\ \text{to cost of} \\ \text{services}}}$$

where $k = 1, \cdots, N$

We wish to focus attention explicitly on the flow of materials through the economy. By definition of the Leontief input coefficients (now related to materials flow), we have:

$C_{kj} X_j$ (physical) quantity transferred from k to j

$C_{jk} X_k$ quantity transferred from j to k

Hence, material flows *from* the environment to all other sectors are given by:

(12)
$$\sum_{k=1}^{N} C_{0k} X_k = \sum_{j=1}^{L} r_j^{m} = \sum_{j=1}^{L} \sum_{k=1}^{N} a_{jk}^{m} X_k$$
$$ = \sum_{j=1}^{L} \sum_{k=1}^{N} b_{jk}^{m} Y_k$$

using equation (1), as modified.[18] Obvi-

ously, comparing the first and third terms,

(13)
$$\underbrace{C_{0k}}_{\substack{\text{total material} \\ \text{flow (0 to } k)}} = \underbrace{\sum_{j=1}^{L} a_{jk}^{m}}_{\substack{\text{all raw materials} \\ \text{(0 to } k)}}$$

Flows into and out of the environmental sector must be in balance:

(14)
$$\underbrace{\sum_{k=1}^{N} C_{0k} X_k}_{\substack{\text{sum of all raw} \\ \text{material flows}}} = \underbrace{\sum_{k=1}^{N} C_{k0} X_0 + C_{f0} X_0}_{\substack{\text{sum of all return} \\ \text{(waste) flows}}}$$

Material flows to and from the final sector must also balance:

(15)
$$\underbrace{\sum_{k=1}^{N} C_{kf} X_f}_{\substack{\text{sum of all} \\ \text{final goods}}}$$
$$ = \underbrace{\sum_{k=1}^{N} C_{fk} X_k}_{\substack{\text{sum of all} \\ \text{materials} \\ \text{recycled}}} + \underbrace{C_{f0} X_0}_{\substack{\text{waste residuals} \\ \text{(plus accumulation[19])}}}$$

---

[18] Ignoring, for convenience, any materials flow from the environment *directly* to the final consumption sector.

[19] For convenience, we can treat accumulation in the final sector as a return flow to the environment. In truth, structures actually *become* part of our environment, although certain disposal costs may be deferred.

Of course, by definition, X_f is the sum of the final demands:

$$(16) \qquad X_f = \sum_{j=1}^{N} Y_j$$

Substituting (16) into the left side of (15) and (2a) into the right side of (15), we obtain an expression for the waste flow in terms of final demands:

$$(17) \quad C_{f0}X_0 = \sum_{j=1}^{N}\sum_{k=1}^{N}(C_{jf} - C_{fj}A_{jk})Y_k$$

The treatment could be simplified slightly if we assumed that there is no recycling per se. Thus, in the context of the model, we could suppose that all residuals return to the environmental sector,[20] where some of them (e.g., waste paper) become "raw materials." They would then be indistinguishable from new raw materials, however, and price differentials between the two would be washed out. In principle, this is an important distinction to retain.

III. *Inclusion of Externalities*

The physical flow of materials between various intermediate (production) sectors and the final (consumption) sector tends to be accompanied by, and correlated with, a (reverse) flow of dollars.[21] However, the physical flow of materials from and back to the environment is only partly reflected by actual dollar flows, namely, land rents and payments for raw materials. There are three classes of physical exchange for which there exist no counterpart economic transactions. These are: (1) private use for production inputs of "common property" resources, notably air, streams, lakes, and the ocean; (2) private use of the assimila-tive capacity of the environment to "dispose of" or dilute wastes and residuals; (3) inadvertent or unwanted material inputs to productive processes—diluents and pollutants.

All these goods (or "bads") are physically transferred at zero price, not because they are not scarce relative to demand—they often are in developed economies—or because they confer no service or disservice on the user—since they demonstrably do so—but because there exist no social institutions that permit the resources in question to be "owned," and exchanged in the market.

The allocation of resources corresponding to a Pareto optimum cannot be attained without subjecting the above-mentioned nonmarket and involuntary exchanges to the moderation of a market or a surrogate thereof. In principle, the influence of a market might be simulated, to a first approximation, by introducing a set of shadow (or virtual) prices.[22] These may well be zero, where supply truly exceeds demand, or negative (i.e., costs) in some instances; they will be positive in others. The exchanges are, of course, real.

The Walras-Cassel model can be generalized to handle these effects in the following way:

1. One can introduce a set of R common-property resources or services of raw materials $\{r_1^{cp}, \cdots, r_R^{cp}\}$ as a subset of the set $\{r_j\}$; these will have corresponding virtual prices $\{v_j^{cp}\}$, which would constitute an "income" from the environment. Such resources include the atmosphere; streams, lakes, and oceans; landscape; wildlife and biological diversity; and the indispensable assimilative capacity of the environment (its ability to accept and neutralize or recycle residuals).[23]

[20] In calculating actual quantities, we would (by convention) ignore the weight of oxygen taken free from the atmosphere in combustion and return as CO_2. However, such inputs will be treated explicitly later.

[21] To be precise, the dollar flow governs and is governed by a combined flow of materials and services (value added).

[22] A similar concept exists in mechanics where the forces producing "reaction" (to balance action and reaction) are commonly described as "virtual forces."

[23] Economists have previously suggested generalization of the Walras-Cassel model to take account of public goods. One of the earliest appears to be Schles-

2. One can introduce a set of S environmental *disservices* imposed on consumers of material resources, by forcing them to accept unwanted inputs $\{r_1^u, \cdots, r_s^u\}$ (pollutants, contaminants, etc.); these disservices would have negative value, giving rise to *negative* virtual prices $\{u_j\}$.[24]

The matrix coefficients $\{a_{ij}\}$ and $\{b_{ij}\}$ can be further partitioned to take account of this additional refinement, and equations analogous to (9), (10), and (11) can be generalized in the obvious way. Equation (6) carries over unchanged, but (7) must be appropriately generalized to take account of the altered situation. Actually, (7) breaks up into several groups of equations:

$$(18) \qquad \overset{m}{r_k} = \overset{m}{G_k}(p_1, \cdots, p_N)$$
$$k = 1, \cdots, L$$

$$(19) \qquad \overset{s}{r_k} = \overset{s}{G_k}(p_1, \cdots, p_N)$$
$$k = 1, \cdots, P$$

However, as we have noted at the outset, the supplies of common-property resources and environmental services or disservices are *not* regulated directly by market prices of other goods and services. In the case of common-property resources, the supplies are simply constants fixed by nature or otherwise determined by accident or noneconomic factors.

The total value of these services performed by the environment cannot be

calculated but it is suggestive to consider the situation if the natural reservoir of air, water, minerals, etc., were very much smaller, as if the earth were a submarine or "spaceship" (i.e., a vehicle with no assimilative and/or regenerative capacity). In such a case, all material resources would have to be recycled,[25] and the cost of all goods would necessarily reflect the cost of reprocessing residuals and wastes for reuse. In this case, incidentally, the ambient level of unrecovered wastes continuously circulating through the resource inventory of the system (i.e., the spaceship) would in general be nonzero because of the difficulty of 100 percent efficient waste-removal of air and water. However, although the quantity of waste products in constant circulation may fluctuate within limits, it cannot be allowed to increase monotonically with time, which means that as much material must be recycled, on the average, as is discarded. The value of common resources plus the assimilation services performed by the environment, then, is only indirectly a function of the ambient level of untreated residuals per se, or the disutility caused thereby, which depend on the cost efficiency of the available treatment technology. Be this as it may, of course, the bill of goods produced in a spaceship economy would certainly be radically different from that we are familiar with. For this reason, no standard economic comparison between the two situations is meaningful. The measure of worth we are seeking is actually the difference between the total welfare "produced" by a spaceship economy, where 100 percent of all residuals are promptly recycled, vis-à-vis the existing welfare output on earth, where resource inventories are substantial and

---

inger [23]. We are indebted to Otto Eckstein for calling our attention to this key reference.

[24] The notion of introducing the possibility of negative prices in general equilibrium theory has apparently been discussed before, although we are not aware of any systematic development of the idea in the published literature. In this connection, it is worth pointing out the underlying similarity of negative prices and effluent taxes—which have been, and still are being considered as an attractive alternative to subsidies and federal standard-setting as a means of controlling air and water pollution. Such taxes would, of course, be an explicit attempt to rectify an imbalance caused by a market failure.

[25] Any consistent deviation from this 100 per cent rule implies an accumulation of waste products, on the average, which, by definition, is inconsistent with maintaining an equilibrium.

complete recycling need not be contemplated for a very long time to come.

This welfare difference might well be very large, although we possess no methodological tools for quantifying it. In any case, the resource inventory and assimilative capacity of the environment probably contribute very considerably to our standard of living.

If these environmental contributions were paid for, the overall effect on prices would presumably to be push them generally upward. However, the major *differential* effect of undervaluing the environmental contribution is that goods produced by high residual-producing processes, such as papermaking, are substantially underpriced vis-à-vis goods which involve more economical uses of basic resources. This is, however, not socially disadvantageous per se: that is, it causes no misallocation of resources unless, or until, the large resource inventory and/or the assimilative capacity of the environment are used up. When this happens, however, as it now has in most highly industrialized regions, either a market must be allowed to operate or some other form of decision rule must be introduced to permit a rational choice to be made, e.g., between curtailing or controlling the production of residuals or tolerating the effects (disservice) thereof.

It appears that the natural inventory of most common resources used as inputs (e.g., air as an input to combustion and respiratory processes) is still ample,[26] but the assimilative capacity of the environment has already been exceeded in many areas, with important external costs resulting. This suggests a compromise treatment. If an appropriate price could be charged to the producers of the residuals and used to compensate the inadvertent recipients—with the price determined by appropriate Pareto preference criteria—there would be no particular analytic purpose in keeping books on the exchange of the other environmental benefits mentioned, although they are quantitatively massive. We will, therefore, in the remainder of the discussion omit the common-property variables $\{r_j^{cp}\}$ and the corresponding virtual-price variables $\{v_j^{cp}\}$ defined previously, retaining only the terms $\{r_j^u\}$ and $\{u_{jk}\}$. The variable $\{r_j^u\}$ represents a physical quantity of the jth unwanted input. There are S such terms, by assumption, whose magnitudes are proportional to the levels of consumption of basic raw materials, subject to the existing technology. However, residuals production is not immutable: it can be increased or decreased by investment, changes in materials processing technology, raw material substitutions, and so forth.

At first glance it might seem entirely reasonable to assert that the *supplies* of unwanted residuals received will be functions of the (negative) prices (i.e., compensation) paid for them, in analogy with (7). Unfortunately, this assertion immediately introduces a theoretical difficulty, since the assumption of unique coefficients $\{a_{ij}\}$ and $\{C_{ij}\}$[27] is not consistent with the possibility of factor or process substitution or joint-production, as stated earlier. To permit such substitutions, one would have to envision a very large collection of alternative sets of coefficients: one complete set of a's and C's for each specific combination of factors and processes. Maximization of any objective function (such as GNP) would involve solving the entire system of equations as many times as there are combinations of factors and pro-

[26] Water is an exception in arid regions; in humid regions, however, water "shortages" are misnomers: they are really consequences of excessive use of watercourses as cheap means of waste disposal. But some ecologists have claimed that oxygen depletion may be a very serious long-run problem; see Cole [10].

[27] Or $\{b_{ij}\}$ and $\{A_{ij}\}$.

$$
(21) \qquad [r] = \begin{bmatrix} r^m \\ r^s \\ r^u \end{bmatrix} = M \left\{ \begin{bmatrix} a^m \\ \cdots \\ a^s \\ \cdots \\ a^u \end{bmatrix} X \right\} N = \begin{bmatrix} b^m \\ \cdots \\ b^s \\ \cdots \\ b^u \end{bmatrix} Y
$$

$$
\underbrace{\qquad\qquad}_{N}
$$

cesses, and picking out that set of solutions which yields the largest value. Alternatively, if the a's and C's are assumed to be continuously variable functions (of each other), the objective function could also, presumably, be parameterized. However, as long as the a's and C's are uniquely given, the supply of the kth unwanted residual is only marginally under the control of the producer, since it will be produced in strict relationship to the composition of the bill of final goods $\{Y_j\}$.

Hence, for the present model it is only correct to assume

$$
(20) \qquad r_k^u = G_k^u(Y_1, \cdots, Y_N)
$$

This limitation does not affect the existence of an equilibrium solution for the system of equations; it merely means that the shadow prices $\{u_{jk}\}$ which would emerge from such a solution for given coefficients $\{a_{ij}\}$, $\{b_{ij}\}$, and $\{C_{ij}\}$ might be considerably higher than the real economic optimum, since the latter could only be achieved by introducing factor and process changes.

Of course, the physical inputs are also related to the physical outputs of goods, as in (21).

Written out in full detail (21) is equivalent to:

$$
(22) \quad \text{raw materials} \quad r_k^m = \sum_{j=1}^{N} a_{kj}^m X_j = \sum_{j=1}^{N} b_{kj}^m Y_j
$$

$$
k = 1, \cdots, L
$$

$$
(23) \quad \text{labor and technical services} \quad r_k^s = \sum_{j=1}^{N} a_{kj}^s X_j = \sum_{j=1}^{N} b_{kj}^s Y_j
$$

$$
k = 1, \cdots, P
$$

$$
(24) \quad \text{unwanted inputs} \quad r_k^u = \sum_{j=1}^{N} a_{kj}^u X_j = \sum_{j=1}^{N} b_{kj}^u Y_j
$$

$$
k = 1, \cdots, S
$$

where, of course,

$$
(25) \qquad L + P + S = M
$$

The corresponding matrix equation for the prices of goods, in terms of production costs, is

$$
(26) \qquad [p_1, \cdots, p_N] = [v^m \vdots v^s \vdots u] \begin{bmatrix} b^m \\ \cdots \\ b^s \\ \cdots \\ b^u \end{bmatrix}
$$

Written out in the standard form, we obtain

$$
p_k = \underbrace{\sum_{j=1}^{L} b_{jk}^m v_j^m}_{\substack{\text{cost of raw} \\ \text{materials}}} + \underbrace{\sum_{j=1}^{P} b_{jk}^s v_j^s}_{\substack{\text{cost of labor} \\ \text{and technical} \\ \text{services}}}
$$

$$
(27) \qquad + \underbrace{\sum_{j=1}^{S} b_{jk}^u v_j^u}_{\substack{\text{cost (compensa-} \\ \text{tion) for pro-} \\ \text{viding environ-} \\ \text{mental disser-} \\ \text{vices}}}
$$

$$
k = 1, \cdots, N
$$

Evidently, the coefficients b_{jk}^u are empirically determined by the structure of the regional economy and its geography. It is assumed that a single overall (negative) price for each residual has meaning, even though each productive sector—and even each consumer—has his own individual utility function. Much the same assumption is conventionally made, and accepted, in the case of positive real prices.

All of the additional variables now fit into the general framework of the original Walras-Cassel analysis. Indeed, we have $2N+2M-1$ variables (r_i, Y_i, p_i, v_i) (allowing an arbitrary normalization factor for the price level) and $2N+2M-1$ independent equations.[28] If solutions exist for the Walras-Cassel system of equations, the arguments presumably continue to hold true for the generalized model. In any case, a discussion of such mathematical questions would carry us too far from our main theme.

IV. *Concluding Comments*

The limited economics literature currently available which is devoted to environmental pollution problems has generally taken a partial equilibrium view of the matter, as well as treated the pollution of particular environmental media, such as air and water, as separate problems.[29] This no doubt reflects the propensity of the theoretical literature to view externalities as exceptional and minor. Clearly, the partial equilibrium approach in particular is very convenient theoretically and empirically for it permits external damage and control cost functions to be defined for each particular case without reference to broader interrelationships and adjustments in the economy.

We have argued in this paper that the production of residuals is an inherent and general part of the production and consumption process and, moreover, that there are important trade-offs between the gaseous, liquid, or solid forms that these residuals may take. Further, we have argued that under conditions of intensive economic and population development the environmental media which can receive and assimilate residual wastes are not free goods but natural resources of great value with respect to which voluntary exchange cannot operate because of their common property characteristics. We have also noted, in passing, that the assimilative capacity of environmental media can sometimes be altered and that therefore the problem of achieving Pareto optimality reaches beyond devising appropriate shadow prices and involves the planning and execution of investments with public goods aspects.

We have exhibited a formal mathematical framework for tracing residuals flows in the economy and related it to the general equilibrium model of resources allocation, altered to accommodate recycle and containing unpriced sectors to represent the environment. This formulation, in contrast to the usual partial equilibrium treatments, implies knowledge of all preference and production functions including relations between residuals discharge and external cost and all possible factor and process substitutions. While we feel that it represents reality with greater fidelity than the usual view, it also implies a central planning problem of impossible difficulty, both from the standpoint of data collection and computation.

What, if any, help can the general interdependency approach we have outlined offer in dealing with pollution problems effectively and reasonably efficiently? A minimal contribution is its warning that partial equilibrium approaches, while more

[28] There is one redundant equation in the system, which expresses the identity between gross product and gross income for the system as a whole (sometimes called "Walras law").

[29] See, for example, the essays in Wolozin [29].

tractable, may lead to serious errors. Second, in projecting waste residuals for an economy—a regional economy would usually be the most relevant for planning and control—the inter-industry materials flow model can provide a much more conceptually satisfying and accurate tool for projecting future residuals production than the normal aggregative extrapolations.[30] The latter not only treat gaseous, liquid, and solid wastes separately, but do not take account of input-output relations and the fact that the materials account for the region must balance.

We think that in the next few years it will be possible to make improved regional projections of residuals along the lines sketched above. Undoubtedly, there will also be further progress in empirically estimating external costs associated with residuals discharge and in estimating control costs via various alternative measures. On the basis of this kind of information, a control policy can be devised. However, this approach will still be partial. Interrelations between the regional and national economy must be treated simplistically and to be manageable, the analysis must confine itself to a specific projected bill of goods.

The basic practical question which remains to be answered is whether an iterated series of partial equilibrium treatments—e.g., focusing on one industry or region at a time, *ceteris paribus*—would converge toward the general equilibrium solution, or not. We know of no theoretical test of convergence which would be applicable in this case but, in the absence of such a criterion, would be willing to admit the possible relevance of an empirical sensitivity test more or less along the following lines: take a major residuals-producing industry (such as electric power) and parametrize its cost structure in terms of emission control levels, allowing all technically feasible permutations of factor (fuel) inputs and processes. It would be a straightforward, but complicated, operations research problem to determine the minimum cost solution as a function of the assumed (negative) price of the residuals produced. If possible industry patterns—factor and process combinations—exist which would permit a high level of emission control at only a small increase in power production cost, then it might be possible to conclude that for a significant range of (negative) residuals prices the effect on power prices —and therefore on the rest of the economy —would not be great. Such a conclusion would support the convergence hypothesis. If, on the other hand, electric power prices are very sensitive to residuals prices, then one would at least have to undertake a deeper study of consumer preference functions to try to determine what residuals prices would actually be if a market mechanism existed. If people prove to have a strong antipathy to soot and sulfur dioxide, for instance, resulting in a high (negative) price for these unwanted inputs, then one would be forced to suspect that the partial equilibrium approach is probably not convergent to the general equilibrium solution and that much more elaborate forms of analysis will be required.

[30] Some efforts to implement these concepts are already underway. Walter Isard and his associates have prepared an input-output table for Philadelphia which includes coefficients representing waterborne wastes (unpublished). The recent study of waste management in the New York Metropolitan region by the Regional Plan Association took a relatively broad view of the waste residuals problem [31]. Relevant data on several industries are being gathered. Richard Frankel's not yet published study of thermal power in which the range of technical options for controlling residuals, and their costs, is being explored is notable in this regard. His and other salient studies are described in Ayres and Kneese [2].

REFERENCES

1. R. U. AYRES, "Stability of Biosystems in Sea Water," Tech. Rept. No. 142, Hudson Laboratories, Columbia University, New York 1967.

2. —— AND A. V. KNEESE, "Environmental Pollution," in U.S. Congress, Joint Economic Committee, *Federal Programs for the Development of Human Resources*, Vol. 2, Washington 1968.

3. W. J. BAUMOL, *Welfare Economics and the Theory of the State*. Cambridge 1967.

4. K. E. BOULDING, "The Economics of the Coming Spaceship Earth," in H. Jarrett, ed., *Environmental Quality in a Growing Economy*, Baltimore 1966, pp. 3–14.

5. J. H. BOYD, "Collective Facilities in Water Quality Management," Appendix to Kneese and Bower [17].

6. J. W. BUCHANAN AND WM. C. STUBBLEBINE, "Externality," *Economica*, Nov. 1962, *29*, 371–84.

7. —— AND G. TULLOCK, "Public and Private Interaction under Reciprocal Externality," in J. Margolis, ed., *The Public Economy of Urban Communities*, Baltimore 1965, pp. 52–73.

8. G. CASSEL, *The Theory of Social Economy*. New York 1932.

9. R. H. COASE, "The Problem of Social Cost," *Jour. Law and Econ.*, Oct. 1960, *3*, 1–44.

10. L. COLE, "Can the World be Saved?" Paper presented at the 134th Meeting of the American Association for the Advancement of Science, December 27, 1967.

11. R. K. DAVIS, *The Range of Choice in Water Management*. Baltimore 1968.

12. O. A. DAVIS AND A. WHINSTON, "Externalities, Welfare, and the Theory of Games," *Jour. Pol. Econ.*, June 1962, *70*, 241–62.

13. R. DORFMAN, P. SAMUELSON AND R. M. SOLOW, *Linear Programming and Economic Analysis*. New York 1958.

14. I. FISHER, *Nature of Capital and Income*. New York 1906.

15. A. C. HARBERGER, "Monopoly and Resources Allocation," *Am. Econ. Rev.*, Proc., May 1954, *44*, 77–87.

16. K. W. KAPP, *The Social Costs of Private Enterprise*. Cambridge 1950.

17. A. V. KNEESE AND B. T. BOWER, *Managing Water Quality: Economics, Technology, Institutions*. Baltimore 1968.

18. F. H. KNIGHT, *Risk, Uncertainty, and Profit*. Boston and New York 1921.

19. H. LEIBENSTEIN, "Allocative Efficiency vs. 'X-Efficiency,'" *Am. Econ. Rev.*, June 1966, *56*, 392–415.

20. W. P. LOWRY, "The Climate of Cities," *Sci. Am.*, Aug. 1967, *217*, 15–23.

21. E. J. MISHAN, "Reflections on Recent Developments in the Concept of External Effects," *Canadian Jour. Econ. Pol. Sci.*, Feb. 1965, *31*, 1–34.

22. A. C. PIGOU, *Economics of Welfare*. London 1952.

23. K. SCHLESINGER, "Über die Produktionsgleichungen der ökonomischen Wertlehre," *Ergebnisse eines mathematischen Kolloquiums*, No. 6. Vienna, F. Denticke, 1933.

24. D. SCHWARTZMAN, "The Burden of Monopoly," *Jour. Pol. Econ.*, Dec. 1960, *68*, 627–30.

25. T. SCITOVSKY, "Two Concepts of External Economies," *Jour. Pol. Econ.*, Apr. 1954, *62*, 143–51.

26. F. SMITH, *The Economic Theory of Industrial Waste Production and Disposal*, draft of a doctoral dissertation, Northwestern Univ. 1967.

27. R. TURVEY, "On Divergencies between Social Cost and Private Cost," *Economica*, Nov. 1962, *30*, 309–13.

28. L. WALRAS, *Elements d'economie politique pure*, Jaffé translation. London 1954.

29. H. WOLOZIN, ed., *The Economics of Air Pollution*. New York 1966.

30. CONSERVATION FOUNDATION, *Implications of Rising Carbon Dioxide Content of the Atmosphere*. New York 1963.

31. REGIONAL PLAN ASSOCIATION, *Waste Management*, a Report of the Second Regional Plan. New York 1968.

Impact of Population Growth

Complacency concerning this component of man's predicament is unjustified and counterproductive.

Paul R. Ehrlich and John P. Holdren

The interlocking crises in population, resources, and environment have been the focus of countless papers, dozens of prestigious symposia, and a growing avalanche of books. In this wealth of material, several questionable assertions have been appearing with increasing frequency. Perhaps the most serious of these is the notion that the size and growth rate of the U.S. population are only minor contributors to this country's adverse impact on local and global environments (1, 2). We propose to deal with this and several related misconceptions here, before persistent and unrebutted repetition entrenches them in the public mind— if not the scientific literature. Our discussion centers around five theorems which we believe are demonstrably true and which provide a framework for realistic analysis:

1) Population growth causes a *disproportionate* negative impact on the environment.

2) Problems of population size and growth, resource utilization and depletion, and environmental deterioration must be considered jointly and on a global basis. In this context, population control is obviously not a panacea —it is necessary but not alone sufficient to see us through the crisis.

3) Population density is a poor measure of population pressure, and redistributing population would be a dangerous pseudosolution to the population problem.

4) "Environment" must be broadly construed to include such things as the

Dr. Ehrlich is professor of biology at Stanford University, Palo Alto, California, and Dr. Holdren is a physicist at the Lawrence Radiation Laboratory, University of California, Livermore. This article is adapted from a paper presented before the President's Commission on Population Growth and the American Future on 17 November 1970.

physical environment of urban ghettos, the human behavioral environment, and the epidemiological environment.

5) Theoretical solutions to our problems are often not operational and sometimes are not solutions.

We now examine these theorems in some detail.

Population Size and Per Capita Impact

In an agricultural or technological society, each human individual has a negative impact on his environment. He is responsible for some of the simplification (and resulting destabilization) of ecological systems which results from the practice of agriculture (3). He also participates in the utilization of renewable and nonrenewable resources. The total negative impact of such a society on the environment can be expressed, in the simplest terms, by the relation

$$I = P \cdot F$$

where P is the population, and F is a function which measures the per capita impact. A great deal of complexity is subsumed in this simple relation, however. For example, F increases with per capita consumption if technology is held constant, but may decrease in some cases if more benign technologies are introduced in the provision of a constant level of consumption. (We shall see in connection with theorem 5 that there are limits to the improvements one should anticipate from such "technological fixes.")

Pitfalls abound in the interpretation of manifest increases in the total impact I. For instance, it is easy to mistake changes in the composition of

resource demand or environmental impact for absolute per capita increases, and thus to underestimate the role of the population multiplier. Moreover, it is often assumed that population size and per capita impact are independent variables, when in fact they are not. Consider, for example, the recent article by Coale (1), in which he disparages the role of U.S. population growth in environmental problems by noting that since 1940 "population has increased by 50 percent, but per capita use of electricity has been multiplied several times." This argument contains both the fallacies to which we have just referred.

First, a closer examination of very rapid increases in many kinds of consumption shows that these changes reflect a shift among alternatives within a larger (and much more slowly growing) category. Thus the 760 percent increase in electricity consumption from 1940 to 1969 (4) occurred in large part because the electrical *component* of the energy budget was (and is) increasing much faster than the budget itself. (Electricity comprised 12 percent of the U.S. energy consumption in 1940 versus 22 percent today.) The total energy use, a more important figure than its electrical component in terms of resources and the environment, increased much less dramatically—140 percent from 1940 to 1969. Under the simplest assumption (that is, that a given increase in population size accounts for an exactly proportional increase in consumption), this would mean that 38 percent of the increase in energy use during this period is explained by population growth (the actual population increase from 1940 to 1969 was 53 percent). Similar considerations reveal the imprudence of citing, say, aluminum consumption to show that population growth is an "unimportant" factor in resource use. Certainly, aluminum consumption has swelled by over 1400 percent since 1940, but much of the increase has been due to the substitution of aluminum for steel in many applications. Thus a fairer measure is combined consumption of aluminum and steel, which has risen only 117 percent since 1940. Again, under the simplest assumption, population growth accounts for 45 percent of the increase.

The "simplest assumption" is not valid, however, and this is the second flaw in Coale's example (and in his thesis). In short, he has failed to recognize that per capita consumption of energy and resources, and the associated per capita impact on the environment, are themselves functions of the population size. Our previous equation is more accurately written

$$I = P \cdot F(P)$$

displaying the fact that impact can increase faster than linearly with population. Of course, whether $F(P)$ is an increasing or decreasing function of P depends in part on whether diminishing returns or economies of scale are dominant in the activities of importance. In populous, industrial nations such as the United States, most economies of scale are already being exploited; we are on the diminishing returns part of most of the important curves.

As one example of diminishing returns, consider the problem of providing nonrenewable resources such as minerals and fossil fuels to a growing population, even at fixed levels of per capita consumption. As the richest supplies of these resources and those nearest to centers of use are consumed, we are obliged to use lower-grade ores, drill deeper, and extend our supply networks. All these activities increase our per capita use of energy and our per capita impact on the environment. In the case of partly renewable resources such as water (which is effectively nonrenewable when groundwater supplies are mined at rates far exceeding natural recharge), per capita costs and environmental impact escalate dramatically when the human population demands more than is locally available. Here the loss of free-flowing rivers and other economic, esthetic, and ecological costs of massive water-movement projects represent increased per capita diseconomies directly stimulated by population growth.

Diminishing returns are also operative in increasing food production to meet the needs of growing populations. Typically, attempts are made both to overproduce on land already farmed and to extend agriculture to marginal land. The former requires disproportionate energy use in obtaining and distributing water, fertilizer, and pesticides. The latter also increases per capita energy use, since the amount of energy invested per unit yield increases as less desirable land is cultivated.

Similarly, as the richest fisheries stocks are depleted, the yield per unit effort drops, and more and more energy per capita is required to maintain the supply (5). Once a stock is depleted it may not recover—it may be nonrenewable.

Population size influences per capita impact in ways other than diminishing returns. As one example, consider the oversimplified but instructive situation in which each person in the population has links with every other person—roads, telephone lines, and so forth. These links involve energy and materials in their construction and use. Since the number of links increases much more rapidly than the number of people (6), so does the per capita consumption associated with the links.

Other factors may cause much steeper positive slopes in the per capita impact function, $F(P)$. One such phenomenon is the *threshold effect*. Below a certain level of pollution trees will survive in smog. But, at some point, when a small increment in population produces a small increment in smog, living trees become dead trees. Five hundred people may be able to live around a lake and dump their raw sewage into the lake, and the natural systems of the lake will be able to break down the sewage and keep the lake from undergoing rapid ecological change. Five hundred and five people may overload the system and result in a "polluted" or eutrophic lake. Another phenomenon capable of causing near-discontinuities is the *synergism*. For instance, as cities push out into farmland, air pollution increasingly becomes a mixture of agricultural chemicals with power plant and automobile effluents. Sulfur dioxide from the city paralyzes the cleaning mechanisms of the lungs, thus increasing the residence time of potential carcinogens in the agricultural chemicals. The joint effect may be much more than the sum of the individual effects. Investigation of synergistic effects is one of the most neglected areas of environmental evaluation.

Not only is there a connection between population size and per capita damage to the environment, but the cost of maintaining environmental quality at a given level escalates disproportionately as population size increases. This effect occurs in part because costs increase very rapidly as one tries to reduce contaminants per unit

volume of effluent to lower and lower levels (diminishing returns again!). Consider municipal sewage, for example. The cost of removing 80 to 90 percent of the biochemical and chemical oxygen demand, 90 percent of the suspended solids, and 60 percent of the resistant organic material by means of secondary treatment is about 8 cents per 1000 gallons (3785 liters) in a large plant (7). But if the volume of sewage is such that its nutrient content creates a serious eutrophication problem (as is the case in the United States today), or if supply considerations dictate the reuse of sewage water for industry, agriculture, or groundwater recharge, advanced treatment is necessary. The cost ranges from two to four times as much as for secondary treatment (17 cents per 1000 gallons for carbon absorption; 34 cents per 1000 gallons for disinfection to yield a potable supply). This dramatic example of diminishing returns in pollution control could be repeated for stack gases, automobile exhausts, and so forth.

Now consider a situation in which the limited capacity of the environment to absorb abuse requires that we hold man's impact in some sector constant as population doubles. This means *per capita effectiveness* of pollution control in this sector must double (that is, effluent per person must be halved). In a typical situation, this would yield doubled per capita costs, or quadrupled total costs (and probably energy consumption) in this sector for a doubling of population. Of course, diminishing returns and threshold effects may be still more serious: we may easily have an eightfold increase in control costs for a doubling of population. Such arguments leave little ground for the assumption, popularized by Barry Commoner (2, 8) and others, that a 1 percent rate of population growth spawns only 1 percent effects.

It is to be emphasized that the possible existence of "economies of scale" does not invalidate these arguments. Such savings, if available at all, would apply in the case of our sewage example to a change in the amount of effluent to be handled at an installation of a given type. For most technologies, the United States is already more than populous enough to achieve such economies and is doing so. They are accounted for in our example by citing figures for the largest treatment plants of each type. Population growth, on the other hand, forces us into quantitative *and* qualitative changes in how we handle each unit volume of effluent—what fraction and what kinds of material we remove. Here economies of scale do not apply at all, and diminishing returns are the rule.

Global Context

We will not deal in detail with the best example of the global nature and interconnections of population resource and environmental problems—namely, the problems involved in feeding a world in which 10 to 20 million people starve to death annually (*9*), and in which the population is growing by some 70 million people per year. The ecological problems created by high-yield agriculture are awesome (*3, 10*) and are bound to have a negative feedback on food production. Indeed, the Food and Agriculture Organization of the United Nations has reported that in 1969 the world suffered its first absolute decline in fisheries yield since 1950. It seems likely that part of this decline is attributable to pollution originating in terrestrial agriculture.

A second source of the fisheries decline is, of course, overexploitation of fisheries by the developed countries. This problem, in turn, is illustrative of the situation in regard to many other resources, where similarly rapacious and shortsighted behavior by the developed nations is compromising the aspirations of the bulk of humanity to a decent existence. It is now becoming more widely comprehended that the United States alone accounts for perhaps 30 percent of the nonrenewable resources consumed in the world each year (for example, 37 percent of the energy, 25 percent of the steel, 28 percent of the tin, and 33 percent of the synthetic rubber) (*11*). This behavior is in large part inconsistent with American rhetoric about "developing" the countries of the Third World. *We* may be able to afford the technology to mine lower grade deposits when we have squandered the world's rich ores, but the underdeveloped countries, as their needs grow and their means remain meager, will not be able to do so. Some observers argue that the poor countries are today economically dependent on our use of their resources, and indeed that economists in these countries complain that world demand for their raw materials is too low (*1*).

This proves only that their economists are as shortsighted as ours.

It is abundantly clear that the entire context in which we view the world resource pool and the relationships between developed and underdeveloped countries must be changed, if we are to have any hope of achieving a stable and prosperous existence for all human beings. It cannot be stated too forcefully that the developed countries (or, more accurately, the overdeveloped countries) are the principal culprits in the consumption and dispersion of the world's nonrenewable resources (*12*) as well as in appropriating much more than their share of the world's protein. Because of this consumption, and because of the enormous negative impact on the global environment accompanying it, the population growth in these countries must be regarded as the most serious in the world today.

In relation to theorem 2 we must emphasize that, even if population growth were halted, the present population of the world could easily destroy civilization as we know it. There is a wide choice of weapons—from unstable plant monocultures and agricultural hazes to DDT, mercury, and thermonuclear bombs. If population size were reduced and per capita consumption remained the same (or increased), we would still quickly run out of vital, high-grade resources or generate conflicts over diminishing supplies. Racism, economic exploitation, and war will not be eliminated by population control (of course, they are unlikely to be eliminated without it).

Population Density and Distribution

Theorem 3 deals with a problem related to the inequitable utilization of world resources. One of the commonest errors made by the uninitiated is to assume that population density (people per square mile) is the critical measure of overpopulation or underpopulation. For instance, Wattenberg states that the United States is not very crowded by "international standards" because Holland has 18 times the population density (*13*). We call this notion "the Netherlands fallacy." The Netherlands actually requires large chunks of the earth's resources and vast areas of land not within its borders to maintain itself. For example, it is the second largest per capita importer of protein in the world, and it imports 63 percent of its cereals, including 100 percent of its corn and rice. It also imports all of its cotton, 77 percent of its wool, and all of its iron ore, antimony, bauxite, chromium, copper, gold, lead, magnesite, manganese, mercury, molybdenum, nickel, silver, tin, tungsten, vanadium, zinc, phosphate rock (fertilizer), potash (fertilizer), asbestos, and diamonds. It produces energy equivalent to some 20 million metric tons of coal and consumes the equivalent of over 47 million metric tons (*14*).

A certain preoccupation with density as a useful measure of overpopulation is apparent in the article by Coale (*1*). He points to the existence of urban problems such as smog in Sydney, Australia, "even though the total population of Australia is about 12 million in an area 80 percent as big as the United States," as evidence that environmental problems are unrelated to population size. His argument would be more persuasive if problems of population *distribution* were the only ones with environmental consequences, and if population distribution were unrelated to resource distribution and population size. Actually, since the carrying capacity of the Australian continent is far below that of the United States, one would *expect* distribution problems—of which Sydney's smog is one symptom—to be encountered at a much lower total population there. Resources, such as water, are in very short supply, and people cluster where resources are available. (Evidently, it cannot be emphasized enough that carrying capacity includes the availability of a wide variety of resources in addition to space itself, and that population pressure is measured relative to the carrying capacity. One would expect water, soils, or the ability of the environment to absorb wastes to be the limiting resource in far more instances than land area.)

In addition, of course, many of the most serious environmental problems are essentially independent of the way in which population is distributed. These include the global problems of weather modification by carbon dioxide and particulate pollution, and the threats to the biosphere posed by man's massive inputs of pesticides, heavy metals, and oil (*15*). Similarly, the problems of resource depletion and ecosystem simplification by agriculture depend on how many people there are and their patterns of consumption, but not in any major way on how they are distributed.

Naturally, we do not dispute that smog and most other familiar urban ills are serious problems, or that they are related to population distribution. Like many of the difficulties we face, these problems will not be cured simply by stopping population growth; direct and well-conceived assaults on the problems themselves will also be required. Such measures may occasionally include the redistribution of population, but the considerable difficulties and costs of this approach should not be underestimated. People live where they do not because of a perverse intention to add to the problems of their society but for reasons of economic necessity, convenience, and desire for agreeable surroundings. Areas that are uninhabited or sparsely populated today are presumably that way because they are deficient in some of the requisite factors. In many cases, the remedy for such deficiencies—for example, the provision of water and power to the wastelands of central Nevada—would be extraordinarily expensive in dollars, energy, and resources and would probably create environmental havoc. (Will we justify the rape of Canada's rivers to "colonize" more of our western deserts?)

Moving people to more "habitable" areas, such as the central valley of California or, indeed, most suburbs, exacerbates another serious problem—the paving-over of prime farmland. This is already so serious in California that, if current trends continue, about 50 percent of the best acreage in the nation's leading agricultural state will be destroyed by the year 2020 (16). Encouraging that trend hardly seems wise.

Whatever attempts may be made to solve distribution-related problems, they will be undermined if population growth continues, for two reasons. First, population growth and the aggravation of distribution problems are correlated—part of the increase will surely be absorbed in urban areas that can least afford the growth. Indeed, barring the unlikely prompt reversal of present trends, most of it will be absorbed there. Second, population growth puts a disproportionate drain on the very financial resources needed to combat its symptoms. Economist Joseph Spengler has estimated that 4 percent of national income goes to support our 1 percent per year rate of population growth in the United States (17). The 4 percent figure now amounts to about $30 billion per year. It seems safe to conclude that the faster we grow the less likely it is that we will find the funds either to alter population distribution patterns or to deal more comprehensively and realistically with our problems.

Meaning of Environment

Theorem 4 emphasizes the comprehensiveness of the environment crisis. All too many people think in terms of national parks and trout streams when they say "environment." For this reason many of the suppressed people of our nation consider ecology to be just one more "racist shuck" (18). They are apathetic or even hostile toward efforts to avert further environmental and sociological deterioration, because they have no reason to believe they will share the fruits of success (19). Slums, cockroaches, and rats are ecological problems, too. The correction of ghetto conditions in Detroit is neither more nor less important than saving the Great Lakes—both are imperative.

We must pay careful attention to sources of conflict both within the United States and between nations. Conflict within the United States blocks progress toward solving our problems; conflict among nations can easily "solve" them once and for all. Recent laboratory studies on human beings support the anecdotal evidence that crowding may increase aggressiveness in human males (20). These results underscore long-standing suspicions that population growth, translated through the inevitable uneven distribution into physical crowding, will tend to make the solution of all of our problems more difficult.

As a final example of the need to view "environment" broadly, note that human beings live in an epidemiological environment which deteriorates with crowding and malnutrition—both of which increase with population growth. The hazard posed by the prevalence of these conditions in the world today is compounded by man's unprecedented mobility: potential carriers of diseases of every description move routinely and in substantial numbers from continent to continent in a matter of hours. Nor is there any reason to believe that modern medicine has made widespread plague impossible (21). The Asian influenza epidemic of 1968 killed relatively few people only because the virus happened to be nonfatal to people in otherwise good health, not because of public health measures. Far deadlier viruses, which easily could be scourges without precedent in the population at large, have on more than one occasion been confined to research workers largely by good luck [for example, the Marburgvirus incident of 1967 (22) and the Lassa fever incident of 1970 (21, 23)].

Solutions: Theoretical and Practical

Theorem 5 states that theoretical solutions to our problems are often not operational, and sometimes are not solutions. In terms of the problem of feeding the world, for example, technological fixes suffer from limitations in scale, lead time, and cost (24). Thus potentially attractive theoretical approaches—such as desalting seawater for agriculture, new irrigation systems, high-protein diet supplements—prove inadequate in practice. They are too little, too late, and too expensive, or they have sociological costs which hobble their effectiveness (25). Moreover, many aspects of our technological fixes, such as synthetic organic pesticides and inorganic nitrogen fertilizers, have created vast environmental problems which seem certain to erode global productivity and ecosystem stability (26). This is not to say that important gains have not been made through the application of technology to agriculture in the poor countries, or that further technological advances are not worth seeking. But it must be stressed that even the most enlightened technology cannot relieve the necessity of grappling forthrightly and promptly with population growth [as Norman Borlaug aptly observed on being notified of his Nobel Prize for development of the new wheats (27)].

Technological attempts to ameliorate the environmental impact of population growth and rising per capita affluence in the developed countries suffer from practical limitations similar to those just mentioned. Not only do such measures tend to be slow, costly, and insufficient in scale, but in addition they most often *shift* our impact rather than remove it. For example, our first generation of smog-control devices increased emissions of oxides of nitrogen while reducing those of hydrocarbons and carbon monoxide. Our unhappiness about eutrophication has led to the replacement of phosphates in detergents with compounds like NTA—nitrilotriacetic acid—which has carcinogenic breakdown products and apparently enhances teratogenic effects of heavy metals (28). And our distaste for lung diseases apparently induced by sulfur dioxide inclines us to accept the hazards of radioactive waste disposal, fuel reprocessing, routine low-level emissions of radiation, and an apparently small but finite risk of catastrophic accidents associated with nuclear fission power plants. Similarly, electric automobiles would simply shift part of the environmental burden of personal transportation from the vicinity of highways to the vicinity of power plants.

We are not suggesting here that electric cars, or nuclear power plants, or substitutes for phosphates are inherently bad. We argue rather that they, too, pose environmental costs which must be weighed against those they eliminate. In many cases the choice is not obvious, and in *all* cases there will be some environmental impact. The residual per capita impact, after all the best choices have been made, must then be multiplied by the population engaging in the activity. If there are too many people, even the most wisely managed technology will not keep the environment from being overstressed.

In contending that a change in the way we use technology will invalidate these arguments, Commoner (2, 8) claims that our important environmental problems began in the 1940's with the introduction and rapid spread of certain "synthetic" technologies: pesticides and herbicides, inorganic fertilizers, plastics, nuclear energy, and high-compression gasoline engines. In so arguing, he appears to make two unfounded assumptions. The first is that man's pre-1940 environmental impact was innocuous and, without changes for the worse in technology, would have remained innocuous even at a much larger population size. The second assumption is that the advent of the new technologies was independent of the attempt to meet human needs and desires in a growing population. Actually, man's record as a simplifier of ecosystems and plunderer of resources can be traced from his

probable role in the extinction of many Pleistocene mammals (29), through the destruction of the soils of Mesopotamia by salination and erosion, to the deforestation of Europe in the Middle Ages and the American dustbowls of the 1930's, to cite only some highlights. Man's contemporary arsenal of synthetic technological bludgeons indisputably magnifies the potential for disaster, but these were evolved in some measure to *cope* with population pressures, not independently of them. Moreover, it is worth noting that, of the four environmental threats viewed by the prestigious Williamstown study (15) as globally significant, three are associated with pre-1940 technologies which have simply increased in scale [heavy metals, oil in the seas, and carbon dioxide and particulates in the atmosphere, the latter probably due in considerable part to agriculture (30)]. Surely, then, we can anticipate that supplying food, fiber, and metals for a population even larger than today's will have a profound (and destabilizing) effect on the global ecosystem under *any* set of technological assumptions.

Conclusion

John Platt has aptly described man's present predicament as "a storm of crisis problems" (31). Complacency concerning any component of these problems—sociological, technological, economic, ecological—is unjustified and counterproductive. It is time to admit that there are no monolithic solutions to the problems we face. Indeed, population control, the redirection of technology, the transition from open to closed resource cycles, the equitable distribution of opportunity and the ingredients of prosperity must *all* be accomplished if there is to be a future worth having. Failure in any of these areas will surely sabotage the entire enterprise.

In connection with the five theorems elaborated here, we have dealt at length with the notion that population growth in industrial nations such as the United States is a minor factor, safely ignored. Those who so argue often add that, anyway, population control would be the slowest to take effect of all possible attacks on our various problems, since the inertia in attitudes and in the age structure of the popu-

lation is so considerable. To conclude that this means population control should be assigned low priority strikes us as curious logic. Precisely because population is the most difficult and slowest to yield among the components of environmental deterioration, we must start on it at once. To ignore population today because the problem is a tough one is to commit ourselves to even gloomier prospects 20 years hence, when most of the "easy" means to reduce per capita impact on the environment will have been exhausted. The desperate and repressive measures for population control which might be contemplated then are reason in themselves to proceed with foresight, alacrity, and compassion today.

References and Notes

1. A. J. Coale, *Science* **170**, 132 (1970).
2. B. Commoner, *Saturday Rev.* **53**, 50 (1970); *Humanist* **30**, 10 (1970).
3. For a general discussion, see P. R. Ehrlich and A. H. Ehrlich, *Population, Resources, Environment* (Freeman, San Francisco, 1970), chap. 7. More technical treatments of the relationship between complexity and stability may be found in R. H. MacArthur, *Ecology* **36**, 533 (1955); D. R. Margalef, *Gen. Syst.* **3**, 3671 (195 → E. G. Leigh, Jr., *Proc. Nat. Acad. Sci. U.S.* **53**, 777 (1965); and O. T. Loucks, "Evolution of diversity, efficiency, and stability of a community," paper delivered at AAAS meeting, Dallas, Texas, 30 Dec. 1968.
4. The figures used in this paragraph are all based on data in *Statistical Abstract of the United States 1970* (U.S. Department of Commerce) (Government Printing Office, Washington, D.C., 1970).
5. A dramatic example of this effect is given in R. Payne's analysis of the whale fisheries [*N.Y. Zool. Soc. Newsl.* (Nov. 1968)]. The graphs from Payne's paper are reproduced in Ehrlich and Ehrlich (3).
6. If N is the number of people, then the number of links is $N(N-1)/2$, and the number of links per capita is $(N-1)/2$.
7. These figures and the others in this paragraph are from *Cleaning Our Environment: The Chemical Basis for Action* (American Chemical Society, Washington, D.C., 1969), pp. 95–162.
8. In his unpublished testimony before the President's Commission on Population Growth and the American Future (17 Nov. 1970), Commoner acknowledged the operation of diminishing returns, threshold effects, and so on. Since such factors apparently do not account for *all* of the increase in per capita impact on the environment in recent decades, however, Commoner drew the unwarranted conclusion that they are negligible.
9. R. Dumont and B. Rosier, *The Hungry Future* (Praeger, New York, 1969), pp. 34–35.
10. L. Brown, *Sci. Amer.* **223**, 160 (1970); P. R. Ehrlich, *War on Hunger* **4**, 1 (1970).
11. These figures are based on data from the *United Nations Statistical Yearbook 1969* (United Nations, New York, 1969), with estimates added for the consumption by Mainland China when none were included.
12. The notion that dispersed resources, because they have not left the planet, are still available to us, and the hope that mineral supplies can be extended indefinitely by the application of vast amounts of energy to common rock have been the subject of lively debate elsewhere. See, for example, the articles by P. Cloud, T. Lovering, A. Weinberg, *Texas Quart.* **11**, 103, 127, 90 (Summer 1968); and *Resources and Man* (National Academy of Sciences) (Freeman, San Francisco, 1969). While the pessimists seem to have had the better of this argument, the entire matter is

academic in the context of the rate problem we face in the next 30 years. Over that time period, at least, cost, lead time, and logistics will see to it that industrial economies and dreams of development stand or fall with the availability of high-grade resources.

13. B. Wattenberg, *New Republic* **162**, 18 (4 Apr. and 11 Apr. 1970).
14. These figures are from (*II*), from the *FAO Trade Yearbook*, the *FAO Production Yearbook* (United Nations, New York, 1968), and from G. Borgstrom, *Too Many* (Collier-Macmillan, Toronto, Ont., 1969).
15. *Man's Impact on the Global Environment, Report of the Study of Critical Environmental Problems* (M.I.T. Press, Cambridge, Mass., 1970).
16. *A Model of Society, Progress Report of the Environmental Systems Group* (Univ. of California Institute of Ecology, Davis, April 1969).
17. J. J. Spengler, in *Population: The Vital Revolution*, R. Freedman, Ed. (Doubleday, New York, 1964), p. 67.
18. R. Chrisman, *Scanlan's* **1**, 46 (August 1970).
19. A more extensive discussion of this point is given in an article by P. R. Ehrlich and A. H. Ehrlich, in *Global Ecology: Readings Toward a Rational Strategy for Man*, J. P. Holdren and P. R. Ehrlich, Eds. (Harcourt, Brace, Jovanovich, New York, in press).
20. J. L. Freedman, A. Levy, J. Price, R. Welte, M. Katz, P. R. Ehrlich, in preparation.
21. J. Lederberg, *Washington Post* (15 Mar. and 22 Mar. 1970).
22. C. Smith, D. Simpson, E. Bowen, I. Zlotnik, *Lancet* **1967-II**, 1119, 1128 (1967).
23. Associated Press wire service, 2 Feb. 1970.
24. P. R. Ehrlich and J. P. Holdren, *BioScience* **19**, 1065 (1969).
25. See L. Brown [*Seeds of Change* (Praeger, New York, 1970)] for a discussion of unemployment problems exacerbated by the Green Revolution.
26. G. Woodwell, *Science* **168**, 429 (1970).
27. *New York Times*, 22 Oct. 1970, p. 18; *Newsweek* **76**, 50 (2 Nov. 1970).
28. S. S. Epstein, *Environment* **12**, No. 7, 2 (Sept. 1970); *New York Times* service, 17 Nov. 1970.
29. G. S. Krantz, *Amer. Sci.* **58**, 164 (Mar.–Apr. 1970).
30. R. A. Bryson and W. M. Wendland, in *Global Effects of Environmental Pollution*, S. F. Singer, Ed. (Springer-Verlag, New York, 1970).
31. J. Platt, *Science* **166**, 1115 (1969).

GROWTH
IN
THE
WORLD
SYSTEM

In the circumference of a circle the beginning and end are common.

HERACLITUS, 500 B.C.

We have discussed food, nonrenewable resources, and pollution absorption as separate factors necessary for the growth and maintenance of population and industry. We have looked at the rate of growth in the demand for each of these factors and at the possible upper limits to the supply. By making simple extrapolations of the demand growth curves, we have attempted to estimate, roughly, how much longer growth of each of these factors might continue at its present rate of increase. Our conclusion from these extrapolations is one that many perceptive people have already realized—that the short doubling times of many of man's activities, combined with the immense quantities being doubled, will bring us close to the limits to growth of those activities surprisingly soon.

Extrapolation of present trends is a time-honored way of looking into the future, especially the very near future, and especially if the quantity being considered is not much in-fluenced by other trends that are occurring elsewhere in the system. Of course, none of the five factors we are examining here is independent. Each interacts constantly with all the others. We have already mentioned some of these interactions. Population cannot grow without food, food production is increased by growth of capital, more capital requires more resources, discarded resources become pollution, pollution interferes with the growth of both population and food.

Furthermore, over long time periods each of these factors also feeds back to influence itself. The rate at which food production increases in the 1970's, for example, will have some effect on the size of the population in the 1980's, which will in turn determine the rate at which food production must increase for many years thereafter. Similarly, the rate of resource consumption in the next few years will influence both the size of the capital base that must be maintained and the amount of resources left in the earth. Existing capital and available resources will then interact to determine future resource supply and demand.

The five basic quantities, or levels—population, capital, food, nonrenewable resources, and pollution—are joined by still other interrelationships and feedback loops that we have not yet discussed. Clearly it is not possible to assess the long-term future of any of these levels without taking all the others into account. Yet even this relatively simple system has such a complicated structure that one cannot intuitively understand how it will behave in the future, or how a change in one variable might ultimately affect each of the others. To achieve such understanding, we must extend our intuitive capabilities so that we can follow the complex, interrelated behavior of many variables simultaneously.

Reprinted with permission from Dennis Meadows. The full publication can be accessed at https://www.dartmouth.edu/library/digital/publishing/meadows/ltg/

In this chapter we describe the formal world model that we have used as a first step toward comprehending this complex world system. The model is simply an attempt to bring together the large body of knowledge that already exists about cause-and-effect relationships among the five levels listed above and to express that knowledge in terms of interlocking feedback loops. Since the world model is so important in understanding the causes of and limits to growth in the world system, we shall explain the model-building process in some detail.

In constructing the model, we followed four main steps:

1. We first listed the important causal relationships among the five levels and traced the feedback loop structure. To do so we consulted literature and professionals in many fields of study dealing with the areas of concern—demography, economics, agronomy, nutrition, geology, and ecology, for example. Our goal in this first step was to find the most basic structure that would reflect the major interactions between the five levels. We reasoned that elaborations on this basic structure, reflecting more detailed knowledge, could be added after the simple system was understood.

2. We then quantified each relationship as accurately as possible, using global data where it was available and characteristic local data where global measurements had not been made.

3. With the computer, we calculated the simultaneous operation of all these relationships over time. We then tested the effect of numerical changes in the basic assumptions to find the most critical determinants of the system's behavior.

4. Finally, we tested the effect on our global system of the various policies that are currently being proposed to enhance or change the behavior of the system.

These steps were not necessarily followed serially, because often new information coming from a later step would lead us back to alter the basic feedback loop structure. There is not one inflexible world model; there is instead an evolving model that is continuously criticized and updated as our own understanding increases.

A summary of the present model, its purpose and limitations, the most important feedback loops it contains, and our general procedure for quantifying causal relationships follows.

THE PURPOSE OF THE WORLD MODEL

In this first simple world model, we are interested only in the broad behavior modes of the population-capital system. By *behavior modes* we mean the tendencies of the variables in the system (population or pollution, for example) to change as time progresses. A variable may increase, decrease, remain constant, oscillate, or combine several of these characteristic modes. For example, a population growing in a limited environment can approach the ultimate carrying capacity of that environment in several possible ways. It can adjust smoothly to an equilibrium below the environmental limit by means of a gradual decrease in growth rate, as shown below. It can over-

shoot the limit and then die back again in either a smooth or an oscillatory way, also as shown below. Or it can overshoot

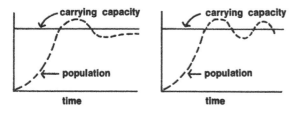

the limit and in the process decrease the ultimate carrying capacity by consuming some necessary nonrenewable resource, as diagramed below. This behavior has been noted in many natural systems. For instance, deer or goats, when natural enemies are absent, often overgraze their range and cause erosion or destruction of the vegetation.[28]

A major purpose in constructing the world model has been to determine which, if any, of these behavior modes will be most characteristic of the world system as it reaches the limits to growth. This process of determining behavior modes is "prediction" only in the most limited sense of the word. The output graphs reproduced later in this book show values for world population, capital, and other variables on a time scale that begins in the year 1900 and continues until 2100. These graphs are *not* exact predictions of the values of the variables at any particular year in the future. They are indications of the system's behavioral tendencies only.

The difference between the various degrees of "prediction" might be best illustrated by a simple example. If you throw a ball straight up into the air, you can predict with certainty what its general behavior will be. It will rise with decreasing velocity, then reverse direction and fall down with increasing velocity until it hits the ground. You know that it will not continue rising forever, nor begin to orbit the earth, nor loop three times before landing. It is this sort of elemental understanding of behavior modes that we are seeking with the present world model. If one wanted to predict exactly how high a thrown ball would rise or exactly where and when it would hit the ground, it would be necessary to make a detailed calculation based on precise information about the ball, the altitude, the wind, and the force of the initial throw. Similarly, if we wanted to predict the size of the earth's population in 1993 within a few percent, we would need a very much more complicated model than the one described here. We would also need information about the world system more precise and comprehensive than is currently available.

Because we are interested at this point only in broad behavior modes, this first world model need not be extremely detailed. We thus consider only one general population, a population that statistically reflects the average characteristics of the global population. We include only one class of pollutants—the long-lived, globally distributed family of pollutants, such as lead, mercury, asbestos, and stable pesticides and radioisotopes—

whose dynamic behavior in the ecosystem we are beginning to understand. We plot one generalized resource that represents the combined reserves of all nonrenewable resources, although we know that each separate resource will follow the general dynamic pattern at its own specific level and rate.

This high level of aggregation is necessary at this point to keep the model understandable. At the same time it limits the information we can expect to gain from the model. Questions of detail cannot be answered because the model simply does not yet contain much detail. National boundaries are not recognized. Distribution inequalities of food, resources, and capital are included implicitly in the data but they are not calculated explicitly nor graphed in the output. World trade balances, migration patterns, climatic determinants, and political processes are not specifically treated. Other models can, and we hope will, be built to clarify the behavior of these important subsystems.*

Can anything be learned from such a highly aggregated model? Can its output be considered meaningful? In terms of exact predictions, the output is not meaningful. We cannot forecast the precise population of the United States nor the GNP of Brazil nor even the total world food production for the year 2015. The data we have to work with are certainly not sufficient for such forecasts, even if it were our purpose to make them. On the other hand, it is vitally important to gain some understanding of the causes of growth in human society, the limits to growth, and the behavior of our socio-economic systems when the limits are reached. Man's knowledge of the

* We have built numerous submodels ourselves in the course of this study to investigate the detailed dynamics underlying each sector of the world model. A list of those studies is included in the appendix.

Figure 23 POPULATION GROWTH AND CAPITAL GROWTH FEEDBACK LOOPS

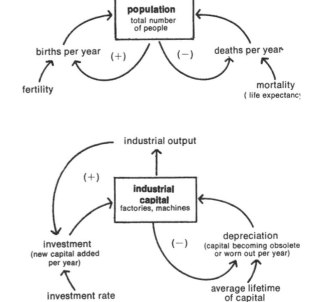

The central feedback loops of the world model govern the growth of population and of industrial capital. The two positive feedback loops involving births and investment generate the exponential growth behavior of population and capital. The two negative feedback loops involving deaths and depreciation tend to regulate this exponential growth. The relative strengths of the various loops depend on many other factors in the world system.

behavior modes of these systems is very incomplete. It is currently not known, for example, whether the human population will continue growing, or gradually level off, or oscillate around some upper limit, or collapse. We believe that the aggregated world model is one way to approach such questions. The model utilizes the most basic relationships among people, food, investment, depreciation, resources, output—relationships that are the same the world over, the same in any part of human society or in society as a whole. In fact, as we indicated at the beginning of this book, there are advantages to considering such questions with as broad a space-time horizon as possible. Questions of detail, of individual nations, and of short-term pressures can be asked much more sensibly when the overall limits and behavior modes are understood.

THE FEEDBACK LOOP STRUCTURE

In chapter I we drew a schematic representation of the feedback loops that generate population growth and capital growth. They are reproduced together in figure 23.

A review of the relationships diagrammed in figure 23 may be helpful. Each year the population is increased by the total number of births and decreased by the total number of deaths that have taken place during that year. The absolute number of births per year is a function of the average fertility of the population and of the size of the population. The number of deaths is related to the average mortality and the total population size. As long as births exceed deaths, the population grows. Similarly, a given amount of industrial capital, operating at constant efficiency, will be able to produce a certain amount of output each year. Some of that output will be more factories, machines, etc., which are investments to increase the stock of capital goods. At the same time some capital equipment will depreciate or be discarded each year. To keep industrial capital growing, the investment rate must exceed the

Figure 24 FEEDBACK LOOPS OF POPULATION, CAPITAL, AGRICULTURE, AND POLLUTION

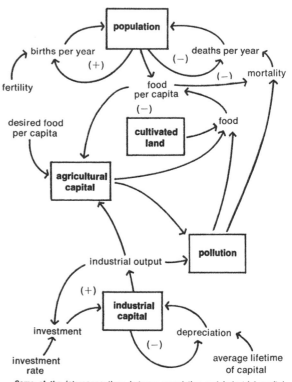

Some of the interconnections between population and industrial capital operate through agricultural capital, cultivated land, and pollution. Each arrow indicates a causal relationship, which may be immediate or delayed, large or small, positive or negative, depending on the assumptions included in each model run.

depreciation rate.

In all our flow diagrams, such as figure 23, the arrows simply indicate that one variable has some influence on another. The *nature* and *degree* of influence are not specified, although of course they must be quantified in the model equations. For simplicity, we often omit noting in the flow diagrams that several of the causal interactions occur only after a delay. The delays are included explicitly in the model calculations.

Population and capital influence each other in many ways, some of which are shown in figure 24. Some of the output of industrial capital is agricultural capital—tractors, irrigation ditches, and fertilizers, for example. The amount of agricultural capital and land area under cultivation strongly influences the amount of food produced. The food per capita (food produced divided by the population) influences the mortality of the population. Both industrial and agricultural activity can cause pollution. (In the case of agriculture, the pollution consists largely of pesticide residues, fertilizers that cause eutrophication, and salt deposits from improper irrigation.) Pollution may affect the mortality of the population directly and also indirectly by decreasing agricultural output.[29]

There are several important feedback loops in figure 24. If everything else in the system remained the same, a population *increase* would decrease food per capita, and thus increase mortality, increase the number of deaths, and eventually lead to a population decrease. This negative feedback loop is diagramed below.

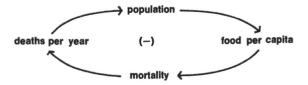

Another negative feedback loop (shown below) tends to counterbalance the one shown above. If the food per capita *decreases* to a value lower than that desired by the population, there will be a tendency to increase agricultural capital, so that future food production and food per capita can *increase*.

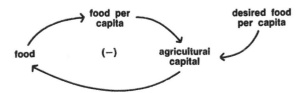

Other important relationships in the world model are illustrated in figure 25. These relationships deal with population, industrial capital, service capital, and resources.

Industrial output includes goods that are allocated to service capital—houses, schools, hospitals, banks, and the equipment they contain. The output from this service capital divided by the population gives the average value of services per capita. Services per capita influence the level of health services and thus the mortality of the population. Services also include education and research into birth control methods as well as distribution of birth control information and devices. Services per capita are thus related to fertility.

Figure 25 FEEDBACK LOOPS OF POPULATION, CAPITAL, SERVICES, AND RESOURCES

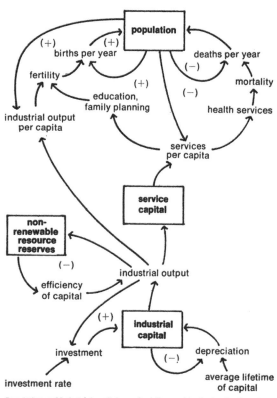

Population and industrial capital are also influenced by the levels of service capital (such as health and education services) and of nonrenewable resource reserves.

A changing industrial output per capita also has an observable effect (though typically after a long delay) on many social factors that influence fertility.

Each unit of industrial output consumes some nonrenewable resource reserves. As the reserves gradually diminish, more capital is necessary to extract the same amount of resource from the earth, and thus the efficiency of capital decreases (that is, more capital is required to produce a given amount of finished goods).

The important feedback loops in figure 25 are shown below.

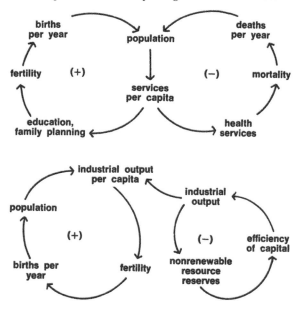

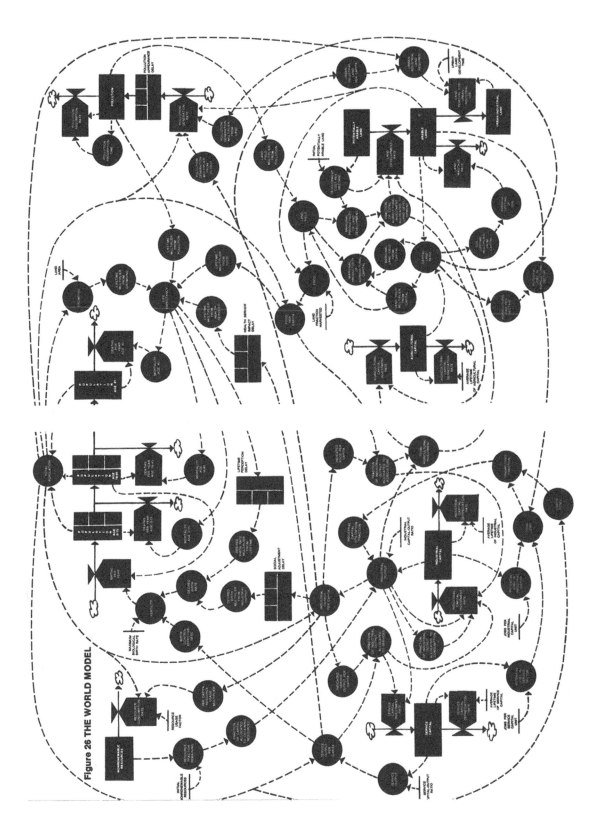

Figure 26 THE WORLD MODEL

Figure 26 THE WORLD MODEL

The entire world model is represented here by a flow diagram in formal System Dynamics notation. Levels, or physical quantities that can be measured directly, are indicated by rectangles ▮, *rates that influence those levels by valves* ▶◀, *and auxiliary variables that influence the rate equations by circles* ●. *Time delays are indicated by sections within rectangles* ▮▮▮ *Real flows of people, goods, money, etc. are shown by solid arrows* ━━▶ *and causal relationships by broken arrows* ─ ─ ─ ─ ▶ *Clouds* ☁ *represent sources or sinks that are not important to the model behavior.*

The relationships shown in figures 24 and 25 are typical of the many interlocking feedback loops in the world model. Other loops include such factors as the area of cultivated land and the rate at which it is developed or eroded, the rate at which pollution is generated and rendered harmless by the environment, and the balance between the labor force and the number of jobs available. The complete flow diagram for the world model, incorporating all these factors and more, is shown in figure 26.

QUANTITATIVE ASSUMPTIONS

Each of the arrows in figure 26 represents a general relationship that we know is important or potentially important in the population-capital system. The structure is, in fact, sufficiently general that it might also represent a single nation or even a single city (with the addition of migration and trade flows across boundaries). To apply the model structure of figure 26 to a nation, we would quantify each relationship in the structure with numbers characteristic of that nation. To represent the world, the data would have to reflect average characteristics of the whole world.

Most of the causal influences in the real world are nonlinear.

That is, a certain change in a causal variable (such as an increase of 10 percent in food per capita) may affect another variable (life expectancy, for example) differently, depending on the point within the possible range of the second variable at which the change takes place. For instance, if an increase in food per capita of 10 percent has been shown to increase life expectancy by 10 years, it may not follow that an increase of food per capita by 20 percent will increase life expectancy by 20 years. Figure 27 shows the nonlinearity of the relationship between food per capita and life expectancy. If there is little food, a small increase may bring about a large increase in life expectancy of a population. If there is already sufficient food, a further increase will have little or no effect. Nonlinear relationships of this sort have been incorporated directly into the world model.*

The current state of knowledge about causal relationships in the world ranges from complete ignorance to extreme accuracy. The relationships in the world model generally fall in the middle ground of certainty. We do know something about the direction and magnitude of the causal effects, but we rarely have fully accurate information about them. To illustrate how we operate on this intermediate ground of knowledge, we present here three examples of quantitative relationships from the world model. One is a relationship between economic variables that is relatively well understood; another involves sociopsychological variables that are well studied but difficult to quantify; and the third one relates biological variables that

---

* The data in figure 27 have not been corrected for variations in other factors, such as health care. Further information on statistical treatment of such a relationship and on its incorporation into the model equations will be presented in the technical report.

Figure 27 NUTRITION AND LIFE EXPECTANCY

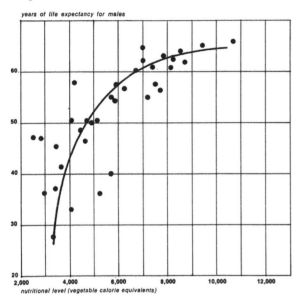

Life expectancy of a population is a nonlinear function of the nutrition that the population receives. In this graph nutritional level is given in vegetable calorie equivalents. Calories obtained from animal sources, such as meat or milk, are multiplied by a conversion factor (roughly 7, since about 7 calories of vegetable feed are required to produce 1 calorie of animal origin). Since food from animal sources is of greater value in sustaining human life, this measure takes into account both quantity and quality of food. Each point on the graph represents the average life expectancy and nutritional level of one nation in 1953.

SOURCE: M. Cépède, F. Houtart, and L. Grond, *Population and Food* (New York: Sheed and Ward, 1964).

are, as yet, almost totally unknown. Although these three examples by no means constitute a complete description of the world model, they illustrate the reasoning we have used to construct and quantify it.

Per capita resource use

As the world's population and capital plant grow, what will happen to the demand for nonrenewable resources? The amount of resources consumed each year can be found by multiplying the population times the per capita resource usage rate. Per capita resource usage rate is not constant, of course. As a population becomes more wealthy, it tends to consume more resources per person per year. The flow diagram expressing the relationship of population, per capita resource usage rate, and wealth (as measured by industrial output per capita) to the resource usage rate is shown below.

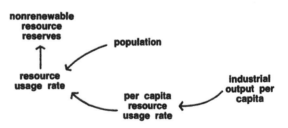

The relationship between wealth (industrial output per capita) and resource demand (per capita resource usage rate) is expressed by a nonlinear curve of the form shown in figure 28. In figure 28 resource use is defined in terms of the world average resource consumption per capita in 1970, which is set

Figure 28 INDUSTRIAL OUTPUT PER CAPITA AND RESOURCE USAGE

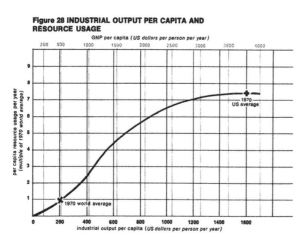

The postulated model relationship between resources consumed per person and industrial output per person is S-shaped. In nonindustrialized societies resource consumption is very low, since most production is agricultural. As industrialization increases, nonrenewable resource consumption rises steeply, and then becomes nearly level at a very high rate of consumption. Point × indicates the 1970 world average resource consumption rate; point + indicates the 1970 US average consumption rate. The two horizontal scales give the resource consumption relationship in terms of both industrial output per capita and GNP per capita.

equal to 1. Since world average industrial output per capita in 1970 was about $230,[30] we know that the curve goes through the point marked by an ×. In 1970 the United States had an average industrial output per capita of about $1,600, and the average citizen consumed approximately seven times the world average per capita resource usage.[31] The point on the curve that would represent the US level of consumption is marked by

a +. We assume that, as the rest of the world develops economically, it will follow basically the US pattern of consumption—a sharp upward curve as output per capita grows, followed by a leveling off. Justification for that assumption can be found in the present pattern of world steel consumption (see figure 29). Although there is some variation in the steel consumption curve from the general curve of figure 28, the overall pattern is consistent, even given the differing economic and political structures represented by the various nations.

Additional evidence for the general shape of the resource consumption curve is shown by the history of US consumption of steel and copper plotted in figure 30. As the average individual income has grown, the resource usage in both cases has risen, at first steeply and then less steeply. The final plateau represents an average saturation level of material possessions. Further income increases are spent primarily on services, which are less resource consuming.

The S-shaped curve of resource usage shown in figure 28 is included in the world model only as a representation of apparent *present* policies. The curve can be altered at any time in the model simulation to test the effects of system changes (such as recycling of resources) that would either increase or decrease the amount of nonrenewable resources each person consumes. Actual model runs shown later in this book will illustrate the effects of such policies.

Desired birth rate

The number of births per year in any population equals the number of women of reproductive age times the average fertility (the average number of births per woman per year). There may be numerous factors influencing the fertility of a

Figure 29 WORLD STEEL CONSUMPTION AND GNP PER CAPITA

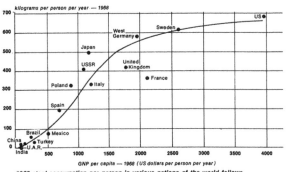

1968 steel consumption per person in various nations of the world follows the general S-shaped pattern shown in figure 28.

SOURCES: Steel consumption from UN Department of Economic and Social Affairs, *Statistical Yearbook 1969* (New York: United Nations, 1970). GNP per capita from *World Bank Atlas* (Washington, DC: International Bank for Reconstruction and Development, 1970).

population. In fact the study of fertility determinants is a major occupation of many of the world's demographers. In the world model we have identified three major components of fertility—maximum biological birth rate, birth control effectiveness, and desired birth rate. The relationship of these components to fertility is expressed in the diagram below.

Figure 30 US COPPER AND STEEL CONSUMPTION AND GNP PER CAPITA

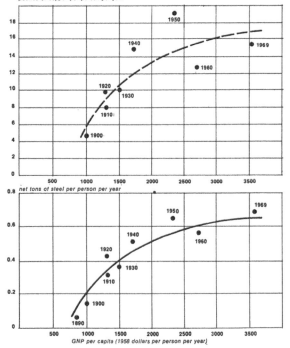

Per capita copper and steel consumption in the United States underwent a period of rapid increase as total productivity rose, followed by a period of much slower increase after consumption reached a relatively high rate.

SOURCES: Copper and steel consumption from *Metal Statistics* (Somerset, NJ: American Metal Market Company, 1970). Historical population and GNP from US Department of Commerce, *US Economic Growth* (Washington, DC: Government Printing Office, 1969).

Figure 31 BIRTH RATES AND GNP PER CAPITA

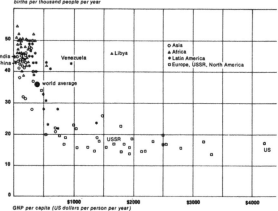

Birth rates in the world's nations show a regular downward trend as GNP per capita increases. More than one-half of the world's people are represented in the upper left-hand corner of the graph, where GNP per capita is less than $500 per person per year and birth rates range from 40 to 50 per thousand persons per year. The two major exceptions to the trend, Venezuela and Libya, are oil-exporting nations, where the rise in income is quite recent and income distribution is highly unequal.

SOURCE: US Agency for International Development, *Population Program Assistance* (Washington, DC: Government Printing Office, 1970).

The *maximum biological birth rate* is the rate at which women would bear children if they practiced no method of birth control throughout their entire reproductive lifetimes. This rate is biologically determined, depending mainly on the general health of the population. The *desired birth rate* is the rate that would result if the population practiced "perfect"

birth control and had only planned and wanted children. *Birth control effectiveness* measures the extent to which the population is able to achieve the desired birth rate rather than the maximum biological one. Thus "birth control" is defined very broadly to include any method of controlling births actually practiced by a population, including contraception, abortion, and sexual abstinence. It should be emphasized that perfect birth control effectiveness does *not* imply low fertility. If desired birth rate is high, fertility will also be high.

These three factors influencing fertility are in turn influenced by other factors in the world system. Figure 31 suggests that industrialization might be one of the more important of these factors.

The relation between crude birth rates and GNP per capita of all the nations in the world follows a surprisingly regular pattern. In general, as GNP rises, the birth rate falls. This appears to be true, despite differences in religious, cultural, or political factors. Of course, we cannot conclude from this figure that a rising GNP per capita directly causes a lower birth rate. Apparently, however, a number of social and educational changes that ultimately lower the birth rate are associated with increasing industrialization. These social changes typically occur only after a rather long delay.

Where in the feedback loop structure does this inverse relationship between birth rate and per capita GNP operate? Most evidence would indicate that it does not operate through the maximum biological birth rate. If anything, rising industrialization implies better health, so that the number of births possible might increase as GNP increases. On the other hand, birth control effectiveness would also increase, and this effect certainly contributes to the decline in births shown in figure 31.

Figure 32 FAMILIES WANTING FOUR OR MORE CHILDREN AND GNP PER CAPITA

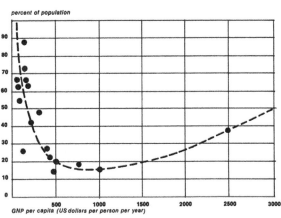

Respondents to family planning surveys in seventeen different countries indicated how many children they would like to have. The percentage of respondents desiring large families (four or more children) shows a relationship to average GNP per capita comparable to the trend shown in figure 31.

SOURCE: Bernard Berelson et al., *Family Planning and Population Programs* (Chicago: University of Chicago Press, 1965).

We suggest, however, that the major effect of rising GNP is on the *desired* birth rate. Evidence for this suggestion is shown in figure 32. The curve indicates the percentage of respondents to family planning surveys wanting more than four children as a function of GNP per capita. The general shape of the curve is similar to that of figure 31, except for the slight increase in desired family size at high incomes.

The economist J. J. Spengler has explained the general response of desired birth rate to income in terms of the economic and social changes that occur during the process of

Figure 33 DESIRED FAMILY SIZE

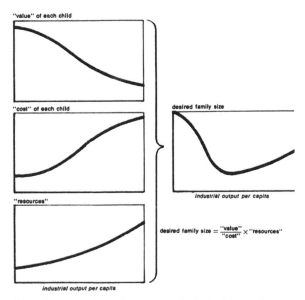

Schematic representation of the economic determinants of family size follows a rough cost-benefit analysis. The resultant curve summarizes the balance between value and cost of children and resources available for child-raising, all as a function of increasing industrialization. This composite curve is similar to the curves in figures 31 and 32.

industrialization.[32] He believes that each family, consciously or unconsciously, weighs the value and cost of an additional child against the resources the family has available to devote to that child. This process results in a general attitude about family size that shifts as income increases, as shown in figure 33.

The "value" of a child includes monetary considerations, such as the child's labor contribution to the family farm or business and the eventual dependence on the child's support when the parents reach old age. As a country becomes industrialized, child labor laws, compulsory education, and social security provisions all reduce the potential monetary value of a child. "Value" also includes the more intangible values of a child as an object of love, a carrier of the family name, an inheritor of the family property, and a proof of masculinity. These values tend to be important in any society, and so the reward function always has a positive value. It is particularly important in poor societies, where there are almost no alternative modes of personal gratification.

The "cost" of a child includes the actual financial outlays necessary to supply the child's needs, the opportunity costs of the mother's time devoted to child care, and the increased responsibility and decreased freedom of the family as a whole. The cost of children is very low in a traditional society. No additional living space is added to house a new child, little education or medical care is available, clothing and food requirements are minimal. The mother is generally uneducated and assigns no value to her time. The family has little freedom to do anything that a child would hinder, and the extended family structure is there to provide child care if it should become necessary, for example, for a parent to leave home to find a job.

As family income increases, however, children are given more than the basic food and clothing requirements. They receive better housing and medical care, and education becomes both necessary and expensive. Travel, recreation, and alternative employment for the mother become possibilities that are not compatible with a large family. The extended family structure tends to disappear with industrialization, and substitute child care is costly.

The "resources" that a family has to devote to a child generally increase with income. At very high income, the value and cost curves become nearly invariant with further increases in income, and the resource curve becomes the dominant factor in the composite desired birth rate. Thus, in rich countries, such as the United States, desired family size becomes a direct function of income. It should be noted that "resources" is partially a psychological concept in that present actual income must be modified by an expectation of future income in planning family size.

We have summarized all these social factors by a feedback loop link between industrial output per capita and desired birth rate. The general shape of the relationship is shown on the right side of figure 33. We do not mean to imply by this link that rising income is the only determinant of desired family size, or even that it is a direct determinant. In fact we include a delay between industrial output per capita and desired family size to indicate that this relationship requires a social adjustment, which may take a generation or two to complete. Again, this relationship may be altered by future policies or social changes. As it stands it simply reflects the historical behavior of human society. Wherever economic development has taken place, birth rates have fallen. Where industrialization has not occurred, birth rates have remained high.

Pollution effect on lifetime

We have included in the world model the possibility that

pollution will influence the life expectancy of the world's population. We express this relationship by a "lifetime multiplier from pollution," a function that multiplies the life expectancy otherwise indicated (from the values of food and medical services) by the contribution to be expected from pollution. If pollution were severe enough to lower the life expectancy to 90 percent of its value in the absence of pollution, the multiplier would equal 0.9. The relationship of pollution to life expectancy is diagramed below.

There are only meager global data on the effect of pollution on life expectancy. Information is slowly becoming available about the toxicity to humans of specific pollutants, such as mercury and lead. Attempts to relate statistically a given concentration of pollutant to the mortality of a population have been made only in the field of air pollution.[83]

Although quantitative evidence is not yet available, there is little doubt that a relationship does indeed exist between pollution and human health. According to a recent Council on Environmental Quality report:

Serious air pollution episodes have demonstrated how air pollution can severely impair health. Further research is spawning a growing body of evidence which indicates that even the long-term effects of exposure to low concentrations of pollutants can damage health and cause chronic disease and premature death, especially for the most vulnerable—the aged and those already suffering from respiratory diseases. Major illnesses linked to air pollution include emphysema, bronchitis, asthma, and lung cancer.[84]

What will be the effect on human lifetime as the present level of global pollution increases? We cannot answer this question accurately, but we do know that there will be *some* effect. We would be more in error to ignore the influence of pollution on life expectancy in the world model than to include it with our best guess of its magnitude. Our approach to a "best guess" is explained below and illustrated in figure 34.

If an increase in pollution by a factor of 100 times the present global level would have absolutely no effect on lifetime, the straight line A in figure 34 would be the correct representation of the relationship we seek. Life expectancy would be unrelated to pollution. Curve A is very unlikely, of course, since we know that many forms of pollution are damaging to the human body. Curve B or any similar curve that rises above curve A is even more unlikely since it indicates that additional pollution will increase average lifetime. We can expect that the relationship between pollution and lifetime is negative, although we do not know what the exact shape or slope of a curve expressing it will be. Any one of the curves labeled C, or any other negative curve, might represent the correct function.

Our procedure in a case like this is to make several different estimates of the probable effect of one variable on another and then to test each estimate in the model. If the model behavior is very sensitive to small changes in a curve, we know we must obtain more information before including it. If (as in this case) the behavior mode of the entire model is not substantially altered by changes in the curve, we make a conservative guess of its shape and include the corresponding values in our calculation. Curve C" in figure 34 is the one we believe most accurately depicts the relationship between life expectancy

Figure 34 THE EFFECT OF POLLUTION ON LIFETIME

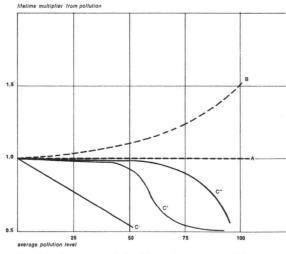

lifetime multiplier from pollution

average pollution level

The relationship between level of pollution and average human lifetime might follow many different curves. Curve A indicates that pollution has no effect on lifetime (normal life expectancy is multiplied by 1.0). Curve B represents an enhancement of lifetime as pollution increases (normal life expectancy is multiplied by a number greater than 1.0). The curves C, C', and C" reflect differing assumptions about deleterious effects of pollution on lifetime. The relationship used in the world model is shaped like curve C".

and pollution. This curve assumes that an increase in global pollution by a factor of 10 would have almost no effect on lifetime but an increase by a factor of 100 would have a great effect.

The usefulness of the world model

The relationships discussed above comprise only three of the hundred or so causal links that make up the world model. They have been chosen for presentation here as examples of the kind of information inputs we have used and the way in which we have used them. In many cases the information available is not complete. Nevertheless, we believe that the model based on this information is useful even at this preliminary stage for several reasons.

First, we hope that by posing each relationship as a hypothesis, and emphasizing its importance in the total world system, we may generate discussion and research that will eventually improve the data we have to work with. This emphasis is especially important in the areas in which different sectors of the model interact (such as pollution and human lifetime), where interdisciplinary research will be necessary.

Second, even in the absence of improved data, information now available is sufficient to generate valid basic behavior modes for the world system. This is true because the model's feedback loop structure is a much more important determinant of overall behavior than the exact numbers used to quantify the feedback loops. Even rather large changes in input data do not generally alter the *mode* of behavior, as we shall see in the following pages. Numerical changes may well affect the *period* of an oscillation or the *rate* of growth or the *time* of a collapse,* but they will not affect the fact that the basic mode is oscillation or growth or collapse.* Since we intend to use the

* The importance of structure rather than numbers is a most difficult concept to present without extensive examples from the observation and modeling of dynamic systems. For further discussion of this point, see chapter 6 of J. W. Forrester's *Urban Dynamics* (Cambridge, Mass.: MIT Press, 1969).

world model only to answer questions about behavior modes, not to make exact predictions, we are primarily concerned with the correctness of the feedback loop structure and only secondarily with the accuracy of the data. Of course when we do begin to seek more detailed, short-term knowledge, exact numbers will become much more important.

Third, if decision-makers at any level had access to precise predictions and scientifically correct analyses of alternate policies, we would certainly not bother to construct or publish a simulation model based on partial knowledge. Unfortunately, there is no perfect model available for use in evaluating today's important policy issues. At the moment, our only alternatives to a model like this, based on partial knowledge, are mental models, based on the mixture of incomplete information and intuition that currently lies behind most political decisions. A dynamic model deals with the same incomplete information available to an intuitive model, but it allows the organization of information from many different sources into a feedback loop structure that can be exactly analyzed. Once all the assumptions are together and written down, they can be exposed to criticism, and the system's response to alternative policies can be tested.

WORLD MODEL BEHAVIOR

Now we are at last in a position to consider seriously the questions we raised at the beginning of this chapter. As the world system grows toward its ultimate limits, what will be its most likely behavior mode? What relationships now existent will change as the exponential growth curves level off? What will the world be like when growth comes to an end?

There are, of course, many possible answers to these ques-

tions. We will examine several alternatives, each dependent on a different set of assumptions about how human society will respond to problems arising from the various limits to growth.

Let us begin by assuming that there will be in the future no great changes in human values nor in the functioning of the global population-capital system as it has operated for the last one hundred years. The results of this assumption are shown in figure 35. We shall refer to this computer output as the "standard run" and use it for comparison with the runs based on other assumptions that follow. The horizontal scale in figure 35 shows time in years from 1900 to 2100. With the computer we have plotted the progress over time of eight quantities:

- **━━━** population (total number of persons)
- **━ ━ ━** industrial output per capita (dollar equivalent per person per year)
- **━━━** food per capita (kilogram-grain equivalent per person per year)
- **•••••••** pollution (multiple of 1970 level)
- **━•━•━** nonrenewable resources (fraction of 1900 reserves remaining)
- B   crude birth rate (births per 1000 persons per year)
- D   crude death rate (deaths per 1000 persons per year)
- S   services per capita (dollar equivalent per person per year)

Each of these variables is plotted on a different vertical scale. We have deliberately omitted the vertical scales and we have made the horizontal time scale somewhat vague because we want to emphasize the general behavior modes of these computer outputs, not the numerical values, which are only approxi-

Figure 35 WORLD MODEL STANDARD RUN

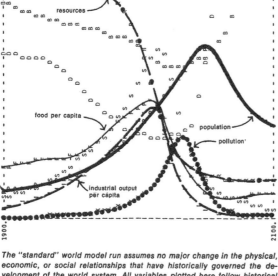

The "standard" world model run assumes no major change in the physical, economic, or social relationships that have historically governed the development of the world system. All variables plotted here follow historical values from 1900 to 1970. Food, industrial output, and population grow exponentially until the rapidly diminishing resource base forces a slowdown in industrial growth. Because of natural delays in the system, both population and pollution continue to increase for some time after the peak of industrialization. Population growth is finally halted by a rise in the death rate due to decreased food and medical services.

mately known. The scales are, however, exactly equal in all the computer runs presented here, so results of different runs may be easily compared.

All levels in the model (population, capital, pollution, etc.) begin with 1900 values. From 1900 to 1970 the variables plotted in figure 35 (and numerous other variables included in the model but not plotted here) agree generally with their historical values to the extent that we know them. Population rises from 1.6 billion in 1900 to 3.5 billion in 1970. Although the birth rate declines gradually, the death rate falls more quickly, especially after 1940, and the rate of population growth increases. Industrial output, food, and services per capita increase exponentially. The resource base in 1970 is still about 95 percent of its 1900 value, but it declines dramatically thereafter, as population and industrial output continue to grow.

The behavior mode of the system shown in figure 35 is clearly that of overshoot and collapse. In this run the collapse occurs because of nonrenewable resource depletion. The industrial capital stock grows to a level that requires an enormous input of resources. In the very process of that growth it depletes a large fraction of the resource reserves available. As resource prices rise and mines are depleted, more and more capital must be used for obtaining resources, leaving less to be invested for future growth. Finally investment cannot keep up with depreciation, and the industrial base collapses, taking with it the service and agricultural systems, which have become dependent on industrial inputs (such as fertilizers, pesticides, hospital laboratories, computers, and especially energy for mechanization). For a short time the situation is especially serious because population, with the delays inherent in the age structure and the process of social adjustment, keeps rising. Population finally decreases when the death rate is driven upward by lack of food and health services.

The exact timing of these events is not meaningful, given the great aggregation and many uncertainties in the model. It is significant, however, that growth is stopped well before the year 2100. We have tried in every doubtful case to make the most optimistic estimate of unknown quantities, and we have also ignored discontinuous events such as wars or epidemics, which might act to bring an end to growth even sooner than our model would indicate. In other words, the model is biased to allow growth to continue longer than it probably can continue in the real world. *We can thus say with some confidence that, under the assumption of no major change in the present system, population and industrial growth will certainly stop within the next century, at the latest.*

The system shown in figure 35 collapses because of a resource crisis. What if our estimate of the global stock of resources is wrong? In figure 35 we assumed that in 1970 there was a 250-year supply of all resources, at 1970 usage rates. The static reserve index column of the resource table in chapter II will verify that this assumption is indeed optimistic. But let us be even more optimistic and assume that new discoveries or advances in technology can *double* the amount of resources economically available. A computer run under that assumption is shown in figure 36.

The overall behavior mode in figure 36—growth and collapse—is very similar to that in the standard run. In this case the primary force that stops growth is a sudden increase in the level of pollution, caused by an overloading of the natural absorptive capacity of the environment. The death rate rises abruptly from pollution and from lack of food. At the same time resources are severely depleted, in spite of the doubled amount available, simply because a few more years of exponential growth in industry are sufficient to consume those extra resources.

Is the future of the world system bound to be growth and then collapse into a dismal, depleted existence? Only if we make the initial assumption that our present way of doing things will not change. We have ample evidence of mankind's

Figure 36 WORLD MODEL WITH NATURAL RESOURCE RESERVES DOUBLED

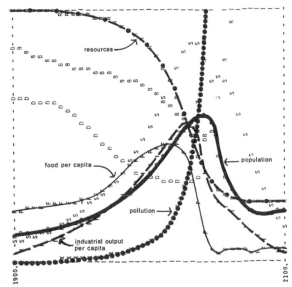

To test the model assumption about available resources, we doubled the resource reserves in 1900, keeping all other assumptions identical to those in the standard run. Now industrialization can reach a higher level since resources are not so quickly depleted. The larger industrial plant releases pollution at such a rate, however, that the environmental pollution absorption mechanisms become saturated. Pollution rises very rapidly, causing an immediate increase in the death rate and a decline in food production. At the end of the run resources are severely depleted in spite of the doubled amount initially available.

ingenuity and social flexibility. There are, of course, many likely changes in the system, some of which are already taking place. The Green Revolution is raising agricultural yields in nonindustrialized countries. Knowledge about modern methods of birth control is spreading rapidly. Let us use the world model as a tool to test the possible consequences of the new technologies that promise to raise the limits to growth.

1

ON THE LOGIC OF WEALTH DISTRIBUTION AND RISK DISTRIBUTION

In advanced modernity the social production of *wealth* is systematically accompanied by the social production of *risks*. Accordingly, the problems and conflicts relating to distribution in a society of scarcity overlap with the problems and conflicts that arise from the production, definition and distribution of techno-scientifically produced risks.

This change from the logic of wealth distribution in a society of scarcity to the logic of risk distribution in late modernity is connected historically to (at least) two conditions. First, it occurs – as is recognizable today – where and to the extent that *genuine material need* can be objectively reduced and socially isolated through the development of human and technological productivity, as well as through legal and welfare-state protections and regulations. Second, this categorical change is likewise dependent upon the fact that in the course of the exponentially growing productive forces in the modernization process, hazards and potential threats have been unleashed to an extent previously unknown.[1]

To the extent that these conditions occur, one historical type of thinking and acting is relativized or overridden by another. The concepts of 'industrial' or 'class society', in the broadest sense of Marx or Weber, revolved around the issue of how socially produced wealth could be distributed in a socially unequal and *also* 'legitimate' way. This overlaps with the new *paradigm of risk society* which is based on the solution of a similar and yet quite different problem. How can the risks and hazards systematically produced as part of modernization be prevented, minimized, dramatized, or channeled? Where they do finally see the light of day in the shape of 'latent side effects', how can they be limited and distributed away so that they neither hamper the modernization process nor exceed the limits of that which is 'tolerable' – ecologically, medically, psychologically and socially?

We are therefore concerned no longer exclusively with making nature useful, or with releasing mankind from traditional constraints, but also and essentially with problems resulting from techno-economic development itself. Modernization is becoming *reflexive*; it is becoming its own theme. Questions of the development and employment of technologies (in the realms of nature, society and the personality) are being eclipsed by questions of the political and economic 'management' of the risks of actually or potentially utilized technologies – discovering, administering, acknowledging, avoiding or concealing such hazards with respect to specially defined horizons of relevance. The promise of security grows with the risks and destruction and must be reaffirmed over and over again to an alert and critical public through cosmetic or real interventions in the techno-economic development.

Both 'paradigms' of inequality are systematically related to definite periods of modernization. The distribution of socially produced wealth and related conflicts occupy the foreground so long as obvious material need, the 'dictatorship of scarcity', rules the thought and action of people (as today in large parts of the so-called Third World). Under these conditions of 'scarcity society', the modernization process takes place with the claim of opening the gates to hidden sources of social wealth with the keys of techno-scientific development. These promises of emancipation from undeserved poverty and dependence underlie action, thought and research in the categories of social inequality, from the class through the stratified to the individualized society.

In the welfare states of the West a double process is taking place now. On the one hand, the struggle for one's 'daily bread' has lost its urgency as a cardinal problem overshadowing everything else, compared to material subsistence in the first half of this century and to a Third World menaced by hunger. For many people problems of 'overweight' take the place of hunger. This development, however, withdraws the legitimizing basis from the modernization process, the struggle against obvious scarcity, for which one was prepared to accept a few (no longer completely) unseen side effects.

Parallel to that, the knowledge is spreading that the sources of wealth are 'polluted' by growing 'hazardous side effects'. This is not at all new, but it has remained unnoticed for a long time in the efforts to overcome poverty. This dark side is also gaining importance through the over-development of productive forces. In the modernization process, more and more *destructive* forces are also being unleashed, forces before which the human imagination stands in awe. Both sources feed a growing critique of modernization, which loudly and contentiously determines public discussions.

In systematic terms, sooner or later in the continuity of modernization the social positions and conflicts of a 'wealth-distributing' society begin to be joined by those of a 'risk-distributing' society. In West Germany we have faced the beginning of this transition since the early 1970s at the latest – that is my thesis. That means that two types of topics and conflicts overlap here. We do not *yet* live in a risk society, but we also no longer live *only* within the distribution conflicts of scarcity societies. To the extent that this transition occurs, there will be a real transformation of society which will lead us out of the previous modes of thought and action.

Can the concept of risk carry the theoretical and historical significance which is demanded of it here? Is this not a primeval phenomenon of human action? Are not risks already characteristic of the industrial society period, against which they are being differentiated here? It is also true that risks are not an invention of modernity. Anyone who set out to discover new countries and continents – like Columbus – certainly accepted 'risks'. But these were *personal* risks, not global dangers like those that arise for all of humanity from nuclear fission or the storage of radioactive waste. In that earlier period, the word 'risk' had a note of bravery and adventure, not the threat of self-destruction of all life on Earth.

Forests have also been dying for some centuries now – first through being transformed into fields, then through reckless overcutting. But the death of forests today occurs *globally*, as the *implicit* consequence of industrialization – with quite different social and political consequences. Heavily wooded countries like Norway and Sweden, which hardly have any pollutant-intensive industries of their own, are also affected. They have to settle up the pollution accounts of other highly industrialized countries with dying trees, plants and animal species.

It is reported that sailors who fell into the Thames in the early nineteenth century did not drown, but rather choked to death inhaling the foul-smelling and poisonous fumes of this London sewer. A walk through the narrow streets of a medieval city would also have been like running the gauntlet for the nose. 'Excrement piles up everywhere, in the streets, at the turnpikes, in the carriages . . . The façades of Parisian houses are decomposing from urine . . . the socially organized constipation threatens to pull all of Paris into the process of putrescent decomposition' (Corbin 1984: 41ff.). It is nevertheless striking that hazards in those days assaulted the nose or the eyes and were thus perceptible to the senses, while the risks of civilization today typically *escape perception* and are localized in the sphere of *physical and chemical formulas* (e.g. toxins in foodstuffs or the nuclear threat).

Another difference is directly connected to this. In the past, the hazards could be traced back to an *under*supply of hygienic technology. Today they have their basis in industrial *over*production. The risks and hazards of today thus differ in an essential way from the superficially similar ones in the Middle Ages through the global nature of their threat (people, animals and plants) and through their *modern* causes. They are risks *of modernization*. They are a *wholesale product* of industrialization, and are systematically intensified as it becomes global.

The concept of risk is directly bound to the concept of reflexive modernization. *Risk* may be defined as a *systematic way of dealing with hazards and insecurities induced and introduced by modernization itself*. Risks, as opposed to older dangers, are consequences which relate to the threatening force of modernization and to its globalization of doubt. They are *politically reflexive*.

Risks, in this meaning of the word, are certainly as old as that development itself. The immiseration of large parts of the population – the 'poverty risk' – kept the nineteenth century holding its breath. 'Threats to skills' and 'health risks' have long been a theme of automation processes and the related social conflicts, protections (and research). It did take some time and struggle to establish social welfare state norms and minimize or limit these kinds of risk politically. Nevertheless, the ecological and high-tech risks that have upset the public for some years now, which will be the focus of what follows, have a new quality. In the afflictions they produce they are no longer tied to their place of origin – the industrial plant. By their nature they endanger *all* forms of life on this planet. The normative bases of their calculation – the concept of accident and insurance, medical precautions, and so on – do not fit the basic dimensions of these modern threats. Atomic plants, for example, are not privately insured or insurable. Atomic accidents are accidents no more (in the limited sense of the word 'accident'). They outlast generations. The affected even include those not yet alive at the time or in the place where the accident occurred but born years later and long distances away.

This means that the calculation of risk as it has been established so far by science and legal institutions *collapses*. Dealing with these consequences of modern productive and destructive forces in the normal terms of risk is a false but nevertheless very effective way of legitimizing them. Risk scientists normally do so as if there is not the gap of a century between the local accidents of the nineteenth century and the often creeping, catastrophic potentials at the end of the twentieth century. Indeed, if you distinguish between calculable and non-calculable threats, under the surface of risk calculation new kinds of *industrialized, decision-produced incalculabilities and threats* are spreading within the globalization of high-risk industries, whether for warfare or welfare purposes. Max Weber's concept of 'rationalization' no longer grasps this late modern reality, produced by successful rationalization. *Along with the growing capacity of technical options [Zweckrationalität] grows the incalculability of their consequences.* Compared to these global consequences, the hazards of primary industrialization indeed belonged to a different age. The dangers of highly developed nuclear and chemical productive forces abolish the foundations and categories according to which we have thought and acted to this point, such as space and time, work and leisure time, factory and nation state, indeed even the borders between continents. To put it differently, in the risk society the unknown and unintended consequences come to be a dominant force in history and society.[2]

The social architecture and political dynamics of such potentials for self-endangerment in civilization will occupy the center of these discussions. The argument can be set out in five theses:

(1) Risks such as those produced in the late modernity differ essentially from wealth. By risks I mean above all radioactivity, which completely evades human perceptive abilities, but also toxins and pollutants in the air, the water and foodstuffs, together with the accompanying short- and long-term effects on plants, animals and people. They induce systematic

and often *irreversible* harm, generally remain *invisible*, are based on *causal interpretations*, and thus initially only exist in terms of the (scientific or anti-scientific) *knowledge* about them. They can thus be changed, magnified, dramatized or minimized within knowledge, and to that extent they are particularly *open to social definition and construction*. Hence the mass media and the scientific and legal professions in charge of defining risks become key social and political positions.

(2) Some people are more affected than others by the distribution and growth of risks, that is, *social risk positions* spring up. In some of their dimensions these follow the inequalities of class and strata positions, but they bring a fundamentally different distributional logic into play. Risks of modernization sooner or later also strike those who produce or profit from them. They contain a *boomerang effect*, which breaks up the pattern of class and national society. Ecological disaster and atomic fallout ignore the borders of nations. Even the rich and powerful are not safe from them. These are hazards not only to health, but also to legitimation, property and profit. *Connected* to the recognition of modernization risks are *ecological devaluations and expropriations*, which frequently and systematically enter into contradiction to the profit and property interests which advance the process of industrialization. Simultaneously, risks produce *new international inequalities*, firstly between the Third World and the industrial states, secondly among the industrial states themselves. They undermine the order of national jurisdictions. In view of the universality and supra-nationality of the circulation of pollutants, the life of a blade of grass in the Bavarian Forest ultimately comes to depend on the making and keeping of international agreements. Risk society in this sense is a world risk society.

(3) Nevertheless, the diffusion and commercialization of risks do not break with the logic of capitalist development completely, but instead they raise the latter to a new stage. There are always losers but also winners in risk definitions. The space between them varies in relation to different issues and power differentials. Modernization risks from the winners' points of view are *big business*. They are the insatiable demands long sought by economists. Hunger can be sated, needs can be satisfied, but *civilization* risks are a *bottomless barrel of demands*, unsatisfiable, infinite, self-producible. One could say along with Luhmann that with the advent of risks, the economy becomes 'self-referential', independent of the surrounding satisfaction of human needs. But that means: with the economic exploitation of the risks it sets free, industrial society produces the hazards and the political potential of the risk society.

(4) One can *possess* wealth, but one can only be *afflicted* by risks; they are, so to speak, *ascribed* by civilization. [Bluntly, one might say: in class and stratification positions being determines consciousness, while in risk positions *consciousness determines being*.] Knowledge gains a new political significance. Accordingly the political potential of the risk society

must be elaborated and analyzed in a sociological theory of the origin and diffusion of *knowledge about risks*.

(5) Socially recognized risks, as appears clearly in the discussions of forest destruction, contain a peculiar political explosive: what *was* until now *considered unpolitical becomes political – the elimination of the causes in the industrialization process itself*. Suddenly the public and politics extend their rule into the private sphere of plant management – into product planning and technical equipment. What is at stake in the public dispute over the definition of risks is revealed here in an exemplary fashion: not just secondary health problems for nature and mankind, but the *social, economic and political consequences of these side effects* – collapsing markets, devaluation of capital, bureaucratic checks on plant decisions, the opening of new markets, mammoth costs, legal proceedings and loss of face. In smaller or larger increments – a smog alarm, a toxic spill, etc. – what thus emerges in risk society is the *political potential of catastrophes*. Averting and managing these can include a *reorganization of power and authority*. Risk society is a *catastrophic* society. In it the exceptional condition threatens to become the norm.

Scientific Definition and Distributions of Pollutants

The debate on pollutant and toxic elements in air, water and foodstuffs, as well as on the destruction of nature and the environment in general, is still being conducted exclusively or dominantly in the terms and formulas of *natural* science. It remains unrecognized that a social, cultural and political meaning is inherent in such scientific 'immiseration formulas'. There exists accordingly a danger that an environmental discussion conducted exclusively in chemical, biological and technological terms will inadvertently include human beings in the picture only as *organic material*. Thus the discussion runs the risk of making the same mistake for which it has long and justly reproached the prevailing optimism with respect to industrial progress; it runs the risk of atrophying into a discussion of nature *without* people, without asking about matters of social and cultural significance. Particularly the debates over the last few years, in which all arguments critical of technology and industry were once again deployed, have remained at heart *technocratic* and *naturalistic*. They exhausted themselves in the invocation and publication of the pollutant levels in the air, water and foodstuffs, in relative figures of population growth, energy consumption, food requirements, raw material shortages and so on. They did so with a passion and a singlemindedness as if there had never been people such as a certain Max Weber, who apparently wasted his time showing that without including structures of social power and distribution, bureaucracies, prevailing norms and rationalities, such a debate is either meaningless or absurd, and probably both. An understanding has crept in, according to which modernity is reduced to the frame of reference of technology and nature in the manner of perpetrator and victim. The social, cultural and political risks of modernization remain hidden by this very approach, and from this way of thinking (which is also that of the political environmental movement).

Let us illustrate this with an example. The Rat der Sachverständigen für Umweltfragen (Council of Experts on Environmental Issues) determines in a report that 'in mother's milk beta-hexachlorocyclohexane, hexachlorobenzol and DDT are often found in significant concentrations' (1985: 33). These toxic substances are contained in pesticides and herbicides that have by now been taken off the market. According to the report their origin is undetermined (33). At another point it is stated: 'The exposure of the population to lead is not dangerous on average' (35). What is concealed behind that statement? Perhaps by analogy the following distribution. Two men have two apples. One eats both of them. Thus they have eaten *on average* one each. Transferred to the distribution of foodstuffs on the global scale this statement would mean: 'on average' all the people in the world have enough to eat. The cynicism here is obvious. In one part of the Earth people are dying of hunger, while in the other the consequences of overeating have become a major item of expense. It may be, of course, that this statement about pollutants and toxins is *not* cynical, that the *average* exposure is also the *actual* exposure of *all* groups in the population. But do we know that? In order to defend this statement, is it not a prerequisite that we know what other poisons the people are forced to inhale and ingest? It is astonishing how *as a matter of course* one inquires about 'the average'. A person who inquires about the average already excludes many socially unequal risk positions. But that is exactly what that person cannot know. Perhaps there are groups and living conditions for which the levels of lead and the like that are 'on average harmless' constitute a *mortal danger*?

The next sentence of the report reads: 'Only in the vicinity of industrial emitters are dangerous concentrations of lead sometimes found in children.' What is characteristic is not just the absence of any social differentiations in this and other reports on pollutants and toxins. It is also characteristic *how* differentiations are made – along *regional* lines with regard to emission sources and according to *age* differences – both criteria that are rooted in *biological* (or more generally, natural scientific) thinking. This cannot be blamed on the expert committees. It only reflects the general state of scientific and social thought with regard to environmental problems. These are generally viewed as matters of nature and technology, or of economics and medicine. What is astonishing about that is that the industrial pollution of the environment and the destruction of nature, with their multifarious effects on the health and social life of people, which only arise in highly developed societies, are characterized by a *loss of social thinking*. This loss becomes caricature – this absence seems to strike no one, not even sociologists themselves.

People inquire about and investigate the distribution of pollutants, toxins, contamination of water, air, and foodstuffs. The results are presented to an alarmed public on multi-colored 'environmental maps', differentiated along regional lines. To the extent that the state of the environment is to be presented in this way, this mode of presentation and consideration is obviously appropriate. As soon as *consequences for people* are to be drawn from it, however, the underlying thought *short-circuits*. Either one implies broadly that *all* people are *equally* affected in the identified pollution centers – independent of their income, education, occupation and the associated eating, living and recreational opportunities and habits (which would have to be proved). Or one ultimately excludes people and the extent of their affliction entirely and speaks only about pollutants and their distributions and effects on the region.

The pollution debate conducted in terms of natural science correspondingly moves between the false conclusion of social afflictions based on biological ones, and a view of nature which excludes the selective affliction of people as well as the social and cultural meaning connected to it. At the same time what is not taken into consideration is that *the same* pollutants can have quite *different* meanings for *different* people, according to age, gender, eating habits, type of work, information, education and so on.

What is particularly aggravating is that investigations which start from individual pollutants can *never* determine the concentration of pollutants *in people*. What may seem 'insignificant' for a single product, is perhaps extremely significant when collected in the 'consumer reservoirs' which people have become in the advanced stage of total marketing. We are in the presence here of a *category error*. A pollution analysis oriented to nature and products is incapable of answering questions about safety, at least as long as the 'safety' or 'danger' has anything to do with the people who swallow or breathe the stuff. What is known is that the taking of several medications can nullify or amplify the effect of each individual one. Now people obviously do not (yet) live by medications alone. They also breathe the pollutants in the air, drink those in the water, eat those in the vegetables, and so on. In other words, the insignificances can add up quite significantly. Do they thereby become more and more insignificant – as is usual for sums according to the rules of mathematics?

Thomas E. Graedel, Braden R. Allenby, and Peter B. Linhart (1993)

Implementing Industrial Ecology

Industrial ecology (IE), a new ensemble concept — elements of which have been recognized for years, arises from the perception that human economic activity is causing unacceptable changes in basic environmental support systems. The IE concept is needed to guide industrial activities over the next several decades if we are to move in the direction of achieving sustainable development practices. As applied to manufacturing, this systems-oriented concept suggests that industrial design and manufacturing processes are not performed in isolation from their surroundings, but rather are influenced by them and, in turn, have influence on them.

The essence of IE is embodied in the following definition:

> *The concept of industrial ecology is one in which economic systems are viewed not in isolation from their surrounding systems, but in concert with them. As applied to industrial operations, it requires a systems view in which one seeks to optimize the total*

The authors are with AT&T Bell Laboratories, Murray Hill, NJ 07974.

> *materials cycle from virgin material, to finished material, to component, to product, to waste product, and to ultimate disposal. Factors to be optimized include resources, energy, and capital.*

A more extensive definition uses a biological analogy to describe the perspective from which IE views an industrial system. A full consideration of industrial ecology could well include the entire scope of economic activity, but we limit our discussion to aspects of the problem that are specific to manufacturing industries.

A Systems Description

Traditional biological ecology is defined as *the scientific study of the interactions that determine the distribution and abundance of organisms.* The relationship between this concept and that of industrial activities has been discussed by Frosch and Gallopoulos [1]:

> *"In a biological ecosystem, some of the organisms use sunlight, water, and minerals to grow, while others consume the first, alive or dead, along with*

---

minerals and gases, and produce wastes of their own. These wastes are in turn food for other organisms, some of which may convert the wastes into the minerals used by the primary producers, and some of which consume each other in a complex network of processes in which everything produced is used by some organism for its own metabolism. Similarly, in the industrial ecosystem, each process and network of processes must be viewed as a dependent and interrelated part of a larger whole. The analogy between the industrial ecosystem concept and the biological ecosystem is not perfect, but much could be gained if the industrial system were to mimic the best features of the biological analogue.

It is instructive to think of the materials cycles involved with the earliest of Earth's life forms. At that time, the potentially usable resources were so large and the amount of life so small, that the existence of life forms had essentially no impact on available resources. This individual component process might be described as *linear.* We term this pattern Type I ecology; schematically, it takes the form of Fig. 1(a).

An aspect of biological ecology that is implied in this definition is that the totality of the ecosystems is sustainable over the long term, although individual components of the systems may undergo transitory periods of expansion or decay. In the larger picture represented by an ecosystem in which proximal resources are limited, the resulting life forms are interlinked and form the complex system of biological communities. This evolution has produced the efficiently operating system with which we are familiar. In this system, the flows of material within the proximal domain may be quite large, but the flows into and out of that domain (i.e., from resources and to waste) are quite small. Schematically, such a Type II system might be expressed as in Fig. 1(b).

This system is much more efficient than the previous one, but it clearly is not sustainable over the long term, because the flows are all in one direction, that is, the system is "running down." To be ultimately sustainable, biological ecosystems have evolved over the long term to be almost completely cyclical in nature, with "resources" and "waste" being undefined, since waste to one component of the system represents resources to another. This Type III system, in which complete cyclicity has been achieved, may be pictured as in Fig. 2. Note that the exception to the cyclicity of the overall system is that energy (in the form of solar radiation) is available as an external resource. It is also important to recognize that the cycles within the system tend to function on widely differing temporal and spatial scales, a

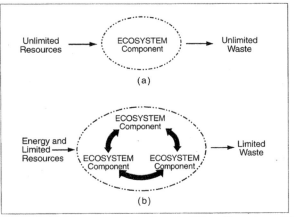

Fig. 1. (a) Linear materials flows in "Type I" ecology. (b) Quasicyclic materials flows in "Type I" ecology.

behavior that greatly complicates analysis and understanding of the system.

The ideal anthropogenic use of the materials and resources available for industrial processes (broadly defined to include agriculture, the urban infrastructure, etc.) would be one that is similar to the ensemble biological model. Many uses of materials have been and continue to be essentially dissipative, however. That is, the materials are degraded, dispersed, and lost to the economic system in the course of a single normal use [2], mimicking the Type I unconstrained resource diagram above. This pattern can be associated with the maturation of the Industrial Revolution which, in concert with exponential increases in human population and agricultural production, took place in a context of global plenty. Many present day industrial processes and products still remain inherently dissipative.

Not all present day technological processes are completely dissipative, however. Where specific materials are sufficiently precious, some demonstrate ecological Type II behavior, at least in part. We present a few examples later.

On the broadest of product and time scales, many indications demonstrate today that the flows in the ensemble of industrial ecosystems are so large that resource limitations are setting in: the rapid changes in atmospheric ozone, increases in atmospheric carbon dioxide, and the filling of available waste disposal sites being examples. Accordingly, industrial systems (and other anthropogenic systems) are and will be increasingly under pressure to evolve from linear (Type I) to semicyclic (Type II) modes of operation. For the past decade or two, industrial organizations have largely been in the position of responding to legislation enacted as a consequence of a real or perceived environmental crisis. Such a mode of operation is essentially unplanned, and imposes significant economic

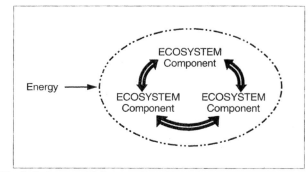

Fig. 2. Cyclic materials flows in "Type III" ecology.

costs. Industrial ecology, in its implementation, is intended to accomplish the evolution of manufacturing from Type I to Type II behavior through understanding of the interplay of processes and flows and by optimizing the ensemble of considerations.

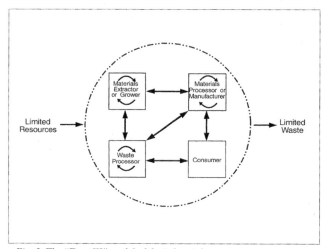

Fig. 3. The "Type III" model of the industrial ecosystem.

The central domain of industrial ecology is conveniently pictured with four central nodes: the materials extractor or grower, the materials processor or manufacturer, the consumer, and the waste processor. To the extent that they perform operations within the nodes in a cyclic manner, or organize to encourage cyclic flow of materials within the entire industrial ecosystem, they evolve into modes of operation that are more efficient, have less disruptive impact on external support systems, and are more like Type III ecological behavior. The schematic model of such an industrial ecosystem is shown in Fig. 3. Note that the flows within the nodes and within the industrial ecological system as a whole are much larger than the external resource and waste flows.

Resource Flows

Applied industrial ecology is the study of driving factors influencing the specification of flows of selected materials between economic processes. Examples of these flows and some of the associated perspectives and constraints, drawn from the manufacturing sector, are illustrated and discussed below.

▼ Industrial Production of Materials

The resource flows involved in the production of materials are shown in Fig. 4(a). Beginning with virgin materials, the flows proceed through cycles of extraction, separation and/or refining, and physical and chemical preparation to produce finished materials. A typical example might be the extraction from the earth of copper ore and the eventual production of copper wire.

Fig. 4(b) shows, in addition to the central flows (Mx), the materials flows which occur away from the central spine. To the right are the wastes consigned to disposal (MxD), typically large fractions in the early stages of the process and smaller fractions at later stages. To the left are flows of recycled material (MxR). Near the center are recycled flows that occur during the production process itself (MxC). A complementary flow occurs for any chemicals used in processing the material.

The flows of materials and processes, as shown in Fig. 4, occur within what has traditionally been termed "heavy industry," and the flows are generally under the sole control of the "materials supplier."

▼ Industrial Manufacture of Products

The resource flows involved in the manufacture of products from the finished materials produced by the materials supplier are shown in Fig. 5(a). An important distinction between this figure and Fig. 4 is that several finished materials are generally involved. A typical example is the production of a cable connector from metal and plastic.

Fig. 5(b) shows the materials flows that occur away from the central spine. As in Fig. 4(b), waste flows, recycled material flows, and in-process recycle flows are indicated. It is the processes of Fig. 5(b) that are addressed by industrial physical designers to the extent that materials flows are addressed at all in the design process.

It is instructive to examine Fig. 5(b) from the perspective of constraints to optimizing the industrial ecology of the process. At the forming step, material may enter from three types of streams: the virgin materials streams ($P1x$), the process recycle streams ($P1C$), and the recycled material streams ($P1R$). Unrecycled waste exits in disposal stream $P2D$. Optimization involves decreasing or eliminating $P2D$, increasing $P1C$, and utilizing whatever external recycled material is available ($P1R$), within the constraints of customer preference and existing price structures. The use of any recycled material stream thus involves a tradeoff between insuring its purity and suitability against the cost of using virgin finished materials. A similar analysis applies to each step of the process.

The flows of materials and processes as shown in Fig. 5 are within the "manufacturing" community, and the materials flows are generally under the sole control of the "manufacturer."

▼ The Customer Product Cycle

The resource flows involved in the customer portion of the industrial ecology cycle are shown in Fig. 6. Optimization opportunities here include the need to avoid resource dissipation, especially in the handling of obsolete products. To the extent that customers favor the recycle stream $C3R$ at the expense of the waste disposal stream $C3D$, industrial ecology is optimized.

Although exceptions exist, the situation under existing economic and governmental constraints is that customer decisions are made independently of either the material supplier (who may also be a material recycler) or the manufacturer.

▼ The Industrial Ecology Cycle

When Figs. 4-6 are combined, they result in the ensemble industrial ecology cycle shown in Fig. 7. One might view this figure as a potential diagram in which the energy expended to achieve a given flow increases as one moves upwards in the diagram, since it takes much less energy to recycle materials from one of the lower stages to an intermediate stage than to begin at a higher stage. Ayres [2] puts the point differently: that the embedded information in materials increases at the expense of energy invested as one approaches the bottom of the materials flow chain. Moreover, the waste associated with preliminary processing stages, such as the refining of ore, if often far greater per unit of output than the waste produced in the final product manufacturing stages.

The industrial ecology cycle is one in which, at present, three essentially decoupled participants are acting in their own interests. Just as the relationships between finished materials suppliers and manufacturers have been markedly changed by just-in-time (JIT) manufacturing techniques, so those relationships have the potential to be changed by industrial ecology manufacturing techniques. As with JIT, accomplishing IE will require close cooperation among equipment designers, process designers, process engineers, and suppliers. This process may begin to come about as corporations begin to implement design for environment (DFE) procedures.

Table I
Raw Material, Utility, and Environmental Summary for Hot Drink Containers

Item	Paper cup	Polyfoam cup
Per cup		
Raw Materials:		
Wood and bark (g)	33	0
Petroleum (g)	4.1	3.2
Finished weight (g)	10	11.5
Cost	2.5x	x
Per metric ton of material		
Utilities:		
Steam (kg)	9000-12 000	5000
Power (kWh)	980	120-180
Cooling water (m³)	50	154
Water effluent:		
Volume (m³)	50-190	0.5-2
Susp. solids (kg)	35-60	Trace
BOD (kg)	30-50	0.07
Organochlorides (kg)	5-7	0
Metal salts (kg)	1-20	20
Air emissions:		
Chlorine (kg)	0.5	0
Sulfides (kg)	2.0	0
Particulates (kg)	5-15	0.1
Pentane (kg)	0	35-50
Recycle potential		
Primary user	Possible	Easy
After use	Low	High
Ultimate disposal		
Heat recovery (MJ/kg)	20	40
Mass to landfill (g)	10.1	1.5
Biodegradable	Yes	No

Preliminary Examples of Industrial Ecology Studies

▼ Types of IE Assessments

Industrial ecology is usually pictured in one of two ways. The first is material-specific; that is, it selects a particular material or group of materials and analyzes the ways in which they flow through the industrial ecosystem. An example of such analyses is that recently performed by many industries to assess and minimize the use of chlorofluorocarbon compounds. Such an analysis in manufacturing operations is general-

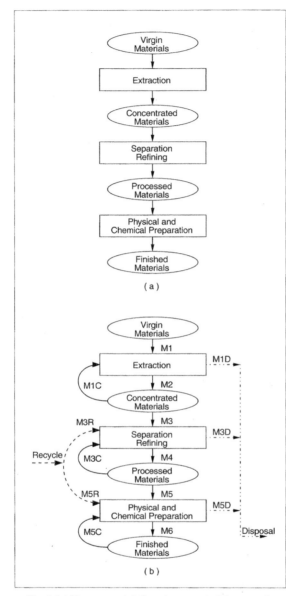

Fig. 4. (a) Linear materials flows in raw materials processing.
(b) Cyclic materials flows in raw materials processing.

an analysis is particularly appropriate at the initial design stage of a product, when decisions on alternate materials or processes can often be made prior to the investment of large amounts of capital for equipment or process development, an action that often locks in a particular material or process for the long term.

In order to contrast these two approaches, each of which has its place in IE, we present below a few brief examples of preliminary IE analyses.

▼ The Platinum Group Metals

Examples of the opportunities presented by IE are contained in Frosch and Gallopoulos's analysis of the cycles of the platinum group metals [3]. The flows are shown in Fig. 8, which is a materials-specific IE diagram and represents the current state of things rather than a concerted effort to design the system in accordance with IE principles. Fig. 8 demonstrates that the platinum group metals are recovered efficiently from jewelry and other fabricated objects, two uses that constitute about 60% of consumption. Industrial catalysts and chemicals, also efficiently recycled, account for another 6%. The fastest growing use for the metals is in automobile catalytic converters, an application marked by low recycling rates. The infrastructure is only now being set up to collect the millions of converters that enter automotive scrap yards each year and to recover the approximately two grams of platinum in each converter.

▼ Paper versus Polystyrene

A second attempt to apply IE principles to an existing situation is the comparison of Hocking [4] on the analysis of relative merit of paper versus polystyrene foam as the material of construction for hot drink containers. He points out that the instinctive supposition is that the paper cup is superior in environmental merit. However, a paper cup requires much more raw material (Table I) and, in addition, consumes as much energy as a polystyrene cup to produce. The organic chemicals needed in the papermaking process, together with the large relative amounts of water effluent, provide substantial external impacts compared with the only significant emittant associated only with the polystyrene cup: the pentane used as a blowing agent. The polystyrene cup is easier to recycle and approximately as easy to incinerate. Landfill degradation tends to favor the paper cup, but recent studies indicate that even "biodegradable" materials remain undegraded over very long periods, rendering that advantage relatively unimportant. The conclusion from Hocking's analysis is that "it would appear that polystyrene foam cups should be given a much more even-handed assessment as regards their environmen-

ly made while products are in their manufacturing cycle, because any modifications to materials or processes tend to be costly, difficult, and constrained by product design.

The second type of industrial ecology analysis is product-specific; that is, it selects a particular product and analyzes the ways in which its different component materials flows can be modified or redirected in order to optimize product-environment interaction. Such

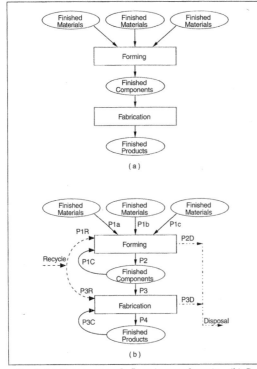

Fig. 5. (a) Linear materials flows in manufacturing. (b) Cyclic materials flows in manufacturing.

tal impact relative to paper cups than they have received during the past few years."

The implications of studies such as Hocking's, still in their infancy, are much larger than whether foam cups or paper cups are preferable. Indeed, they have potential application to all industrial operations, both those dealing with components entering the manufacturing operation and with products leaving the manufacturing cycle. It is worth noting that Hocking's assessment has not faced the really crucial issue in implementing IE in an industrial setting: assigning dollar values to each of the relative comparisons made in his analysis. It is with that issue that regulatory, legal, capital, and other corporate factors come into play. Nonetheless, his study demonstrates that comprehensive analyses of competitive processes can produce results which may be contrary to the "conventional wisdom."

BMW's DFD Roadster

One aspect of IE is the recycling flow after use (Flow *C3R* in Fig. 6). This aspect has been addressed in detail by Bavarian Motor Works in Landshut, Germany, where a design for disassembly (DFD) plant has been constructed. The initial vehicle designed according to DFD principles is the Z1 roadster [5].

The Z1 roadster has doors, bumpers, and front, rear, and side panels made of recyclable thermoplastic. It avoids the use of adhesives and screws, using instead "pop-in, pop-out" two-way fasteners. Some twenty different kinds of plastics, with varying properties, are used to construct the vehicle, and recycling is limited by this diversity. BMW hopes to reduce the types of plastics to five over the longer term.

The Z1 roadster is a prototype of the IE approach, if a limited one. It applies IE to an expensive product where economics is not a significant factor in design choices, and deals with a product for which many of the materials and assembly techniques had been well researched over the last decade or more.

Approaches to Industrial Ecology

▼ *Constraints and Drivers*

How can industrial ecology be studied and optimized, so that its best students become its most efficient practitioners? One needs to recognize that human institutions of various kinds are involved in promoting or constraining desirable material flows. For example:

▼ The search for engineering excellence can often promote cyclic behavior within the manufacturing node by designing processes to promote materials reuse. In the present instance, the high value that engineers place on efficiency is in harmony with ecological objectives.

▼ The social goal of avoiding toxic wastes may promote process changes to reduce the quantity of wastes or (better) to substitute materials or components that result in less toxic or nontoxic wastes. Alternatively, the use of potentially toxic materials can be designed with

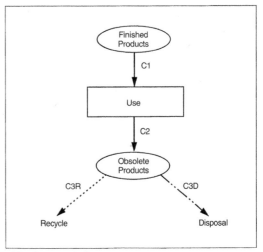

Fig. 6. Materials flows in customer use.

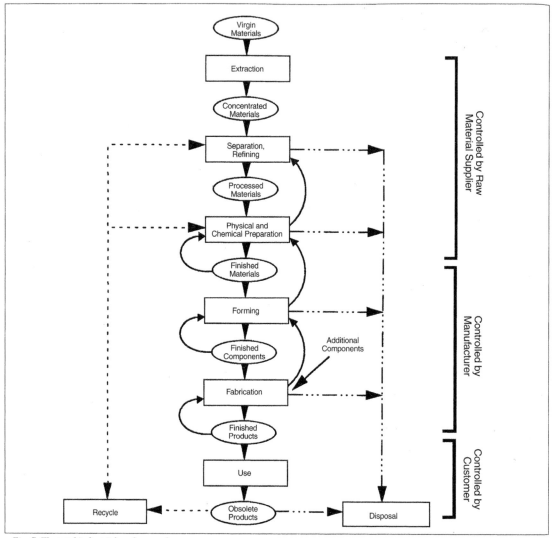

Fig. 7. The total industrial ecology cycle.

proper controls and endpoint treatment to neutralize the waste.

▼ It may be difficult to raise capital to alter a process in order to improve its cyclic nature unless such a change has private benefits (profit) as well as public ones.

▼ Certain government regulations conceived in the absence of ecological thinking may make reuse of materials so difficult that enhanced waste flow is *de facto* encouraged.

▼ The price system, by failing to include relevant externalities in prices and costs, may preclude adoption of industrial ecology by manufacturers and producers (see below).

▼ Consumers' desire for a high "standard of living" may encourage extended product use or, alternatively, may promote early product disposal.

▼ Rapid technological evolution and obsolescence contributes to an increased waste stream.

Inherent in many of these examples is the realization that closing the loops in the industrial ecosystem is often difficult or impossible when driven by exhortation alone. Many of the significant factors are "externalities" from a free market standpoint. In at least certain cases, however, it may be possible to use prices as incentives for ecologically sound behavior. To some extent this happens naturally, as in the increasing cost of waste disposal, but it may also be stimulated artificially by carbon dioxide taxes, limited numbers of permits to pollute, and so forth. Alternatively, regulations such as the Clean Air Act can be promulgated to dictate modes of industrial operation.

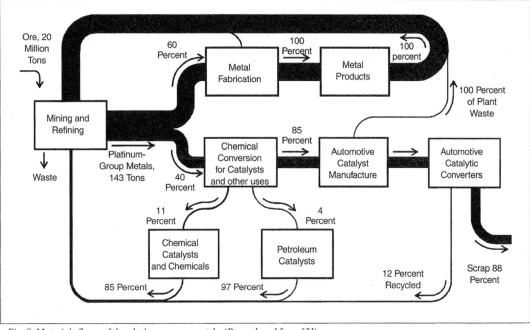

Fig. 8. Materials flows of the platinum group metals. (Reproduced from [3]).

▼ *Formulating the Industrial Ecology Model*

It is a characteristic lesson of ecology that subsystems cannot be studied in isolation without the danger of serious distortion. In physics and astronomy, a system may be only weakly coupled to other systems; in ecology, including industrial ecology, the coupling is usually much stronger. One cannot put the whole world into one model, however. In a typical industrial ecology problem, the modeler "draws a circle" around the subsystem under study (this system should be as large as feasible), and represents the rest of the world as realistically as possible using a few parameters on the boundary of the circle. It is the recognition of these parameters that distinguishes industrial ecology from "industry as usual." (It should be noted that it is often difficult to quantify some aspects of situations, for example, where considerations of social structures or quality of life influence recycling or waste disposal options. These aspects of the problem will require cooperative thinking among industrial, economic, and governmental spheres.) An important facet of industrial ecology is thus that IE does not consider industrial systems in isolation, nor can it be implemented in isolation.

Industrial ecology problems must ultimately be posed in the framework of mathematical models. Exploring the full range of complexity of these models is outside the scope of this paper, but it is useful to comment on possible approaches.

The mathematical tools used for industrial ecology problems should be selected to fit particular situations. In many cases, the most natural tools seem to be the input-output models of economics (called Leontief models when applied to an entire economy).

Some questions in industrial ecology can be formulated as problems in linear programming. For example, the profit of a firm can be approximated as a linear function of the quantity of each of its outputs, and the (resource and other) constraints on the firm, some of which may express ecological considerations, can be approximated as linear inequalities [6].

Many situations are strongly nonlinear, however. Some may involve thresholds or a sudden "tipping" from one mode of operation into another. For example, the availability of a particular raw material or the capacity of certain waste disposal sites may change abruptly at a certain level of use. In such situations, certain nonlinear programming techniques may be appropriate tools.

The challenge in implementing analyses for industrial ecology is not in performing the mathematical operations in a model, it is in the insight and information that must be brought to defining the magnitudes of the interactions among the components of the industrial ecology cycle.

Balancing Constraints and Aspirations

It is now becoming widely recognized that the achievement of a cultural and economic dynamic state compatible with the long term carrying capacity of Earth is required to balance fundamental environmental constraints and legitimate human aspirations [7]. Industrial ecology is a vital component of that achievement, and one that can be accomplished only when formulated from a solid theoretical understanding of its component parts. This formulation can be expressed mathematically in various ways, in which the external objectives and constraints deal with such factors as optimizing internal recycling during manufacture, minimizing waste disposal requirements, and minimizing the incorporation of materials whose toxic properties or resource limitations argue against their use. As manufacturing begins a transition from the traditional linear behavior to the cyclic behavior of industrial ecology, it is a transition in which those industries whose perspective of IE goals is clear will be those that are best prepared to implement and profit by IE principles.

References

[1] R.A. Frosch and N. Gallopoulos, "Towards and industrial ecology," paper presented to the Royal Society, London, Feb. 21, 1990.
[2] R. Ayres, "Industrial metabolism," in *Technology and Environment*, J.H. Ausubel and H.E. Sladovich, Eds. Washington, DC: National Academy Press, Washington, DC, 1989.
[3] R.A. Frosch and N. Gallopoulos, "Strategies for manufacturing," *Sci. Amer.*, vol. 261, no. 3, p. 144, 1989.
[4] M.B. Hocking, "Paper versus polystyrene: A complex choice," *Science*, vol. 251, p. 504, 1991.
[5] B. Nussbaum and J. Templeman, "Built to last - Until it's time to take it apart," *Bus. Week*, p. 102, Sept. 17, 1990.
[6] R. Fourer, D.M. Gay, andB.W. Kernighan, "A modeling language for mathematical programming," *Manag. Sci.*, vol. 36, p. 519, May 1990.
[7] World Commission on Environment and Development, *Our Common Future*. Oxford, U.K.: Oxford Univ. Press, 1987.   T&S

Metabolism and Colonization. Modes of Production and the Physical Exchange between Societies and Nature

MARINA FISCHER-KOWALSKI AND HELMUT HABERL[1]

ABSTRACT   *Written by a sociologist and a biologist, this paper attempts an interdisciplinary approach to describing the basic exchange relations between human societies and their natural environments. One type of exchange relation is termed 'metabolism' and related to the biological metabolism member organisms of societies require. A historical overview (part 1) demonstrates this exchange relation in terms of mass throughput per inhabitant to have grown in the course of human cultural evolution—without necessarily increasing the quality of life of those concerned—to the twentyfold it now amounts to in industrial societies (as is demonstrated empirically for Austria in part 3). A strategy of 'contraction of physical metabolism' (reduction of physical growth irrespective of 'economic' growth) of industrial societies is proposed as a strategic means of survival and possible ways to this goal are discussed quantitatively. The other exchange relation termed 'colonization' refers to treatments of natural environments that purposively change some components to render better exploitability (for the purpose of social metabolism), while still relying upon their basic self-regenerating qualities. It is sketched how colonization strategies developed historically, and it is demonstrated empirically that industrial societies now use about 50% of the available plant biomass (the energetic basis of all animal life) upon their territories for human purposes (part 4). Part 5 classifies different 'paradigms' for judging the 'harmfulness' of social interventions into the environment and outlines the logic of an information system that would enable society to generate an awareness of its own interventions into nature. On the whole the paper presents a theoretical as well as an empirical attempt to view societies as physical systems (among other physical systems on this planet) and confront sociology with the paradigmatic task to analyze the social regulation of these physical processes.*

1. Introduction

Well over a decade ago Catton and Dunlap (1978) suggested it would require a new paradigm within sociology to successfully deal with the relationship between society and nature. They accused the social sciences and sociology in particular to follow a 'human exceptionalism paradigm' (HEP) that would not allow human beings and society to be viewed as 'one among other' forms of life on this planet, as part of the 'web of nature'. They outlined some general features of what they called the 'new environmental paradigm' (NEP) required for a fundamentally different approach. As far as we can see, such a paradigmatic change has not taken place (Dunlap and Mertig, 1991; Devall, 1991). And we can well understand how highly uncomfortable such a paradigmatic change for sociology would be: As uncomfortable (if not irreconcilable) as a 'no growth'-perspective is for modern economics.

Excerpt from Metabolism and Colonization. M. Fischer-Kowalski and H. Haberl (1993). Modes of Production and the Physical Exchange between Societies and Nature: Innovation 6(4): 415–429, 439–442, Taylor & Francis, reprinted by permission of the publisher.

The vantage point for a paradigmatic change is a preconception that something is wrong with the relationship between society and nature, that society behaves in a way which destroys the natural basis it rests upon. There is no agreement, however, how radical this 'wrongness' ought to be conceived: Whether it is a matter of minor amendments or of imminent danger to survival; whether it is something that will more or less regulate itself or whether it will require massive intervention; or, finally, whether it cannot be regulated by society at all. All these conceptions do exist, among sociologists as among other people, and there seems to be no professional consensus among sociologists about the means and ways to approach truth.

Mainstream sociology tends to view society as a system of communication (see Luhmann, 1986) and disregards its material and physical properties. It is similar with neo-classical economics: The economy is regarded as a system of stocks and flows of money, and it is only the monetary side of reality that is addressed by economic theory. At best, physical concepts are discussed as tools for the development of monetarization[2].

These kinds of theories are not very helpful in conceptualizing the relationships between societies and their natural environments. To do so, societies must be looked upon as forms of organization that are supposed to serve one ultimate purpose: Sustaining the species humankind. These forms of organization vary considerably over time and space, as is well known, but in all cases they have to provide for the basics of nutrition and biological reproduction under specific environmental circumstances. It is not self-evident that they are able to do so; many societies are known to have collapsed or even died out because they could not.

Procuring sufficient nutrition, housing and material welfare, societies have evolved by organizing a continuous flow of energy and materials from their natural environments, transforming them in various ways useful for human life and discharging them again into natural environments. According to physics, the energy and materials intake will equal the amount of output, only the quality will have changed: In terms of thermodynamics entropy will have increased, thereby providing the differentials that biological and social life depend upon. We suggest to include these physical processes into the conception of society. This permits looking at societies from a physical perspective: As systems of material stocks and flows regulated by biological as well as cultural, technological and economic processes. These systems are sustained in specific environments. If the environment changes (possibly as a consequence of the system's behaviour, or for endogenous reasons such as meteorites or the like) the system will change. In contrast to ecosystems these changes do not occur by biological evolution, but by something we choose to call 'cultural evolution' (Harris, 1991, p.25). Cultural evolution does not depend upon changes in the composition of genetic information (all humans are, whatever their social organization may be, genetically very much alike, and their genetic outfit has practically not been altered since the advent of homo sapiens), but on the composition of culturally available information incorporated in human brains[3] and social organizations.

2. The question for sociology: how does society regulate its exchange processes with its natural environment?

Societies, just as ecosystems, may be characterized by their energy density. Energy density means the amount of energy taken in and being transformed by the system per calculation unit (space or organism). The minimal energy density of societies is determined biologically by the metabolism of its member organisms. Since humans are pretty big mammals, they require a fairly high energy flow (not per kilogram mass, since in this respect mice, for example, have a higher turnover than humans, but per organism), even at a minimum level of reproduction[4]. The energy density of contemporary industrial societies, of course, is a multiple of this.

Ecosystems with a high energy density seem to follow one of two alternative patterns:

(1) 'Closed cycle systems' have fast and rather closed cycles for all important nutrients. They manage to gain control over most environmental conditions necessary for their existence. An example for this pattern are tropical rainforests. They are able to continuously recycle their own nutrients. They can practically go on forever as long as none of their highly interdependent parts are destroyed by external forces (such as men chopping thousands of trees). They create and reproduce the environment they need: Their micro-atmospere, their water circulation and their mineral cycles. Of course, they depend on external ('environmental') circumstances such as the intensity of sunshine and the macro-climate. But these circumstances are independent of the system. Thus the 'carrying capacity' of the environment for such systems is basically determined by the size of the planet (respectively, in the case of rainforests, by its terrestrial space).

(2) 'Flow systems', for example floodplains and updwelling regions in oceans, depend upon an external nutrient supply that they cannot themselves reproduce. They can only exist as long as this nutrient supply is sustained (and the system's offproducts are being removed or can be deposited). Thus, flow systems depend on their environment to remain relatively constant, although they contribute to change it: They extract nutrients from it and discharge outputs to it. Here the notion of 'carrying capacity' refers to many processes within the environment determining its ability to regenerate, processes out of the system's reach of control.

So far societies obviously belong to the second type, the 'flow type' of systems. Locally several collapses have occurred, but globally that seems to have worked well, if judged by the size of the human population. How could it work?

Strategy one: small is beautiful

One strategy is to remain small in relation to the relevant natural environment. If the system is small enough, the effects its intake and its output have upon the environment remain negligible (see Figure 1). 'Size' has to be operationalized; it corresponds to energy density or, to use an organismic term referring to matter rather than to energy, to the dimension of the system's metabolism. In the case of societies this metabolism is a function of the size of the population and their mode of production. Minimally it is the sum of the metabolisms of the human organisms that are sustained by a society.

This strategy was employed by the hunter-and-gatherer societies that make up for most of human history (for a time span of 30.000 years or more). Living in bands of well below 100 members, the prehistoric (and the few contemporary examples surviving) paleolithic societies had a very low population density (even in favorable surroundings less than two persons per square mile, which is much less than one person per square kilomete; see Harris, 1990, p.24), and their societies' metabolism basically equalled the sum of the biological metabolisms of their members. As can be judged from anthropological findings (Angel, 1975), they were well nourished, and it probably took them relatively few working hours to secure their subsistence (Harris, 1991, p.75; Lee, 1969). This was possible because of almost zero growth of the population. Overall population growth rates for the time between 40.000 to 10.000 b.c. are an estimated 0.001% per year. Had they stood at the 0.5% rate of later agricultural societies, the world population would

Figure 1. Small system's metabolism in large environment: hunter-and-gatherer societies.

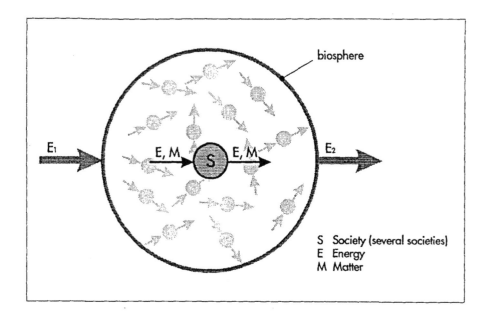

have arrived at a stunning $6*10^{23}$ at the end of this period, which would have been 10^{14} times more than it is now (Harris, 1990, p.24).

There are many indications that the stability of the size of the population was not due to 'natural causes' but was effectively culturally regulated by these societies, mainly by means of (preferrably female) infanticide (Hassan, 1973, and extensively Harris, 1990, p.26 ff.)[5].

Nevertheless, mankind expanded during this period to an estimated number of 5 million people and spread from its African origins all over the planet, even to reach Australia (as the only placental mammal). And despite their negligible size, human societies depleted their local and regional natural environments to a degree that forced them to change their mode of production. For whichever reason—because of climatic change (the end of a glacial period, the Wurm, occurred at about 10,000 BC) or as a result of improved hunting techniques—many large mammals were extinct by that time and hunters had to shift towards what Flannery (1969) termed 'Broadspectrum-Hunting'. This did not only imply an intensification of effort on the part of the hunters, but also an accelerated exhaustion of the environment.

Strategy two: terrestrial colonization

Middle Eastern societies were the first to develop a new mode of production based upon the domestication of animals and agriculture. This implies a new mode of intervention into the environment: Colonization. This is a type of intervention qualitatively different from the exchange relations we described as 'metabolism' above. By their metabolism societies extract materials ('nutrients') from the environment and discharge residual products ('feces'). These nutrients do or do not regenerate by processes beyond the reach

Figure 2. Metabolism supported by colonization: agricultural societies

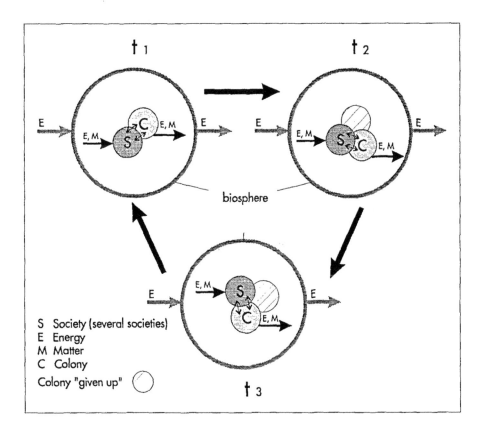

of intentional social regulation[6]. By 'colonizing' parts of the environment, society intentionally changes some elements of the environment so that it is rendered more exploitable for social needs. At the same time society intentionally contributes to organizing its regeneration. The most simple form of this is sketched in Figure 2, where this regeneration is achieved by leaving colonies out of use for some time period.

Historically and regionally one may find various different examples of creating and treating 'colonies'. Slash-and-burn methods in tropical rainforests (where the ashes of burnt trees serve as fertilizers), fertilization by animal manure and fallow periods in rainforest agriculture, and flooding systems in irrigation agricultures. But 'colonization' does not only refer to space (and the problem of maintaining soil fertility), but also to the domestication and breeding of animals and the cultivation of plants. Colonization seems to be the core invention of the neolithic 'revolution', preceding changes in what we would consider changes in 'metabolistic' techniques such as making use of metals for tools. The problem common to all 'colonization' is how to maintain the basic self-regenerating quality of colonies despite intensified exploitation, how to organize the investment of human labour and resources into this regeneration process and how to set limits to 'exploitation beyond reproduction'[7]. Harris (1990) describes the critical role of the cultural regulation of food-taboos and prescriptions: He analyzes that the prohibition of eating pork in the Middle East corresponded to pigs becoming direct nutritional competitors of humans as a consequence of deforestation in this region (before, pigs had been able to feed on themselves in the woods). Had there not been established a general religious taboo (Third Book of Moses) on pigs, their further breeding would have required those grains (and other increasingly scarce resources like shadow and water-pools) needed for the supply of humans. Similarly, the cultural creation of the 'holy cow' in India (and, even more so, religiously supported complete vegetarianism) can be explained as a means to protect the environment from depletion and, respectively, to secure the protein supply of humans at costs which the environment can bear (Harris, 1990, p.180). Under specific conditions even extensive ritual cannibalism as with the Aztecs may occur and can be interpreted as 'environmental protection' supplying protein to humans without doing so at the expense of the environment (Harris 1990, p.128).

Colonized environments allow for a much higher population density than their natural predecessors. At the same time, they seem to stimulate population growth. Under hunter-and-gatherer conditions the nutrient supply largely depended on the natural reproduction of game and plants; there was not much that could be gained by the intensification of labour. Under agricultural conditions extra labour, particularly child labour, may render returns beyond costs, and so people tend to have more children—or so at least Harris (1990, p.95) argues. During the neolithic period the human population increased by 0.1% annually, during the Ancient Empires by 0.5%, rendering a world population above 200 millions by the end of this period (Hassan, 1981, cited in Harris, 1991, p.96)[8]. This population increase occurs, although the average nutritional and health status of the people is worse than before.

The innovation of colonization also creates new preconditions for the metabolism of societies. The latter become sedentary; centralized political power and centralized infrastructures are established. As a consequence, gigantic masses of stone are moved and processed for the construction of buildings, storehouses, fortresses, city walls, dams, roads and the like. Millions of cubic meters of soil are being ploughed and exposed to erosion. In the same vein, human intervention into regional water households increases: Irrigation systems are established, the water supply for urban settlements is organized. The technological innovations in metallurgy and shipping cause whole regions like the Mediterranean to be stripped of their woods, with the accompanying climatic changes.

Thus, the invention of agriculture and animal husbandry permitted more people in larger societies to exist. Living conditions, however, were more miserable on average (in terms of nutritional status, workload and social inequality) and interventions into the environment intensified. The metabolism of societies had increased in size and become diversified: Whereas before it had been concentrated on (and more or less confined to) human nutrition, there were now added means of construction, metals and a lot more use of water and wood. As a consequence of this, despite the lower nutritional level (and the scarcity of meat in the average diet), the physical metabolism in terms of mass throughput per person living in the societies under this new mode of production was probably several times the biological minimum.

Nevertheless, periodic famines were a constant feature of agrarian societies and a recurring motive for colonial expansions. The most far-reaching and successful expansion was the expansion of the Old World into the New World beginning in the sixteenth century. Crosby (1990) attributes the success of this expansion not so much to the technological superiority of the Europeans in shipping and armament, but to their 'ecological sample-bag' inherited from the very beginnings of this civilization in the Middle East. According to him, the most important elements of this sample bag that accounted for superiority were domestic animals and microbes. The rich collection of domestic animals (such as goats, sheep, pigs, horses, hens etc.), in combination with the sustained ability to digest milk (which was genetically specific for Europeans), permitted long journeys to far away new settlements without dying from malnutrition; the collections of microbes bred in urban settlements and able to quickly decimate or even extinguish the virgin populations the conquerors came in contact with supplied the best of all weapons. Thus, the environmental depletion of Europe caused by population pressure and the agrarian mode of production could (at least temporarily) be resolved by huge migration processes lasting for centuries and resulting in the colonization of an ever increasing amount of land all over the planet.

Parallel to this colonial expansion a new mode of production developed gradually within Europe, for which this expansion and the accompanying improvements in the means of transportation (and increased temptations of trading) were of key importance. We are not going to discuss the reasons for this transition to industrialism here in detail, but we will elaborate on what this means in relation to the environment.

The whole terrestrial sphere of the planet becomes colonized: National states formally take possession of all land and inland waters (and even of parts of the sea, with the sole terrestrial exception being the antarctic which so far has been spared by means of international contracts). Large parts of these colonies are colonies only in a formal sense, though: They are no longer 'no-man's-land', but neither are they maintained in a way to preserve their capacity for self-regeneration as was (at least to a limited extent) the case with the traditional agricultural colonies. They are simply treated as inexhaustible 'mines of nature' and exploited accordingly. For a long period the strategies of 'colonization' in the more specific sense are expanded to ever increasing territories, but do not change in quality.

Strategy three: exploitation/deposition of subterrestrial resources

Major qualitative changes occur with regard to society's metabolism, though. In the first phase human and animal labour is substituted by machines. In contrast to animals, machines do not compete with human nutrition by consuming edible biomass but require large amounts of subterrestrial resources. In view of the constant scarcity of life-sustaining resources of the agricultural era (the supply of animal power competing with humans for nutrition), this may be looked upon as an environmentally benefactory innovation: No other living system that human societies might depend upon is vitally interested in subterrestrial resources like metal ores or fossil fuels. But unfortunately this induces new problems, not at the input side, but at the output side of social metabolism. As a consequence the atmospheric dimension of human metabolism grows considerably: Whereas before, this metabolism of society did not amount to much more than the biological minimum (required by human breathing and the breathing of the animals which humans kept for their own subsistence) plus fire for warming and cooking, now vast amounts of medium- and long-term deposits of carbon, in combination with other substances, were released into the atmosphere. Short term environmental consequences of this were felt quickly—such as the large scale deforestation of Western Europe which enforced the use of coal and, in turn, caused regional smogs detrimental to agricultural productivity (Bowlus, 1988)—and long term consequences will be felt for centuries to come.

At the same time industrialization strongly enhanced society's metabolism with respect to water: Water as a source of energy and cooling, water as a means of industrial processing and water supplies for the steadily growing urban settlements—in addition to agricultural uses.

With the advent of the 'petrochemical age', due to the affluence of subterrestrial reserves of (ancient) biomass, society's metabolism multiplied once more: Quantitatively, as a consequence of mass production and mass consumption of new goods, and qualitatively, by the production and emission of 'artificial' substances that enter all natural cycles.

Fairly late in the process of industrialization[9] changes in the modes of colonization occurred as well, which led to a sharp increase of biomass productivity per acre—the so-called 'green revolution'. This innovation consisted of the industrial production of fertilizers and the use of petrochemical products to defeat nutritional competitors.

During the transition to the industrial mode of production (historically in Europe, North America and Japan, and contemporarily in most of the rest of the world) the traditional cultural regulations of population growth broke down. However insufficient they had been, what followed was (and is) a veritable population explosion. Population growth rates throughout the 19th century Europe exceeded 0.7% per year (from 187 millions in 1800 to 410 millions in 1900) despite a population export that resulted in growth rates of 1.8% per year for America (Braudel, 1985, p.34). In the early phases of industrial development, theories like that of Malthus made it seem inevitable that most members of industrial societies would have to be kept just at the mimimum level of reproduction, otherwise they would bear more children than the labour market could absorb and soil could feed (Sieferle, 1990, p.90). In later stages the mechanism proved to be different: The industrial mode of living very much changed the cost-benefit relation of having children, i.e. lowered the benefits and increased the costs, so that it became culturally established to have very few children or even no children at all (Heinsohn *et al.*, 1979). This cultural change was effective even before the invention of elegant technologies that permitted non-reproductive coital intercourse. Thus, in the centres of industrialism the problem of population growth seems to be under cultural and technological control. This is not the case at the peripheries, and it is still a matter of sheer hope that a 'demographic transition' will take place there.

Not only had the cultural controls of population growth broken down, but the same collapse had affected the cultural checks and balances concerning the treatment of 'natural colonies'. Social arrangements against their 'exploitation beyond reproduction' or attention and care for their self-regenerating processes were neglected even for some traditional colonies of agriculture. Other realms of the natural environment which societies strongly acted upon, such as the atmosphere or water cycles, or the evolution of 'wild' species, were not treated with the care even an 'exploited colony' receives, but simply used as if they had an infinite capacity of regeneration. To include them in 'colonization strategies', i.e. to treat them in ways agricultural societies had learned to treat arable soil, is exactly what some 'environmental policies' now seem to try.

Fancy strategy four: global closed cycle system

Now let's come back to the system's approach from above. One possible solution for an energy-intensive flow system that keeps growing[10] and growing within a limited environment is to become all-encompassing, to incorporate its environment. This is equivalent to the creation of a mega-social system that culturally, economically and technically manages and controls the former 'natural environment' as a part of itself, or at least as a 'colony' that it exploits but takes cares to regenerate (Figure 3). This, of course, means the transformation into a closed cycle system as described above, able to create and to reproduce its own nutrients.

All possible damages inflicted upon the 'natural colonies' would have to be treated as damages within society. All of nature would be 'posessed' by specific social subjects for

Figure 3. Global control: The whole biosphere as a colony of industrial mega-society

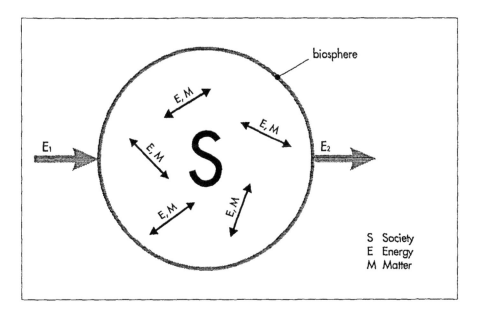

specific purposes (not neccessarily for profit purposes, however), and damages caused by one subject would be damages to someone else's posessions to be persecuted within social rules. All 'natural resources' would be represented within the economy as (not neccessarily private) capital stocks. The mega-social system would have to establish all mechanisms for distribution and exchange, and not only on a contemporary, but also on an inter-temporal (i.e., intergenerational) basis. This would mean the full mastery over nature, the perfection of the project of the Enlightenment.

Even if this vision were technically feasible (which environmental economists, e.g. Immler (1989), and technicians tend to assume, whereas biologists tend to deny it), it would be a huge social task. For the time being the (fully developed) industrial mode of production feeds about 20% of the planet's human population properly, and its side effects keep about 25% starving (UNDP, 1992). Mechanisms of exchange and distribution are not solved satisfactorily, not within societies and, let alone, between and among societies. And there is no agreement to be reached that this state of affairs is on its way to improvement—many would claim the contrary. Can one reasonably entrust the improvement of nature to the management of the industrial world?

We are very sceptical. So it seems worthwhile to consider another, more modest, basic model of relationships between societies and natural systems, pinpointed in Figure 4.

Fancy strategy five: multifold conviviality

The story of Figure 4 reads as follows: Let us assume human societies (not one unified social system) to share the biosphere with one another and with many other (natural) systems, systems of different qualitative and evolutionary status. Among these systems various exchange relationships exist. They have co-evolved, they are (to a varying degree) interdependent, and they all are to a certain extent self-organized, adaptive and contingent (Nicolis and Prigogine, 1987).

Figure 4. Multifold conviviality

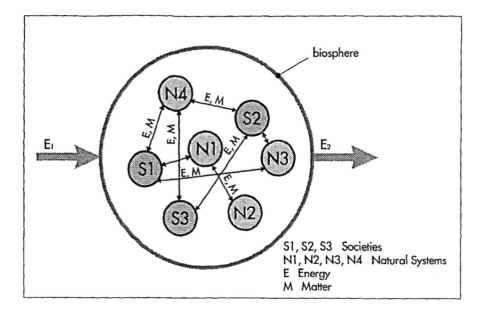

Let us use the metaphor of a neighbourhood upon which one depends: In order to survive, and even more in order to get along well, you have to know a lot about your neighbours—much more than just what you might get from them. You will have presumptions about their habits, about the language they speak, about their character, you will learn on whom and on what they depend, and you will try out what you must do or not do in order to get from them what you need, or to avoid getting from them what you don't like. You might find out you could intimidate and dominate some of them, probably with the help of others, but you would better not foster the illusion you could control them all. And if you are tempted to such an illusion, remind yourself of your limited knowledge and the contingency of the game. And be careful not to make real enemies, because to fight them will keep you busy all the time and distract your energies from more rewarding enterprises. Be loyal to your allies, but never rely completely upon them. Be aware: If you become a real nuisance for your whole neighbourhood, they'll extinguish you. They learn about you as well as you learn about them. So institutionalize ambassadors to warn against deviation from fair terms of trade, don't load too much responsibility upon your back (so let them live their way if possible) and keep proper records of your neighbours (particularly on your own part in the various deals, since you'll harvest what you've sowed) so you won't be taken by surprise.

Such a 'communicative' (or anthropomorphic) view of exchanges not only among societies but also between societies and natural systems supposedly appears very strange to natural scientists accustomed to an objectivistic paradigm of nature. It should not appear all that strange to social scientists, though. But is it not sheer romanticism? Friendly illusionary utopics? We do not know. The less friendly global control vision does not look very realistic, either. And, maybe, both visions share some common first steps.

So let us take a look at some features of the metabolism and at some parameters for colonization strategies of a contemporary industrial society.

3. The metabolism of an industrial society

Just a few months ago, the debate on sustainable development and possible limits of growth culminated at the UN Conference on Environment and Development (UNCED) in Rio de Janeiro. The variety of conceptions reaches from the 'Beyond the Limits' perspective of Meadows *et al.* (1992) to the point of view according to which 'only economic growth can guarantee sustainability' (see, for example, Schmidheiny/BCSD 1992).

Obviously, if there are limits to socio-economic growth, they are physical, not monetary limits. The 'size' of the social system vis-à-vis its natural environment is determined by the size of its metabolism in physical terms. The most basic physical terms are energy and matter.

Most industrial societies publish calculations of their energy-metabolism, or of what we would consider part of their energy-metabolism: The use of energy-carriers for combustion (in industry, transport and households) and the use of hydroelectric energy, in terms of Joule per year extracted. This does not include the nutritional energy extracted from biomass on which humans live (in competition against other species). Both components of energy metabolism have about the same size, as we will show for Austria below.

It is absolutely uncommon for industrial societies to calculate their metabolism in terms of matter, i.e. mass throughput per time unit. In economic input-output statistics a pragmatic consensus has been established over the decades on how to define a national economy's boundary and what monetary flows (and stocks) you have to consider in order to be able to calculate overall national product. An analogous problem has to be solved if one attempts to calculate the physical flows (and stocks) of a society in a specified period of time. In order to determine what amount of materials (in tons) flows 'through' a society per year, or is being stocked 'within' a society, you have to be able to tell at what point some material 'enters' and at what point it 'leaves' society. This is by no means trivial. Do the waters of a river crossing a national border or snow that falls 'enter' this society and then 'leave' it again? Is the crumb of arable land part of society's material 'stock', and if so, to the depth of how many centimeters? Many questions of this kind open up.

The most reasonable solution for these problems, so it seemed to us, was to look for the physical counterparts of economic processes. Only matter that is intentionally processed within a society's economy is what counts. At the 'environmental origin' of these processes one usually has to deal with some conceptual and even more so operational problem: How much material is turned over and put aside in mining? How much earth is turned over by ploughs (for a discussion of the problem see Bringezu, 1993)? How do you deal with the amount of water stored by dams for hydroelectric purposes? Do you calculate agricultural products including or excluding the water they contain (e.g. grass-hay)?

Steurer (1992) has made a first attempt to handle these problems empirically for Austrian society and to calculate the material annual throughput for 1988. He defines the 'boundaries' of the Austrian society in a twofold manner:

1   As a boundary between Austrian society and other societies; This conforms to the ordinary definitions of the national state: Products of another society that cross the Austrian state border are considered to 'enter' Austrian society as imports (not in monetary units, but in tons). This, of course, includes many raw materials sold as goods—but it does not include the extra materials that have entered the production process in the country of origin. In the same manner exports 'leave' Austrian society.

2   As a boundary between the Austrian economy and its natural environment. If ground-water, for example, is extracted and used for agricultural, urban or industrial water supply, these amounts are considered a throughput. It is the same with the amount of hay cattle consumes, with the mass of air that is used for combustion processes[11] (and the oxygen transformed into CO_2), and with the amount of metal ores (together with a lot of other material), petroleum or gas that are extracted.

Steurer's calculations are based upon the available variety of economic statistics. Some problems of both a conceptual and an empirical nature would require further research, but the results as presented in Figure 5 serve well as an overview of the whole structure of a modern industrial society's metabolism.

Austria has 7,8 million inhabitants. Austrian society's material throughput per year amounts to almost 4 billion tons, which is a rate of nearly 500 tons per inhabitant per year, or *1.3 tons per day*.

It is amazing to see that the overall structure of an industrial society's metabolism very much resembles that of an ecosystem: The most important throughput is water. Water amounts to 88% of the total metabolism; three quarters of this water are used by industry (mainly for cooling purposes) and for the production of electricity; only 5% are used for irrigation in agriculture and the remaining 20% by households (and small crafts) (see Steurer, 1992, p.6). The total throughput of water amounts to 423 tons per inhabitant per year, or 1160 liters per day (as compared to a biological minimum of two liters for drinking, and maybe another 50 liters for cooking and washing).

The next largest segment of the metabolism is air: 8% of Austrian society's throughput consists of air (respectively its oxygen part) for combustion and for technical processing[12] (mainly in the production of iron). Per inhabitant this makes for 38 tons per year or 105 kg per day. Human (and livestock) respiration is not included, as it reasonably should be. An adult human needs about 13 kg of air per day for breathing. Another conceptual problem arises with regard to the net oxygen production of forests. In Austria forests produce about 20% of the amount of oxygen used by combustion. But the economic function of forests so far is not the recycling of air, since this is just the way they function naturally. Thus oxygen production by forests was not included.

Only 4% of material input into society consists of solid stuff; almost one third of this is put on stock (see also Figure 6). 75% of solid input materials are extracted from the natural environment within national boundaries; one fourth is being imported, of which raw materials (mainly: energy carriers, timber and metals) account for three quarters. It seems remarkable that, in terms of mass input, even highly developed industrial societies live on their 'national territories'. They make use of local water, local air and even mostly local solid raw materials. The major exception from this 'territoriality' are fossil fuels: They are imported from other territories.

It seems also remarkable that industrial societies, on the level of national economies, overwhelmingly live upon direct extractions from natural environments. This even holds true if water and air are disregarded, as may be gathered from Figure 6: 87% of all material goods flowing into the economy per year are 'primary products'. Construction materials (road stone, sand etc.) constitute the largest fraction: They make up for almost one third. The second largest fraction is food: Animal food and grains amount to almost one quarter of total input[13]. A total of 36 million tons per year of biomass for nutrition is consumed; this amounts to 12.6 kg per inhabitant per day, corresponds to roughly 45.000 kcal/per day and is about twenty times higher than the biological metabolism.

The extraction (or import) of timber (for construction, manufacturing and combustion)

Figure 5. Material throughput of the Austrian Society (1988)

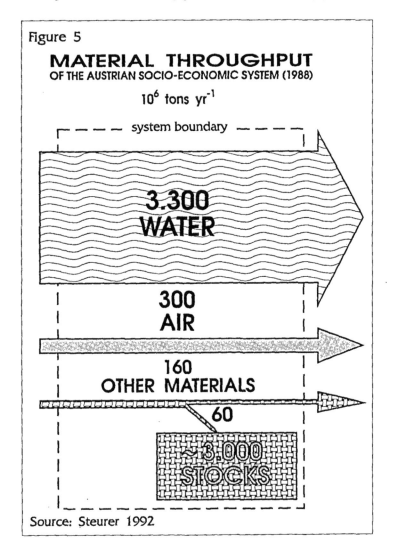

Figure 5

MATERIAL THROUGHPUT
OF THE AUSTRIAN SOCIO-ECONOMIC SYSTEM (1988)
10^6 tons yr^{-1}

— — — system boundary — — — —

3.300 WATER

300 AIR

160 OTHER MATERIALS

60

~3.000 STOCKS

Source: Steurer 1992

amounts to 13%. Thus the direct extraction of biomass altogether accounts for 36% of the solid matter of the inputs a society uses.

The next largest fraction (14%) is made up by fossil fuels; another 4% by metal ores. Thus, roughly one fifth of the annual inputs of an industrial society is extracted from our planet's long term subterrestrial deposits and not regenerable within time spans human societies may count upon.

Let us now look at the 'use'-side of the inputs. A large fraction (52 million tons or about one third of the total) is put on stock and is added to the already huge amount of buildings, roads, dams and the like. (One should be aware that these stocks, as a rule, prevent or reduce biomass production, see section 4 of this paper). Almost half of the material input (approximately 75 million tons) is put to short term uses, i.e. nutrition and materials for combustion. Its residual products' are immediately emitted into the

Figure 6. Materials balance of Austria (1988), solid materials only

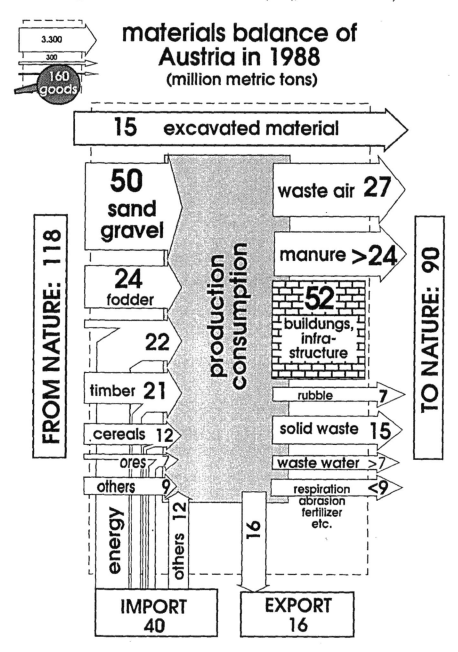

environment. The rest, amounting to about one fifth, circulates (or is stored) within the system somewhat longer (Steuer, 1992, estimates a period of 5–10 years) or gets exported to other social systems.

If we apply Meadows' and Randers' (1992) argument on the possibility of decreasing society's inputs from nature by increasing the longevity of its products, one may see that it can apply only to a small fraction of the inputs. However, longevity may be a culturally important, though quantitatively only small contribution to the possible 'contraction' of social metabolism: It can work only on one fifth of the inputs. If we triple the life-span (or the recycling periods) of these goods, we arrive at a theoretical potential reduction of 7% of annual inputs.

A large contribution might be expected from changes in nutrition: An increase in vegetarianism could mean a 10–20% deduction from annual inputs. Quantitatively almost equally important would be more efficient uses of energy: Increasing energy efficiency and reducing transportation (particularly road traffic) would render another 10% reduction of material input. And, finally, one could imagine cutting the net growth of construction and think of recycling construction materials: Maybe this would amount to another cut of 10–20%. All these measures taken together might amount to cutting down half of society's (solid) inputs from the environment. This would automatically be accompanied by a reduction of oxygen and water inputs. These calculations are very tentative, of course, but they may render an impression of possible changes that is more reliable than the usual sweeping estimates (e.g. Meadows' and Rander's 'factor 8').

Now let us take a look at the 'output'-side of Figure 6 that Steurer has carefully cross-checked against the Austrian emission- and waste-statistics. The only physical 'output' to the natural environment that may be considered a (albeit, problematic) social contribution to its regeneration are an estimated 24 million tons of fertilizers per year (animal manure). About 50% of the input matter is emitted to the environment in the form of 'wastes' more or less detrimental to it (such as carbon in the form of CO_2, sewage or solid wastes to be deposited). The rest is either exported as goods or put on stock.

How wasteful this economy works can be seen more clearly by material input-output analyses of single economic branches, as Payer (1991) has demonstrated[14]: In manufacturing industries ten tons of output in the form of goods correspond to 1–4 tons of output in the form of waste.

The material outputs are of course not equally 'harmful' to the environment. Moreover, they can be harmful in different functional contexts, and they may have a different weight in proportion to the natural cycles themselves. As Ayres (1991) showed, human activities currently influence the natural turnover of the carbon-, nitrogen-, sulfur- and phosphorous-cycle (the key cycles of the biomass productivity of this planet) to the extent of between 5 and several hundred per cent.

We will go into the problem of how to judge environmental 'harmfulness' within section 5 of this contribution. But so far we have tried to argue that the sheer size of society's metabolism in terms of matter is a meaningful global parameter for the effect the system will have on its environment. And if 'limits to growth' are to be discussed, the only relevant limits are of a physical nature. Monetary values, national incomes and profits may grow as much as they like, as far as the environment is concerned.

•••

References

Angel, J. L. (1975) 'Paleoecology, Paleodemography and Health', in: Polgar St. (ed.) *Population, Ecology and Social Evolution*, Den Haag, Mouton.

Ayres, R. U. (1991): 'Industrial Metabolism. Theory and Policy', in: Ayres, R. U., Simonis, U. E. (eds) *Industrial Metabolism*, New York (forthcoming).

Bringezu, St. (1993) 'Stoffströme und Strukturwandel', Paper presented at the seminar 'Ökologie und Ökonomie', IFF-Vienna (unpublished manuscript, Wuppertal Institut für Klima, Umwelt, Energie GMbH).

Birdsell, J. (1968) 'Some Predictions for the Pleistocene Based on Equilibrium Systems Among Recent Hunter-Gatherers', in: Lee, R., De Vore, I. (eds) *Man the Hunter*, Chicago, Aldine.

Bowles, Ch. R. (1988) 'Die Umweltkrise im Europa des 14. Jahrhunderts', in: Sieferle, R. P. (ed.) *Fortschritte der Naturzerstörung*, Frankfurt, Suhrkamp.

Braudel, F. (1985) *Sozialgeschichte des 15.-18.Jahrhunderts*, vol.1, *Der Alltag*, München, Kindler.

Catton, W. R. Jr. and Dunlap, R. E.(1978) 'Environmental Sociology: A New Paradigm', *The American Sociologist*, 13, pp.41–49.

Corniere, P. (1986) 'Natural Resource Accounts in France. An Example: Inland Waters', in: *Information and Natural Resources*, Paris, OECD.

Cronon, W. (1983) *Changes in the Land; Indians, Colonists, and the Ecology of New England*, New York, Hill and Wang.

Crosby, A. W. (1990) *Die Früchte des Weißen Mannes. Ökologischer Imperialismus 900-1900*, Frankfurt/New York, Campus.

Diamond, J. M. (1987) 'Human Use of World Resources', *Nature*, 328, pp.479–480.

Devall, B. (1990) 'Deep Ecology and Radical Environmentalism', *Society and Natural Resources*, 4(3), pp.247–258.

Duby, G. (1984) *Krieger und Bauern. Die Entwicklung der mittelalterlichen Wirtschaft und Gesellschaft bis um 1200*, Frankfurt, Suhrkamp.

Dunlap, R. E. and Mertig, A. G. (1990) 'The Evolution of the U.S. Environmental Movement from 1970 to 1990: An Overview', *Society and Natural Resources*, 4(3), pp.209–218.

Fischer-Kowalski, M., Haberl, H., Payer, H., Steurer, A., Zangerl-Weisz, H. (1991a) *Verursacherbezogene Umweltindikatoren—Kurzfassung*, Research Report IFF-Soziale Ökologie no. 10, Wien.

Fischer-Kowalski, M., Haberl, H., Wenzl, P., Zangerl-Weisz, H. (1991b) '"Emissions" and "Purposeful Interventions into Life Processes"—Indicators for the Austrian Environmental Accounting System', Paper presented at the Conference on 'Ecologic Bioprocessing' of the Österreichische Gesellschaft für Bioprozeßtechnik (GBPT), Graz, October 1991.

Fischer-Kowalski, M., Haberl, H. (1992) 'Purposive Interventions into Life Processes—an Attempt to Describe the Structural Dimensions of the Man-Animal-Relationship', Paper presented at the Conference on 'Science and the Human-Animal-Relationship' in Amsterdam, Research Report IFF-Soziale Ökologie no. 23, Wien.

Flannery, K. (1969) 'Origins and Ecological Effects of Early Domestication in Iran and the Near East', in: Ucko, P., Dimbleby, G. W. (eds) *The Domestication and Exploitation of Plants and Animals*, Chicago, Aldine.

Friend, A. (1988) 'Natural Resource Accounting: A Canadian Perspective', in: Ahmad, Y. J., El Serafy, S., Lutz, E. (eds) *Environmental and Resource Accounting and their Relevance to the Measurement of Sustainable Development*, Washington D.C., World Bank.

Galtung, J. (1975) *Strukturelle Gewalt. Beiträge zur Friedens- und Konfliktforschung*, Reinbek, Rowohlt.

Goldsmith, E. and Hildyard, N. (1984) *The Social and Environmental Effects of Large Dams*, Vol. 1, Overview, Wadebridge Ecological Centre.

Haberl, H. (1991) *Gezielte Eingriffe in Lebensprozesse*, Research Report IFF-Soziale Ökologie no. 11, Wien.

Harris, M. (1990) *Kannibalen und Könige. Die Wachstumsgrenzen der Hochkulturen*, Darmstadt, Klett & Kotta.

Harris, M. (1991) *Cultural Anthropology*, (3rd edn), New York, Harper & Collins.

Hassan, F. (1975) 'Size, Density and Growth Rate of Hunting-Gathering Populations', in: Polgar, St. (ed.) *Population, Ecology and Social Evolution*, Den Haag, Mouton.

Heinsohn, G., Knieper, R., Steiger, O. (1979) *Menschenproduktion. Allgemeine Bevölkerungslehre der Neuzeit*, Frankfurt, Suhrkamp.

Immler, H. (1989) *Vom Wert der Natur. Zur ökologischen Reform von Wirtschaft und Gesellschaft*, Opladen, Westdeutscher Verlag.

Kahlert, J. (1990) *Alltagstheorien in der Umweltpädagogik*, Weinheim, Deutscher Studienverlag.

Korab, R. (1991) 'Ökologische Orientierungen: Naturwahrnehmung als Sozialer Prozeß', in: Pellert, A. (ed.) *Vernetzung und Widerspruch. Zur Neuorganisation der Wissenschaft*, München/Wien, Profil Verlag.

Lee, R. (1968) 'Problems in the Study of Hunters and Gatherers', in: Lee, R., De Vore, I. (eds) *Man the Hunter*, Chicago, Aldine.

Luhmann, N. (1986) *Ökologische Kommunikation*, Opladen, Westdeutscher Verlag.

Max-Neef, M. A. (1991) 'Speculations and Reflections on the Future', Official document No.1 prepared for the Preparatory Committee of the Santiago Encounter, March 13–15, Santiago de Chile.

Meadows, D., Meadows, D., Randers, J. (1992) *Beyond the Limits; Global Collapse or a Sustainable Future*, London, Earthscan Publications.

Moscovici, S. (1990) *Versuch einer Menschlichen Geschichte der Natur*, Frankfurt, Suhrkamp.

Nicolis, G., Prigogine, I. (1987) *Die Erforschung des Komplexen. Auf dem Weg zu einem Neuen Verständnis der Naturwissenschaften*, München-Zürich.

Odum, E. P. (1983) *Grundlagen der Ökologie*, Band 1, Grundlagen, (2nd edn), Stuttgart, Thieme.

Odum, E. P. (1991) *Prinzipien der Ökologie: Lebensräume, Stoffkreisläufe, Wachstumsgrenzen*, Heidelberg, Verlag Spektrum der Wissenschaft.

Oechsle, M. (1988) *Der Ökologische Naturalismus. Zum Verhältnis von Natur und Gesellschaft im Ökologischen Diskurs*, Frankfurt/New York, Campus.

Payer, H. (1991) *Indikatoren für die Materialintensität der Österreichischen Wirtschaft*, Research Report IFF-Soziale Ökologie no. 14, Wien.

Pearce, D., Markandya, A., Barbier, E. B. (1990) *Blueprint for a Green Economy*, London.

Prigogine, I., Stengers, S. (1990) *Dialog mit der Natur*, München.

Schmidheiny, S./ BCSD (1992) *Kurswechsel. Globale Unternehmerische Perspektiven für Entwicklung und Umwelt*, München, Artemis & Winkler.

Sieferle, R. P. (1990) *Bevölkerungswachstum und Naturzerstörung*, Frankfurt, Suhrkamp.

Steurer, A. (1992) *Stoffstrombilanz Österreich 1988*, Research Report IFF-Soziale Ökologie no. 6, Wien.

UNDP—United Nations Development Programme (1992) *Human Development Report 1992*, New York/Oxford, Oxford University Press.

Vitousek, P. M., Ehrlich, P. R., Ehrlich, A. H., Matson, P. A. (1986) 'Human Appropriation of the Products of Photosynthesis', *BioScience*, 36(6), pp.368–373.

Wahlert, G. and Wahlert, H. (1981) *Was Darwin noch nicht Wissen Konnte: Die Naturgeschichte der Biosphäre*, München, Dtv.

Wenzl, P., Zangerl-Weisz, H. (1991) *Gentechnik als Gezielter Eingriff in Rebensprozesse; Vorüberlegungen für Verursacherbezogene Umweltindikatoren*, Research Report IFF-Soziale Ökologie no. 12, Wien.

Wright, D. H. (1990) 'Human Impacts on Energy Flow Through Natural Ecosystems, and Implications for Species Endangerment', *Ambio*, 19(4), pp.189–194.

Notes

1. This article is to be seen as an outcome of a team research process. We particularly wish to acknowledge the contributions of Thomas Hartmann-Macho, Harald Payer and Anton Steurer (all IFF-Soziale Ökologie, Vienna) and the careful reviewing by Stefan Bringezu (Wuppertal-Institute).

2. It has, however, to be said that attempts in this direction are becoming somewhat more common. See, for example, Pearce *et al.* (1990) and Ayres (1991).

3. Cultural evolution may nevertheless, and typically does, induce biological evolution by changing the conditions for selection in the environment. The increase in number and improvement of hunting techniques in early hunter-and-gatherer societies have probably contributed to the extinction of the pleistocene megafauna in the Old and New World (Harris, 1990, p.38). The development of urban agglomerations supported the evolution of various microbes that consequently caused epidemical catastrophes (Crosby, 1990, p.193), AIDS being one of the latest examples of such a process.

4. The average 2200 kcal per organism and day needed, have to be extracted from biomass (since humans are heterotrophic organisms that cannot live on sunlight and anorganic matter). The built-up of biomass can be described in terms of energy: about 0.4% of photosynthetically active sunlight is incorporated into plants as human edible matter (Harris, 1991, p.83). If humans choose to eat animal meat, this ratio is decreased by 10–1, making 0.04% photosynthetically active sunlight available for human nutrition.

5. Birdsell (1968) calculated infanticide-rates from data gathered on Australian aborigines still living by a hunter-and-gatherer mode of production. He arrived at an estimate of 50%.

6. In the biological cycle 'feces' serve as fertilizers of soil and, thereby, contribute to the regeneration of nutrients. This is a nice invention of evolution, but has little to do with the intentions of human beings: they have to get rid of their feces, whether this contributes to fertilization or not. It is possible, however, to socially organize the use of feces for intentional fertilization. This was done, for example, during the Chinese cultural revolution when big cities were organized to contribute to soil fertility by transporting feces to the country-side. More 'civilized' societies, inspired by the British example of WC, socially organize the deletion of human feces to surface waters, thereby 'overfertilized'.

7. In hunter-and-gatherer societies, the protection of the environment from 'exploitation beyond reproduction' had also probably to be achieved through cultural means. Duby (1984, p.70) for example, reports that there existed during the 6th century in Europe strong taboos against the cutting of trees which the Church had difficulties to overcome despite an abundance of undernourished people.

8. There was no steady growth of population, though. In the centres of the large irrigation cultures such as China, India or Egypt, there were periods of population growth, periods of population reduction and periods of stagnation. For centuries the standard of living in China, North-India, Mesopotamia or Egypt remained constant, and just a little above or below the level of impoverishment. If the population density of some region increased above a certain level, the level of living fell below misery; this led to wars, famines and a reduction of the population (Harris, 1990, p.198). Similarly the European population was substantially reduced in the course of the collapse of the Roman Empire

and the peoples' migration between the 2nd and the 7th century (Duby, 1984, p.20) and then again during the 14th century as a consequence of famines and plagues (Bowlus, 1988).

9. Important innovations in the agriculture of the 16th to 18th century were crop rotation with lucerne (bearing symbiotic bacteria able to find atmospheric nitrogen), enhanced husbandry of livestock and the introduction of the potato (Wahlert and Wahlert, 1981). In 1840 Justus von Liebig discovered the principles of chemical fertilization. The technology for the production of soluble nitrogen from atmospheric nitrogen was invented by Fritz Haber at the beginning of the 20th century. The practical application of these inventions on a mass scale is a fairly young phenomenon.

10. 'Growing' with respect to the size of its metabolism, which is both a function of its size of population and its mode of production. The per capita metabolism of Western Europeans in terms of mass and energy is at least 20 times higher than the more or less biological minimum required by an Australian aborigine.

11. The biological respiration process of humans is not included in this calculation, though. Whereas about 110 kg (or 80 kubic meters) of air per day are put through combustion processes (per inhabitant) humans need about 10 kubic meters (or 13.7 kg) per day for their respiration.

12. But not air used for cooling purposes and as pressurized air. These amounts seemed too difficult to calculate.

13. Animal meat, if 'produced' in Austria, was regarded not as a primary product (except for deer from hunting, which is quantitatively negligible) but as secondary, primary product being the vegetable basis of animal feeding (see Steurer, 1992, p.8).

14. The need for a supplementary system of physical accounting connected to the traditional SNA has gained understanding during the last years. Ambitious attempts have been started by the Norwegian, the French and the Canadian governments (see Corniere, 1986; Friend, 1988; OECD, 1988). The model of material balances presented here is to be seen as a contribution to that discussion in progress.

•••

Conclusion

Looking Forward: Legacy Readings and Contemporary Socio-Environmental Research

William R. Burnside, Kathryn J. Fiorella, Meghan L. Avolio, Steven M. Alexander, and Simone Pulver

Water sources are drying from anthropogenic changes, disease is spilling over from wildlife to human societies, and Indigenous communities are being displaced so others can harvest minerals and timber. The year is 2025. Or is it 1525, 1825, or 2125? Human societies have interacted intensely with local and distant environments for centuries and, in most places, millennia. With continuing growth in population, economy, and technology, changes in climate, oceans, and land use now dwarf those in past centuries and pose existential threats on a planetary scale. The underlying "great acceleration" in human activity has spurred study and soul searching.

Yet socio-environmental research – structured inquiry about the reciprocal relationships between society and environment – has a deep and rich past as well. Just as prior land use has shaped current landscapes, the pieces in this volume shape how we research and teach about socio-environmental issues today – they leave a legacy. Our goal in convening and contextualizing these readings is to illuminate, through this landscape view, the nature of and historic coherence behind the diversity of topics, views, and voices. We hope linking that legacy to contemporary approaches helps inform and motivate you to pioneer future work.

Present-day efforts to understand changing socio-environmental terrains are now flourishing. Diversifying frameworks and approaches aim at re-bridging disciplinary divides and understanding the complex linkages between and within social and environmental domains. Research varies from the expansive and conceptual to the focused and methodological. It is a stimulating and energizing mix, and the will to action is building.

Imagine you happen upon a party hosted by the socio-environmental research community, past and present. It is dark and cold outside, and the warm glow and laughter are tempting. The door is open just enough. It is a big gathering, and even those invited recognize but a few faces. It is daunting, even for insiders. In concluding, we aim to be gracious hosts, to welcome and introduce you to key socio-environmental approaches, frameworks, and fields active now and to link them to legacy readings on which they build.

Continuity and Change in Socio-Environmental Research

Juxtaposing the readings in this volume highlights both the continuities and changes in socio-environmental relationships and their study over centuries. One insight is the striking continuity of underlying themes. In exploring the terrain of socio-environmental questions and issues, research has concentrated over time on enduring themes, carving the equivalent of central channels in a delta. A focus on population, shared resources, technology, and systems has characterized socio-environmental research since the 1700s. This is unsurprising given both the fundamental role these play for societies of any size and the challenges they present as societies expand.

Continuity also characterizes societies' search for principles to understand and guide their relationships with environments. Debates around spiritual, utilitarian, and precautionary framings of socio-environmental relationships characterize the legacy selections from Humboldt and Darwin in the 1800s, Semple and Boas in the early 1900s, and Rappaport and Hardin in the 1960s.

Such continuities in socio-environmental challenges and research are countered with associated changes in both. Since the 1700s, socio-environmental scholarship evolved from integrated, as researchers with broad expertise recognized society and environment as interdependent, to de-integrated as scholars marked turf and academic disciplines fractured and emphasized depth over breadth. Now calls for reintegration within and across disciplines reflect the expected pendulum swing back and the growing recognition, as from climate change, pandemic disease, and other catastrophic examples, of the perils of studying society and environment in isolation. Returning to our delta, imagine water together in a few initial, braided channels that then separates into a growing number of distinct channels and, ultimately, reweaves and remixes.

Over the centuries, the temporal and spatial scales of many challenges have changed dramatically as well. Pre-industrial concerns, such as deforestation and associated freshwater loss, often directly implicated both society and environment in part since problems arose in roughly the same place and time. Growing globalization and time lags now obscure many linkages, as when deforestation for agriculture in one region alters climatic patterns that generate water scarcity in other regions, requiring researchers to reframe systems and problems more expansively.

As socio-environmental issues have grown in scale and associated complexity, research has also evolved to recognize and focus increasingly on tradeoffs, synergies, and scenarios. For example, different policies or actions may favor either agricultural livelihoods or biodiversity in crop-producing regions, a tradeoff, or may promote both, a synergy. Researchers increasingly focus on such options and strive to integrate diverse sectors and considerations, particularly those aspects and actors historically ignored.

Contemporary Approaches to Socio-Environmental Research

If the legacy readings constitute the central channels of a delta that have separated and rewoven over centuries, its end streams are today's approaches to socio-environmental research. We conclude by linking the legacy readings to this expanding diversity of contemporary scholarship. This effort is inevitably illustrative rather than exhaustive.

We looked for contemporary approaches offering distinctive insights, adding something new to the existing legacy of scholarship. We also prioritized those that have become or are becoming centers of gravity, often anchored by one or more academic journals, a professional association, and/or a research institute. Some, like sustainability science and political ecology, are more established; others, like planetary health and infrastructure studies, are relatively new. Finally, we strove to profile a diversity of approaches, including ones allied with the social sciences, natural sciences, and humanities. We recognize the limits of these filters, the difficulties of characterizing a dynamic research terrain, and our restricted knowledge and urge you to link forward other approaches to legacy readings.

As with early scholarship, the contemporary socio-environmental research discussed often transcends disciplines. Although some can be subsumed within or even comprise fields, researchers vary in their embrace of given labels and methods. Moreover, contemporary approaches do not sit in isolation, and some of the most interesting work is occurring at the boundaries between two or more.

We group contemporary approaches into six broad categories reflecting both similarities and shared histories: sustainability science and allied systems frameworks; research on the ecology of economic and industrial systems; approaches that emphasize power and justice in socio-environmental relationships; approaches that focus on a specific research context, such as cities, land, or food; approaches that emphasize tools for integration; and planetary-scale approaches.

Sustainability Science and Allied Systems Frameworks

Sustainability science is a growing umbrella field for science and technology-based, use-oriented socio-environmental research (Komiyama and Takeuchi 2006; Bettencourt and Kaur 2011; Kates 2011). Building on the 1987 Brundtland Report's definition of sustainable development as "the kind of development that meets the needs of the present without compromising the ability of future generations to meet their own needs" (WCED 1987), sustainability science focuses on understanding interactions between society and environment to inform equitable living and development within the Earth's constraints (see Kates et al. 2001; Clark and Dickson 2003; Clark and Harley 2020). It grew from international collaborations around global change (e.g., see National Research Council 1999) and the developing concept and embrace of sustainability as an idea and ideal (Clark and Munn 1986; Kates et al. 2001; Steffen et al. 2020). Today, journals, degrees, and research programs are proliferating, and the UN's Sustainable Development Goals and initiatives such as its Decade of Ocean Science for Sustainable Development are focusing research efforts.

Yet sustainability science's emphasis emerged from and is enriched by earlier scholarship (Bettencourt and Kaur 2011; Kates 2011). **Darwin (1859, Part I)** noted our mutual interdependence in the web of life and **Humboldt (1814, Part I)** the concept of the Earth as a system (Steffen et al. 2020) whose unity defied reductionism and specialization (Kates 2012). The ability of societies, as part of this system, to modify and even despoil environments (**Marsh 1864, Part I**; see also Clark et al. 2004) and the challenges as population, consumption, and technology grow (**Ehrlich and Holdren 1971, Part VI**) suggest limits (**Meadows et al. 1972, Part VI**) and the imperative to develop sustainably.

Several contemporary systems-oriented frameworks and approaches are closely associated with sustainability science. Emerging in the late 1990s and early 2000s, these prioritize visualizing socio-environmental systems with conceptual diagrams (Pulver et al. 2018). Their genesis and development reflect both common efforts to address global environmental change and cross-fertilization among often overlapping research communities.

In the 1990s, Berkes and Folke (1998) catalyzed work on **social–ecological systems** – ecosystems linked to associated social systems – as an alternative to conventional approaches to managing natural resources. Integrating insights from adaptive management (e.g., Holling 1978; Walters 1986) and institutional approaches to common-pool resource management (e.g., **Ostrom 1990, Part III**), they extracted lessons from a set of case studies of successful management in the face of environmental change (Berkes and Folke 1998; Berkes, Colding and Folke 2003). In turn, these initiatives informed research on the sustainability, resilience, vulnerability, and other properties of social–ecological systems. Ostrom (2007, 2009), a contributor to the 2003 volume, leveraged decades of work on common-pool resources, institutions, and collective action to propose the **social–ecological systems framework,** a diagnostic approach widely used (Partelow 2018) for organizing and comparing variables that influence the sustainability of social–ecological systems. Ostrom's work on common-pool resources was influenced by **Gordon (1954, Part III)** (e.g., Ostrom 1999).

Allied research considers the resilience and vulnerability of social–ecological systems. **Resilience**, a pioneering systems approach originating in ecology (**Holling 1973, Part IV**), considers an ecosystem's ability to absorb shocks while maintaining function. Research on social–ecological resilience considers the associated capacities to adapt, transform, learn, and innovate (i.e., Carpenter et al. 2001; Holling 2001; Holling and Gunderson 2002; see Folke 2006 for an in-depth history). Today, a resilience approach is being applied broadly and at different scales, as on the resilience of households to food insecurity (e.g., Melketo et al. 2021). Regulation of system components is key, and **Rappaport (1967, Part II)** informed understanding of the role of ritual in such regulation (Berkes 2012). **Vulnerability** is a parallel but distinct framework and focus (Turner 2010). With origins in the social sciences and, in particular, work on hazards and risk (e.g., **White 1942, Part II; Sen 1981, Part III; Beck 1986, Part VI**), this research considers the vulnerability of socio-environmental systems to risks, such as climate change operating at different scales (Turner et al. 2003).

A focus parallel to that on social–ecological systems originated in the 2000s from a systems ecology tradition. **Coupled human-and-natural systems, or "CHANS,"** are systems of reciprocally interacting human and natural components. CHANS research emphasizes feedback loops, nonlinearities, emergent behavior, threshold responses, and interactions within and across scales (see Liu et al. 2007a, b, 2021) and integrates knowledge from multiple disciplines (Alberti et al. 2011). It reflects legacies of **Marsh (1864, Part I)**, **Moran (1981, Part II)**, and others considering society and environment issues as reciprocal and complex, as well as **Ellis and Swift (1988, Part IV)**, who showcased how proper coupling between resources and users could promote sustainability.

The Ecology of Economic and Industrial Systems

Ecological economics, established as a new interdisciplinary field in the 1980s, treats the economy as a subset of the Earth's natural systems (Van den Bergh 2001). In this view, healthy ecosystems and their resources are natural capital. While the recognition of this value dates back at least to **Ricardo (1817, Part I)**, ecological economists in the 1970s and 1980s emphasized the "capital stock" of natural ecosystems on which economies and societies rely (see Costanza et al. 2017 for an in-depth history). The need to maintain that stock in the face of economic activity, and thus to sustain societies more broadly, is a central challenge and idea (see **Jevons 1865, Part I; Daly 1974, Part III**). Later, **ecosystem services** research emerged to quantify the societal benefits and services provided by healthy ecosystems (**Ehrlich and Mooney 1983** and **Odum 1969, Part IV**; Costanza et al. 1997; Daily 1997). The valuation of those ecosystems and their assets is the focus of the **natural-capital** approach, which promotes this value in decision-making (e.g., Fenichel and Abbott 2014). Both ecosystem-services and natural-capital approaches were galvanized in the early 2000s by the Millennium Ecosystem Assessment, an international effort to document the status and trends of the Earth's ecosystems (MEA 2006). Indeed, central to ecosystem services and natural capital approaches are the benefits to people from ecosystems.

Relatedly, **industrial ecology** uses an ecological lens to analyze industrial systems. Emerging from production engineering in the 1990s (see **Graedel et al. 1993, Part VI**), it applies notions of ecological cyclicity and efficiency to industrial systems, such as those used to refine natural resources and produce finished goods. A popular associated methodology is life cycle assessment (LCA), which traces the life cycle and environmental load of such products, following flows of energy and materials from their natural sources through manufacture, product use, and disposal or recycling (Roy et al. 2009). A newer iteration, social life cycle assessment (S-LCA), adds a social component (Dreyer et al. 2006). More recently, research in industrial ecology and ecological economics has coalesced around the idea of a **circular economy**, defined by the Ellen MacArthur Foundation as "an industrial economy that is restorative or regenerative by intention and design" (Geissdoerfer et al. 2017). A circular economy seeks to replace linear cradle-to-grave conceptions with the closed loops and circularity seen in many biophysical cycles. Attendant concerns also focus on slowing and minimizing material and energy throughput. Effectively, a circular economy attempts to physically internalize the externalities considered by **Ayres and Kneese (1969, Part VI)**.

Considering Power and Inequality

As a counterpoint to these perspectives, **political ecology** focuses on the power relations that underlie socio-environmental system management and use. Political ecologists are critical of the idea of sustainability and sustainable development (e.g., **Lélé 1991, Part III**), advocating for a more holistic recognition of the factors sustaining livelihoods (e.g., **Chambers and Conway 1992, Part II**). The field emerged in the 1970s and 1980s in opposition to "apolitical" narratives about the drivers of environmental degradation. These included ecoscarcity narratives that blame environmental degradation on increasing populations (i.e., the **Malthus 1798, Part I** and **Hardin 1968, Part III** traditions) and efficiency narratives that blame degradation on the failure to correctly value and account for nature in economic systems, such as with the environmental externalities **(Ayers and Kneese 1969, Part VI)** and ecosystem services **(Ehrlich and Mooney 1983, Part IV)** traditions. The political ecology critique of both is that they ignore power and politics as central to explanations of socio-environmental change and the distribution of associated costs and benefits (Robbins 2019). As an interdisciplinary field, political ecology's origins are rooted in early critiques of capitalism and markets, as by **Marx (1867, Part I)** and **Polanyi (1944, Part III)**, and in the disciplines of anthropology and geography (see early readings in **Part II**). One of the field's founding texts is **Blaikie and Brookfield (1987, Part II)** (Watts 2000). Although initially focused on rural natural resource-dependent communities in the Global South, a political ecology lens has been applied widely, from waste and water infrastructures in urban systems to disaster recovery to novel ecosystems and interspecies ethics.

Political ecology's focus on power and politics is mirrored in **environmental justice** research, which considers the unequal distribution of environmental benefits and burdens across communities. Environmental justice scholarship arose in the 1970s alongside the associated social movement in the United States (Mohai et al. 2009; Pellow 2017). Environmental justice research analyzes the disproportionate exposure of low-income communities and communities of color to pollution and other environmental harms and how communities contest these burdens. Robert **Bullard's 1990** book *Dumping in Dixie* **(Part III)**, encapsulates these interconnections, as does more recent research (e.g., Agyeman et al. 2016). **Climate justice studies**, which emerged out of environmental justice research and activism in the 2000s, brings an ethics and justice perspective to climate change. It investigates the reciprocal linkages by which marginalized and vulnerable communities, who have contributed least to the problem of climate change, bear the brunt of its impacts (see **Boff 1995, Part V**) and are often further marginalized by mitigation and adaptation strategies (Schlosberg and Collins 2014).

Contextual Approaches and Fields

A growing body of contemporary socio-environmental research focuses on a specific research context, such as cities, land, or food.

Urban research has a deep history, as cities have long been sites for investigating reciprocal relationships between society and environment. The modern veneer of many cities today hides their ancient origins; cities evolved in environments worldwide along with intensifying agricultural production (**Childe 1936** and **Mumford 1956, Part II**). As cities expanded and became more distant from food production, the growing physical separation marked a metabolic rift considered by **Marx** (**1867, Part I**) and still echoing today. The concentration of people and resources manifested in urban pollution and waste, whose distribution afflicts communities unequally (see **Bullard 1990, Part III**). Building on these legacies and other early work (e.g., Park 1915), contemporary urban research is increasingly integrative (National Research Council 2010), making new linkages and combining ideas and methods from ecology, sociology, anthropology, geography, urban planning, civil engineering, and public health. As urban ecosystems grew, so did the subfield of **urban ecology**, which began with ecologists studying ecological processes in green areas within cities (e.g., Kunick 1987) or along urban–rural gradients (e.g., **McDonnell and Pickett 1990, Part IV**). Today, transdisciplinary teams focus on how complex feedbacks between society and nature shape urban ecosystems (Zhou et al. 2021) and consider the ecology in, of, and for cities (Pickett et al. 2016). Expanding beyond the urban research tradition and linking to discussions of environmental injustice, political ecology, anthropology, and the environmental humanities, current research on **infrastructure** delves into the ways in which material media embody historical legacies of exclusions and privilege, channel action and organization, and co-constitute lived realities of society and environment (Simone 2004; Larkin 2013; Amin 2014).

Moving out in scale, **land system science** and its relative, land change science, focus on the terrestrial components of the Earth system. They investigate interactions between society and environment as affected by and reflected in land (Lambin and Meyfroidt 2011; Verburg et al. 2013). Land change science seeks to understand the socio-environmental dynamics that moderate changed land uses and covers (Rindfuss et al. 2004), including reciprocal interactions with natural resources (e.g., **Melville 1990, Part V**). Land system science uses this understanding, including of land-use legacies (see **Gómez-Pompa and Kaus 1992, Part V**), to transform land systems through governance and stakeholder engagement (Verburg et al. 2015). Both share roots in geography starting at least with **Humboldt (1814, Part I)** (Turner et al. 2016). Contemporary research is often associated with the Global Land Programme (https://glp.earth/).

Closely linked to land (and aquatic) systems, **food systems research** integrates the many food-related elements and activities that affect food provision and security, encompassing dynamics from production to consumption (see Ericksen 2008; Ingram 2011; Timmermans et al. 2014). Highly interdisciplinary, food systems research combines nutrition, agriculture, ecology, fisheries, public health, economics, and policy, among others. Prominent initiatives include international global environmental-change research programs (e.g., see Seitzinger et al. 2015) and the EAT-Lancet commission (https://eatforum.org/eat-lancet-commission), which explicitly integrates health and broader sustainability elements of food systems (see also Downs et al. 2020). **Hong (1793, Part I)** and **Malthus (1798, Part I)** highlighted early on the challenge of sufficient food production for growing populations, while **Boserup (1965, Part VI)** argued that more mouths catalyze more technology to feed them. This debate continues today, as does recognition and research showing how food access reflects power to command food (see **Sen 1981, Part III** and

Chambers and Conway 1992, Part II), how changes in crops affect agricultural systems **(Ho 1959, Part V)**, and how sovereignty over the means of production may be instrumental to food security.

Integrating Sectors and Systems

Nexus approaches examine interactions and interdependencies among multiple sectors, such as among food, water, and energy (the "food–water–energy nexus," or FEW). Focusing on such an intersection illuminates synergies and tradeoffs (Liu et al. 2018), as when crop irrigation and hydraulic fracking compete for groundwater. Nexus approaches were catalyzed in the 1980s by the growing realization of linked issues surrounding natural resources (Scott et al. 2015), such as insecurity around water, energy, and food in India, as considered by **Shiva (1988, Part V)**. A nexus can also be among resources and situations, such as water, land, and flood disasters, as recognized by **Gilbert White (1942, Part II)**, or between land and poverty **(Blaikie and Brookfield 1987, Part II)**. The local or regional interactions considered by nexus approaches originally can scale up, and researchers are increasingly focusing there. Nexus analyses are supported by a number of methods and tools, including integrated assessment models.

    Integrated assessment models (IAMs) combine social and environmental data in a single analysis by coupling multiple subsystem models. Integrated assessment models were developed for climate-change research and aim to be science-based decision support tools (Weyant 2017), as for predicting how fossil-fuel policies could affect energy use, greenhouse-gas production, and associated warming (Risbey et al. 1996). While early integrated assessments linked economic and atmospheric models (e.g., Nordhaus 2018), they have since become more comprehensive, adding energy, agriculture, governance, education, and health subsystem models (e.g., Kling et al. 2017). Integrated assessment models build on the legacy of the world-systems models developed in the 1970s by Donella Meadows and her collaborators (**Meadows et al. 1972, Part VI**). Future directions for climate integrated assessment models include increasing resolution in spatial and temporal scales and integrating impacts, adaptation, and vulnerability (Graham et al. 2020). They continue to influence work of the IPCC and to feed growing public and international interest in the scenarios they explore.

The Ultimate Stage: Our Planet in the Anthropocene

Planetary perspectives take an especially wide lens, foregrounding the global scale of resource use, constraints, and impacts and, in turn, the implications of this framing. They carry forward some of the earliest concerns of socio-environmental research (see **Part I**) – about food, population, resources, and limits – that reappear across disciplines and decades throughout these legacy readings. The notion of the Anthropocene, a new Earth-system epoch dominated by humans, embodies the planetary lens (Steffen et al. 2007; Zalasiewicz et al. 2010). This conceptualization gained momentum around 2000 with the backing of atmospheric chemist Paul Crutzen, ecologist Eugene Stoermer, and historian John McNeil (Crutzen and Stoermer 2000; McNeill 2001). Informed by the idea that the environment provisions ecosystem services (**Ehrlich and Mooney 1983, Part IV**), planetary perspectives grapple with global scales of change (Steffen et al. 2020) and also imply associated bounds. Rockström et al. (2009a, b) proposed the idea of *planetary boundaries*, global thresholds of pollution, biodiversity loss, and other phenomena whose crossing could breach the "safe operating space" for humanity. Researchers consider the uncertainty around these limits and risks from approaching them, such as from overuse of nitrogen in agricultural fertilizers, echoing **Hutchinson (1948, Part IV)**. In explicitly addressing the health impacts of approaching or exceeding these boundaries, the new field of **planetary health** addresses the ways environmental change is negatively impacting human health (Whitmee et al. 2015). Planetary health reflects the legacy of concerns raised by **Carson (1962, Part IV)** and **Norgaard (1994, Part III)** in highlighting the reciprocal effects of pesticide use. The field also grapples with the ways environmental changes may alter food production and nutrition, such as those observed within fisheries **(Pauly 1995, Part IV)**. The sense of associated responsibility has inspired an activist focus among ecologists on **Earth stewardship** (Chapin et al. 2011), which recognizes the value of local and traditional knowledge (see **Gadgil et al. 1993, Part IV**) and builds on powerful articulations by **Leopold (1949, Part V)** and others (Rozzi et al. 2015).

Anthropocene narratives in the environmental humanities parallel planetary perspectives in the natural sciences. Humanists are grappling with the ethical, philosophical, and spiritual implications of the Anthropocene (e.g., Neimanis et al. 2015, echoing **White 1967** and **Boff 1995, Part V**), and with the associated imperative for a more sustainable future (e.g., Merchant 2020). The Anthropocene ignites questions about the exceptionalism of humans (see **Leopold 1949, Part V; Catton and Dunlap 1978, Part III; Cajete 1994, Part V**), the benefits of technological and societal advancements **(Sahlins 1972, Part II)**, the systems responsible for environmental harm **(Gadgil et al. 1993, Part IV; Shiva 1988, Part V)**, the grief of a transformed Earth system **(Marsh 1864, Part I; Boff 1995, Part V)**, and the politics of knowledge production and meaning making through narrative **(Cronon 1992, Part V)**.

Challenges Going Forward

Even as socio-environmental research diversifies and matures, challenges remain. One is the continuing barrier to integrating disciplines and their research processes. The quantitative approaches favored in the natural and some social sciences remain too often at odds, both methodologically and epistemologically, with the rhetorical theory favored in the humanities and other social science fields. Visual frameworks, such as the growing varieties of box-and-arrow models, have attempted to bridge this, but they often simply form a third way. The clear value of each of these approaches makes this prospect tantalizing and worthwhile. Another challenge is recognizing the tradeoffs between specificity and generality. Small-scale, place-based studies are increasingly overshadowed by large, international efforts, however worthwhile. There is also the challenge of growing feedbacks from research itself. As more and more research aspires to policy relevance and actionability, scholars will need to reflect on how their findings and narratives may affect the systems they study.

In the face of these challenges and of the intensifying Anthropocene, consider what the persistence and ever-growing vibrancy of socio-environmental research suggests. Its existence is an answer to the question, posed by Richard York in the introductory commentary to Part I, "Is a better world possible?" Addressing that question repeatedly and with reference to ideals – wise use, equity, justice, holism, and even understanding – is a continuing expression of optimism.

Conducting Socio-Environmental Research in Light of Legacy Readings

We believe the legacy of socio-environmental research can enrich your study and work deeply. The readings here illustrate powerful ways to conceptualize socio-environmental elements and relationships – framings that shed light where it has been missing, highlight key connections, and transcend the specifics. Recognizing in today's research concerns and debates voiced decades and centuries ago suggests many of these are timeless and fundamental. Newly discovered natural resources, for example, will likely still be subject to shortages, require wise governance, and generate externalities experienced unevenly. Those enjoying the spoils will avert the spoilage, and affected landscapes and waterscapes will co-develop with technology and broader culture, generating ethics and narratives. Collectively, we hope readers feel empowered by this understanding and these examples, with the caveat that no one approach works for all systems or situations and that the conceptualizations emerged from the human imagination. There is ample room for creativity, for new ways to conceptualize and study socio-environmental relationships and problems.

The legacy readings also showcase qualities of successful socio-environmental scholarship. Tracing the evolution of concerns and ideas exposes you to how scholars responded to events in ways that resonated. Reading within disciplines suggests how strengths and weaknesses are tested and mature under one lens, while reading across them shows how different lenses provided new insights and even changed the fields involved. The readings showcase a real and stimulating diversity of systems, settings, worldviews, and voices. The more deeply and broadly you read, the more you will notice gaps. They deserve your attention. We hope the vantage points gained from this volume suggest new areas, insights, and approaches in your own work.

Finally, the legacy readings can catalyze and enrich collaborations. Research around society and environment is, like its focus, complex and usually interdisciplinary. Projects are enhanced by interdisciplinary research teams, and productive groups are increasingly diverse. Yet most training continues to be relatively disciplinary. In convening foundational readings, this volume juxtaposes ideas, places, topics, and voices usually considered in relative isolation. Reading how researchers from different fields view environments and issues will provide vocabulary and knowledge that can help bridge disciplinary divides, including shared "boundary concepts" (Star and Griesemer 1989) useful in knitting together interdisciplinary teams. We hope you leave excited by the wealth of possible synergies with other disciplines and the collective ability to make a lasting impact.

Think back, and recall the evening party with which we started, the gathering of socio-environmental scholars past and present. It is later now, and the buzz of names and ideas spills out into the night. As you leave, perhaps with some new friends, the landscape looks more familiar. You see things you had missed that mark different paths. The stories are alive in your mind. Some new stories echoed older ones – like variations on basic melodies – you heard almost as whispers from interior rooms. Despite the cool air, you feel warmed by being part of this community, by this conversation. As you walk, now with surer footing, the voices flood back. It is time to add yours.

References

Agyeman, Julian, David Schlosberg, Luke Craven, and Caitlin Matthews. "Trends and Directions in Environmental Justice: From Inequity to Everyday Life, Community, and Just Sustainabilities." *Annual Review of Environment and Resources* 41 (2016): 321–340.

Alberti, Marina, Heidi Asbjornsen, Lawrence A. Baker, Nicholas Brozovic, Laurie E. Drinkwater, Scott A. Drzyzga et al. "Research on Coupled Human and Natural Systems (CHANS): Approach, Challenges, and Strategies." *Bulletin of the Ecological Society of America* 92, no 2 (2011): 218–228.

Amin, Ash. "Lively Infrastructure." *Theory, Culture and Society* 31, no. 7/8 (2014): 137–161.

Berkes, Fikret. *Sacred Ecology.* New York: Routledge, 2012.

Berkes, Fikret, and Carl Folke, eds. *Linking Social and Ecological Systems: Management Practices and Social Mechanisms for Building Resilience.* Cambridge: Cambridge University Press, 1998.

Berkes, Fikret, Johan Colding, and Carl Folke, eds. *Navigating Social-Ecological Systems: Building Resilience for Complexity and Change.* Cambridge: Cambridge University Press, 2003.

Bettencourt, Luís M. A. and Jasleen Kaur. "Evolution and structure of sustainability science." *Proceedings of the National Academy of Sciences* 108, no. 49 (2011): 19540–19545.

Bettencourt, Luís M. A., José Lobo, Dirk Helbing, Christian Kühnert, and Geoffrey B. West. "Growth, Innovation, Scaling, and the Pace of Life in Cities." *Proceedings of the National Academy of Sciences* 104, no. 17 (2007): 7301–7306.

Carpenter, Steve, Brian Walker, J. Marty Anderies, and Nick Abel (2001). "From Metaphor to Measurement: Resilience of What to What?" *Ecosystems* 4: 765–781.

Chapin, F. Stuart, Mary E. Power, Steward T. A. Pickett, Amy Freitag, Julie A. Reynolds, Robert B. Jackson et al. "Earth Stewardship: Science for Action to Sustain the Human-Earth System." *Ecosphere* 2, no. 8 (2011): 1–20.

Clark, William C., and Nancy M. Dickson. "Sustainability Science: The Emerging Research Program." *Proceedings of the National Academy of Sciences* 100, no. 14 (2003): 8059–8061.

Clark, William C., and Alicia G. Harley. "Sustainability Science: Toward a Synthesis." *Annual Review of Environment and Resources* 45 (2020): 331–386.

Clark, William C., and Robert E. Munn. *Sustainable Development of the Biosphere.* Cambridge: Cambridge University Press, 1986.

Clark, William C., Paul J. Crutzen, and Hans J. Schellnhuber. "Science for Global Sustainability: Toward a New Paradigm." In *Earth System Analysis for Sustainability*, edited by Hans J. Schellnhuber, Paul J. Crutzen, William C. Clark, Martin Clauseen, and Hermann Held, 1–28. Cambridge, MA: MIT Press, 2004.

Costanza, Robert, Ralph d'Arge, Rudolf De Groot, Stephen Farber, Monica Grasso, Bruce Hannon et al. "The Value of the World's Ecosystem Services and Natural Capital." *Nature* 387, no. 6630 (1997): 253–260.

Costanza, Robert, Rudolf De Groot, Leon Braat, Ida Kubiszewski, Lorenzo Fioramonti, Paul Sutton et al. "Twenty Years of Ecosystem Services: How Far Have We Come and How Far Do We Still Need to Go?" *Ecosystem Services* 28 (2017): 1–16.

Crutzen, Paul, and Eugene Stoermer. "The Anthropocene." *Global Change Newsletter* 41 (2000): 17–18.

Daily, Gretchen C. *Nature's Services: Societal Dependence on Natural Ecosystems.* Washington, DC: Island Press, 1997.

Downs, Shauna M., Selena Ahmed, Jessica Fanzo, and Anna Herforth. "Food Environment Typology: Advancing an Expanded Definition, Framework, and Methodological Approach for Improved Characterization of Wild, Cultivated, and Built Food Environments toward Sustainable Diets." *Foods* 9, no. 4 (2020): 532.

Dreyer, Louise, Michael Hauschild, and Jens Schierbeck. "A Framework for Social Life Cycle Impact Assessment." *International Journal of Life Cycle Assessment* 11, no. 2 (2006): 88–97.

Ericksen, Polly J. "Conceptualizing Food Systems for Global Environmental Change Research." *Global Environmental Change* 18, no. 1 (2008): 234–245.

Fenichel, Eli P., and Joshua K. Abbott. "Natural Capital: From Metaphor to Measurement." *Journal of the Association of Environmental and Resource Economists* 1, no. 1/2 (2014): 1–27.

Folke, Carl. "Resilience: The Emergence of a Perspective for Social–Ecological Systems Analyses." *Global Environmental Change* 16, no. 3 (2006): 253–267.

Geissdoerfer, Martin, Paulo Savaget, Nancy M. P. Bocken, and Erik Jan Hultink. "The Circular Economy: A new sustainability paradigm?" *Journal of Cleaner Production* 143 (2017): 757–768.

Graham, Neal T., Mohamad I. Hejazi, Min Chen, Evan G. R. Davies, James A. Edmonds, Son H. Kim et al. "Humans Drive Future Water Scarcity Changes across All Shared Socioeconomic Pathways." *Environmental Research Letters* 15, no. 1 (2020): 014007.

Holling, Crawford S. *Adaptive Environmental Assessment and Management.* Hoboken, NJ: John Wiley and Sons, 1978.

Holling, Crawford S. "Understanding the Complexity of Economic, Ecological, and Social Systems." *Ecosystems* 4, no. 5 (2001): 390–405.

Holling, Crawford S., and Lance H. Gunderson. *Panarchy: Understanding Transformations in Human and Natural Systems*. Washington, DC: Island Press, 2002.

Ingram, John. "A Food Systems Approach to Researching Food Security and Its Interactions with Global Environmental Change." *Food Security* 3, no. 4 (2011): 417–431.

Kates, Robert W. "What Kind of a Science Is Sustainability Science?" *Proceedings of the National Academy of Sciences* 108, no. 49 (2011): 19449–19450.

Kates, Robert W. "From the Unity of Nature to Sustainability Science: Ideas and Practice." In *Sustainability Science: The Emerging Paradigm and the Urban Environment*, edited by Michael P. Weinstein and Robert E. Turner, 3–19. New York: Springer, 2012.

Kates, Robert W., William C. Clark, Robert Corell, J. Michael Hall, Carlo C. Jaeger, Ian Lowe et al. "Sustainability Science." *Science* 292, no. 5517 (2001): 641–642.

Kling, Catherine. L., Raymond W. Arritt, Gray Calhoun, and David A. Keiser (2017). "Integrated Assessment Models of the Food, Energy, and Water Nexus: A Review and an Outline of Research Needs." *Annual Review of Resource Economics* 9 (2017): 143–163.

Komiyama, Hiroshi, and Kazuhiko Takeuchi. "Sustainability Science: Building a New Discipline." *Sustainability Science* 1 (2006): 1–6.

Kunick, Wolfram. "Woody Vegetation in Settlements." *Landscape and Urban Planning* 14 (1987): 57–78.

Lambin, Eric F., and Patrick Meyfroidt. "Global Land Use Change, Economic Globalization, and the Looming Land Scarcity." *Proceedings of the National Academy of Sciences* 108, no. 9 (2011): 3465–3472.

Larkin, Brian. "The Politics and Poetics of Infrastructure." *Annual Review of Anthropology* 42 (2013): 327–343.

Liu, Jianguo, Thomas Dietz, Stephen R. Carpenter, Carl Folke, Marina Alberti, Charles L. Redman et al. "Coupled Human and Natural Systems." *AMBIO: A Journal of the Human Environment* 36, no. 8 (2007a): 639–649.

Liu, Jianguo, Thomas Dietz, Stephen R. Carpenter, Marina Alberti, Carl Folke, Emilio Moran et al. "Complexity of Coupled Human and Natural Systems." *Science* 317, no. 5844 (2007b): 1513–1516

Liu, Jianguo, Vanessa Hull, H. Charles J. Godfray, David Tilman, Peter Gleick, Holger Hoff et al. "Nexus Approaches to Global Sustainable Development." *Nature Sustainability* 1, no. 9 (2018): 466–476.

Liu, Jianguo, Thomas Dietz, Stephen R. Carpenter, William W. Taylor, Marina Alberti, Peter Deadman et al. "Coupled Human and Natural Systems: The Evolution and Applications of an Integrated Framework." *Ambio* 50, no. 10 (2021): 1778–1783.

McNeill, John Robert. *Something New under the Sun: An Environmental History of the Twentieth-Century World*. New York: W.W. Norton and Company, 2001.

Melketo, Tagesse, Martin Schmidt, Michelle Bonatti, Stefan Sieber, Klaus Müller, and Marcos Lana. "Determinants of Pastoral Household Resilience to Food Insecurity in Afar Region, Northeast Ethiopia." *Journal of Arid Environments* 188 (2021): 104454.

Merchant, Carolyn. *The Anthropocene and the Humanities: From Climate Change to a New Age of Sustainability*. New Haven, CT: Yale University Press, 2020.

Millennium Ecosystem Assessment (MEA). *Ecosystems and Human Well-being: Current State and Trends. Volume 1.* Washington, DC: Island Press, 2006.

Mohai, Paul, David Pellow, and J. Timmons Roberts. "Environmental Justice." *Annual Review of Environment and Resources* 34 (2009): 405–430.

National Research Council. *Global Environmental Change: Research Pathways for the Next Decade.* Washington, DC: National Academies Press, 1999.

National Research Council. *Pathways to Urban Sustainability: Research and Development on Urban Systems: Summary of a Workshop.* Washington, DC: National Academies Press, 2010.

Neimanis, Astrida, Cecilia Åsberg, and Johan Hedrén. "Four Problems, Four Directions for Environmental Humanities: Toward Critical Posthumanities for the Anthropocene." *Ethics and the Environment* 20, no. 1 (2015): 67–97.

Nordhaus, William. "Evolution of Modeling of the Economics of Global Warming: Changes in the DICE Model, 1992–2017." *Climatic Change* 148, no. 4 (2018): 623–640.

Ostrom, Elinor. "Coping with Tragedies of the Commons." *Annual Review of Political Science* 2, no. 1 (1999): 493–535.

Ostrom, Elinor. "A Diagnostic Approach for Going beyond Panaceas." *Proceedings of the National Academy of Sciences* 104, no. 39 (2007): 15181–15187.

Ostrom, Elinor. "A General Framework for Analyzing Sustainability of Social-Ecological Systems." *Science* 325, no. 5939 (2009): 419–422.

Park, Robert E. "The City: Suggestions for the Investigation of Human Behavior in the City Environment." *American Journal of Sociology* 20, no. 5 (1915): 577–612.

Partelow, Stefan. "A Review of the Social-Ecological Systems Framework: Applications, Methods, Modifications, and Challenges." *Ecology and Society* 23, no. 4 (2018): 36.

Pellow, David Naguib. *What Is Critical Environmental Justice?* Hoboken, NJ: John Wiley and Sons, 2017.

Pickett, Steward T. A., Mary L. Cadenasso, Daniel L. Childers, Mark J. McDonnell, and Weiqi Zhou. "Evolution and Future of Urban Ecological Science: Ecology in, of, and for the City." *Ecosystem Health and Sustainability* 2, no. 7 (2016): e01229.

Pulver, Simone, Nicola Ulibarri, Kathryn L. Sobocinski, Steven M. Alexander, Michelle L. Johnson, Paul F. McCord, and Jampel Dell'Angelo. "Frontiers in Socio-Environmental Research." *Ecology and Society* 23, no. 3 (2018).

Rindfuss, Ronald R., Stephen J. Walsh, Billie L. Turner, Jefferson Fox, and Vinod Mishra. "Developing a Science of Land Change: Challenges and Methodological Issues." *Proceedings of the National Academy of Sciences* 101, no. 39 (2004): 13976–13981.

Risbey, James, Milind Kandlikar, and Anand Patwardhan. "Assessing Integrated Assessments." *Climatic Change* 34, no. 3 (1996): 369–395.

Robbins, Paul. *Political Ecology: A Critical Introduction.* Hoboken, NJ: John Wiley & Sons, 2019.

Rockström, Johan, Will Steffen, Kevin Noone, Åsa Persson, F. Stuart Chapin, Eric F. Lambin et al. "A Safe Operating Space for Humanity." *Nature* 461, no. 7263 (2009a): 472–475.

Rockström, Johan, Will Steffen, Kevin Noone, Åsa Persson, F. Stuart Chapin III, Eric Lambin et al. "Planetary Boundaries: Exploring the Safe Operating Space for Humanity." *Ecology and Society* 14, no. 2 (2009b).

Roy, Poritosh, Daisuke Nei, Takahiro Orikasa, Qingyi Xu, Hiroshi Okadome, Nobutaka Nakamura, and Takeo Shiina. "A Review of Life Cycle Assessment (LCA) on some Food Products." *Journal of Food Engineering* 90, no. 1 (2009): 1–10.

Rozzi, Ricardo, F. Stuart Chapin III, J. Baird Callicott, Steward T. A. Pickett, Mary E. Power, Juan J. Armesto, and Roy H. May Jr., eds. *Earth Stewardship: Linking Ecology and Ethics in Theory and Practice. Volume 2*. Dordrecht: Springer, 2015.

Schlosberg, David, and Lisette B. Collins. "From Environmental to Climate Justice: Climate Change and the Discourse of Environmental Justice." *Wiley Interdisciplinary Reviews: Climate Change* 5, no. 3 (2014): 359–374.

Scott, Christopher A., Mathew Kurian, and James L. Wescoat. "The Water–Energy–Food Nexus: Enhancing Adaptive Capacity to Complex Global Challenges." In *Governing the Nexus: Water, Soil and Waste Resources Considering Global Change*, edited by Mathew Kurian and Reza Ardakanian, pp. 15–38. Cham: Springer, 2015.

Seitzinger, Sybil P., Owen Gaffney, Guy Brasseur, Wendy Broadgate, Phillipe Ciais, Martin Claussen et al. "International Geosphere–Biosphere Programme and Earth System Science: Three Decades of Co-evolution." *Anthropocene* 12 (2015): 3–16.

Simone, AbdouMaliq. "People As Infrastructure: Intersecting Fragments in Johannesburg." *Public Culture* 16, no. 3 (2004): 407–429.

Star, Susan L., and James R. Griesemer. "Institutional Ecology 'Translations' and Boundary Objects: Amateurs and Professionals in Berkeley's Museum of Vertebrate Zoology, 1907–39." *Social Studies of Science* 19, no. 3 (1989): 387–420.

Steffen, Will. *Sustainability or Collapse? An Integrated History and Future of People on Earth*. Cambridge: MIT Press, 2007.

Steffen, Will, Katherine Richardson, Johan Rockström, Hans Joachim Schellnhuber, Opha P. Dube, Sébastien Dutreuil et al. "The Emergence and Evolution of Earth System Science." *Nature Reviews Earth and Environment* 1, no. 1 (2020): 54–63.

Timmermans, Antoine. J. M., Jane Ambuko, Walter Belik, and Jikun Huang. *Food Losses and Waste in the Context of Sustainable Food Systems*. Rome: CFS Committee on World Food Security, High Level Panel of Experts, 2014.

Turner II, Billie L. "Vulnerability and Resilience: Coalescing or Paralleling Approaches for Sustainability Science?" *Global Environmental Change* 20, no. 4 (2010): 570–576.

Turner II, Billie L., Roger E. Kasperson, Pamela A. Matson, James J. McCarthy, Robert W. Corell, Lindsey Christensen et al. "A Framework for Vulnerability Analysis in Sustainability Science." *Proceedings of the National Academy of Sciences* 100, no. 14 (2003): 8074–8079.

Turner II, Billie L., Jackie Geoghegan, Deborah Lawrence, Claudia Radel, Birgit Schmook, Colin Vance et al. "Land System Science and the Social–Environmental System: The Case of Southern Yucatán Peninsular Region (SYPR) project." *Current Opinion in Environmental Sustainability* 19 (2016): 18–29.

Van den Bergh, Jeroen C. "Ecological Economics: Themes, Approaches, and Differences with Environmental Economics." *Regional Environmental Change* 2, no. 1 (2001): 13–23.

Verburg, Peter H., Karl-Heinz Erb, Ole Mertz, and Giovana Espindola. "Land System Science: Between Global Challenges and Local Realities." *Current Opinion in Environmental Sustainability* 5, no. 5 (2013): 433–437.

Verburg, Peter H., Neville Crossman, Erle C. Ellis, Andreas Heinimann, Patrick Hostert, Ole Mertz et al. "Land System Science and Sustainable Development of the Earth System: A Global Land Project Perspective." *Anthropocene* 12 (2015): 29–41.

Walters, Carl J. *Adaptive Management of Renewable Resources*. New York: Macmillan Publishers Ltd., 1986.

Watts, Michael. "Political Ecology." In *A Companion to Economic Geography*, edited by Eric S. Sheppard and Trevor J. Barnes, 257–274. Oxford: Blackwell, 2000.

Weyant, John. "Some Contributions of Integrated Assessment Models of Global Climate Change." *Review of Environmental Economics and Policy* 11, no. 1 (2017): 115–137.

Whitmee, Sarah, Andy Haines, Chris Beyrer, Frederick Boltz, Anthony G. Capon, Braulio Ferreira de Souza Dias et al. "Safeguarding Human Health in the Anthropocene Epoch: Report of The Rockefeller Foundation–Lancet Commission on Planetary Health." *The Lancet* 386, no. 10007 (2015): 1973–2028.

World Commission on Environment and Development (WCED). *Our Common Future*. Oxford: Oxford University Press, 1987.

Zalasiewicz, Jan, Mark Williams, Will Steffen, and Paul Crutzen. "The New World of the Anthropocene." *Environmental Science & Technology*, 44, no. 7 (2010): 2228–2231.

Zhou, Weiqi, Steward T. A. Pickett, and Timon McPhearson. "Conceptual Frameworks Facilitate Integration for Transdisciplinary Urban Science." *npj Urban Sustainability* 1, no. 1 (2021): 1–11.